**RULES FOR DERIVATIVES**
**(continued)**

***Constant Times a Function***   Let $k$ be a real number. Then the derivative of $y = k \cdot f(x)$ is
$$y' = k \cdot f'(x).$$

***Sum or Difference Rule***   If $f(x) = u(x) \pm v(x)$, then $f'(x) = u'(x) \pm v'(x)$.

***Product Rule***   If $f(x) = u(x) \cdot v(x)$, then
$$f'(x) = u(x) \cdot v'(x) + v(x) \cdot u'(x).$$

***Quotient Rule***   If $f(x) = \dfrac{u(x)}{v(x)}$, and $v(x) \neq 0$, then
$$f'(x) = \frac{v(x) \cdot u'(x) - u(x) \cdot v'(x)}{[v(x)]^2}.$$

***Chain Rule***   If $y$ is a function of $u$, say $y = f(u)$, and if $u$ is a function of $x$, say $u = g(x)$, then $y = f(u) = f[g(x)]$, and
$$\frac{dy}{dx} = \frac{dy}{du} \cdot \frac{du}{dx}.$$

***Chain Rule (alternate form)***   If $y = f[g(x)]$, then $y' = f'[g(x)] \cdot g'(x)$.

***Generalized Power Rule***   Let $u$ be a function of $x$, and let $y = u^n$, for any real number $n$. Then $y' = n \cdot u^{n-1} \cdot u'$.

4.3   ***Logarithmic Function***   If $f(x) = \ln |g(x)|$, then
$$f'(x) = \frac{d}{dx} \ln |g(x)| = \frac{g'(x)}{g(x)}.$$

4.3   ***Exponential Function***   If $f(x) = e^{g(x)}$, then
$$f'(x) = \frac{d}{dx} e^{g(x)} = g'(x) \, e^{g(x)}.$$

---

3.2   **FIRST DERIVATIVE TEST**

Let $c$ be a critical number for a function $f$. Suppose that $f$ is differentiable on $(a, b)$, and that $c$ is the only critical number for $f$ in $(a, b)$.

1. $(c)$ is a relative maximum of $f$ if the derivative $f'(x)$ is positive in the interval $(a, c)$ and negative in the interval $(c, b)$.

2. $f(c)$ is a relative minimum of $f$ if the derivative $f'(x)$ is negative in the interval $(a, c)$ and positive in the interval $(c, b)$.

---

3.4   **SECOND DERIVATIVE TEST**

Let $f''$ exist on some open interval containing $c$, and let $f'(c) = 0$.

1. If $f''(c) > 0$, then $f(c)$ is a relative minimum.

2. If $f''(c) < 0$, then $f(c)$ is a relative maximum.

3. If $f''(c) = 0$, then the test gives no information about extrema.

# CALCULUS WITH APPLICATIONS

## FOURTH EDITION

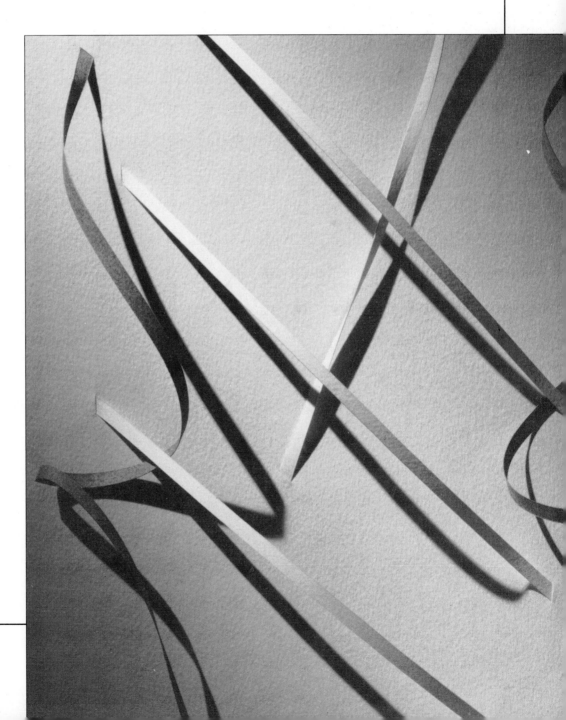

# CALCULUS WITH APPLICATIONS
## FOURTH EDITION

*Margaret L. Lial*
*American River College*

*Charles D. Miller*

SCOTT, FORESMAN AND COMPANY
Glenview, Illinois   London, England

## ■■ TO THE STUDENT

If you want further help with this course, you may want to obtain a copy of the *Student's Solutions Manual* that accompanies this textbook. This manual provides detailed step-by-step solutions to the odd-numbered exercises in the textbook and can help you study and understand the course material. Your college bookstore either has this manual or can order it for you.

**Cover art:** Vasily Kandinsky, ''Extended,'' May–June 1926. Collection of the Solomon R. Guggenheim Museum, New York.

Photo: David Heald

**Library of Congress Cataloging-in-Publication Data**

Lial, Margaret L.
    Calculus with applications/Margaret L. Lial, Charles D. Miller.—4th ed.
        p.    cm.
      Includes index.
      1. Calculus.    I. Miller, Charles David.    II. Title.
    QA303.L482   1989
    515—dc19                                       88-29047
                                                         CIP

    ISBN 0-673-38251-6

Portions of this book appear in *Calculus with Applications, Brief Version,* © 1989, and *Finite Mathematics and Calculus with Applications,* Third Edition, © 1989, 1985, 1980 Scott, Foresman and Company.

    2 3 4 5 6—VHJ—93 92 91 90 89

# PREFACE

*Calculus with Applications* is a solid, applications-oriented text intended for students majoring in business, management, economics, or the life or social sciences. The fourth edition of this book represents an extensive revision of the previous edition. Many new examples, exercises, and applications have been added, and portions of the text have been completely rewritten. In response to comments from those who have used the most recent edition of the book, greater emphasis has been given to explanations of topics from business and economics. Interesting new applications in these areas add to the rich diversity of applications provided for students with varied career interests.

This text is designed to be useful to the student and helpful for the instructor. The student will find explanations and examples that are clear, direct, and to the point. The exercises are carefully graded, and numerous examples that correspond to the exercises are provided. Abundant applied problems and Extended Applications show students how mathematics is used in the careers they are exploring. The instructor will find a comprehensive instructional text carefully planned to provide options for varying courses.

*Calculus with Applications* can be used for either a one- or two-term (full-year) calculus course. A prerequisite of three to four semesters of high school algebra or the equivalent is assumed. In addition, a course in trigonometry is helpful if Chapter 10 on trigonometric functions is to be included in the course. For those students who have not studied algebra recently, a review of algebra is given at the beginning of the book. This review may be used either in class or individually for reference.

## KEY FEATURES

**APPLICATIONS** Examples and exercise sets include a wide range of practical applications to motivate student interest. An Index of Applications, grouped by discipline, includes those applications given in the exercises, examples, and Extended Applications, and is located after the table of contents.

**EXAMPLES** Approximately 630 worked-out examples clearly illustrate the techniques and concepts presented and prepare students for success with the exercises. For clarity, the end of each example is indicated with the symbol ▬

**EXERCISES** More than 3900 exercises, including about 3160 drill problems and 750 applications, feature a wide range of difficulty from routine to challenging problems. A new format is used for exercises in this edition. Routine exercises are given first, followed by a clearly marked applications section arranged by discipline, with a descriptive heading for each application.

**EXTENDED APPLICATIONS**   Special lengthy applications are included at the end of most chapters. Designed to help motivate students in the study of mathematics, the Extended Applications are derived from current literature in various fields and show how the course material can actually be applied.

**FORMAT**   Important rules, definitions, theorems, and equations are enclosed in boxes and highlighted with a title in the margin, for ease of study and review. A second color is used to annotate equations, illuminate troublesome areas, and clarify concepts and processes. A new format in the exercise sets highlights the many diverse applications included in the book.

**STUDY AIDS**   Each chapter ends with a list of key words presented in the chapter and an extensive set of review exercises. The key words are numbered according to the section in which each appeared, for extra studying help.

**FLEXIBILITY**   *Calculus with Applications* has been designed to be as flexible as possible. An algebra review is included at the beginning of the book. Chapter 1, on functions and graphs, can be skipped by well-prepared students. After covering the core ideas of calculus in Chapters 2 through 5, instructors can choose among the remaining chapters, each of which is independent of the others. For example, the text can be tailored for classes with mainly business students by covering Chapter 1 (if necessary), Chapters 2 through 5, and Chapter 7, omitting the examples from other disciplines, and assigning the applied exercises designated for business and economics. If time permits, selected topics from the remaining chapters can be included.

## ▬ CONTENT FEATURES

An **intuitive and visual approach** is used. New topics are introduced with an appeal to past experience. See, for example, the introduction to maxima and minima in Section 3.2.

**New motivational material** in this edition includes **historical references** throughout and more attention to the background and development of new topics.

**Graphing techniques and curve sketching** are emphasized.

The **derivative is introduced early.** An intuitive approach employs graphs and tables wherever possible.

The chapter on trigonometric functions emphasizes **derivatives and integrals involving trigonometric functions.**

**Comprehensive coverage and logical sequencing** of topics give students a thorough introduction to the calculus.

**Expanded discussions of topics from business and economics** are included. See, for example, marginal analysis in Section 2.4, elasticity of demand in Section 3.6, and consumers' and producers' surplus in Section 5.5. In addition, more material on the mathematics of finance is covered in Section 4.6.

## SUPPLEMENTS

The **Instructor's Guide and Solutions Manual** gives a lengthy set of test questions for each chapter, organized by section, plus answers to all the questions. It also provides complete solutions to all of the even-numbered text exercises.

The **Instructor's Answer Manual** gives the answers to every text exercise, collected in one convenient location.

The **Student's Solutions Manual,** available for purchase by students, provides detailed, worked-out solutions to all of the odd-numbered text exercises, plus a chapter test for each chapter. Answers to the chapter test questions are given at the back of the book.

The **Scott, Foresman Test Generator for Mathematics** enables instructors to select questions by section or chapter or to use a ready-made test for each chapter. Instructors may generate tests in multiple-choice or open-response formats, scramble the order of questions while printing, and produce multiple versions of each test (up to nine with Apple, up to 25 with IBM). The system features a preview option that allows instructors to view questions before printing, to regenerate variables, and to skip certain questions if desired.

**Computer Applications for Finite Mathematics and Calculus** by Donald R. Coscia is a softbound textbook packaged with two diskettes (in Apple II and IBM-PC versions) with programs and exercises keyed to the text. The programs allow students to solve meaningful problems without the difficulties of extensive computation. This book bridges the gap between the text and the computer by providing additional explanations and exercises for solution using a microcomputer. Appropriate exercises in *Calculus with Applications* are identified by the heading "For the Computer."

A set of two-color **transparencies** showing charts, figures, and portions of examples is available and can be used to accompany a lecture.

## ACKNOWLEDGMENTS

Many instructors helped in the preparation of this revision. Special thanks go to Louis F. Hoelzle, Bucks County Community College. In addition, substantial contributions were made by Chris Burditt, Napa Valley College; Linda B. Holden, Indiana University; John Hornsby, University of New Orleans; and Thomas B. Muenzenberger, Kansas State University.

The following instructors reviewed the manuscript and made many helpful suggestions for improving this edition: Charles C. Clever, South Dakota State University; Jule M. Connolly, Wake Forest University; B. Jan Davis, University of Southern Mississippi; Joseph B. Dennin, Fairfield University; Charles W. Johnson, Louisiana State University at Shreveport; Samuel C. Jordan, Jackson State University; Thomas J. Miles, Central Michigan University; William C. Ramaley, Fort Lewis College; and James A. Verhanovitz, Delta College.

The following instructors answered a questionnaire and offered their opinions on the third edition of this textbook and on the applied calculus course itself.

Blanche Abramov, Pace University, Pleasantville-Briarcliff Campus
William J. Adams, Pace University, Pace Plaza
Stephen Andrilli, La Salle University
Thomas Arbutiski, Community College of Allegheny County
Richard Armstrong, Saint Louis Community College at Florissant Valley
James Arnold, University of Wisconsin–Milwaukee
Howard B. Beckwith, California State University, Long Beach
Ronald Bensema, Joliet Junior College
Fattaneh Cauley, Pepperdine University
Charles C. Clever, South Dakota State University
Jule M. Connolly, Wake Forest University
Ronald J. Czochor, Glassboro State College
Richard M. Davitt, University of Louisville
Duane E. Deal, Ball State University
Bruce Edwards, University of Florida
Ronald Faulstich, Moraine Valley Community College
Howard Frisinger, Colorado State University
Susan S. Garstka, Moraine Valley Community College
Gerald K. Goff, Oklahoma State University
Warren B. Gordon, City University of New York, Baruch College
Ronald Hatton, Sacramento City College
Linda B. Holden, Indiana University
Donald Jessup, University of Nevada at Reno
Charles W. Johnson, Louisiana State University at Shreveport
David B. Johnson, Diablo Valley College
Robert Kleinberg, University of Texas at Austin
Dennis Lewandowski, Michigan Technological University
Stanley M. Lukawecki, Clemson University
Walter E. Mientka, University of Nebraska–Lincoln
Thomas B. Muenzenberger, Kansas State University
Fredric R. Plachy, Clayton State College
Richard L. Poss, Saint Norbert College
Joyce Riseberg, Montgomery College
Stephen B. Rodi, Austin Community College
Thomas Spradley, American River College
Nathaniel H. Stevens, City University of New York, Lehman College
Philip W. Stoddard, Hartnell College
Ann Thorne, College of Du Page
Karen E. Zak, Washington College

Thanks also to Bill Poole, Linda Youngman, and Janet Tilden, editors at Scott, Foresman, who contributed a great deal to the final book.

# CONTENTS

## 9   PROBABILITY AND CALCULUS   596

## 10   THE TRIGONOMETRIC FUNCTIONS   631

## 11   SEQUENCES AND SERIES   672

# ≡ TABLES  733

# INDEX OF APPLICATIONS

## SOCIAL SCIENCES

# Review of Algebra

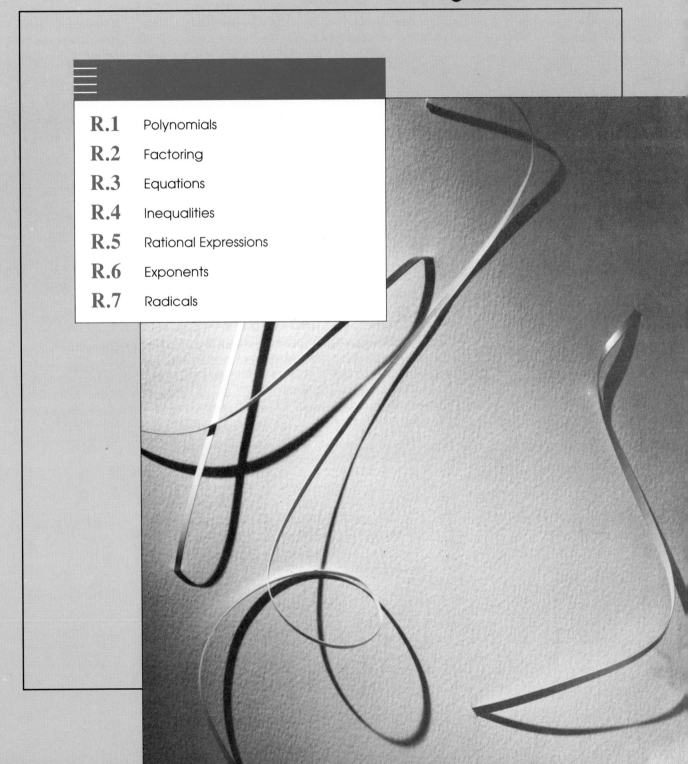

The study of calculus is impossible without a thorough knowledge of elementary algebra. The word *algebra* is derived from the Arabic word *al-jabr,* which appeared in the title of a book by Arab mathematician al-Khowarizmi in the early ninth century A.D. The study of algebra in modern times has grown to include many abstract ideas, but in this text we will be concerned primarily with the topics of elementary algebra found in al-Khowarizmi's book.

This algebra review is designed for self-study; you can study it all at once or use it as a reference when needed throughout the course. Since this is a review, answers to all exercises are given in the answer section at the back of the book.

## R.1  POLYNOMIALS

A **polynomial** is an expression of the form

$$a_n x^n + a_{n-1} x^{n-1} + \cdots + a_1 x + a_0,$$

where $a_0, a_1, a_2, \ldots, a_n$ are real numbers, $n$ is a natural number, and $a_n \neq 0$. Examples of polynomials include

$$5x^4 + 2x^3 + 6x, \qquad 8m^3 + 9m^2 - 6m + 3, \qquad 10p, \qquad \text{and} \qquad -9.$$

An expression such as $9p^4$ is a **term;** the number 9 is the **coefficient,** $p$ is the **variable,** and 4 is the **exponent.** The expression $p^4$ means $p \cdot p \cdot p \cdot p$, while $p^2$ means $p \cdot p$, and so on. Terms having the same variable and the same exponent, such as $9x^4$ and $-3x^4$, are **like terms.** Terms that do not have both the same variable and the same exponent, such as $m^2$ and $m^4$, are **unlike terms.**

Polynomials can be added or subtracted by using the **distributive property,** shown below.

If $a$, $b$, and $c$ are real numbers, then

$$a(b + c) = ab + ac \qquad \text{and} \qquad (b + c)a = ba + ca.$$

Only like terms may be added or subtracted. For example,

$$12y^4 + 6y^4 = (12 + 6)y^4 = 18y^4,$$

and

$$-2m^2 + 8m^2 = (-2 + 8)m^2 = 6m^2,$$

but the polynomial $8y^4 + 2y^5$ cannot be further simplified. To subtract polynomials, use the fact that $-(a + b) = -a - b$. In the next example, we show how to add and subtract polynomials.

■■ EXAMPLE 1

Add or subtract as indicated.

**(a)** $(8x^3 - 4x^2 + 6x) + (3x^3 + 5x^2 - 9x + 8)$

Combine like terms.

$(8x^3 - 4x^2 + 6x) + (3x^3 + 5x^2 - 9x + 8)$
$= (8x^3 + 3x^3) + (-4x^2 + 5x^2) + (6x - 9x) + 8$
$= 11x^3 + x^2 - 3x + 8$

**(b)** $(-4x^4 + 6x^3 - 9x^2 - 12) + (-3x^3 + 8x^2 - 11x + 7)$
$= -4x^4 + 3x^3 - x^2 - 11x - 5$

**(c)** $(2x^2 - 11x + 8) - (7x^2 - 6x + 2)$
$= (2x^2 - 11x + 8) + (-7x^2 + 6x - 2)$
$= -5x^2 - 5x + 6$ ■

The distributive property is used also when multiplying polynomials, as shown in the next example.

■■ EXAMPLE 2

Multiply.

**(a)** $8x(6x - 4)$

$$8x(6x - 4) = 8x(6x) - 8x(4)$$
$$= 48x^2 - 32x$$

**(b)** $(3p - 2)(5p + 1)$

$$(3p - 2)(5p + 1) = (3p - 2)(5p) + (3p - 2)(1)$$
$$= 3p(5p) - 2(5p) + 3p(1) - 2(1)$$
$$= 15p^2 - 10p + 3p - 2$$
$$= 15p^2 - 7p - 2$$ ■

When two binomials are multiplied, as in Example 2(b), the FOIL method (First, Outer, Inner, Last) is used as a shortcut. This method is shown below.

■■ EXAMPLE 3

Find $(2m - 5)(m + 4)$ using the FOIL method.

$$(2m - 5)(m + 4) = \overset{\text{F}}{(2m)(m)} + \overset{\text{O}}{(2m)(4)} + \overset{\text{I}}{(-5)(m)} + \overset{\text{L}}{(-5)(4)}$$
$$= 2m^2 + 8m - 5m - 20$$
$$= 2m^2 + 3m - 20$$ ■

## R.1 EXERCISES

*Perform the indicated operations.*

1. $(2x^2 - 6x + 11) + (-3x^2 + 7x - 2)$
2. $(-4y^2 - 3y + 8) - (2y^2 - 6y - 2)$
3. $-3(4q^2 - 3q + 2) + 2(-q^2 + q - 4)$
4. $2(3r^2 + 4r + 2) - 3(-r^2 + 4r - 5)$
5. $(.613x^2 - 4.215x + .892) - .47(2x^2 - 3x + 5)$
6. $.83(5r^2 - 2r + 7) - (7.12r^2 + 6.423r - 2)$
7. $-9m(2m^2 + 3m - 1)$
8. $(6k - 1)(2k - 3)$
9. $(5r - 3s)(5r + 4s)$
10. $(9k + q)(2k - q)$
11. $\left(\dfrac{2}{5}y + \dfrac{1}{8}z\right)\left(\dfrac{3}{5}y + \dfrac{1}{2}z\right)$
12. $\left(\dfrac{3}{4}r - \dfrac{2}{3}s\right)\left(\dfrac{5}{4}r + \dfrac{1}{3}s\right)$
13. $(.012x - .17)(.3x + .54)$
14. $(6.2m - 3.4)(.7m + 1.3)$
15. $(3p - 1)(9p^2 + 3p + 1)$
16. $(2p - 1)(3p^2 - 4p + 5)$
17. $(2m + 1)(4m^2 - 2m + 1)$
18. $(k + 2)(12k^3 - 3k^2 + k + 1)$
19. $(m - n + k)(m + 2n - 3k)$
20. $(r - 3s + t)(2r - s + t)$

## R.2 FACTORING

Multiplication of polynomials relies on the distributive property. The reverse process, where a polynomial is written as a product of other polynomials, is called **factoring.** For example, one way to factor the number 18 is to write it as the product $9 \cdot 2$. When 18 is written as $9 \cdot 2$, both 9 and 2 are called **factors** of 18. It is true that $18 = 36 \cdot 1/2$, but 36 and 1/2 are not considered factors of 18; only integers are used as factors. The number 18 also can be written with three integer factors as $2 \cdot 3 \cdot 3$. The integer factors of 18 are $\pm 1, \pm 2, \pm 3, \pm 6, \pm 9, \pm 18$.

To factor the algebraic expression $15m + 45$, first note that both $15m$ and 45 can be divided by 15. In fact, $15m = 15 \cdot m$ and $45 = 15 \cdot 3$. Thus, the distributive property can be used to write

$$15m + 45 = \mathbf{15} \cdot m + \mathbf{15} \cdot 3 = \mathbf{15}(m + 3).$$

Both 15 and $m + 3$ are factors of $15m + 45$. Since 15 divides into all terms of $15m + 45$ (and is the largest number that will do so), 15 is the **greatest common factor** for the polynomial $15m + 45$. The process of writing $15m + 45$ as $15(m + 3)$ is called **factoring out** the greatest common factor.

### EXAMPLE 1

Factor out the greatest common factor.

**(a)** $12p - 18q$

Both $12p$ and $18q$ are divisible by 6. Therefore,

$$12p - 18q = \mathbf{6} \cdot 2p - \mathbf{6} \cdot 3q = 6(2p - 3q).$$

**(b)** $8x^3 - 9x^2 + 15x$

Each of these terms is divisible by $x$.

$$8x^3 - 9x^2 + 15x = (8x^2) \cdot x - (9x) \cdot x + 15 \cdot x$$
$$= x(8x^2 - 9x + 15) \quad \text{or} \quad (8x^2 - 9x + 15)x \quad \blacksquare$$

A polynomial that has no greatest common factor (other than 1) may still be factorable. For example, the polynomial $x^2 + 5x + 6$ can be factored as $(x + 2)(x + 3)$. To see that this is correct, find the product $(x + 2)(x + 3)$; you should get $x^2 + 5x + 6$.

There are two different ways to factor a polynomial of three terms such as $x^2 + 5x + 6$, depending on whether the coefficient of $x^2$ is 1 or a number other than 1. If the coefficient is 1, proceed as shown in the following example.

### ▰▰ EXAMPLE 2

Factor $y^2 + 8y + 15$.

Since the coefficient of $y^2$ is 1, factor by finding two numbers whose *product* is 15 and whose *sum* is 8. Use trial and error to find these numbers. Begin by listing all pairs of integers having a product of 15. As you do this, also form the sum of each pair of numbers.

| *Products* | *Sums* |
|:---:|:---:|
| $15 \cdot 1 = 15$ | $15 + 1 = 16$ |
| $\mathbf{5 \cdot 3 = 15}$ | $\mathbf{5 + 3 = 8}$ |
| $(-1) \cdot (-15) = 15$ | $-1 + (-15) = -16$ |
| $(-5) \cdot (-3) = 15$ | $-5 + (-3) = -8$ |

The numbers 5 and 3 have a product of 15 and a sum of 8. Thus, $y^2 + 8y + 15$ factors as

$$y^2 + 8y + 15 = (y + 5)(y + 3).$$

The answer also can be written as $(y + 3)(y + 5)$.  $\blacksquare$

If the coefficient of the squared term is *not* 1, work as shown in the next example.

### ▰▰ EXAMPLE 3

Factor $2x^2 + 9xy - 5y^2$.

The factors of $2x^2$ are $2x$ and $x$; the possible factors of $-5y^2$ are $-5y$ and $y$, or $5y$ and $-y$. Try various combinations of these factors until one works (if, indeed, any work). For example, try the product $(2x + 5y)(x - y)$.

$$(2x + 5y)(x - y) = 2x^2 - 2xy + 5xy - 5y^2$$
$$= 2x^2 + 3xy - 5y^2$$

This product is not correct, so try another combination.

$$(2x - y)(x + 5y) = 2x^2 + 10xy - xy - 5y^2$$
$$= 2x^2 + 9xy - 5y^2$$

Since this combination gives the correct polynomial,

$$2x^2 + 9xy - 5y^2 = (2x - y)(x + 5y). \quad \blacksquare$$

Four special factorizations occur so often that they are listed here for future reference.

**SPECIAL FACTORIZATIONS**

| | |
|---|---|
| $x^2 - y^2 = (x + y)(x - y)$ | **Difference of two squares** |
| $x^2 + 2xy + y^2 = (x + y)^2$ | **Perfect square** |
| $x^3 - y^3 = (x - y)(x^2 + xy + y^2)$ | **Difference of two cubes** |
| $x^3 + y^3 = (x + y)(x^2 - xy + y^2)$ | **Sum of two cubes** |

A polynomial that cannot be factored is called a **prime polynomial.**

**EXAMPLE 4**

Factor each of the following.

**(a)** $64p^2 - 49q^2 = (8p)^2 - (7q)^2 = (8p + 7q)(8p - 7q)$

**(b)** $x^2 + 36$ is a prime polynomial.

**(c)** $x^2 + 12x + 36 = (x + 6)^2$

**(d)** $9y^2 - 24yz + 16z^2 = (3y - 4z)^2$

**(e)** $y^3 - 8 = y^3 - 2^3 = (y - 2)(y^2 + 2y + 4)$

**(f)** $m^3 + 125 = m^3 + 5^3 = (m + 5)(m^2 - 5m + 25)$

**(g)** $8k^3 - 27z^3 = (2k)^3 - (3z)^3 = (2k - 3z)(4k^2 + 6kz + 9z^2)$ $\quad \blacksquare$

## R.2 EXERCISES

*Factor each of the following. If a polynomial cannot be factored, write* prime. *Factor out the greatest common factor as necessary.*

**1.** $8a^3 - 16a^2 + 24a$

**2.** $3y^3 + 24y^2 + 9y$

**3.** $25p^4 - 20p^3q + 100p^2q^2$

**4.** $60m^4 - 120m^3n + 50m^2n^2$

**5.** $m^2 + 9m + 14$

**6.** $x^2 + 4x - 5$

**7.** $z^2 + 9z + 20$

**8.** $b^2 - 8b + 7$

**9.** $a^2 - 6ab + 5b^2$

**10.** $s^2 + 2st - 35t^2$

**11.** $y^2 - 4yz - 21z^2$

**12.** $6a^2 - 48a - 120$

**13.** $3m^3 + 12m^2 + 9m$

**14.** $2x^2 - 5x - 3$

**15.** $3a^2 + 10a + 7$

**16.** $2a^2 - 17a + 30$

**17.** $15y^2 + y - 2$

**18.** $21m^2 + 13mn + 2n^2$

**19.** $24a^4 + 10a^3b - 4a^2b^2$

**20.** $32z^5 - 20z^4a - 12z^3a^2$

**21.** $x^2 - 64$

**22.** $9m^2 - 25$

**23.** $121a^2 - 100$

**24.** $9x^2 + 64$

**25.** $z^2 + 14zy + 49y^2$

**26.** $m^2 - 6mn + 9n^2$

**27.** $9p^2 - 24p + 16$

**28.** $a^3 - 216$

**29.** $8r^3 - 27s^3$

**30.** $64m^3 + 125$

## ▦ R.3 EQUATIONS

Equations that can be written in the form $ax + b = 0$, where $a$ and $b$ are real numbers, with $a \neq 0$, are **linear equations.** Examples of linear equations include $5y + 9 = 16$, $8x = 4$, and $-3p + 5 = -8$. Equations that are *not* linear include absolute value equations such as $|x| = 4$. The following properties are used to solve linear equations.

**PROPERTIES OF REAL NUMBERS**

> For all real numbers $a$, $b$, and $c$:
>
> $a(b + c) = ab + ac$ 　　　　　　　　 **Distributive property**
>
> **If $a = b$, then $a + c = b + c$.** 　　 **Addition property of equality**
> (The same number may be added to both sides of an equation.)
>
> **If $a = b$, then $ac = bc$.** 　　　　　 **Multiplication property of equality**
> (The same number may be multiplied on both sides of an equation.)

The following example shows how these properties are used to solve linear equations. Of course, the solutions should always be checked by substitution in the original equation.

## ▦ EXAMPLE 1

Solve $2x - 5 + 8 = 3x + 2(2 - 3x)$.

| | |
|---|---|
| $2x - 5 + 8 = 3x + 4 - 6x$ | Distributive property |
| $2x + 3 = -3x + 4$ | Combine like terms |
| $5x + 3 = 4$ | Add $3x$ to both sides |
| $5x = 1$ | Add $-3$ to both sides |
| $x = \dfrac{1}{5}$ | Multiply on both sides by $\dfrac{1}{5}$ |

Check in the original equation.  ▦

Although equations involving absolute value are not linear equations, sometimes they may be reduced to linear equations by using the following property.

**EQUATIONS WITH ABSOLUTE VALUE**

If $k \geq 0$, $|ax + b| = k$ is equivalent to

$$ax + b = k \quad \text{or} \quad -(ax + b) = k.$$

**EXAMPLE 2**

Solve each equation.

**(a)** $|2x - 5| = 7$

Rewrite $|2x - 5| = 7$ as follows.

$$
\begin{array}{llll}
2x - 5 = 7 & \quad \text{or} \quad & -(2x - 5) = 7 \\
2x = 12 & & -2x + 5 = 7 \\
x = 6 & & -2x = 2 \\
& & x = -1
\end{array}
$$

Check both answers by substituting them into the original equation.

**(b)** $|3x + 1| = |x - 1|$

Use a variation of the property given above to rewrite the equation.

$$
\begin{array}{llll}
3x + 1 = x - 1 & \quad \text{or} \quad & -(3x + 1) = x - 1 \\
2x = -2 & & -3x - 1 = x - 1 \\
x = -1 & & -4x = 0 \\
& & x = 0
\end{array}
$$

Be sure to check both answers.

An equation with 2 as the highest exponent is a *quadratic equation*. A **quadratic equation** has the form $ax^2 + bx + c = 0$, where $a$, $b$, and $c$ are real numbers and $a \neq 0$. A quadratic equation written in the form $ax^2 + bx + c = 0$ is said to be in **standard form.**

The simplest way to solve a quadratic equation, but one that is not always applicable, is by factoring. This method depends on the **zero-factor property.**

**ZERO-FACTOR PROPERTY**

If $a$ and $b$ are real numbers, with $ab = 0$, then

$$a = 0, \; b = 0, \; \text{or both.}$$

**EXAMPLE 3**

Solve $6r^2 + 7r = 3$.

First write the equation in standard form.

$$6r^2 + 7r - 3 = 0$$

Now factor $6r^2 + 7r - 3$ to get

$$(3r - 1)(2r + 3) = 0.$$

By the zero-factor property, the product $(3r - 1)(2r + 3)$ can equal 0 only if

$$3r - 1 = 0 \qquad \text{or} \qquad 2r + 3 = 0.$$

Solve each of these equations separately to find that the solutions are 1/3 and $-3/2$. Check these solutions by substituting them in the original equation. ■

If a quadratic equation cannot be solved easily by factoring, use the *quadratic formula* (the derivation of the quadratic formula is given in most algebra books).

**QUADRATIC FORMULA**

The solutions of the quadratic equation $ax^2 + bx + c = 0$, where $a \neq 0$, are given by

$$x = \frac{-b \pm \sqrt{b^2 - 4ac}}{2a}.$$

■ **EXAMPLE 4**

Solve $x^2 - 4x - 5 = 0$ by the quadratic formula.

The equation is already in standard form (it has 0 alone on one side of the equals sign), so the values of $a$, $b$, and $c$ from the quadratic formula are easily identified. The coefficient of the squared term gives the value of $a$; here, $a = 1$. Also, $b = -4$ and $c = -5$. (Be careful to use the correct signs.) Substitute these values into the quadratic formula.

$$x = \frac{-(-4) \pm \sqrt{(-4)^2 - 4(1)(-5)}}{2(1)} \qquad \text{Let } a = 1, b = -4, c = -5$$

$$x = \frac{4 \pm \sqrt{16 + 20}}{2} \qquad\qquad (-4)^2 = (-4)(-4) = 16$$

$$x = \frac{4 \pm 6}{2} \qquad\qquad \sqrt{16 + 20} = \sqrt{36} = 6$$

The $\pm$ sign represents the two solutions of the equation. To find all of the solutions, first use $+$ and then use $-$.

$$x = \frac{4 + 6}{2} = \frac{10}{2} = 5 \qquad \text{or} \qquad x = \frac{4 - 6}{2} = \frac{-2}{2} = -1$$

The two solutions are 5 and $-1$. ■

### ■ EXAMPLE 5

Solve $x^2 + 1 = 4x$.

First, add $-4x$ on both sides of the equals sign in order to get 0 alone on the right side.

$$x^2 - 4x + 1 = 0$$

Now identify the letters $a$, $b$, and $c$. Here $a = 1$, $b = -4$, and $c = 1$. Substitute these numbers into the quadratic formula.

$$x = \frac{-(-4) \pm \sqrt{(-4)^2 - 4(1)(1)}}{2(1)}$$

$$= \frac{4 \pm \sqrt{16 - 4}}{2}$$

$$= \frac{4 \pm \sqrt{12}}{2}$$

Simplify the solutions by writing $\sqrt{12}$ as $\sqrt{4 \cdot 3} = \sqrt{4} \cdot \sqrt{3} = 2\sqrt{3}$. Substituting $2\sqrt{3}$ for $\sqrt{12}$ gives

$$x = \frac{4 \pm 2\sqrt{3}}{2}$$

$$= \frac{2(2 \pm \sqrt{3})}{2} \qquad \text{Factor } 4 \pm 2\sqrt{3}$$

$$= 2 \pm \sqrt{3}.$$

The two solutions are $2 + \sqrt{3}$ and $2 - \sqrt{3}$.

The exact values of the solutions are $2 + \sqrt{3}$ and $2 - \sqrt{3}$. Decimal approximations of these solutions (to the nearest thousandth) are

$$2 + \sqrt{3} \approx 2 + 1.732 = 3.732$$

and $\qquad\qquad 2 - \sqrt{3} \approx 2 - 1.732 = .268$ ■

Sometimes the quadratic formula will give a result with a negative number under the radical sign, such as $2 \pm \sqrt{-3}$. A solution of this type is not a real number. Since this text deals only with real numbers, such solutions cannot be used.

When an equation includes fractions, first eliminate all denominators by multiplying both sides of the equation by a **common denominator,** a number that can be divided (with no remainder) by each denominator in the equation. When an equation involves fractions with variable denominators, it is *necessary* to check all solutions in the original equation to be sure that no solution will lead to a zero denominator.

### ▰▰ EXAMPLE 6

Solve each equation.

**(a)** $\dfrac{r}{10} - \dfrac{2}{15} = \dfrac{3r}{20} - \dfrac{1}{5}$

The denominators are 10, 15, 20, and 5. Each of these numbers can be divided into 60, so 60 is a common denominator. Multiply both sides of the equation by 60 and use the distributive property. (If a common denominator cannot be found easily, all the denominators in the problem can be multiplied together to produce one.)

$$\frac{r}{10} - \frac{2}{15} = \frac{3r}{20} - \frac{1}{5}$$

$$60\left(\frac{r}{10} - \frac{2}{15}\right) = 60\left(\frac{3r}{20} - \frac{1}{5}\right)$$

$$60\left(\frac{r}{10}\right) - 60\left(\frac{2}{15}\right) = 60\left(\frac{3r}{20}\right) - 60\left(\frac{1}{5}\right)$$

$$6r - 8 = 9r - 12$$

Add $-6r$ and 12 to both sides.

$$6r - 8 + (-6r) + 12 = 9r - 12 + (-6r) + 12$$

$$4 = 3r$$

Multiply both sides by 1/3 to get

$$r = \frac{4}{3}.$$

**(b)** $\dfrac{3}{x^2} - 12 = 0$

Begin by multiplying both sides of the equation by $x^2$ to get $3 - 12x^2 = 0$. This equation could be solved by using the quadratic formula with $a = -12$, $b = 0$, and $c = 3$. Another method, which works well for the type of quadratic equation in which $b = 0$, is shown below.

$$3 - 12x^2 = 0$$

$$3 = 12x^2 \qquad \text{Add } 12x^2$$

$$\frac{1}{4} = x^2 \qquad \text{Multiply by } \frac{1}{12}$$

$$\pm\frac{1}{2} = x \qquad \text{Take square roots}$$

There are two solutions, $-1/2$ and $1/2$.

(c) $\dfrac{2}{k} - \dfrac{3k}{k + 2} = \dfrac{k}{k^2 + 2k}$

Factor $k^2 + 2k$ as $k(k + 2)$. The common denominator for all the fractions is $k(k + 2)$. Multiplying both sides by $k(k + 2)$ gives the following.

$$2(k + 2) - 3k(k) = k$$
$$2k + 4 - 3k^2 = k$$
$$-3k^2 + k + 4 = 0 \qquad \text{Add } -k; \text{ rearrange terms}$$
$$3k^2 - k - 4 = 0 \qquad \text{Multiply by } -1$$
$$(3k - 4)(k + 1) = 0 \qquad \text{Factor}$$

$$3k - 4 = 0 \qquad \text{or} \qquad k + 1 = 0$$

$$k = \dfrac{4}{3} \qquad\qquad k = -1$$

Verify that the solutions are 4/3 and $-1$.  ▆

## R.3 EXERCISES

*Solve each equation.*

**1.** $.2m - .5 = .1m + .7$

**2.** $\dfrac{5}{6}k - 2k + \dfrac{1}{3} = \dfrac{2}{3}$

**3.** $3r + 2 - 5(r + 1) = 6r + 4$

**4.** $2[m - (4 + 2m) + 3] = 2m + 2$

**5.** $|3x + 2| = 9$

**6.** $|4 - 7x| = 15$

**7.** $|2x + 8| = |x - 4|$

**8.** $|5x + 2| = |8 - 3x|$

*Solve each of the following equations by factoring or by using the quadratic formula. If the solutions involve square roots, give both the exact and the approximate solutions.*

**9.** $(y - 5)(y + 4) = 0$

**10.** $x^2 + 5x + 6 = 0$

**11.** $r^2 - 5r - 6 = 0$

**12.** $a^2 + 5a = 24$

**13.** $x^2 = 3 + 2x$

**14.** $m^2 + 16 = 8m$

**15.** $2k^2 - k = 10$

**16.** $6x^2 - 5x = 4$

**17.** $m(m - 7) = -10$

**18.** $9x^2 - 16 = 0$

**19.** $z(2z + 7) = 4$

**20.** $12y^2 - 48y = 0$

*Use the quadratic formula to solve each equation. If the solutions involve square roots, give both the exact and the approximate solutions.*

**21.** $3x^2 - 5x + 1 = 0$

**22.** $2m^2 = m + 4$

**23.** $p^2 + p - 1 = 0$

**24.** $k^2 - 10k = -20$

**25.** $2x^2 + 12x + 5 = 0$

**26.** $2r^2 - 7r + 5 = 0$

**27.** $6k^2 - 11k + 4 = 0$

**28.** $x^2 + 3x = 10$

**29.** $2x^2 = 3x + 5$

**30.** $2x^2 - 7x + 30 = 0$

**31.** $3k^2 + k = 6$

**32.** $5m^2 + 5m = 0$

*Solve each of the following equations.*

**33.** $\dfrac{3x - 2}{7} = \dfrac{x + 2}{5}$

**34.** $\dfrac{x}{3} - 7 = 6 - \dfrac{3x}{4}$

**35.** $\dfrac{m}{2} - \dfrac{1}{m} = \dfrac{6m + 5}{12}$

**36.** $-\dfrac{3k}{2} + \dfrac{9k - 5}{6} = \dfrac{11k + 8}{k}$

**37.** $\dfrac{4}{x - 3} - \dfrac{8}{2x + 5} + \dfrac{3}{x - 3} = 0$

**38.** $\dfrac{5}{2p + 3} - \dfrac{3}{p - 2} = \dfrac{4}{2p + 3}$

**39.** $\dfrac{2}{m} + \dfrac{m}{m + 3} = \dfrac{3m}{m^2 + 3m}$

**40.** $\dfrac{2y}{y - 1} = \dfrac{5}{y} + \dfrac{10 - 8y}{y^2 - y}$

**41.** $\dfrac{p}{p + 4} + \dfrac{2p - 4}{p^2 + 3p - 4} = \dfrac{2}{p - 1}$

**42.** $\dfrac{1}{x - 2} - \dfrac{3x}{x - 1} = \dfrac{2x + 1}{x^2 - 3x + 2}$

**43.** $\dfrac{5}{a} + \dfrac{-7}{a + 1} = \dfrac{a^2 - 2a + 4}{a^2 + a}$

**44.** $\dfrac{2b^2 + 5b - 8}{b^2 + 2b} + \dfrac{5}{b + 2} = -\dfrac{3}{b}$

---

# R.4  INEQUALITIES

To write that one number is greater than or less than another number, use the following symbols.

**INEQUALITY SYMBOLS**

| | |
|---|---|
| $<$ means *is less* than | $\leq$ means *is less than or equal to* |
| $>$ means *is greater than* | $\geq$ means *is greater than or equal to* |

An equation states that two expressions are equal; an **inequality** states that they are unequal. A **linear inequality** is an inequality that can be simplified to the form $ax < b$. (Properties introduced in this section are given only for $<$, but they are equally valid for $>$, $\leq$, or $\geq$.) Linear equalities are solved with the following properties.

**PROPERTIES OF INEQUALITY**

For all real numbers $a$, $b$, and $c$:

1. If $a < b$, then $a + c < b + c$.
2. If $a < b$ and if $c > 0$, then $ac < bc$.
3. If $a < b$ and if $c < 0$, then $ac > bc$.

Pay careful attention to property 3; it says that if both sides of an inequality are multiplied by a negative number, the direction of the inequality symbol must be reversed.

## EXAMPLE 1
Solve $4 - 3y \leq 7 + 2y$.

Use the properties of inequality.

$$4 - 3y + (-4) \leq 7 + 2y + (-4) \qquad \text{Add } -4 \text{ to both sides}$$
$$-3y \leq 3 + 2y$$

Remember that *adding* the same number to both sides never changes the direction of the inequality symbol.

$$-3y + (-2y) \le 3 + 2y + (-2y) \qquad \text{Add } -2y \text{ to both sides}$$
$$-5y \le 3$$

Multiply both sides by $-1/5$. Since $-1/5$ is negative, change the direction of the inequality symbol.

$$-\frac{1}{5}(-5y) \ge -\frac{1}{5}(3)$$

$$y \ge -\frac{3}{5} \quad \blacksquare$$

The solution $y \ge -3/5$ in Example 1 represents an interval on the number line. **Interval notation** often is used for writing intervals. With interval notation $y \ge -3/5$ is written as $[-3/5, +\infty)$. This is an example of a **half-open interval,** since one endpont, $-3/5$, is included. The **open interval** $(2, 5)$ corresponds to $2 < x < 5$, with neither endpoint included. The **closed interval** $[2, 5]$ includes both endpoints and corresponds to $2 \le x \le 5$.

The **graph** of an interval shows all points on a number line that correspond to the numbers in the interval. To graph the interval $[-3/5, +\infty)$, for example, use a bracket at $-3/5$, since $-3/5$ is part of the solution. To show that the solution includes all real numbers greater than or equal to $-3/5$, draw a heavy arrow pointing to the right (the positive direction). See Figure 1.

**FIGURE 1**

### ▰ EXAMPLE 2

Solve $-2 < 5 + 3m < 20$. Graph the solution.

The inequality $-2 < 5 + 3m < 20$ says that $5 + 3m$ is *between* $-2$ and 20. Solve this inequality with an extension of the properties given above. Work as follows, first adding $-5$ to each part.

$$-2 + (-5) < 5 + 3m + (-5) < 20 + (-5)$$
$$-7 < 3m < 15$$

Now multiply each part by $1/3$.

$$-\frac{7}{3} < m < 5$$

A graph of the solution is given in Figure 2; here parentheses are used to show that $-7/3$ and 5 are *not* part of the graph.   ▰

**FIGURE 2**

A **quadratic inequality** has the form $ax^2 + bx + c > 0$ (or $<$, or $\le$, or $\ge$). The highest exponent is 2. The next few examples show how to solve quadratic inequalities.

### ▬ EXAMPLE 3

Solve the quadratic inequality $x^2 - x - 12 < 0$.

This inequality is solved with values of $x$ that make $x^2 - x - 12$ negative ($< 0$). The quantity $x^2 - x - 12$ changes from positive to negative or from negative to positive at the point where it equals 0. For this reason, first solve the *equation* $x^2 - x - 12 = 0$.

$$x^2 - x - 12 = 0$$
$$(x - 4)(x + 3) = 0$$
$$x = 4 \quad \text{or} \quad x = -3$$

**FIGURE 3**

Locating $-3$ and 4 on a number line, as shown in Figure 3, determines three intervals A, B, and C. Decide which intervals include numbers that make $x^2 - x - 12$ negative by substituting a number from each interval in the polynomial. For example,

choose $-4$ from interval A: $(-4)^2 - (-4) - 12 = 8 > 0$;

choose $\phantom{-}0$ from interval B: $0^2 - 0 - 12 = -12 < 0$;

choose $\phantom{-}5$ from interval C: $5^2 - 5 - 12 = 8 > 0$.

**FIGURE 4**

Only numbers in interval B satisfy the given inequality, so the solution is $(-3, 4)$. A graph of this solution is shown in Figure 4. ▬

### ▬ EXAMPLE 4

Solve the quadratic inequality $r^2 + 3r \geq 4$.

First solve the equation $r^2 + 3r = 4$.

$$r^2 + 3r = 4$$
$$r^2 + 3r - 4 = 0$$
$$(r - 1)(r + 4) = 0$$
$$r = 1 \quad \text{or} \quad r = -4$$

These two solutions determine three intervals on the number line: $(-\infty, -4)$, $(-4, 1)$, and $(1, +\infty)$. Testing a number from each interval and the endpoints of the intervals shows that the solution includes the numbers less than or equal to $-4$, together with the numbers greater than or equal to 1. This solution is written in interval notation as

**FIGURE 5**

$$(-\infty, -4] \cup [1, +\infty).^*$$

A graph of the solution is given in Figure 5. ▬

Inequalities with fractions can be solved in a manner similar to the way quadratic inequalities are solved.

---

*The symbol $\cup$ indicates the *union* of two sets, which includes all elements in either set.

■■ EXAMPLE 5

Solve $\dfrac{2x - 3}{x} \geq 1$.

As with quadratic inequalities, first solve the corresponding equation.

$$\frac{2x - 3}{x} = 1$$

$$2x - 3 = x$$

$$x = 3$$

The solution, $x = 3$, determines the intervals on the number line where the fraction may change from greater than 1 to less than 1. This change also may occur on either side of a number that makes the denominator equal 0. Here, the $x$-value that makes the denominator 0 is

$$x = 0.$$

Test each of the three intervals determined by the numbers 0 and 3.

$$\text{For } (-\infty, 0), \text{ choose } -1: \quad \frac{2(-1) - 3}{-1} = 5 \geq 1.$$

$$\text{For } (0, 3), \quad \text{choose} \quad 1: \quad \frac{2(1) - 3}{1} = -1 \not\geq 1.$$

$$\text{For } (3, +\infty), \text{ choose} \quad 4: \quad \frac{2(4) - 3}{4} = \frac{5}{4} \geq 1.$$

The symbol $\not\geq$ means "is *not* greater than or equal to." Testing the endpoints 0 and 3 shows that the solution is $(-\infty, 0) \cup [3, +\infty)$. ■■

■■ EXAMPLE 6

Solve $\dfrac{x^2 - 3x}{x^2 - 9} < 4$.

Begin by solving the corresponding equation.

$$\frac{x^2 - 3x}{x^2 - 9} = 4$$

$$x^2 - 3x = 4x^2 - 36 \qquad \text{Multiply by } x^2 - 9$$

$$0 = 3x^2 + 3x - 36 \qquad \text{Get 0 on one side}$$

$$0 = x^2 + x - 12 \qquad \text{Multiply by } \frac{1}{3}$$

$$0 = (x + 4)(x - 3) \qquad \text{Factor}$$

$$x = -4 \quad \text{or} \quad x = 3$$

Now set the denominator equal to 0 and solve that equation.

$$x^2 - 9 = 0$$
$$(x - 3)(x + 3) = 0$$
$$x = 3 \quad \text{or} \quad x = -3$$

The intervals determined by the three (different) solutions are $(-\infty, -4)$, $(-4, -3)$, $(-3, 3)$, and $(3, +\infty)$. Testing a number from each interval and the endpoints of the intervals in the given inequality shows that the solution is

$$(-\infty, -4) \cup (-3, 3) \cup (3, +\infty). \quad \blacksquare$$

## R.4 EXERCISES

*Solve each inequality and graph the solution.*

**1.** $-3p - 2 \geq 1$

**2.** $6k - 4 < 3k - 1$

**3.** $m - (4 + 2m) + 3 < 2m + 2$

**4.** $-2(3y - 8) \geq 5(4y - 2)$

**5.** $3p - 1 < 6p + 2(p - 1)$

**6.** $x + 5(x + 1) > 4(2 - x) + x$

**7.** $-7 < y - 2 < 4$

**8.** $8 \leq 3r + 1 \leq 13$

**9.** $-4 \leq \dfrac{2k - 1}{3} \leq 2$

**10.** $-1 \leq \dfrac{5y + 2}{3} \leq 4$

**11.** $\dfrac{3}{5}(2p + 3) \geq \dfrac{1}{10}(5p + 1)$

**12.** $\dfrac{8}{3}(z - 4) \leq \dfrac{2}{9}(3z + 2)$

*Solve each of the following quadratic inequalities. Graph each solution.*

**13.** $(m + 2)(m - 4) < 0$

**14.** $(t + 6)(t - 1) \geq 0$

**15.** $y^2 - 3y + 2 < 0$

**16.** $2k^2 + 7k - 4 > 0$

**17.** $q^2 - 7q + 6 \leq 0$

**18.** $2k^2 - 7k - 15 \leq 0$

**19.** $6m^2 + m > 1$

**20.** $10r^2 + r \leq 2$

**21.** $2y^2 + 5y \leq 3$

**22.** $3a^2 + a > 10$

**23.** $x^2 \leq 25$

**24.** $p^2 - 16p > 0$

*Solve the following inequalities.*

**25.** $\dfrac{m - 3}{m + 5} \leq 0$

**26.** $\dfrac{r + 1}{r - 1} > 0$

**27.** $\dfrac{k - 1}{k + 2} > 1$

**28.** $\dfrac{a - 5}{a + 2} < -1$

**29.** $\dfrac{2y + 3}{y - 5} \leq 1$

**30.** $\dfrac{a + 2}{3 + 2a} \leq 5$

**31.** $\dfrac{7}{k + 2} \geq \dfrac{1}{k + 2}$

**32.** $\dfrac{5}{p + 1} > \dfrac{12}{p + 1}$

**33.** $\dfrac{3x}{x^2 - 1} < 2$

**34.** $\dfrac{8}{p^2 + 2p} > 1$

**35.** $\dfrac{z^2 + z}{z^2 - 1} \geq 3$

**36.** $\dfrac{a^2 + 2a}{a^2 - 4} \leq 2$

## ▥ R.5 RATIONAL EXPRESSIONS

Many algebraic fractions are **rational expressions,** which are quotients of polynomials with nonzero denominators. Examples include

$$\frac{8}{x-1}, \qquad \frac{3x^2 + 4x}{5x - 6}, \qquad \text{and} \qquad \frac{2 + \dfrac{1}{y}}{y} = \frac{2y + 1}{y^2}.$$

Methods for working with rational expressions are summarized below.

**PROPERTIES OF RATIONAL EXPRESSIONS**

▥ For all mathematical expressions $P$, $Q$, $R$, and $S$, with $Q$ and $S \neq 0$:

$$\frac{P}{Q} = \frac{PS}{QS} \qquad\qquad \textbf{Fundamental property}$$

$$\frac{P}{Q} \cdot \frac{R}{S} = \frac{PR}{QS} \qquad\qquad \textbf{Multiplication}$$

$$\frac{P}{Q} + \frac{R}{Q} = \frac{P + R}{Q} \qquad\qquad \textbf{Addition}$$

$$\frac{P}{Q} - \frac{R}{Q} = \frac{P - R}{Q} \qquad\qquad \textbf{Subtraction}$$

$$\frac{P}{Q} \div \frac{R}{S} = \frac{P}{Q} \cdot \frac{S}{R} \quad (R \neq 0) \qquad \textbf{Division}$$

### ▬ EXAMPLE 1

Write each rational expression in lowest terms.

**(a)** $\dfrac{8x + 16}{4} = \dfrac{8(x + 2)}{4} = \dfrac{4 \cdot 2(x + 2)}{4} = 2(x + 2)$

The answer also could be written as $2x + 4$, if desired.

**(b)** $\dfrac{k^2 + 7k + 12}{k^2 + 2k - 3} = \dfrac{(k + 4)(k + 3)}{(k - 1)(k + 3)} = \dfrac{k + 4}{k - 1}$

The answer cannot be further reduced. ▬

### ▬ EXAMPLE 2

Perform each operation.

**(a)** $\dfrac{3y + 9}{6} \cdot \dfrac{18}{5y + 15}$

Factor where possible, then multiply numerators and denominators and reduce to lowest terms.

$$\frac{3y + 9}{6} \cdot \frac{18}{5y + 15} = \frac{3(y + 3)}{6} \cdot \frac{18}{5(y + 3)}$$

$$= \frac{3 \cdot 18(y + 3)}{6 \cdot 5(y + 3)}$$

$$= \frac{3 \cdot 6 \cdot 3(y + 3)}{6 \cdot 5(y + 3)} = \frac{3 \cdot 3}{5} = \frac{9}{5}$$

**(b)** $\dfrac{m^2 + 5m + 6}{m + 3} \cdot \dfrac{m}{m^2 + 3m + 2} = \dfrac{(m + 2)(m + 3)}{m + 3} \cdot \dfrac{m}{(m + 2)(m + 1)}$

$$= \frac{m(m + 2)(m + 3)}{(m + 3)(m + 2)(m + 1)} = \frac{m}{m + 1}$$

**(c)** $\dfrac{9p - 36}{12} \div \dfrac{5(p - 4)}{18}$

$$= \frac{9p - 36}{12} \cdot \frac{18}{5(p - 4)} \qquad \text{Invert and multiply}$$

$$= \frac{9(p - 4)}{12} \cdot \frac{18}{5(p - 4)} = \frac{27}{10}$$

**(d)** $\dfrac{4}{5k} - \dfrac{11}{5k}$

As mentioned above, when two rational expressions have the same denominators, we subtract by subtracting the numerators.

$$\frac{4}{5k} - \frac{11}{5k} = \frac{4 - 11}{5k} = -\frac{7}{5k}$$

**(e)** $\dfrac{7}{p} + \dfrac{9}{2p} + \dfrac{1}{3p}$

These three fractions cannot be added until their denominators are the same. A common denominator into which $p$, $2p$, and $3p$ all divide is $6p$. Use the fundamental property to rewrite each rational expression with a denominator of $6p$.

$$\frac{7}{p} + \frac{9}{2p} + \frac{1}{3p} = \frac{6 \cdot 7}{6 \cdot p} + \frac{3 \cdot 9}{3 \cdot 2p} + \frac{2 \cdot 1}{2 \cdot 3p}$$

$$= \frac{42}{6p} + \frac{27}{6p} + \frac{2}{6p}$$

$$= \frac{42 + 27 + 2}{6p} = \frac{71}{6p} \quad \blacksquare$$

## R.5 EXERCISES

*Write each rational expression in lowest terms.*

**1.** $\dfrac{7z^2}{14z}$  **2.** $\dfrac{25p^3}{10p^2}$  **3.** $\dfrac{8k + 16}{9k + 18}$  **4.** $\dfrac{3(t + 5)}{(t + 5)(t - 3)}$

**5.** $\dfrac{8x^2 + 16x}{4x^2}$  **6.** $\dfrac{36y^2 + 72y}{9y}$  **7.** $\dfrac{m^2 - 4m + 4}{m^2 + m - 6}$  **8.** $\dfrac{r^2 - r - 6}{r^2 + r - 12}$

**9.** $\dfrac{x^2 + 3x - 4}{x^2 - 1}$  **10.** $\dfrac{z^2 - 5z + 6}{z^2 - 4}$  **11.** $\dfrac{8m^2 + 6m - 9}{16m^2 - 9}$  **12.** $\dfrac{6y^2 + 11y + 4}{3y^2 + 7y + 4}$

*Perform the indicated operations.*

**13.** $\dfrac{9k^2}{25} \cdot \dfrac{5}{3k}$  **14.** $\dfrac{15p^3}{9p^2} \div \dfrac{6p}{10p^2}$  **15.** $\dfrac{a + b}{2p} \cdot \dfrac{12}{5(a + b)}$

**16.** $\dfrac{a - 3}{16} \div \dfrac{a - 3}{32}$  **17.** $\dfrac{2k + 8}{6} \div \dfrac{3k + 1\,2}{2}$  **18.** $\dfrac{9y - 18}{6y + 12} \cdot \dfrac{3y + 6}{15y - 30}$

**19.** $\dfrac{4a + 12}{2a - 10} \div \dfrac{a^2 - 9}{a^2 - a - 20}$  **20.** $\dfrac{6r - 18}{9r^2 + 6r - 24} \cdot \dfrac{12r - 16}{4r - 12}$

**21.** $\dfrac{k^2 - k - 6}{k^2 + k - 12} \cdot \dfrac{k^2 + 3k - 4}{k^2 + 2k - 3}$  **22.** $\dfrac{m^2 + 3m + 2}{m^2 + 5m + 4} \div \dfrac{m^2 + 5m + 6}{m^2 + 10m + 24}$

**23.** $\dfrac{2m^2 - 5m - 12}{m^2 - 10m + 24} \div \dfrac{4m^2 - 9}{m^2 - 9m + 18}$  **24.** $\dfrac{6n^2 - 5n - 6}{6n^2 + 5n - 6} \cdot \dfrac{12n^2 - 17n + 6}{12n^2 - n - 6}$

**25.** $\dfrac{a + 1}{2} - \dfrac{a - 1}{2}$  **26.** $\dfrac{3}{p} + \dfrac{1}{2}$  **27.** $\dfrac{2}{y} - \dfrac{1}{4}$

**28.** $\dfrac{1}{6m} + \dfrac{2}{5m} + \dfrac{4}{m}$  **29.** $\dfrac{1}{m - 1} + \dfrac{2}{m}$  **30.** $\dfrac{6}{r} - \dfrac{5}{r - 2}$

**31.** $\dfrac{8}{3(a - 1)} + \dfrac{2}{a - 1}$  **32.** $\dfrac{2}{5(k - 2)} + \dfrac{3}{4(k - 2)}$  **33.** $\dfrac{2}{x^2 - 2x - 3} + \dfrac{5}{x^2 - x - 6}$

**34.** $\dfrac{2y}{y^2 + 7y + 12} - \dfrac{y}{y^2 + 5y + 6}$  **35.** $\dfrac{3k}{2k^2 + 3k - 2} - \dfrac{2k}{2k^2 - 7k + 3}$  **36.** $\dfrac{4m}{3m^2 + 7m - 6} - \dfrac{m}{3m^2 - 14m + 8}$

## R.6 EXPONENTS

Recall that $a^2 = a \cdot a$, while $a^3 = a \cdot a \cdot a$, and so on. In this section a more general meaning is given to the symbol $a^n$.

**DEFINITION OF EXPONENT**

If $n$ is a natural number, then
$$a^n = a \cdot a \cdot a \ldots a,$$
where $a$ appears as a factor $n$ times.

In the expression $a^n$, $n$ is the **exponent** and $a$ is the **base.** This definition can be extended by defining $a^n$ for zero and negative integer values of $n$.

**ZERO AND NEGATIVE EXPONENTS**

If $a$ is any nonzero real number, and if $n$ is a positive integer, then

$$a^0 = 1 \quad \text{and} \quad a^{-n} = \frac{1}{a^n}.$$

(The symbol $0^0$ is meaningless.)

### EXAMPLE 1

(a) $6^0 = 1$

(b) $(-9)^0 = 1$

(c) $3^{-2} = \dfrac{1}{3^2} = \dfrac{1}{9}$

(d) $9^{-1} = \dfrac{1}{9^1} = \dfrac{1}{9}$

(e) $\left(\dfrac{3}{4}\right)^{-1} = \dfrac{1}{(3/4)^1} = \dfrac{1}{3/4} = \dfrac{4}{3}$

The following properties rely on the definitions of exponents given above.

**PROPERTIES OF EXPONENTS**

For any integers $m$ and $n$, and any real numbers $a$ and $b$ for which the following exist,

1. $a^m \cdot a^n = a^{m+n}$;

2. $\dfrac{a^m}{a^n} = a^{m-n}$;

3. $(a^m)^n = a^{mn}$;

4. $(ab)^m = a^m \cdot b^m$;

5. $\left(\dfrac{a}{b}\right)^m = \dfrac{a^m}{b^m}$.

### EXAMPLE 2

Use the properties of exponents to simplify each of the following. Leave answers with positive exponents. Assume that all variables represent positive real numbers.

(a) $7^4 \cdot 7^6 = 7^{4+6} = 7^{10}$      Property 1

(b) $\dfrac{9^{14}}{9^6} = 9^{14-6} = 9^8$      Property 2

(c) $\dfrac{r^9}{r^{17}} = r^{9-17} = r^{-8} = \dfrac{1}{r^8}$      Property 2

(d) $(2m^3)^4 = 2^4 \cdot (m^3)^4 = 16m^{12}$      Properties 3 and 4

**(e)** $(3x)^4 = 3^4 \cdot x^4$                 Property 4

**(f)** $\left(\dfrac{9}{7}\right)^6 = \dfrac{9^6}{7^6}$                 Property 5

**(g)** $\dfrac{2^{-3} \cdot 2^5}{2^4 \cdot 2^{-7}} = \dfrac{2^2}{2^{-3}} = 2^{2-(-3)} = 2^5$

**(h)** $2^{-1} + 3^{-1} = \dfrac{1}{2} + \dfrac{1}{3} = \dfrac{5}{6}$ ■

For *even* values of *n,* the expression $a^{1/n}$ is defined to be the **positive *n*th root** of *a*. For example, $a^{1/2}$ denotes the positive second root, or **square root,** of *a*, while $a^{1/4}$ is the positive fourth root of *a*. When *n* is *odd*, there is only one *n*th root, which has the same sign as *a*. For example, $a^{1/3}$, the **cube root** of *a*, has the same sign as *a*. By definition, if $b = a^{1/n}$, then $b^n = a$.

### ■ EXAMPLE 3

(A calculator will be helpful here.)

**(a)** $121^{1/2} = 11$, since 11 is positive and $11^2 = 121$.

**(b)** $625^{1/4} = 5$, since $5^4 = 625$.

**(c)** $256^{1/4} = 4$

**(d)** $64^{1/6} = 2$

**(e)** $27^{1/3} = 3$

**(f)** $(-32)^{1/5} = -2$

**(g)** $128^{1/7} = 2$

**(h)** $(-49)^{1/2}$ is not a real number. ■

In the following definition, the domain of an exponent is extended to include all rational numbers.

**DEFINITION OF $a^{m/n}$**

■ For all real numbers *a* for which the indicated roots exist, and for any rational number *m/n,*

$$a^{m/n} = (a^{1/n})^m.$$

### ■ EXAMPLE 4

**(a)** $27^{2/3} = (27^{1/3})^2 = 3^2 = 9$

**(b)** $32^{2/5} = (32^{1/5})^2 = 2^2 = 4$

**(c)** $64^{4/3} = (64^{1/3})^4 = 4^4 = 256$

**(d)** $25^{3/2} = (25^{1/2})^3 = 5^3 = 125$ ▬

All the properties for integer exponents given in this section also apply to any rational exponent on a nonnegative real-number base.

## ▬ EXAMPLE 5

**(a)** $\dfrac{27^{1/3} \cdot 27^{5/3}}{27^3} = \dfrac{27^{1/3+5/3}}{27^3} = \dfrac{27^2}{27^3} = 27^{2-3} = 27^{-1} = \dfrac{1}{27}$

**(b)** $m^{2/3}(m^{7/3} + 2m^{1/3}) = m^{2/3+7/3} + 2m^{2/3+1/3} = m^3 + 2m$ ▬

In calculus, it is often necessary to factor expressions involving fractional exponents, as shown in the next example.

## ▬ EXAMPLE 6

Factor $(x^2 + 5)(3x - 1)^{-1/2}(2) + (3x - 1)^{1/2}(2x)$.

There is a common factor of 2. Also, $(3x - 1)^{-1/2}$ and $(3x - 1)^{1/2}$ have a common factor. Always factor out the quantity to the *smallest* exponent. Here $-1/2 < 1/2$, so the common factor is $2(3x - 1)^{-1/2}$ and the factored form is

$$2(3x - 1)^{-1/2}[(x^2 + 5) + (3x - 1)x] = 2(3x - 1)^{-1/2}(4x^2 - x + 5).$$ ▬

## ▬ R.6 EXERCISES

*Evaluate each expression. Write all answers without exponents.*

**1.** $8^{-2}$

**2.** $3^{-4}$

**3.** $6^{-3}$

**4.** $5^0$

**5.** $(-12)^0$

**6.** $2^{-1} + 4^{-1}$

**7.** $-2^{-4}$

**8.** $(-2)^{-4}$

**9.** $-(-3)^{-2}$

**10.** $-(-3^{-2})$

**11.** $\left(\dfrac{5}{8}\right)^2$

**12.** $\left(\dfrac{6}{7}\right)^3$

**13.** $\left(\dfrac{1}{2}\right)^{-3}$

**14.** $\left(\dfrac{1}{5}\right)^{-3}$

**15.** $\left(\dfrac{2}{7}\right)^{-2}$

**16.** $\left(\dfrac{4}{3}\right)^{-3}$

*Simplify each expression. Assume that all variables represent positive real numbers. Write answers with only positive exponents.*

**17.** $\dfrac{7^5}{7^9}$

**18.** $\dfrac{3^{-4}}{3^2}$

**19.** $\dfrac{2^{-5}}{2^{-2}}$

**20.** $\dfrac{6^{-1}}{6}$

**21.** $4^{-3} \cdot 4^6$

**22.** $\dfrac{8^9 \cdot 8^{-7}}{8^{-3}}$

**23.** $\dfrac{10^8 \cdot 10^{-10}}{10^4 \cdot 10^2}$

**24.** $\left(\dfrac{5^{-6} \cdot 5^3}{5^{-2}}\right)^{-1}$

**25.** $\dfrac{x^4 \cdot x^3}{x^5}$

**26.** $\dfrac{y^9 \cdot y^7}{y^{13}}$

**27.** $\dfrac{(4k^{-1})^2}{2k^{-5}}$

**28.** $\dfrac{(3z^2)^{-1}}{z^5}$

**29.** $\dfrac{2^{-1}x^3y^{-3}}{xy^{-2}}$

**30.** $\dfrac{5^{-2}m^2y^{-2}}{5^2m^{-1}y^{-2}}$

**31.** $\left(\dfrac{a^{-1}}{b^2}\right)^{-3}$

**32.** $\left(\dfrac{2c^2}{d^3}\right)^{-2}$

*Evaluate each expression, assuming that $a = 2$ and $b = -3$.*

**33.** $a^{-1} + b^{-1}$

**34.** $b^{-2} - a$

**35.** $\dfrac{2b^{-1} - 3a^{-1}}{a + b^2}$

**36.** $\dfrac{3a^2 - b^2}{b^{-3} + 2a^{-1}}$

**37.** $\left(\dfrac{a}{3}\right)^{-1} + \left(\dfrac{b}{2}\right)^{-2}$

**38.** $\left(\dfrac{2b}{5}\right)^2 - 3\left(\dfrac{a^{-1}}{4}\right)$

*Write each number without exponents.*

**39.** $81^{1/2}$

**40.** $27^{1/3}$

**41.** $8^{2/3}$

**42.** $1000^{2/3}$

**43.** $32^{2/5}$

**44.** $-125^{2/3}$

**45.** $\left(\dfrac{4}{9}\right)^{1/2}$

**46.** $\left(\dfrac{64}{27}\right)^{1/3}$

**47.** $16^{-5/4}$

**48.** $625^{-1/4}$

**49.** $\left(\dfrac{27}{64}\right)^{-1/3}$

**50.** $\left(\dfrac{121}{100}\right)^{-3/2}$

*Simplify each expression. Write all answers with only positive exponents. Assume that all variables represent positive real numbers.*

**51.** $2^{1/2} \cdot 2^{3/2}$

**52.** $27^{2/3} \cdot 27^{-1/3}$

**53.** $\dfrac{4^{2/3} \cdot 4^{5/3}}{4^{1/3}}$

**54.** $\dfrac{3^{-5/2} \cdot 3^{3/2}}{3^{7/2} \cdot 3^{-9/2}}$

**55.** $\dfrac{7^{-1/3} \cdot 7r^{-3}}{7^{2/3} \cdot (r^{-2})^2}$

**56.** $\dfrac{12^{3/4} \cdot 12^{5/4} \cdot y^{-2}}{12^{-1} \cdot (y^{-3})^{-2}}$

**57.** $\dfrac{6k^{-4} \cdot (3k^{-1})^{-2}}{2^3 \cdot k^{1/2}}$

**58.** $\dfrac{8p^{-3} \cdot (4p^2)^{-2}}{p^{-5}}$

**59.** $\dfrac{a^{4/3} \cdot b^{1/2}}{a^{2/3} \cdot b^{-3/2}}$

**60.** $\dfrac{x^{1/3} \cdot y^{2/3} \cdot z^{1/4}}{x^{5/3} \cdot y^{-1/3} \cdot z^{3/4}}$

**61.** $\dfrac{k^{-3/5} \cdot h^{-1/3} \cdot t^{2/5}}{k^{-1/5} \cdot h^{-2/3} \cdot t^{1/5}}$

**62.** $\dfrac{m^{7/3} \cdot n^{-2/5} \cdot p^{3/8}}{m^{-2/3} \cdot n^{3/5} \cdot p^{-5/8}}$

**63.** $\dfrac{k^{3/2} \cdot k^{-1/2}}{k^{1/4} \cdot k^{3/4}}$

**64.** $\dfrac{m^{2/5} \cdot m^{3/5} \cdot m^{-4/5}}{m^{1/5} \cdot m^{-6/5}}$

**65.** $\dfrac{x^{-2/3} \cdot x^{4/3}}{x^{1/2} \cdot x^{-3/4}}$

**66.** $\dfrac{-4a^{1/2} \cdot a^{2/3}}{a^{-5/6}}$

**67.** $\dfrac{8y^{2/3}y^{-1}}{2^{-1}y^{3/4} \cdot y^{-1/6}}$

**68.** $\dfrac{9 \cdot k^{1/3} \cdot k^{-1/2} \cdot k^{-1/6}}{k^{-2/3}}$

*Factor each expression.*

**69.** $(x^2 + 2)(x^2 - 1)^{-1/2}(x) + (x^2 - 1)^{1/2}(2x)$

**70.** $(3x - 1)(5x + 2)^{1/2}(15) + (5x + 2)^{1/2}(5)$

**71.** $(2x + 5)^2\left(\dfrac{1}{2}\right)(x^2 - 4)^{-1/2}(2x) + (x^2 - 4)^{1/2}(2)(2x + 5)$

**72.** $(4x^2 + 1)^2(2x - 1)^{-1/2} + (2x - 1)^{1/2}(2)(4x^2 + 1)$

# R.7  RADICALS

We have defined $a^{1/n}$ as the positive $n$th root of $a$ for appropriate values of $a$ and $n$. An alternate notation for $a^{1/n}$ uses radicals.

**RADICALS**

> If $n$ is an even natural number and $a > 0$, or $n$ is an odd natural number, then
> $$a^{1/n} = \sqrt[n]{a}.$$

The symbol $\sqrt[n]{\phantom{a}}$ is a **radical sign,** the number $a$ is the **radicand,** and $n$ is the **index** of the radical. The familiar symbol $\sqrt{a}$ is used instead of $\sqrt[2]{a}$.

## EXAMPLE 1

(a)  $\sqrt[4]{16} = 16^{1/4} = 2$

(b)  $\sqrt[5]{-32} = -2$

(c)  $\sqrt[3]{1000} = 10$

(d)  $\sqrt[6]{\dfrac{64}{729}} = \dfrac{2}{3}$

With $a^{1/n}$ written as $\sqrt[n]{a}$, $a^{m/n}$ also can be written using radicals.

$$a^{m/n} = (\sqrt[n]{a})^m \quad \text{or} \quad a^{m/n} = \sqrt[n]{a^m}$$

The following properties of radicals depend on the definitions and properties of exponents.

**PROPERTIES OF RADICALS**

> For all real numbers $a$ and $b$ and natural numbers $m$ and $n$ such that $\sqrt[n]{a}$ and $\sqrt[n]{b}$ are real numbers:
>
> 1. $\left(\sqrt[n]{a}\right)^n = a$
> 2. $\sqrt[n]{a^n} = \begin{cases} |a| & \text{if } n \text{ is even} \\ a & \text{if } n \text{ is odd} \end{cases}$
> 3. $\sqrt[n]{a} \cdot \sqrt[n]{b} = \sqrt[n]{ab}$
> 4. $\dfrac{\sqrt[n]{a}}{\sqrt[n]{b}} = \sqrt[n]{\dfrac{a}{b}} \quad (b \neq 0)$
> 5. $\sqrt[m]{\sqrt[n]{a}} = \sqrt[mn]{a}$

Property 3 can be used to simplify certain radicals. For example, since $48 = 16 \cdot 3$,

$$\sqrt{48} = \sqrt{16 \cdot 3} = \sqrt{16} \cdot \sqrt{3} = 4\sqrt{3}.$$

## ▰ EXAMPLE 2

**(a)** $\sqrt{1000} = \sqrt{100 \cdot 10} = \sqrt{100} \cdot \sqrt{10} = 10\sqrt{10}$

**(b)** $\sqrt{128} = \sqrt{64 \cdot 2} = 8\sqrt{2}$

**(c)** $\sqrt{108} = 6\sqrt{3}$

**(d)** $\sqrt[3]{54} = \sqrt[3]{27 \cdot 2} = \sqrt[3]{27} \cdot \sqrt[3]{2} = 3\sqrt[3]{2}$

**(e)** $\sqrt{288m^5} = \sqrt{144 \cdot m^4 \cdot 2m} = 12m^2\sqrt{2m}$

**(f)** $2\sqrt{18} - 5\sqrt{32} = 2\sqrt{9 \cdot 2} - 5\sqrt{16 \cdot 2}$

$$= 2\sqrt{9} \cdot \sqrt{2} - 5\sqrt{16} \cdot \sqrt{2}$$

$$= 2(3)\sqrt{2} - 5(4)\sqrt{2} = -14\sqrt{2} \quad ▰$$

The next example shows how to *rationalize* (remove all radicals from) the denominator in an expression containing radicals.

## ▰ EXAMPLE 3

Simplify each of the following expressions by rationalizing the denominator.

**(a)** $\dfrac{4}{\sqrt{3}}$

To rationalize the denominator, multiply by $\sqrt{3}/\sqrt{3}$ (or 1) so that the denominator of the product is a rational number.

$$\frac{4}{\sqrt{3}} \cdot \frac{\sqrt{3}}{\sqrt{3}} = \frac{4\sqrt{3}}{3}$$

**(b)** $\dfrac{1}{1 - \sqrt{2}}$

The best approach here is to multiply both numerator and denominator by the number $1 + \sqrt{2}$. The expressions $1 + \sqrt{2}$ and $1 - \sqrt{2}$ are conjugates.* Doing so gives

$$\frac{1}{1 - \sqrt{2}} = \frac{1(1 + \sqrt{2})}{(1 - \sqrt{2})(1 + \sqrt{2})} = \frac{1 + \sqrt{2}}{1 - 2} = -1 - \sqrt{2}. \quad ▰$$

Sometimes it is advantageous to rationalize the *numerator* of a rational expression.

---

*If $a$ and $b$ are real numbers, the **conjugate** of $a + b$ is $a - b$.

### ▬ EXAMPLE 4

Rationalize the numerator of $\dfrac{\sqrt{2} - \sqrt{3}}{5}$.

Multiply numerator and denominator by the conjugate of the numerator, $\sqrt{2} + \sqrt{3}$.

$$\frac{\sqrt{2} - \sqrt{3}}{5} \cdot \frac{\sqrt{2} + \sqrt{3}}{\sqrt{2} + \sqrt{3}} = \frac{2 - 3}{5(\sqrt{2} + \sqrt{3})} = \frac{-1}{5(\sqrt{2} + \sqrt{3})} \quad ▬$$

## ▬ R.7 EXERCISES

*Simplify each expression. Assume that all variables represent positive real numbers.*

**1.** $\sqrt[3]{125}$

**2.** $\sqrt[4]{1296}$

**3.** $\sqrt[5]{-3125}$

**4.** $\sqrt{50}$

**5.** $\sqrt{2000}$

**6.** $\sqrt{32y^5}$

**7.** $7\sqrt{2} - 8\sqrt{18} + 4\sqrt{72}$

**8.** $4\sqrt{3} - 5\sqrt{12} + 3\sqrt{75}$

**9.** $2\sqrt{5} - 3\sqrt{20} + 2\sqrt{45}$

**10.** $\sqrt{50} - 8\sqrt{8} + 4\sqrt{18}$

**11.** $6\sqrt{27} - 3\sqrt{12} + 5\sqrt{48}$

**12.** $3\sqrt{28} - 4\sqrt{63} + \sqrt{112}$

**13.** $3\sqrt[3]{16} - 4\sqrt[3]{2}$

**14.** $\sqrt[3]{2} - \sqrt[3]{16} + 2\sqrt[3]{54}$

**15.** $2\sqrt[3]{3} + 4\sqrt[3]{24} - \sqrt[3]{81}$

**16.** $\sqrt[3]{32} - 5\sqrt[3]{4} + 2\sqrt[3]{108}$

**17.** $\sqrt{2x^3y^2z^4}$

**18.** $\sqrt{98r^3s^4t^{10}}$

**19.** $\sqrt[3]{16z^5x^8y^4}$

**20.** $\sqrt[4]{x^8y^7z^{11}}$

**21.** $\sqrt{a^3b^5} - 2\sqrt{a^7b^3} + \sqrt{a^3b^9}$

**22.** $\sqrt{p^7q^3} - \sqrt{p^5q^9} + \sqrt{p^9q}$

*Rationalize each denominator. Assume that all radicands represent positive real numbers.*

**23.** $\dfrac{5}{\sqrt{7}}$

**24.** $\dfrac{-2}{\sqrt{3}}$

**25.** $\dfrac{-3}{\sqrt{12}}$

**26.** $\dfrac{4}{\sqrt{8}}$

**27.** $\dfrac{3}{1 - \sqrt{5}}$

**28.** $\dfrac{5}{2 - \sqrt{6}}$

**29.** $\dfrac{-2}{\sqrt{3} - \sqrt{2}}$

**30.** $\dfrac{1}{\sqrt{10} + \sqrt{3}}$

**31.** $\dfrac{1}{\sqrt{r} - \sqrt{3}}$

**32.** $\dfrac{5}{\sqrt{m} - \sqrt{5}}$

**33.** $\dfrac{y - 5}{\sqrt{y} - \sqrt{5}}$

**34.** $\dfrac{z - 11}{\sqrt{z} - \sqrt{11}}$

**35.** $\dfrac{\sqrt{x} + \sqrt{x + 1}}{\sqrt{x} - \sqrt{x + 1}}$

**36.** $\dfrac{\sqrt{p} + \sqrt{p^2 - 1}}{\sqrt{p} - \sqrt{p^2 - 1}}$

*Rationalize each numerator. Assume that all radicands represent positive real numbers.*

**37.** $\dfrac{1 + \sqrt{2}}{2}$

**38.** $\dfrac{1 - \sqrt{3}}{3}$

**39.** $\dfrac{\sqrt{x}}{1 + \sqrt{x}}$

**40.** $\dfrac{\sqrt{p}}{1 - \sqrt{p}}$

**41.** $\dfrac{\sqrt{x} + \sqrt{x + 1}}{\sqrt{x} - \sqrt{x + 1}}$

**42.** $\dfrac{\sqrt{p} + \sqrt{p^2 - 1}}{\sqrt{p} - \sqrt{p^2 - 1}}$

# 1 Functions and Graphs

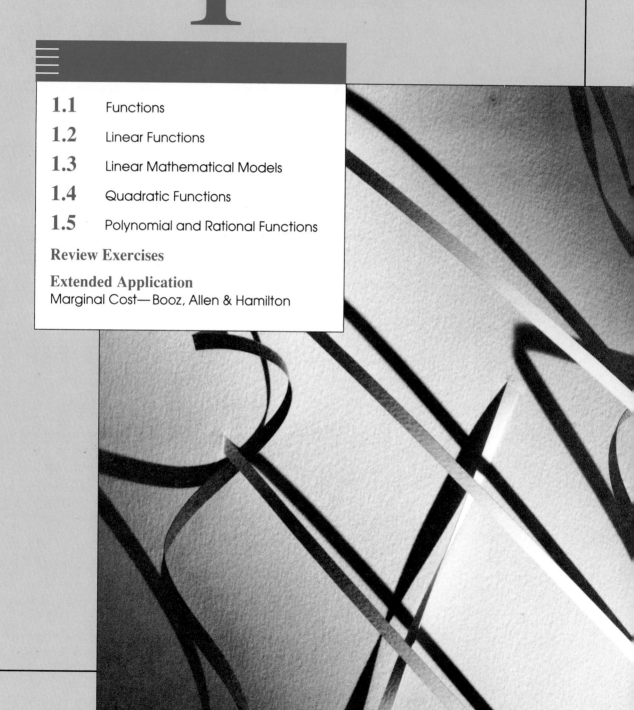

Figure 1(a) shows the speed in miles per hour of a Porsche 928 over elapsed time in seconds as the car accelerates from rest.* For example, the figure shows that 15 seconds after starting, the car is going 90 miles per hour. Figure 1(b) shows the variation in blood pressure for a typical person.† (Systolic and diastolic pressures are the upper and lower limits in the periodic changes in pressure that produce the pulse. The length of time between peaks is called the period of the pulse.) After .8 seconds, the blood pressure is the same as its starting value, 80 millimeters. Figures 1(a) and 1(b) illustrate *functions,* which are rules or procedures that yield just one value of one variable from any given value of another variable.

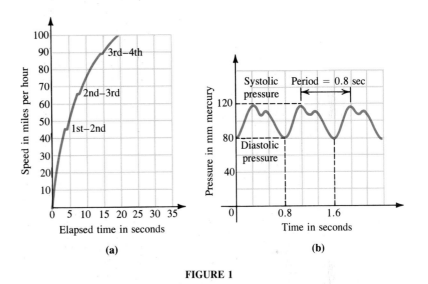

**FIGURE 1**

Functions are useful in describing many situations involving two variables because each value of one variable corresponds to only one value of the other variable. We can find the exact speed of the Porsche after 10 seconds because the car cannot have two different speeds at the same moment. Similarly, a person can have only one blood pressure at a given instant. A function used in this way, as a mathematical description of a real-world situation, is called a **mathematical model.** Constructing a mathematical model requires a solid understanding of the situation to be modeled, along with a good knowledge of the possible mathematical ideas that can be used to construct the model. In this chapter we discuss a variety of functions useful in modeling.

---

*From *Road & Track,* April and May 1978. Reprinted with permission.

†From *Calculus for the Life Sciences* by Rodolfo De Sapio. Copyright © 1976, 1978 by W. H. Freeman and Company. Reprinted by permission.

## ≡ 1.1  FUNCTIONS

The cost of leasing an office in an office building depends on the number of square feet of space in the office. Typical associations between the area of an office in square feet and the monthly rent in dollars are shown in Figure 2.

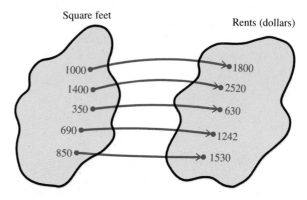

**FIGURE 2**

Another association might be set up between investments in a mutual fund and the corresponding earnings, assuming an annual return of 12%. Typical associations are shown in Figure 3.

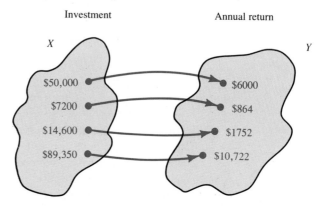

**FIGURE 3**

In this second example, we could use a formula to show how the numbers in set $X$ are used to obtain the numbers in set $Y$. If $x$ is a dollar amount from set $X$, then the corresponding annual return $y$ in dollars from set $Y$ can be found with the formula

$$y = .12 \times x \qquad \text{or} \qquad y = .12x.$$

In this example, $x$, the investment in the mutual fund, is called the **independent variable.** Because the annual return *depends* on the amount of the investment, $y$ is called the **dependent variable.** When a specific number, say 2000, is substituted for $x$, then $y$ takes on one specific value—in this case, $.12 \times 2000 = 240$. The variable $y$ is said to be a function of $x$.

This pair of numbers, one for $x$ and one for $y$, can be written as an **ordered pair** $(x, y)$, where the order of the numbers is important. Using ordered pairs, the information shown in Figure 3 can be expressed as the set

$$\{(50{,}000, 6000), (7200, 864), (14{,}600, 1752), (89{,}350, 10{,}722)\}.$$

Alternatively, the ordered pairs can be given in a table, as shown in Figure 4.

The set of points in a plane that correspond to the ordered pairs of a function is called the **graph** of the function. The function defined by $y = .12x$ that contains the ordered pairs shown in the table also can be illustrated with a graph, as shown in Figure 4.

| $x$ | $y$ |
|---|---|
| 7,200 | 864 |
| 14,600 | 1,752 |
| 50,000 | 6,000 |
| 89,350 | 10,722 |

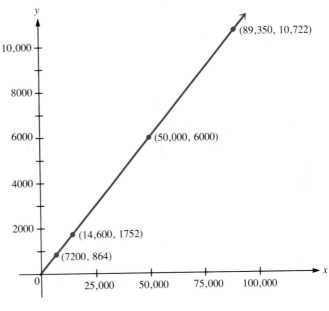

**FIGURE 4**

The following formal definition of a function summarizes the discussion above.

**DEFINITION OF FUNCTION**

A **function** is a rule that assigns to each element of a set $X$ exactly one element from a set $Y$.

### ▬ EXAMPLE 1

Which of the following are functions?

**(a)**

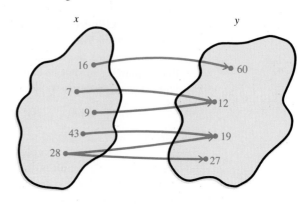

**FIGURE 5**

Figure 5 shows that an *x*-value of 28 corresponds to *two* *y*-values, 19 and 27. In a function, each *x* must correspond to exactly one *y*, so this correspondence is not a function.

**(b)** The optical reader at the checkout counter in many stores that converts codes to prices

For each code, the reader produces exactly one price, so this is a function.

**(c)** The $x^2$ key on a calculator

This correspondence between input and output is a function because the calculator produces just one $x^2$ value for each *x* value entered.

**(d)**

| $x$ | 1 | 1 | 2 | 2 | 3 | 3 |
|---|---|---|---|---|---|---|
| $y$ | 3 | $-3$ | 5 | $-5$ | 8 | $-8$ |

Since at least one *x*-value corresponds to more than one *y*-value, this table does not define a function.

**(e)** The set of ordered pairs with first elements mothers and second elements their children

Here the mother is the independent variable and the child is the dependent variable. For a given mother, there may be several children, so this correspondence is not a function.

**(f)** The set of ordered pairs with first elements children and second elements their birth mothers

In this case the child is the independent variable and the mother is the dependent variable. Since each child has only one birth mother, this is a function. ▬

As shown in Example 1, there are many ways to define functions. Almost every function we will use in this book will be defined by an equation, such as the equation $y = .12x$ discussed earlier in this section.

As we have seen, a function is a correspondence between the elements of two sets. These sets are given special names.

**DOMAIN AND RANGE**

> ▰ The set of all possible values for the independent variable in a function is called the **domain** of the function; the set of all possible values for the dependent variable is the **range.**

The domain and range may or may not be the same set. For example, the domain of the function in Figure 2 is a set of positive real numbers, and the range is a set of decimals representing dollars and cents. For the function in Figure 3, both the domain and range are a set of positive decimal numbers.

▰ EXAMPLE 2

Decide whether each of the following equations represents a function. (Assume that $x$ represents the independent variable here, an assumption we shall make throughout this book.) Give the domain and range of any functions.

**(a)** $y = -4x + 11$

For a given value of $x$, calculating $-4x + 11$ produces exactly one value of $y$. (For example, if $x = -7$, then $y = -4(-7) + 11 = 39$.) Since one value of the independent variable leads to exactly one value of the dependent variable, $y = -4x + 11$ is a function. Both $x$ and $y$ may take on any real-number values, so both the domain and range are the set of all real numbers.

**(b)** $y^2 = x$

Suppose $x = 36$. Then $y^2 = x$ becomes $y^2 = 36$, from which $y = 6$ or $y = -6$. Since one value of the independent variable can lead to two values of the dependent variable, $y^2 = x$ does not represent a function. ▰

The following *agreement on domains* is customary.

**AGREEMENT ON DOMAINS**

> ▰ Unless otherwise stated, assume that the domain of all functions defined by an equation is the largest set of real numbers that are meaningful replacements for the independent variable.

For example, suppose

$$y = \frac{-4x}{2x - 3}.$$

Any real number can be used for $x$ except $x = 3/2$, which makes the denominator equal 0. By the agreement on domains, the domain of this function is the set of all real numbers except 3/2.

From now on, we will write domains and ranges in **interval notation.** With this notation, the set of real numbers less than 4 is written $(-\infty, 4)$. The symbol $\infty$ (the symbol for infinity) is not a real number; it shows that the interval includes *all* real numbers less than 4.

A square bracket is used to indicate that a number is included in the interval. For example, $2 \le x < 5$ is written $[2, 5)$, indicating that 2 is included but 5 is not. Similarly, $x \ge 10$ is written $[10, +\infty)$. Using interval notation, the set of all real numbers is the interval $(-\infty, +\infty)$. Parentheses and brackets also are used to graph intervals on the number line. Refer to the Review of Algebra for examples of graphs of intervals.

## ▬ EXAMPLE 3

Find the domain and range for each of the functions defined as follows.

**(a)** $y = x^2$

Any number may be squared, so the domain is the set of all real numbers, written $(-\infty, +\infty)$. Since $x^2 \ge 0$ for every value of $x$, the range is $[0, +\infty)$.

**(b)** $y = \sqrt{6 - x}$

For $y$ to be a real number, $6 - x$ must be nonnegative. This happens only when $6 - x \ge 0$, or $6 \ge x$, making the domain $(-\infty, 6]$. The range is $[0, +\infty)$ because $\sqrt{6 - x}$ is always nonnegative.

**(c)** $y = \sqrt{2x^2 + 5x - 12}$

The domain includes only those values of $x$ satisfying $2x^2 + 5x - 12 \ge 0$. Consult the Review of Algebra for a discussion of the methods for solving a quadratic inequality. Using these methods and interval notation produces the domain

$$(-\infty, -4] \cup [3/2, +\infty).*$$

As in part (b), the range is $[0, +\infty)$.

**(d)** $y = \dfrac{1}{x + 3}$

Since the denominator cannot be zero, $x \ne -3$ and the domain is

$$(-\infty, -3) \cup (-3, +\infty).$$

Because the numerator can never be zero, $y \ne 0$. There are no other restrictions on $y$, so the range is $(-\infty, 0) \cup (0, +\infty)$. ▬

*Function Notation*    The letters $f$, $g$, and $h$ frequently are used to represent functions. For example, $f$ might be used to name the function defined by $y = 5 - 3x$. For a given value of $x$ in the domain of a function $f$, there is exactly one corresponding value of $y$ in the range. To emphasize that $y$ is obtained by

---

*The *union* of sets $A$ and $B$, written $A \cup B$, is defined as the set of all elements in $A$ or $B$ or both.

applying function $f$ to the element $x$, replace $y$ with the symbol $f(x)$, read "$f$ of $x$" or "$f$ at $x$." Here $x$ is the independent variable; either $y$ or $f(x)$ represents the dependent variable.

Using $f(x)$ to replace $y$ in the equation $y = 5 - 3x$ gives

$$f(x) = 5 - 3x.$$

If 2 is chosen as a value of $x$, $f(x)$ becomes

$$f(2) = 5 - 3 \cdot 2$$
$$f(2) = -1.$$

In a similar manner,

$$f(-4) = 5 - 3(-4) = 17, \qquad f(0) = 5, \qquad f(-6) = 23,$$

and so on.

To understand how a function works, think of a function $f$ as a machine—for example, a calculator or computer—that takes an input $x$ from the domain and uses it to produce an output $f(x)$, as shown in Figure 6.

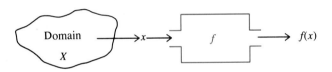

**FIGURE 6**

## ■ EXAMPLE 4

Let $g(x) = -x^2 + 4x - 5$. Find each of the following.

**(a)** $g(3)$

Replace $x$ with 3.

$$g(3) = -3^2 + 4 \cdot 3 - 5 = -9 + 12 - 5 = -2$$

**(b)** $g(a)$

Replace $x$ with $a$ to get

$$g(a) = -a^2 + 4a - 5.$$

This replacement of one variable with another is important in later chapters.

**(c)** $\begin{aligned} g(x + h) &= -(x + h)^2 + 4(x + h) - 5 \\ &= -(x^2 + 2xh + h^2) + 4(x + h) - 5 \\ &= -x^2 - 2xh - h^2 + 4x + 4h - 5 \end{aligned}$

**(d)** $g\left(\dfrac{2}{r}\right) = -\left(\dfrac{2}{r}\right)^2 + 4\left(\dfrac{2}{r}\right) - 5 = -\dfrac{4}{r^2} + \dfrac{8}{r} - 5$ ■

*Graphs*   As mentioned earlier, it often is useful to draw a graph of the ordered pairs produced by a function. To graph a function, we use a **Cartesian coordinate system,** as shown in Figure 7. The horizontal number line, or **x-axis,** represents the elements from the domain of the function, and the vertical or **y-axis** represents the elements from the range. The point where the number lines cross is the zero point on both of these number lines; this point is called the **origin.**

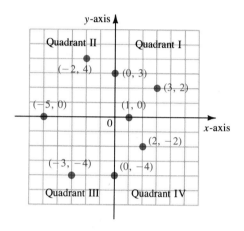

**FIGURE 7**

The name "Cartesian" honors René Descartes (1596–1650), a brilliant but sickly man. According to legend, Descartes was lying in bed when he noticed an insect crawling on the ceiling and realized that if he could determine the distance from the bug to each of two perpendicular walls, he could describe its position at any given moment. The same idea could be used to locate a point in a plane.

Each point in a Cartesian coordinate system corrresponds to an ordered pair of real numbers. Several points and their corresponding ordered pairs are shown in Figure 7. In the point $(-2, 4)$, for example, $-2$ is the **x-coordinate** and 4 is the **y-coordinate.** From now on, instead of referring to "the point corresponding to the ordered pair $(-2, 4)$," we will say "the point $(-2, 4)$."

The x-axis and y-axis divide the plane into four parts or **quadrants.** For example, quadrant I includes points whose x- and y-coordinates are both positive. The quadrants are numbered as shown in Figure 7. The points of the axes themselves belong to no quadrant.

## ▰▰ EXAMPLE 5

Let $f(x) = 3 - 2x$, with domain $\{-2, -1, 0, 1, 2, 3, 4\}$. Graph the ordered pairs produced by this function.

If $x = -2$, then $f(-2) = 3 - 2(-2) = 7$, giving the corresponding ordered pair $(-2, 7)$. Using additional values of $x$ in the equation gives the ordered pairs listed below.

| $x$ | $-2$ | $-1$ | $0$ | $1$ | $2$ | $3$ | $4$ |
|---|---|---|---|---|---|---|---|
| $y$ | $7$ | $5$ | $3$ | $1$ | $-1$ | $-3$ | $-5$ |
| Ordered Pair | $(-2, 7)$ | $(-1, 5)$ | $(0, 3)$ | $(1, 1)$ | $(2, -1)$ | $(3, -3)$ | $(4, -5)$ |

These ordered pairs are graphed in Figure 8.

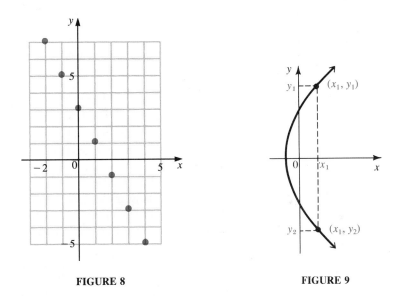

**FIGURE 8**                    **FIGURE 9**

If a graph is to represent a function, each value of $x$ from the domain must lead to exactly one value of $y$. In the graph in Figure 9, the domain value $x_1$ leads to *two* $y$-values, $y_1$ and $y_2$. Since the given $x$-value corresponds to two different $y$-values, this is not the graph of a function. This example suggests the **vertical line test** for the graph of a function.

**VERTICAL LINE TEST**

If a vertical line intersects a graph in more than one point, the graph is not the graph of a function.

# EXAMPLE 6

Use the vertical line test to decide which of the graphs in Figure 10 are graphs of functions.

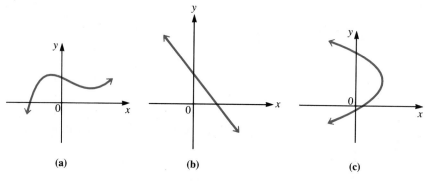

**(a)**                     **(b)**                     **(c)**

**FIGURE 10**

**(a)** Every vertical line intersects this graph in at most one point, so this is the graph of a function.

**(b)** Again, each vertical line intersects the graph in at most one point, showing that this is the graph of a function.

**(c)** It is possible for a vertical line to intersect the graph in part (c) twice. This is not the graph of a function.  ▬

# EXAMPLE 7

An overnight delivery service charges $25 for a package weighing up to 2 pounds. For each additional pound there is an additional charge of $3. Let $D(x)$ represent the cost to send a package weighing $x$ pounds. Graph $D(x)$ for $x$ in the interval $(0, 6]$.

For $x$ in the interval $(0, 2]$, $y = 25$. For $x$ in $(2, 3]$, $y = 25 + 3 = 28$. For $x$ in $(3, 4]$, $y = 28 + 3 = 31$, and so on. The graph is shown in Figure 11.  ▬

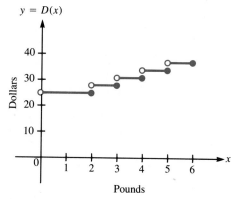

**FIGURE 11**

The function discussed in Example 7 is called a **step function.** Many real-life situations are best modeled by step functions. Additional examples are given in the exercises.

## 1.1 EXERCISES

*Which of the following rules define y as a function of x?*

**1.** 

**2.** 
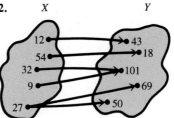

**3.**

| x | y |
|----|----|
| 3 | 9 |
| 2 | 4 |
| 1 | 1 |
| 0 | 0 |
| −1 | 1 |
| −2 | 4 |
| −3 | 9 |

**4.**

| x | y |
|----|----|
| 9 | 3 |
| 4 | 2 |
| 1 | 1 |
| 0 | 0 |
| 1 | −1 |
| 4 | −2 |
| 9 | −3 |

**5.** $y = x^3$    **6.** $y = \sqrt{x}$    **7.** $x = |y|$    **8.** $x = y^4 - 1$

*List the ordered pairs obtained from each equation, given $\{-2, -1, 0, 1, 2, 3\}$ as domain. Graph each set of ordered pairs. Give the range.*

**9.** $y = x - 1$    **10.** $y = 2x + 3$    **11.** $y = -4x + 9$    **12.** $y = -6x + 12$

**13.** $2x + y = 9$    **14.** $3x + y = 16$    **15.** $2y - x = 5$    **16.** $6x - y = -3$

**17.** $y = x(x + 1)$    **18.** $y = (x - 2)(x - 3)$    **19.** $y = x^2$    **20.** $y = -2x^2$

**21.** $y = \dfrac{1}{x + 3}$    **22.** $y = \dfrac{-2}{x + 4}$    **23.** $y = \dfrac{3x - 3}{x + 5}$    **24.** $y = \dfrac{2x + 1}{x + 3}$

*Write each expression in interval notation. Graph each interval.*

**25.** $x < 0$    **26.** $x \geq -3$    **27.** $1 \leq x < 2$

**28.** $-5 < x \leq -4$    **29.** $-9 > x$    **30.** $6 \leq x$

*Using the variable x, write each interval as an inequality.*

**31.** $(-4, 3)$    **32.** $[2, 7)$    **33.** $(-\infty, -1]$    **34.** $(3, +\infty)$

**35.**

**36.**

**37.**

**38.**

*Give the domain of each function defined as follows.*

**39.** $f(x) = 2x$

**40.** $f(x) = x + 2$

**41.** $f(x) = x^4$

**42.** $f(x) = (x - 2)^2$

**43.** $f(x) = \sqrt{16 - x^2}$

**44.** $f(x) = |x - 1|$

**45.** $f(x) = (x - 3)^{1/2}$

**46.** $f(x) = (3x + 5)^{1/2}$

**47.** $f(x) = \dfrac{2}{x^2 - 4}$

**48.** $f(x) = \dfrac{-8}{x^2 - 36}$

**49.** $f(x) = -\sqrt{\dfrac{2}{x^2 + 9}}$

**50.** $f(x) = -\sqrt{\dfrac{5}{x^2 + 36}}$

**51.** $f(x) = \sqrt{x^2 - 4x - 5}$

**52.** $f(x) = \sqrt{15x^2 + x - 2}$

*Give the domain and the range of each function.*

**53.**

**54.**

**55.**

**56.**

*For each function, find* **(a)** $f(4)$, **(b)** $f(-3)$, **(c)** $f(-1/2)$, **(d)** $f(a)$, *and*
**(e)** $f(2/m)$.

**57.** $f(x) = 3x + 2$

**58.** $f(x) = 5x - 6$

**59.** $f(x) = -x^2 + 5x + 1$

**60.** $f(x) = (x + 3)(x - 4)$

**61.** $f(x) = \dfrac{2x + 1}{x - 2}$

**62.** $f(x) = \dfrac{3x - 5}{2x + 3}$

*Use each graph to find* **(a)** $f(-2)$, **(b)** $f(0)$, **(c)** $f(1/2)$, *and* **(d)** $f(4)$.

**63.**

**64.**

**65.**

**66.**

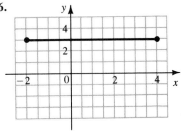

*Let* $f(x) = 6x - 2$ *and* $g(x) = x^2 - 2x + 5$ *to find the following values.*

**67.** $f(m - 3)$

**68.** $f(2r - 1)$

**69.** $g(r + h)$

**70.** $g(z - p)$

**71.** $g\left(\dfrac{3}{q}\right)$

**72.** $g\left(-\dfrac{5}{z}\right)$

*Decide whether each graph represents a function.*

**73.**

**74.**

**75.**

**76.**

**77.**

**78.**

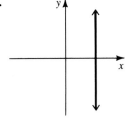

*For each function defined as follows, find*  **(a)** $f(x + h)$,  **(b)** $f(x + h) - f(x)$,

*and*  **(c)** $\dfrac{f(x + h) - f(x)}{h}$.

**79.** $f(x) = x^2 - 4$       **80.** $f(x) = 8 - 3x^2$       **81.** $f(x) = 6x + 2$

**82.** $f(x) = 4x - 11$       **83.** $f(x) = \dfrac{1}{x}$       **84.** $f(x) = -\dfrac{1}{x^2}$

---

## ▆ APPLICATIONS

### BUSINESS AND ECONOMICS

**Saw Rental**  **85.** A chain-saw rental firm charges $7 per day or fraction of a day to rent a saw, plus a fixed fee of $4 for resharpening the blade. Let $S(x)$ represent the cost of renting a saw for $x$ days.

Find each of the following.

**(a)** $S\left(\dfrac{1}{2}\right)$       **(e)** $S(4)$

**(b)** $S(1)$       **(f)** $S\left(4\dfrac{1}{10}\right)$

**(c)** $S\left(1\dfrac{1}{4}\right)$       **(g)** $S\left(4\dfrac{9}{10}\right)$

**(d)** $S\left(3\dfrac{1}{2}\right)$

**(h)** A portion of the graph of $y = S(x)$ is shown here. Explain how the graph could be continued.

**(i)** What is the independent variable?

**(j)** What is the dependent variable?

**Car Rental**  **86.** To rent a midsized car from one agency costs $40 per day or fraction of a day. If you pick up the car in Boston and drop it off in Utica, there is a fixed $40 charge. Let $C(x)$ represent the cost of renting the car for $x$ days and taking it from Boston to Utica.

Find each of the following.

**(a)** $C\left(\dfrac{3}{4}\right)$   **(b)** $C\left(\dfrac{9}{10}\right)$   **(c)** $C(1)$   **(d)** $C\left(1\dfrac{5}{8}\right)$   **(e)** $C\left(2\dfrac{1}{9}\right)$

**(f)** Graph the function defined by $y = C(x)$ for $0 < x \le 5$.

**(g)** What is the independent variable?

**(h)** What is the dependent variable?

## 1.2 LINEAR FUNCTIONS

Many practical situations can be described (or at least approximated) with a *linear function*. Some examples are the relationships between Fahrenheit and Celsius temperatures, between price and supply of some consumer goods, and between demand and supply of some commodities.

**LINEAR FUNCTION**

A function $f$ is **linear** if its equation can be written as

$$f(x) = ax + b$$

for real numbers $a$ and $b$.

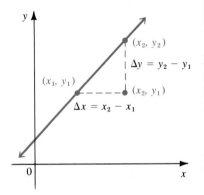

**FIGURE 12**

As the name implies, every linear function has a graph that is a straight line. An important characteristic of a straight line is its *slope,* a numerical measure of the steepness of the line. To find this measure, start with the line through the two distinct points $(x_1, y_1)$ and $(x_2, y_2)$, as shown in Figure 12. (Assume $x_1 \neq x_2$.) The difference

$$x_2 - x_1$$

is called the *change in x* and written with the symbol $\Delta x$ (read "delta x"), where $\Delta$ is the Greek letter delta. In the same way, the *change in y* is

$$\Delta y = y_2 - y_1.$$

The **slope** of a nonvertical line is defined as the quotient of the change in $y$ and the change in $x$.

**SLOPE OF A LINE**

The slope $m$ of the nonvertical line through the distinct points $(x_1, y_1)$ and $(x_2, y_2)$ is

$$m = \frac{\Delta y}{\Delta x} = \frac{y_2 - y_1}{x_2 - x_1}.$$

Similar triangles can be used to show that the slope does not depend on which pair of points on the line is used to calculate slope. That is, the same slope will be obtained for any two points on the line.

The slope of a line can be found only if the line is nonvertical, because $x_2 \neq x_1$ for a nonvertical line, so that the denominator $x_2 - x_1 \neq 0$. The slope of a vertical line is not defined.

## EXAMPLE 1

Find the slope of the line through each of the following pairs of points.

**(a)** $(-4, 8)$ and $(2, -3)$

Choosing $x_1 = -4$, $y_1 = 8$, $x_2 = 2$, and $y_2 = -3$ gives $\Delta y = -3 - 8 = -11$ and $\Delta x = 2 - (-4) = 6$. By definition, the slope is

$$m = \frac{\Delta y}{\Delta x} = -\frac{11}{6}.$$

**(b)** $(2, 7)$ and $(2, -4)$

A sketch shows that the line through $(2, 7)$ and $(2, -4)$ is vertical. As mentioned above, the slope of a vertical line is undefined. (An attempt to use the definition of slope here would produce a zero denominator.)

**(c)** $(5, -3)$ and $(-2, -3)$

By the definition of slope,

$$m = \frac{-3 - (-3)}{-2 - 5} = 0. \quad \blacksquare$$

Drawing a graph of the line in Example 1(c) shows that it is horizontal, which suggests that the slope of a horizontal line is 0.

The slope of a horizontal line is 0.

The slope of a vertical line is not defined.

Figure 13 shows lines of various slopes. As suggested by the figure, a line with a positive slope goes up from left to right, and a line with a negative slope goes down.

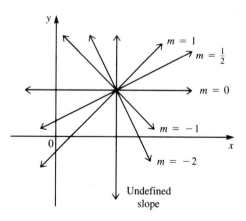

**FIGURE 13**

### EXAMPLE 2

Graph the line that passes through $(-1, 5)$ and has slope $-5/3$.

First, locate the point $(-1, 5)$, as shown in Figure 14. Since the slope of this line is $-5/3$, a change of 3 units horizontally produces a change of $-5$

units vertically, giving a second point, (2, 0), which is used to complete the graph.

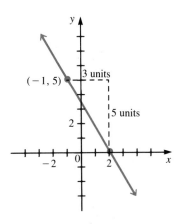

**FIGURE 14**

***Equations of a Line*** The slope of a line, together with a point on the line, can be used to find an equation of the line. The procedure for finding the equation depends on whether or not the line is vertical. The vertical line through the point $(a, 0)$ passes through all points of the form $(a, y)$, making the equation $x = a$.

**VERTICAL LINE**

An equation of the vertical line through the point $(a, 0)$ is $x = a$.

Let $m$ be the slope of a nonvertical line. Assume that the fixed point $(x_1, y_1)$ is on the line. Let $(x, y)$ represent any other point on the line. The point $(x, y)$ can be on the line if and only if the slope of the line through $(x_1, y_1)$ and $(x, y)$ is $m$—that is, if

$$\frac{y - y_1}{x - x_1} = m.$$

Multiplying both sides by $x - x_1$ gives

$$y - y_1 = m(x - x_1).$$

This result is summarized below.

**POINT-SLOPE FORM**

The line with slope $m$ passing through the point $(x_1, y_1)$ has an equation

$$y - y_1 = m(x - x_1).$$

This equation is called the **point-slope form** of the equation of a line.

### ▬ EXAMPLE 3

Write an equation of each line.

**(a)** Through $(-4, 1)$, with slope $-3$

Use the point-slope form of the equation of a line, with $x_1 = -4$, $y_1 = 1$, and $m = -3$.

$$y - 1 = -3[x - (-4)]$$
$$y - 1 = -3(x + 4)$$
$$y - 1 = -3x - 12$$
$$y = -3x - 11 \quad \text{or} \quad 3x + y = -11$$

**(b)** Through $(-3, 2)$ and $(2, -4)$

First, find the slope with the definition of slope:

$$m = \frac{-4 - 2}{2 - (-3)} = \frac{-6}{5}.$$

Either $(-3, 2)$ or $(2, -4)$ can be used for $(x_1, y_1)$. Choosing $x_1 = -3$ and $y_1 = 2$ gives

$$y - 2 = \frac{-6}{5}[x - (-3)]$$
$$5(y - 2) = -6(x + 3)$$
$$5y - 10 = -6x - 18$$
$$5y = -6x - 8 \quad \text{or} \quad 6x + 5y = -8.$$

Verify that using $(2, -4)$ instead of $(-3, 2)$ leads to the same result. ▬

Any value of $x$ where a graph crosses the $x$-axis is called an **$x$-intercept** for the graph. Any value of $y$ where the graph crosses the $y$-axis is called a **$y$-intercept** for the graph. The graph in Figure 15 has $x$-intercepts $x_1$, $x_2$, and $x_3$ and $y$-intercept $y_1$. As suggested by the graph, $x$-intercepts can be found by letting $y = 0$ and $y$-intercepts by letting $x = 0$.

**FIGURE 15**

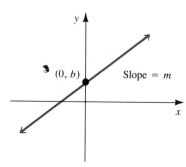

**FIGURE 16**

Figure 16 shows a line with $y$-intercept $b$; the line goes through $(0, b)$. If the slope of the line is $m$, then by the point-slope form, an equation of the line is

$$y - y_1 = m(x - x_1)$$
$$y - b = m(x - 0)$$
$$y = mx + b.$$

This result, which shows both the slope and the $y$-intercept, is the **slope-intercept form** of the equation of a line. Reversing these steps shows that any equation of the form $y = mx + b$ has a graph that is a line with slope $m$ that passes through the point $(0, b)$.

**SLOPE-INTERCEPT FORM**

The line with slope $m$ passing through the point $(0, b)$ has an equation
$$y = mx + b.$$
This equation is called the slope-intercept form of the equation of a line.

This result, together with the fact that vertical lines have equations of the form $x = a$, shows that every line has an equation of the form $ax + by + c = 0$, where $a$ and $b$ are not both 0. Conversely, assuming $b \neq 0$ and solving $ax + by + c = 0$ for $y$ gives $y = (-a/b)x - c/b$. By the result above, this equation is a line with slope $-a/b$ and $y$-intercept $-c/b$. If $b = 0$, solve for $x$ to get $x = -c/a$, a vertical line. In any case, the equation $ax + by + c = 0$ has a straight line for its graph.

If $a$ and $b$ are not both 0, then the equation $ax + by + c = 0$ has a line for its graph. Also, any line has an equation of the form $ax + by + c = 0$.

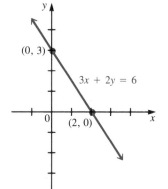

**FIGURE 17**

## ▰ EXAMPLE 4

Graph $3x + 2y = 6$.

By the work above, this equation has a line for its graph. Two distinct points on the line are enough to locate the graph. The intercepts often provide the necessary points. To find the $x$-intercept, let $y = 0$.

$$3x + 2(0) = 6$$
$$3x = 6$$
$$x = 2$$

The $x$-intercept is 2. Let $x = 0$ to find that the $y$-intercept is 3. These two intercepts were used to get the graph shown in Figure 17.

Alternatively, solve $3x + 2y = 6$ for $y$ to get

$$3x + 2y = 6$$
$$2y = -3x + 6$$
$$y = -\frac{3}{2}x + 3$$

By the slope-intercept form, the graph of this equation is the line with $y$-intercept 3 and slope $-3/2$. (This means that the line goes down 3 units for each 2 units it goes to the right.) ▬

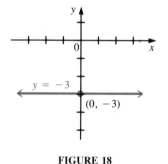

**FIGURE 18**

▬ EXAMPLE 5

Graph $y = -3$.

For a line to have an $x$-intercept, there must be a value of $x$ that makes $y = 0$. Here, however, $y = -3 \neq 0$. This means the line has no $x$-intercept, a situation that can happen only if the line is parallel to the $x$-axis (see Figure 18). ▬

Example 5 suggests the following generalization.

**HORIZONTAL LINE**

> An equation of the horizontal line through $(0, a)$ is $y = a$.

*Parallel and Perpendicular Lines*    One application of slope involves deciding whether two lines are parallel. Since two parallel lines are equally "steep," they should have the same slope. Also, two lines with the same "steepness" are parallel.

**PARALLEL LINES**

> Two nonvertical lines are parallel if and only if they have the same slope.

▬ EXAMPLE 6

Find the equation of the line that passes through the point $(3, 5)$ and is parallel to the line $2x + 5y = 4$.

The slope of $2x + 5y = 4$ can be found by writing the equation in slope-intercept form.

$$2x + 5y = 4$$
$$y = -\frac{2}{5}x + \frac{4}{5}$$

This result shows that the slope is $-2/5$. Since the lines are parallel, $-2/5$ is also the slope of the line whose equation is needed. This line passes through

(3, 5). Substituting $m = -2/5$, $x_1 = 3$, and $y_1 = 5$ into the point-slope form gives

$$y - y_1 = m(x - x_1)$$

$$y - 5 = -\frac{2}{5}(x - 3)$$

$$5(y - 5) = -2(x - 3)$$

$$5y - 25 = -2x + 6$$

$$2x + 5y = 31. \quad \blacksquare$$

As mentioned above, two nonvertical lines are parallel if and only if they have the same slope. Two lines having slopes with a product of $-1$ are perpendicular. A proof of this fact, which depends on similar triangles from geometry, is given as Exercise 57 in this section.

**PERPENDICULAR LINES**

Two lines, neither of which is vertical, are perpendicular if and only if their slopes have a product of $-1$.

### ▰ EXAMPLE 7

Find the slope of the line $L$ perpendicular to the line having the equation $5x - y = 4$.

To find the slope, write $5x - y = 4$ in slope-intercept form:

$$y = 5x - 4.$$

The slope is 5. Since the lines are perpendicular, if line $L$ has slope $m$, then

$$5m = -1$$

$$m = -\frac{1}{5}. \quad \blacksquare$$

*Supply and Demand*  Linear functions often are good choices for **supply and demand curves.** Typically, as the price of an item increases, the demand for the item decreases, while the supply increases. For example, several years ago the price of gasoline increased rapidly. As the price continued to escalate, most buyers became more and more prudent in their use of gasoline in order to restrict their demand to an affordable amount. Consequently, the overall demand for gasoline decreased and the supply increased. The demand decreased to a point where there was an oversupply of gasoline. This caused prices to fall until supply and demand were approximately balanced. Many other factors were involved in the situation, but the relationship between price, supply, and demand was nonetheless typical. Some commodities, however, such as cosmetics, medicine, and dog food, are exceptions to these typical relationships.

# ▰▰ EXAMPLE 8

Suppose an economist studies the supply and demand for a product over a number of years and concludes that the price $p$ and the demand $x$, in appropriate units and for an appropriate domain, are related by the equation

$$p = 60 - \frac{3}{4}x, \qquad \text{Demand}$$

while the price $p$ and the supply $x$ are related by

$$p = \frac{3}{4}x. \qquad \text{Supply}$$

**(a)** Find the demand at a price of $45 and at a price of $18.

Start with the demand function

$$p = 60 - \frac{3}{4}x$$

and replace $p$ with 45.

$$45 = 60 - \frac{3}{4}x$$

Solve this equation to find that

$$x = 20.$$

Thus, at a price of $45, 20 units will be demanded.

Substitute 18 for $p$ to find the demand when the price is $18.

$$p = 60 - \frac{3}{4}x$$

$$18 = 60 - \frac{3}{4}x$$

$$-42 = -\frac{3}{4}x$$

$$x = 56$$

When the price is lowered from $45 to $18, the demand increases from 20 to 56.

**(b)** Find the supply at a price of $60 and at a price of $12.

Substitute 60 for $p$ in the supply equation

$$p = \frac{3}{4}x$$

to find that the supply is 80 units. Similarly, replacing $p$ with 12 in the supply equation gives a supply of 16 units. If the price decreases from $60 to $12, the supply also decreases, from 80 units to 16 units.

**(c)** Graph both functions on the same axes.

Economists consider $p$ to be the dependent variable in the supply and demand functions, so the results of part (a) are written as the ordered pairs (20, 45) and (56, 18). The line through the corresponding points is the graph of $p = 60 - (3/4)x$, shown in black in Figure 19.

Use the ordered pairs (80, 60) and (16, 12) from the work in part (b) to get the supply graph shown in color in Figure 19.   ■

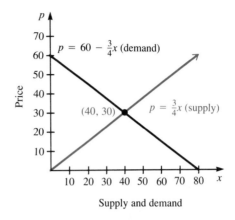

**FIGURE 19**

In Example 8 supply and demand are determined by

$$p = \frac{3}{4}x \quad \text{and} \quad p = 60 - \frac{3}{4}x,$$

respectively. To find a price where supply and demand are equal, solve the equation

$$\frac{3}{4}x = 60 - \frac{3}{4}x,$$

getting

$$x = 40.$$

Supply and demand will be equal when $x = 40$. This happens at a price of

$$p = \frac{3}{4}x$$

$$p = \frac{3}{4}(40) = 30,$$

or \$30. (Find the same result by using the demand function.) If the price of the item is more than \$30, the supply will exceed the demand. At a price less than \$30, the demand will exceed the supply. Only at a price of \$30 will demand and supply be equal. For this reason, \$30 is called the *equilibrium price*. When the price is \$30, demand and supply both equal 40 units, the *equilibrium supply* or *equilibrium demand*.

Generalizing, the **equilibrium price** of a commodity is the price at the point where the supply and demand graphs for that commodity cross. The **equilibrium demand** is the demand at the same point, and the **equilibrium supply** is the supply at this point.

## 1.2 EXERCISES

*Find the slope of each line that has a slope.*

**1.** Through $(4, 5)$ and $(-1, 2)$

**2.** Through $(5, -4)$ and $(1, 3)$

**3.** Through $(8, 4)$ and $(8, -7)$

**4.** Through $(1, 5)$ and $(-2, 5)$

**5.** $y = 2x$

**6.** $y = 3x - 2$

**7.** $5x - 9y = 11$

**8.** $4x + 7y = 1$

**9.** $x = -6$

**10.** The $x$-axis

**11.** The line parallel to $2y - 4x = 7$

**12.** The line perpendicular to $6x = y - 3$

**13.** Through $(-1.978, 4.806)$ and $(3.759, 8.125)$

**14.** Through $(11.72, 9.811)$ and $(-12.67, -5.009)$

*Write an equation in the form $ax + by = c$ for each line.*

**15.** Through $(1, 3)$, $m = -2$

**16.** Through $(2, 4)$, $m = -1$

**17.** Through $(6, 1)$, $m = 0$

**18.** Through $(-8, 1)$, with undefined slope

**19.** Through $(4, 2)$ and $(1, 3)$

**20.** Through $(8, -1)$ and $(4, 3)$

**21.** Through $(0, 3)$ and $(4, 0)$

**22.** Through $(-3, 0)$ and $(0, -5)$

**23.** $x$-intercept 3, $y$-intercept $-2$

**24.** $x$-intercept $-2$, $y$-intercept 4

**25.** Vertical, through $(-6, 5)$

**26.** Horizontal, through $(8, 7)$

**27.** Through $(-1.76, 4.25)$, with slope $-5.081$

**28.** Through $(5.469, 11.08)$, with slope 4.723

*Graph each line.*

**29.** Through $(-1, 3)$ $m = 3/2$

**30.** Through $(-2, 8)$, $m = -1$

**31.** Through $(3, -4)$, $m = -1/3$

**32.** Through $(-2, -3)$, $m = -3/4$

**33.** $3x + 5y = 15$    **34.** $2x - 3y = 12$

**35.** $4x - y = 8$    **36.** $x + 3y = 9$

**37.** $x + 2y = 0$    **38.** $3x - y = 0$

**39.** $x = -1$    **40.** $y + 2 = 0$

**41.** $y = -3$    **42.** $x = 5$

*Write an equation for each line in Exercises 43–50.*

**43.** Through $(-1, 4)$ parallel to $x + 3y = 5$

**44.** Through $(2, -5)$, parallel to $y - 4 = 2x$

**45.** Through $(3, -4)$, perpendicular to $x + y = 4$

**46.** Through $(-2, 6)$, perpendicular to $2x - 3y = 5$

**47.** $x$-intercept $-2$, parallel to $y = 2x$

**48.** $y$-intercept 3, parallel to $x + y = 4$

**49.** The line with $y$-intercept 2 and perpendicular to $3x + 2y = 6$

**50.** The line with $x$-intercept $-2/3$ and perpendicular to $2x - y = 4$

**51.** Do the points $(4, 3)$, $(2, 0)$, and $(-18, -12)$ lie on the same line? (*Hint:* Find the equation of the line through two of the points.)

**52.** Find $k$ so that the line through $(4, -1)$ and $(k, 2)$ is
   **(a)** parallel to $3y + 2x = 6$,
   **(b)** perpendicular to $2y - 5x = 1$.

**53.** Use slopes to show that the quadrilateral with vertices at $(1, 3)$, $(-5/2, 2)$, $(-7/2, 4)$, and $(2, 1)$ is a parallelogram.

**54.** Use slopes to show that the square with vertices at $(-2, 3)$, $(4, 3)$, $(4, -1)$, and $(-2, -1)$ has diagonals that are perpendicular.

**55.** Use similar triangles from geometry to show that the slope of a line is the same, no matter which two distinct points on the line are chosen to compute it.

**56.** Suppose that $(0, b)$ and $(x_1, y_1)$ are distinct points on the line $y = mx + b$. Show that $(y_1 - b)/x_1$ is the slope of the line, and that $m = (y_1 - b)/x_1$.

**57.** To show that two perpendicular lines, neither of which is vertical, have slopes with a product of $-1$, go through the following steps. Let line $L_1$ have equation $y = m_1x + b_1$, and let $L_2$ have equation $y = m_2x + b_2$. Assume that $L_1$ and $L_2$ are perpendicular, and use right triangle $MPN$ shown in the figure. Prove each of the following statements.
   **(a)** $MQ$ has length $m_1$.
   **(b)** $QN$ has length $-m_2$.
   **(c)** Triangles $MPQ$ and $PNQ$ are similar.
   **(d)** $m_1/1 = 1/-m_2$ and $m_1m_2 = -1$.

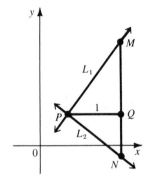

## ▤ APPLICATIONS

### BUSINESS AND ECONOMICS

**Supply and Demand** **58.** Let the supply and demand functions for sugar be given by the following equations.

$$\text{Supply: } p = 1.4x - .6 \qquad \text{Demand: } p = -2x + 3.2$$

   **(a)** Graph these equations on the same axes.
   **(b)** Find the equilibrium demand.
   **(c)** Find the equilibrium price.

**Supply and Demand** **59.** Let the supply and demand functions for a product be given by the following equations.

$$\text{Supply: } p = \frac{2}{5}x \qquad \text{Demand: } p = 100 - \frac{2}{5}x$$

   **(a)** Graph these equations on the same axes.
   **(b)** Find the equilibrium demand.
   **(c)** Find the equilibrium price.

**Cost Analysis** **60.** A company finds that it can make a total of 20 small trailers for $13,900 and 10 for $7500. Let $y$ be the total cost to produce $x$ trailers. Assume that the relationship between the number of trailers produced and the cost is linear. Find the slope of the line, and give an equation for the line in the form $y = mx + b$.

**Rental Car Cost** **61.** To rent a midsized car costs $27 per day or fraction of a day. If the car is picked up in Lansing and dropped off in West Lafayette, there is a fixed $25 dropoff charge. Let $C(x)$ represent the cost of renting the car for $x$ days, taking it from Lansing to West Lafayette.

Find each of the following.

(a) $C(3/4)$    (b) $C(9/10)$    (c) $C(1)$    (d) $C\left(1\dfrac{5}{8}\right)$    (e) $C(2.4)$

(f) Graph $y = C(x)$.

(g) Is $C$ a function?

(h) Is $C$ a linear function?

## LIFE SCIENCES

**Pollution** **62.** When a certain industrial pollutant is introduced into a river, the reproduction of catfish declines. In a given period of time, depositing three tons of the pollutant results in a fish population of 37,000. Also, 12 tons of pollutant produce a fish population of 28,000. Let $y$ be the fish population when $x$ tons of pollutant are introduced into the river. Assuming a linear relationship, find the slope and an equation of the line in the form $y = mx + b$.

**Human Growth** **63.** A person's tibia bone goes from ankle to knee. A male with a tibia 40 cm in length has a height of 177 cm, while a tibia 43 cm in length corresponds to a height of 185 cm.

(a) Write a linear equation showing how the height of a male, $h$, relates to the length of his tibia, $t$.

(b) Estimate the heights of males having tibias of length 38 cm and 45 cm.

(c) Estimate the length of the tibia for a male whose height is 190 cm.

**Human Growth** **64.** The radius bone goes from the wrist to the elbow. A female whose radius bone is 24 cm long is 167 cm tall, while a radius of 26 cm corresponds to a height of 174 cm.

(a) Write a linear equation showing how the height of a female, $h$, corresponds to the length of her radius bone, $r$.

(b) Estimate the heights of females having radius bones of length 23 cm and 27 cm.

(c) Estimate the length of a radius bone for a height of 170 cm.

## SOCIAL SCIENCES

**Voting Trends** **65.** According to research done by the political scientist James March, if the Democrats win 45% of the two-party vote for the House of Representatives, they win 42.5% of the seats. If the Democrats win 55% of the vote, they win 67.5% of the seats. Let $y$ be the percent of seats won and $x$ the percent of the two-party vote. Assuming a linear relationship, find the slope and an equation of the line in the form $y = mx + b$.

**Voting Trends**   **66.** If the Republicans win 45% of the two-party vote, they win 32.5% of the seats (see Exercise 65). If they win 60% of the vote, they get 70% of the seats. Let *y* represent the percent of the seats and *x* the percent of the vote. Assume the relationship is linear and find the slope and an equation of the line in the form *y* = *mx* + *b*.

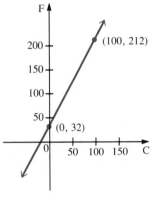

**FIGURE 20**

## 1.3  LINEAR MATHEMATICAL MODELS

Throughout this book, we construct mathematical models that describe real-world situations. In this section we look at some situations that lead to linear functions as mathematical models.

***Temperature***   One of the most common linear relationships found in everyday situations deals with temperature. Recall that water freezes at 32° Fahrenheit and 0° Celsius, while it boils at 212° Fahrenheit and 100° Celsius. The ordered pairs (0, 32) and (100, 212) are graphed in Figure 20 on axes showing Fahrenheit (*F*) as a function of Celsius (*C*). The line joining them is the graph of the function.

### EXAMPLE 1

Derive an equation relating *F* and *C*.

   To derive the required linear equation, first find the slope using the given ordered pairs, (0, 32) and (100, 212).

$$m = \frac{212 - 32}{100 - 0} = \frac{9}{5}$$

The *F*-intercept of the graph is 32, so by the slope-intercept form the equation of the line is

$$F = \frac{9}{5}C + 32.$$

With simple algebra this equation can be rewritten to give *C* in terms of *F*:

$$C = \frac{5}{9}(F - 32). \quad \blacksquare$$

***Sales Analysis***   It is common to compare the change in sales of two companies by comparing the rates at which these sales change. If the sales of the two companies can be approximated by linear functions, we can use the work of the last section to find the rate of change of the dependent variable compared to the change in the independent variable.

■■ EXAMPLE 2

The chart below shows sales in two different years for two different companies.

| Company | Sales in 1985 | Sales in 1988 |
|---------|---------------|---------------|
| A | $10,000 | $16,000 |
| B | 5000 | 14,000 |

A study of company records suggests that the sales of both companies have increased linearly (that is, the sales can be closely approximated by a linear function).

**(a)** Find a linear equation describing the sales for Company A.

To find a linear equation describing the sales, let $x = 0$ represent 1985, so that 1988 corresponds to $x = 3$. Then, by the chart above, the line representing the sales for Company A passes through the points $(0, 10,000)$ and $(3, 16,000)$. The slope of the line through these points is

$$\frac{16,000 - 10,000}{3 - 0} = 2000.$$

Using the point-slope form of the equation of a line gives

$$y - 10,000 = 2000(x - 0)$$
$$y = 2000x + 10,000$$

as an equation describing the sales of Company A.

**(b)** Find a linear equation describing the sales for Company B.

Since the sales for Company B also have increased linearly, they can be described by a line through $(0, 5000)$ and $(3, 14,000)$. Using the same procedure as in part (a) gives

$$y = 3000x + 5000$$

as the equation describing the sales of Company B.  ■

**Average Rate of Change**    Notice that the sales for Company A in Example 2 increased from $10,000 to $16,000 over the period from 1985 to 1988, representing a total increase of $6000 in three years.

$$\text{Average rate of change in sales} = \frac{\$6000}{3} = \$2000 \text{ per year}$$

This is the same as the slope found in part (a) of the example. Verify that the average annual rate of change in sales for Company B over the three-year period also agrees with the slope found in part (b). Management needs to watch the rate of change in sales closely in order to be aware of any unfavorable trends. If the rate of change is decreasing, then sales growth is slowing down, and this trend may require some response.

For $y = f(x)$, as $x$ changes from $x$ to $x + \Delta x$, the **average change in $y$ with respect to $x$** is given by

$$\frac{\text{Change in } y}{\text{Change in } x} = \frac{f(x + \Delta x) - f(x)}{(x + \Delta x) - x}$$

$$= \frac{f(x + \Delta x) - f(x)}{\Delta x}$$

$$= \frac{\Delta y}{\Delta x}.$$

For data that can be modeled with a linear function, the average rate of change, which is the same as the slope of the line, is constant.

### ■ EXAMPLE 3

Suppose that a researcher has concluded that a dosage of $x$ grams of a certain stimulant causes a rat to gain

$$y = 2x + 50$$

grams of weight, for appropriate values of $x$. If the researcher administers 30 grams of the stimulant, how much weight will the rat gain? What is the average rate of change of weight gain?

Let $x = 30$. The rat will gain

$$y = 2(30) + 50 = 110$$

grams of weight.

The average rate of change of weight gain with respect to the amount of stimulant is given by the slope of the line. The slope of $y = 2x + 50$ is 2, so that when the dose is varied by 1 gram the difference in weight gain is 2 grams.  ■

*Cost Analysis*   The cost of manufacturing an item commonly consists of two parts. The first is a **fixed cost** for designing the product, setting up a factory, training workers, and so on. Within broad limits, the fixed cost is constant for a particular product and does not change as more items are made. The second part is a *cost per item* for labor, materials, packing, shipping, and so on. The total value of this second cost *does* depend on the number of items made.

### ■ EXAMPLE 4

Suppose that the cost of producing clock-radios can be approximated by

$$C(x) = 12x + 100,$$

where $C(x)$ is the cost in dollars to produce $x$ radios. The cost to produce 0 radios is

$$C(0) = 12(0) + 100 = 100,$$

or $100. This sum, $100, is the fixed cost.

Once the company has invested the fixed cost into the clock-radio project, what then will be the additional cost per radio? As an example, we first find the cost of a total of 5 radios:

$$C(5) = 12(5) + 100 = 160,$$

or $160. The cost of 6 radios is

$$C(6) = 12(6) + 100 = 172,$$

or $172.

The sixth radio itself costs $172 − $160 = $12 to produce. In the same way, the 81st radio costs $C(81) − C(80) = \$1072 − \$1060 = \$12$ to produce. In fact, the $(n + 1)$st radio costs

$$C(n + 1) − C(n) = [12(n + 1) + 100] − [12n + 100]$$
$$= 12,$$

or $12, to produce. The number 12 is also the slope of the graph of the cost function $C(x) = 12x + 100$. ▬

In economics, **marginal cost** is the rate of change of cost $C(x)$ at a level of production $x$. The marginal cost is equal to the slope of the cost function. It approximates the cost of producing an additional item. In the clock-radio example, the marginal cost of each radio is $12. With *linear functions,* the marginal cost, which is equal to the slope of the cost function, is constant. Marginal cost is important to management in making decisions in areas such as cost control, pricing, and production planning.

The work in Example 4 can be generalized. Suppose the total cost to make $x$ items is given by the cost function $C(x) = mx + b$. The fixed cost is found by letting $x = 0$:

$$C(0) = m \cdot 0 + b = b;$$

thus, the fixed cost is $b$ dollars. The additional cost of the $(n + 1)$st item is

$$C(n + 1) − C(n) = [m(n + 1) + b] − [mn + b]$$
$$= mn + m + b − mn − b$$
$$= m.$$

This is exactly equal to the slope of the line $C(x) = mx + b$, which represents the marginal cost.

**COST FUNCTION**

▬ In a cost function of the form $C(x) = mx + b$, $m$ represents the marginal cost per item and $b$ the fixed cost. Conversely, if the fixed cost of producing an item is $b$ and the marginal cost is $m$, then the **cost function** $C(x)$ for producing $x$ items is $C(x) = mx + b$.

If $C(x)$ is the total cost to manufacture $x$ items, then the **average cost** per item is given by

$$\overline{C}(x) = \frac{C(x)}{x}.$$

In Example 4, the average cost per clock-radio is

$$\overline{C}(x) = \frac{C(x)}{x} = \frac{12x + 100}{x} = 12 + \frac{100}{x}.$$

The second term, $100/x$, shows that as more and more items are produced and the fixed cost is spread over a larger number of items, the average cost per item decreases.

### ▰▰ EXAMPLE 5

The marginal cost for raising a certain type of frog for laboratory study is $12 per unit of frogs, while the cost to produce 100 units is $1500.

**(a)** Find the cost function $C(x)$, given that it is linear.

Since the cost function is linear, it can be expressed in the form $C(x) = mx + b$. The marginal cost is $12 per unit, which gives the value for $m$, leading to $C(x) = 12x + b$. To find $b$, use the fact that the cost of producing 100 units of frogs is $1500, or $C(100) = 1500$. Substituting $x = 100$ and $C(x) = 1500$ into $C(x) = 12x + b$ gives

$$C(x) = 12x + b$$
$$1500 = 12 \cdot 100 + b$$
$$1500 = 1200 + b$$
$$300 = b.$$

The model is given by $C(x) = 12x + 300$, where the fixed cost is $300.

**(b)** Find the average cost per item to produce 50 units and 300 units.

The average cost per item is

$$\overline{C}(x) = \frac{C(x)}{x} = \frac{12x + 300}{x} = 12 + \frac{300}{x}.$$

If 50 units of frogs are produced, the average cost is

$$\overline{C}(50) = 12 + \frac{300}{50} = 18,$$

or $18 per unit. Producing 300 units of frogs will lead to an average cost of

$$\overline{C}(300) = 12 + \frac{300}{300} = 13,$$

or $13 per unit. ▰

*Break-Even Analysis*   The **revenue** $R(x)$ from selling $x$ units of a product is the product of the price per unit $p$ and the number of units sold (demand) $x$, so that

$$R(x) = px.$$

The corresponding **profit** $P(x)$ is the difference between revenue $R(x)$ and cost $C(x)$. That is,

$$P(x) = R(x) - C(x).$$

A company can make a profit only if the revenue received from its customers exceeds the cost of producing its goods and services. The number of units at which revenue just equals cost is the **break-even point.**

## ■■■ EXAMPLE 6

A firm producing poultry feed finds that the total cost $C(x)$ of producing $x$ units is given by

$$C(x) = 20x + 100.$$

Management plans to charge $24 per unit for the feed.

**(a)**  How many units must be sold for the firm to break even?

The firm will break even (no profit and no loss), as long as revenue just equals cost, or $R(x) = C(x)$. From the given information, since $R(x) = px$ and $p = \$24$,

$$R(x) = 24x.$$

Substituting for $R(x)$ and $C(x)$ in the equation $R(x) = C(x)$ gives

$$24x = 20x + 100,$$

from which $x = 25$. The firm breaks even by selling 25 units. The graphs of $C(x) = 20x + 100$ and $R(x) = 24x$ are shown in Figure 21. The break-even point (where $x = 25$) is shown on the graph. If the company produces more than 25 units (if $x > 25$), it makes a profit. If it produces less than 25 units, it loses money.

**(b)**  What is the profit if 100 units of feed are sold?

Use the formula for profit $P(x)$.

$$P(x) = R(x) - C(x)$$
$$= 24x - (20x + 100)$$
$$= 4x - 100$$

Then $P(100) = 4(100) - 100 = 300$. The firm will make a profit of $300 from the sale of 100 units of feed.  ■■

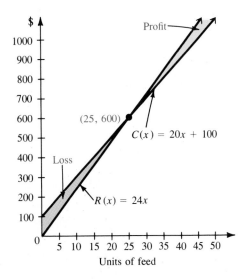

**FIGURE 21**

## 1.3 EXERCISES

*Write a cost function for each situation. Identify all variables used.*

**1.** A chain-saw rental firm charges $12 plus $1 per hour.

**2.** A trailer-hauling service charges $45 plus $2 per mile.

**3.** A parking garage charges 50¢ plus 35¢ per half-hour.

**4.** For a one-day rental, a car rental firm charges $44 plus 28¢ per mile.

*Assume that each situation can be expressed as a linear cost function, and find the appropriate cost function.*

**5.** Fixed cost, $100; 50 items cost $1600 to produce

**6.** Fixed cost, $400; 10 items cost $650 to produce

**7.** Fixed cost, $1000; 40 items cost $2000 to produce

**8.** Fixed cost, $8500; 75 items cost $11,875 to produce

**9.** Marginal cost, $50; 80 items cost $4500 to produce

**10.** Marginal cost, $120; 100 items cost $15,800 to produce

**11.** Marginal cost, $90; 150 items cost $16,000 to produce

**12.** Marginal cost, $120; 700 items cost $96,500 to produce

# ▦ APPLICATIONS

## BUSINESS AND ECONOMICS

**Sales Analysis**    **13.** Suppose the sales of a particular brand of electric guitar satisfy the relationship

$$S(x) = 300x + 2000,$$

where $S(x)$ represents the number of guitars sold in year $x$, with $x = 0$ corresponding to 1987.

Find the sales in each of the following years.

(a) 1989    (b) 1990    (c) 1991    (d) 1987

(e) Find the annual rate of change of sales.

**Sales Analysis**    **14.** Assume that the sales of a certain appliance dealer are approximated by a linear function. Suppose that sales were \$850,000 in 1982 and \$1,262,500 in 1987. Let $x = 0$ represent 1982.

(a) Find an equation giving the dealer's yearly sales.

(b) What were the dealer's approximate sales in 1985?

(c) Estimate sales in 1990.

**Sales Analysis**    **15.** Assume that the sales of a certain automobile parts company are approximated by a linear function. Suppose that sales were \$200,000 in 1981 and \$1,000,000 in 1988. Let $x = 0$ represent 1981 and $x = 7$ represent 1988.

(a) Find the equation giving the company's yearly sales.

(b) Find the approximate sales in 1983.

(c) Estimate the sales in 1990.

**Marginal Cost**    **16.** In deciding whether to set up a new manufacturing plant, company analysts have established that a reasonable function for the total cost to product $x$ items is

$$C(x) = 500,000 + 4.75x.$$

(a) Find the total cost to produce 100,000 items.

(b) Find the marginal cost of the items to be produced in this plant.

**Marginal Cost**    **17.** The manager of a local restaurant has found that his cost function for producing coffee is $C(x) = .097x$, where $C(x)$ is the total cost in dollars of producing $x$ cups. (He is ignoring the cost of the coffee pot and the cost of labor.)

Find the total cost of producing the following numbers of cups of coffee.

(a) 1000 cups    (b) 1001 cups

(c) Find the marginal cost of the 1001st cup.

(d) What is the marginal cost for *any* cup?

**Average Cost**    *Let $C(x)$ be the total cost in dollars to manufacture $x$ items. Find the average cost per item in Exercises 18 and 19.*

**18.** $C(x) = 500,000 + 4.75x$

     (a) $x = 1000$    (b) $x = 5000$    (c) $x = 10,000$

**19.** $C(x) = 800 + 20x$

     (a) $x = 10$    (b) $x = 50$    (c) $x = 200$

**Break-Even Analysis**      **20.** The cost to produce $x$ units of squash is $C(x) = 100x + 6000$, while the revenue is $R(x) = 500\, x$. Find the break-even point.

**Break-Even Analysis**      **21.** The cost to produce $x$ units of wire is $C(x) = 50x + 5000$, while the revenue is $R(x) = 60x$. Find the break-even point and the revenue at the break-even point.

**Break-Even Analysis**      *Suppose that you are the manager of a firm. You are considering the manufacture of a new product, so you ask the accounting department to produce cost estimates and the sales department to produce estimates for revenue and sales. After you receive the data, you must decide whether to go ahead with production of the new product. Analyze the data in Exercises 22–25 (find a break-even point), and then decide what you would do.*

**22.** $C(x) = 105x + 6000$; $R(x) = 250x$; not more than 400 units can be sold

**23.** $C(x) = 85x + 900$; $R(x) = 105x$; not more than 38 units can be sold

**24.** $C(x) = 1000x + 5000$; $R(x) = 900x$ (*Hint:* What does a negative break-even point mean?)

**25.** $C(x) = 70x + 500$; $R(x) = 60x$

**Break-Even Analysis**      **26.** The graph on the left shows the productivity of U.S. and Japanese workers in appropriate units over a 35-year period. Estimate the break-even point (the point at which workers in the two countries produced the same amounts).*

**Break-Even Analysis**      **27.** The graph on the right gives U.S. imports and exports in billions of dollars over a five-year period. Estimate the break-even point.*

**EXERCISE 26**

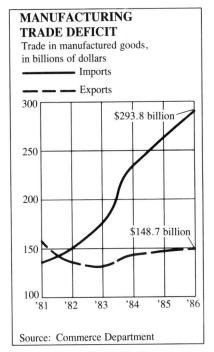

**EXERCISE 27**

*Figures, "Manufacturing Productivity Index" and "Manufacturing Trade Deficit" as appeared in *The Sacramento Bee,* December 21, 1987. Reprinted by permission of The Associated Press.

## LIFE SCIENCES

**Bacterial Growth**    **28.** Let $N(x) = -5x + 100$ represent the number of bacteria (in thousands) present in a certain tissue culture at time $x$, measured in hours, after an antibacterial spray is introduced into the environment.

Find the number of bacteria present at each of the following times.

(a) $x = 0$     (b) $x = 6$     (c) $x = 20$

(d) What is the hourly rate of change in the number of bacteria? Interpret the negative sign in the answer.

**Medical Expenses**    **29.** Over a recent three-year period, medical expenses in the U.S. rose in a linear pattern from 10.3% of the gross national product in year 0 to 11.2% in year 3.

(a) Assuming that this change in medical expenses continues to be linear, write a linear function describing the percent of the gross national product devoted to medical expenses, $y$, in terms of the year, $x$.

(b) Give the average rate of change of medical expenses from year 0 to 3. Compare it with the slope of the line in part (a). What do you find?

**Population**    **30.** The number of children in the U.S. from five to thirteen years old decreased from 31.2 million in 1980 to 30.3 million in 1986.

(a) Write a linear function describing this population $y$ in terms of year $x$ for the given period.

(b) What was the average rate of change in this population over the period from 1980 to 1986?

## SOCIAL SCIENCES

**Stimulus Effect**    **31.** In psychology, the just-noticeable-difference (JND) for some stimulus is defined as the amount by which the stimulus must be increased so that a person will perceive it as having barely been increased. For example, suppose a research study indicates that a line 40 centimeters in length must be increased to 42 cm before a subject thinks that it is longer. In this case, the JND would be $42 - 40 = 2$ cm. In a particular experiment, the JND is given by

$$y = .03x,$$

where $x$ represents the original length of the line and $y$ the JND.

Find the JND for lines having the following lengths.

(a) 10 cm     (b) 20 cm     (c) 50 cm     (d) 100 cm

(e) Find the rate of change in the JND with respect to the original length of the line.

**Passage of Time**    **32.** Most people are not very good at estimating the passage of time. Some people's estimations are too fast, and those of others are too slow. One psychologist has constructed a mathematical model for actual time as a function of estimated time: if $y$ represents actual time and $x$ estimated time, then

$$y = mx + b,$$

where $m$ and $b$ are constants that must be determined experimentally for each person. Suppose that for a particular person, $m = 1.25$ and $b = -5$. Find $y$ for each of the following values of $x$.

(a) 30 min     (b) 60 min     (c) 120 min     (d) 180 min

**Passage of Time**   **33.** In Exercise 32, suppose that for another person, $m = .85$ and $b = 1.2$. Find $y$ for each of the following values of $x$.

(a) 15 min    (b) 30 min    (c) 60 min    (d) 120 min

For this same person, find $x$ for each of the following values of $y$.

(e) 60 min    (f) 90 min

## PHYSICAL SCIENCES

**Temperature**   **34.** Use the formulas for conversion between Fahrenheit and Celsius derived in Example 1 to convert each temperature.

(a) 58° F to Celsius

(b) −40° F to Celsius

(c) 50° C to Fahrenheit

**Temperature**   **35.** Use the formulas for conversion between Fahrenheit and Celsius derived in Example 1 to convert each temperature.

(a) 98.6° F to Celsius

(b) −40° C to Fahrenheit

(c) 20° C to Fahrenheit

## GENERAL

**Class Size**   **36.** Let $R(x) = -8x + 240$ represent the number of students present in a large business calculus class, where $x$ represents the number of hours of study required weekly.

Find the number of students present at each of the following levels of required study.

(a) $x = 0$    (b) $x = 5$    (c) $x = 10$

(d) What is the rate of change of the number of students in the class with respect to the number of hours of study? Interpret the negative sign in the answer.

(e) The professor in charge of the class likes to have exactly 16 students. How many hours of study must he require in order to have exactly 16 students?

# ▬ 1.4   QUADRATIC FUNCTIONS

By definition, a linear function is defined by

$$f(x) = ax + b,$$

for real numbers $a$ and $b$. Another type of function that is very useful in applications is a *quadratic function,* in which the independent variable is squared. A quadratic function is an especially good model for many situations with a maximum or a minimum function value. Quadratic functions also may be used to describe supply and demand curves, cost, revenue, and profit, as well as other quantities.

**QUADRATIC FUNCTION**

A **quadratic function** is defined by
$$f(x) = ax^2 + bx + c,$$
where $a$, $b$, and $c$ are real numbers, with $a \neq 0$.

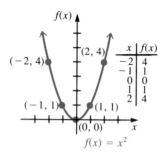

| $x$ | $f(x)$ |
|----|----|
| $-2$ | 4 |
| $-1$ | 1 |
| 0 | 0 |
| 1 | 1 |
| 2 | 4 |

$f(x) = x^2$

**FIGURE 22**

The simplest quadratic function has $f(x) = x^2$, with $a = 1$, $b = 0$, and $c = 0$. This function can be graphed by choosing several values of $x$ and then finding the corresponding values of $f(x)$. Plot the resulting ordered pairs $(x, f(x))$, and draw a smooth curve through them, as in Figure 22. This graph is called a **parabola.** Every quadratic function has a parabola as its graph. The lowest (or highest) point on a parabola is the **vertex** of the parabola. The vertex of the parabola in Figure 22 is $(0, 0)$.

If the graph in Figure 22 were folded in half along the $y$-axis, the two halves of the parabola would match exactly. This means that the graph of a quadratic function is *symmetric* with respect to a vertical line through the vertex; this line is the **axis** of the parabola.

There are many real-world instances of parabolas. For example, cross sections of spotlight reflectors or radar dishes form parabolas. Also, a projectile thrown in the air follows a parabolic path.

## EXAMPLE 1
Graph $y = x^2 - 4$.

Each value of $y$ will be 4 less than the corresponding value of $y$ in $y = x^2$. The graph of $y = x^2 - 4$ has the same shape as that of $y = x^2$ but is 4 units lower. See Figure 23. The vertex of the parabola (on this parabola, the *lowest* point) is at $(0, -4)$. The $x$-intercepts can be found by letting $y = 0$ to get

$$0 = x^2 - 4,$$

from which $x = 2$ and $x = -2$ are the $x$-intercepts. The axis of the parabola is the vertical line $x = 0$.  ▬

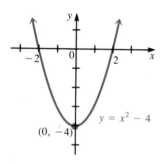

$y = x^2 - 4$

$(0, -4)$

**FIGURE 23**

## EXAMPLE 2
Graph $y = (x - 4)^2$.

If we choose values of $x$ and find the corresponding values of $y$, we see that this parabola is located 4 units to the right of the graph of $y = x^2$. The vertex is at $(4, 0)$ and is the only $x$-intercept. Setting $x$ equal to 0 gives the $y$-intercept, $y = 16$. Plot additional points (with $x$-values close to the vertex) as needed. For example, the points $(2, 4)$ and $(6, 4)$ are on the graph. The axis is the vertical line $x = 4$. See Figure 24.  ▬

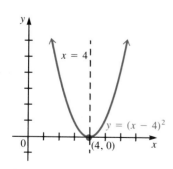

$x = 4$

$y = (x - 4)^2$

$(4, 0)$

**FIGURE 24**

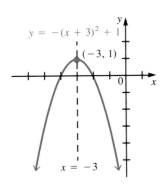

$y = -(x + 3)^2 + 1$

$(-3, 1)$

$x = -3$

**FIGURE 25**

### ◼◼ EXAMPLE 3

Graph $y = -(x + 3)^2 + 1$.

This parabola is shifted 3 units to the left and 1 unit up compared to $y = x^2$. Because of the minus sign in front of the squared quantity, the graph opens downward. The vertex, $(-3, 1)$, is the highest point on the graph. The axis is the line $x = -3$. See Figure 25.  ◼◼

Since the vertex of a parabola is the highest or lowest point on the parabola, it is the most important point for graphing or for applications with quadratic models. The vertex and axis of a parabola can be found quickly if the equation of the parabola is in the form

$$y = a(x - h)^2 + k,$$

for real numbers $a$, $h$, and $k$, with $a \neq 0$. An equation not given in this form can be converted by a process called **completing the square.** This process is explained next.

### ◼◼ EXAMPLE 4

Graph $y = -3x^2 - 2x + 1$.

To rewrite $-3x^2 - 2x + 1$ in the form $a(x - h)^2 + k$, first factor $-3$ from $-3x^2 - 2x$ to get

$$y = -3\left(x^2 + \frac{2}{3}x\right) + 1.$$

Half the coefficient of $x$ is 1/3, and $(1/3)^2 = 1/9$. Add and subtract 1/9 inside the parentheses as follows:

$$y = -3\left(x^2 + \frac{2}{3}x + \frac{1}{9} - \frac{1}{9}\right) + 1.$$

Using the distributive property and simplifying gives

$$y = -3\left(x^2 + \frac{2}{3}x + \frac{1}{9}\right) - 3\left(-\frac{1}{9}\right) + 1$$

$$= -3\left(x^2 + \frac{2}{3}x + \frac{1}{9}\right) + \frac{4}{3}.$$

Factor to get

$$y = -3\left(x + \frac{1}{3}\right)^2 + \frac{4}{3}.$$

This result shows that the axis is the vertical line

$$x + \frac{1}{3} = 0 \qquad \text{or} \qquad x = -\frac{1}{3}$$

and that the vertex is $(-1/3, 4/3)$. Use these results and plot additional ordered pairs as needed to get the graph in Figure 26.  ◼◼

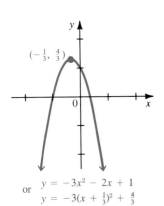

$\left(-\frac{1}{3}, \frac{4}{3}\right)$

or $\quad y = -3x^2 - 2x + 1$
$\quad\quad y = -3(x + \frac{1}{3})^2 + \frac{4}{3}$

**FIGURE 26**

The examples in this section suggest the following result.

PARABOLAS

> When the equation of a parabola is written in the form
> $$y = a(x - h)^2 + k$$
> the vertex is $(h, k)$. The axis is the vertical line $x = h$. The parabola opens upward if $a > 0$ and downward if $a < 0$.

Since the graph of a quadratic function opens upward if $a > 0$, the vertex in that case will be the lowest point on the graph, and the $y$-value of the vertex, $k$, gives the *minimum* function value. Similarly, if $a < 0$, the vertex is the highest point on the graph and $k$ is the *maximum* function value. The fact that the vertex of a parabola of the form $y = ax^2 + bx + c$ is the highest or lowest point on the graph can be used in applications to find a maximum or a minimum value.

### ■ EXAMPLE 5

Ms. Tilden owns and operates Aunt Emma's Pie Shop. She has hired a consultant to analyze her business operations. The consultant tells her that her profit $P(x)$ from the sale of $x$ units of pies is given by

$$P(x) = 120x - x^2.$$

How many units of pies must be sold in order to maximize the profit? What is the maximum possible profit?

The profit function can be rewritten as $P(x) = -x^2 + 120x + 0$. Complete the square to rewrite $P(x)$ as

$$P(x) = -(x - 60)^2 + 3600.$$

The graph of $P$ is a parabola that has vertex $(60, 3600)$ and opens downward. Since the parabola opens downward, the vertex leads to *maximum* profit. Figure 27 shows the portion of the profit function in quadrant I. (Why is quadrant I the only one of interest here?) The maximum profit of $3600 is reached when 60 units of pies are sold. In this case, profit increases as more pies are sold up to 60 units and then decreases as more pies are sold past this point. ■

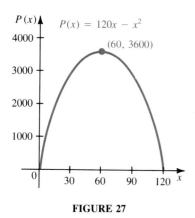

**FIGURE 27**

---

### ≡ 1.4 EXERCISES

**1.** Graph the functions in parts (a)–(d) on the same coordinate system.

(a) $f(x) = 2x^2$    (b) $f(x) = 3x^2$    (c) $f(x) = \frac{1}{2}x^2$    (d) $f(x) = \frac{1}{3}x^2$

(e) How does the coefficient affect the shape of the graph?

2. Graph the functions in parts (a)–(d) on the same coordinate system.

   (a) $y = \frac{1}{2}x^2$   (b) $y = -\frac{1}{2}x^2$   (c) $y = 4x^2$   (d) $y = -4x^2$

   (e) What effect does the minus sign have on the graph?

3. Graph the functions in parts (a)–(d) on the same coordinate system.

   (a) $f(x) = x^2 + 2$   (b) $f(x) = x^2 - 1$   (c) $f(x) = x^2 + 1$   (d) $f(x) = x^2 - 2$

   (e) How do these graphs differ from the graph of $f(x) = x^2$?

4. Graph the functions in parts (a)–(d) on the same coordinate system.

   (a) $f(x) = (x - 2)^2$   (b) $f(x) = (x + 1)^2$   (c) $f(x) = (x + 3)^2$   (d) $f(x) = (x - 4)^2$

   (e) How do these graphs differ from the graph of $f(x) = x^2$?

*Graph each parabola and give its vertex and axis.*

5. $y = (x - 2)^2$               6. $y = (x + 4)^2$               7. $y = (x + 3)^2 - 4$

8. $y = (x - 5)^2 - 4$          9. $y = -2(x + 3)^2 + 2$       10. $y = -3(x - 2)^2 + 1$

11. $y = -\frac{1}{2}(x + 1)^2 - 3$   12. $y = \frac{2}{3}(x - 2)^2 - 1$   13. $y = x^2 - 2x + 3$

14. $y = x^2 + 6x + 5$          15. $y = -x^2 - 4x + 2$        16. $y = -x^2 + 6x - 6$

17. $y = 2x^2 - 4x + 5$         18. $y = -3x^2 + 24x - 46$     19. $y = -\frac{1}{3}x^2 + 2x + 4$

20. $y = \frac{5}{2}x^2 + 10x + 8$   21. $y = 3x^2 + 6x + 2$        22. $y = 2x^2 + 12x - 16$

*For each function in Exercises 23–26, find several points on the graph and then sketch the graph.*

23. $y = .14x^2 + .56x - .3$                    24. $y = .82x^2 + 3.24x - .4$

25. $y = -.09x^2 - 1.8x + .5$                   26. $y = -.35x^2 + 2.8x - .3$

27. Let $x$ be in the interval $[0, 1]$. Use a graph to suggest that the product $x(1 - x)$ is always less than or equal to 1/4. For what values of $x$ does the product equal 1/4?

# ≡ APPLICATIONS

## BUSINESS AND ECONOMICS

**Minimizing Cost**   28. George Duda runs a sandwich shop. By studying data concerning his past costs, he has found that the cost of operating his shop is given by

$$C(x) = 2x^2 - 20x + 360,$$

where $C(x)$ is the daily cost in dollars to make $x$ units of sandwiches. Find the number of units George must sell to minimize the cost. What is the minimum cost?

**Maximizing Revenue**   29. The revenue of a charter bus company depends on the number of unsold seats. If the revenue $R(x)$ is given by

$$R(x) = 5000 + 50x - x^2,$$

where $x$ is the number of unsold seats, find the maximum revenue and the number of unsold seats that corresponds to maximum revenue.

**Maximizing Revenue**    **30.** A charter flight charges a fare of $200 per person plus $4 per person for each unsold seat on the plane. The plane holds 100 passengers. Let $x$ represent the number of unsold seats.

(a) Find an expression for the total revenue received for the flight. (*Hint:* Multiply the number of people flying, $100 - x$, by the price per ticket.)

(b) Graph the expression from part (a).

(c) Find the number of unsold seats that will produce the maximum revenue.

(d) What is the maximum revenue?

**Maximizing Revenue**    **31.** The demand for a certain type of cosmetic is given by

$$p = 500 - x,$$

where $p$ is the price in dollars when $x$ units are demanded.

(a) Find the revenue $R(x)$ that would be obtained at a price of $p$. (*Hint:* Revenue = Demand × Price)

(b) Graph the revenue function $R(x)$.

(c) From the graph of the revenue function, estimate the price that will produce maximum revenue.

(d) What is the maximum revenue?

## LIFE SCIENCES

**Maximizing Population**    **32.** The number of mosquitoes $M(x)$, in millions, in a certain area of Kentucky depends on the June rainfall $x$, in inches, approximately as follows.

$$M(x) = 10x - x^2$$

Find the rainfall that will produce the maximum number of mosquitoes.

**Maximizing Chlorophyll Production**    **33.** For the months of June through October, the percent of maximum possible chlorophyll production in a leaf is approximated by $C(x)$, where

$$C(x) = 10x + 50.$$

Here $x$ is time in months with $x = 1$ representing June. From October through December, $C(x)$ is approximately by

$$C(x) = -20(x - 5)^2 + 100,$$

with $x$ as above. Find the percent of maximum possible chlorophyll production in each of the following months.

(a) June  (b) July  (c) September  (d) October  (e) November  (f) December

**Maximizing Chlorophyll Production**    **34.** Use your results from Exercise 33 to sketch a graph of $y = C(x)$, from June through December. In what month is chlorophyll production at a maximum?

## PHYSICAL SCIENCES

**Maximizing the Height of an Object**    **35.** If an object is thrown upward with an initial velocity of 32 ft/sec, then its height after $t$ sec is given by

$$h = 32t - 16t^2.$$

Find the maximum height attained by the object. Find the number of seconds it takes the object to hit the ground.

**Maximizing the Height of an Object**

**36.** If an object is thrown upward with an initial velocity of 64 ft/sec, then its height after $t$ sec is given by

$$h = -16t^2 + 64t.$$

Find the maximum height attained by the object. Find the number of seconds it takes the object to hit the ground.

## GENERAL

**Maximizing Area** **37.** Glenview Community College wants to construct a rectangular parking lot on land bordered on one side by a highway. It has 320 ft of fencing to use along the other three sides. What should be the dimensions of the lot if the enclosed area is to be a maximum? (*Hint:* Let $x$ represent the width of the lot, and let $320 - 2x$ represent the length. Graph the area parabola, $A = x(320 - 2x)$, and investigate the vertex.)

**Maximizing Area** **38.** What would be the maximum area that could be enclosed by the college's 320 ft of fencing if it decided to close the entrance by enclosing all four sides of the lot? (See Exercise 37.)

**Number Analysis** **39.** Find two numbers whose sum is 20 and whose product is a maximum. (*Hint:* Let $x$ and $20 - x$ be the two numbers, and write an equation for the product.)

**Number Analysis** **40.** Find two numbers whose sum is 45 and whose product is a maximum.

**Parabolic Arch** **41.** An arch is shaped like a parabola. It is 30 m wide at the base and 15 m high. How wide is the arch 10 m from the ground?

**Parabolic Culvert** **42.** A culvert is shaped like a parabola, 18 cm across the top and 12 cm deep. How wide is the culvert 8 cm from the top?

## 1.5 POLYNOMIAL AND RATIONAL FUNCTIONS

In the previous sections we discussed linear and quadratic functions and their graphs. Both of these functions are special types of *polynomial functions*.

**POLYNOMIAL FUNCTION**

> A **polynomial function** of degree $n$, where $n$ is a nonnegative integer, is defined by
> $$f(x) = a_n x^n + a_{n-1} x^{n-1} + \ldots + a_1 x + a_0,$$
> where $a_n, a_{n-1}, \ldots, a_1$, and $a_0$ are real numbers, with $a_n \neq 0$.

For $n = 1$, a polynomial function takes the form

$$f(x) = a_1 x + a_0,$$

a linear function. A linear function, therefore, is a polynomial function of degree 1. (Note, however, that a linear function of the form $f(x) = a_0$ for a real number $a_0$ is a polynomial function of degree 0.) A polynomial function of degree 2 is a quadratic function.

Accurate graphs of polynomial functions of degree 3 or more require methods of calculus that we will discuss later. In this section, we will use point plotting to get reasonable sketches of the graphs of polynomial functions.

The simplest polynomial functions of higher degree are those of the form $f(x) = x^n$. To graph $f(x) = x^3$, for example, find several ordered pairs that satisfy $y = x^3$, then plot the points and connect them with a smooth curve. The graph of $f(x) = x^3$ is shown as a black curve in Figure 28. This same figure also shows the graph of $f(x) = x^5$ in color.

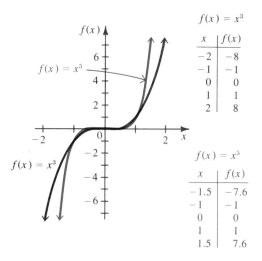

$f(x) = x^3$

| $x$ | $f(x)$ |
|-----|--------|
| $-2$ | $-8$ |
| $-1$ | $-1$ |
| $0$ | $0$ |
| $1$ | $1$ |
| $2$ | $8$ |

$f(x) = x^5$

| $x$ | $f(x)$ |
|-----|--------|
| $-1.5$ | $-7.6$ |
| $-1$ | $-1$ |
| $0$ | $0$ |
| $1$ | $1$ |
| $1.5$ | $7.6$ |

**FIGURE 28**

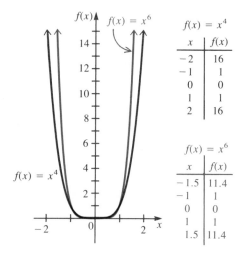

$f(x) = x^4$

| $x$ | $f(x)$ |
|-----|--------|
| $-2$ | $16$ |
| $-1$ | $1$ |
| $0$ | $0$ |
| $1$ | $1$ |
| $2$ | $16$ |

$f(x) = x^6$

| $x$ | $f(x)$ |
|-----|--------|
| $-1.5$ | $11.4$ |
| $-1$ | $1$ |
| $0$ | $0$ |
| $1$ | $1$ |
| $1.5$ | $11.4$ |

**FIGURE 29**

The graphs of $f(x) = x^4$ and $f(x) = x^6$ can be sketched in a similar manner. Figure 29 shows $f(x) = x^4$ as a black curve and $f(x) = x^6$ in color. These graphs have symmetry about the $y$-axis, as does the graph of $f(x) = ax^2$ for a nonzero real number $a$. As with the graph of $f(x) = ax^2$, the value of $a$ in $f(x) = ax^n$ affects the direction of the graph. When $a > 0$, the graph has the same general appearance as the graph of $f(x) = x^n$. However, if $a < 0$, the graph is rotated $180°$ about the $x$-axis.

### ▰ EXAMPLE 1

Graph $f(x) = 8x^3 - 12x^2 + 2x + 1$.

Letting $x$ take values from $-3$ through 3 leads to the values of $f(x)$ given in the following table.

| $x$ | $-3$ | $-2$ | $-1$ | $0$ | $1$ | $2$ | $3$ |
|-----|------|------|------|-----|-----|-----|-----|
| $f(x)$ | $-329$ | $-115$ | $-21$ | $1$ | $-1$ | $21$ | $115$ |

Plot the points with reasonable $y$-values. You will probably need to calculate additional ordered pairs between these points to get a good idea of how the graph should look. See Figure 30. ▰

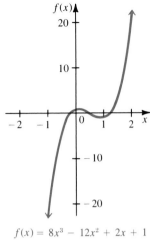

$$f(x) = 8x^3 - 12x^2 + 2x + 1$$

**FIGURE 30**

$$f(x) = 3x^4 - 14x^3 + 54x - 3$$

**FIGURE 31**

### ■ EXAMPLE 2

Graph $f(x) = 3x^4 - 14x^3 + 54x - 3$.

Complete a table of ordered pairs.

| $x$ | $-3$ | $-2$ | $-1$ | $0$ | $1$ | $2$ | $3$ |
|------|------|------|------|------|------|------|------|
| $f(x)$ | $456$ | $49$ | $-40$ | $-3$ | $40$ | $41$ | $24$ |

The graph is shown in Figure 31.   ■

As suggested by the graphs above, the domain of a polynomial function is the set of all real numbers. The range of a polynomial function of odd degree is also the set of all real numbers. Some typical graphs of polynomial functions of odd degree are shown in Figure 32. These graphs suggest that for every polynomial function $f$ of odd degree there is at least one real value of $x$ for which $f(x) = 0$. Such a value of $x$ is called a **real zero** of $f$; these values are also the $x$-intercepts of the graph.

Degree 3;
three real zeros

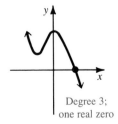

Degree 3;
one real zero

Degree 5;
three real zeros

**FIGURE 32**

Polynomial functions of even degree have a range that takes either the form $(-\infty, k]$ or the form $[k, +\infty)$ for some real number $k$. Figure 33 shows two typical graphs of polynomial functions of even degree.

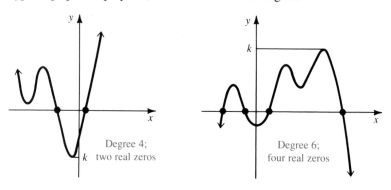

Degree 4; two real zeros

Degree 6; four real zeros

**FIGURE 33**

*Rational Functions*   Many situations require mathematical models that are quotients. A common model for such situations is a *rational function.*

**RATIONAL FUNCTION**

A function defined by

$$f(x) = \frac{p(x)}{q(x)},$$

where $p(x)$ and $q(x)$ are polynomial functions and $q(x) \neq 0$, is called a **rational function.**

Since any values of $x$ such that $q(x) = 0$ are excluded from the domain, a rational function usually has a graph with one or more breaks.

### ▇▇ EXAMPLE 3

Graph $y = \dfrac{2}{1 + x}$.

This function is undefined for $x = -1$, since $-1$ leads to a 0 denominator. For this reason, the graph of this function will not intersect the vertical line $x = -1$. Since $x$ can take on any value except $-1$, the values of $x$ can approach $-1$ as closely as desired from either side of $-1$.

*x approaches* $-1$

| $x$ | $-.5$ | $-.8$ | $-.9$ | $-.99$ | $-1.01$ | $-1.1$ | $-1.2$ | $-1.5$ |
|---|---|---|---|---|---|---|---|---|
| $1 + x$ | .5 | .2 | .1 | .01 | $-.01$ | $-.1$ | $-.2$ | $-.5$ |
| $y = \dfrac{2}{1 + x}$ | 4 | 10 | 20 | 200 | $-200$ | $-20$ | $-10$ | $-4$ |

$|y|$ *gets larger and larger*

The table above suggests that as $x$ gets closer and closer to $-1$ from either side, the sum $1 + x$ gets closer and closer to 0, and $|2/(1 + x)|$ gets larger and larger. Thus, the graph of the function approaches the vertical line $x = -1$ without ever touching the line.

As $|x|$ gets larger and larger, $y = 2/(1 + x)$ gets closer and closer to 0, as shown in the table below.

| $x$ | $-101$ | $-11$ | $-2$ | $0$ | $9$ | $99$ |
|---|---|---|---|---|---|---|
| $1 + x$ | $-100$ | $-10$ | $-1$ | $1$ | $10$ | $100$ |
| $y = \dfrac{2}{1 + x}$ | $-.02$ | $-.2$ | $-2$ | $2$ | $.2$ | $.02$ |

The graph of the function approaches the horizontal line $y = 0$. Using the information from the tables and plotting a few additional points (shown with the figure) gives the graph in Figure 34.  ▬

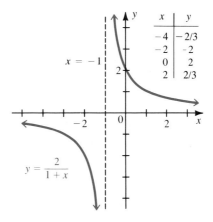

| $x$ | $y$ |
|---|---|
| $-4$ | $-2/3$ |
| $-2$ | $-2$ |
| $0$ | $2$ |
| $2$ | $2/3$ |

$x = -1$

$y = \dfrac{2}{1 + x}$

**FIGURE 34**

In Example 3, the vertical line $x = -1$ and the horizontal line $y = 0$ are *asymptotes,* defined below.

**ASYMPTOTES**

▬ If a number $k$ makes the denominator 0 in a rational function, then the line $x = k$ is a **vertical asymptote.***

If the values of $y$ approach a number $k$ as $|x|$ gets larger and larger, the line $y = k$ is a **horizontal asymptote.**

---

*Actually, we should make sure that $x = k$ does not also make the numerator 0. If both the numerator and denominator are 0, then there may be no vertical asymptote at $k$.

### ■ EXAMPLE 4

Graph $y = \dfrac{3x + 2}{2x + 4}$.

The value $x = -2$ makes the denominator 0, so the line $x = -2$ is a vertical asymptote. To find a horizontal asymptote, find $y$ as $x$ gets larger and larger, as in the following chart.

| $x$ | $y = \dfrac{3x + 2}{2x + 4}$ | *Ordered Pair* |
|---|---|---|
| 10 | $\dfrac{32}{24} = 1.33$ | (10, 1.33) |
| 20 | $\dfrac{62}{44} = 1.41$ | (20, 1.41) |
| 100 | $\dfrac{302}{204} = 1.48$ | (100, 1.48) |
| 100,000 | $\dfrac{300{,}002}{200{,}004} = 1.49998$ | (100,000, 1.49998) |

The chart suggests that as $x$ gets larger and larger, $(3x + 2)/(2x + 4)$ gets closer and closer to 1.5, or 3/2, making the line $y = 3/2$ a horizontal asymptote. Use a calculator to show that as $x$ gets more negative and takes on the values $-10$, $-100$, $-1000$, $-100,000$, and so on, the graph again approaches the line $y = 3/2$. Using these asymptotes and plotting several points leads to the graph of Figure 35. ■

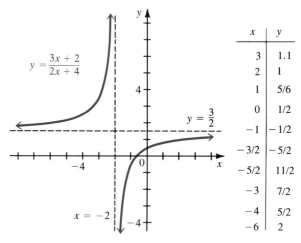

| $x$ | $y$ |
|---|---|
| 3 | 1.1 |
| 2 | 1 |
| 1 | 5/6 |
| 0 | 1/2 |
| $-1$ | $-1/2$ |
| $-3/2$ | $-5/2$ |
| $-5/2$ | 11/2 |
| $-3$ | 7/2 |
| $-4$ | 5/2 |
| $-6$ | 2 |

**FIGURE 35**

In Example 4 above, $y = 3/2$ was a horizontal asymptote for the rational function $y = (3x + 2)/(2x + 4)$. An equation for the horizontal asymptote also can be found by solving $y = (3x + 2)/(2x + 4)$ for $x$. To do this, first multiply both sides of the equation by $2x + 4$. This gives

$$y(2x + 4) = 3x + 2$$

or
$$2xy + 4y = 3x + 2.$$

Collect all terms containing $x$ on one side of the equation:

$$2xy - 3x = 2 - 4y.$$

Factor out $x$ on the left and solve for $x$.

$$x(2y - 3) = 2 - 4y$$

$$x = \frac{2 - 4y}{2y - 3}$$

This form of the equation shows that $y$ cannot take on the value 3/2. This means that the line $y = 3/2$ is a horizontal asymptote.

In many situations involving environmental pollution, much of the pollutant can be removed from the air or water at a fairly reasonable cost, but the last small part of the pollutant can be very expensive to remove. Cost as a function of the percentage of pollutant removed from the environment can be calculated for various percentages of removal, with a curve fitted through the resulting data points. This curve then leads to a mathematical model of the situation. Rational functions are often a good choice for these **cost-benefit models.**

## ▬ EXAMPLE 5

Suppose a cost-benefit model is given by

$$y = \frac{18x}{106 - x},$$

where $y$ is the cost (in thousands of dollars) of removing $x$ percent of a certain pollutant. The domain of $x$ is the set of all numbers from 0 to 100 inclusive; any amount of pollutant from 0% to 100% can be removed. Find the cost to remove the following amounts of the pollutant: 100%, 95%, 90%, and 80%. Graph the function.

Removal of 100% of the pollutant would cost

$$y = \frac{18(100)}{106 - 100} = 300,$$

or $300,000. Check that 95% of the pollutant can be removed for $155,000, 90% for $101,000, and 80% for $55,000. Using these points, as well as others that could be obtained from the function above, gives the graph shown in Figure 36 (on the next page). ▬

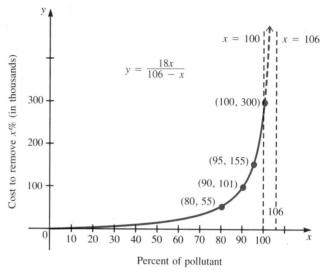

**FIGURE 36**

*Sketch the graph of each polynomial function by point plotting.*

**1.** $f(x) = (x + 1)^3$

**2.** $f(x) = x^3 + 1$

**3.** $f(x) = x^3 - 7x - 6$

**4.** $f(x) = x^3 + x^2 - 4x - 4$

**5.** $f(x) = x^4 - 5x^2 + 6$

**6.** $f(x) = x^3 - 3x^2 - x + 3$

**7.** $f(x) = 6x^3 + 11x^2 - x - 6$

**8.** $f(x) = x^4 - 2x^2 - 8$

**9.** $f(x) = x^4 + x^3 - 2$

**10.** $f(x) = 6x^4 - x^3 - 23x^2 + 4x + 12$

**11.** $f(x) = 8x^4 - 2x^3 - 47x^2 - 52x - 15$

**12.** $f(x) = x^3 - 4x^2 - 5x + 18$

**13.** $f(x) = x^3 + 3x^2 - 9x - 11$

**14.** $f(x) = -x^3 + 4x^2 + 3x - 8$

**15.** $f(x) = -x^3 + 6x^2 - x - 14$

**16.** $f(x) = 2x^3 + 4x - 1$

**17.** $f(x) = 2x^3 + 4x + 1$

**18.** $f(x) = x^4 - 4x^3 + 3x - 2$

**19.** $f(x) = 2x^4 - 3x^3 + 4x^2 + 5x - 1$

**20.** $f(x) = -x^4 - 2x^3 + 3x^2 + 3x + 5$

*Find the horizontal and vertical asymptotes for each of the following rational functions. Draw the graph of each function.*

**21.** $y = \dfrac{-4}{x - 4}$

**22.** $y = \dfrac{-1}{x + 3}$

**23.** $y = \dfrac{2}{3 + 2x}$

**24.** $y = \dfrac{4}{5 + 3x}$

**25.** $y = \dfrac{3x}{x - 1}$

**26.** $y = \dfrac{4x}{3 - 2x}$

**27.** $y = \dfrac{x + 1}{x - 4}$

**28.** $y = \dfrac{x - 3}{x + 5}$

**29.** $y = \dfrac{1 - 2x}{5x + 20}$

**30.** $y = \dfrac{6 - 3x}{4x + 12}$

**31.** $y = \dfrac{-x - 4}{3x + 6}$

**32.** $y = \dfrac{-x + 8}{2x + 5}$

## ≡ APPLICATIONS

### BUSINESS AND ECONOMICS

**Average Cost**   **33.** Suppose the average cost per unit $C(x)$, in dollars, to produce $x$ units of margarine is given by

$$C(x) = \frac{500}{x + 30}.$$

(a) Find $C(10)$, $C(20)$, $C(50)$, $C(75)$, and $C(100)$.

(b) Which of the intervals $(0, +\infty)$ and $[0, +\infty)$ would be a more reasonable domain for $C$? Why?

(c) Graph $y = C(x)$.

**Cost Analysis**   **34.** In a recent year, the cost per ton, $y$, to build an oil tanker of $x$ thousand deadweight tons is approximated by

$$y = \frac{110,000}{x + 225}.$$

(a) Find $y$ for $x = 25$, $x = 50$, $x = 100$, $x = 200$, $x = 300$, and $x = 400$.

(b) Graph the function.

**Tax Rates**   *Exercises 35 and 36 refer to the* Laffer curve, *originated by the economist Arthur Laffer. It has been a center of controversy. An idealized version of this curve is shown here.*

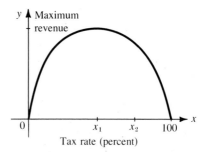

*According to this curve, increasing a tax rate, say from $x_1$ percent to $x_2$ percent on the graph, can actually lead to a decrease in government revenue. All economists agree on the endpoints—0 revenue at tax rates of both 0% and 100%—but there is much disagreement on the location of the rate $x_1$ that produces maximum revenue.*

**35.** Suppose an economist studying the Laffer curve produced the rational function

$$y = \frac{60x - 6000}{x - 120},$$

where $y$ is government revenue in millions from a tax rate of $x$ percent, with the function valid for $50 \le x \le 100$.

Find the revenue from the following tax rates.

(a) 50%      (b) 60%      (c) 80%      (d) 100%

(e) Graph the function.

**36.** Suppose the economist in Exercise 35 studies a different tax, this time producing

$$y = \frac{80x - 8000}{x - 110},$$

where $y$ is the government revenue in tens of millions of dollars for a tax rate of $x$ percent, with the function valid for $55 \leq x \leq 100$.

Find the revenue from the following tax rates.

**(a)** 55%   **(b)** 60%   **(c)** 70%   **(d)** 90%   **(e)** 100%

**(f)** Graph the function.

**Cost-Benefit Model**   **37.** Suppose a cost-benefit model is given by

$$y = \frac{6.7x}{100 - x}.$$

where $y$ is the cost in thousands of dollars of removing $x$ percent of a given pollutant.

Find the cost of removing each of the following percents of pollutants.

**(a)** 50%   **(b)** 70%   **(c)** 80%   **(d)** 90%   **(e)** 95%   **(f)** 98%   **(g)** 99%

**(h)** Is it possible, according to this function, to remove *all* the pollutant?

**(i)** Graph the function.

**Cost-Benefit Model**   **38.** Suppose a cost-benefit model is given by

$$y = \frac{6.5x}{102 - x},$$

where $y$ is the cost in thousands of dollars of removing $x$ percent of a certain pollutant.

Find the cost of removing each of the following percents of pollutants.

**(a)** 0%   **(b)** 50%   **(c)** 80%   **(d)** 90%   **(e)** 95%   **(f)** 99%   **(g)** 100%

**(h)** Graph the function.

## LIFE SCIENCES

**Cardiac Output**   **39.** A technique for measuring cardiac output depends on the concentration of a dye after a known amount is injected into a vein near the heart. In a normal heart, the concentration of the dye at time $x$ (in seconds) is given by the function

$$g(x) = -.006x^4 + .140x^3 - .053x^2 + 1.79x.$$

Graph $g(x)$.

**Population Variation**   **40.** During the early part of the twentieth century, the deer population of the Kaibab Plateau in Arizona experienced a rapid increase, because hunters had reduced the number of natural predators. The increase in population depleted the food resources and eventually caused the population to decline. For the period from 1905 to 1930, the deer population was approximated by

$$D(x) = -.125x^5 + 3.125x^4 + 4000,$$

where $x$ is time in years from 1905.

**(a)** Use a calculator to find enough points to graph $D(x)$.

**(b)** From the graph, over what period of time (from 1905 to 1930) was the population increasing? relatively stable? decreasing?

**Alcohol Concentration**  **41.** The polynomial function
$$A(x) = -.015x^3 + 1.058x$$

gives the approximate alcohol concentration (in tenths of a percent) in an average person's bloodstream $x$ hr after drinking about eight ounces of 100-proof whiskey. The function is approximately valid for $x$ in the interval [0, 8].

**(a)** Graph $A(x)$.

**(b)** Using the graph you drew for part (a), estimate the time of maximum alcohol concentration.

**(c)** In one state, a person is legally drunk if the blood alcohol concentration exceeds .15%. Use the graph from part (a) to estimate the period in which this average person is legally drunk.

**Drug Dosage**  **42.** To calculate the drug dosage for a child, pharmacists may use the formula
$$d(x) = \frac{Dx}{x + 12},$$

where $x$ is the child's age in years and $D$ is the adult dosage. Let $D = 70$, the adult dosage of the drug Naldecon.

**(a)** What is the vertical asymptote for this function?

**(b)** What is the horizontal asymptote for this function?

**(c)** Graph $d(x)$.

## PHYSICAL SCIENCES

**Electronics**  **43.** In electronics, the circuit gain is given by
$$G(R) = \frac{R}{r + R},$$

where $R$ is the resistance of a temperature sensor in the circuit and $r$ is constant. Let $r = 1000$ ohms.

**(a)** Find any vertical asymptotes of the function.

**(b)** Find any horizontal asymptotes of the function.

**(c)** Graph $G(R)$.

**Oil Pressure**  **44.** The pressure of the oil in a reservoir tends to drop with time. By taking sample pressure readings, petroleum engineers have found that the *change* in pressure in a particular oil reservoir is given by
$$P(t) = t^3 - 25t^2 + 200t,$$

where $t$ is time in years from the date of the first reading.

**(a)** Graph $P(t)$.

**(b)** From what time period is the *change* in pressure (the drop in pressure) increasing? decreasing?

### GENERAL

**Antique Cars**  **45.** Antique car fans often enter their cars in a *concours d'élègance* in which a maximum of 100 points can be awarded to a particular car. Points are awarded for the general attractiveness of the car. Based on a recent article in *Business Week*, we constructed the following mathematical model for the cost, in thousands of dollars, of restoring a car so that it will win $x$ points.

$$C(x) = \frac{10x}{49(101 - x)}$$

Find the cost of restoring a car so that it will win the following numbers of points.

**(a)** 99    **(b)** 100

### FOR THE COMPUTER

*Approximate maximum or minimum values of polynomial functions on given intervals can be found as follows. First, evaluate the function at the left endpoint of the given interval. Then add .1 to the value of x and reevaluate the polynomial. Keep doing this until the right endpoint of the interval is reached. Then identify the approximate maximum and minimum value for the polynomial on the interval. Use this procedure in Exercises 46–50.*

**46.** $y = x^3 + 4x^2 - 8x - 8$,  $[-3.8, -3]$

**47.** $y = x^3 + 4x^2 - 8x - 8$,  $[.3, 1]$

**48.** $y = 2x^3 - 5x^2 - x + 1$,  $[-1, 0]$

**49.** $y = x^4 - 7x^3 + 13x^2 + 6x - 28$,  $[-2, -1]$

**50.** $y = x^4 - 7x^3 + 13x^2 + 6x - 28$,  $[2, 3]$

*Get a table of ordered pairs for each of the following over the given interval at increments of .5; then graph the function.*

**51.** $f(x) = x^3 + 3x^2 - 2x + 1$,  $[-3, 1.5]$

**52.** $f(x) = -3x^4 - 2x^3 + x^2 + x$,  $[-3, 1.5]$

**53.** $f(x) = \dfrac{-2x^2}{x^2 - 10}$,  $[-6, 2]$

**54.** $f(x) = \dfrac{5x + 4}{2x^2 - 1}$,  $[-4, 6]$

*Find the horizontal asymptotes for each of the following. Use this information together with some ordered pairs to sketch the graph of each function.*

**55.** $f(x) = \dfrac{-2x^2 + x - 1}{2x + 3}$

**56.** $f(x) = \dfrac{3x + 2}{x^2 - 4}$

**57.** $f(x) = \dfrac{2x^2 - 5}{x^2 - 1}$

**58.** $f(x) = \dfrac{4x^2 - 1}{x^2 + 1}$

| ≣ | **KEY WORDS** | ≣ |
|---|---|---|

*To understand the concepts presented in this chapter, you should know the meaning and use of the following words. For easy reference, the section in the chapter where a word (or expression) was first used is given with each item.*

**1.1**  function
  mathematical model
  independent variable
  dependent variable
  ordered pair
  domain
  range
  interval notation
  Cartesian coordinate system
  *x*-axis
  *y*-axis
  origin
  *x*-coordinate
  *y*-coordinate
  quadrant
  vertical line test
  step function
**1.2**  linear function
  slope
  point-slope form

  slope-intercept form
  supply and demand curves
  equilibrium price
**1.3**  average rate of change
  fixed cost
  marginal cost
  cost function
  break-even point
**1.4**  quadratic function
  parabola
  vertex
  axis
  completing the square
**1.5**  polynomial function
  real zero
  rational function
  vertical asymptote
  horizontal asymptote
  cost-benefit model

---

**CHAPTER 1        REVIEW EXERCISES**

*List the ordered pairs obtained from each of the following if the domain of x for each exercise is* $\{-3, -2, -1, 0, 1, 2, 3\}$. *Graph each set of ordered pairs. Give the range.*

**1.** $2x - 5y = 10$      **2.** $3x + 7y = 21$      **3.** $y = (2x + 1)(x - 1)$      **4.** $y = (x + 4)(x + 3)$

**5.** $y = -2 + x^2$      **6.** $y = 3x^2 - 7$      **7.** $y = \dfrac{2}{x^2 + 1}$      **8.** $y \dfrac{-3 + x}{x + 10}$

**9.** $y + 1 = 0$      **10.** $y = 3$

*In Exercises 11–14, find*  **(a)** $f(6)$,   **(b)** $f(-2)$,   **(c)** $f(-4)$,   *and*   **(d)** $f(r + 1)$.

**11.** $f(x) = 4x - 1$      **12.** $f(x) = 3 - 4x$

**13.** $f(x) = -x^2 + 2x - 4$      **14.** $f(x) = 8 - x - x^2$

**15.** Let $f(x) = 5x - 3$ and $g(x) = -x^2 + 4x$. Find each of the following.

  **(a)** $f(-2)$      **(b)** $g(3)$      **(c)** $g(-4)$      **(d)** $f(5)$

  **(e)** $g(-k)$      **(f)** $g(3m)$      **(g)** $g(k - 5)$      **(h)** $f(3 - p)$

*Graph each of the following.*

**16.** $y = 6 - 2x$      **17.** $y = 4x + 3$      **18.** $2x + 7y = 14$      **19.** $3x - 5y = 15$

**20.** $y = 1$      **21.** $x + 2 = 0$      **22.** $x + 3y = 0$      **23.** $y = 2x$

*Find the slope for each line that has a slope.*

**24.** Through $(4, -1)$ and $(3, -3)$

**25.** Through $(-2, 5)$ and $(4, 7)$

**26.** Through the origin and $(0, 7)$

**27.** Through the origin and $(11, -2)$

**28.** $4x - y = 7$

**29.** $2x + 3y = 15$

**30.** $3y - 1 = 14$

**31.** $x + 4 = 9$

*Find an equation in the form $ax + by = c$ for each line.*

**32.** Through $(8, 0)$, with slope $-1/4$

**33.** Through $(5, -1)$, with slope $2/3$

**34.** Through $(2, -3)$ and $(-3, 4)$

**35.** Through $(5, -2)$ and $(1, 3)$

**36.** Slope 0, through $(-2, 5)$

**37.** Undefined slope, through $(-1, 4)$

**38.** Through $(2, -1)$, parallel to $3x - y = 1$

**39.** Through $(0, 5)$, perpendicular to $8x + 5y = 3$

**40.** Through $(2, -10)$, perpendicular to a line with undefined slope

**41.** Through $(3, -5)$, parallel to $y = 4$

**42.** Through $(-7, 4)$, perpendicular to $y = 8$

*Graph each line.*

**43.** Through $(2, -4)$, $m = 3/4$

**44.** Through $(0, 5)$, $m = -2/3$

**45.** Through $(-4, 1)$, $m = 3$

**46.** Through $(-3, -2)$, $m = -1$

*Graph each of the following.*

**47.** $y = x^2 - 4$

**48.** $y = 6 - x^2$

**49.** $y = 3(x + 1)^2 - 5$

**50.** $y = -\dfrac{1}{4}(x - 2)^2 + 3$

**51.** $y = x^2 - 4x + 2$

**52.** $y = -3x^2 - 12x - 1$

**53.** $f(x) = x^3 + 5$

**54.** $f(x) = 1 - x^4$

**55.** $f(x) = 2x^3 - 11x^2 - 2x + 2$

**56.** $f(x) = x^3 - 3x^2 - 4x - 2$

**57.** $f(x) = x^4 - 4x^3 - 5x^2 + 14x - 15$

**58.** $f(x) = x^4 + x^3 - 7x^2 - x + 6$

**59.** $f(x) = \dfrac{8}{x}$

**60.** $f(x) = \dfrac{2}{3x - 1}$

**61.** $f(x) = \dfrac{4x - 2}{3x + 1}$

**62.** $f(x) = \dfrac{6x}{(x - 1)(x + 2)}$

▬ APPLICATIONS

## BUSINESS AND ECONOMICS

**Supply and Demand** **63.** The supply and demand for a certain commodity are related by the equations

$$\text{Supply: } p = 6x + 3 \quad \text{and} \quad \text{Demand: } p = 19 - 2x,$$

where $p$ represents the price at a supply or demand, respectively, of $x$ units.

Find the supply and the demand at each of the following prices.

(a) 10     (b) 15     (c) 18

(d) Graph both the supply and the demand functions on the same axes.

(e) Find the equilibrium price.

(f) Find the equilibrium supply and the equilibrium demand.

**Supply   64.** For a particular product, 72 units will be supplied at a price of 6, while 104 units will be supplied at a price of 10. Write a supply function for this product. Assume it is a linear function.

**Cost**   *Find the linear cost functions in Exercises 65–67.*

**65.** Eight units cost $300; fixed cost is $60

**66.** Twelve units cost $445; 50 units cost $1585

**67.** Thirty units cost $1500; 120 units cost $5640

**Cost   68.** The cost of producing $x$ units of a product is $C(x)$, where

$$C(x) = 20x + 100.$$

The product sells for $40 per unit.

(a) Find the break-even point.

(b) What revenue will the company receive if it sells just that number of units?

## LIFE SCIENCES

**Fever   69.** A certain viral infection causes a fever that typically lasts 6 days. A model of the fever (in °F) on day $x, 1 \leq x \leq 6$, is

$$F(x) = -\frac{2}{3}x^2 + \frac{14}{3}x + 96.$$

According to the model, on what day should the maximum fever occur? What is the maximum fever?

**Pollution   70.** The cost to remove $x$ percent of a pollutant is

$$y = \frac{7x}{100 - x},$$

in thousands of dollars.

Find the cost of removing each of the following percents of pollution.

(a) 80%     (b) 50%     (c) 90%

(d) Graph the function.

(e) Can all of the pollutant be removed?

**Medicine Dosage   71.** Different rules exist for determining medicine dosages for children and adults. Two methods for finding dosages for children are given here. If the child's age in years is $A$, the adult dosage is $d$, and the child's dosage is $c$, then

$$c = \frac{A}{A + 12}d \quad \text{or} \quad c = \frac{A + 1}{24}d.$$

(a) At what age are the two dosages equal?

(b) Carefully graph $c = f(A) = \dfrac{A}{A + 12}$ and $c = g(A) = \dfrac{A + 1}{24}$ for the interval $2 \leq A \leq 13$ on the same axes by plotting points. Give the intervals where $f(A) > g(A)$ and where $g(A) > f(A)$.

(c) From the graphs in part (b), at what ages do the functions $f$ and $g$ appear to differ the most?

## GENERAL

**Postage**    **72.** Assume that it costs 30¢ to mail a letter weighing one ounce or less, with each additional ounce, or portion of an ounce, costing 27¢. Let $C(x)$ represent the cost to mail a letter weighing $x$ oz.

Find the costs of mailing letters of the following weights.

(a) 3.4 oz    (b) 1.02 oz    (c) 5.9 oz    (d) 10 oz

(e) Graph $C$.

(f) Give the domain and range of $C$.

**EXTENDED APPLICATION**

## MARGINAL COST—BOOZ, ALLEN & HAMILTON

Booz, Allen & Hamilton Inc. is a large management consulting firm.* One service they provide to client companies is profitability studies showing ways in which the client can increase profit levels. The client company requesting the analysis presented in this case is a large producer of a staple food. The company buys from farmers and then processes the food in its mills, resulting in a finished product. The company sells some food at retail under its own brands and some in bulk to other companies who use the product in the manufacture of convenience foods.

The client company has been reasonably profitable in recent years, but the management retained Booz, Allen & Hamilton to see whether its consultants could suggest ways of increasing company profits. The management of the company had long operated with the philosophy of trying to process and sell as much of its product as possible, since, they felt, this would lower the average processing cost per unit sold. The consultants found, however, that the client's fixed mill costs were quite low, and that, in fact, processing extra units made the cost per unit start to increase. (There are several reasons for this: the company must run three shifts, machines break down more often, and so on.)

In this application, we shall discuss the marginal cost of two of the company's products. The marginal cost (approximate cost of producing an extra unit) of production for product A was found by the consultants to be approximated by the linear function

$$y = .133x + 10.09,$$

where $x$ is the number of units produced (in millions) and $y$ is the marginal cost. (Here the marginal cost is not a constant, as it was in the examples in the text.)

For example, at a level of production of 3.1 million units, an additional unit of product A would cost about

$$y = .133(3.1) + 10.09 \approx \$10.50.†$$

---

*This case was supplied by John R. Dowdle of the Chicago office of Booz, Allen & Hamilton Inc. Reprinted by permission.

†The symbol $\approx$ means *approximately equal to.*

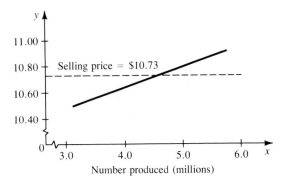

**FIGURE 37**

At a level of production of 5.7 million units, an extra unit costs $10.85. Figure 37 shows a graph of the marginal cost function from $x = 3.1$ to $x = 5.7$, the domain over which the function above was found to apply.

The selling price for product A is $10.73 per unit, so that, as shown in Figure 37, the company was losing money on many units of the product that it sold. Since the selling price could not be raised if the company were to remain competitive, the consultants recommended that production of product A be cut.

For product B, the Booz, Allen & Hamilton consultants found a marginal cost function given by

$$y = .0667x + 10.29,$$

with $x$ and $y$ as defined above. Verify that at a production level of 3.1 million units, the marginal cost is about $10.50, while at a production level of 5.7 million units, the marginal cost is about $10.67. Since the selling price of this product is $9.65, the consultants again recommended a cutback in production.

The consultants ran similar cost analyses of other products made by the company and then issued their recommendation: the company should reduce total production by 2.1 million units. The analysts predicted that this would raise profits for the products under discussion from $8.3 million annually to $9.6 million—which is very close to what actually happened when the client took the advice.

**EXERCISES**

1. At what level of production, $x$, was the marginal cost of a unit of product A equal to the selling price?

2. Graph the marginal cost function for product B from $x = 3.1$ million units to $x = 5.7$ million units.

3. Find the number of units for which marginal costs equals the selling price for product B.

4. For product C, the marginal cost of production is

$$y = .133x + 9.46.$$

   **(a)** Find the marginal costs at production levels of 3.1 million units and 5.7 million units.

   **(b)** Graph the marginal cost function.

   **(c)** For a selling price of $9.57, find the production level for which the cost equals the selling price.

# 2 The Derivative

In Chapter 1 we examined linear functions in detail and considered many of their practical applications. These functions are particularly useful because the graph of a linear function is a straight line and the slope of a line represents a rate of change of one variable (dependent) with respect to another (independent). For instance, if the dependent variable is cost and the independent variable is the number of units of an item sold, the rate of change is called the marginal cost. Similar interpretations are given to the rate of change if the dependent variable is profit or revenue.

If the function that describes a situation is linear, the rate of change remains constant no matter what value of the independent variable is used. If the function is not linear, however, the rate of change varies. Variable rates of change cannot be expressed using algebra; a branch of calculus, called *differential calculus,* is used to find such rates of change. Rates of change are important because they can be used in many different practical applications to find the optimum values of functions that describe profits, costs, nutritional values, and other quantities.

The basic concept of differential calculus is the *derivative,* introduced in Section 3 of this chapter. Since the definition of the derivative involves the idea of a *limit,* limits are discussed first.

## ▤ 2.1 LIMITS

The idea of a limit is fundamental to the study of calculus. In everyday usage, we talk about limiting one's activities, stretching a spring to its limit, or meeting the weight limit for a boxing class. In each of these situations, the use of the word *limit* suggests a boundary that may be reached, not reached, or exceeded. A mathematical limit has characteristics similar to those of a physical limit.

As mentioned in Chapter 1, we can find the asymptotes for a rational function by using tables of values to decide what happens to $f(x)$ as $x$ gets close to some number, say $a$. If the values of $f(x)$ get closer and closer to a number $L$ as the values of $x$ approach $a$, we say that $L$ is the *limit* of $f(x)$ as $x$ approaches $a$. This is written as

$$f(x) \rightarrow L \text{ as } x \rightarrow a,$$

or more commonly, as

$$\lim_{x \to a} f(x) = L.$$

### ■ EXAMPLE 1

Find $\lim\limits_{x \to 1} f(x)$ if $f(x) = \dfrac{x^2 - 4}{x - 2}$.

To find this limit, choose several values of $x$ close to 1 and on either side of 1. The results for several different values of $x$ are shown in the following table.

| | $x$ approaches 1 from the left | | | | ↓ | $x$ approaches 1 from the right | | | | | |
|---|---|---|---|---|---|---|---|---|---|---|---|
| $x$ | .8 | .9 | .99 | .9999 | 1 | 1.0000001 | 1.0001 | 1.001 | 1.01 | 1.05 | 1.1 |
| $f(x)$ | 2.8 | 2.9 | 2.99 | 2.9999 | 3 | 3.0000001 | 3.0001 | 3.001 | 3.01 | 3.05 | 3.1 |

$f(x)$ approaches 3 ↑ $f(x)$ approaches 3

The table suggests that as $x$ approaches 1 from *either direction*, $f(x)$ approaches 3, so

$$\lim_{x \to 1} f(x) = 3. \quad ■$$

The phase "$x$ approaches 1 from the left" is written $x \to 1^-$. Similarly, "$x$ approaches 1 from the right" is written $x \to 1^+$. These expressions are used to write **one-sided limits:** the **limit from the left,** such as

$$\lim_{x \to 1^-} f(x) = 3,$$

and the **limit from the right,** such as

$$\lim_{x \to 1^+} f(x) = 3.$$

A **two-sided limit,** such as

$$\lim_{x \to 1} f(x) = 3,$$

exists only if the two one-sided limits are the same—that is, if $f(x)$ approaches the same value as $x$ approaches a given value from *either* side.

One might ask "Why not just evaluate $f(1)$, since $f(1) = 3$?" In many cases limits can be evaluated by direct substitution, but in other cases they cannot. It may happen that the function value does not equal the limit, or perhaps the function is not defined at the particular value of $x$ that is of interest. It is important to remember that in evaluating a limit at some value of $x$, only function values *near* that number are of interest. The examples in this section should illustrate these ideas.

### EXAMPLE 2

Find $\lim\limits_{x\to2} \dfrac{x^2 - 4}{x - 2}$.

The value $x = 2$ is not in the domain of $f(x) = (x^2 - 4)/(x - 2)$, since 2 makes the denominator equal 0. However, $f$ may still have a limit as $x$ approaches 2. Start with a table of values.

| $x$ | 1.8 | 1.9 | 1.99 | 1.9999 | 2 | 2.0000001 | 2.00001 | 2.001 | 2.05 | 2.1 |
|---|---|---|---|---|---|---|---|---|---|---|
| $f(x)$ | 3.8 | 3.9 | 3.99 | 3.9999 | 4 | 4.0000001 | 4.00001 | 4.001 | 4.05 | 4.1 |

*x approaches 2 from the left*  ↓  *x approaches 2 from the right*

*f(x) approaches 4*  ↑  *f(x) approaches 4*

The table suggests that the values of $f(x)$ get closer and closer to 4 as the values of $x$ get closer and closer to 2 from either side, so

$$\lim_{x\to2^-} f(x) = 4 \quad \text{and} \quad \lim_{x\to2^+} f(x) = 4.$$

Thus,

$$\lim_{x\to2} \frac{x^2 - 4}{x - 2} = 4. \ \blacksquare$$

You may have noticed that, if $x \neq 2$,

$$\frac{x^2 - 4}{x - 2} = \frac{(x + 2)(x - 2)}{x - 2} = x + 2.$$

The values in the table could have been found more easily if we had evaluated $x + 2$ instead of $(x^2 - 4)/(x - 2)$; the limit also could have been found in this way.

As the values of $x$ approach 2, the values of the expression $x + 2$ approach $2 + 2 = 4$. Thus,

$$\lim_{x\to2} \frac{x^2 - 4}{x - 2} = \lim_{x\to2} (x + 2) = 4.$$

A graph of $f(x) = (x^2 - 4)/(x - 2)$ is shown in Figure 1. Compare it with the graph of $g(x) = x + 2$.

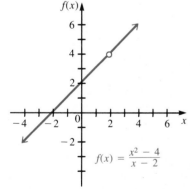

$f(x) = \dfrac{x^2 - 4}{x - 2}$

**FIGURE 1**

### EXAMPLE 3

Let $g(x) = \dfrac{x^2 + 4}{x - 2}$ and find $\lim\limits_{x\to2} g(x)$.

Make a table of values as in the previous examples.

| | x approaches 2 from the left | | | | | x approaches 2 from the right | | |
|---|---|---|---|---|---|---|---|---|
| x | 1.8 | 1.9 | 1.99 | 1.999 | 2 | 2.001 | 2.01 | 2.05 |
| g(x) | −36.2 | −76.1 | −796 | −7996 | | 8004 | 804 | 164 |

Limit does
not exist.

| g(x) gets smaller and smaller | | g(x) gets larger and larger |
|---|---|---|

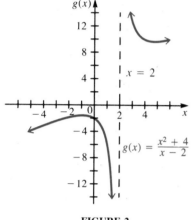

**FIGURE 2**

Both the table above and the graph in Figure 2 suggest that as $x \to 2$ from the left, $g(x)$ gets more and more negative, hence, smaller and smaller. This is indicated by writing

$$\lim_{x \to 2^-} g(x) = -\infty.$$

The symbol $-\infty$ does not represent a real number; it simply indicates that as $x \to 2^-$, $g(x)$ gets smaller without bound. In the same way, the behavior of the function as $x \to 2$ from the right is indicated by writing

$$\lim_{x \to 2^+} g(x) = +\infty.$$

Since there is no real number that $g(x)$ approaches as $x \to 2$ (from either side),

$$\lim_{x \to 2} \frac{x^2 + 4}{x - 2} \quad \text{does not exist.} \quad \blacksquare$$

The preceding examples suggest the following intuitive definition of limit.

**LIMIT**

> If the values $f(x)$ of a function $f$ can be made as close as desired to a single number $L$ for all values of $x$ sufficiently close to the number $a$, with $x \neq a$, then $L$ is the **limit** of $f(x)$ as $x$ approaches $a$, written
>
> $$\lim_{x \to a} f(x) = L.$$

This definition is *intuitive* because the expression "as close as desired" has not been defined. A more formal definition of limit would be needed for proving the rules for limits given later in this section.

Several observations about limits are in order. First, the definition of limit implies that the values of a function cannot approach two different limits at the same time, so that if a limit exists, it is unique. Second, a limit may not always exist, as seen in Example 3. Third, sometimes a limit does not exist because the function values do not approach some unique number from both the left and the right of an $x$-value. For example, in Figure 3, the values of $f(x)$ do not approach some fixed number at $x = -4$ and at $x = 3$, so that both $\lim_{x \to -4} f(x)$

and $\lim_{x \to 3} f(x)$ do not exist. Finally, a limit can exist even at a point where the function is not defined, since limits give the behavior of the function *near* some $x$-value, not *at* the $x$-value. In Figure 3, $\lim_{x \to 4} f(x) = 1$, even though $f(x)$ is not defined (because of the hole in the graph).

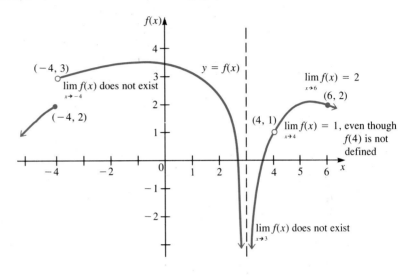

**FIGURE 3**

These observations are summarized below.

**EXISTENCE OF A LIMIT**

The limit of a function $f$ as $x \to a$ exists only if

$$\lim_{x \to a^-} f(x) = L = \lim_{x \to a^+} f(x)$$

for some real number $L$.

If $f(x)$ becomes infinitely large or infinitely small as $x$ approaches the number $a$ from either side, then the limit does not exist. Furthermore, if $L \ne M$, and

$$\lim_{x \to a^-} f(x) = L \quad \text{and} \quad \lim_{x \to a^+} f(x) = M,$$

then

$$\lim_{x \to a} f(x) \text{ does not exist.}$$

***Rules for Limits*** As shown by the examples above, tables and graphs can be used to find limits. However, it is more efficient to use the rules for limits given on the next page. (Proofs of these rules require a formal definition of limit, which we have not given.)

**RULES FOR LIMITS**

Let $a$, $k$, $n$, $A$, and $B$ be real numbers, and let $f$ and $g$ be functions such that

$$\lim_{x \to a} f(x) = A \quad \text{and} \quad \lim_{x \to a} g(x) = B.$$

1. If $k$ is a constant, then $\lim_{x \to a} k = k$ and $\lim_{x \to a} k \cdot f(x) = k \cdot \lim_{x \to a} f(x)$.

2. $\lim_{x \to a} [f(x) \pm g(x)] = \lim_{x \to a} f(x) \pm \lim_{x \to a} g(x) = A \pm B$

   (The limit of a sum or difference is the sum or difference of the limits.)

3. If $p(x)$ is a polynomial, then $\lim_{x \to a} p(x) = p(a)$.

4. $\lim_{x \to a} [f(x) \cdot g(x)] = [\lim_{x \to a} f(x)] \cdot [\lim_{x \to a} g(x)] = A \cdot B$
   (The limit of a product is the product of the limits.)

5. $\lim_{x \to a} \dfrac{f(x)}{g(x)} = \dfrac{\lim\limits_{x \to a} f(x)}{\lim\limits_{x \to a} g(x)} = \dfrac{A}{B}$ if $B \neq 0$

   (The limit of a quotient is the quotient of the limits, provided the limit of the denominator is not zero.)

6. For any real number $n$, $\lim_{x \to a} [f(x)]^n = [\lim_{x \to a} f(x)]^n = A^n$.

7. $\lim_{x \to a} f(x) = \lim_{x \to a} g(x)$ if $f(x) = g(x)$ for all $x \neq a$.

This list may seem imposing, but most limit problems have solutions that agree with your common sense. Algebraic techniques and tables of values can also help resolve questions about limits, as some of the following examples will show.

### ■ EXAMPLE 4

Find each limit.

**(a)** $\lim\limits_{x \to -2} 5$

By Rule 1, $\lim\limits_{x \to -2} 5 = 5$.

**(b)** $\lim\limits_{x \to 1} (x + 3)$

$$\lim_{x \to 1} (x + 3) = \lim_{x \to 1} x + \lim_{x \to 1} 3 \qquad \text{Rule 2}$$

$$= 1 + 3 = 4 \qquad \text{Rules 3 and 1}$$

**(c)** $\lim\limits_{x \to -1} 5(3x^2 + 2)$

$$\lim\limits_{x \to -1} 5(3x^2 + 2) = 5 \lim\limits_{x \to -1} (3x^2 + 2) \qquad \text{Rule 1}$$
$$= 5\,[3(-1)^2 + 2] \qquad \text{Rule 3}$$
$$= 25$$

**(d)** $\lim\limits_{x \to 4} \dfrac{x}{x + 2}$

$$\lim\limits_{x \to 4} \frac{x}{x + 2} = \frac{\lim\limits_{x \to 4} x}{\lim\limits_{x \to 4} (x + 2)} \qquad \text{Rule 5}$$
$$= \frac{4}{4 + 2} = \frac{2}{3}$$

**(e)** $\lim\limits_{x \to 9} \sqrt{4x - 11}$

As $x \to 9$, the expression $4x - 11$ approaches $4 \cdot 9 - 11 = 25$. Using Rule 6, with $n = 1/2$, gives

$$\lim\limits_{x \to 9} (4x - 11)^{1/2} = [\lim\limits_{x \to 9} (4x - 11)]^{1/2} = \sqrt{25} = 5. \quad \blacksquare$$

As Example 4 suggests, the rules for limits actually mean that many limits can be found simply by evaluation. The next examples illustrate some exceptions.

**■ EXAMPLE 5**

Find $\lim\limits_{x \to 2} \dfrac{x^2 + x - 6}{x - 2}$.

Rule 5 cannot be used here, since

$$\lim\limits_{x \to 2} (x - 2) = 0.$$

We can, however, simplify the function by rewriting the fraction as

$$\frac{x^2 + x - 6}{x - 2} = \frac{(x + 3)(x - 2)}{x - 2} = x + 3.$$

Now Rule 7 can be used.

$$\lim\limits_{x \to 2} \frac{x^2 + x - 6}{x - 2} = \lim\limits_{x \to 2} (x + 3) = 2 + 3 = 5 \quad \blacksquare$$

■ EXAMPLE 6

Find $\lim\limits_{x \to 4} \dfrac{\sqrt{x} - 2}{x - 4}$.

As $x \to 4$, the numerator approaches 0 and the denominator also approaches 0, giving the meaningless expression 0/0. Algebra can be used to rationalize the numerator by multiplying both the numerator and the denominator by $\sqrt{x} + 2$. This gives

$$\frac{\sqrt{x} - 2}{x - 4} \cdot \frac{\sqrt{x} + 2}{\sqrt{x} + 2} = \frac{\sqrt{x} \cdot \sqrt{x} - 2\sqrt{x} + 2\sqrt{x} - 4}{(x - 4)(\sqrt{x} + 2)}$$

$$= \frac{x - 4}{(x - 4)(\sqrt{x} + 2)} = \frac{1}{\sqrt{x} + 2}$$

if $x \neq 4$. Now use rules for limits.

$$\lim_{x \to 4} \frac{\sqrt{x} - 2}{x - 4} = \lim_{x \to 4} \frac{1}{\sqrt{x} + 2} = \frac{1}{\sqrt{4} + 2} = \frac{1}{2 + 2} = \frac{1}{4} \quad ■$$

■ EXAMPLE 7

Find $\lim\limits_{x \to 1} \dfrac{x + 1}{x^2 - 1}$.

Again, rule 5 cannot be used since $\lim\limits_{x \to 1} x^2 - 1 = 0$. If $x \neq 1$, the function can be rewritten as

$$\frac{x + 1}{x^2 - 1} = \frac{x + 1}{(x + 1)(x - 1)} = \frac{1}{x - 1}.$$

Then

$$\lim_{x \to 1} \frac{x + 1}{x^2 - 1} = \lim_{x \to 1} \frac{1}{x - 1}$$

by Rule 7. None of the rules can be used to find

$$\lim_{x \to 1} \frac{1}{x - 1},$$

but a table of values would show that the values of $1/(x - 1)$ increase as $x \to 1^+$ and the values of $1/(x - 1)$ decrease as $x \to 1^-$. Therefore,

$$\lim_{x \to 1} \frac{1}{x - 1} \quad \text{does not exist.} \quad ■$$

***Limits at Infinity*** Sometimes it is useful to examine the behavior of the values of $f(x)$ as $x$ gets larger and larger (or smaller and smaller). For example, suppose a small pond normally contains 12 units of dissolved oxygen in a fixed volume of water. Suppose also that at time $t = 0$ a quantity of organic waste is introduced into the pond, with the oxygen concentration $t$ weeks later given by

$$f(t) = \frac{12t^2 - 15t + 12}{t^2 + 1}.$$

As time goes on, what will be the ultimate concentration of oxygen? Will it return to 12 units? After 2 weeks, the pond contains

$$f(2) = \frac{12 \cdot 2^2 - 15 \cdot 2 + 12}{2^2 + 1} = \frac{30}{5} = 6$$

units of oxygen, and after 4 weeks, it contains

$$f(4) = \frac{12 \cdot 4^2 - 15 \cdot 4 + 12}{4^2 + 1} \approx 8.5$$

$$f(t) = \frac{12t^2 - 15t + 12}{t^2 + 1}$$

**FIGURE 4**

units. Choosing several values of $t$ and finding the corresponding values of $f(t)$ leads to the graph in Figure 4.

The graph suggests that as time goes on, the oxygen level gets closer and closer to the original 12 units. Consider a table of values as $t$ gets larger and larger.

| $t$ | 10 | 100 | 1000 | 10,000 |
|---|---|---|---|---|
| $f(t)$ | 10.5 | 11.85 | 11.985 | 11.9985 |

The table suggests that

$$\lim_{t \to +\infty} f(t) = 12,$$

where $t \to +\infty$ means that $t$ increases without bound. (Similarly, $x \to -\infty$ means that $x$ *decreases* without bound; that is, $x$ becomes more and more negative.) Thus, the oxygen concentration will approach 12, but it will never be *exactly* 12.

The graphs of $f(x) = 1/x$ (in black) and $g(x) = 1/x^2$ (in color) shown in Figure 5 lead to examples of such **limits at infinity.** The graphs and table of values indicate that $\lim_{x \to +\infty} (1/x) = 0$, $\lim_{x \to -\infty} (1/x) = 0$, $\lim_{x \to +\infty} (1/x^2) = 0$, and $\lim_{x \to -\infty} (1/x^2) = 0$, suggesting the following rule.

**LIMITS AT INFINITY**

> For any positive real number $n$,
> $$\lim_{x \to +\infty} \frac{1}{x^n} = 0 \quad \text{and} \quad \lim_{x \to -\infty} \frac{1}{x^n} = 0.$$

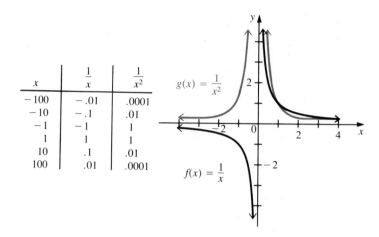

| $x$ | $\dfrac{1}{x}$ | $\dfrac{1}{x^2}$ |
|---|---|---|
| $-100$ | $-.01$ | $.0001$ |
| $-10$ | $-.1$ | $.01$ |
| $-1$ | $-1$ | $1$ |
| $1$ | $1$ | $1$ |
| $10$ | $.1$ | $.01$ |
| $100$ | $.01$ | $.0001$ |

**FIGURE 5**

The rules for limits given earlier in this section remain unchanged when $a$ is replaced with $+\infty$ or $-\infty$.

## ■ EXAMPLE 8
Find each limit.

**(a)** $\displaystyle\lim_{x\to+\infty} \frac{8x + 6}{3x - 1}$

We can use the rule $\displaystyle\lim_{x\to+\infty} 1/x^n = 0$ to find this limit by first dividing numerator and denominator by $x$, as follows.

$$\lim_{x\to+\infty} \frac{8x + 6}{3x - 1} = \lim_{x\to+\infty} \frac{\dfrac{8x}{x} + \dfrac{6}{x}}{\dfrac{3x}{x} - \dfrac{1}{x}} = \lim_{x\to+\infty} \frac{8 + 6 \cdot \dfrac{1}{x}}{3 - \dfrac{1}{x}} = \frac{8 + 0}{3 - 0} = \frac{8}{3}$$

**(b)** $\displaystyle\lim_{x\to-\infty} \frac{4x^2 - 6x + 3}{2x^2 - x + 4}$

Divide each term of the numerator and denominator by $x^2$, the highest power of $x$.

$$\lim_{x\to-\infty} \frac{4x^2 - 6x + 3}{2x^2 - x + 4} = \lim_{x\to-\infty} \frac{4 - 6 \cdot \dfrac{1}{x} + 3 \cdot \dfrac{1}{x^2}}{2 - \dfrac{1}{x} + 4 \cdot \dfrac{1}{x^2}}$$

$$= \frac{4 - 0 + 0}{2 - 0 + 0} = \frac{4}{2} = 2$$

**(c)** $\displaystyle\lim_{x\to+\infty} \frac{3x + 2}{4x^3 - 1} = \lim_{x\to+\infty} \frac{3 \cdot \dfrac{1}{x^2} + 2 \cdot \dfrac{1}{x^3}}{4 - \dfrac{1}{x^3}} = \frac{0 + 0}{4 - 0} = \frac{0}{4} = 0$

Here, the highest power of $x$ is $x^3$, which is used to divide each term in numerator and denominator.

**(d)** $\displaystyle\lim_{x\to+\infty} \frac{3x^2 + 2}{4x - 3} = \lim_{x\to+\infty} \frac{3 + \dfrac{2}{x^2}}{\dfrac{4}{x} - \dfrac{3}{x^2}} = \frac{3 + 0}{0 - 0} = \frac{3}{0}$

Division by 0 is undefined, so this function has no limit. ▬

The method used in Example 8 is a useful way to rewrite expressions with fractions so that the rules for limits at infinity can be used.

**FINDING LIMITS AT INFINITY**

If $f(x) = \dfrac{p(x)}{q(x)}$, for polynomials $p(x)$ and $q(x)$, $q(x) \neq 0$, $\displaystyle\lim_{x\to-\infty} f(x)$ and $\displaystyle\lim_{x\to+\infty} f(x)$ can be found as follows.

**1.** Divide $p(x)$ *and* $q(x)$ by the highest power of $x$ in either polynomial.

**2.** Use the rules for limits, including the rules for limits at infinity,

$$\lim_{x\to+\infty} \frac{1}{x^n} = 0 \quad \text{and} \quad \lim_{x\to-\infty} \frac{1}{x^n} = 0,$$

to find the limit of the result from Step 1.

Limits at infinity can be used to find horizontal asymptotes when graphing rational functions.

▬ **EXAMPLE 9**

Graph $f(x) = \dfrac{3x^2}{x^2 + 5}$.

There is no vertical asymptote, because $x^2 + 5 \neq 0$ for any value of $x$. Find any horizontal asymptote by calculating $\displaystyle\lim_{x\to+\infty} f(x)$ and $\displaystyle\lim_{x\to-\infty} f(x)$. First, divide both the numerator and the denominator of $f(x)$ by $x^2$.

$$\lim_{x\to+\infty} \frac{3x^2}{x^2 + 5} = \lim_{x\to+\infty} \frac{\dfrac{3x^2}{x^2}}{\dfrac{x^2}{x^2} + \dfrac{5}{x^2}} = \frac{3}{1 + 0} = 3$$

Verify that the limit of $f(x)$ as $x \rightarrow -\infty$ is also 3. Thus, the horizontal asymptote is $y = 3$. Plot a few points (several are needed near the origin) including the intercept at the origin, and use the fact that the graph approaches the line $y = 3$ as $x \rightarrow +\infty$ and also as $x \rightarrow -\infty$, to get the graph that is shown in Figure 6. ■

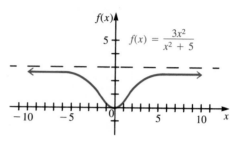

**FIGURE 6**

## 2.1 EXERCISES

*Decide whether each limit exists. If a limit exists, find its value.*

**1.** $\lim\limits_{x \to 3} f(x)$

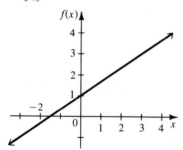

**2.** $\lim\limits_{x \to 2} F(x)$

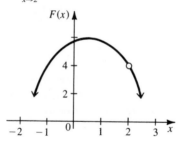

**3.** $\lim\limits_{x \to -2} f(x)$

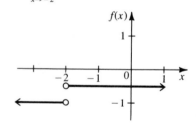

**4.** $\lim\limits_{x \to 3} g(x)$

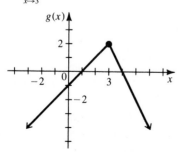

**5.** $\lim\limits_{x \to 0} f(x)$

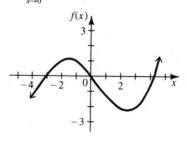

**6.** $\lim\limits_{x \to 1} g(x)$

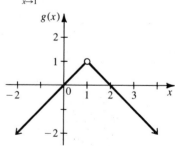

**7.** $\lim_{x \to 3} F(x)$

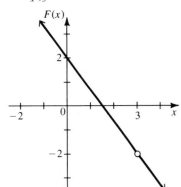

**8.** $\lim_{x \to +\infty} f(x)$

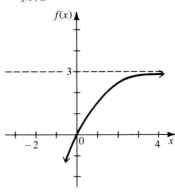

**9.** $\lim_{x \to -\infty} g(x)$

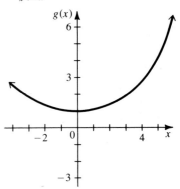

*Complete the tables and use the results to find the indicated limits. (You will need a calculator with a $\sqrt{x}$ key or a computer for Exercises 13 and 14.)*

**10.** If $f(x) = 2x^2 - 4x + 3$, find $\lim_{x \to 1} f(x)$.

| x | .9 | .99 | .999 | 1.001 | 1.01 | 1.1 |
|---|----|-----|------|-------|------|-----|
| f(x) | | | 1.000002 | 1.000002 | | |

**11.** If $k(x) = \dfrac{x^3 - 2x - 4}{x - 2}$, find $\lim_{x \to 2} k(x)$.

| x | 1.9 | 1.99 | 1.999 | 2.001 | 2.01 | 2.1 |
|---|-----|------|-------|-------|------|-----|
| k(x) | | | | | | |

**12.** If $f(x) = \dfrac{2x^3 + 3x^2 - 4x - 5}{x + 1}$, find $\lim_{x \to -1} f(x)$.

| x | -1.1 | -1.01 | -1.001 | -.999 | -.99 | -.9 |
|---|------|-------|--------|-------|------|-----|
| f(x) | | | | | | |

**13.** If $h(x) = \dfrac{\sqrt{x} - 2}{x - 1}$, find $\lim_{x \to 1} h(x)$.

| x | .9 | .99 | .999 | 1.001 | 1.01 | 1.1 |
|---|----|-----|------|-------|------|-----|
| h(x) | | | | | | |

**14.** If $f(x) = \dfrac{\sqrt{x} - 3}{x - 3}$, find $\lim_{x \to 3} f(x)$.

| x | 2.9 | 2.99 | 2.999 | 3.001 | 3.01 | 3.1 |
|---|-----|------|-------|-------|------|-----|
| f(x) | | | | | | |

*Use the properties of limits to help decide whether the following limits exist. If a limit exists, find its value.*

**15.** $\lim\limits_{x \to 2} (2x^3 + 5x^2 + 2x + 1)$

**16.** $\lim\limits_{x \to -1} (4x^3 - x^2 + 3x - 1)$

**17.** $\lim\limits_{x \to 3} \dfrac{5x - 6}{2x + 1}$

**18.** $\lim\limits_{x \to -2} \dfrac{2x + 1}{3x - 4}$

**19.** $\lim\limits_{x \to 1} \dfrac{2x^2 - 6x + 3}{3x^2 - 4x + 2}$

**20.** $\lim\limits_{x \to 2} \dfrac{-4x^2 + 6x - 8}{3x^2 + 7x - 2}$

**21.** $\lim\limits_{x \to 3} \dfrac{x^2 - 9}{x - 3}$

**22.** $\lim\limits_{x \to -2} \dfrac{x^2 - 4}{x + 2}$

**23.** $\lim\limits_{x \to -2} \dfrac{x^2 - x - 6}{x + 2}$

**24.** $\lim\limits_{x \to 5} \dfrac{x^2 - 3x - 10}{x - 5}$

**25.** $\lim\limits_{x \to 3} \sqrt{x^2 - 4}$

**26.** $\lim\limits_{x \to 3} \sqrt{x^2 - 5}$

**27.** $\lim\limits_{x \to 4} \dfrac{-6}{(x - 4)^2}$

**28.** $\lim\limits_{x \to -2} \dfrac{3x}{(x + 2)^3}$

**29.** $\lim\limits_{x \to 0} \dfrac{x^3 - 4x^2 + 8x}{2x}$

**30.** $\lim\limits_{x \to 0} \dfrac{-x^5 - 9x^3 + 8x^2}{5x}$

**31.** $\lim\limits_{x \to 0} \dfrac{[1/(x + 3)] - 1/3}{x}$

**32.** $\lim\limits_{x \to 0} \dfrac{[-1/(x + 2)] + 1/2}{x}$

**33.** $\lim\limits_{x \to 25} \dfrac{\sqrt{x} - 5}{x - 25}$

**34.** $\lim\limits_{x \to 36} \dfrac{\sqrt{x} - 6}{x - 36}$

**35.** $\lim\limits_{x \to 5} \dfrac{\sqrt{x} - \sqrt{5}}{x - 5}$

**36.** $\lim\limits_{x \to 8} \dfrac{\sqrt{x} - \sqrt{8}}{x - 8}$

**37.** $\lim\limits_{h \to 0} \dfrac{(x + h)^2 - x^2}{h}$

**38.** $\lim\limits_{h \to 0} \dfrac{(x + h)^3 - x^3}{h}$

*Decide whether the following limits exist. If a limit exists, find its value.*

**39.** $\lim\limits_{x \to +\infty} \dfrac{3x}{5x - 1}$

**40.** $\lim\limits_{x \to +\infty} \dfrac{5x}{3x - 1}$

**41.** $\lim\limits_{x \to -\infty} \dfrac{2x + 3}{4x - 7}$

**42.** $\lim\limits_{x \to -\infty} \dfrac{8x + 2}{2x - 5}$

**43.** $\lim\limits_{x \to +\infty} \dfrac{x^2 + 2x}{2x^2 - 2x + 1}$

**44.** $\lim\limits_{x \to +\infty} \dfrac{x^2 + 2x - 5}{3x^2 + 2}$

**45.** $\lim\limits_{x \to +\infty} \dfrac{3x^3 + 2x - 1}{2x^4 - 3x^3 - 2}$

**46.** $\lim\limits_{x \to +\infty} \dfrac{2x^2 - 1}{3x^4 + 2}$

**47.** $\lim\limits_{x \to +\infty} (\sqrt{x^2 + 4} - x)$

**48.** $\lim\limits_{x \to +\infty} (x - \sqrt{x^2 - 9})$

---

## ▤ APPLICATIONS

### BUSINESS AND ECONOMICS

**Production**   **49.** The graph on the next page shows the profit from the daily production of $x$ thousand kilograms of an industrial chemical.

Use the graph to find the following limits.

**(a)** $\lim\limits_{x \to 6} P(x)$      **(b)** $\lim\limits_{x \to 10} P(x)$      **(c)** $\lim\limits_{x \to 15} P(x)$

Find each of the following values and compare with the corresponding limits found in parts (a)–(c).

**(d)** $P(6)$      **(e)** $P(10)$      **(f)** $P(15)$

**(g)** Use the graph to estimate the number of units of the chemical that must be produced before the second shift is as profitable as the first.

**EXERCISE 49**

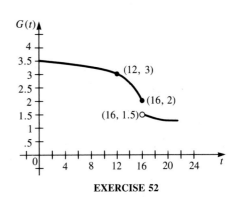

**EXERCISE 52**

**Average Cost**    **50.** The cost for manufacturing a particular videotape is

$$c(x) = 15,000 + 6x,$$

where $x$ is the number of tapes produced. The average cost per tape, denoted by $\bar{c}(x)$, is found by dividing $c(x)$ by $x$. Find the following.

(a) $\bar{c}(1000)$    (b) $\bar{c}(100,000)$    (c) $\lim\limits_{x \to 10,000} \bar{c}(x)$    (d) $\lim\limits_{x \to +\infty} \bar{c}(x)$

**Employee Productivity**    **51.** A company training program has determined that, on the average, a new employee can do $P(s)$ pieces of work per day after $s$ days of on-the-job training, where

$$P(s) = \frac{75s}{s + 8}.$$

Find the following.

(a) $P(1)$    (b) $P(11)$    (c) $\lim\limits_{s \to 11} P(s)$    (d) $\lim\limits_{s \to +\infty} P(s)$

**Consumer Demand**    **52.** When the price of an essential commodity (such as gasoline) rises rapidly, consumption drops slowly at first. If the price continues to rise, however, a "tipping" point may be reached, at which consumption takes a sudden, substantial drop. Suppose the accompanying graph shows the consumption of gasoline, $G(t)$, in millions of gallons, in a certain area. We assume that the price is rising rapidly. Here $t$ is time in months after the price began rising. Use the graph to find the following.

(a) $\lim\limits_{t \to 12} G(t)$    (b) $\lim\limits_{t \to 16} G(t)$    (c) $G(16)$

(d) The tipping point (in months)

## LIFE SCIENCES

**Drug Concentration**    **53.** The concentration of a drug in a patient's bloodstream $h$ hours after it was injected is given by

$$A(h) = \frac{.17h}{h^2 + 2}.$$

Find the following.

**(a)** $A(.5)$   **(b)** $A(1)$   **(c)** $\lim\limits_{h\to 1} A(h)$   **(d)** $\lim\limits_{h\to +\infty} A(h)$

## SOCIAL SCIENCES

**Legislative Voting**   **54.** Members of a legislature often must vote repeatedly on the same bill. As time goes on, members may change their votes. Suppose that $p_0$ is the probability that an individual legislator favors an issue before the first roll call vote, and suppose that $p$ is the probability of a change in position from one vote to the next. Then the probability that the legislator will vote "yes" on the $n$th roll call is given by

$$p_n = \frac{1}{2} + \left(p_0 - \frac{1}{2}\right)(1 - 2p)^n.*$$

For example, the chance of a "yes" on the third roll call vote is

$$p_3 = \frac{1}{2} + \left(p_0 - \frac{1}{2}\right)(1 - 2p)^3.$$

Suppose that there is a chance of $p_0 = .7$ that Congressman Stephens will favor the budget appropriation bill before the first roll call, but only a probability of $p = .2$ that he will change his mind on the subsequent vote. Find and interpret the following.

**(a)** $p_2$   **(b)** $p_4$   **(c)** $p_8$   **(d)** $\lim\limits_{n\to +\infty} p_n$

## FOR THE COMPUTER

*Use a table of values to find the following limits.*

**55.** $\lim\limits_{x\to 1.3} \dfrac{x^3 - 2x^2 + 5x - 8}{2x^4 - 3x^2}$

**56.** $\lim\limits_{x\to 5.2} \dfrac{x^4 - 6x^3 - 10x + 5}{4x^3 + 6x^2 - 12x - 6}$

**57.** $\lim\limits_{x\to 0} \dfrac{x^3 - 4x^2 + 2x}{x^4 + 5x^3 + 4x}$

**58.** $\lim\limits_{x\to 1} \dfrac{3.1x^3 + 5.2\sqrt{x}}{-2x^3 + 12.3x - 2\sqrt{x}}$

*If* $\lim\limits_{x\to a} f(x) = L$, *how close must x be to a for f(x) to be within .01 of L in the limits in Exercises 59 and 60?*

**59.** $\lim\limits_{x\to -1} (3x - 4) = -7$

**60.** $\lim\limits_{x\to -2} \dfrac{x^2 - 4}{x + 2} = -4$

## ▦ 2.2 RATES OF CHANGE

One of the main applications of calculus is determining how one variable changes in relation to another. A person in business wants to know how profit changes with respect to advertising, while a person in medicine wants to know how a patient's reaction to a drug changes with respect to the dose. For example, suppose a manager is testing two different advertising campaigns in

---

*See John W. Bishir and Donald W. Drewes, *Mathematics in the Behavioral and Social Sciences* (New York: Harcourt Brace Jovanovich, 1970), p. 538.

different parts of the country for three months, with the intention of canceling the least effective of the two after that time. The graphs in Figure 7 show sales as a function of time (in months) since the two campaigns began. The graphs suggest that although campaign 1 consistently did better than campaign 2 for the first three months, its *rate of change in sales* is decreasing after that time. The rate of change in sales for campaign 2, however, is increasing at time $t = 3$ months.

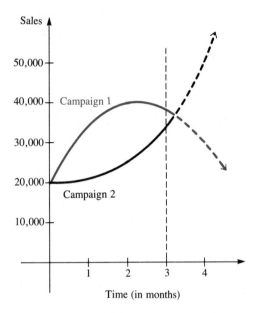

**FIGURE 7**

The **average rate of change** of sales with respect to time in months on some interval is the change in sales divided by the change in time on that interval. On the interval [2, 4], for campaign 1,

$$\text{Average Rate of Change} = \frac{\text{Change in Sales}}{\text{Change in Time}}$$

$$= \frac{30,000 - 40,000}{4 - 2} = -5000.$$

For campaign 2, on [2, 4],

$$\text{Average Rate of Change} = \frac{50,000 - 25,000}{4 - 2} = 12,500.$$

The negative result for campaign 1 indicates that sales were *decreasing* at the rate of 5000 per month over the two-month period, while the positive result for campaign 2 indicates *increasing* sales at the rate of 12,500 over the same two-month period. These rates show that the second campaign will be the most productive in the future, and campaign 1 should be discontinued.

### ■ EXAMPLE 1

The graph in Figure 8 shows the profit $P(x)$, in hundreds of thousands of dollars, from a highly popular new computer program $x$ months after its introduction on the market. The graph shows that profit increases until the program has been on the market for 25 months, after which profit decreases as the popularity of the program decreases. Here the average rate of change of *profit* with respect to *time* on some interval is defined as the change in profit divided by the change in time on the interval.

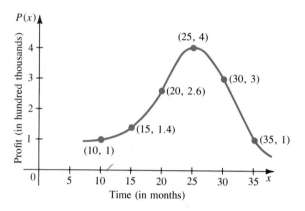

**FIGURE 8**

**(a)** On the interval from 15 to 25 months the average rate of change of profit with respect to time is given by

$$\frac{4 - 1.4}{25 - 15} = \frac{2.6}{10} = .26.$$

On the average, each month from 15 months to 25 months shows an increase in profit of .26 hundred thousand dollars, or $26,000.

**(b)** From 25 months to 30 months, the average rate of change is

$$\frac{3 - 4}{30 - 25} = \frac{-1}{5} = -.20,$$

with each month in this interval showing an average *decline* in profits of .20 hundred thousand dollars, or $20,000.

**(c)** From 10 months to 35 months, the average rate of change of profit is

$$\frac{1 - 1}{35 - 10} = \frac{0}{25} = 0,$$

or $0. ■

**AVERAGE RATE OF CHANGE**

The average rate of change of $f(x)$ with respect to $x$ for a function $f$ as $x$ changes from $a$ to $b$, where $a < b$, is

$$\frac{f(b) - f(a)}{b - a}.$$

As Example 1(c) suggests, finding the average rate of change of a function over a large interval can lead to answers that are not very helpful.

The results often are more useful if the average rate of change is found over a fairly narrow interval, with the usefulness generally increasing as the width of the interval decreases. In fact, the most useful result comes from taking the limit of the average rate of change as the width of the interval approaches 0.

As an example, suppose that for a certain firm, the profit $P(x)$, in thousands of dollars, is related to the volume of production by the mathematical model

$$P(x) = 16x - x^2,$$

where $x$ represents the number of units produced. Assume that the company can produce any nonnegative number of units. (A *unit* here can represent any number of a product; $x$ might represent thousands or tens of thousands of an item.) Naturally, the company wants profits to be as large as possible. A graph of the profit function is shown in Figure 9. The graph shows that an increase in production from $x = 1$ to $x = 2$ will increase profits more than an increase in production from $x = 6$ to $x = 7$. An increase from $x = 7$ to $x = 8$ produces

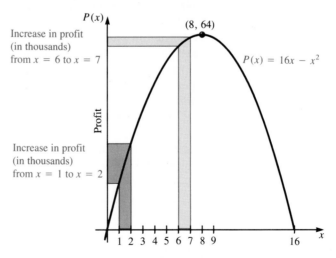

**FIGURE 9**

very little increase in profits, while an increase from $x = 8$ to $x = 9$ actually produces a decline in total profit (perhaps because of additional production costs).

As part of an analysis of the production line, the manager wants to know the rate of change of profit when exactly 1 unit is produced. By taking smaller and smaller intervals near $x = 1$, this rate of change can be approximated as follows. Using the profit function $P(x) = 16x - x^2$, the profit from producing 1 unit is

$$P(1) = 16(1) - 1^2 = 15,$$

or \$15,000. The profit from producing 4 units is

$$P(4) = 16(4) - 4^2 = 48,$$

or \$48,000. The average rate of change in profit for this 3-unit increase in production is found by dividing the change in profits by the change in production to get

$$\text{Average Rate of Change} = \frac{P(4) - P(1)}{4 - 1}$$

$$= \frac{48 - 15}{3}$$

$$= 11,$$

or \$11,000 per unit. The profit from producing 2 units is

$$P(2) = 16(2) - 2^2 = 28,$$

or \$28,000. A 1-unit change in production (from 1 to 2 units) increases profit by $28 - 15 = 13$, or \$13,000, per unit.

An increase in production from 1 unit to 4 units leads to an average rate of increase in profit of 11 thousand dollars per unit, while an increase from 1 unit to 2 units leads to an average rate of increase in profit of 13 thousand dollars per unit. Suppose production changes from 1 unit to $1 + h$ units, where $h$ is a small number; what will happen to the average rate of change in profit? To find out, make a table for some selected values of $h$.

**Average Rate of Change in Profit (in Thousands of Dollars) for a Production Level of 1**

| $h$ | $-.1$ | $-.001$ | $.001$ | $.1$ |
|---|---|---|---|---|
| $1 + h$ | $.9$ | $.999$ | $1.001$ | $1.1$ |
| $P(1 + h)$ | $13.59$ | $14.985999$ | $15.013999$ | $16.39$ |
| $P(1)$ | $15$ | $15$ | $15$ | $15$ |
| Change in Profit $= P(1 + h) - P(1)$ | $-1.41$ | $-.014001$ | $.013999$ | $1.39$ |
| $\dfrac{\text{Average Rate of}}{\text{Change in Profit}} = \dfrac{P(1 + h) - P(1)}{h}$ | $14.1$ | $14.001$ | $13.999$ | $13.9$ |

Average rate of change approaches 14

The numbers in the bottom row of the table are found by dividing the change in profit (row 5) by the change in production (row 1); that is, by evaluating the quotient

$$\frac{P(1 + h) - P(1)}{(1 + h) - 1} = \frac{P(1 + h) - P(1)}{h}$$

for the different values of $h$.

The table suggests that as $h$ approaches 0 (or as production gets closer and closer to $1 + 0 = 1$), the average rate of change of profit approaches the limit 14. This limit,

$$\lim_{h \to 0} \frac{P(1 + h) - P(1)}{h},$$

is called the *instantaneous rate of change,* or just the *rate of change,* of the profit at a production level of $x = 1$.

The rate of change in profit varies as the level of production varies. At the exact instant when production is 1 unit, however, the rate of change in profit is 14 thousand dollars per unit.

**INSTANTANEOUS RATE OF CHANGE**

> The **instantaneous rate of change** for a function $f$ when $x = x_0$ is
>
> $$\lim_{h \to 0} \frac{f(x_0 + h) - f(x_0)}{h},$$
>
> provided this limit exists.

**Velocity**   A good example of an instantaneous rate of change comes from *velocity.* Suppose a car leaves a city at time $t = 0$ and travels due west. Let $s(t)$ represent the position of the car (its distance from the city in kilometers) at time $t$. Suppose $s(t)$ is given by

$$s(t) = 10t^2 + 30t.$$

Since velocity gives the rate of change of distance with respect to time, the average rate of change, or **average velocity,** during the second hour of driving (between time $t = 1$ and $t = 2$) is given by the quotient of the change in distance and the change in time:

$$\frac{s(2) - s(1)}{2 - 1} = \frac{(10 \cdot 2^2 + 30 \cdot 2) - (10 \cdot 1^2 + 30 \cdot 1)}{1}$$

$$= \frac{100 - 40}{1} = 60,$$

or 60 kilometers per hour. To find the instantaneous rate of change of distance at a particular time $t$, use the definition given above. The **instantaneous velocity** at $t = 2$ is

$$\lim_{h \to 0} \frac{s(2 + h) - s(2)}{h}.$$

In this example,

$$s(2 + h) = 10(2 + h)^2 + 30(2 + h)$$
$$= 10(4 + 4h + h^2) + 30(2 + h)$$
$$= 40 + 40h + 10h^2 + 60 + 30h$$
$$= 100 + 70h + 10h^2.$$

Also, $s(2) = 100$. Now find the limit.

$$\lim_{h \to 0} \frac{s(2 + h) - s(2)}{h} = \lim_{h \to 0} \frac{(100 + 70h + 10h^2) - 100}{h}$$
$$= \lim_{h \to 0} \frac{70h + 10h^2}{h}$$
$$= \lim_{h \to 0} (70 + 10h) = 70$$

At the instant that the car is 2 hours out of town, its velocity is 70 kilometers per hour. Even though the car averaged 60 kilometers per hour from $t = 1$ to $t = 2$, the speedometer reading probably varied considerably over that time period. The actual speedometer reading at $t = 2$ hours is the instantaneous velocity at $t = 2$, which is 70 kilometers per hour.

To generalize the results of the car example, suppose an object is moving in a straight line, with its position (distance from some fixed point) given by $s(t)$, where $t$ represents time. The quotient

$$\frac{s(t + h) - s(t)}{h}$$

represents the average rate of change of the distance, or the average velocity of the object. The instantaneous velocity at time $t$ (often called just the velocity at time $t$) is the limit of the quotient above as $h$ approaches 0.

**VELOCITY**

If $v(t)$ represents the velocity at time $t$ of an object moving in a straight line with position $s(t)$, then

$$v(t) = \lim_{h \to 0} \frac{s(t + h) - s(t)}{h},$$

provided this limit exists.

### ▬ EXAMPLE 2

The distance in feet of an object from a starting point is given by $s(t) = 2t^2 - 5t + 40$, where $t$ is time in seconds.

**(a)** Find the average velocity of the object from 2 seconds to 4 seconds.

The average velocity is

$$\frac{s(4) - s(2)}{4 - 2} = \frac{52 - 38}{2} = \frac{14}{2} = 7$$

feet per second.

**(b)** Find the instantaneous velocity at 4 seconds.

For $t = 4$, the instantaneous velocity is

$$\lim_{h \to 0} \frac{s(4 + h) - s(4)}{h}$$

feet per second. Since

$$
\begin{aligned}
s(4 + h) &= 2(4 + h)^2 - 5(4 + h) + 40 \\
&= 2(16 + 8h + h^2) - 20 - 5h + 40 \\
&= 32 + 16h + 2h^2 - 20 - 5h + 40 \\
&= 2h^2 + 11h + 52,
\end{aligned}
$$

and 

$$s(4) = 2(4)^2 - 5(4) + 40 = 52,$$

the instantaneous velocity at $t = 4$ is

$$\lim_{h \to 0} \frac{2h^2 + 11h}{h} = \lim_{h \to 0} (2h + 11) = 11. \quad \blacksquare$$

## ▰▰ EXAMPLE 3

The velocity of blood cells is of interest to physicians; a slower velocity than normal might indicate a constriction, for example. Suppose the position of a red blood cell in a capillary is given by

$$s(t) = 1.2t + 5,$$

where $s(t)$ gives the position of a cell in millimeters from some initial point and $t$ is time in seconds. Find the velocity of this cell at time $t$.

Evaluate the limit given above. To find $s(t + h)$, substitute $t + h$ for the variable $t$ in $s(t) = 1.2t + 5$.

$$s(t + h) = 1.2(t + h) + 5$$

Now use the definition of velocity.

$$
\begin{aligned}
v(t) &= \lim_{h \to 0} \frac{s(t + h) - s(t)}{h} \\
&= \lim_{h \to 0} \frac{1.2(t + h) + 5 - (1.2t + 5)}{h} \\
&= \lim_{h \to 0} \frac{1.2t + 1.2h + 5 - 1.2t - 5}{h} = \lim_{h \to 0} \frac{1.2h}{h} = 1.2
\end{aligned}
$$

The velocity of the blood cell is a constant 1.2 millimeters per second. ▬

▬ EXAMPLE 4

A company determines that the cost in dollars of manufacturing $x$ units of a certain item is

$$C(x) = 100 + 5x - x^2.$$

**(a)** Find the average rate of change of cost per item for manufacturing between 1 and 5 items.

Use the formula for average rate of change. The cost to manufacture 1 item is

$$C(1) = 100 + 5(1) - (1)^2 = 104,$$

or $104. The cost to manufacture 5 items is

$$C(5) = 100 + 5(5) - (5)^2 = 100,$$

or $100. The average rate of change of cost is

$$\frac{C(5) - C(1)}{5 - 1} = \frac{100 - 104}{4}$$

$$= -1.$$

Thus, on the average, cost is decreasing at the rate of $1 per item when production is increased from 1 to 5 items.

**(b)** Find the instantaneous rate of change of cost with respect to the number of items produced when just 1 item is produced.

The instantaneous rate of change for $x = 1$ is given by

$$\lim_{h \to 0} \frac{C(1 + h) - C(1)}{h}$$

$$= \lim_{h \to 0} \frac{[100 + 5(1 + h) - (1 + h)^2] - [100 + 5(1) - 1^2]}{h}$$

$$= \lim_{h \to 0} \frac{[100 + 5 + 5h - 1 - 2h - h^2 - 104]}{h}$$

$$= \lim_{h \to 0} \frac{3h - h^2}{h} \qquad\qquad \text{Combine terms}$$

$$= \lim_{h \to 0} (3 - h) \qquad\qquad \text{Divide by } h$$

$$= 3. \qquad\qquad\qquad \text{Calculate the limit}$$

When 1 item is manufactured, the cost is increasing at the rate of $3 per item. Thus, the instantaneous rate of change of cost represents the approximate cost of manufacturing an additional item. As mentioned in Chapter 1, this rate of change of cost is called the marginal cost.

**(c)** Find the instantaneous rate of change of cost when 5 items are made.

The instantaneous rate of change for $x = 5$ is given by

$$\lim_{h \to 0} \frac{C(5 + h) - C(5)}{h}$$

$$= \lim_{h \to 0} \frac{[100 + 5(5 + h) - (5 + h)^2] - [100 + 5(5) - 5^2]}{h}$$

$$= \lim_{h \to 0} \frac{[100 + 25 + 5h - 25 - 10h - h^2] - 100}{h}$$

$$= \lim_{h \to 0} \frac{-5h - h^2}{h}$$

$$= \lim_{h \to 0} (-5 - h)$$

$$= -5.$$

When 5 items are manufactured, the cost is *decreasing* at the rate of $5 per item—that is, the marginal cost when $x = 5$ is $-$5. Notice that as the number of items produced goes up, the marginal cost goes down, as might be expected. ■

## 2.2 EXERCISES

*Find the average rate of change for each function over the given interval.*

**1.** $y = x^2 + 2x$ between $x = 0$ and $x = 3$

**2.** $y = -4x^2 - 6$ between $x = 2$ and $x = 5$

**3.** $y = 2x^3 - 4x^2 + 6x$ between $x = -1$ and $x = 1$

**4.** $y = -3x^3 + 2x^2 - 4x + 1$ between $x = 0$ and $x = 1$

**5.** $y = \sqrt{x}$ between $x = 1$ and $x = 4$

**6.** $y = \sqrt{3x - 2}$ between $x = 1$ and $x = 2$

**7.** $y = \dfrac{1}{x - 1}$ between $x = -2$ and $x = 0$

**8.** $y = \dfrac{-5}{2x - 3}$ between $x = 2$ and $x = 4$

*Use the properties of limits to find the following limits for $s(t) = t^2 + 5t + 2$.*

**9.** $\lim\limits_{h \to 0} \dfrac{s(6 + h) - s(6)}{h}$

**10.** $\lim\limits_{h \to 0} \dfrac{s(1 + h) - s(1)}{h}$

**11.** $\lim\limits_{h \to 0} \dfrac{s(10 + h) - s(10)}{h}$

*Use the properties of limits to find the following limits for $s(t) = t^3 + 2t + 9$.*

**12.** $\lim\limits_{h \to 0} \dfrac{s(1 + h) - s(1)}{h}$

**13.** $\lim\limits_{h \to 0} \dfrac{s(4 + h) - s(4)}{h}$

APPLICATIONS

### BUSINESS AND ECONOMICS

**Catalog Sales**   **14.** The graph shows the total sales in thousands of dollars from the distribution of *x* thousand catalogs. Find and interpret the average rate of change of sales with respect to the number of catalogs distributed for the following changes in *x*.

(a) 10 to 20
(b) 10 to 40
(c) 20 to 30
(d) 30 to 40

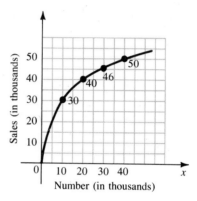

**EXERCISE 14**

**Sales**   **15.** The graph shows annual sales (in units) of a typical product. Sales increase slowly at first to some peak, hold steady for a while, and then decline as the product goes out of style. Find and interpret the average annual rate of change in sales for the following changes in years.

(a) 1 to 3
(b) 2 to 4
(c) 3 to 6
(d) 5 to 7
(e) 7 to 9
(f) 8 to 11
(g) 9 to 10
(h) 10 to 12

**EXERCISE 15**

**Stock Prices**  **16.** The graph shows the monthly closing stock price in dollars per share for National Healthcare from November 1985 to September 1987.* Estimate and interpret the average monthly rate of change of stock prices between the following months.

    **(a)** January 1986 and January 1987

    **(b)** January 1986 and July 1986

    **(c)** August 1986 and January 1987

    **(d)** July 1987 and September 1987

**National Healthcare's Fall**

Monthly closing stock price, in dollars per share

**EXERCISE 16**

**Long-Distance Phone Costs**  **17.** The graph shows the federal phone system long-distance costs from 1963 to 1987.* Estimate and interpret the average annual rate of change of costs during the following periods.

    **(a)** 1985–87　　**(b)** 1982–85　　**(c)** 1965–75　　**(d)** 1975–85

**Gasoline Prices**  **18.** The graph shows the average retail gasoline price for August in the years from 1977 through 1987.* Estimate and interpret the average rate of change of prices during the following periods.

    **(a)** 1977 to 1981　　**(b)** 1981 to 1985　　**(c)** 1985 to 1986　　**(d)** 1986 to 1987

**Government Phone Bill**

Federal phone system long-distance costs (Cents per minute)

Source: General Services Administration

**EXERCISE 17**

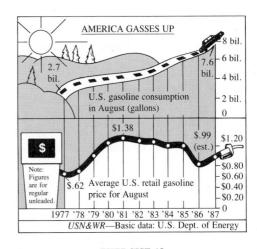

**EXERCISE 18**

**Profit** 19. Profit figures for a small firm over a 7-year period are shown in the following table.

| Year | 1981 | 1982 | 1983 | 1984 | 1985 | 1986 | 1987 |
|---|---|---|---|---|---|---|---|
| Profit (in Dollars) | 5000 | 6000 | 6500 | 6800 | 7200 | 7500 | 8000 |

(a) Find the average rate of change of profit from 1981 to 1987.

(b) Find the average rate of change of profit from 1982 to 1986.

**Profit** 20. Suppose that the total profit in hundreds of dollars from selling $x$ items is given by

$$P(x) = 2x^2 - 5x + 6.$$

Find the average rate of change of profit for the following changes in $x$.

(a) 2 to 4    (b) 2 to 3

(c) Use $P(x) = 2x^2 - 5x + 6$ to complete the following table.

| $h$ | 1 | .1 | .01 | .001 | .0001 |
|---|---|---|---|---|---|
| $2 + h$ | 3 | 2.1 | 2.01 | 2.001 | 2.0001 |
| $P(2 + h)$ | 9 | 4.32 | 4.0302 | 4.003002 | 4.00030002 |
| $P(2)$ | 4 | 4 | 4 | 4 | 4 |
| $P(2 + h) - P(2)$ | 5 | .32 | .0302 | .003002 | .00030002 |
| $\dfrac{P(2 + h) - P(2)}{h}$ | 5 | 3.2 | _____ | _____ | _____ |

(d) Use the bottom row of the chart to find $\lim\limits_{h \to 0} \dfrac{P(2 + h) - P(2)}{h}$.

(e) Use the rule for limits from Section 1 of this chapter to find

$$\lim_{h \to 0} \frac{P(2 + h) - P(2)}{h}.$$

(f) What is the instantaneous rate of change of profit with respect to the number of items produced when $x = 2$? (This number, called the *marginal profit* at $x = 2$, is the approximate profit from producing the third item.)

**Profit** 21. Redo the chart in Exercise 20. This time, change the second line to $4 + h$, the third line to $P(4 + h)$, and so on. Then find

$$\lim_{h \to 0} \frac{P(4 + h) - P(4)}{h}.$$

(As in Exercise 20, this result is the marginal profit, the approximate profit from producing the fifth item.)

**Revenue** 22. The revenue from producing $x$ units of an item is

$$R = 10x - .002x^2.$$

(a) Find the marginal revenue when 1000 units are produced.

(b) Find the additional revenue if production is increased from 1000 to 1001 units.

(c) Find the average rate of change of revenue when production is increased from 1000 to 3000 units.

**Demand** 23. Suppose customers in a hardware store are willing to buy $N(p)$ boxes of nails at $p$ dollars per box, as given by

$$N(p) = 80 - 5p^2, \quad 1 \le p \le 4.$$

(a) Find the average rate of change of demand for a change in price from $2 to $3.

(b) Find the instantaneous rate of change of demand when the price is $2.

(c) Find the instantaneous rate of change of demand when the price is $3.

## LIFE SCIENCES

**Bacteria Population**  **24.** The graph shows the population in millions of bacteria $t$ minutes after a bactericide is introduced into a culture. Find and interpret the average rate of change of population with respect to time for the following time intervals.

(a) 1 to 2
(b) 1 to 3
(c) 2 to 3
(d) 2 to 5
(e) 3 to 4
(f) 4 to 5

**EXERCISE 24**

## SOCIAL SCIENCES

**Memory**  **25.** In an experiment, teenagers memorize a list of 15 words and then listen to rock music for certain periods of time. After $t$ minutes of listening to rock music, the typical subject can remember approximately

$$R(t) = -.03t^2 + 15$$

of the words, where $0 \le t \le 20$. Find the instantaneous rate at which the typical student remembers the words at the following times.

(a) $t = 5$    (b) $t = 15$

## PHYSICAL SCIENCES

**Temperature**  **26.** The graph shows the temperature $T$ in degrees Celsius as a function of the altitude $h$ in feet when an inversion layer is over Southern California. (An inversion layer is formed when air at a higher altitude, say 3000 ft, is warmer than air at sea level, even though air normally is cooler with increasing altitude.) Estimate and interpret the average rate of change in temperature for the following changes in altitude.

(a) 1000 to 3000 ft
(b) 1000 to 5000 ft
(c) 3000 to 9000 ft
(d) 1000 to 9000 ft

**EXERCISE 26**

**Velocity** **27.** A car is moving along a straight test track. The position in feet of the car, $s(t)$, at various times $t$ is measured, with the following results.

| $t$ (seconds) | 0 | 2 | 6 | 10 | 12 | 18 | 20 |
|---|---|---|---|---|---|---|---|
| $s(t)$ (feet) | 0 | 10 | 14 | 18 | 30 | 36 | 40 |

Find and interpret the average velocities for the following changes in $t$.

(a) 0 to 6 sec      (b) 2 to 10 sec      (c) 2 to 12 sec

(d) 6 to 10 sec      (e) 6 to 18 sec      (f) 12 to 20 sec

**Velocity** *In Exercises 28–29, tell which graph, (a) or (b), represents velocity and which represents position.*

**28. (a)**

**(b)**

**29. (a)**

**(b)**

**Velocity** **30.** The distance of a particle from some fixed point is given by

$$s(t) = t^2 + 5t + 2,$$

where $t$ is time measured in seconds.

Find the average velocity of the particle over each of the following intervals.

(a) 4 to 6 sec      (b) 4 to 5 sec

(c) Complete the following table.

| $h$ | 1 | .1 | .01 | .001 | .0001 |
|---|---|---|---|---|---|
| $4 + h$ | 5 | 4.1 | 4.01 | 4.001 | 4.0001 |
| $s(4 + h)$ | 52 | 39.31 | 38.1301 | 38.013001 | 38.00130001 |
| $s(4)$ | 38 | 38 | 38 | 38 | 38 |
| $s(4 + h) - s(4)$ | 14 | 1.31 | .1301 | .013001 | .00130001 |
| $\dfrac{s(4 + h) - s(4)}{h}$ | 14 | 13.1 | ____ | ____ | ____ |

**(d)** Find $\lim\limits_{h \to 0} \dfrac{s(4 + h) - s(4)}{h}$, and give the instantaneous velocity of the particle when $x = 4$.

## 2.3 DEFINITION OF THE DERIVATIVE

In the previous section, the formula

$$\lim_{h \to 0} \frac{f(x_0 + h) - f(x_0)}{h}$$

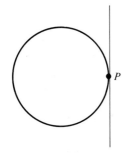

**FIGURE 10**

was used to calculate the instantaneous rate of change of a function $f$ at the point where $x = x_0$. The same formula can be used to find the slope of a line *tangent* to the graph of a function $f$ at the point $x = x_0$. (Remember that the slope of a line is really the rate of change of $y$ with respect to $x$.)

In geometry, a tangent line to a circle is defined as a line that touches the circle at only one point, as at point $P$ in Figure 10. The semicircle in Figure 11 is the graph of a function. Comparing the various lines in Figure 12 to the tangent at point $P$ in Figure 11 suggests that the lines in Figure 12 at $P_1$ and $P_3$ are tangent to the curve, while the lines at $P_2$ and $P_5$ are not. The tangent lines just touch the curve, while the other lines pass through it. To decide about the line at $P_4$, we need to define the idea of a tangent line to the graph of a function more carefully, as follows.

**FIGURE 12**

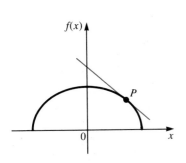

**FIGURE 11**

To determine the slope of a line tangent to the graph of a function $f$ at a given point, let $R$ be a fixed point with coordinates $(x_0, f(x_0))$ on the graph of a function $y = f(x)$, as in Figure 13. Choose a different point $S$ on the graph and draw the line through $R$ and $S$; this line is called a **secant line.** If $S$ has coordinates $(x_0 + h, f(x_0 + h))$, then by the definition of slope, the slope of the secant line $RS$ is

$$\text{Slope of Secant} = \frac{\Delta y}{\Delta x} = \frac{f(x_0 + h) - f(x_0)}{x_0 + h - x_0} = \frac{f(x_0 + h) - f(x_0)}{h}.$$

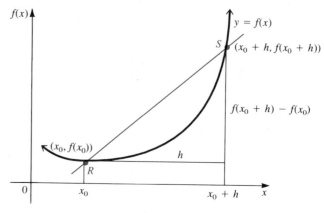

**FIGURE 13**

This slope corresponds to the average rate of change of $y$ with respect to $x$ over the interval from $x_0$ to $x_0 + h$. As $h$ approaches 0, point $S$ will slide along the curve, getting closer and closer to the fixed point $R$. See Figure 14, which shows successive positions $S_1$, $S_2$, $S_3$, and $S_4$ of the point $S$. If the slopes of the corresponding secant lines approach a limit as $h$ approaches 0, then this limit is defined to be the slope of the tangent line at point $R$.

**TANGENT LINE**

> The **tangent line** to the graph of $y = f(x)$ at the point $(x_0, f(x_0))$ is the line through this point having slope
> $$\lim_{h \to 0} \frac{f(x_0 + h) - f(x_0)}{h},$$
> provided this limit exists.

The slope of this line at a point is also called the **slope of the curve** at the point and corresponds to the instantaneous rate of change of $y$ with respect to $x$ at the point.

In certain applications of mathematics it is necessary to determine the equation of a line tangent to the graph of a function at a given point, as in the next example.

### ■ EXAMPLE 1

Find the slope of the tangent line to the graph of $f(x) = x^2 + 2$ at $x = -1$. Find the equation of the tangent line.

Use the definition that was given above, with $f(x) = x^2 + 2$ and $x_0 = -1$.

The slope of the tangent line is

$$\text{Slope of Tangent} = \lim_{h \to 0} \frac{f(x_0 + h) - f(x_0)}{h}.$$

$$= \lim_{h \to 0} \frac{[(-1 + h)^2 + 2] - [(-1)^2 + 2]}{h}$$

$$= \lim_{h \to 0} \frac{[1 - 2h + h^2 + 2] - [1 + 2]}{h}$$

$$= \lim_{h \to 0} \frac{-2h + h^2}{h}$$

$$= \lim_{h \to 0} (-2 + h) = -2.$$

The slope of the tangent line at $(-1, f(-1)) = (-1, 3)$ is $-2$. The equation of the tangent line can be found with the point-slope form of the equation of a line from Chapter 1.

$$y - y_1 = m(x - x_1)$$

$$y - 3 = -2[x - (-1)]$$

$$y - 3 = -2(x + 1)$$

$$y - 3 = -2x - 2$$

$$y = -2x + 1$$

Figure 15 shows a graph of $f(x) = x^2 + 2$, along with a graph of the tangent line at $x = -1$. ▬

**FIGURE 14**

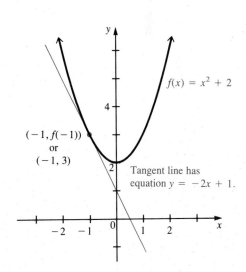

**FIGURE 15**

**The Derivative**    The special limit

$$\lim_{h \to 0} \frac{f(x_0 + h) - f(x_0)}{h}$$

has now been used in two different ways—as the instantaneous rate of change of $y$ with respect to $x$ and also as the slope of a tangent line. This particular limit is so important that it is given a special name, the *derivative*.

**DERIVATIVE**

The **derivative** of the function $f$ at $x$, written $f'(x)$, is defined as

$$f'(x) = \lim_{h \to 0} \frac{f(x + h) - f(x)}{h},$$

provided this limit exists.

The alternative notation $y'$ is sometimes used for the derivative of $y = f(x)$.

Notice that the derivative is a function of $x$ since $f'(x)$ varies as $x$ varies. This differs from the slope of the tangent, given above, in which a specific value of $x$, namely $x_0$, is used.

This new function given by $y' = f'(x)$ has as domain all the points at which the specified limit exists, and its value at $x$ is $f'(x)$. The function $f'$ is called the derivative of $f$ with respect to $x$. If $x$ is a value in the domain of $f$, and if $f'(x)$ exists, then $f$ is **differentiable** at $x$.

In summary, the derivative of a function $f$ is a new function $f'$. The mathematical process that produces $f'$ is **differentiation**. This new function has several interpretations. Two of them are given here.

1. The function $f'$ represents the *slope* of the graph of $f(x)$ at any point $x$. If the derivative is evaluated at the point $x = x_0$, then it represents the slope of the curve at that point.

2. The function $f'$ represents the *instantaneous rate of change* of $y$ with respect to $x$. This instantaneous rate of change could be interpreted as marginal cost, revenue, or profit (if the original function represented cost, revenue, or profit) or velocity (if the original function described displacement along a line). From now on we will use "rate of change" to mean "instantaneous rate of change."

The next few examples show how to use the definition to find the derivative of a function by means of a four-step procedure summarized after Example 5.

**EXAMPLE 2**

Let $f(x) = x^2$.

**(a)** Find the derivative.

By definition, for all values of $x$ where the following limit exists, the derivative is given by

$$f'(x) = \lim_{h \to 0} \frac{f(x + h) - f(x)}{h}.$$

Use the following sequence of steps to evaluate this limit.

*Step 1*   Find $f(x + h)$.

Replace $x$ with $x + h$ in the equation for $f(x)$. Simplify the result.

$$f(x) = x^2$$

$$f(x + h) = (x + h)^2$$

$$= x^2 + 2xh + h^2$$

(Note that $f(x + h) \neq f(x) + h$, since $f(x) + h = x^2 + h$.)

*Step 2*   Find $f(x + h) - f(x)$.

Since $f(x) = x^2$,

$$f(x + h) - f(x) = (x^2 + 2xh + h^2) - x^2 = 2xh + h^2.$$

*Step 3*   Find and simplify the quotient $\dfrac{f(x + h) - f(x)}{h}$.

$$\frac{f(x + h) - f(x)}{h} = \frac{2xh + h^2}{h}$$

$$= \frac{h(2x + h)}{h}$$

$$= 2x + h$$

*Step 4*   Finally, find the limit as $h$ approaches 0.

$$f'(x) = \lim_{h \to 0} \frac{f(x + h) - f(x)}{h}$$

$$= \lim_{h \to 0} (2x + h)$$

$$= 2x + 0$$

$$= 2x$$

**(b)** Calculate and interpret $f'(3)$.

$$f'(3) = 2 \cdot 3 = 6$$

The number 6 is the slope of the tangent line to the graph of $f(x) = x^2$ at the point where $x = 3$, that is, at $(3, f(3)) = (3, 9)$. See Figure 16. ▬

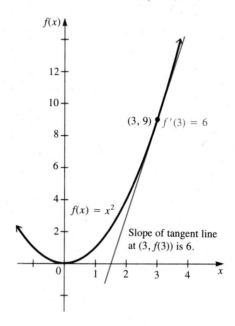

**FIGURE 16**

In Example 2(b), do not confuse $f(3)$ and $f'(3)$. The value $f(3)$ is found by substituting 3 for $x$ in $f(x)$; doing so gives $f(3) = 3^2 = 9$. On the other hand, $f'(3)$ is the slope of the tangent line to the curve at $x = 3$; as Example 2 shows, $f'(3) = 2 \cdot 3 = 6$.

▬ **EXAMPLE 3**

Let $f(x) = 2x^3 + 4x$. Find $f'(x)$, $f'(2)$, and $f'(-3)$.

Go through the four steps used above.

*Step 1* Find $f(x + h)$ by replacing $x$ with $x + h$.

$$\begin{aligned}
f(x + h) &= 2(x + h)^3 + 4(x + h) \\
&= 2(x^3 + 3x^2h + 3xh^2 + h^3) + 4(x + h) \\
&= 2x^3 + 6x^2h + 6xh^2 + 2h^3 + 4x + 4h
\end{aligned}$$

*Step 2*  $f(x + h) - f(x) = 2x^3 + 6x^2h + 6xh^2 + 2h^3 + 4x + 4h - 2x^3 - 4x$

$$= 6x^2h + 6xh^2 + 2h^3 + 4h$$

*Step 3*  $\dfrac{f(x + h) - f(x)}{h} = \dfrac{6x^2h + 6xh^2 + 2h^3 + 4h}{h}$

$$= \dfrac{h(6x^2 + 6xh + 2h^2 + 4)}{h}$$

$$= 6x^2 + 6xh + 2h^2 + 4$$

*Step 4*  Now use the rules for limits to get

$$f'(x) = \lim_{h \to 0} \frac{f(x + h) - f(x)}{h}$$

$$= \lim_{h \to 0} (6x^2 + 6xh + 2h^2 + 4)$$

$$= 6x^2 + 6x(0) + 2(0)^2 + 4$$

$$= 6x^2 + 4.$$

Use this result to find $f'(2)$ and $f'(-3)$.

$$f'(2) = 6 \cdot 2^2 + 4$$

$$= 28$$

$$f'(-3) = 6 \cdot (-3)^2 + 4$$

$$= 58$$ ▬

## ▬ EXAMPLE 4

Let $f(x) = \dfrac{4}{x}$. Find $f'(x)$.

*Step 1*  $f(x + h) = \dfrac{4}{x + h}$

*Step 2*  $f(x + h) - f(x) = \dfrac{4}{x + h} - \dfrac{4}{x}$

$$= \frac{4x - 4(x + h)}{x(x + h)} \qquad \text{Find a common denominator}$$

$$= \frac{4x - 4x - 4h}{x(x + h)} \qquad \text{Simplify the numerator}$$

$$= \frac{-4h}{x(x + h)}$$

*Step 3* $\dfrac{f(x + h) - f(x)}{h} = \dfrac{\dfrac{-4h}{x(x + h)}}{h}$

$= \dfrac{-4h}{x(x + h)} \cdot \dfrac{1}{h}$   Invert and multiply

$= \dfrac{-4}{x(x + h)}$

*Step 4*  $f'(x) = \lim_{h \to 0} \dfrac{f(x + h) - f(x)}{h} = \lim_{h \to 0} \dfrac{-4}{x(x + h)}$

$= \dfrac{-4}{x(x + 0)}$

$= \dfrac{-4}{x(x)} = \dfrac{-4}{x^2}$ ▬

Notice that in Example 4 neither $f(x)$ nor $f'(x)$ is defined when $x = 0$.

## ▬ EXAMPLE 5

Let $f(x) = \sqrt{x}$. Find $f'(x)$.

*Step 1*  $f(x + h) = \sqrt{x + h}$

*Step 2*  $f(x + h) - f(x) = \sqrt{x + h} - \sqrt{x}$

*Step 3*  $\dfrac{f(x + h) - f(x)}{h} = \dfrac{\sqrt{x + h} - \sqrt{x}}{h}$

At this point, in order to be able to divide by $h$, multiply both numerator and denominator by $\sqrt{x + h} + \sqrt{x}$; that is, rationalize the *numerator*.

$\dfrac{f(x + h) - f(x)}{h} = \dfrac{\sqrt{x + h} - \sqrt{x}}{h} \cdot \dfrac{\sqrt{x + h} + \sqrt{x}}{\sqrt{x + h} + \sqrt{x}}$

$= \dfrac{(\sqrt{x + h})^2 - (\sqrt{x})^2}{h(\sqrt{x + h} + \sqrt{x})}$

$= \dfrac{x + h - x}{h(\sqrt{x + h} + \sqrt{x})}$

$= \dfrac{1}{\sqrt{x + h} + \sqrt{x}}$

*Step 4*  $f'(x) = \lim_{h \to 0} \dfrac{1}{\sqrt{x + h} + \sqrt{x}} = \dfrac{1}{\sqrt{x} + \sqrt{x}} = \dfrac{1}{2\sqrt{x}}$ ▬

**FINDING $f'(x)$ FROM THE DEFINITION OF DERIVATIVE**

The four steps used when finding the derivative $f'(x)$ for a function $y = f(x)$ are summarized here.

1. Find $f(x + h)$.
2. Find and simplify $f(x + h) - f(x)$.
3. Divide by $h$ to get $\dfrac{f(x + h) - f(x)}{h}$.
4. Let $h \to 0$; $f'(x) = \lim\limits_{h \to 0} \dfrac{f(x + h) - f(x)}{h}$ if this limit exists.

### EXAMPLE 6

The cost in dollars to manufacture $x$ graphic calculators is given by $C(x) = -.005x^2 + 20x + 150$. Find the rate of change of cost with respect to the number manufactured when 100 calculators are made and when 1000 calculators are made.

The rate of change of cost is given by the derivative of the cost function,

$$C'(x) = \lim_{h \to 0} \frac{C(x + h) - C(x)}{h}.$$

Going through the steps for finding $C'(x)$ gives

$$C'(x) = -.01x + 20.$$

When $x = 100$,

$$C'(100) = -.01(100) + 20 = 19.$$

This rate of change of cost per calculator gives the marginal cost at $x = 100$, which means the approximate cost of producing the 101st calculator is $19. The actual cost of manufacturing the 101st item is found as follows:

$$
\begin{aligned}
\text{Cost of Making 101st Calculator} = \Delta C &= C(101) - C(100) \\
&= [-.005(101)^2 + 20(101) + 150] \\
&\quad - [-.005(100)^2 + 20(100) + 150] \\
&= 2118.995 - 2100 \\
&= 18.995,
\end{aligned}
$$

or $19. Here, the value of $C'(x)$ at $x = 100$ agrees almost exactly with the actual cost of the next item.

When 1000 calculators are made, the marginal cost is

$$C'(1000) = -.01(1000) + 20 = 10,$$

or $10, and the actual cost of the 1001st calculator is

$$C(1001) - C(1000) = [-.005(1001)^2 + 20(1001) + 150]$$
$$-[-.005(1000)^2 + 20(1000) + 150]$$
$$= 15{,}159.995 - 15{,}150$$
$$= 9.995,$$

or $10. Again, the value of $C'(x)$ at a particular value of $x$ gives a very good approximation of the cost of the next item. ■

■■ **EXAMPLE 7**

A sales representative for a textbook publishing company frequently makes a four-hour drive from her home in a large city to a university in another city. If $s(t)$ represents her distance (in miles) from home $t$ hours into the trip, then $s(t)$ is given by

$$s(t) = -5t^3 + 30t^2.$$

**(a)** How far from home will she be after 1 hour? after 1½ hours?

Her distance from home after 1 hour is

$$s(1) = -5(1)^3 + 30(1)^2 = 25,$$

or 25 miles. After $1\frac{1}{2}$ (or $\frac{3}{2}$) hours, her distance from home is

$$s\left(\frac{3}{2}\right) = -5\left(\frac{3}{2}\right)^3 + 30\left(\frac{3}{2}\right)^2 = \frac{405}{8} = 50.625,$$

or 50.625 miles.

**(b)** How far apart are the two cities?

Since the trip takes 4 hours and the distance is given by $s(t)$, the university city is $s(4) = 160$ miles from her home.

**(c)** How fast is she driving 1 hour into the trip? $1\frac{1}{2}$ hours into the trip?

Velocity (or speed) is the instantaneous rate of change in position with respect to time. We need to find the value of the derivative $s'(t)$ at $t = 1$ and $t = 1\frac{1}{2}$. Going through the four steps for finding the derivative gives $s'(t) = -15t^2 + 60t$ as the velocity of the car at any time $t$. At $t = 1$, the velocity is

$$s'(1) = -15(1)^2 + 60(1) = 45,$$

or 45 miles per hour. At $t = 1\frac{1}{2}$, the velocity is

$$s'\left(\frac{3}{2}\right) = -15\left(\frac{3}{2}\right)^2 + 60\left(\frac{3}{2}\right) \approx 56,$$

about 56 miles per hour.

**(d)** Does she ever exceed the speed limit of 65 miles per hour on the trip?

To find the maximum velocity, notice that the graph of the velocity function $s'(t) = -15t^2 + 60t$ is a parabola opening downward. The maximum velocity will occur at the vertex. Verify that the vertex of this parabola is (2, 60). Thus, her maximum velocity during the trip is 60 miles per hour, so she never exceeds the speed limit. ■

***Existence of the Derivative*** The definition of the derivative included the phrase "provided this limit exists." If the limit used to define the derivative does not exist, then of course the derivative does not exist. For example, a derivative cannot exist at a point where the function itself is not defined. If there is no function value for a particular value of $x$, there can be no tangent line for that value. This was the case in Example 4—there was no tangent line (and no derivative) when $x = 0$.

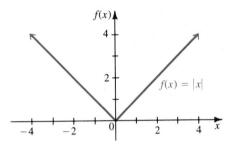

**FIGURE 17**

Derivatives also do not exist at "corners" or "sharp points" on a graph. For example, the function graphed in Figure 17 is the **absolute value function,** defined as

$$f(x) = \begin{cases} x \text{ if } x \geq 0 \\ -x \text{ if } x < 0, \end{cases}$$

and written $f(x) = |x|$. By the definition of derivative, the derivative at any value of $x$ is given by

$$\lim_{h \to 0} \frac{f(x + h) - f(x)}{h},$$

provided this limit exists. To find the derivative at 0 for $f(x) = |x|$, replace $x$ with 0 and $f(x)$ with $|0|$ to get

$$\lim_{h \to 0} \frac{|0 + h| - |0|}{h} = \lim_{h \to 0} \frac{|h|}{h}.$$

For this limit to exist, the limit from the right and the limit from the left must both exist and be the same number. For the limit from the right, let $h > 0$; for the limit from the left, let $h < 0$. If $h > 0$, then $|h| = h$, and

$$\frac{|h|}{h} = \frac{h}{h} = 1,$$

while if $h < 0$, then $|h| = -h$, and

$$\frac{|h|}{h} = \frac{-h}{h} = -1.$$

For positive values of $h$, the quotient $|h|/h$ is 1, while it is $-1$ for negative values of $h$. The two one-sided limits are not the same, and so

$$\lim_{h \to 0} \frac{|h|}{h} \quad \text{does not exist,}$$

and there is no derivative at 0.

The derivative of $f(x) = |x|$ does exist for all values of $x$ other than 0. In fact, $f'(x) = 1$ if $x > 0$ and $f'(x) = -1$ if $x < 0$. A graph of $f'(x)$ is shown in Figure 18.

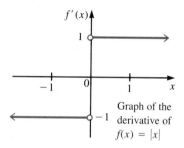

**FIGURE 18**

A graph of the function $f(x) = x^{1/3}$ is shown in Figure 19. As the graph suggests, the tangent line is vertical when $x = 0$. Since a vertical line has an undefined slope, the derivative of $f(x) = x^{1/3}$ cannot exist when $x = 0$.

Figure 20 shows various ways that a derivative can fail to exist.

**FIGURE 19**

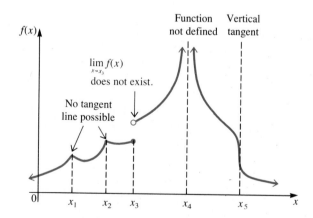

**FIGURE 20**

## ▤ 2.3 EXERCISES

*Find the slope of the tangent line to each curve when x has the given value. (Hint for Exercise 5: In Step 3, multiply numerator and denominator by $\sqrt{16 + h} + \sqrt{16}$.)*

**1.** $f(x) = -4x^2 + 11x; \quad x = -2$     **2.** $f(x) = 6x^2 - 4x; \quad x = -1$     **3.** $f(x) = -2/x; \quad x = 4$

**4.** $f(x) = 6/x; \quad x = -1$            **5.** $f(x) = \sqrt{x}; \quad x = 16$         **6.** $f(x) = -3\sqrt{x}; \quad x = 1$

*Find the equation of the tangent line to each curve when x has the given value.*

**7.** $f(x) = x^2 + 2x; \quad x = 3$          **8.** $f(x) = 6 - x^2; \quad x = -1$     **9.** $f(x) = 5/x; \quad x = 2$

**10.** $f(x) = -3/(x + 1); \quad x = 1$     **11.** $f(x) = 4\sqrt{x}; \quad x = 9$       **12.** $f(x) = \sqrt{x}; \quad x = 25$

*Estimate the slope of the tangent line to each curve at the given point (x, y).*

**13.**

**14.**

**15.**

**16.**

**17.**

**18.**

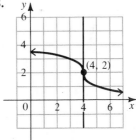

*Find $f'(x)$, $f'(2)$, $f'(0)$, and $f'(-3)$ for each of the following.*

**19.** $f(x) = -4x^2 + 11x$        **20.** $f(x) = 6x^2 - 4x$         **21.** $f(x) = 8x + 6$

**22.** $f(x) = -9x - 5$          **23.** $f(x) = x^3 + 3x$          **24.** $f(x) = 2x^3 - 14x$

**25.** $f(x) = -\dfrac{2}{x}$             **26.** $f(x) = \dfrac{6}{x}$             **27.** $f(x) = \dfrac{4}{x - 1}$

**28.** $f(x) = \dfrac{3}{x + 2}$          **29.** $f(x) = \sqrt{x}$            **30.** $f(x) = -3\sqrt{x}$

*Find the x-values where the following do not have derivatives.*

**31.**

**32.**

**33.**

**34.**

**35.**

**36.**

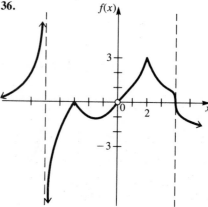

---

### APPLICATIONS

### BUSINESS AND ECONOMICS

**Demand**   **37.** Suppose the demand for a certain item is given by

$$D(p) = -2p^2 + 4p + 6,$$

where $p$ represents the price of the item.

(a) Find $D'(p)$, the rate of change of demand with respect to price.

(b) Find and interpret $D'(10)$.

**Profit**    **38.** The profit from the expenditure of $x$ thousand dollars on advertising is given by
$$P(x) = 1000 + 32x - 2x^2.$$
Find the marginal profit at the following expenditures. In each case, decide whether the firm should increase the expenditure.

(a) $x = 8$    (b) $x = 6$    (c) $x = 12$    (d) $x = 20$

**Revenue**    **39.** The revenue generated from the sale of $x$ picnic tables is given by
$$R(x) = 20x - \frac{x^2}{500}.$$

(a) Find the marginal revenue when $x = 1000$ units.

(b) Estimate the revenue from the sale of the 1001st item by finding $R'(1000)$.

(c) Determine the actual revenue from the sale of the 1001st item.

**Cost**    **40.** The cost of producing $x$ tacos is $C(x) = 1000 + .24x^2$, $0 \le x \le 30,000$.

(a) Find the marginal cost, $C'(x)$.

(b) Find and interpret $C'(100)$.

(c) Find the exact cost to produce the 101st taco.

(d) Compare the answers to (b) and (c). How are they related?

## LIFE SCIENCES

**Bacteria Population**    **41.** A biologist has estimated that if a bactericide is introduced into a culture of bacteria, the number of bacteria, $B(t)$, present at time $t$ (in hours) is given by
$$B(t) = 1000 + 50t - 5t^2 \text{ million.}$$
Find the rate of change of the number of bacteria with respect to time for the following values of $t$.

(a) 2    (b) 3    (c) 4    (d) 5    (e) 6

**Bacteria Population**    **42.** When does the population of bacteria in Exercise 41 start to decline?

**Shellfish Population**    **43.** In one research study, the population of a certain shellfish in an area at time $t$ was closely approximated by the graph below. Estimate the derivative at each of the marked points.

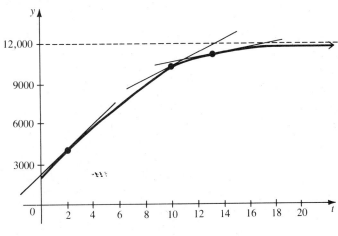

**EXERCISE 43**

## PHYSICAL SCIENCES

**Temperature**  **44.** The graph shows the temperature in degrees Celsius as a function of the altitude $h$ in feet when an inversion layer is over Southern California. (See Exercise 26 in the previous section.) Estimate the derivatives of $T(h)$ at the marked points.

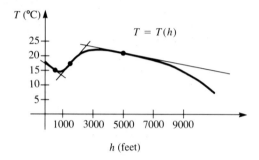

**EXERCISE 44**

## FOR THE COMPUTER

*Find the slope of the tangent line to the graph of each of the following at the given value of x.*

**45.** $f(x) = \dfrac{x^2 + 4.9}{3x^2 - 1.7x}$   at 4.9

**46.** $f(x) = \dfrac{3x - 7.2}{x^2 + 5x + 2.3}$   at $-6.2$

**47.** $f(x) = \sqrt{x^3 + 4.8}$   at 8.17

**48.** $f(x) = \sqrt{6x^2 + 3.5}$   at 3.49

**49.** $f(x) = \dfrac{2.8}{\sqrt{x} + 1.2}$   at 2.65

**50.** $f(x) = \dfrac{-3.7}{4.1 + 2\sqrt{x}}$   at 1.23

---

## ▤ 2.4  TECHNIQUES FOR FINDING DERIVATIVES

In the previous section, the derivative of a function was defined as a special limit. The mathematical process of finding this limit, called *differentiation*, resulted in a new function that was interpreted in several different ways. Using the definition to calculate the derivative of a function is a very involved process even for simple functions. In this section we develop rules that make the calculation of derivatives much easier. Keep in mind that even though the process of finding a derivative will be greatly simplified with these rules, *the interpretation of the derivative will not change.*

Several alternative notations for the derivative are used. In the previous section the symbols $y'$ and $f'(x)$ were used to represent the derivative of $y = f(x)$. Sometimes it is important to show that the derivative is taken with respect to a particular variable; for example, if $y$ is a function of $x$, the notation

$$\frac{dy}{dx} \qquad \text{or} \qquad D_x y$$

can be used for the derivative of $y$ with respect to $x$. The $dy/dx$ notation for the derivative is sometimes referred to as *Leibniz notation*, named after one of the co-inventors of the calculus, Gottfried Wilhelm von Leibniz (1646–1716). (The other was Sir Isaac Newton, 1642–1727.)

Another notation is used to write a derivative without functional symbols such as $f$ or $f'$. With this notation, the derivative of $y = 2x^3 + 4x$, for example, which was found in the last section to be $y' = 6x^2 + 4$, would be written

$$\frac{d}{dx}[2x^3 + 4x] = 6x^2 + 4,$$

or

$$D_x[2x^3 + 4x] = 6x^2 + 4.$$

Either $\dfrac{d}{dx}[f(x)]$ or $D_x[f(x)]$ represents the derivative of the function $f$ with respect to $x$.

**NOTATIONS FOR THE DERIVATIVE**

The derivative of $y = f(x)$ may be written in any of the following ways:

$$f'(x), \qquad y', \qquad \frac{dy}{dx}, \qquad \frac{d}{dx}[f(x)], \qquad \text{or} \qquad D_x[f(x)].$$

A variable other than $x$ often may be used as the independent variable. For example, if $y = f(t)$ gives population growth as a function of time, then the derivative of $y$ with respect to $t$ could be written

$$f'(t), \qquad \frac{dy}{dt}, \qquad \frac{d}{dt}[f(t)], \qquad \text{or} \qquad D_t[f(t)].$$

Now we will use the definition

$$f'(x) = \lim_{h \to 0} \frac{f(x + h) - f(x)}{h}$$

to develop some rules for finding derivatives more easily than by the four-step process given in the previous section.

The first rule tells how to find the derivative of a constant function defined by $f(x) = k$, where $k$ is a constant real number. Since $f(x + h)$ is also $k$, by definition $f'(x)$ is

$$f'(x) = \lim_{h \to 0} \frac{f(x + h) - f(x)}{h}$$

$$= \lim_{h \to 0} \frac{k - k}{h} = \lim_{h \to 0} \frac{0}{h} = \lim_{h \to 0} 0 = 0,$$

establishing the following rule.

**CONSTANT RULE**

> If $f(x) = k$, where $k$ is any real number, then
>
> $$f'(x) = 0.$$
>
> (The derivative of a constant is 0.)

Figure 21 illustrates this constant rule geometrically; it shows a graph of the horizontal line $y = k$. At any point $P$ on this line, the tangent line at $P$ is the line itself. Since a horizontal line has a slope of 0, the slope of the tangent line is 0. This agrees with the result above: the derivative of a constant is 0.

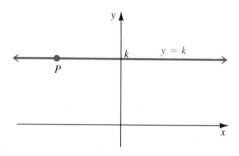

**FIGURE 21**

## ▰ EXAMPLE 1

**(a)** If $f(x) = 9$, then $f'(x) = 0$.

**(b)** If $y = \pi$, then $y' = 0$.

**(c)** If $y = 2^3$, then $dy/dx = 0$.  ▰

Functions of the form $y = x^n$, where $n$ is a real number, are very common in applications. To get a rule for finding the derivative of such a function, we can use the definition to work out the derivatives for various special values of $n$. This was done in the previous section in Example 2 to show that for $f(x) = x^2$, $f'(x) = 2x$.

For $f(x) = x^3$, the derivative is found as follows.

$$\begin{aligned}
f'(x) &= \lim_{h \to 0} \frac{f(x + h) - f(x)}{h} \\
&= \lim_{h \to 0} \frac{(x + h)^3 - x^3}{h} \\
&= \lim_{h \to 0} \frac{(x^3 + 3x^2h + 3xh^2 + h^3) - x^3}{h}
\end{aligned}$$

The binomial theorem (discussed in most intermediate and college algebra texts) was used to expand $(x + h)^3$ in the last step. Now, the limit can be determined.

$$f'(x) = \lim_{h \to 0} \frac{3x^2h + 3xh^2 + h^3}{h}$$

$$= \lim_{h \to 0} (3x^2 + 3xh + h^2)$$

$$= 3x^2$$

The results in the table below were found in a similar way, using the definition of the derivative. (These results are modifications of some of the examples and exercises from the previous section.)

| Function | $n$ | Derivative |
|---|---|---|
| $y = x^2$ | 2 | $y' = 2x = 2x^1$ |
| $y = x^3$ | 3 | $y' = 3x^2$ |
| $y = x^4$ | 4 | $y' = 4x^3$ |
| $y = x^{-1}$ | $-1$ | $y' = -1 \cdot x^{-2} = \dfrac{-1}{x^2}$ |
| $y = x^{1/2}$ | 1/2 | $y' = \dfrac{1}{2}x^{-1/2} = \dfrac{1}{2x^{1/2}}$ |

These results suggest the following rule.

**POWER RULE**

If $f(x) = x^n$ for any real number $n$, then

$$f'(x) = nx^{n-1}.$$

(The derivative of $f(x) = x^n$ is found by multiplying by the exponent $n$ and decreasing the exponent on $x$ by 1.)

While the power rule is true for every real-number value of $n$, a proof is given here only for positive integer values of $n$. This proof follows the steps used above in finding the derivative of $y = x^3$.

For any real numbers $p$ and $q$, by the binomial theorem,

$$(p + q)^n = p^n + np^{n-1}q + \frac{n(n-1)}{2}p^{n-2}q^2 + \cdots + npq^{n-1} + q^n.$$

Replacing $p$ with $x$ and $q$ with $h$ gives

$$(x + h)^n = x^n + nx^{n-1}h + \frac{n(n-1)}{2}x^{n-2}h^2 + \cdots + nxh^{n-1} + h^n,$$

from which

$$(x + h)^n - x^n = nx^{n-1}h + \frac{n(n-1)}{2}x^{n-2}h^2 + \cdots + nxh^{n-1} + h^n.$$

Dividing each term by $h$ yields

$$\frac{(x + h)^n - x^n}{h} = nx^{n-1} + \frac{n(n - 1)}{2}x^{n-2}h + \cdots + nxh^{n-2} + h^{n-1}.$$

Use the definition of derivative, and the fact that each term except the first contains $h$ as a factor and thus approaches 0 as $h$ approaches 0, to get

$$f'(x) = \lim_{h \to 0} \frac{(x + h)^n - x^n}{h}$$

$$= nx^{n-1} + \frac{n(n - 1)}{2}x^{n-2}0 + \cdots + nx0^{n-2} + 0^{n-1}$$

$$f'(x) = nx^{n-1}.$$

This shows that the derivative of $f(x) = x^n$ is $f'(x) = nx^{n-1}$, proving the power rule for positive integer values of $n$.

**POWER RULE**

> If $f(x) = x^n$ for any real number $n$, then
> $$f'(x) = nx^{n-1}.$$

## ■ EXAMPLE 2

**(a)** If $y = x^6$, find $y'$.

$$y' = 6x^{6-1} = 6x^5$$

**(b)** If $y = x = x^1$, find $y'$.

$$y' = 1x^{1-1} = x^0 = 1$$

**(c)** If $y = 1/x^3$, find $dy/dx$.

Use a negative exponent to rewrite this equation as $y = x^{-3}$; then

$$\frac{dy}{dx} = -3x^{-3-1} = -3x^{-4} \quad \text{or} \quad \frac{-3}{x^4}.$$

**(d)** Find $D_x(x^{4/3})$.

$$D_x(x^{4/3}) = \frac{4}{3}x^{4/3-1} = \frac{4}{3}x^{1/3}$$

**(e)** If $y = \sqrt{x}$, find $dy/dx$.

Rewrite this as $y = x^{1/2}$; then

$$\frac{dy}{dx} = \frac{1}{2}x^{1/2-1} = \frac{1}{2}x^{-1/2} \quad \text{or} \quad \frac{1}{2x^{1/2}} \quad \text{or} \quad \frac{1}{2\sqrt{x}}. \quad ■$$

The next rule shows how to find the derivative of the product of a constant and a function.

**CONSTANT TIMES A FUNCTION**

Let $k$ be a real number. If $f'(x)$ exists, then
$$D_x[kf(x)] = kf'(x).$$
(The derivative of a constant times a function is the constant times the derivative of the function.)

This rule is proved with the definition of the derivative and rules for limits.

$$D_x[kf(x)] = \lim_{h \to 0} \frac{kf(x + h) - kf(x)}{h}$$

$$= \lim_{h \to 0} k \frac{[f(x + h) - f(x)]}{h}$$

$$= k \lim_{h \to 0} \frac{f(x + h) - f(x)}{h}$$

$$= kf'(x)$$

## ▆ EXAMPLE 3

**(a)** If $y = 8x^4$, find $y'$.

$$y' = 8(4x^3) = 32x^3$$

**(b)** If $y = -\frac{3}{4}x^{12}$, find $dy/dx$.

$$\frac{dy}{dx} = -\frac{3}{4}(12x^{11}) = -9x^{11}$$

**(c)** Find $D_x(-8x)$.

$$D_x(-8x) = -8(1) = -8$$

**(d)** Find $D_x(10x^{3/2})$.

$$D_x(10x^{3/2}) = 10\left(\frac{3}{2}x^{1/2}\right) = 15x^{1/2}$$

**(e)** If $y = \frac{6}{x}$, find $\frac{dy}{dx}$.

Rewrite this as $y = 6x^{-1}$; then

$$\frac{dy}{dx} = 6(-1x^{-2}) = -6x^{-2} \quad \text{or} \quad \frac{-6}{x^2}. \quad \blacksquare$$

The final rule in this section is for the derivative of a function that is a sum or difference of terms.

**SUM OR DIFFERENCE RULE**

> If $f(x) = u(x) \pm v(x)$, and if $u'(x)$ and $v'(x)$ exist, then
> $$f'(x) = u'(x) \pm v'(x).$$
> (The derivative of a sum or difference of functions is the sum or difference of the derivatives.)

The proof of the sum part of this rule is as follows: If $f(x) = u(x) + v(x)$, then

$$f'(x) = \lim_{h \to 0} \frac{[u(x + h) + v(x + h)] - [u(x) + v(x)]}{h}$$

$$= \lim_{h \to 0} \frac{[u(x + h) - u(x)] + [v(x + h) - v(x)]}{h}$$

$$= \lim_{h \to 0} \left[ \frac{u(x + h) - u(x)}{h} + \frac{v(x + h) - v(x)}{h} \right]$$

$$= \lim_{h \to 0} \frac{u(x + h) - u(x)}{h} + \lim_{h \to 0} \frac{v(x + h) - v(x)}{h}$$

$$= u'(x) + v'(x).$$

A similar proof can be given for the difference of two functions.

## ▰ EXAMPLE 4

Find the derivative of each function.

**(a)** $y = 6x^3 + 15x^2$

Let $f(x) = 6x^3$ and $g(x) = 15x^2$; then $y = f(x) + g(x)$. Since $f'(x) = 18x^2$ and $g'(x) = 30x$,

$$\frac{dy}{dx} = 18x^2 + 30x.$$

**(b)** $p(t) = 12t^4 - 6\sqrt{t} + \dfrac{5}{t}$

Rewrite $p(t)$ as $p(t) = 12t^4 - 6t^{1/2} + 5t^{-1}$; then

$$p'(t) = 48t^3 - 3t^{-1/2} - 5t^{-2}.$$

Also, $p'(t)$ may be written as $p'(t) = 48t^3 - 3/\sqrt{t} - 5/t^2$.

**(c)** $f(x) = 5\sqrt[3]{x^2} + 4x^{-2} + 7$

Rewrite $f(x)$ as $f(x) = 5x^{2/3} + 4x^{-2} + 7$.

Then
$$D_x f(x) = \frac{10}{3} x^{-1/3} - 8x^{-3},$$

or
$$D_x f(x) = \frac{10}{3\sqrt[3]{x}} - \frac{8}{x^3}. \ \blacksquare$$

The rules developed in this section make it possible to find the derivative of a function more directly, so that applications of the derivative can be dealt with more effectively. The following examples illustrate some business applications.

***Marginal Analysis*** In previous sections we discussed the concepts of marginal cost, marginal revenue, and marginal profit. These concepts are summarized here.

In business and economics the rates of change of such variables as cost, revenue, and profits are important considerations. Economists use the word marginal to refer to rates of change: for example, *marginal cost* refers to the rate of change of cost. Since the derivative of a function gives the rate of change of the function, a marginal cost (or revenue, or profit) function is found by taking the derivative of the cost (or revenue, or profit) function. Roughly speaking, the marginal cost at some level of production $x$ is the cost to produce the $(x + 1)$st item. (Similar statements could be made for revenue or profit.)

To see why it is reasonable to say that the marginal cost function is approximately the cost of producing one more unit, look at Figure 22, where $C(x)$ represents the cost of producing $x$ units of some item. Then the cost of producing $x + 1$ units is $C(x + 1)$. The cost of the $x + 1$st unit is therefore $C(x + 1) - C(x)$. This quantity is shown on the graph in Figure 22.

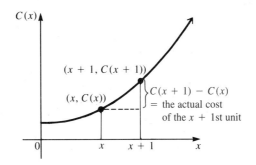

**FIGURE 22**

Now if $C(x)$ is the cost function, then the marginal cost $C'(x)$ represents the slope of the tangent line at any point $(x, C(x))$. The graph in Figure 23 shows the cost function $C(x)$ and the tangent line at a point $(x, C(x))$. Remember what it means for a line to have a given slope. If the slope of the line is $C'(x)$, then

$$\frac{\Delta y}{\Delta x} = C'(x) = \frac{C'(x)}{1},$$

and beginning at any point on the line and moving 1 unit to the right requires moving $C'(x)$ units up to get back to the line again. The vertical distance from the horizontal line to the tangent line shown in Figure 23 is therefore $C'(x)$.

Superimposing the graphs from Figures 22 and 23 as in Figure 24 shows that $C'(x)$ is indeed very close to $C(x + 1) - C(x)$. The two values are closest when $C'(x)$ is very large, so that 1 unit is relatively small.

**FIGURE 23**

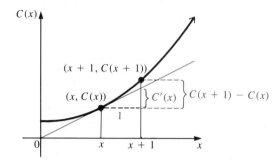

**FIGURE 24**

### ■ EXAMPLE 5

Suppose that the total cost in hundreds of dollars to produce $x$ thousand barrels of a beverage is given by

$$C(x) = 4x^2 + 100x + 500.$$

Find the marginal cost for the following values of $x$.

**(a)** $x = 5$

To find the marginal cost, first find $C'(x)$, the derivative of the total cost function.

$$C'(x) = 8x + 100$$

When $x = 5$,

$$C'(5) = 8(5) + 100 = 140.$$

After 5 thousand barrels of the beverage have been produced, the cost to produce one thousand more barrels will be *approximately* 140 hundred dollars, or $14,000.

The *actual* cost to produce one thousand more barrels is $C(6) - C(5)$:

$$C(6) - C(5) = (4 \cdot 6^2 + 100 \cdot 6 + 500) - (4 \cdot 5^2 + 100 \cdot 5 + 500)$$
$$= 1244 - 1100$$
$$= 144,$$

144 hundred dollars, or $14,400.

**(b)** $x = 30$

After 30 thousand barrels have been produced, the cost to produce one thousand more barrels will be approximately

$$C'(30) = 8(30) + 100 = 340,$$

or $34,000. Notice that the cost to produce an additional thousand barrels of beverage has increased by approximately $20,000 at a production level of 30,000 barrels compared to a production level of 5000 barrels. Management must be careful to keep track of marginal costs. If the marginal cost of producing an extra unit exceeds the revenue received from selling it, then the company will lose money on that unit.  ▬

***Demand Functions***  The demand function, defined by $p = f(x)$, relates the number of units $x$ of an item that consumers are willing to purchase to the price $p$. (Demand functions were also discussed in Chapter 1.) The total revenue $R(x)$ is related to price per unit and the amount demanded (or sold) by the equation

$$R(x) = xp = x \cdot f(x).$$

▬ **EXAMPLE 6**

The demand function for a certain product is given by

$$p = \frac{50,000 - x}{25,000}.$$

Find the marginal revenue when $x = 10,000$ units and $p$ is in dollars.

From the given function for $p$, the revenue function is given by

$$R(x) = xp$$

$$= x\left(\frac{50,000 - x}{25,000}\right)$$

$$= \frac{50,000\,x - x^2}{25,000}$$

$$= 2x - \frac{1}{25,000}\,x^2.$$

The marginal revenue is

$$R'(x) = 2 - \frac{2}{25,000}\,x.$$

When $x = 10,000$, the marginal revenue is

$$R'(10,000) = 2 - \frac{2}{25,000}\,(10,000) = 1.2,$$

or \$1.20 per unit. Thus, the next item sold (at sales of 10,000) will produce additional revenue of about \$1.20 per unit. ▰

In economics the demand function is written in the form $p = f(x)$, as shown above. From the perspective of a consumer, it is probably more reasonable to think of the quantity demanded as a function of price. Mathematically, these two viewpoints are equivalent. In Example 6, the demand function could have been written from the consumer's viewpoint as

$$x = 50{,}000 - 25{,}000p.$$

### ▰ EXAMPLE 7

Suppose that the cost function for the product in Example 6 is given by

$$C(x) = 2100 + .25x, \qquad \text{where } 0 \leq x \leq 30{,}000.$$

Find the marginal profit from the production of the following numbers of units.

**(a)** 15,000

From Example 6, the revenue from the sale of $x$ units is

$$R(x) = 2x - \frac{1}{25{,}000} x^2.$$

Since profit, $P$, is given by $P = R - C$,

$$
\begin{aligned}
P(x) &= R(x) - C(x) \\[2mm]
&= \left( 2x - \frac{1}{25{,}000} x^2 \right) - (2100 + .25x) \\[2mm]
&= 2x - \frac{1}{25{,}000} x^2 - 2100 - .25x \\[2mm]
&= 1.75x - \frac{1}{25{,}000} x^2 - 2100.
\end{aligned}
$$

The marginal profit from the sale of $x$ units is

$$P'(x) = 1.75 - \frac{2}{25{,}000} x = 1.75 - \frac{1}{12{,}500} x.$$

At $x = 15{,}000$ the marginal profit is

$$P'(15{,}000) = 1.75 - \frac{1}{12{,}500} (15{,}000) = .55,$$

or \$.55 per unit.

**(b)** 21,875

When $x = 21,875$, the marginal profit is

$$P'(21,875) = 1.75 - \frac{1}{12,500}(21,875) = 0.$$

**(c)** 25,000

When $x = 25,000$, the marginal profit is

$$P'(25,000) = 1.75 - \frac{1}{12,500}(25,000) = -.25,$$

or $-\$.25$ per unit.

As shown by parts (b) and (c), if more than 21,875 units are sold, the marginal profit is negative. This indicates that increasing production beyond that level will *reduce* profit. ▬

The final example shows a medical application of the derivative as the rate of change of a function.

### ▬ EXAMPLE 8

A tumor has the approximate shape of a cone (see Figure 25). The radius of the base of the tumor is fixed by the bone structure at 2 centimeters, but the tumor is growing along the height of the cone (the volume of the tumor therefore being a function of its height). The formula for the volume of a cone is $V = \frac{1}{3}\pi r^2 h$, where $r$ is the radius of the base and $h$ is the height of the cone. Find the rate of change in the volume of the tumor with respect to the height.

The symbol $dV/dh$ (instead of $V'$) emphasizes that the rate of change of volume is to be found with respect to the height. For this tumor, $r$ is fixed at 2 centimeters. By substituting 2 for $r$,

$$V = \frac{1}{3}\pi r^2 h$$

becomes

$$V = \frac{1}{3}\pi 2^2 \cdot h \quad \text{or} \quad V = \frac{4}{3}\pi h.$$

Since $4\pi/3$ is a constant,

$$\frac{dV}{dh} = \frac{4\pi}{3} \approx 4.2 \text{ cubic centimeters per centimeter.}$$

For each additional centimeter that the tumor grows in height, its volume will increase approximately 4.2 cubic centimeters. ▬

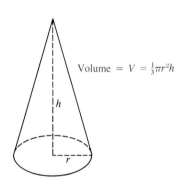

Volume $= V = \frac{1}{3}\pi r^2 h$

**FIGURE 25**

## 2.4 EXERCISES

*Find the derivative of each function defined as follows.*

**1.** $f(x) = 9x^2 - 8x + 4$

**2.** $f(x) = 10x^2 + 4x - 9$

**3.** $y = 10x^3 - 9x^2 + 6x$

**4.** $y = 3x^3 - x^2 - 12x$

**5.** $y = x^4 - 5x^3 + 9x^2 + 5$

**6.** $y = 3x^4 + 11x^3 + 2x^2 - 4x$

**7.** $f(x) = 6x^{1.5} - 4x^5$

**8.** $f(x) = -2x^{2.5} + 8x^5$

**9.** $y = -15x^{3.2} + 2x^{1.9}$

**10.** $y = 18x^{1.6} - 4x^{3.1}$

**11.** $y = 24t^{3/2} + 4t^{1/2}$

**12.** $y = -24t^{5/2} - 6t^{1/2}$

**13.** $y = 8\sqrt{x} + 6x^{3/4}$

**14.** $y = -100\sqrt{x} - 11x^{2/3}$

**15.** $g(x) = 6x^{-5} - x^{-1}$

**16.** $y = -2x^{-4} + x^{-1}$

**17.** $y = -4x^{-2} + 3x^{-3}$

**18.** $g(x) = 8x^{-4} - 9x^{-2}$

**19.** $y = 10x^{-2} + 3x^{-4} - 6x$

**20.** $y = x^{-5} - x^{-2} + 5x^{-1}$

**21.** $f(t) = \dfrac{6}{t} - \dfrac{8}{t^2}$

**22.** $f(t) = \dfrac{4}{t} + \dfrac{2}{t^3}$

**23.** $y = \dfrac{9}{x^4} - \dfrac{8}{x^3} + \dfrac{2}{x}$

**24.** $y = \dfrac{3}{x^6} + \dfrac{1}{x^5} - \dfrac{7}{x^2}$

**25.** $f(x) = 12x^{-1/2} - 3x^{1/2}$

**26.** $r(x) = -30x^{-1/2} + 5x^{1/2}$

**27.** $p(x) = -10x^{-1/2} + 8x^{-3/2}$

**28.** $h(x) = x^{-1/2} - 14x^{-3/2}$

**29.** $y = \dfrac{6}{\sqrt[4]{x}}$

**30.** $y = \dfrac{-2}{\sqrt[3]{x}}$

**31.** $y = \dfrac{-5t}{\sqrt[3]{t^2}}$

**32.** $g(t) = \dfrac{9t}{\sqrt{t^3}}$

*Find each of the following.*

**33.** $\dfrac{dy}{dx}$ if $y = 8x^{-5} - 9x^{-4}$

**34.** $\dfrac{dy}{dx}$ if $y = -3x^{-2} - 4x^{-5}$

**35.** $D_x\left[9x^{-1/2} + \dfrac{2}{x^{3/2}}\right]$

**36.** $D_x\left[\dfrac{8}{\sqrt[4]{x}} - \dfrac{3}{\sqrt{x^3}}\right]$

**37.** $f'(-2)$ if $f(x) = 6x^2 - 4x$

**38.** $f'(3)$ if $f(x) = 9x^3 - 8x^2$

**39.** $f'(4)$ if $f(t) = 2\sqrt{t} - \dfrac{3}{\sqrt{t}}$

**40.** $f'(8)$ if $f(t) = -5\sqrt[3]{t} + \dfrac{6}{\sqrt[3]{t}}$

*In Exercises 41–46 find the slope of the tangent line to the graph of the given function at the given value of x. Find the equation of the tangent line in Exercises 41 and 42.*

**41.** $y = x^4 - 5x^3 + 2$; $x = 2$

**42.** $y = -2x^5 - 7x^3 + 8x^2$; $x = 1$

**43.** $y = 3x^{3/2} - 2x^{1/2}$; $x = 9$

**44.** $y = -2x^{1/2} + x^{3/2}$; $x = 4$

**45.** $y = 5x^{-1} - 2x^{-2}$; $x = 3$

**46.** $y = -x^{-3} + x^{-2}$; $x = 1$

## APPLICATIONS

### BUSINESS AND ECONOMICS

**Revenue** **47.** Assume that a demand equation is given by $x = 5000 - 100p$. Find the marginal revenue for the following production levels (values of $x$). (*Hint:* Solve the demand equation for $p$ and use $R(x) = xp$.)

**(a)** 1000 units **(b)** 2500 units **(c)** 3000 units

**Profit**   **48.** Suppose that for the situation in Exercise 47 the cost of producing $x$ units is given by $C(x) = 3000 - 20x + .03x^2$. Find the marginal profit for each of the following production levels.

(a) 500 units      (b) 815 units      (c) 1000 units

**Demand**   **49.** If the price of a product is given by

$$P(x) = \frac{1000}{x} + 1000,$$

where $x$ represents the demand for the product, find the rate of change of price when the demand is 10.

**Sales**   **50.** Often sales of a new product grow rapidly at first and then level off with time. This is the case with the sales represented by the function

$$S(t) = 100 - 100t^{-1},$$

where $t$ represents time in years. Find the rate of change of sales for the following values of $t$.

(a) $t = 1$      (b) $t = 10$

**Profit**   **51.** The profit in dollars from the sale of $x$ thousand expensive cassette recorders is

$$P(x) = x^3 - 5x^2 + 7x.$$

Find the marginal profit for the following values of $x$.

(a) $x = 1$      (b) $x = 2$      (c) $x = 5$      (d) $x = 10$

(e) Interpret the results from parts (a)–(d).

**Cost**   **52.** The total cost to produce $x$ handcrafted weathervanes is

$$C(x) = 100 + 8x - x^2 + 4x^3.$$

Find the marginal cost for the following values of $x$.

(a) $x = 0$      (b) $x = 4$      (c) $x = 6$      (d) $x = 8$

(e) Interpret the results from parts (a)–(d).

**Profit**   **53.** An analyst has found that a company's costs and revenues for one product are given by

$$C(x) = 2x \quad \text{and} \quad R(x) = 6x - \frac{x^2}{1000},$$

respectively, where $x$ is the number of items produced.

(a) Find the marginal cost function.

(b) Find the marginal revenue function.

(c) Using the fact that profit is the difference between revenue and costs, find the marginal profit function.

(d) What value of $x$ makes marginal profit equal 0?

(e) Find the profit when the marginal profit is 0.

(As we shall see in the next chapter, this process is used to find *maximum* profit.)

## LIFE SCIENCES

**Blood Sugar Level**   **54.** Insulin affects the glucose, or blood sugar, level of some diabetics according to the function

$$G(x) = -.2x^2 + 450,$$

where $G(x)$ is the blood sugar level one hour after $x$ units of insulin are injected.

(This mathematical model is only approximate, and it is valid only for values of $x$ less than about 40.)

(a) Find $G(0)$.     (b) Find $G(25)$.

**Blood Sugar Level** 55. Using function $G$ from Exercise 54, find $dG/dx$ for the following values of $x$. Interpret your answers.

(a) $x = 10$     (b) $x = 25$

**Blood Vessel Volume** 56. A short length of blood vessel has a cylindrical shape. The volume of a cylinder is given by $V = \pi r^2 h$. Suppose an experimental device is set up to measure the volume of blood in a blood vessel of fixed length 80 mm.

(a) Find $dV/dr$.

Suppose a drug is administered that causes the blood vessel to expand. Evaluate $dV/dr$ for the following values of $r$.

(b) 4 mm     (c) 6 mm     (d) 8 mm

**Insect Mating Patterns** 57. In an experiment testing methods of sexually attracting male insects to sterile females, equal numbers of males and females of a certain species are permitted to intermingle. Assume that

$$M(t) = 4t^{3/2} + 2t^{1/2}$$

approximates the number of matings observed among the insects in an hour, where $t$ is the temperature in degrees Celsius. (This formula is valid only for certain temperature ranges.) Find each of the following.

(a) $M(16)$     (b) $M(25)$     (c) The rate of change of $M$ when $t = 16$

## PHYSICAL SCIENCES

**Acid Concentration** 58. Suppose $P(t) = 100/t$ represents the percent of acid in a chemical solution after $t$ days of exposure to an ultraviolet light source.
Find the percent of acid in the solution after the following numbers of days.

(a) 1 day     (b) 100 days

(c) Find and interpret $P'(100)$.

**Velocity** We saw earlier in this chapter that the velocity of a particle moving in a straight line is given by

$$\lim_{h \to 0} \frac{s(t + h) - s(t)}{h},$$

where $s(t)$ gives the position of the particle at time $t$. This limit is actually the derivative of $s(t)$, so the velocity of the particle is given by $s'(t)$. If $v(t)$ represents velocity at time $t$, then $v(t) = s'(t)$. For each of the position functions in Exercises 59–64, find (a) $v(t)$ and (b) the velocity when $t = 0$, $t = 5$, and $t = 10$.

59. $s(t) = 6t + 5$     60. $s(t) = 9 - 2t$
61. $s(t) = 11t^2 + 4t + 2$     62. $s(t) = 25t^2 - 9t + 8$
63. $s(t) = 4t^3 + 8t^2$     64. $s(t) = -2t^3 + 4t^2 - 1$

**Velocity** 65. If a rock is dropped from a 144-ft building, its position (in feet above the ground) is given by $s(t) = -16t^2 + 144$, where $t$ is the time in seconds since it was dropped.

(a) What is its velocity 1 second after being dropped? 2 seconds after being dropped?

**(b)** When will it hit the ground?

**(c)** What is its velocity upon impact?

**Velocity** **66.** A ball is thrown vertically upward from the ground at a velocity of 64 feet per second. Its distance from the ground at $t$ seconds is given by $s(t) = -16t^2 + 64t$.

**(a)** How fast is the ball moving 2 seconds after being thrown? 3 seconds after being thrown?

**(b)** How long after the ball is thrown does it reach its maximum height?

**(c)** How high will it go?

## ▤ 2.5 DERIVATIVES OF PRODUCTS AND QUOTIENTS

In the previous section we developed several rules for finding derivatives. We develop two additional rules in this section, again using the definition of the derivative.

The derivative of a sum of two functions is found from the sum of the derivatives. What about products? Is the derivative of a product equal to the product of the derivatives? For example, if

$$u(x) = 2x + 3 \quad \text{and} \quad v(x) = 3x^2;$$

then $\quad u'(x) = 2 \quad \text{and} \quad v'(x) = 6x.$

Let $f(x)$ be the product of $u$ and $v$; that is, $f(x) = (2x + 3)(3x^2) = 6x^3 + 9x^2$. By the rules of the preceding section, $f'(x) = 18x^2 + 18x$. On the other hand, $u'(x) = 2$ and $v'(x) = 6x$, with the product $u'(x) \cdot v'(x) = 2(6x) = 12x \neq f'(x)$. In this example, the derivative of a product is *not* equal to the product of the derivatives, nor is this usually the case.

The rule for finding derivatives of products is given below.

**PRODUCT RULE**

> ▤ If $f(x) = u(x) \cdot v(x)$, and if $u'(x)$ and $v'(x)$ both exist, then
> $$f'(x) = u(x) \cdot v'(x) + v(x) \cdot u'(x).$$
> (The derivative of a product of two functions is the first function times the derivative of the second, plus the second function times the derivative of the first.)

To sketch the method used to prove the product rule, let

$$f(x) = u(x) \cdot v(x).$$

Then $f(x + h) = u(x + h) \cdot v(x + h)$, and, by definition, $f'(x)$ is given by

$$f'(x) = \lim_{h \to 0} \frac{f(x + h) - f(x)}{h}$$

$$= \lim_{h \to 0} \frac{u(x + h) \cdot v(x + h) - u(x) \cdot v(x)}{h}.$$

Now subtract and add $u(x + h) \cdot v(x)$ in the numerator, giving

$$f'(x) = \lim_{h \to 0} \frac{u(x + h) \cdot v(x + h) - u(x + h) \cdot v(x) + u(x + h) \cdot v(x) - u(x) \cdot v(x)}{h}$$

$$= \lim_{h \to 0} \frac{u(x + h)[v(x + h) - v(x)] + v(x)[u(x + h) - u(x)]}{h}$$

$$= \lim_{h \to 0} u(x + h) \left[ \frac{v(x + h) - v(x)}{h} \right] + \lim_{h \to 0} v(x) \left[ \frac{u(x + h) - u(x)}{h} \right]$$

$$= \lim_{h \to 0} u(x + h) \cdot \lim_{h \to 0} \frac{v(x + h) - v(x)}{h} + \lim_{h \to 0} v(x) \cdot \lim_{h \to 0} \frac{u(x + h) - u(x)}{h}. \quad (*)$$

If $u'$ and $v'$ both exist, then

$$\lim_{h \to 0} \frac{u(x + h) - u(x)}{h} = u'(x) \quad \text{and} \quad \lim_{h \to 0} \frac{v(x + h) - v(x)}{h} = v'(x).$$

The fact that $u'$ exists can be used to prove

$$\lim_{h \to 0} u(x + h) = u(x),$$

and since no $h$ is involved in $v(x)$,

$$\lim_{h \to 0} v(x) = v(x).$$

Substituting these results into equation (*) gives

$$f'(x) = u(x) \cdot v'(x) + v(x) \cdot u'(x),$$

the desired result.

### ▬ EXAMPLE 1

Let $f(x) = (2x + 3)(3x^2)$. Use the product rule to find $f'(x)$.

Here $f$ is given as the product of $u(x) = 2x + 3$ and $v(x) = 3x^2$. By the product rule and the fact that $u'(x) = 2$ and $v'(x) = 6x$,

$$\begin{aligned} f'(x) &= u(x) \cdot v'(x) + v(x) \cdot u'(x) \\ &= (2x + 3)(6x) + (3x^2)(2) \\ &= 12x^2 + 18x + 6x^2 = 18x^2 + 18x. \end{aligned}$$

This result is the same as that found at the beginning of the section. ▬

### ▬ EXAMPLE 2

Find the derivative of $y = (\sqrt{x} + 3)(x^2 - 5x)$.

Let $u(x) = \sqrt{x} + 3 = x^{1/2} + 3$, and $v(x) = x^2 - 5x$. Then

$$\begin{aligned} y' &= u(x) \cdot v'(x) + v(x) \cdot u'(x) \\ &= (x^{1/2} + 3)(2x - 5) + (x^2 - 5x)\left( \frac{1}{2} x^{-1/2} \right). \end{aligned}$$

Simplify by multiplying and combining terms.

$$y' = (2x)(x^{1/2}) + 6x - 5x^{1/2} - 15 + (x^2)\left(\frac{1}{2}x^{-1/2}\right) - (5x)\left(\frac{1}{2}x^{-1/2}\right)$$

$$= 2x^{3/2} + 6x - 5x^{1/2} - 15 + \frac{1}{2}x^{3/2} - \frac{5}{2}x^{1/2}$$

$$= \frac{5}{2}x^{3/2} + 6x - \frac{15}{2}x^{1/2} - 15 \quad \blacksquare$$

We could have found the derivatives above by multiplying out the original functions. The product rule then would not have been needed. In the next section, however, we shall see products of functions where the product rule is essential.

What about *quotients* of functions? To find the derivative of the quotient of two functions, use the next result.

**QUOTIENT RULE**

> If $f(x) = \dfrac{u(x)}{v(x)}$, if all indicated derivatives exist, and if $v(x) \neq 0$, then
>
> $$f'(x) = \frac{v(x) \cdot u'(x) - u(x) \cdot v'(x)}{[v(x)]^2}$$
>
> (The derivative of a quotient is the denominator times the derivative of the numerator, minus the numerator times the derivative of the denominator, all divided by the square of the denominator.)

The proof of the quotient rule is similar to that of the product rule and is left for the exercises.

■ **EXAMPLE 3**

Find $f'(x)$ if $f(x) = \dfrac{2x - 1}{4x + 3}$.

Let $u(x) = 2x - 1$, with $u'(x) = 2$. Also, let $v(x) = 4x + 3$, with $v'(x) = 4$. Then, by the quotient rule,

$$f'(x) = \frac{v(x) \cdot u'(x) - u(x) \cdot v'(x)}{[v(x)]^2}$$

$$= \frac{(4x + 3)(2) - (2x - 1)(4)}{(4x + 3)^2}$$

$$= \frac{8x + 6 - 8x + 4}{(4x + 3)^2}$$

$$= \frac{10}{(4x + 3)^2}. \quad \blacksquare$$

■■■ EXAMPLE 4

Find $D_x\left(\dfrac{x^{-1} - 2}{x^{-2} + 4}\right)$.

Use the quotient rule.

$$D_x\left(\frac{x^{-1} - 2}{x^{-2} + 4}\right) = \frac{(x^{-2} + 4)(-x^{-2}) - (x^{-1} - 2)(-2x^{-3})}{(x^{-2} + 4)^2}$$

$$= \frac{-x^{-4} - 4x^{-2} + 2x^{-4} - 4x^{-3}}{(x^{-2} + 4)^2}$$

$$= \frac{x^{-4} - 4x^{-3} - 4x^{-2}}{(x^{-2} + 4)^2}$$

When derivatives are used to solve practical problems, the work is easier if the derivatives are first simplified. This derivative, for example, can be simplified if the negative exponents are removed by multiplying numerator and denominator by $x^4$ and using $(x^{-2} + 4)^2 = x^{-4} + 8x^{-2} + 16$.

$$\frac{x^{-4} - 4x^{-3} - 4x^{-2}}{(x^{-2} + 4)^2} = \frac{1 - 4x - 4x^2}{x^4(x^{-4} + 8x^{-2} + 16)}$$

$$= \frac{1 - 4x - 4x^2}{1 + 8x^2 + 16x^4} \quad ■■$$

■■■ EXAMPLE 5

Find $D_x\left(\dfrac{(3 - 4x)(5x + 1)}{7x - 9}\right)$.

This function has a product within a quotient. Instead of multiplying the factors in the numerator first (which is an option), we can use the quotient rule together with the product rule, as follows. Use the quotient rule first to get

$$D_x\left(\frac{(3 - 4x)(5x + 1)}{7x - 9}\right)$$

$$= \frac{(7x - 9)[D_x(3 - 4x)(5x + 1)] - [(3 - 4x)(5x + 1)D_x(7x - 9)]}{(7x - 9)^2}$$

Now use the product rule to find $D_x(3 - 4x)(5x + 1)$ in the numerator.

$$= \frac{(7x - 9)[(3 - 4x)5 + (5x + 1)(-4)] - (3 + 11x - 20x^2)(7)}{(7x - 9)^2}$$

$$= \frac{(7x - 9)(15 - 20x - 20x - 4) - (21 + 77x - 140x^2)}{(7x - 9)^2}$$

$$= \frac{(7x - 9)(11 - 40x) - 21 - 77x + 140x^2}{(7x - 9)^2}$$

$$= \frac{-280x^2 + 437x - 99 - 21 - 77x + 140x^2}{(7x - 9)^2}$$

$$= \frac{-140x^2 + 360x - 120}{(7x - 9)^2} \quad \blacksquare$$

***Average Cost*** Suppose $y = C(x)$ gives the total cost to manufacture $x$ items. As mentioned earlier, the average cost per item is found by dividing the total cost by the number of items. The rate of change of average cost, called the *marginal average cost,* is the derivative of the average cost.

**AVERAGE COST**

If the total cost to manufacture $x$ items is given by $C(x)$, then the **average cost per item** is

$$\overline{C}(x) = \frac{C(x)}{x}.$$

The **marginal average cost** is the derivative of the average cost function, $\overline{C}'(x)$.

A company naturally would be interested in making the average cost as small as possible. The next chapter will show that this can be done by using the derivative of $C(x)/x$. This derivative often can be found by means of the quotient rule, as in the next example.

**▬ EXAMPLE 6**

The total cost in thousands of dollars to manufacture $x$ electrical generators is given by $C(x)$, where

$$C(x) = x^3 + 15x^2 + 1000.$$

**(a)** Find the average cost per generator.

The average cost is given by the total cost divided by the number of items, or

$$\overline{C}(x) = \frac{C(x)}{x} = \frac{x^3 + 15x^2 + 1000}{x}.$$

**(b)** Find the marginal average cost.

The marginal average cost is the derivative of the average cost function.

Using the quotient rule,

$$\frac{d}{dx}\left[\overline{C}(x)\right] = \frac{x(3x^2 + 30x) - (x^3 + 15x^2 + 1000)(1)}{x^2}$$

$$= \frac{3x^3 + 30x^2 - x^3 - 15x^2 - 1000}{x^2}$$

$$= \frac{2x^3 + 15x^2 - 1000}{x^2}. \quad \blacksquare$$

### ▄▄ EXAMPLE 7

Suppose the cost in dollars of manufacturing $x$ hundred items is given by

$$C(x) = 3x^2 + 7x + 12.$$

**(a)** Find the average cost.

The average cost is

$$\overline{C}(x) = \frac{C(x)}{x} = \frac{3x^2 + 7x + 12}{x} = 3x + 7 + \frac{12}{x}.$$

**(b)** Find the marginal average cost.

The marginal average cost is

$$\frac{d}{dx}(\overline{C}(x)) = \frac{d}{dx}\left(3x + 7 + \frac{12}{x}\right) = 3 - \frac{12}{x^2}.$$

**(c)** Find the marginal cost.

The marginal cost is

$$\frac{d}{dx}(C(x)) = \frac{d}{dx}(3x^2 + 7x + 12) = 6x + 7.$$

**(d)** Average cost is generally minimized when the marginal average cost is zero. Find the level of production that minimizes average cost.

Set the derivative $\overline{C}'(x) = 0$ and solve for $x$.

$$3 - \frac{12}{x^2} = 0$$

$$\frac{3x^2 - 12}{x^2} = 0$$

$$3x^2 - 12 = 0$$

$$x^2 = 4$$

$$x = 2$$

Since $x$ is in hundreds, production of 200 items will minimize average cost. ▄▄

## 2.5 EXERCISES

*Use the product rule to find the derivative of each of the following. (Hint for Exercise 11: $a^2 = a \cdot a$.)*

**1.** $y = (2x - 5)(x + 4)$

**2.** $y = (3x + 7)(x - 1)$

**3.** $f(x) = (8x - 2)(3x + 9)$

**4.** $f(x) = (4x + 1)(7x + 12)$

**5.** $y = (3x^2 + 2)(2x - 1)$

**6.** $y = (5x^2 - 1)(4x + 3)$

**7.** $y = (t^2 + t)(3t - 5)$

**8.** $y = (2t^2 - 6t)(t + 2)$

**9.** $y = (9x^2 + 7x)(x^2 - 1)$

**10.** $y = (2x^2 - 4x)(5x^2 + 4)$

**11.** $y = (2x - 5)^2$

**12.** $y = (7x - 6)^2$

**13.** $k(t) = (t^2 - 1)^2$

**14.** $g(t) = (3t^2 + 2)^2$

**15.** $y = (x + 1)(\sqrt{x} + 2)$

**16.** $y = (2x - 3)(\sqrt{x} - 1)$

**17.** $g(x) = (5\sqrt{x} - 1)(2\sqrt{x} + 1)$

**18.** $g(x) = (-3\sqrt{x} + 6)(4\sqrt{x} - 2)$

*Use the quotient rule to find the derivative of each of the following.*

**19.** $y = \dfrac{x + 1}{2x - 1}$

**20.** $y = \dfrac{3x - 5}{x - 4}$

**21.** $f(x) = \dfrac{7x + 1}{3x + 8}$

**22.** $f(x) = \dfrac{6x - 11}{8x + 1}$

**23.** $y = \dfrac{2}{3x - 5}$

**24.** $y = \dfrac{-4}{2x - 11}$

**25.** $y = \dfrac{5 - 3t}{4 + t}$

**26.** $y = \dfrac{9 - 7t}{1 - t}$

**27.** $y = \dfrac{x^2 + x}{x - 1}$

**28.** $y = \dfrac{x^2 - 4x}{x + 3}$

**29.** $f(t) = \dfrac{t - 2}{t^2 + 1}$

**30.** $f(t) = \dfrac{4t + 11}{t^2 - 3}$

**31.** $y = \dfrac{3x^2 + x}{2x^2 - 1}$

**32.** $y = \dfrac{-x^2 + 6x}{4x^2 + 1}$

**33.** $g(x) = \dfrac{x^2 - 4x + 2}{x + 3}$

**34.** $k(x) = \dfrac{x^2 + 7x - 2}{x - 2}$

**35.** $p(t) = \dfrac{\sqrt{t}}{t - 1}$

**36.** $r(t) = \dfrac{\sqrt{t}}{2t + 3}$

**37.** $y = \dfrac{5x + 6}{\sqrt{x}}$

**38.** $y = \dfrac{9x - 8}{\sqrt{x}}$

**39.** Find an equation of the line tangent to the graph of $f(x) = x/(x - 2)$ at $(3, 3)$.

**40.** Following the steps used to prove the product rule for derivatives, prove the quotient rule for derivatives.

## APPLICATIONS

### BUSINESS AND ECONOMICS

**Average Cost** **41.** The total cost to produce $x$ units of perfume is
$$C(x) = (3x + 2)(3x + 4).$$
Find the average cost for each of the following production levels.
**(a)** 10 units  **(b)** 20 units  **(c)** $x$ units
**(d)** Find the marginal average cost function.

**Average Profit** **42.** The total profit from selling $x$ units of self-help books is
$$P(x) = (5x - 6)(2x + 3).$$
Find the average profit from each of the following sales levels.
**(a)** 8 units  **(b)** 15 units  **(c)** $x$ units
**(d)** Find the marginal average profit function.

**Fuel Efficiency**    **43.** Suppose you are the manager of a trucking firm, and one of your drivers reports that she has calculated that her truck burns fuel at the rate of

$$G(x) = \frac{1}{200}\left(\frac{800 + x^2}{x}\right)$$

gallons per mile when traveling at $x$ miles per hour on a smooth, dry road.

(a) If the driver tells you that she wants to travel 20 miles per hour, what should you tell her? (*Hint:* Take the derivative of $G$ and evaluate it for $x = 20$. Then interpret your results.)

(b) If the driver wants to go 40 miles per hour, what should you say? (*Hint:* Find $G'(40)$.)

**Employee Training**    **44.** A company that manufactures bicycles has determined that a new employee can assemble $M(d)$ bicycles per day after $d$ days of on-the-job training, where

$$M(d) = \frac{200d}{3d + 10}.$$

(a) Find $M'(d)$.

(b) Find and interpret $M'(2)$ and $M'(5)$.

## LIFE SCIENCES

**Muscle Reaction**    **45.** When a certain drug is injected into a muscle, the muscle responds by contracting. The amount of contraction, $s$, in millimeters, is related to the concentration of the drug, $x$, in milliliters, by

$$s(x) = \frac{x}{m + nx},$$

where $m$ and $n$ are constants.

(a) Find $s'(x)$.

(b) Evaluate $s'(x)$ when $x = 50$, $m = 10$, and $n = 3$.

**Bacteria Population**    **46.** Assume that the total number (in millions) of bacteria present in a culture at a certain time $t$ is given by

$$N(t) = (t - 10)^2(2t) + 50.$$

(a) Find $N'(t)$.

Find the rate at which the population of bacteria is changing at each of the following times.

(b) $t = 8$    (c) $t = 11$

(d) The answer in part (b) is negative, and the answer in part (c) is positive. What does this mean in terms of the population of bacteria?

## SOCIAL SCIENCES

**Memory Retention**    **47.** Some psychologists contend that the number of facts of a certain type that are remembered after $t$ hr is given by

$$f(t) = \frac{kt}{at - m},$$

where $k$, $m$, and $a$ are constants. Find $f'(t)$ for each of the following times if $a = 99$, $k = 90$, and $m = 90$.

(a) $t = 1$    (b) $t = 10$

## 2.6 THE CHAIN RULE

Many of the most useful functions for modeling are created by combining simpler functions. Viewing complex functions as combinations of simpler functions often makes them easier to understand and use.

***Composition of Functions***   Suppose a function $f$ assigns to each element $x$ in set $X$ some element $y = f(x)$ in set $Y$. Suppose also that a function $g$ takes each element in set $Y$ and assigns to it a value $z = g[f(x)]$ in set $Z$. By using both $f$ and $g$, an element $x$ in $X$ is assigned to an element $z$ in $Z$, as illustrated in Figure 26.

**FIGURE 26**

The result of this process is a new function called the *composition* of functions $g$ and $f$ and defined as follows.

**COMPOSITE FUNCTION**

> Let $f$ and $g$ be functions. The **composite function,** or **composition,** of $g$ and $f$ is the function whose values are given by $g[f(x)]$ for all $x$ in the domain of $f$ such that $f(x)$ is in the domain of $g$.

## EXAMPLE 1

Let $f(x) = 2x - 1$ and $g(x) = \sqrt{3x + 5}$. Find each of the following.

**(a)** $g[f(4)]$

Find $f(4)$ first.

$$f(4) = 2 \cdot 4 - 1 = 8 - 1 = 7$$

Then

$$g[f(4)] = g[7] = \sqrt{3 \cdot 7 + 5} = \sqrt{26}.$$

**(b)** $f[g(4)]$

Since $g(4) = \sqrt{3 \cdot 4 + 5} = \sqrt{17}$,

$$f[g(4)] = 2 \cdot \sqrt{17} - 1 = 2\sqrt{17} - 1.$$

**(c)** $f[g(-2)]$ does not exist since $-2$ is not in the domain of $g$. ■

■■■ **EXAMPLE 2**

Let $f(x) = 4x + 1$ and $g(x) = 2x^2 + 5x$. Find each of the following.

**(a)** $g[f(x)]$

Using the given functions,

$$
\begin{aligned}
g[f(x)] &= g[4x + 1] \\
&= 2(4x + 1)^2 + 5(4x + 1) \\
&= 2(16x^2 + 8x + 1) + 20x + 5 \\
&= 32x^2 + 16x + 2 + 20x + 5 \\
&= 32x^2 + 36x + 7.
\end{aligned}
$$

**(b)** $f[g(x)]$

By the definition above, with $f$ and $g$ interchanged,

$$
\begin{aligned}
f[g(x)] &= f[2x^2 + 5x] \\
&= 4(2x^2 + 5x) + 1 \\
&= 8x^2 + 20x + 1. \quad ■■
\end{aligned}
$$

As Example 2 shows, it is not always true that $f[g(x)] = g[f(x)]$. (In fact, it is rare to find two functions $f$ and $g$ such that $f[g(x)] = g[f(x)]$. The domain of both composite functions given in Example 2 is the set of all real numbers.

**The Chain Rule** A leaking oil well off the Gulf Coast is spreading a circular film of oil over the water surface. At any time $t$, in minutes, after the beginning of the leak, the radius of the circular oil slick is given by

$$
r(t) = 4t, \qquad \text{with} \qquad \frac{dr}{dt} = 4,
$$

where $dr/dt$ is the rate of change in radius over time. The area of the oil slick is given by

$$
A(r) = \pi r^2, \qquad \text{with} \qquad \frac{dA}{dr} = 2\pi r,
$$

where $dA/dr$ is the rate of change in area per unit change in radius.

As these derivatives show, the radius is increasing 4 times as fast as the time $t$, and the area is increasing $2\pi r$ times as fast as the radius $r$. It seems reasonable, then, that the area is increasing $2\pi r \cdot 4 = 8\pi r$ times as fast as time. That is,

$$
\frac{dA}{dt} = \frac{dA}{dr} \cdot \frac{dr}{dt} = 2\pi r \cdot 4 = 8\pi r.
$$

Since $r = 4t$,

$$\frac{dA}{dt} = 8\pi(4t) = 32\pi t.$$

To check this, use the fact that $r = 4t$ and $A = \pi r^2$ to get the same result:

$$A = \pi(4t)^2 = 16\pi t^2, \quad \text{with} \quad \frac{dA}{dt} = 32\pi t.$$

The product used above,

$$\frac{dA}{dt} = \frac{dA}{dr} \cdot \frac{dr}{dt},$$

is an example of the **chain rule,** which is used to find the derivative of a composite function.

**CHAIN RULE**

> If $y$ is a function of $u$, say $y = f(u)$, and if $u$ is a function of $x$, say $u = g(x)$, then $y = f(u) = f[g(x)]$, and
>
> $$\frac{dy}{dx} = \frac{dy}{du} \cdot \frac{du}{dx}.$$

One way to remember the chain rule is to pretend that $dy/du$ and $du/dx$ are fractions, with $du$ "canceling out." The proof of the chain rule requires advanced concepts and therefore is not given here.

### EXAMPLE 3

Find $dy/dx$ if $y = (3x^2 - 5x)^{1/2}$.
   Let $y = u^{1/2}$, and $u = 3x^2 - 5x$. Then

$$\frac{dy}{dx} = \frac{dy}{du} \cdot \frac{du}{dx}$$

$$= \frac{1}{2}u^{-1/2} \cdot (6x - 5).$$

Replacing $u$ with $3x^2 - 5x$ gives

$$\frac{dy}{dx} = \frac{1}{2}(3x^2 - 5x)^{-1/2}(6x - 5) = \frac{6x - 5}{2(3x^2 - 5x)^{1/2}}.$$

The following alternative version of the chain rule is stated in terms of composite functions.

**CHAIN RULE
(ALTERNATE FORM)**

If $y = f[g(x)]$, then
$$y' = f'[g(x)] \cdot g'(x).$$
(To find the derivative of $f[g(x)]$, find the derivative of $f(x)$, replace each $x$ with $g(x)$, and then multiply the result by the derivative of $g(x)$.)

## EXAMPLE 4
Use the chain rule to find $D_x\sqrt{15x^2 + 1}$.

Write $\sqrt{15x^2 + 1}$ as $(15x^2 + 1)^{1/2}$. Let $f(x) = x^{1/2}$ and $g(x) = 15x^2 + 1$. Then $\sqrt{15x^2 + 1} = f[g(x)]$ and
$$D_x(15x^2 + 1)^{1/2} = f'[g(x)] \cdot g'(x).$$

Here $f'(x) = \frac{1}{2}x^{-1/2}$, with $f'[g(x)] = \frac{1}{2}[g(x)]^{-1/2} = \frac{1}{2}(15x^2 + 1)^{-1/2}$, and

$$D_x\sqrt{15x^2 + 1} = \frac{1}{2}[g(x)]^{-1/2} \cdot g'(x)$$

$$= \frac{1}{2}(15x^2 + 1)^{-1/2} \cdot (30x)$$

$$= \frac{15x}{(15x^2 + 1)^{1/2}}.$$

While the chain rule is essential for finding the derivatives of some of the functions discussed later, the derivatives of the algebraic functions discussed so far can be found by the following *generalized power rule*, a special case of the chain rule.

**GENERALIZED
POWER RULE**

Let $u$ be a function of $x$, and let $y = u^n$, for any real number $n$. Then
$$y' = n \cdot u^{n-1} \cdot u'.$$
(The derivative of $y = u^n$ is found by decreasing the exponent on $u$ by 1 and multiplying the result by the exponent $n$ and by the derivative of $u$ with respect to $x$.)

## EXAMPLE 5

**(a)** Use the generalized power rule to find the derivative of $y = (3 + 5x)^2$. Let $u = 3 + 5x$, and $n = 2$. Then $u' = 5$. By the generalized power rule,

$$y' = \frac{dy}{dx} = n \cdot u^{n-1} \cdot u'$$

$$n \qquad u \qquad n-1 \qquad u'$$
$$\downarrow \qquad \downarrow \qquad \downarrow \qquad \downarrow$$

$$= 2 \cdot (3 + 5x)^{2-1} \cdot \frac{d}{dx}(3 + 5x)$$

$$= 2(3 + 5x)^{2-1} \cdot 5 = 10(3 + 5x)$$

$$= 30 + 50x.$$

**(b)** Find $y'$ if $y = (3 + 5x)^{-3/4}$.

Use the generalized power rule, with $u = 3 + 5x$, $n = -3/4$, and $u' = 5$.

$$y' = -\frac{3}{4}(3 + 5x)^{-3/4-1} \cdot 5 = -\frac{15}{4}(3 + 5x)^{-7/4}$$

This result could not have been found by any of the rules given in previous sections. ■

### ■ EXAMPLE 6

Find the derivative of each function.

**(a)** $y = 2(7x^2 + 5)^4$

Let $u = 7x^2 + 5$. Then $u' = 14x$, and

$$n \qquad u \quad n-1 \qquad u'$$
$$\downarrow \qquad \downarrow \quad \downarrow \qquad \downarrow$$

$$y' = 2 \cdot 4(7x^2 + 5)^{4-1} \cdot \frac{d}{dx}(7x^2 + 5)$$

$$= 2 \cdot 4(7x^2 + 5)^3(14x)$$

$$= 112x(7x^2 + 5)^3.$$

**(b)** $y = \sqrt{9x + 2}$

Write $y = \sqrt{9x + 2}$ as $y = (9x + 2)^{1/2}$. Then

$$y' = \frac{1}{2}(9x + 2)^{-1/2}(9) = \frac{9}{2}(9x + 2)^{-1/2}.$$

The derivative also can be written as

$$y' = \frac{9}{2(9x + 2)^{1/2}} \quad \text{or} \quad y' = \frac{9}{2\sqrt{9x + 2}}. \quad ■$$

Sometimes both the generalized power rule and either the product or quotient rule are needed to find a derivative, as the next examples show.

### ■ EXAMPLE 7

Find the derivative of $y = 4x(3x + 5)^5$.
  Write $4x(3x + 5)^5$ as the product

$$(4x) \cdot (3x + 5)^5.$$

To find the derivative of $(3x + 5)^5$, let $u = 3x + 5$, with $u' = 3$. Now use the product rule and the generalized power rule.

$$\begin{array}{ll}
& \overbrace{\text{Derivative of } (3x + 5)^5} \qquad \overbrace{\text{Derivative of } 4x} \\
y' = & 4x[5(3x + 5)^4 \cdot 3] + (3x + 5)^5(4) \\
= & 60x(3x + 5)^4 + 4(3x + 5)^5 \\
= & 4(3x + 5)^4[15x + (3x + 5)^1] \qquad \text{Factor out the greatest} \\
& \qquad\qquad\qquad\qquad\qquad\qquad\qquad \text{common factor, } 4(3x + 5)^4 \\
= & 4(3x + 5)^4(18x + 5) \quad ■
\end{array}$$

### ■ EXAMPLE 8

Find $D_x \dfrac{(3x + 2)^7}{x - 1}$.

  Use the quotient rule and the generalized power rule.

$$\begin{aligned}
D_x \frac{(3x + 2)^7}{x - 1} &= \frac{(x - 1)[7(3x + 2)^6 \cdot 3] - (3x + 2)^7(1)}{(x - 1)^2} \\[2mm]
&= \frac{21(x - 1)(3x + 2)^6 - (3x + 2)^7}{(x - 1)^2} \\[2mm]
&= \frac{(3x + 2)^6[21(x - 1) - (3x + 2)]}{(x - 1)^2} \qquad \text{Factor out the} \\
&\qquad\qquad\qquad\qquad\qquad\qquad\qquad\qquad \text{greatest common} \\
&\qquad\qquad\qquad\qquad\qquad\qquad\qquad\qquad \text{factor, } (3x + 2)^6 \\[2mm]
&= \frac{(3x + 2)^6[21x - 21 - 3x - 2]}{(x - 1)^2} \qquad \text{Simplify inside} \\
&\qquad\qquad\qquad\qquad\qquad\qquad\qquad\qquad \text{brackets} \\[2mm]
&= \frac{(3x + 2)^6(18x - 23)}{(x - 1)^2} \quad ■
\end{aligned}$$

  Some applications requiring the use of the chain rule or the generalized power rule are illustrated in the next examples.

### ■ EXAMPLE 9

The revenue realized by a small city from the collection of fines from parking tickets is given by

$$R(n) = \frac{8000n}{n + 2},$$

where $n$ is the number of work hours each day that can be devoted to parking patrol. At the outbreak of a flu epidemic, 30 work hours are used daily in parking patrol, but during the epidemic that number is decreasing at the rate of 6 work hours per day. How fast is revenue from parking fines decreasing during the epidemic?

We want to find $dR/dt$, the change in revenue with respect to time. By the chain rule,

$$\frac{dR}{dt} = \frac{dR}{dn} \cdot \frac{dn}{dt}.$$

First find $dR/dn$, as follows.

$$\frac{dR}{dn} = \frac{(n + 2)(8000) - 8000n(1)}{(n + 2)^2} = \frac{16000}{(n + 2)^2}$$

Since $n = 30$, $dR/dn = 15.625$. Also, $dn/dt = -6$. Thus,

$$\frac{dR}{dt} = (15.625)(-6) = -93.75.$$

Revenue is being lost at the rate of about $94 per day.  ▬

▬ **EXAMPLE 10**

Suppose a sum of $500 is deposited in an account with an interest rate of $r$ percent per year compounded monthly. At the end of 10 years, the balance in the account is given by

$$A = 500\left(1 + \frac{r}{1200}\right)^{120}.$$

Find the rate of change of $A$ with respect to $r$ if $r = 5$, 7, or 9.

First find $dA/dr$ using the generalized power rule.

$$\frac{dA}{dr} = (120)(500)\left(1 + \frac{r}{1200}\right)^{119}\left(\frac{1}{1200}\right)$$

$$= 50\left(1 + \frac{r}{1200}\right)^{119}$$

If $r = 5$,

$$\frac{dA}{dr} = 50\left(1 + \frac{5}{1200}\right)^{119}$$

$$\approx 82.01,$$

or $82.01 per percentage point. If $r = 7$,

$$\frac{dA}{dr} = 50\left(1 + \frac{7}{1200}\right)^{119}$$

$$\approx 99.90,$$

or $99.90 per percentage point. If $r = 9$,

$$\frac{dA}{dr} = 50\left(1 + \frac{9}{1200}\right)^{119}$$
$$\approx 121.66,$$

or $121.66 per percentage point. ■

The chain rule can be used to develop the formula for **marginal revenue product,** an economic concept that approximates the change in revenue when a manufacturer hires an additional employee. Start with $R = px$, where $R$ is total revenue from the daily production of $x$ units and $p$ is the price per unit. The demand function is $p = f(x)$, as before. Also, $x$ can be considered a function of the number of employees, $n$. Since $R = px$, and $x$ and therefore $p$ depend on $n$, $R$ can also be considered a function of $n$. To find an expression for $dR/dn$, use the product rule for derivatives on the function $R = px$ to get

$$\frac{dR}{dn} = p \cdot \frac{dx}{dn} + x \cdot \frac{dp}{dn}. \tag{1}$$

By the chain rule,

$$\frac{dp}{dn} = \frac{dp}{dx} \cdot \frac{dx}{dn}.$$

Substituting for $dp/dn$ in equation (1) gives

$$\frac{dR}{dn} = p \cdot \frac{dx}{dn} + x\left(\frac{dp}{dx} \cdot \frac{dx}{dn}\right)$$

$$= \left(p + x \cdot \frac{dp}{dx}\right)\frac{dx}{dn}. \qquad \text{Factor out } \frac{dx}{dn}$$

The equation for $dR/dn$ gives the marginal revenue product.

### ■ EXAMPLE 11

Find the marginal revenue product $dR/dn$ (in dollars) when $n = 20$ if the demand function is $p = 600/\sqrt{x}$ and $x = 5n$.

As shown above,

$$\frac{dR}{dn} = \left(p + x \cdot \frac{dp}{dx}\right)\frac{dx}{dn}.$$

Find $dp/dx$ and $dx/dn$. From

$$p = \frac{600}{\sqrt{x}} = 600x^{-1/2},$$

we have the derivative

$$\frac{dp}{dx} = -300x^{-3/2}.$$

Also, from $x = 5n$,

$$\frac{dx}{dn} = 5.$$

Then, by substitution,

$$\frac{dR}{dn} = \left[\frac{600}{\sqrt{x}} + x\,(-300x^{-3/2})\right]5 = \frac{1500}{\sqrt{x}}.$$

If $n = 20$, then $x = 100$ and

$$\frac{dR}{dn} = \frac{1500}{\sqrt{100}} = 150.$$

This means that hiring an additional employee when production is at a level of 20 items will produce an increase in revenue of $150.   ▬

# 2.6 EXERCISES

*Let $f(x) = 4x^2 - 2x$ and $g(x) = 8x + 1$. Find each of the following.*

**1.** $f[g(2)]$          **2.** $f[g(-5]$          **3.** $g[f(2)]$

**4.** $g[f(-5)]$        **5.** $f[g(k)]$          **6.** $g[f(5z)]$

*Find $f[g(x)]$ and $g[f(x)]$ in each of the following.*

**7.** $f(x) = 8x + 12;\ \ g(x) = 3x - 1$          **8.** $f(x) = -6x + 9;\ \ g(x) = 5x + 7$

**9.** $f(x) = -x^3 + 2;\ \ g(x) = 4x$          **10.** $f(x) = 2x;\ \ g(x) = 6x^2 - x^3$

**11.** $f(x) = \dfrac{1}{x};\ \ g(x) = x^2$          **12.** $f(x) = \dfrac{2}{x^4};\ \ g(x) = 2 - x$

**13.** $f(x) = \sqrt{x + 2};\ \ g(x) = 8x^2 - 6$          **14.** $f(x) = 9x^2 - 11x;\ \ g(x) = 2\sqrt{x + 2}$

**15.** $f(x) = \dfrac{1}{x - 5};\ \ g(x) = \dfrac{2}{x}$          **16.** $f(x) = \dfrac{8}{x - 6};\ \ g(x) = \dfrac{4}{3x}$

**17.** $f(x) = \sqrt{x + 1};\ \ g(x) = \dfrac{-1}{x}$          **18.** $f(x) = \dfrac{8}{x};\ \ g(x) = \sqrt{3 - x}$

*Write each function as the composition of two functions. (There may be more than one way to do this.)*

**19.** $y = (3x - 7)^{1/3}$          **20.** $y = (5 - x)^{2/5}$

**21.** $y = \sqrt{9 - 4x}$          **22.** $y = -\sqrt{13 + 7x}$

**23.** $y = \dfrac{\sqrt{x} + 3}{\sqrt{x} - 3}$

**24.** $y = \dfrac{2}{\sqrt{x} + 5}$

**25.** $y = (x^{1/2} - 3)^2 + (x^{1/2} - 3) + 5$

**26.** $y = (x^2 + 5x)^{1/3} - 2(x^2 + 5x)^{2/3} + 7$

*Find the derivative of each function defined as follows.*

**27.** $y = (2x + 9)^2$

**28.** $y = (8x - 3)^2$

**29.** $f(x) = 6(5x - 1)^3$

**30.** $f(x) = -8(3x + 2)^3$

**31.** $k(x) = -2(12x^2 + 5)^3$

**32.** $g(x) = 5(3x^2 - 5)^3$

**33.** $y = 9(x^2 + 5x)^4$

**34.** $y = -3(x^2 - 5x)^4$

**35.** $s(t) = 12(2t + 5)^{3/2}$

**36.** $s(t) = 45(3t - 8)^{3/2}$

**37.** $y = -7(4x^2 + 9x)^{3/2}$

**38.** $y = 11(5x^2 + 6x)^{3/2}$

**39.** $f(t) = 8\sqrt{4t + 7}$

**40.** $g(t) = -3\sqrt{7t - 1}$

**41.** $y = -2\sqrt{x^2 + 4x}$

**42.** $y = 4\sqrt{2x^2 + 3}$

**43.** $r(t) = 4t(2t + 3)^2$

**44.** $m(t) = -6t(5t - 1)^2$

**45.** $y = (x + 2)(x - 1)^2$

**46.** $y = (3x + 1)^2(x + 4)$

**47.** $f(x) = 5(x + 3)^2(x - 1)^2$

**48.** $g(x) = -9(x + 4)^2(2x - 3)^2$

**49.** $y = (5x + 1)^2\sqrt{2x}$

**50.** $y = (3x + 5)^2\sqrt{4x}$

**51.** $y = \dfrac{1}{(3x - 4)^2}$

**52.** $y = \dfrac{-5}{(2x + 1)^2}$

**53.** $p(t) = \dfrac{(2t + 3)^2}{4t - 1}$

**54.** $r(t) = \dfrac{(5t - 6)^2}{3t + 4}$

**55.** $y = \dfrac{x^2 + 4x}{(3x + 2)^2}$

**56.** $y = \dfrac{3x^2 - x}{(2x - 1)^2}$

**57.** $y = (x^{1/2} + 1)(x^{1/2} - 1)^{1/2}$

**58.** $y = (3 - x^{2/3})(x^{2/3} + 2)^{1/2}$

---

### ▰ APPLICATIONS

#### BUSINESS AND ECONOMICS

**Demand**   **59.** Suppose the demand for a certain brand of vacuum cleaner is given by

$$D(p) = \frac{-p^2}{100} + 500,$$

where $p$ is the price in dollars. If the price, in terms of the cost $c$, is expressed as

$$p(c) = 2c - 10,$$

find the demand in terms of the cost.

**Revenue**   **60.** Assume that the total revenue from the sale of $x$ television sets is given by

$$R(x) = 1000\left(1 - \frac{x}{500}\right)^2.$$

Find the marginal revenue for the following values of $x$.

**(a)** $x = 100$     **(b)** $x = 150$     **(c)** $x = 200$     **(d)** $x = 400$

**(e)** Find the average revenue from the sale of $x$ sets.

**(f)** Find the marginal average revenue.

**Interest** 61. A sum of $1500 is deposited in an account with an interest rate of $r$ percent per year, compounded daily. At the end of 5 years, the balance in the account is given by

$$A = 1500\left(1 + \frac{r}{36500}\right)^{1825}.$$

Find the rate of change of $A$ with respect to $r$ for the following interest rates.

(a) $r = 6$    (b) $r = 8$    (c) $r = 9$

**Demand** 62. Suppose a demand function is given by

$$x = 30\left(5 - \frac{p}{\sqrt{p^2 + 1}}\right),$$

where $x$ is the demand for a product and $p$ is the price per item in dollars. Find the rate of change in the demand for the product (i.e., find $dx/dp$).

**Depreciation** 63. A certain truck depreciates according to the formula

$$V = \frac{6000}{1 + .3t + .1t^2},$$

where $t$ is measured in years and $t = 0$ represents the time of purchase (in years). Find the rate at which the value of the truck is changing at the following times.

(a) $t = 2$    (b) $t = 4$

**Cost** 64. Suppose the cost in dollars of manufacturing $x$ items is given by

$$C = 2000x + 3500,$$

and the demand equation is given by

$$x = \sqrt{15,000 - 1.5p}.$$

(a) Find an expression for the revenue $R$.

(b) Find an expression for the profit $P$.

(c) Find an expression for the marginal profit.

(d) Determine the value of the marginal profit for $p = \$25$.

**Marginal Revenue Product** 65. Find the marginal revenue product for a manufacturer with 8 workers if the demand function is $p = 300/x^{1/3}$ and if $x = 8n$.

**Marginal Revenue Product** 66. Suppose the demand function for a product is $p = 200/x^{1/2}$. Find the marginal revenue product if there are 25 employees and if $x = 15n$.

## LIFE SCIENCES

**Fish Population** 67. Suppose the population $P$ of a certain species of fish depends on the number $x$ (in hundreds) of a smaller fish that serves as its food supply, so that

$$P(x) = 2x^2 + 1.$$

Suppose, also, that the number $x$ (in hundreds) of the smaller species of fish depends upon the amount $a$ (in appropriate units) of its food supply, a kind of plankton. Suppose

$$x = f(a) = 3a + 2.$$

Find $P[f(a)]$, the relationship between the population $P$ of the large fish and the amount $a$ of plankton available.

**Oil Pollution**  68. An oil well off the Gulf Coast is leaking, with the leak spreading oil over the surface as a circle. At any time $t$, in minutes, after the beginning of the leak, the radius of the circular oil slick on the surface is $r(t) = t^2$ feet. Let $A(r) = \pi r^2$ represent the area of a circle of radius $r$. Find and interpret $A[r(t)]$.

**Thermal Inversion**  69. When there is a thermal inversion layer over a city (as happens often in Los Angeles), pollutants cannot rise vertically but are trapped below the layer and must disperse horizontally. Assume that a factory smokestack begins emitting a pollutant at 8 A.M. Assume that the pollutant disperses horizontally, forming a circle. If $t$ represents the time, in hours, since the factory began emitting pollutants ($t = 0$ represents 8 A.M.), assume that the radius of the circle of pollution is $r(t) = 2t$ miles. Let $A(r) = \pi r^2$ represent the area of a circle of radius $r$. Find and interpret $A[r(t)]$.

**Bacteria Population**  70. The total number of bacteria (in millions) present in a culture is given by

$$N(t) = 2t(5t + 9)^{1/2} + 12,$$

where $t$ represents time in hours after the beginning of an experiment. Find the rate of change of the population of bacteria with respect to time for each of the following.

**(a)** $t = 0$    **(b)** $t = 7/5$    **(c)** $t = 8$    **(d)** $t = 11$.

**Calcium Usage**  71. To test an individual's use of calcium, a researcher injects a small amount of radioactive calcium into the person's bloodstream. The calcium remaining in the bloodstream is measured each day for several days. Suppose the amount of the calcium remaining in the bloodstream in milligrams per cubic centimeter $t$ days after the initial injection is approximated by

$$C(t) = \frac{1}{2}(2t + 1)^{-1/2}.$$

Find the rate of change of $C$ with respect to time for each of the following.

**(a)** $t = 0$    **(b)** $t = 4$    **(c)** $t = 6$ (Use a calculator.)    **(d)** $t = 7.5$

**Drug Reaction**  72. The strength of a person's reaction to a certain drug is given by

$$R(Q) = Q\left(C - \frac{Q}{3}\right)^{1/2},$$

where $Q$ represents the quantity of the drug given to the patient and $C$ is a constant.

**(a)** The derivative $R'(Q)$ is called the *sensitivity* to the drug. Find $R'(Q)$.

**(b)** Find $R'(Q)$ if $Q = 87$ and $C = 59$.

## 2.7 CONTINUITY AND DIFFERENTIABILITY

To draw the graphs of most of the functions we have used so far, we selected several values of $x$ and found the corresponding values of $y$. Then we plotted the resulting ordered pairs and drew a smooth curve through them. This process worked reasonably well because most of the functions graphed were *continuous*—the graph could be drawn without lifting pencil from paper. A function that is not continuous at some point is *discontinuous* at the point.

Looking first at graphs having points of discontinuity will clarify the idea of continuity at a point. Studying the reasons why the functions are discontinuous at various points will lead to a reasonable definition of continuity. For example, the function shown in Figure 27(a) is discontinuous at $x = 2$ because of the "hole" in the graph at (2, 3). The function of Figure 27(b) is discontinuous at $x = -3$ because of the "jump" in the graph when $x = -3$.

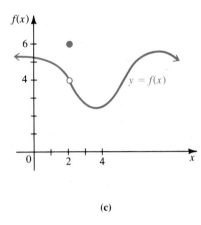

(a)

(b)

(c)

**FIGURE 27**

Finally, the function in Figure 27(c) is discontinuous when $x = 2$. Even though $f(2)$ is defined (here $f(2) = 6$), the value of $f(2)$ is not "close" to the values of $f(x)$ for values of $x$ "close" to 2. That is,

$$\lim_{x \to 2} f(x) \neq f(2).$$

Now that we have seen some different ways for a function to have a point of discontinuity, look again at Figure 27(b). The function shown in Figure 27(b), which is *not* continuous when $x = -3$, *is* continuous when $x = -5$. To see the difference, note first that $f(-5)$ is defined, so that a point on the graph corresponds to $x = -5$. Second, $\lim_{x \to -5} f(x)$ exists. As $x$ approaches $-5$ from either side of $-5$, the values of $f(x)$ approach one particular number, $f(-5)$, which is 0. These examples suggest the following definition.

**DEFINITION OF CONTINUITY AT A POINT**

A function $f$ is **continuous** at $c$ if the following three conditions are satisfied:

**1.** $f(c)$ is defined;

**2.** $\lim_{x \to c} f(x)$ exists;

**3.** $\lim_{x \to c} f(x) = f(x)$.

If $f$ is not continuous at $c$, it is **discontinuous** there.

## ■ EXAMPLE 1

Tell why the functions are discontinuous at the indicated points.

**(a)** $f(x)$ in Figure 28 at $x = 3$

The open circle on the graph of Figure 28 at the point where $x = 3$ means that $f(3)$ is not defined. Because of this, part (1) of the definition fails.

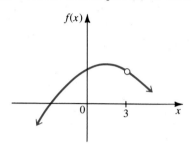

**FIGURE 28**

**(b)** $h(x)$ in Figure 29 at $x = 0$

According to the graph of Figure 29, $h(0) = -1$. Also, as $x$ approaches 0 from the left, $h(x)$ is $-1$. As $x$ approaches 0 from the right, however, $h(x)$ is 1. Since there is no single number approached by the values of $h(x)$ as $x$ approaches 0, $\lim_{x \to 0} h(x)$ does not exist, and part (2) of the definition fails.

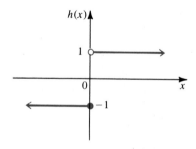

**FIGURE 29**

**(c)** $g(x)$ in Figure 30 at $x = 4$

In Figure 30, the heavy dot above 4 shows that $g(4)$ is defined. In fact, $g(4) = 1$. The graph also shows, however, that

$$\lim_{x \to 4} g(x) = -2,$$

so $\lim_{x \to 4} g(x) \neq g(4)$, and part (3) of the definition fails.

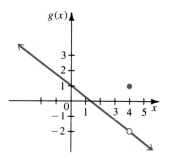

**FIGURE 30**

**(d)** $f(x)$ in Figure 31 at $x = -2$

The function $f$ graphed in Figure 31 is not defined at $-2$, and $\lim\limits_{x \to -2} f(x)$ does not exist. Either of these reasons is sufficient to show that $f$ is not continuous at $-2$. (Function $f$ *is* continuous at any value of $x$ greater than $-2$, however.) ▬

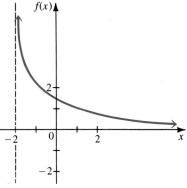

**FIGURE 31**

The following properties of continuous functions may be used to construct

**PROPERTIES OF CONTINUOUS FUNCTIONS**

$f(x) = k$ is continuous for all $x$.

$f(x) = x^n$ and $g(x) = \sqrt[n]{x}$ are continuous for all positive integers $n$ and all real numbers $x$ in the domain of the functions.

If $f(x)$ and $g(x)$ are continuous at a point, then so are $f(x) + g(x)$, $f(x) - g(x)$, and $f(x) \cdot g(x)$.

If $f(x)$ and $g(x)$ are continuous at a point, and if $g(x) \neq 0$, then $f(x)/g(x)$ is continuous at the point.

The next table lists some key functions and tells where each is continuous.

**Continuous Functions**

| *Type of Function* | *Where It Is Continuous* | *Graphic Example* |
|---|---|---|
| **Polynomial Function** $y = a_n x^n + a_{n-1} x^{n-1} + \cdots + a_1 x + a_0,$ where $a_n, a_{n-1}, \cdots, a_1, a_0$ are real numbers, not all 0 | For all $x$ | |
| **Rational Function** $y = \dfrac{p(x)}{q(x)},$ where $p(x)$ and $q(x)$ are polynomials, with $q(x) \neq 0$ | For all $x$ where $q(x) \neq 0$ | |
| **Root Function** $y = \sqrt{ax + b},$ where $a$ and $b$ are real numbers, with $a \neq 0$ and $ax + b \geq 0$ | For all $x$ where $ax + b > 0$ | |

Finally, in addition to continuity at a point, continuity on an open interval is defined as follows.*

**CONTINUITY ON AN OPEN INTERVAL**

If a function $f$ is continuous at each point of an open interval, $f$ is said to be **continuous on that interval.**

---

*In a few places later in the book the idea of continuity on a closed interval is needed. While it is perhaps intuitively clear when a function is continuous on a closed interval, a precise definition requires more work than we have given.

### EXAMPLE 2

Is the function in Figure 32 continuous on the following *x*-intervals?

**(a)** $(-2, -1)$

The function is discontinuous only at $x = -2$, 0, and 2. Thus, it is continuous at every point of the open interval $(-2, -1)$. The function is continuous on the open interval.

**(b)** $(1, 3)$

This interval includes the point of discontinuity at $x = 2$, so that the function is not continuous on the open interval $(1, 3)$. ▪

**FIGURE 32**                                    **FIGURE 33**

### EXAMPLE 3

A trailer rental firm charges a flat $4 to rent a hitch. The trailer itself is rented for $11 per day or fraction of a day. Let $C(x)$ represent the cost of renting a hitch and trailer for *x* days.

**(a)** Graph *C*.

The charge for one day is $4 for the hitch and $11 for the trailer, or $15. In fact, if $0 < x \le 1$, then $C(x) = 15$. To rent the trailer for more than one day, but not more than two days, the charge is $4 + 2 \cdot 11 = 26$ dollars. For any value of *x* satisfying $1 < x \le 2$, $C(x) = 26$. Also, if $2 < x \le 3$, then $C(x) = 37$. These results lead to the graph in Figure 33.

**(b)** Find any points of discontinuity for *C*.

As the graph suggests, *C* is discontinuous at $x = 1$, 2, 3, 4, and all other positive integers. ▪

***Continuity and Differentiability*** As shown earlier in this chapter, a function fails to have a derivative at a point where the function is not defined, where the graph of the function has a ''sharp point,'' or where the graph has a vertical tangent line. (See Figure 34).

The function graphed in Figure 34 is continuous on the interval $(x_1, x_2)$ and has a derivative at each point on this interval. On the other hand, the function is also continuous on the interval $(0, x_2)$ but does *not* have a derivative at each point on the interval (see $x_1$ on the graph).

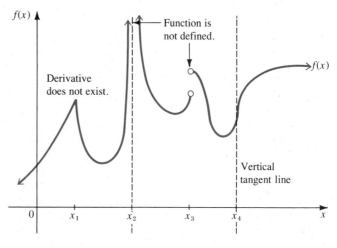

**FIGURE 34**

If the derivative of a function exists at a point, then the function is continuous at that point.

Intuitively, this means that a graph can have a derivative at a point only if the graph is ''smooth'' in the vicinity of the point.

## ▰ EXAMPLE 4

A nova is a star whose brightness suddenly increases and then gradually fades. The cause of the sudden increase in brightness is thought to be an explosion of some kind. The intensity of light emitted by a nova as a function of time is shown in Figure 35.* Notice that although the graph is a continuous curve, it is not differentiable at the point of the explosion. ▰

---

*Reprinted with permission of Macmillan Publishing Company from *Astronomy: The Structure of the Universe* by William J. Kaufmann, III. Copyright © 1977 by William J. Kaufmann, III.

0 ← Point of explosion

**FIGURE 35**

## 2.7 EXERCISES

*Find all points of discontinuity. Tell where the derivatives would fail to exist.*

**1.**

**2.**

**3.**

**4.**

**5.**

**6.**

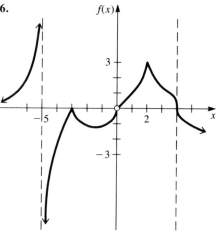

*Decide whether each of the following is continuous at the given values of x.*

**7.** $f(x) = \dfrac{2}{x - 3}$;   $x = 0, 3$

**8.** $f(x) = \dfrac{6}{x}$;   $x = 0, -1$

**9.** $g(x) = \dfrac{1}{x(x - 2)}$;   $x = 0, 2, 4$

**10.** $h(x) = \dfrac{-2}{3x(x + 5)}$;   $x = 0, 3, -5$

**11.** $h(x) = \dfrac{1 + x}{(x - 3)(x + 1)}$;   $x = 0, 3, -1$

**12.** $g(x) = \dfrac{-2x}{(2x + 1)(3x + 6)}$;   $x = 0, -1/2, -2$

**13.** $k(x) = \dfrac{5 + x}{2 + x}$;   $x = 0, -2, -5$

**14.** $f(x) = \dfrac{4 - x}{x - 9}$;   $x = 0, 4, 9$

**15.** $g(x) = \dfrac{x^2 - 4}{x - 2}$;   $x = 0, 2, -2$

**16.** $h(x) = \dfrac{x^2 - 25}{x + 5}$;   $x = 0, 5, -5$

**17.** $p(x) = x^2 - 4x + 11$;   $x = 0, 2, -1$

**18.** $q(x) = -3x^3 + 2x^2 - 4x + 1$;   $x = -2, 3, 1$

**19.** $p(x) = \dfrac{|x + 2|}{x + 2}$;   $x = -2, 0, 2$

**20.** $r(x) = \dfrac{|5 - x|}{x - 5}$;   $x = -5, 0, 5$

*Decide whether the functions graphed in Exercises 20–24 are continuous on the indicated x-intervals.*

**21.** $(-3, 0)$; $(0, 3)$; $(0, 4)$

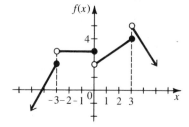

**22.** $(-6, 0)$; $(0, 6)$; $(4, 8)$

**23.** $(-12, 6)$; $(0, 6)$; $(6, 12)$

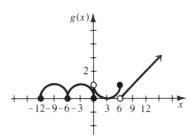

**24.** $(-4, 0)$; $(0, 3)$; $(0, 5)$

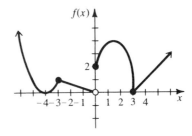

**25.** Let $f(x) = x + 5$ and $g(x) = \sqrt{x}$. Tell whether the following are continuous when $x = -4$, 0, or 4.

(a) $f[g(x)]$    (b) $g[f(x)]$

**26.** Let $f(x) = 2/(3 - x)$ and $g(x) = 4 - x^2$. Tell whether the following are continuous when $x = -3$, $-2$, 0, 2, or 3.

(a) $f[g(x)]$    (b) $g[f(x)]$

---

### ▤ APPLICATIONS

#### BUSINESS AND ECONOMICS

**Cost Analysis** **27.** A company charges $1.20 per pound for a certain fertilizer on all orders not over 100 lb, and $1 per pound for orders over 100 lb. Let $F(x)$ represent the cost for buying $x$ lb of the fertilizer.

(a) Find $F(80)$.    (b) Find $F(150)$.    (c) Graph $y = F(x)$.

(d) Where is $F$ discontinuous?

**Cost Analysis** **28.** The cost to transport a mobile home depends on the distance, $x$, in miles that the home is moved. Let $C(x)$ represent the cost to move a mobile home $x$ miles. One firm charges as follows.

| Cost per Mile | Distance in Miles |
|---|---|
| $2.00 | $0 < x \le 150$ |
| $1.50 | $150 < x \le 400$ |
| $1.25 | $400 < x$ |

(a) Find $C(130)$.    (b) Find $C(210)$.    (c) Find $C(350)$.    (d) Find $C(500)$.

(e) Graph $y = C(x)$.

(f) Where is $C$ discontinuous?

**Car Rental** **29.** Recently, a car rental firm charged $30 per day or portion of a day to rent a car for a period of 1 to 5 days. Days 6 and 7 were then "free," while the charge for days 8 through 12 was again $30 per day. Let $A(t)$ represent the average cost per day to rent the car for $t$ days, where $0 < t \le 12$. Find each of the following.

(a) $A(4)$    (b) $A(5)$    (c) $A(6)$    (d) $A(7)$    (e) $A(8)$

(f) $\lim\limits_{t \to 5} A(t)$    (g) $\lim\limits_{t \to 6} A(t)$

**Prime Interest Rate** **30.** The graph shows the prime interest rate for business loans for the business days of a recent month. Where is the function discontinuous?

Business days

**EXERCISE 30**

## SOCIAL SCIENCES

**Learning Skills** **31.** With certain skills (such as musical skills), learning is rapid at first and then levels off, with sudden insights causing learning to take a jump. A typical graph of such learning is shown in the figure. Where is the function discontinuous?

## PHYSICAL SCIENCES

**Temperature** **32.** Suppose a gram of ice is at a temperature of $-100°C$. The graph on the left shows the temperature of the ice as increasing numbers of calories of heat are applied. It takes 80 calories to melt one gram of ice at $0°C$ into water, and 539 calories to boil one gram of water at $100°C$ into steam. Where is this graph discontinuous?

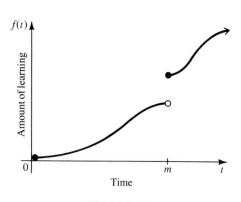

**EXERCISE 31**

**EXERCISE 32**

# ▬ DERIVATIVE SUMMARY

For reference, the definition of derivative and the rules for derivatives developed in this chapter are listed here.

**DEFINITION OF DERIVATIVE**

The derivative of the function $f$ at $x$ is defined as

$$f'(x) = \lim_{h \to 0} \frac{f(x + h) - f(x)}{h}$$

provided this limit exists. Alternative notations for the derivative include $dy/dx$, $D_x y$, $D_x[f(x)]$, $\dfrac{d}{dx}[f(x)]$, and $y'$.

The following rules for derivatives are valid when all the indicated derivatives exist.

**SUMMARY OF RULES FOR DERIVATIVES**

***Constant Rule***   If $f(x) = k$, where $k$ is any real number, then $f'(x) = 0$.

***Power Rule***   If $f(x) = x^n$ for any real number $n$, then $f'(x) = nx^{n-1}$.

***Constant Times a Function***   Let $k$ be a real number. Then the derivative of $y = k \cdot f(x)$ is $y' = k \cdot f'(x)$.

***Sum or Difference Rule***   If $f(x) = u(x) \pm v(x)$, then $f'(x) = u'(x) \pm v'(x)$.

***Product Rule***   If $f(x) = u(x) \cdot v(x)$, then

$$f'(x) = u(x) \cdot v'(x) + v(x) \cdot u'(x).$$

***Quotient Rule***   If $f(x) = \dfrac{u(x)}{v(x)}$, and $v(x) \neq 0$, then

$$f'(x) = \frac{v(x) \cdot u'(x) - u(x) \cdot v'(x)}{[v(x)]^2}.$$

***Chain Rule***   If $y$ is a function of $u$, say $y = f(u)$, and if $u$ is a function of $x$, say $u = g(x)$, then $y = f(u) = f[g(x)]$, and

$$\frac{dy}{dx} = \frac{dy}{du} \cdot \frac{du}{dx}.$$

***Chain Rule (alternate form)***   If $y = f[g(x)]$, then $y' = f'[g(x)] \cdot g'(x)$.

***Generalized Power Rule***   Let $u$ be a function of $x$, and let $y = u^n$, for any real number $n$. Then $y' = n \cdot u^{n-1} \cdot u'$.

| KEY WORDS | To understand the concepts presented in this chapter, you should know the meaning and use of the following words. For easy reference, the section in the chapter where a word (or expression) was first used is given with each item. |

2.1  **limit**
     **limit at infininty**
2.2  **average rate of change**
     **instantaneous rate of change**
     **velocity**
2.3  **tangent line**
     **secant line**
     **slope of a curve**
     **derivative**
     **differentiable**
     **differentiation**

2.5  **product rule**
     **quotient rule**
     **average cost**
     **marginal average cost**
2.6  **chain rule**
     **generalized power rule**
     **marginal revenue product**
2.7  **continuous**
     **discontinuous**
     **continuous on an interval**

## CHAPTER 2     REVIEW EXERCISES

*Decide whether the limits in Exercises 1–18 exist. If a limit exists, find its value.*

**1.** $\lim\limits_{x \to -3} f(x)$

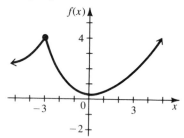

**2.** $\lim\limits_{x \to -1} g(x)$

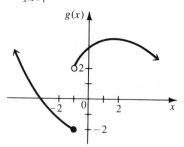

**3.** $\lim\limits_{x \to +\infty} f(x)$

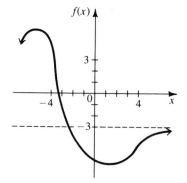

**4.** $\lim\limits_{x \to 0} \dfrac{x^2 - 5}{2x}$

**5.** $\lim\limits_{x \to -1} (2x^2 + 3x + 5)$

**6.** $\lim\limits_{x \to 2} (-x^2 + 4x + 1)$

**7.** $\lim\limits_{x \to 6} \dfrac{2x + 5}{x - 3}$

**8.** $\lim\limits_{x \to 3} \dfrac{2x + 5}{x - 3}$

**9.** $\lim\limits_{x \to 4} \dfrac{x^2 - 16}{x - 4}$

**10.** $\lim\limits_{x \to 2} \dfrac{x^2 + 3x - 10}{x - 2}$

**11.** $\lim\limits_{x \to -4} \dfrac{2x^2 + 3x - 20}{x + 4}$

**12.** $\lim\limits_{x \to 3} \dfrac{3x^2 - 2x - 21}{x - 3}$

**13.** $\lim\limits_{x \to 9} \dfrac{\sqrt{x} - 3}{x - 9}$

**14.** $\lim\limits_{x \to 16} \dfrac{\sqrt{x} - 4}{x - 16}$

**15.** $\lim\limits_{x \to +\infty} \dfrac{x^2 + 5}{5x^2 - 1}$

**16.** $\lim\limits_{x \to +\infty} \dfrac{x^2 + 6x + 8}{x^3 + 2x + 1}$

**17.** $\lim\limits_{x \to -\infty} \left( \dfrac{3}{4} + \dfrac{2}{x} - \dfrac{5}{x^2} \right)$

**18.** $\lim\limits_{x \to -\infty} \left( \dfrac{9}{x^4} + \dfrac{1}{x^2} - 3 \right)$

*Use the graph to find the average rate of change of f on the following intervals.*

**19.** $x = 0$ to $x = 4$

**20.** $x = 2$ to $x = 8$

**21.** $x = 2$ to $x = 4$

**22.** $x = 0$ to $x = 6$

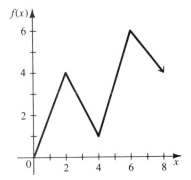

*Find the average rate of change for each of the following on the given interval.*

**23.** $y = 6x^2 + 2$; from $x = 1$ to $x = 4$

**24.** $y = -2x^3 - x^2 + 5$; from $x = -2$ to $x = 6$

**25.** $y = \dfrac{-6}{3x - 5}$; from $x = 4$ to $x = 9$

**26.** $y = \dfrac{x + 4}{x - 1}$; from $x = 2$ to $x = 5$

*Use the definition of the derivative to find the derivative of each of the following.*

**27.** $y = 4x + 3$

**28.** $y = 5x^2 + 6x$

**29.** $y = -x^3 + 7x$

**30.** $y = 11x^2 - x^3$

*Find the slope of the tangent line to the given curve at the given value of x. Find the equation of each tangent line.*

**31.** $y = x^2 - 6x$;   $x = 2$

**32.** $y = 8 - x^2$;   $x = 1$

**33.** $y = \dfrac{3}{x - 1}$;   $x = -1$

**34.** $y = \dfrac{-2}{x + 5}$;   $x = -2$

**35.** $y = (3x^2 - 5x)(2x)$;   $x = -1$

**36.** $y = \dfrac{3}{x^2 - 1}$;   $x = 2$

**37.** $y = \sqrt{6x - 2}$;   $x = 3$

**38.** $y = -\sqrt{8x + 1}$;   $x = 3$

*Find the derivative of each function defined as follows.*

**39.** $y = 5x^2 - 7x - 9$

**40.** $y = x^3 - 4x^2$

**41.** $y = 6x^{7/3}$

**42.** $y = -3x^{-2}$

**43.** $f(x) = x^{-3} + \sqrt{x}$

**44.** $f(x) = 6x^{-1} - 2\sqrt{x}$

**45.** $y = (3t^2 + 7)(t^3 - t)$

**46.** $y = (-5t + 4)(t^3 - 2t^2)$

**47.** $y = 4\sqrt{x}(2x - 3)$

**48.** $y = -3\sqrt{x}(8 - 5x)$

**49.** $g(t) = -3t^{-1/3}(5t + 7)$

**50.** $p(t) = 8t^{3/4}(7t - 2)$

**51.** $y = 12x^{-3/4}(3x + 5)$

**52.** $y = 15x^{-3/5}(x + 6)$

**53.** $k(x) = \dfrac{3x}{x + 5}$

**54.** $r(x) = \dfrac{-8}{2x + 1}$

**55.** $y = \dfrac{\sqrt{x} - 1}{x + 2}$

**56.** $y = \dfrac{\sqrt{x} + 6}{x - 3}$

**57.** $y = \dfrac{x^2 - x + 1}{x - 1}$

**58.** $y = \dfrac{2x^3 - 5x^2}{x + 2}$

**59.** $f(x) = (3x - 2)^4$

**60.** $k(x) = (5x - 1)^6$

**61.** $y = \sqrt{2t - 5}$

**62.** $y = -3\sqrt{8t - 1}$

**63.** $y = 3x(2x + 1)^3$

**64.** $y = 4x^2(3x - 2)^5$

**65.** $r(t) = \dfrac{5t^2 - 7t}{(3t + 1)^3}$

**66.** $s(t) = \dfrac{t^3 - 2t}{(4t - 3)^4}$

**67.** $y = \dfrac{x^2 + 3x - 10}{x - 2}$

**68.** $y = \dfrac{x^2 - x - 6}{x - 3}$

*Find each of the following.*

**69.** $D_x\left[\dfrac{\sqrt{x} + 1}{\sqrt{x} - 1}\right]$

**70.** $D_x\left[\dfrac{2x + \sqrt{x}}{1 - x}\right]$

**71.** $\dfrac{dy}{dt}$ if $y = \sqrt{t^{1/2} + t}$

**72.** $\dfrac{dy}{dx}$ if $y = \dfrac{\sqrt{x} - 1}{x}$

**73.** $f'(1)$ if $f(x) = \dfrac{\sqrt{8 + x}}{x + 1}$

**74.** $f'(-2)$ if $f(t) = \dfrac{2 - 3t}{\sqrt{2 + t}}$

*Decide whether each of the following is continuous at the given values of x.*

**75.** $f(x) = \dfrac{6x + 1}{2x - 3}$;  $x = 3/2, -1/6, 0$

**76.** $f(x) = \dfrac{2}{x(x + 4)}$;  $x = 2, 0, -4$

**77.** $f(x) = \dfrac{-5}{3x(2x - 1)}$;  $x = -5, 0, -1/3, 1/2$

**78.** $f(x) = \dfrac{2 - 3x}{(1 + x)(2 - x)}$;  $x = 2/3, -1, 2, 0$

**79.** $f(x) = \dfrac{x - 6}{x + 5}$;  $x = 6, -5, 0$

**80.** $f(x) = \dfrac{x^2 - 9}{x + 3}$;  $x = 3, -3, 0$

**81.** $f(x) = x^2 + 3x - 4$;  $x = 1, -4, 0$

**82.** $f(x) = 2x^2 - 5x - 3$;  $x = -1/2, 3, 0$

*Decide whether the functions graphed in Exercise 83 and 84 are continuous on the given x-intervals.*

**83.** $(-3, 0); (-3, 3); (0, 4)$

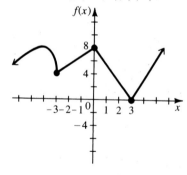

**84.** $(-4, 2); (-4, 6); (2, 6)$

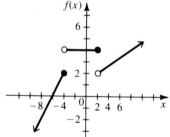

*Identify the x-values where the derivative of f does not exist and where f is discontinuous.*

**85.**

**86.**

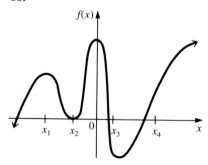

## ▦ APPLICATIONS

### BUSINESS AND ECONOMICS

**Profit** 87. Suppose the profit in cents from selling $x$ pounds of potatoes is given by

$$P(x) = 15x + 25x^2.$$

Find the marginal profit for each of the following.

(a) $x = 6$    (b) $x = 20$    (c) $x = 30$

**Average Profit** 88. In Exercise 87, find the average profit function and the marginal average profit function.

**Sales** 89. The sales of a company are related to its expenditures on research by

$$S(x) = 1000 + 50\sqrt{x} + 10x,$$

where $S(x)$ gives sales in millions when $x$ thousand dollars is spent on research. Find and interpret $dS/dx$ for each of the following.

(a) $x = 9$    (b) $x = 16$    (c) $x = 25$

**Profit** 90. Suppose that the profit in hundreds of dollars from selling $x$ units of a product is given by

$$P(x) = \frac{x^2}{x - 1}, \text{ where } x > 1.$$

Find and interpret the marginal profit for each of the following.

(a) $x = 4$    (b) $x = 12$    (c) $x = 20$

**Costs** 91. A company finds that its costs are related to the amount spent on training programs by

$$T(x) = \frac{1000 + 50x}{x + 1},$$

where $T(x)$ is costs in thousands of dollars when $x$ hundred dollars are spent on training. Find and interpret $T'(x)$ for each of the following.

(a) $x = 9$    (b) $x = 19$

**Revenue** 92. Waverly Products has found that its revenue is related to advertising expenditures by the function

$$R(x) = 5000 + 16x - 3x^2,$$

where $R(x)$ is the revenue in dollars when $x$ hundred dollars are spent on advertising.

(a) Find the marginal revenue function.

(b) Find and interpret the marginal revenue when $x = 10$.

**Cost Analysis** 93. A company charges $1.50 per pound when a certain chemical is bought in lots of 125 lb or less, with a price per pound of $1.35 if more than 125 lb are purchased. Let $C(x)$ represent the cost of $x$ lb. Find each of the following.

(a) $C(100)$    (b) $C(125)$    (c) $C(140)$

(d) Graph $y = C(x)$.

(e) Where is $C$ discontinuous?

**Average Cost** 94. Use the information in Exercise 93 to find the average cost per pound if the following number of pounds are bought.

(a) 100    (b) 125    (c) 140    (d) 200

# 3 Applications of the Derivative

The graph in Figure 1 shows the population of deer on the Kaibab Plateau on the North Rim of the Grand Canyon between 1905 and 1930. In the first two decades of the twentieth century, hunters were effective in reducing the population of predators. With fewer predators, the deer increased in numbers and depleted available food resources. As shown by the graph, the deer population peaked in about 1925 and then declined rapidly.

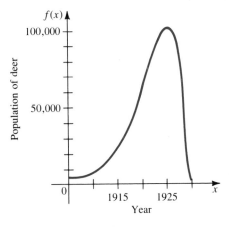

**FIGURE 1**

Given a graph like the one in Figure 1, often we can locate maximum and minimum values simply by looking at the graph. It is difficult to get *exact* values or *exact* locations of maxima and minima from a graph, however, and many functions are difficult to graph. In Chapter 1 we saw how to find exact maximum and minimum values for quadratic functions by identifying the vertex. A more general approach is to use the derivative of a function to determine precise maximum and minimum values of the function. The procedure for doing this is described in the first part of this chapter, which begins with a discussion of increasing and decreasing functions.

## 3.1   INCREASING AND DECREASING FUNCTIONS

Suppose that an amusement park has hired an engineering firm to develop a new roller coaster. The ride must include one very steep drop that actually passes through an underground tunnel. An engineer has developed a design that gives the height $f$, in feet, of the track above the ground $t$ seconds into the ride as

$$f(t) = h = -3t^4 + 28t^3 - 84t^2 + 96t.$$

Meanwhile, a new safety regulation has been passed requiring that such rides be no more than 70 feet high. The engineer is particularly worried about the stretch of track shown in Figure 2. Will it be too high?

**FIGURE 2**

The derivative can be used to answer this question. Remember that the derivative of a function at a point gives the slope of the tangent line to the function at that point. Recall also that a line with positive slope rises from left to right and a line with negative slope falls from left to right.

Now, picture one of the cars on the roller coaster. When the car is on level ground or parallel to level ground, its floor is horizontal, but as the car moves up the slope, its floor tilts upward. When the car reaches a peak, its floor is again horizontal, but it then begins to tilt downward (very steeply) as the car rolls downhill.

If we think of the roller coaster track as a graph, the floor of the car as it moves from left to right along the track represents the tangent line at each point. Using this analogy, we can see that the slope of the tangent will be *positive* when the car travels uphill and *h* is *increasing,* and the slope of the tangent will be *negative* when the car travels downhill and *h* is *decreasing.* (In this case it is also true that the slope will be zero at "peaks and valleys.") See Figure 3.

$$f(t) = h = -3t^4 + 28t^3 - 84t^2 + 96t$$

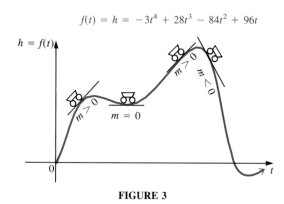

**FIGURE 3**

Thus, on intervals where $f'(t) > 0$, $f(t) = h$ will increase, and on intervals where $f'(t) < 0$, $f(t) = h$ will decrease. We can determine where $f(t)$ "peaks" by finding the intervals on which it increases and decreases. The function

$$f(t) = -3t^4 + 28t^3 - 84t^2 + 96t$$

is differentiated as

$$f'(t) = -12t^3 + 84t^2 - 168t + 96.$$

To determine where $f'(t) > 0$ and where $f'(t) < 0$, first find where $f'(t) = 0$.

$$-12t^3 + 84t^2 - 168t + 96 = 0$$
$$t^3 - 7t^2 + 14t - 8 = 0 \qquad \text{Divide each side by } -12$$

Using synthetic division or long division, we can factor on the left side of the equation to get

$$(t - 1)(t - 2)(t - 4) = 0.$$

Therefore, $f'(t)$ will equal 0 when $t = 1, 2,$ or 4. (These values of $t$ correspond to the two "peaks" and one "valley" on the graph of $f(t)$ in Figure 5.)

The three zeros of $f'(t)$ divide the real number line into four regions, as shown in Figure 4. Choosing a point in each region and evaluating $f'(t)$ there shows that $f'(t)$ is positive between 0 and 1, negative between 1 and 2, positive between 2 and 4, and negative for $t$ greater than 4. Thus, $h = f(t)$ increases from 0 to 1, decreases from 1 to 2, increases from 2 to 4, and decreases thereafter. The graph in Figure 5 shows that the highest point occurs when $t = 4$. At that time, the track is $f(4) = 64$ feet above the ground, safely within the 70-foot limit.

**FIGURE 4**

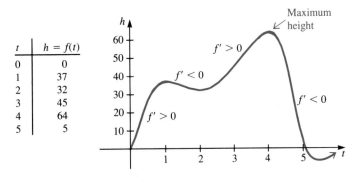

| $t$ | $h = f(t)$ |
|---|---|
| 0 | 0 |
| 1 | 37 |
| 2 | 32 |
| 3 | 45 |
| 4 | 64 |
| 5 | 5 |

**FIGURE 5**

As shown in the example above, a function is *increasing* if the graph goes *up* from left to right and *decreasing* if its graph goes *down* from left to right. Examples of increasing functions are shown in Figure 6 and examples of decreasing functions in Figure 7.

**INCREASING AND DECREASING FUNCTIONS**

Let $f$ be a function defined on some interval. Then for any two numbers $x_1$ and $x_2$ in the interval, $f$ is **increasing** on the interval if

$$f(x_1) < f(x_2) \quad \text{whenever} \quad x_1 < x_2,$$

and $f$ is **decreasing** on the interval if

$$f(x_1) > f(x_2) \quad \text{whenever} \quad x_1 < x_2.$$

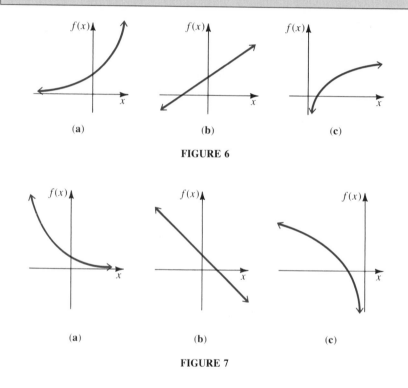

(a)  (b)  (c)

**FIGURE 6**

(a)  (b)  (c)

**FIGURE 7**

### EXAMPLE 1

Where is the function graphed in Figure 8 increasing? Where is it decreasing?

Moving from left to right, the function is increasing up to $-4$, then decreasing from $-4$ to 0, constant (neither increasing nor decreasing) from 0 to 4, increasing from 4 to 6, and decreasing from 6 onward. In interval notation, the function is increasing on $(-\infty, -4)$ and $(4, 6)$, decreasing on $(-4, 0)$ and $(6, +\infty)$, and constant on $(0, 4)$. ∎

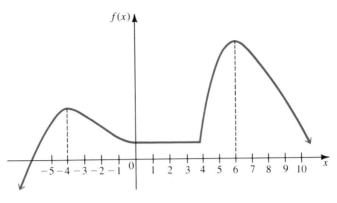

**FIGURE 8**

The following test summarizes the discussion in the opening example.

**TEST FOR INTERVALS WHERE $f(x)$ IS INCREASING AND DECREASING**

Suppose a function $f$ has a derivative at each point in an open interval; then

if $f'(x) > 0$ for each $x$ in the interval, $f$ is *increasing* on the interval;

if $f'(x) < 0$ for each $x$ in the interval, $f$ is *decreasing* on the interval;

if $f'(x) = 0$ for each $x$ in the interval, $f$ is *constant* on the interval.

The derivative $f'(x)$ can change signs from positive to negative (or negative to positive) at points where $f'(x) = 0$, and also at points where $f'(x)$ does not exist. The values of $x$ where this occurs are called *critical numbers*. By definition, the **critical numbers** for a function $f$ are those numbers $c$ in the domain of $f$ for which $f'(c) = 0$ or $f'(c)$ does not exist.

It is shown in more advanced classes that if the critical numbers of a polynomial function are used to determine open intervals on a number line, then the sign of the derivative at any point in the interval will be the same as the sign of the derivative at any other point in the interval. This suggests that the test for increasing and decreasing functions be applied as follows (assuming that no open intervals exist where the function is constant).

**APPLYING THE TEST**

1. Locate on a number line those values of $x$ for which $f'(x) = 0$ or $f'(x)$ does not exist. These points determine several open intervals.
2. Choose a value of $x$ in each of the intervals determined in Step 1. Use these values to decide whether $f'(x) > 0$ or $f'(x) < 0$ in that interval.
3. Use the test above to decide whether $f$ is increasing or decreasing on the interval.

### ▬ EXAMPLE 2

Find the intervals in which the following functions are increasing or decreasing. Locate all points where the tangent line is horizontal. Graph the function.

**(a)** $f(x) = 2x^2 + 6x - 5$

The derivative $f'(x) = 4x + 6$ is 0 when

$$4x + 6 = 0$$

$$x = -\frac{3}{2}.$$

Since there are no values of $x$ where $f'(x)$ fails to exist, the only critical number is $-3/2$. Locate $x = -3/2$ on a number line, as in Figure 9. This point determines two intervals, $(-\infty, -3/2)$ and $(-3/2, +\infty)$.

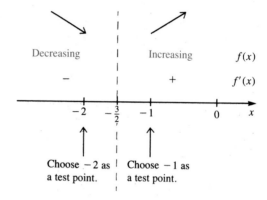

$-\frac{3}{2}$ determines the two intervals.

**FIGURE 9**

Now choose any value of $x$ in the interval $(-3/2, +\infty)$. Choosing $x = -1$ gives

$$f'(-1) = 4 \cdot -1 + 6 = 2,$$

which is positive. Since one value of $x$ in this interval makes $f'(x) > 0$, all values will do so, and therefore $f$ is increasing on $(-3/2, +\infty)$. Use $-2$ as a test point in the interval $(-\infty, -3/2)$:

$$f'(-2) = 4(-2) + 6 = -2,$$

which is negative. This means that $f$ is decreasing on the interval $(-\infty, -3/2)$. The arrows in each interval in Figure 9 indicate where $f$ is increasing or decreasing.

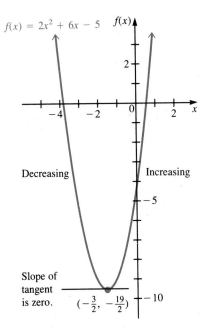

$f(x) = 2x^2 + 6x - 5$

Decreasing

Increasing

Slope of tangent is zero.

$\left(-\frac{3}{2}, -\frac{19}{2}\right)$

**FIGURE 10**

Since a horizontal line has a slope of 0, horizontal tangents can be found by solving the equation $f'(x) = 0$. As shown above, $f'(x) = 0$ when $x = -3/2$.

Up to now our only method of graphing most functions has been by plotting several points that lie on the graph. Now an additional tool is available: the test for determining where a function is increasing or decreasing. (Other tools are discussed in the next few sections.) Using the intervals where $f(x) = 2x^2 + 6x - 5$ is increasing or decreasing, along with the fact (from Chapter 1) that the graph is a parabola, leads to the graph in Figure 10.

**(b)** $f(x) = x^3 + 3x^2 - 9x + 4$

Here $f'(x) = 3x^2 + 6x - 9$. To find the critical numbers, set this derivative equal to 0 and solve the resulting equation by factoring.

$$3x^2 + 6x - 9 = 0$$
$$3(x^2 + 2x - 3) = 0$$
$$3(x + 3)(x - 1) = 0$$
$$x = -3 \quad \text{or} \quad x = 1$$

The tangent line is horizontal at $x = -3$ or $x = 1$. There are no values of $x$ where $f'(x)$ fails to exist. The critical numbers are $-3$ and 1. To determine where the function is increasing or decreasing, locate $-3$ and 1 on a number line, as in Figure 11. (Be sure to place the values on the number line in numerical order.) Choosing the test point $-4$ from the interval on the left gives

$$f'(-4) = 3(-4)^2 + 6(-4) - 9 = 15,$$

which is positive, so $f$ is increasing on $(-\infty, -3)$. Selecting 0 from the middle interval gives $f'(0) = -9$, so $f$ is decreasing on $(-3, 1)$. Finally, choosing 2 in the right-hand region gives $f'(2) = 15$, with $f$ increasing on $(1, +\infty)$. The intervals where $f$ is increasing or decreasing are shown with arrows in Figure 11.

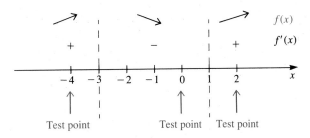

**FIGURE 11**

To graph the function, plot a point in each of the intervals determined by the critical numbers, as well as the points at the critical numbers. Use these points along with the information about where the function is increasing and decreasing to get the graph in Figure 12.

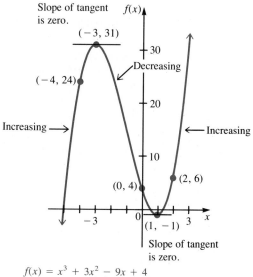

Slope of tangent is zero.

$(-3, 31)$

30

Decreasing

$(-4, 24)$

20

Increasing →

← Increasing

10

$(0, 4)$

$(2, 6)$

$-3$

0

$(1, -1)$

3

$x$

Slope of tangent is zero.

$$f(x) = x^3 + 3x^2 - 9x + 4$$

**FIGURE 12**

**(c)** $f(x) = \dfrac{x - 1}{x + 1}$

Use the quotient rule to find $f'(x)$.

$$f'(x) = \frac{(x + 1)(1) - (x - 1)(1)}{(x + 1)^2}$$

$$= \frac{x + 1 - x + 1}{(x + 1)^2} = \frac{2}{(x + 1)^2}$$

This derivative is never 0, but it fails to exist when $x = -1$. Since $f(x)$ also does not exist at $x = -1$, however, $-1$ is not a critical number. There are no critical numbers. Since the line $x = -1$ is a vertical asymptote for the graph of $f$, it separates the domain into the two intervals $(-\infty, -1)$ and $(-1, +\infty)$. A function can change direction from one side of a vertical asymptote to the other. Use a test point in each of these intervals to find that $f'(x) > 0$ for all $x$ except $-1$. This means that the function $f$ is increasing on both $(-\infty, -1)$ and $(-1, +\infty)$.

Since

$$\lim_{x \to +\infty} \frac{x - 1}{x + 1} = 1,$$

the graph has the line $y = 1$ as a horizontal asymptote. Using all this information and plotting the intercepts and a few points on either side of the vertical asymptote gives the graph in Figure 13. ▬

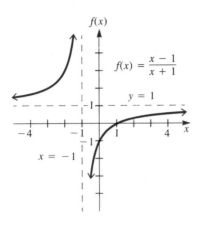

| FIGURE 13 | FIGURE 14 |

It is important to note that the reverse of the test for increasing and decreasing functions is not true—it is possible for a function to be increasing on an interval even though the derivative is not positive at every point in the interval. A good example is given by $f(x) = x^3$, which is increasing on every interval, even though $f'(x) = 0$ when $x = 0$. See Figure 14.

Knowing the intervals where a function is increasing or decreasing can be important in applications, as shown by the next examples.

## ▬ EXAMPLE 3

A company sells an item with the following cost and revenue functions.

$$C(x) = 1.2x - .0006x^2, \qquad 0 \le x \le 1000$$
$$R(x) = 3.6x - .003x^2, \qquad 0 \le x \le 1000$$

Determine any intervals on which the profit function is increasing.

First find the profit function $P(x)$.

$$\begin{aligned} P(x) &= R(x) - C(x) \\ &= (3.6x - .003x^2) - (1.2x - .0006x^2) \\ &= 2.4x - .0024x^2 \end{aligned}$$

To find the interval where this function is increasing, set $P'(x) = 0$.

$$\begin{aligned} P'(x) = 2.4 - .0048x &= 0 \\ x &= 500 \end{aligned}$$

**FIGURE 15**

Use $x = 500$ to determine two intervals on a number line, as shown in Figure 15. Choose $x = 0$ and $x = 1000$ as test points.

$$P'(0) = 2.4 - .0048(0) = 2.4$$
$$P'(1000) = 2.4 - .0048(1000) = -2.4$$

The test points show that the function increases on $(-\infty, 500)$ and decreases on $(500, +\infty)$. See Figure 15. Thus, the profit is increasing on the interval $(0, 500)$ and decreasing on the interval $(500, 1000)$, as shown in Figure 16.

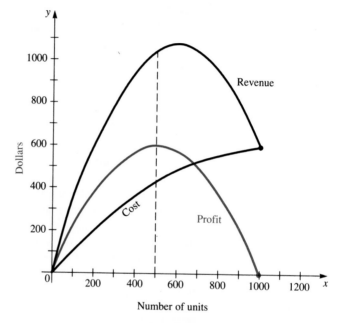

**FIGURE 16**

As the graph in Figure 16 shows, the profit will increase as long as the revenue function increases faster than the cost function. That is, increasing production will produce more profit as long as the marginal revenue is greater than the marginal cost. ▰

▰▰▰ EXAMPLE 4

In the exercises in the previous chapter, the function

$$f(t) = \frac{90t}{99t - 90}$$

gave the number of facts recalled after $t$ hours for $t > 10/11$. Find the intervals in which $f(t)$ is increasing or decreasing.

First find the derivative, $f'(t)$.

$$f'(t) = \frac{(99t - 90)(90) - 90t(99)}{(99t - 90)^2}$$

$$= \frac{8910t - 8100 - 8910t}{(99t - 90)^2} = \frac{-8100}{(99t - 90)^2}$$

Since $(99t - 90)^2$ is positive everywhere in the domain of the function and since the numerator is a negative constant, $f'(t) < 0$ for all $t$ in the domain of $f(t)$. Thus $f(t)$ always decreases and, as expected, the number of words recalled decreases steadily over time.  ▬

## 3.1 EXERCISES

*Find the largest open intervals where the functions graphed as follows are*
**(a)** *increasing, or* **(b)** *decreasing.*

**1.**

**2.**

**3.**

**4.**

**5.**

**6.**

**7.**

**8.**

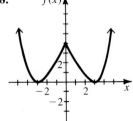

*Find the largest open intervals where the following are* **(a)** *increasing, or* **(b)** *decreasing, and then* **(c)** *graph each function.*

**9.** $f(x) = x^2 + 12x - 6$

**10.** $f(x) = x^2 - 9x + 4$

**11.** $y = 5 + 9x - 3x^2$

**12.** $y = 3 + 4x - 2x^2$

**13.** $f(x) = 2x^3 - 3x^2 - 72x - 4$

**14.** $f(x) = 2x^3 - 3x^2 - 12x + 2$

**15.** $f(x) = 4x^3 - 15x^2 - 72x + 5$

**16.** $f(x) = 4x^3 - 9x^2 - 30x + 6$

**17.** $y = -3x + 6$

**18.** $y = 6x - 9$

**19.** $f(x) = \dfrac{x + 2}{x + 1}$

**20.** $f(x) = \dfrac{x + 3}{x - 4}$

**21.** $y = |x + 4|$

**22.** $y = -|x - 3|$

**23.** $f(x) = -\sqrt{x - 1}$

**24.** $f(x) = \sqrt{5 - x}$

**25.** $y = \sqrt{x^2 + 1}$

**26.** $y = x\sqrt{9 - x^2}$

## ▦ APPLICATIONS

### BUSINESS AND ECONOMICS

**Cost**    **27.** Suppose the total cost $C(x)$ (in dollars) to manufacture a quantity $x$ of weed killer (in hundreds of liters) is given by

$$C(x) = x^3 - 2x^2 + 8x + 50.$$

**(a)** Where is $C(x)$ decreasing?     **(b)** Where is $C(x)$ increasing?

**Housing Starts**    **28.** A county realty group estimates that the number of housing starts per year over the next three years will be

$$H(r) = \frac{300}{1 + .03r^2},$$

where $r$ is the mortgage rate (in percent).

**(a)** Where is $H(r)$ increasing?     **(b)** Where is $H(r)$ decreasing?

**Profit**    **29.** A manufacturer sells video games with the following cost and revenue functions, where $x$ is the number of games sold.

$$C(x) = 4.8x - .0004x^2, \qquad 0 \le x \le 2250$$
$$R(x) = 8.4x - .002x^2, \qquad 0 \le x \le 2250$$

Determine the intervals on which the profit function is increasing.

**Profit**    **30.** A manufacturer of compact disk players has determined that the profit $P(x)$ (in thousands of dollars) is related to the quantity $x$ of players produced (in hundreds) per month by

$$P(x) = \frac{1}{3}x^3 - \frac{7}{2}x^2 + 10x - 2,$$

as long as the number of units produced is fewer than 800 per month.

**(a)** At what production levels is the profit increasing?
**(b)** At what levels is it decreasing?

### LIFE SCIENCES

**Spread of Infection**    **31.** The number of people $P(t)$ (in hundreds) infected $t$ days after an epidemic begins is approximated by

$$P(t) = 2 + 50t - \frac{5}{2}t^2.$$

When will the number of people infected start to decline?

**Alcohol Concentration**   **32.** In Chapter 1 we gave the function

$$A(x) = -.015x^3 + 1.058x$$

as the approximate alcohol concentration (in tenths of a percent) in an average person's bloodstream $x$ hr after drinking eight ounces of 100-proof whiskey. The function applies only for the interval $[0, 8]$.

**(a)** On what time intervals is the alcohol concentration increasing?

**(b)** On what intervals is it decreasing?

**Drug Concentration**   **33.** The percent of concentration of a drug in the bloodstream $x$ hr after the drug is administered is given by

$$K(x) = \frac{4x}{3x^2 + 27}.$$

**(a)** On what time intervals is the concentration of the drug increasing?

**(b)** On what intervals is it decreasing?

**Drug Concentration**   **34.** Suppose a certain drug is administered to a patient, with the percent of concentration of the drug in the bloodstream $t$ hr later given by

$$K(t) = \frac{5t}{t^2 + 1}.$$

**(a)** On what time intervals is the concentration of the drug increasing?

**(b)** On what intervals is it decreasing?

## ▬ 3.2  RELATIVE EXTREMA

Suppose that the manufacturer of a diet soft drink is disappointed by the ineffectiveness of a new series of 30-second television commercials. The company's market research analysts hypothesize that the problem lies in the timing of the commercial's message, "drink sparkling light": either it comes too early in the commercial, before the viewer has become involved, or else it comes too late, after the viewer's attention has faded. After extensive experimentation, the research group finds that the percent of full attention that a viewer devotes to a commercial is a function of time, in seconds, since the commercial began, where

$$\text{Viewer's attention} = f(t) = \frac{-3}{20}(t - 20)^2 + 80, \qquad 0 \le t \le 30.$$

When is the best time to present the commercial's message?

Clearly, the message should be delivered when the viewer's attention is at a maximum. To find this time, find $f'(t)$.

$$f'(t) = \frac{-3}{10}(t - 20) = -.3t + 6$$

The derivative $f'(t)$ is greater than 0 when $-.3t + 6 > 0$, $-3t > -60$, or $t < 20$. Similarly, $f'(t) < 0$ when $-.3t + 6 < 0$, $-3t < -60$ or $t > 20$.

Thus, attention increases for the first 20 seconds and decreases for the last 10 seconds. The message should appear about 20 seconds into the commercial. At that time the viewer will devote $f(20) = 80\%$ of his attention to the commercial.

The maximum level of viewer attention (80%) in the example above is a *relative maximum,* defined as follows.

**RELATIVE MAXIMUM OR MINIMUM**

Let $c$ be a number in the domain of a function $f$. Then $f(c)$ is a **relative maximum** for $f$ if there exists an open interval $(a, b)$ containing $c$ such that

$$f(x) \leq f(c)$$

for all $x$ in $(a, b)$, and $f(c)$ is a **relative minimum** for $f$ if there exists an open interval $(a, b)$ containing $c$ such that

$$f(x) \geq f(c)$$

for all $x$ in $(a, b)$.

A function has a **relative extremum** (plural: **extrema**) at $c$ if it has either a relative maximum or a relative minimum there.

### EXAMPLE 1

Identify the $x$-values of all points where the graph in Figure 17 has relative extrema.

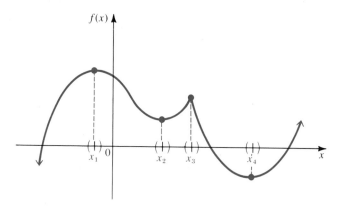

**FIGURE 17**

The parentheses around $x_1$ show an open interval containing $x_1$ such that $f(x) \leq f(x_1)$, so there is a relative maximum of $f(x_1)$ at $x = x_1$. Notice that many other open intervals would work just as well. Similar intervals around $x_2$, $x_3$, and $x_4$ can be used to find a relative maximum of $f(x_3)$ at $x = x_3$ and relative minima of $f(x_2)$ at $x = x_2$ and $f(x_4)$ at $x = x_4$. ■

The function graphed in Figure 18 has relative maxima when $x = x_2$ or $x = x_4$ and relative minima when $x = x_1$ or $x = x_3$. The tangents at the points having $x$-values $x_1$, $x_2$, and $x_3$ are shown in the figure. All three tangents are horizontal and have slope 0. There is no single tangent line at the point where $x = x_4$.

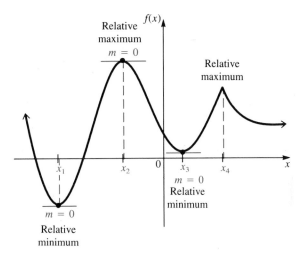

**FIGURE 18**

Since the derivative of a function gives the slope of a tangent line to the graph of the function, to find relative extrema we first identify all values of $x$ where the derivative of the function (the slope of the tangent line) is 0. Next, we identify all $x$-values where the derivative *does not* exist but the function *does* exist. Recall from the preceding section that *critical numbers* are these $x$-values where the derivative is 0 or where the derivative does not exist but the function does exist. A relative extremum *may* exist at a critical number. (A rough sketch of the graph of the function near a critical number often is enough to tell whether an extremum has been found.) These facts about extrema are summarized below.

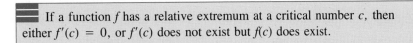

If a function $f$ has a relative extremum at a critical number $c$, then either $f'(c) = 0$, or $f'(c)$ does not exist but $f(c)$ does exist.

Be very careful not to get this result backward. It does *not* say that a function has relative extrema at all critical numbers of the function. For example, Figure 19 shows the graph of $f(x) = x^3$. The derivative, $f'(x) = 3x^2$, is 0 when $x = 0$, so that 0 is a critical number for that function. However, as suggested by the graph of Figure 19, $f(x) = x^3$ has neither a relative maximum nor a relative minimum at $x = 0$ (or anywhere else, for that matter).

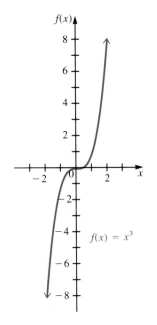

**FIGURE 19**

***First Derivative Test*** Suppose all critical numbers have been found for some function *f*. How is it possible to tell from the equation of the function whether these critical numbers produce relative maxima, relative minima, or neither? One way is suggested by the graph in Figure 20.

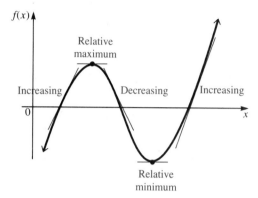

**FIGURE 20**

As shown in Figure 20, on the left of a relative maximum the tangent lines to the graph of a function have positive slope, indicating that the function is increasing. At the relative maximum, the tangent line is horizontal. On the right of the relative maximum the tangent lines have negative slopes, indicating that the function is decreasing. Around a relative minimum the opposite occurs. As shown by the tangent lines in Figure 20, the function is decreasing on the left of the relative minimum, has a horizontal tangent at the minimum, and is increasing on the right of the minimum.

Putting this together with the methods from Section 1 for identifying intervals where a function is increasing or decreasing gives the following *first derivative test* for locating relative extrema.

**FIRST DERIVATIVE TEST**

Let *c* be a critical number for a function *f*. Suppose that *f* is differentiable on $(a, b)$, and that *c* is the only critical number for *f* in $(a, b)$.

**1.** $f(c)$ is a relative maximum of *f* if the derivative $f'(x)$ is positive in the interval $(a, c)$ and negative in the interval $(c, b)$.

**2.** $f(c)$ is a relative minimum of *f* if the derivative $f'(x)$ is negative in the interval $(a, c)$ and positive in the interval $(c, b)$.

The sketches in the following table show how the first derivative test works. Assume the same conditions on *a, b,* and *c* for the table as those given for the first derivative test.

| $f(x)$ has: | Sign of $f'$ in $(a, c)$ | Sign of $f'$ in $(c, b)$ | Sketches |
|---|---|---|---|
| Relative maximum | + | − | |
| Relative minimum | − | + | |
| No relative extrema | + | + | |
| No relative extrema | − | − | |

### ▬▬ EXAMPLE 2

Find all relative extrema for the following functions. Graph each function.

**(a)** $f(x) = 2x^3 - 3x^2 - 72x + 15$

The derivative is $f'(x) = 6x^2 - 6x - 72$. There are no points where $f'(x)$ fails to exist, so the only critical numbers will be found where the derivative equals 0. Setting the derivative equal to 0 gives

$$6x^2 - 6x - 72 = 0$$
$$6(x^2 - x - 12) = 0$$
$$6(x - 4)(x + 3) = 0$$
$$x - 4 = 0 \quad \text{or} \quad x + 3 = 0$$
$$x = 4 \quad \text{or} \quad x = -3.$$

As in the previous section, the critical numbers 4 and $-3$ are used to determine the three intervals $(-\infty, -3)$, $(-3, 4)$, and $(4, +\infty)$ shown on the number line in Figure 21. Any number from each of the three intervals can be used as a test point to find the sign of $f'$ in each interval. Using $-4$, 0, and 5 gives the following information.

$$f'(-4) = 6(-8)(-1) > 0$$
$$f'(0) = 6(-4)(3) < 0$$
$$f'(5) = 6(1)(8) > 0$$

Thus, the derivative is positive on $(-\infty, -3)$, negative on $(-3, 4)$, and positive on $(4, +\infty)$. By Part 1 of the first derivative test, this means that the function has a relative maximum of $f(-3) = 150$ when $x = -3$; by Part 2, $f$ has a relative minimum of $f(4) = -193$ when $x = 4$.

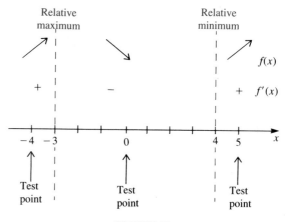

**FIGURE 21**

Using the information found above and plotting the points from the following chart gives the graph in Figure 22. (Be sure to use $f(x)$, not $f'(x)$, to find the $y$-values of the points to plot.)

| $x$ | $-4$ | $-3$ | 0 | 4 | 5 |
|------|------|------|-----|------|------|
| $f(x)$ | 127 | 150 | 15 | $-193$ | $-170$ |

It is not necessary to plot additional points with $x$-values less than $-4$ or greater than 5, since any maximum or minimum must occur at a critical number. Thus, the graph will not have other "turning points."

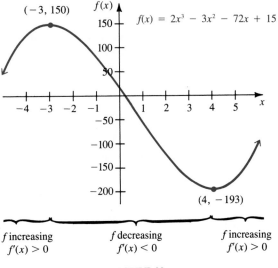

$f(x) = 2x^3 - 3x^2 - 72x + 15$

(−3, 150)

(4, −193)

$f$ increasing
$f'(x) > 0$

$f$ decreasing
$f'(x) < 0$

$f$ increasing
$f'(x) > 0$

**FIGURE 22**

**(b)** $f(x) = 6x^{2/3} - 4x$

Find $f'(x)$:

$$f'(x) = 4x^{-1/3} - 4 = \frac{4}{x^{1/3}} - 4.$$

The derivative fails to exist when $x = 0$, but the function itself is defined when $x = 0$, making 0 a critical number for $f$. To find other critical numbers, set $f'(x) = 0$.

$$f'(x) = 0$$

$$\frac{4}{x^{1/3}} - 4 = 0$$

$$\frac{4}{x^{1/3}} = 4$$

$$4 = 4x^{1/3}$$

$$1 = x^{1/3}$$

$$1 = x$$

The critical numbers 0 and 1 are used to locate the intervals $(-\infty, 0)$, $(0, 1)$, and $(1, +\infty)$ on a number line as in Figure 23. Evaluating $f'(x)$ at the test points $-1$, $1/2$, and 2 and using the first derivative test shows that $f$ has a relative maximum at $x = 1$; the value of this relative maximum is $f(1) = 2$.

Also, $f$ has a relative minimum at $x = 0$; this relative minimum is $f(0) = 0$. Using this information and plotting the points from the following table leads to the graph shown in Figure 24.

| $x$ | $-1$ | 0 | 1/2 | 1 | 2 |
|---|---|---|---|---|---|
| $f(x)$ | 10 | 0 | 1.8 | 2 | 1.5 |

To complete the graph as shown, it would be helpful to also plot the points $(-1/4, 3.4)$ and $(4, -.9)$. Note that the graph has a sharp point at the critical number where the derivative does not exist.

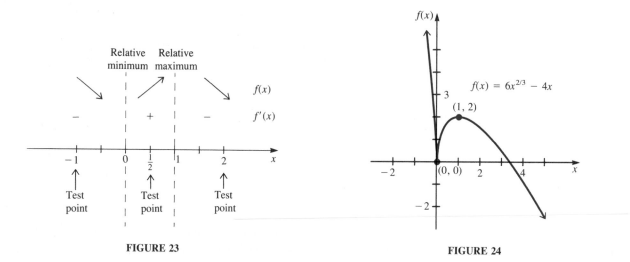

FIGURE 23                                          FIGURE 24

(c)  $f(x) = x^{2/3}$

The derivative is

$$f'(x) = \frac{2}{3}x^{-1/3} = \frac{2}{3x^{1/3}}.$$

The derivative is never 0, but it does fail to exist when $x = 0$. Since $f(0)$ *does* exist but $f'(0)$ does not, 0 is a critical point. Testing a point on either side of 0 gives the results shown in Figure 25. (Use 8 and $-8$ as test points because they are perfect cubes and therefore easy to work with in the equations for $f(x)$ and $f'(x)$.) The function has a relative minimum at 0 with $f(0) = 0^{2/3} = 0$. The function is decreasing on the interval $(-\infty, 0)$ and increasing on $(0, +\infty)$. Using this information and plotting a few points gives the graph in Figure 26. Again, there is a sharp point at the critical number because the derivative does not exist there.  ■

**FIGURE 25**

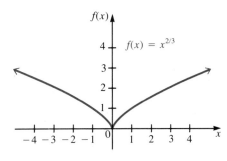

**FIGURE 26**

The next example illustrates the fact that not every critical number yields a relative extremum.

### ■■■ EXAMPLE 3

Find all relative extrema of $f(x) = 4x^3 + 1$ and graph the function.

The derivative is $f'(x) = 12x^2$, which exists for all values of $x$ and is 0 when $x = 0$. The only critical number is 0. Choose $-1$ and 1 as test points. Both numbers make $f'(x)$ positive, so $f$ is increasing for all $x$. Since the value of $f'(x)$ does not change from negative to positive or from positive to negative at the critical number, this function has no extremum. Use this information and plot the points $(-1, -3)$, $(0, 1)$, and $(1, 5)$ to get the graph shown in Figure 27.   ■■

As shown by the next example, a critical number must be in the domain of the function.

### ■■■ EXAMPLE 4

Find all relative extrema of $f(x) = \dfrac{2x}{x - 4}$ and graph $f(x)$.

Find $f'(x)$.

$$f'(x) = \frac{(x - 4)2 - 2x(1)}{(x - 4)^2} = \frac{-8}{(x - 4)^2}$$

Since the numerator can never equal 0, $f'(x)$ is never 0. The derivative fails to exist if

$$(x - 4)^2 = 0$$

or                                           $x = 4,$

but $f(4)$ does not exist, so 4 is not a critical number. The function has no

$f(x)$

$f(x) = 4x^3 + 1$

**FIGURE 27**

critical numbers and thus has no relative extrema. In fact, $x = 4$ is a vertical asymptote of the graph of $f$. To complete the graph, note that

$$\lim_{x \to -\infty} f(x) = 2 \quad \text{and} \quad \lim_{x \to +\infty} f(x) = 2,$$

so that $y = 2$ is a horizontal asymptote. The only intercept is at $(0, 0)$. Plot a few points to get the graph shown in Figure 28. ▬

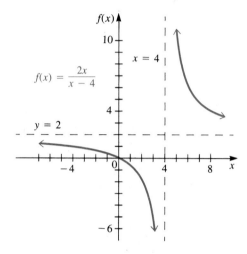

**FIGURE 28**

As mentioned at the beginning of this section, finding the maximum or minimum value of a quantity is important in applications of mathematics. The final example gives specific illustrations.

## ▬ EXAMPLE 5

A small company manufactures and sells bicycles. The production manager has determined that the cost and demand functions for $x(x \geq 0)$ bicycles per week are

$$C(x) = 100 + 30x \quad \text{and} \quad p = 90 - x,$$

where $p$ is the price per bicycle.

**(a)** Find the maximum weekly revenue.

The revenue each week is given by

$$R(x) = xp = x(90 - x) = 90x - x^2.$$

To maximize $R(x) = 90x - x^2$, find $R'(x)$. Then find the critical numbers.

$$R'(x) = 90 - 2x = 0$$
$$90 = 2x$$
$$x = 45$$

Since $R'(x)$ exists for all $x$, 45 is the only critical number. To verify that $x = 45$ will produce a *maximum*, evaluate the derivative on either side of $x = 45$.

$$R'(40) = 10 \quad \text{and} \quad R'(50) = -10$$

This shows that $R(x)$ is increasing up to $x = 45$, then decreasing, so there is a maximum value at $x = 45$ of $R(45) = 2025$. The maximum revenue will be $2025 and will occur when 45 bicycles are produced and sold each week.

**(b)** Find the maximum weekly profit.

Since profit equals revenue minus cost, the profit is given by

$$P(x) = R(x) - C(x)$$
$$= (90x - x^2) - (100 + 30x)$$
$$= -x^2 + 60x - 100.$$

Find the derivative and set it equal to 0 to find the critical numbers. (The derivative exists for all $x$.)

$$P'(x) = -2x + 60 = 0$$
$$60 = 2x$$
$$x = 30$$

The only critical number is 30. Determine whether $x = 30$ produces a maximum by testing a value on each side of 30 in $P'(x)$.

$$P'(20) = 20 \quad \text{and} \quad P'(40) = -20$$

These results show that $P(x)$ increases to $x = 30$ and then decreases, so $x = 30$ produces a maximum value of $P(30) = 800$. Thus, maximum profit of $800 occurs when 30 bicycles are produced and sold each week. Note that this is not the same as the number that should be produced to yield maximum revenue.

**(c)** Find the price the company should charge to realize maximum profit.

As shown in part (b), 30 bicycles per week should be produced and sold to get the maximum profit of $800 per week. Since the price is given by

$$p = 90 - x,$$

if $x = 30$, then $p = 60$. The manager should charge $60 per bicycle and produce and sell 30 bicycles per week to get the maximum profit of $800 per week. Figure 29 shows the graphs of the functions used in this example. Notice that the slopes of the revenue and cost functions are the same at the point where maximum profit occurs. ■

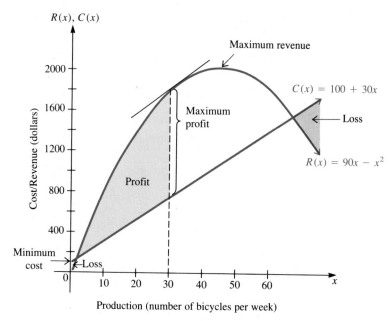

**FIGURE 29**

Additional examples showing how to use the derivative to find extrema in solving applied problems will be given in the next two sections.

## 3.2 EXERCISES

*Find the locations and values of all relative extrema for the functions with graphs as follows. Compare with Exercises 1–8 in the preceding section.*

1.

2.

3.

**4.**

**5.**

**6.**

**7.**

**8.**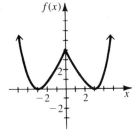

*Find the x-values of all points where the functions defined as follows have any relative extrema. Find the value of any relative extrema.*

**9.** $f(x) = x^2 + 12x - 8$

**10.** $f(x) = x^2 - 4x + 6$

**11.** $f(x) = 8 - 6x - x^2$

**12.** $f(x) = 3 - 4x - 2x^2$

**13.** $f(x) = x^3 + 6x^2 + 9x - 8$

**14.** $f(x) = x^3 + 3x^2 - 24x + 2$

**15.** $f(x) = -\dfrac{4}{3}x^3 - \dfrac{21}{2}x^2 - 5x + 8$

**16.** $f(x) = -\dfrac{2}{3}x^3 - \dfrac{1}{2}x^2 + 3x - 4$

**17.** $f(x) = 2x^3 - 21x^2 + 60x + 5$

**18.** $f(x) = 2x^3 + 15x^2 + 36x - 4$

**19.** $f(x) = x^4 - 18x^2 - 4$

**20.** $f(x) = x^4 - 8x^2 + 9$

**21.** $f(x) = -(8 - 5x)^{2/3}$

**22.** $f(x) = (2 - 9x)^{2/3}$

**23.** $f(x) = 2x + 3x^{2/3}$

**24.** $f(x) = 3x^{5/3} - 15x^{2/3}$

**25.** $f(x) = x - \dfrac{1}{x}$

**26.** $f(x) = x^2 + \dfrac{1}{x}$

**27.** $f(x) = \dfrac{x^2}{x^2 + 1}$

**28.** $f(x) = \dfrac{x^2}{x - 3}$

**29.** $f(x) = \dfrac{x^2 - 2x + 1}{x - 3}$

**30.** $f(x) = \dfrac{x^2 - 6x + 9}{x + 2}$

*Use the derivative to find the vertex of each parabola.*

**31.** $y = -2x^2 + 8x - 1$

**32.** $y = 3x^2 - 12x + 2$

**33.** $y = 2x^2 - 5x + 2$

**34.** $y = -x^2 - 2x + 1$

■ APPLICATIONS

## BUSINESS AND ECONOMICS

**Profit**    *In Exercises 35–37, find* **(a)** *the price p per unit that produces maximum profit;* **(b)** *the number x of units that produces maximum profit; and* **(c)** *the maximum profit P (P = R − C).*

35. $C(x) = 75 + 10x; \qquad p = 70 - 2x$

36. $C(x) = 25x + 5000; \qquad p = 80 - .01x$

37. $C(x) = 75 + 80x - x^2; \qquad p = 100 - 2x$

**Profit**    38. The total profit $P(x)$ (in dollars) from the sale of $x$ units of a certain prescription drug is given by

$$P(x) = -x^3 + 3x^2 + 72x.$$

(a) Find the number of units that should be sold in order to maximize the total profit.

(b) What is the maximum profit?

**Revenue**    39. The demand equation for a certain commodity is

$$p = \frac{1}{3}x^2 - \frac{25}{2}x + 100, \qquad 0 \le x \le 16.$$

Find the values of $x$ and $p$ that maximize revenue.

**Cost**    40. Suppose that the cost function for a product is given by $C(x) = .002x^3 - 9x + 4000$. Find the production level (i.e., value of $x$) that will produce the minimum average cost per unit $\overline{C}(x)$.

## LIFE SCIENCES

**Activity Level**    41. In the summer the activity level of a certain type of lizard varies according to the time of day. A biologist has determined that the activity level is given by the function

$$a(t) = .008t^3 - .27t^2 + 2.02t + 7,$$

where $t$ is the number of hours after 12 noon. When is the activity level highest? When is it lowest? (*Hint:* Use the quadratic formula.)

## SOCIAL SCIENCES

**Class Attendance**    42. A professor has found that the number of students attending a biology class is approximated by

$$S(x) = -x^2 + 20x + 80,$$

where $x$ is the number of hours that the student union is open daily. Find the number of hours that the union should be open so that the number of students attending class is a maximum. Find the maximum number of such students.

## PHYSICAL SCIENCES

**Height**    43. After a great deal of experimentation, two Atlantic Institute of Technology senior physics majors determined that when a bottle of French champagne is shaken several times, held upright, and uncorked, its cork travels according to

$$s(t) = -16t^2 + 64t + 3,$$

where $s$ is its height in feet above the ground $t$ sec after being released. How high will it go?

## 3.3  ABSOLUTE EXTREMA

The function graphed in Figure 30 has a relative minimum at $x_2$ and relative maxima at $x_1$ and $x_3$. In practical situations it is often desirable to know if a function has one function value that is larger than any other or one function value that is smaller than any other. In Figure 30, $f(x_1) \geq f(x)$ for all $x$ in the domain. There is no function value that is smaller than all others, however, because $f(x) \to -\infty$ as $x \to +\infty$ or as $x \to -\infty$.

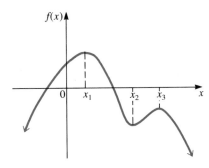

**FIGURE 30**

The largest possible value of a function is called the *absolute maximum* and the smallest possible value of a function is called the *absolute minimum*. As Figure 30 shows, one or both of these may not exist on the domain of the function, $(-\infty, +\infty)$ here. Absolute extrema often coincide with relative extrema, as with $f(x_1)$ in Figure 29. Although a function may have several relative maxima or relative minima, it never has more than one *absolute maximum* or *absolute minimum*.

**ABSOLUTE MAXIMUM OR MINIMUM**

Let $f$ be a function defined on some interval. Let $c$ be a number in the interval. Then $f(c)$ is the **absolute maximum** of $f$ on the interval if

$$f(x) \leq f(c)$$

for every $x$ in the interval, and $f(c)$ is the **absolute minimum** of $f$ on the interval if

$$f(x) \geq f(c)$$

for every $x$ in the interval.

Now look at Figure 31, which shows three functions defined on closed intervals. In each case there is an absolute maximum value and an absolute minimum value. These absolute extrema may occur at the endpoints or at relative extrema. As the graphs in Figure 31 show, an absolute extremum is either the largest or the smallest function value occurring on a closed interval, while a relative extremum is the largest or smallest function value in some (perhaps small) open interval.

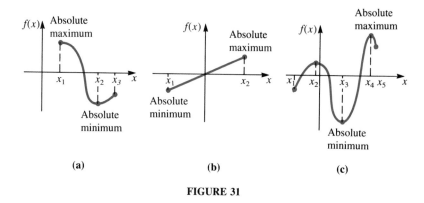

**(a)**          **(b)**          **(c)**

**FIGURE 31**

One of the main reasons for the importance of absolute extrema is given by the following theorem (which is proved in more advanced courses).

**EXTREME VALUE THEOREM**

A function $f$ that is continuous on a closed interval $[a, b]$ will have both an absolute maximum and an absolute minimum on the interval.

As Figure 31 shows, an absolute extremum *may* occur on an open interval. See $x_2$ in part (a) and $x_3$ in part (c), for example. The extreme value theorem says that a function *must* have both an absolute maximum and an absolute minimum on a closed interval. The conditions that $f$ be *continuous* and on a *closed* interval are very important. In Figure 32, $f$ is discontinuous at $x = b$. Since $f(x)$ gets larger and larger as $x \to b$, there is no absolute (or relative) maximum on $(a, b)$. Also, as $x \to a^+$, the values of $f(x)$ get smaller and smaller, but there is no smallest value of $f(x)$ on the *open* interval $(a, b)$, since there is no endpoint.

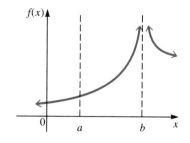

**FIGURE 32**

The extreme value theorem guarantees the existence of absolute extrema. To find these extrema, use the following steps.

**FINDING ABSOLUTE EXTREMA**

> To find absolute extrema for a function $f$ continuous on a closed interval $[a, b]$:
>
> **1.** Find all critical numbers for $f$ in $(a, b)$.
> **2.** Evaluate $f$ for all critical numbers in $(a, b)$.
> **3.** Evaluate $f$ for the endpoints $a$ and $b$ of the interval $[a, b]$.
> **4.** The largest value found in Step 2 or 3 is the absolute maximum for $f$ on $[a, b]$, and the smallest value found is the absolute minimum for $f$ on $[a, b]$.

### ■ EXAMPLE 1

Find the absolute extrema of the function

$$f(x) = 2x^3 - x^2 - 20x - 10$$

on the interval $[-2, 4]$.

First look for critical points in the interval $(-2, 4)$. Set the derivative $f'(x) = 6x^2 - 2x - 20$ equal to 0.

$$6x^2 - 2x - 20 = 0$$
$$2(3x^2 - x - 10) = 0$$
$$2(3x + 5)(x - 2) = 0$$
$$3x + 5 = 0 \quad \text{or} \quad x - 2 = 0$$
$$x = -5/3 \qquad x = 2$$

Since there are no values of $x$ where $f'(x)$ does not exist, the only critical values are $-5/3$ and $2$. Both of these critical values are in the interval $(-2, 4)$. Now evaluate the function at $-5/3$ and $2$, and at the endpoints of its domain, $-2$ and $4$.

| x-value | Value of Function | |
|---------|-------------------|---|
| $-2$ | 10 | |
| $-5/3$ | 11.296 | |
| 2 | $-38$ | ←Absolute minimum |
| 4 | 22 | ←Absolute maximum |

The absolute maximum, 22, occurs when $x = 4$, and the absolute minimum, $-38$, occurs when $x = 2$. A graph of $f$ on $[-2, 4]$ is shown in Figure 33 on the next page. ■

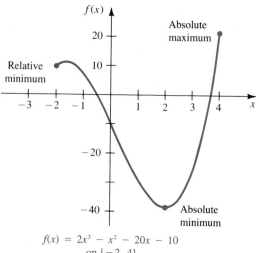

$$f(x) = 2x^3 - x^2 - 20x - 10$$
$$\text{on } [-2, 4]$$

**FIGURE 33**

### ■ EXAMPLE 2

Find the absolute extrema of the function given by

$$f(x) = 4x + \frac{36}{x},$$

on the interval $[1, 6]$.

First determine any critical points in $[1, 6]$.

$$f'(x) = 4 - \frac{36}{x^2} = 0$$

$$\frac{4x^2 - 36}{x^2} = 0$$

$$4x^2 - 36 = 0$$

$$4x^2 = 36$$

$$x^2 = 9$$

$$x = -3 \quad \text{or} \quad x = 3$$

Although $f'(x)$ does not exist at $x = 0$, $f$ is not defined at $x = 0$, so the critical numbers are $-3$ and $3$. Since $-3$ is not in the interval $[1, 6]$, 3 is the only critical number. To find the absolute extrema, evaluate $f(x)$ at $x = 3$ and at the endpoints, where $x = 1$ and $x = 6$.

| *x-value* | *Value of Function* | |
|---|---|---|
| 1 | 40 | ←Absolute maximum |
| 3 | 24 | ←Absolute minimum |
| 6 | 30 | |

Figure 34 shows a graph of $f$ on [1, 6].

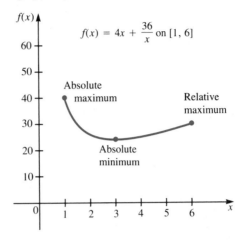

$$f(x) = 4x + \frac{36}{x} \text{ on } [1, 6]$$

Absolute maximum

Relative maximum

Absolute minimum

**FIGURE 34**

Absolute extrema are particularly important in applications of mathematics. In most applications the domain is limited in some natural way that ensures the occurrence of absolute extrema.

#### ■ EXAMPLE 3

A company has found that its weekly profit from the sale of $x$ units of an auto part is given by

$$P(x) = -.02x^3 + 600x - 20,000.$$

Production bottlenecks limit the number of units that can be made per week to no more than 150, while a long-term contract requires that at least 50 units be made each week. Find the maximum possible weekly profit that the firm can make.

Because of the restrictions, the profit function is defined only for the domain [50, 150]. Look first for critical numbers of the function in the open interval (50, 150). Here $P'(x) = -.06x^2 + 600$. Set this derivative equal to 0 and solve for $x$.

$$-.06x^2 + 600 = 0$$
$$-.06x^2 = -600$$
$$x^2 = 10,000$$
$$x = 100 \quad \text{or} \quad x = -100$$

Since $x = -100$ is not in the interval [50, 150], disregard it. Now evaluate

the function at the remaining critical number 100 and at the endpoints of the
domain, 50 and 150.

| x-value | Value of Function | |
|---------|-------------------|---|
| 50 | 7500 | |
| 100 | 20,000 | ←Absolute maximum |
| 150 | 2500 | |

Maximum profit of $20,000 occurs when 100 units are made per week. ■

## 3.3 EXERCISES

*Find the locations of any absolute extrema for the functions with graphs as follows.*

**1.**

**2.**

**3.**

**4.**

**5.** $g(x)$

**6.** $g(x)$

**7.** $f(x)$

**8.** $f(x)$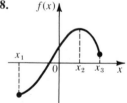

*Find the locations of all absolute extrema for the functions defined as follows, with the
specified domains. A calculator will be helpful for many of these problems.*

**9.** $f(x) = x^2 + 6x + 2$;  $[-4, 0]$

**10.** $f(x) = x^2 - 4x + 1$;  $[-6, 5]$

**11.** $f(x) = 5 - 8x - 4x^2$;  $[-5, 1]$

**12.** $f(x) = 9 - 6x - 3x^2$;  $[-4, 3]$

**13.** $f(x) = x^3 - 3x^2 - 24x + 5$;  $[-3, 6]$

**14.** $f(x) = x^3 - 6x^2 + 9x - 8$;  $[0, 5]$

**15.** $f(x) = \frac{1}{3}x^3 - \frac{1}{2}x^2 - 6x + 3$;  $[-4, 4]$

**16.** $f(x) = \frac{1}{3}x^3 + \frac{3}{2}x^2 - 4x + 1$;  $[-5, 2]$

**17.** $f(x) = x^4 - 32x^2 - 7$;  $[-5, 6]$

**18.** $f(x) = x^4 - 18x^2 + 1$;  $[-4, 4]$

**19.** $f(x) = \frac{1}{1 + x}$;  $[0, 2]$

**20.** $f(x) = \frac{-2}{x + 3}$;  $[1, 4]$

**21.** $f(x) = \dfrac{8 + x}{8 - x}$;  [4, 6]

**22.** $f(x) = \dfrac{1 - x}{3 + x}$;  [0, 3]

**23.** $f(x) = \dfrac{x}{x^2 + 2}$;  [0, 4]

**24.** $f(x) = \dfrac{x - 1}{x^2 + 1}$;  [1, 5]

**25.** $f(x) = (x^2 + 18)^{2/3}$;  [-3, 3]

**26.** $f(x) = (x^2 + 4)^{1/3}$;  [-2, 2]

**27.** $f(x) = (x + 1)(x + 2)^2$;  [-4, 0]

**28.** $f(x) = (x - 3)(x - 1)^3$;  [-2, 3]

**29.** $f(x) = \dfrac{1}{\sqrt{x^2 + 1}}$;  [-1, 1]

**30.** $f(x) = \dfrac{3}{\sqrt{x^2 + 4}}$;  [-2, 2]

---

### ▆ APPLICATIONS

#### BUSINESS AND ECONOMICS

**Average Cost**  **31.** An enterprising (although unscrupulous) business student has managed to get his hands on a copy of the out-of-print solutions manual for an applied calculus text. He plans to duplicate it and sell copies in the dormitories. He figures that demand will be between 100 and 1200 copies, and he wants to minimize his average cost of production. After checking into the cost of paper, duplicating, and the rental of a small van, he estimates that the cost in dollars to produce $x$ hundred manuals is given by $C(x) = x^2 + 200x + 100$. How many should he produce in order to make the *average cost* per unit as small as possible? How much will he have to charge to make a profit?

**Profit**  **32.** The total profit $P(x)$ (in thousands of dollars) from the sale of $x$ hundred thousands of automobile tires is approximated by

$$P(x) = -x^3 + 9x^2 + 120x - 400, \qquad x \ge 5.$$

Find the number of hundred thousands of tires that must be sold to maximize profit. Find the maximum profit.

#### LIFE SCIENCES

**Pollution**  **33.** A marshy region used for agricultural drainage has become contaminated with selenium. It has been determined that flushing the area with clean water will reduce the selenium for a while, but it will then begin to build up again. A biologist has found that the percent of selenium in the soil $x$ months after the flushing begins is given by

$$f(x) = \frac{x^2 + 36}{2x}, \qquad 1 \le x \le 12.$$

When will the selenium be reduced to a minimum? What is the minimum percent?

**34.** The number of salmon swimming upstream to spawn is approximated by

$$S(x) = -x^3 + 3x^2 + 360x + 5000, \qquad 6 \le x \le 20,$$

where $x$ represents the temperature of the water in degrees Celsius. Find the water temperature that produces the maximum number of salmon swimming upstream.

### PHYSICAL SCIENCES

**Gasoline Mileage**  **35.** From information given in a recent business publication we constructed the mathematical model

$$M(x) = -\frac{1}{45}x^2 + 2x - 20, \qquad 30 \le x \le 65,$$

to represent the miles per gallon used by a certain car at a speed of $x$ mph. Find the absolute maximum miles per gallon and the absolute minimum.

**Gasoline Mileage**  **36.** For a certain compact car,

$$M(x) = -.018x^2 + 1.24x + 6.2, \qquad 30 \le x \le 60,$$

represents the miles per gallon obtained at a speed of $x$ mph. Find the absolute maximum miles per gallon and the absolute minimum.

### GENERAL

**Area**  *A piece of wire 12 ft long is cut into two pieces. (See the figure.) One piece is made into a circle and the other piece is made into a square.*

*Let the piece of length x be formed into a circle.*

Radius of circle $= \dfrac{x}{2\pi}$      Area of circle $= \pi\left(\dfrac{x}{2\pi}\right)^2$

Side of square $= \dfrac{12 - x}{4}$      Area of square $= \left(\dfrac{12 - x}{4}\right)^2$

**37.** Where should the cut be made in order to minimize the sum of the areas enclosed by both figures? (*Hint:* Use 3.14 as an approximation for $\pi$. Have a calculator handy.)

**38.** Where should the cut be made in order to make the sum of the areas maximum? (*Hint:* Remember to use the endpoints of a domain when looking for absolute maxima and minima.)

## 3.4 CONCAVITY, THE SECOND DERIVATIVE TEST, AND CURVE SKETCHING

To understand the behavior of a function on an interval, it is important to know the *rate* at which the function is increasing or decreasing. For example, suppose that your friend, a finance major, has made a study of a young company and is trying to get you to invest in its stock. He shows you the following function, which represents the price $P(t)$ of the company's stock since it became available in January two years ago:

$$P(t) = 17 + t^{1/2},$$

where $t$ is the number of months since the stock became available. He points out that the derivative of the function is always positive, so the price of the stock is always increasing. He claims that you cannot help but make a fortune on it. Should you take his advice and invest?

It is true that the price function increases for all $t$. The derivative is

$$P'(t) = \frac{1}{2}t^{-1/2} = \frac{1}{2\sqrt{t}},$$

which is always positive because $\sqrt{t}$ is positive for $t > 0$. The catch lies in *how fast* the function is increasing. The derivative

$$P'(t) = \frac{1}{2\sqrt{t}}$$

tells how fast the price is increasing at any number of months, $t$, since the stock became available. For example, when $t = 1$, $P'(t) = 1/2$, and the price is increasing at the rate of 1/2 dollar, or 50 cents, per month. When $t = 4$, $P'(t) = 1/4$; the stock is increasing at 25 cents per month. At $t = 9$ months, $P'(t) = 1/6$, or about 17 cents per month. By the time you could buy in at $t = 24$ months, the price is increasing at 10 cents per month, and the *rate of increase* looks as though it will continue to decrease.

The rate of increase in $P$ is given by the derivative of $P'(t)$, called the **second derivative** of $P$ and denoted by $P''(t)$. Since $P'(t) = (1/2)t^{-1/2}$,

$$P''(t) = -\frac{1}{4}t^{-3/2} = -\frac{1}{4\sqrt{t^3}}.$$

$P''(t)$ is negative for all positive values of $t$ and therefore confirms the suspicion that the *rate* of increase in price does indeed decrease for all $t \geq 0$. The price of the company's stock will not drop, but the amount of return will certainly not be the "fortune" that your friend predicts. For example, at $t = 24$ months, when you would buy, the price would be $21.90. A year later, it would be $23.00 a share. If you were rich enough to buy 100 shares for $2190, they would be worth $2300 in a year. The increase of $110 is about 5% of the investment—similar to the return that you could get in most savings accounts. The only investors to make a lot of money on this stock would be those who bought early, when the rate of increase was much greater.

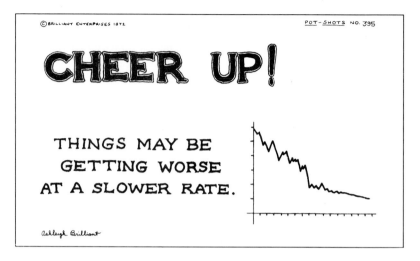

**FIGURE 35**

In Figure 35* the (generally) decreasing function shows that "things are getting worse," but the overall shape of the graph indicates that the *rate* of decrease is slowing down. In other words, the second derivative is negative.

As mentioned earlier, the second derivative of a function $f$, written $f''$, gives the rate of change of the *derivative* of $f$. Before continuing to discuss applications of the second derivative, we need to introduce some additional terminology and notation.

***Higher Derivatives***    If a function $f$ has a derivative $f'$, then the derivative of $f'$, if it exists, is the second derivative of $f$, written $f''(x)$. The derivative of $f''(x)$, if it exists, is called the **third derivative** of $f$, and so on. By continuing this process, we can find **fourth derivatives** and other higher derivatives. For example, if $f(x) = x^4 + 2x^3 + 3x^2 - 5x + 7$, then

$$f'(x) = 4x^3 + 6x^2 + 6x - 5, \qquad \text{First derivative of } f$$

$$f''(x) = 12x^2 + 12x + 6, \qquad \text{Second derivative of } f$$

$$f'''(x) = 24x + 12, \qquad \text{Third derivative of } f$$

and $\qquad f^{(4)}(x) = 24.$ $\qquad\qquad\qquad$ Fourth derivative of $f$

**NOTATION FOR HIGHER DERIVATIVES**

The second derivative of $y = f(x)$ can be written using any of the following notations:

$$f''(x), \qquad y'', \qquad \frac{d^2y}{dx^2}, \qquad \text{or} \qquad D_x^2[f(x)].$$

The third derivative can be written in a similar way. For $n \geq 4$, the $n$th derivative is written $f^{(n)}(x)$.

---

*Ashleigh Brilliant epigrams, Pot-Shots, and Brilliant Thoughts from the book *I May Not Be Totally Perfect, But Parts Of Me Are Excellent, And Other Brilliant Thoughts* used by permission of the author, Ashleigh Brilliant, 117 West Valerio St., Santa Barbara, CA 93101.

### ▰ EXAMPLE 1

Let $f(x) = x^3 + 6x^2 - 9x + 8$.

**(a)** Find $f''(x)$.

To find the second derivative of $f(x)$, find the first derivative, and then take its derivative.

$$f'(x) = 3x^2 + 12x - 9$$
$$f''(x) = 6x + 12$$

**(b)** Find $f''(0)$.

Since $f''(x) = 6x + 12$,

$$f''(0) = 6(0) + 12 = 12. \quad ▰$$

### ▰ EXAMPLE 2

Find $f''(x)$ for the functions defined as follows.

**(a)** $f(x) = \sqrt{x}$

Let $f(x) = \sqrt{x} = x^{1/2}$. Then

$$f'(x) = \frac{1}{2}x^{-1/2},$$

and

$$f''(x) = -\frac{1}{4}x^{-3/2} = -\frac{1}{4} \cdot \frac{1}{x^{3/2}} = \frac{-1}{4x^{3/2}}.$$

**(b)** $f(x) = (x^2 - 1)^2$

Here

$$f'(x) = 2(x^2 - 1)(2x) = 4x(x^2 - 1).$$

Use the product rule to find $f''(x)$.

$$f''(x) = 4x(2x) + (x^2 - 1)(4)$$
$$= 8x^2 + 4x^2 - 4$$
$$= 12x^2 - 4 \quad ▰$$

In the previous chapter we saw that the first derivative of a function represents the rate of change of the function. The second derivative, then, represents the rate of change of the first derivative. If a function describes the position of a particle (along a straight line) at time $t$, then the first derivative gives the velocity of the particle. That is, if $y = s(t)$ describes the position (along a straight line) of the particle at time $t$, then $v(t) = s'(t)$ gives the velocity at time $t$.

The rate of change of velocity is called **acceleration.** Since the second derivative gives the rate of change of the first derivative, the acceleration is the derivative of the velocity. Thus, if $a(t)$ represents the acceleration at time $t$, then

$$a(t) = \frac{d}{dt}v(t) = s''(t).$$

## ▰ EXAMPLE 3

Suppose that a particle is moving along a straight line, with its position at time $t$ given by

$$s(t) = t^3 - 2t^2 - 7t + 9.$$

Find the following.

**(a)** The velocity at any time $t$

The velocity is given by

$$v(t) = s'(t) = 3t^2 - 4t - 7.$$

**(b)** The acceleration at any time $t$

Acceleration is given by

$$a(t) = v'(t) = s''(t) = 6t - 4.$$

**(c)** The time intervals (for $t \geq 0$) when the particle is speeding up or slowing down

When the acceleration is positive, the velocity is increasing, so the particle is speeding up. This happens when $6t - 4 > 0$, or $t > 2/3$. A negative acceleration indicates that the particle is slowing down. This happens when $0 < t < 2/3$. ▰

***Concavity of a Graph*** The first derivative has been used to show where a function is increasing or decreasing and where the extrema occur. The second derivative gives the rate of change of the first derivative; it indicates *how fast* the function is increasing or decreasing. The rate of change of the derivative (the second derivative) affects the *shape* of the graph. Intuitively, we say that a graph is *concave upward* on an interval if it "holds water" and *concave downward* if it "spills water." (See Figure 36.)

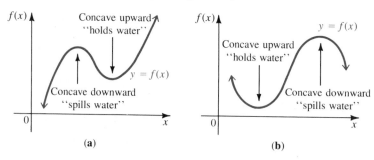

**FIGURE 36**

More precisely, a function is **concave upward** on an interval $(a, b)$ if the graph of the function lies above its tangent line at each point of $(a, b)$. A function is **concave downward** on $(a, b)$ if the graph of the function lies below its tangent line at each point of $(a, b)$. A point where a graph changes concavity is called a **point of inflection.** See Figure 37.

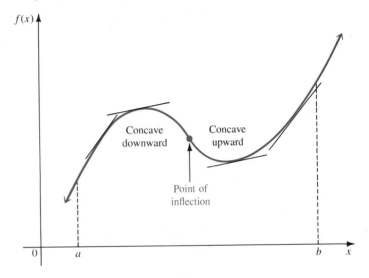

**FIGURE 37**

Just as a function can be either increasing or decreasing on an interval, it can be either concave upward or concave downward on an interval. Examples of various combinations are shown in Figure 38.

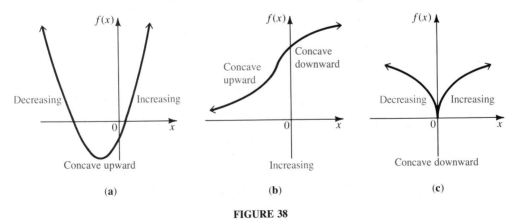

**FIGURE 38**

Figure 39 shows two functions that are concave upward on an interval $(a, b)$. Several tangent lines are shown also. In Figure 39(a), the slopes of the tangent lines (moving from left to right) are first negative, then 0, and then positive. In Figure 39(b), the slopes are all positive, but they get larger.

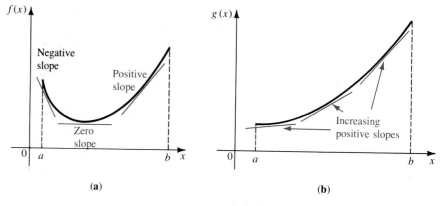

**FIGURE 39**

In both cases, the slopes are *increasing*. The slope at a point on a curve is given by the derivative. Since a function is increasing if its derivative is positive, the slope is increasing if the derivative of the function giving the slopes is positive. Since the derivative of a derivative is the second derivative, a function is concave upward on an interval if its second derivative is positive at each point of the interval.

A similar result is suggested by Figure 40 for functions whose graphs are concave downward. In both graphs, the slopes of the tangent lines are *decreasing* as we move from left to right. Since a function is decreasing if its derivative is negative, a function is concave downward on an interval if its second derivative is negative at each point of the interval. These observations suggest the following test.

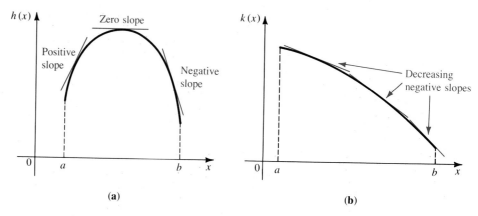

**FIGURE 40**

**TEST FOR CONCAVITY**

> Let $f$ be a function with derivatives $f'$ and $f''$ existing at all points in an interval $(a, b)$. Then $f$ is concave upward on $(a, b)$ if $f''(x) > 0$ for all $x$ in $(a, b)$ and concave downward on $(a, b)$ if $f''(x) < 0$ for all $x$ in $(a, b)$.

### EXAMPLE 4

Find all intervals where $f(x) = x^3 - 3x^2 + 5x - 4$ is concave upward or downward.

The first derivative is $f'(x) = 3x^2 - 6x + 5$, and the second derivative is $f''(x) = 6x - 6$. The function $f$ is concave upward whenever $f''(x) > 0$, or

$$6x - 6 > 0$$
$$6x > 6$$
$$x > 1.$$

Also, $f$ is concave downward if $f''(x) < 0$, or $x < 1$. In interval notation, $f$ is concave upward on $(1, +\infty)$ and concave downward on $(-\infty, 1)$. A graph of $f$ is shown in Figure 41.

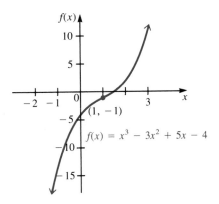

**FIGURE 41**

The graph of the function $f$ in Figure 41 changes from concave downward to concave upward at $x = 1$. As mentioned earlier, a point where the direction of concavity changes is called a point of inflection. This means that the point $(1, f(1))$ or $(1, -1)$ in Figure 41 is a point of inflection. We can locate this point by finding values of $x$ where the second derivative changes from negative to positive, that is, where the second derivative is 0. Setting the second derivative in Example 4 equal to 0 gives

$$6x - 6 = 0$$
$$x = 1.$$

As before, the point of inflection is $(1, f(1))$ or $(1, -1)$.

Example 4 suggests the following result.

At a point of inflection for a function $f$, the second derivative is 0 or does not exist.

Be careful with this statement. Finding a value $f''(x_0) = 0$ does not mean that a point of inflection has been located. For example, if $f(x) = (x - 1)^4$, then $f''(x) = 12(x - 1)^2$, which is 0 at $x = 1$. The graph of $f(x) = (x - 1)^4$ is always concave upward however, so it has no point of inflection. See Figure 42.

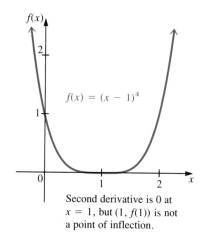

Second derivative is 0 at $x = 1$, but $(1, f(1))$ is not a point of inflection.

**FIGURE 42**

***Curve Sketching*** The test for concavity and the test for increasing and decreasing functions help in sketching the graphs of a variety of functions. This process, called *curve sketching,* uses the following steps.

**CURVE SKETCHING**

To sketch the graph of a function $f$:

1. Find $f'$. Locate any critical points by solving the equation $f'(x) = 0$ and determining where $f'$ does not exist (but $f$ does). Find any relative extrema and determine where $f$ is increasing or decreasing.
2. Find $f''$. Locate potential points of inflection by solving the equation $f''(x) = 0$ and determining where $f''$ does not exist. Determine where $f''$ is concave upward or concave downward.
3. Plot the points at the critical numbers, the inflection points, and other points as needed.

## ■ EXAMPLE 5

Graph each function.

**(a)** $f(x) = 2x^3 - 3x^2 - 12x + 1$

To find the intervals where the function is increasing or decreasing, find the first derivative.

$$f'(x) = 6x^2 - 6x - 12$$

This derivative is 0 when

$$6(x^2 - x - 2) = 0$$
$$6(x - 2)(x + 1) = 0$$
$$x = 2 \quad \text{or} \quad x = -1.$$

These critical numbers divide the number line in Figure 43 into three regions. Testing a number from each region in $f'(x)$ shows that $f$ is increasing on $(-\infty, -1)$ and $(2, +\infty)$ and decreasing on $(-1, 2)$. This is shown with the arrows in Figure 43. By the first derivative test, $f$ has a relative maximum when $x = -1$ and a relative minimum when $x = 2$. The relative maximum is $f(-1) = 8$, while the relative minimum is $f(2) = -19$.

**FIGURE 43**

Now use the second derivative to find the intervals where the function is concave upward or downward. Here

$$f''(x) = 12x - 6,$$

which is 0 when $x = 1/2$. Testing a point with $x$ less than $1/2$, and one with $x$ greater than $1/2$, shows that $f$ is concave downward on $(-\infty, 1/2)$ and concave upward on $(1/2, +\infty)$. The graph has an inflection point at $(1/2, f(1/2)) = (1/2, -11/2)$. This information is summarized in the chart below.

| Interval | $(-\infty, -1)$ | $(-1, 1/2)$ | $(1/2, 2)$ | $(2, \infty)$ |
|---|---|---|---|---|
| *Sign of f'* | + | − | − | + |
| *Sign of f''* | − | − | + | + |
| *f Increasing or Decreasing* | Increasing | Decreasing | Decreasing | Increasing |
| *Concavity of f* | Downward | Downward | Upward | Upward |
| *Shape of Graph* | ⌒ | ⌍ | ⌎ | ⌏ |

Use this information and plot a few points to get the graph in Figure 44.

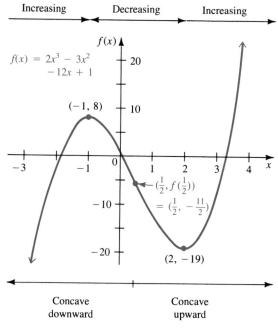

**FIGURE 44**

**(b)** $f(x) = x + 1/x$

Here $f'(x) = 1 - (1/x^2)$, which is 0 when

$$\frac{1}{x^2} = 1$$

$$x^2 = 1$$

$$x = 1 \qquad \text{or} \qquad x = -1.$$

The derivative fails to exist at 0. In fact, the function is undefined at 0, and $x = 0$ is a vertical asymptote. The direction of the graph may change from one side of the asymptote to the other. Evaluating $f'(x)$ in each of the regions determined by the critical numbers and the asymptote shows that $f$ is increasing on $(-\infty, -1)$ and $(1, +\infty)$ and decreasing on $(-1, 0)$ and $(0, 1)$. By the first derivative test, $f$ has a relative maximum when $x = -1$ and a relative minimum when $x = 1$.

The second derivative is

$$f''(x) = \frac{2}{x^3},$$

which is never equal to 0 and does not exist when $x = 0$. (The function itself also does not exist at 0.) Because of this, there may be a change of concavity, but not an inflection point, when $x = 0$. The second derivative is negative when $x$ is negative, making $f$ concave downward on $(-\infty, 0)$. Also, $f''(x) > 0$ when $x > 0$, making $f$ concave upward on $(0, +\infty)$.

| Interval | $(-\infty, -1)$ | $(-1, 0)$ | $(0, 1)$ | $(1, +\infty)$ |
|----------|-----------------|-----------|----------|----------------|
| Sign of $f'$ | + | − | − | + |
| Sign of $f''$ | − | − | + | + |
| $f$ Increasing or Decreasing | Increasing | Decreasing | Decreasing | Increasing |
| Concavity of $f$ | Downward | Downward | Upward | Upward |
| Shape of Graph | | | | |

Use this information and plot points as necessary to get the graph shown in Figure 45. ▬

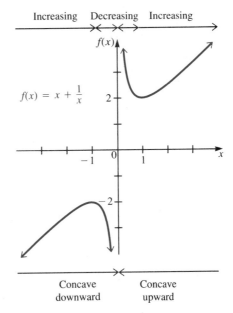

**FIGURE 45**

**Second Derivative Test** The idea of concavity can be used to decide whether a given critical number produces a relative maximum or a relative minimum. This test, an alternative to the first derivative test, is based on the fact that a curve that has a horizontal tangent at a point $c$ and is concave downward on an open interval containing $c$ also has a relative maximum at $c$. A relative minimum occurs when a graph has a horizontal tangent at a point $d$ and is concave upward on an open interval containing $d$. (See Figure 46.)

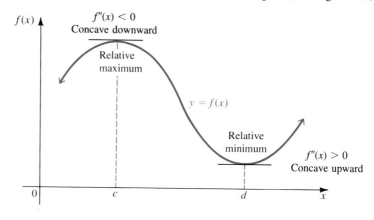

**FIGURE 46**

A function $f$ is concave upward on an interval if $f''(x) > 0$ for all $x$ in the interval, while $f$ is concave downward on an interval if $f''(x) < 0$ for all $x$ in the interval. These ideas lead to the **second derivative test** for relative extrema.

**SECOND DERIVATIVE TEST**

Let $f''$ exist on some open interval containing $c$, and let $f'(x) = 0$.

**1.** If $f''(c) > 0$, then $f(c)$ is a relative minimum.
**2.** If $f''(c) < 0$, then $f(c)$ is a relative maximum.
**3.** If $f''(c) = 0$, then the test gives no information about extrema.

### ■ EXAMPLE 6

Find all relative extrema for

$$f(x) = 4x^3 + 7x^2 - 10x + 8.$$

First, find the points where the derivative is 0. Here $f'(x) = 12x^2 + 14x - 10$. Solve the equation $f'(x) = 0$ to get

$$12x^2 + 14x - 10 = 0$$
$$2(6x^2 + 7x - 5) = 0$$
$$2(3x + 5)(2x - 1) = 0$$
$$3x + 5 = 0 \quad \text{or} \quad 2x - 1 = 0$$
$$3x = -5 \qquad\qquad 2x = 1$$
$$x = -\frac{5}{3} \qquad\qquad x = \frac{1}{2}.$$

Now use the second derivative test. The second derivative is $f''(x) = 24x + 14$. Evaluate $f''(x)$ first at $-5/3$, getting

$$f''\left(-\frac{5}{3}\right) = 24\left(-\frac{5}{3}\right) + 14 = -40 + 14 = -26 < 0,$$

so that by Part 2 of the second derivative test, $-5/3$ leads to a relative maximum of $f(-5/3) = 691/27$. Also, when $x = 1/2$,

$$f''\left(\frac{1}{2}\right) = 24\left(\frac{1}{2}\right) + 14 = 12 + 14 = 26 > 0,$$

with $1/2$ leading to a relative minimum of $f(1/2) = 21/4$.  ▬

The second derivative test works only for those critical numbers $c$ that make $f'(c) = 0$. This test does not work for critical numbers $c$ for which $f'(c)$ does not exist (since $f''(c)$ would not exist either). Also, the second derivative test does not work for critical numbers $c$ that make $f''(c) = 0$. In both of these cases, use the first derivative test.

▬ EXAMPLE 7

The graph in Figure 47 shows the population of catfish in a commercial catfish farm as a function of time. As the graph shows, the population increases rapidly up to a point and then increases at a slower rate. The horizontal dashed line shows that the population will approach some upper limit determined by the capacity of the farm. The point at which the rate of population growth starts to slow is the point of inflection for the graph.

To produce maximum yield of catfish, harvesting should take place at the point of fastest possible growth of the population—here, this is at the point of inflection. The rate of change of the population, given by the first derivative, is increasing up to the inflection point (on the interval where the second derivative is positive) and decreasing past the inflection point (on the interval where the second derivative is negative). This topic is discussed further in the next section.  ▬

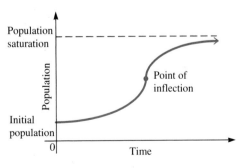

**FIGURE 47**

The *law of diminishing returns* in economics is related to the idea of concavity. The function graphed in Figure 48 gives the output $y$ from a given input $x$. If the input were advertising costs for some product, for example, the output might be the corresponding revenue from sales.

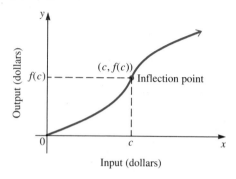

**FIGURE 48**

The graph in Figure 48 shows an inflection point at $(c, f(c))$. For $x < c$, the graph is concave upward, so the rate of change of the slope is increasing. This indicates that the output $y$ is increasing at a faster rate with each additional dollar spent. When $x > c$, however, the graph is concave downward, the rate of change of the slope is decreasing, and the increase in $y$ is smaller with each additional dollar spent. Thus, further input beyond $c$ dollars produces diminishing returns. The point of inflection at $(c, f(c))$ is called the **point of diminishing returns.** Any investment beyond the value $c$ is not considered a good use of capital.

### ▰▰▰ EXAMPLE 8

The revenue $R(x)$ generated from sales of a certain product is related to the amount $x$ spent on advertising by

$$R(x) = \frac{1}{150,000}(600x^2 - x^3), \qquad 0 \le x \le 600,$$

where $x$ and $R(x)$ are in thousands of dollars. Is there a point of diminishing returns for this function? If so, what is it?

Since a point of diminishing returns occurs at an inflection point, look for an $x$-value that makes $R''(x) = 0$. Write the function as

$$R(x) = \frac{600}{150,000}x^2 - \frac{1}{150,000}x^3 = \frac{1}{250}x^2 - \frac{1}{150,000}x^3.$$

Now find $R'(x)$ and then $R''(x)$.

$$R'(x) = \frac{2x}{250} - \frac{3x^2}{150,000} = \frac{1}{125}x - \frac{1}{50,000}x^2$$

$$R''(x) = \frac{1}{125} - \frac{1}{25,000} x$$

Set $R''(x)$ equal to 0 and solve for $x$.

$$\frac{1}{125} - \frac{1}{25,000} x = 0$$

$$-\frac{1}{25,000} x = -\frac{1}{125}$$

$$x = \frac{25,000}{125} = 200$$

Test a number in the interval $(0, 200)$ to see that $R''(x)$ is positive there. Then test a number in the interval $(200, 600)$ to find $R''(x)$ negative in that interval. Since the sign of $R''(x)$ changes from positive to negative at $x = 200$, the graph changes from concave upward to concave downward at that point, and there is a point of diminishing returns at the inflection point $(200, 106^{2/3})$. Any investment in advertising beyond \$200,000 would not pay off. ▬

## 3.4 EXERCISES

*Find $f''$ for each of the following. Then find $f''(0)$, $f''(2)$, and $f''(-3)$.*

**1.** $f(x) = 3x^3 - 4x + 5$

**2.** $f(x) = x^3 + 4x^2 + 2$

**3.** $f(x) = 3x^4 - 5x^3 + 2x^2$

**4.** $f(x) = -x^4 + 2x^3 - x^2$

**5.** $f(x) = 3x^2 - 4x + 8$

**6.** $f(x) = 8x^2 + 6x + 5$

**7.** $f(x) = (x + 4)^3$

**8.** $f(x) = (x - 2)^3$

**9.** $f(x) = \dfrac{2x + 1}{x - 2}$

**10.** $f(x) = \dfrac{x + 1}{x - 1}$

**11.** $f(x) = \dfrac{x^2}{1 + x}$

**12.** $f(x) = \dfrac{-x}{1 - x^2}$

**13.** $f(x) = \sqrt{x + 4}$

**14.** $f(x) = \sqrt{2x + 9}$

**15.** $f(x) = 5x^{3/5}$

**16.** $f(x) = -2x^{2/3}$

*Find $f'''(x)$, the third derivative of $f$, and $f^{(4)}(x)$, the fourth derivative of $f$, for each of the following.*

**17.** $f(x) = -x^4 + 2x^2 + 8$

**18.** $f(x) = 2x^4 - 3x^3 + x^2$

**19.** $f(x) = 4x^5 + 6x^4 - x^2 + 2$

**20.** $f(x) = 3x^5 - x^4 + 2x^3 - 7x$

**21.** $f(x) = \dfrac{x - 1}{x + 2}$

**22.** $f(x) = \dfrac{x + 1}{x}$

**23.** $f(x) = \dfrac{3x}{x - 2}$

**24.** $f(x) = \dfrac{x}{2x + 1}$

*In Exercises 25–42, find the largest open intervals where the functions are concave upward or concave downward. Find any points of inflection.*

**25.**

**26.**

**27.**

**28.**

**29.**

**30.**

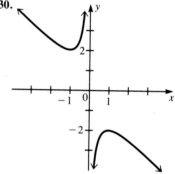

**31.** $f(x) = x^2 + 10x - 9$

**32.** $f(x) = x^2 - 4x + 3$

**33.** $f(x) = -3 + 8x - x^2$

**34.** $f(x) = 8 - 6x - x^2$

**35.** $f(x) = x^3 + 3x^2 - 45x - 3$

**36.** $f(x) = 2x^3 - 3x^2 - 12x + 1$

**37.** $f(x) = -2x^3 + 9x^2 + 168x - 3$

**38.** $f(x) = -x^3 - 12x^2 - 45x + 2$

**39.** $f(x) = \dfrac{3}{x - 5}$

**40.** $f(x) = \dfrac{-2}{x + 1}$

**41.** $f(x) = x(x + 5)^2$

**42.** $f(x) = -x(x - 3)^2$

*Find any critical numbers for f in Exercises 43–60 and then use the second derivative test to decide whether the critical numbers lead to relative maxima or relative minima. If $f''(c) = 0$ for a critical number c, then the second derivative test gives no information. In this case, use the first derivative test instead. Sketch the graph of each function.*

**43.** $f(x) = -x^2 - 10x - 25$

**44.** $f(x) = x^2 - 12x + 36$

**45.** $f(x) = 3x^3 - 3x^2 + 1$

**46.** $f(x) = 2x^3 - 4x^2 + 2$

**47.** $f(x) = -2x^3 - 9x^2 + 108x - 10$

**48.** $f(x) = -2x^3 - 9x^2 + 60x - 8$

**49.** $f(x) = 2x^3 + \dfrac{7}{2}x^2 - 5x + 3$

**50.** $f(x) = x^3 - \dfrac{15}{2}x^2 - 18x - 1$

**51.** $f(x) = (x + 3)^4$

**52.** $f(x) = x^3$

**53.** $f(x) = x^4 - 18x^2 + 5$

**54.** $f(x) = x^4 - 8x^2$

**55.** $f(x) = x + \dfrac{1}{x}$

**56.** $f(x) = 2x + \dfrac{8}{x}$

**57.** $f(x) = \dfrac{x^2 + 25}{x}$

**58.** $f(x) = \dfrac{x^2 + 4}{x}$

**59.** $f(x) = \dfrac{x - 1}{x + 1}$

**60.** $f(x) = \dfrac{x}{1 + x}$

### ▓ APPLICATIONS

#### BUSINESS AND ECONOMICS

**Product Life Cycle**   **61.** The accompanying figure shows the *product life cycle* graph, with typical products marked on it.*

(a) Where would you place home videotape recorders on this graph?

(b) Where would you place rotary-dial telephones?

(c) Which products on the left side of the graph are closest to the left-hand point of inflection? What does the point of inflection mean here?

(d) Which product on the right side of the graph is closest to the right-hand point of inflection? What does the point of inflection mean here?

**Law of Diminishing Returns**   *In Exercises 62 and 63, find the point of diminishing returns for the given functions, where R(x) represents revenue in thousands of dollars and x represents the amount spent on advertising in thousands of dollars.*

**62.** $R(x) = 10,000 - x^3 + 42x^2 + 800x, \quad 0 \le x \le 20$

**63.** $R(x) = \dfrac{4}{27}(-x^3 + 66x^2 + 1050x - 400), \quad 0 \le x \le 25$

---

*Based on ''The Product Life Cycle: A Key to Strategic Marketing Planning'' in *MSU Business Topics* (Winter 1973), p. 30. Reprinted by permission of the publisher, Graduate School of Business Administration, Michigan State University.

## LIFE SCIENCES

**Population Growth**    **64.** When a hardy new species is introduced into an area, the population often increases as shown.

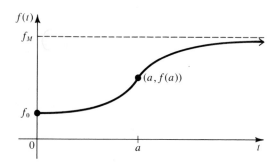

Explain the significance of the following function values on the graph.

**(a)** $f_0$    **(b)** $f(a)$    **(c)** $f_M$

**Bacteria Population**    **65.** Assume that the number of bacteria $R(t)$ (in millions) present in a certain culture at time $t$ (in hours) is given by

$$R(t) = t^2(t - 18) + 96t + 1000.$$

**(a)** At what time before 8 hr will the population be maximized?

**(b)** Find the maximum population.

**Drug Concentration**    **66.** The percent of concentration of a certain drug in the bloodstream $x$ hr after the drug is administered is given by

$$K(x) = \frac{3x}{x^2 + 4}.$$

For example, after 1 hr the concentration is given by

$$K(1) = \frac{3(1)}{1^2 + 4} = \frac{3}{5}\% = .6\% = .006.$$

**(a)** Find the time at which concentration is a maximum.

**(b)** Find the maximum concentration.

**Drug Concentration**    **67.** The percent of concentration of a drug in the bloodstream $x$ hr after the drug is administered is given by

$$K(x) = \frac{4x}{3x^2 + 27}.$$

**(a)** Find the time at which the concentration is a maximum.

**(b)** Find the maximum concentration.

## PHYSICAL SCIENCES

**Chemical Reaction**    **68.** An autocatalytic chemical reaction is one in which the product being formed causes the rate of formation to increase. The rate of a certain autocatalytic reaction is given by

$$V(x) = 12x(100 - x),$$

where $x$ is the quantity of the product present and 100 represents the quantity of chemical present initially. For what value of $x$ is the rate of the reaction a maximum?

**Velocity and Acceleration**     *Each of the functions in Exercise 69–74 gives the displacement at time t of a particle moving along a line. Find the velocity and acceleration functions. Then find the velocity and acceleration at t = 0 and t = 4. Assume that time is measured in seconds and distance is measured in centimeters. Velocity will be in centimeters per second (cm/sec) and acceleration in centimeters per second per second (cm/sec$^2$).*

**69.** $s(t) = 8t^2 + 4t$

**70.** $s(t) = -3t^2 - 6t + 2$

**71.** $s(t) = -5t^3 - 8t^2 + 6t - 3$

**72.** $s(t) = 3t^3 - 4t^2 + 8t - 9$

**73.** $s(t) = \dfrac{-2}{3t + 4}$

**74.** $s(t) = \dfrac{1}{t + 3}$

**Velocity and Acceleration**     **75.** When an object is dropped straight down, the distance in feet that it travels in $t$ seconds is given by

$$s(t) = -16t^2.$$

Find the velocity at each of the following times.

**(a)** After 3 sec     **(b)** After 5 sec     **(c)** After 8 sec

**(d)** Find the acceleration. (The answer here is a constant, the acceleration due to the influence of gravity alone.)

**Projectile Height**     **76.** If an object is thrown directly upward with a velocity of 256 ft/sec, its height above the ground after $t$ sec is given by $s(t) = 256t - 16t^2$. Find the velocity and the acceleration after $t$ sec. What is the maximum height the object reaches? When does it hit the ground?

## FOR THE COMPUTER

*In Exercises 77–80 get a list of values for $f'(x)$ and $f''(x)$ for each function on the given interval as indicated.*

**(a)** *By looking at the sign of $f'(x)$, give the (approximate) intervals where $f(x)$ is increasing and any intervals where $f(x)$ is decreasing.*

**(b)** *Give the (approximate) x-values where any maximums or minimums occur.*

**(c)** *By looking at the sign of $f''(x)$, give the intervals where $f(x)$ is concave upward and where it is concave downward.*

**(d)** *Give the (approximate) x-values of any inflection points.*

**77.** $f(x) = .25x^4 - 2x^3 + 3.5x^2 + 4x - 1$;  $(-5, 5)$ in steps of .5

**78.** $f(x) = 10x^3(x - 1)^2$;  $(-2, 2)$ in steps of .3

**79.** $f(x) = 3.1x^4 - 4.3x^3 + 5.82$;  $(-1, 2)$ in steps of .2

**80.** $f(x) = \dfrac{x}{x^2 + 1}$;  $(-3, 3)$ in steps of .4

## 3.5  APPLICATIONS OF EXTREMA

Finding extrema for realistic problems requires an accurate mathematical model of the problem. For example, suppose the number of units $T(x)$ of an item that can be manufactured daily on a production line can be closely approximated by the mathematical model

$$T(x) = 8x^{1/2} + 2x^{3/2} + 50,$$

where $x$ is the number of employees on the line. Once this mathematical model has been established, it can be used to produce information about the production line. For example, the derivative $T'(x)$ could be used to estimate the change in production resulting from the addition of an extra worker to the line.

In writing the mathematical model itself, it is important to be aware of restrictions on the values of the variables. In the example above, since $x$ represents the number of employees on a production line, $x$ must certainly be restricted to the positive integers, or perhaps to a few common fractional values (we can imagine half-time workers, but not 1/32-time workers).

On the other hand, to apply the tools of calculus to obtain an extremum for some function, the function must be defined and be meaningful at every real-number point in some interval. Because of this, the answer obtained by using a mathematical model of a practical problem might be a number that is not feasible in the setting of the problem.

Usually, the requirement that a continuous function be used instead of a function that can take on only certain selected values is of theoretical interest only. In most cases, calculus gives results that *are* acceptable in a given situation. If not—if, say, the methods of calculus are used on a function $f$ and they lead to the conclusion that $80\sqrt{2}$ units should be produced in order to get the lowest possible cost, it is usually only necessary to calculate $f(80\sqrt{2})$ and then compare this result to various values of $f(x)$ where $x$ is an acceptable number close to $80\sqrt{2}$. The lowest of these values of $f(x)$ then gives minimum cost. In most cases, the result will be very close to the theoretical value.

In this section we give several examples showing applications of calculus to maximum and minimum problems. To solve these examples, go through the following steps.

**SOLVING APPLIED PROBLEMS**

1. Read the problem carefully. Make sure you understand what is given and what is unknown.
2. If possible, sketch a diagram. Label the various parts.
3. Decide on the variable that must be maximized or minimized. Express that variable as a function of *one* other variable.
4. Find the critical points for the function from Step 3. Test each of these for maxima or minima.
5. Check for extrema at any endpoints of the domain of the function from Step 3.

### ■ EXAMPLE 1

Find two numbers $x$ and $y$ for which $2x + y = 30$, such that $xy^2$ is maximized.

First we must decide what is to be maximized and give it a variable. Here, $xy^2$ is to be maximized, so let

$$M = xy^2.$$

Now, express $M$ in terms of just *one* variable. Use the equation $2x + y = 30$ to do that. Solve $2x + y = 30$ for either $x$ or $y$. Solving for $y$ gives

$$2x + y = 30$$
$$y = 30 - 2x.$$

Substitute for $y$ in the expression for $M$ to get

$$M = x(30 - 2x)^2$$
$$= x(900 - 120x + 4x^2)$$
$$= 900x - 120x^2 + 4x^3.$$

Find the critical points for $M$ by finding $M'$, then solving the equation $M' = 0$ for $x$.

$$M' = 900 - 240x + 12x^2 = 0$$
$$12(75 - 20x + x^2) = 0$$
$$(5 - x)(15 - x) = 0$$
$$x = 5 \quad \text{or} \quad x = 15$$

There are two critical numbers, 5 and 15. Use the second derivative test to decide which of these $x$-values will *maximize M*.

$$M'' = -240 + 24x$$
$$M''(5) = -240 + 120 < 0$$
$$M''(15) = -240 + 360 > 0$$

The test shows that $M = xy^2$ is maximized when $x = 5$. Since $y = 30 - 2x = 20$, the values that maximize $xy^2$ are $x = 5$ and $y = 20$. (There are no end-points to check here.)  ■

### ■ EXAMPLE 2

When Power and Money, Inc., charges \$600 for a seminar on management techniques, it attracts 1000 people. For each \$20 decrease in the charge, an additional 100 people will attend the seminar.

**(a)** Find an expression for the total revenue if there are $x$ \$20 decreases in the price.

The price charged will be

$$\text{Price per person} = 600 - 20x,$$

and the number of people in the seminar will be

$$\text{Number of people} = 1000 + 100x.$$

The total revenue, $R(x)$, is given by the product of the price and the number of people attending, or

$$R(x) = (600 - 20x)(1000 + 100x)$$
$$= 600{,}000 + 40{,}000x - 2000x^2.$$

**(b)** Find the value of $x$ that maximizes revenue.

Here $R'(x) = 40{,}000 - 4000x$. Set this derivative equal to 0.

$$40{,}000 - 4000x = 0$$
$$-4000x = -40{,}000$$
$$x = 10$$

Since $R''(x) = -4000$, $x = 10$ leads to maximum revenue.

**(c)** Find the maximum revenue.

Use the function $R(x)$ from part (a):

$$R(10) = 600{,}000 + 40{,}000(10) - 2000(10)^2 = 800{,}000$$

dollars. To get this level of revenue, $1000 + 100(10) = 2000$ people must attend the seminar; each person will pay $600 - 20(10) = \$400$. ▬

▬▬ **EXAMPLE 3**

A truck burns fuel at the rate of

$$G(x) = \frac{1}{200}\left(\frac{800 + x^2}{x}\right)$$

gallons per mile when traveling at $x$ miles per hour on a straight, level road. If fuel costs \$2 per gallon, find the speed that will produce the minimum total cost for a 1000-mile trip. Find the minimum total cost. (This is an extension of Exercise 43 in Section 5 of the previous chapter.)

The total cost of the trip, in dollars, is the product of the number of gallons per mile, the number of miles, and the cost per gallon. If $C(x)$ represents this cost, then

$$C(x) = \left[\frac{1}{200}\left(\frac{800 + x^2}{x}\right)\right](1000)(2)$$
$$= \frac{2000}{200}\left(\frac{800 + x^2}{x}\right)$$
$$= \frac{8000 + 10x^2}{x}$$

To find the value of $x$ that will minimize cost, first find the derivative:

$$C'(x) = \frac{10x^2 - 8000}{x^2}.$$

Set this derivative equal to 0.

$$\frac{10x^2 - 8000}{x^2} = 0$$

$$10x^2 - 8000 = 0$$

$$10x^2 = 8000$$

$$x^2 = 800$$

Take the square root on both sides to get

$$x \approx \pm 28.3 \text{ mph.}$$

Reject $-28.3$ as a speed, leaving $x = 28.3$ as the only critical value. To see if 28.3 leads to a minimum cost, find $C''(x)$.

$$C''(x) = \frac{16{,}000}{x^3}$$

Since $C''(28.3) > 0$, the critical value 28.3 leads to a minimum. (It is not necessary to actually calculate $C''(28.3)$; just check that $C''(28.3) > 0$.)
    The minimum total cost is

$$C(28.3) = \frac{8000 + 10(28.3)^2}{28.3} = 565.69 \text{ dollars.} \quad \blacksquare$$


**■ EXAMPLE 4**

An open box is to be made by cutting a square from each corner of a 12-inch by 12-inch piece of metal and then folding up the sides. The finished box must be at least 1.5 inches deep, but not deeper than 3 inches. What size square should be cut from each corner in order to produce a box of maximum volume?

(a)

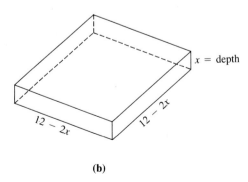

(b)

**FIGURE 49**

Let $x$ represent the length of a side of the square that is cut from each corner, as shown in Figure 49(a). The width of the box is $12 - 2x$, with the length also $12 - 2x$. As shown in Figure 49(b), the depth of the box will be $x$ inches, where $x$ is in the interval $[1.5, 3]$. The volume of the box is given by the product of the length, width, and height. In this example, the volume, $V(x)$, depends on $x$:

$$V(x) = x(12 - 2x)(12 - 2x) = 144x - 48x^2 + 4x^3.$$

The derivative is $V'(x) = 144 - 96x + 12x^2$. Set this derivative equal to 0.

$$12x^2 - 96x + 144 = 0$$
$$12(x^2 - 8x + 12) = 0$$
$$12(x - 2)(x - 6) = 0$$
$$x - 2 = 0 \quad \text{or} \quad x - 6 = 0$$
$$x = 2 \qquad\qquad x = 6$$

Since $x$ must be in the interval $[1.5, 3]$, only the critical number 2 can be used. Find $V(x)$ for $x$ equal to 1.5, 2, and 3 to find the depth that will maximize the volume.

| $x$ | $V(x)$ |
|-----|--------|
| 1.5 | 121.5 |
| 2 | 128 ← Maximum |
| 3 | 108 |

The chart indicates that the box will have maximum volume when $x = 2$ and that the maximum volume will be 128 cubic inches.  ■

For most living things, reproduction is *seasonal*—it can take place only at selected times of the year.* Large whales, for example, reproduce every two years during a relatively short time span of about two months. Shown on the time axis in Figure 50 are the reproductive periods. Let $S$ = number of adults present during the reproductive period and let $R$ = number of adults that return the next season to reproduce.

**FIGURE 50**

*From *Mathematics for the Biosciences* by Michael R. Cullen. Copyright © 1983 PWS Publishers. Reprinted by permission.

If we find a relationship between $R$ and $S$, $R = f(S)$, then we have formed a **spawner-recruit** function or **parent-progeny** function. These functions are notoriously hard to develop because of the difficulty of obtaining accurate counts and because of the many hypotheses that can be made about the life stages. We will simply suppose that the function $f$ takes various forms.

If $R > S$, we can presumably harvest

$$H = R - S = f(S) - S$$

individuals, leaving $S$ to reproduce. Next season, $R = f(S)$ will return and the harvesting process can be repeated, as shown in Figure 51.

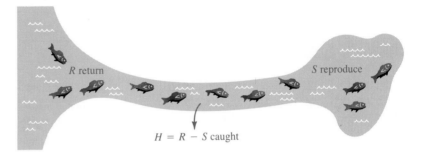

**FIGURE 51**

Let $S_0$ be the number of spawners that will allow as large a harvest as possible without threatening the population with extinction. Then $H(S_0)$ is called the **maximum sustainable harvest.**

### ■■ EXAMPLE 5

Suppose the spawner-recruit function for Idaho rabbits is $f(S) = -.025\,S^2 + 4\,S$, where $S$ is measured in thousands of rabbits. Find $S_0$ and the maximum sustainable harvest, $H(S_0)$.

$S_0$ is the value of $S$ that maximizes $H$. Since

$$\begin{aligned} H(S) &= f(S) - S \\ &= -.025\,S^2 + 4\,S - S \\ &= -.025\,S^2 + 3\,S, \end{aligned}$$

the derivative $H'(S) = -.05\,S + 3$. Set this derivative equal to 0 and solve for $S$.

$$\begin{aligned} 0 &= -.05\,S + 3 \\ .05\,S &= 3 \\ S &= 60 \end{aligned}$$

The number of rabbits needed to sustain the population is $S_0 = 60$ thousand. Note that

$$H''(S) = -.05 < 0,$$

showing that $S = 60$ does indeed produce a *maximum* value of $H$. At $S_0 = 60$, the harvest is

$$H(60) = -.025(60)^2 + 3(60) = 90,$$

or 90 thousand rabbits. These results indicate that after one reproductive season, a population of 60 thousand rabbits will have increased to 150 thousand. Of these, 90 thousand may be harvested, leaving 60 thousand to regenerate the population. Any harvest larger than 90 thousand will threaten the future of the rabbit population, while a harvest smaller than 90 thousand will allow the population to grow larger each season. Thus, 90 thousand is the maximum sustainable harvest for this population. ▬

## 3.5 EXERCISES

## APPLICATIONS

### GENERAL

**Number Analysis**    *Exercises 1–8 involve maximizing and minimizing products and sums of numbers. Solve each problem using steps like those shown in Exercise 1.*

1. Find two numbers $x$ and $y$ such that $x + y = 100$ and the product $P = xy$ is as large as possible.
   (a) Solve $x + y = 100$ for $y$.
   (b) Substitute this result for $y$ into $P = xy$, the equation for the variable that is to be maximized.
   (c) Find $P'$. Solve the equation $P' = 0$.
   (d) Evaluate $x$ and $y$.
   (e) Give the maximum value of $P$.

2. Find two numbers whose sum is 250 and whose product is as large as possible.

3. Find two numbers whose sum is 200 such that the sum of the squares of the two numbers is minimized.

4. Find two numbers whose sum is 30 such that the sum of the squares of the numbers is minimized.

*In Exercises 5–8, use steps (a)–(e) from Exercise 1 to find numbers x and y satisfying the given requirements.*

5. $x + y = 150$ and $x^2y$ is maximized.    6. $x + y = 45$ and $xy^2$ is maximized.
7. $x - y = 10$ and $xy$ is minimized.    8. $x - y = 3$ and $xy$ is minimized.

**Area**    9. A farmer has 1200 m of fencing. He wants to enclose a rectangular field bordering a river, with no fencing needed along the river. (See the sketch.) Let $x$ represent the width of the field.

(a) Write an expression for the length of the field.

(b) Find the area of the field (area = length × width).

(c) Find the value of $x$ leading to the maximum area.

(d) Find the maximum area.

**Area** **10.** Find the dimensions of the rectangular field of maximum area that can be made from 200 m of fencing material. (This fence has four sides.)

**Area** **11.** An ecologist is conducting a research project on breeding pheasants in captivity. She first must construct suitable pens. She wants a rectangular area with two additional fences across its width, as shown in the sketch. Find the maximum area she can enclose with 3600 m of fencing.

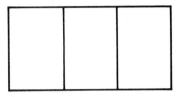

**Travel Time** **12.** A hunter is at a point on a river bank. He wants to get to his cabin, located 3 mi north and 8 mi west. (See the figure.) He can travel 5 mph on the river but only 2 mph on this very rocky land. How far upriver should he go in order to reach the cabin in minimum time? (*Hint:* Distance = Rate × Time.)

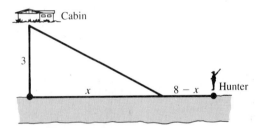

## BUSINESS AND ECONOMICS

**Revenue** **13.** If the price charged for a candy bar is $p(x)$ cents, then $x$ thousand candy bars will be sold in a certain city, where

$$p(x) = 100 - \frac{x}{10}.$$

(a) Find an expression for the total revenue from the sale of $x$ thousand candy bars.

(b) Find the value of $x$ that leads to maximum revenue.

(c) Find the maximum revenue.

**Revenue** **14.** The sale of cassette tapes of "lesser" performers is very sensitive to price. If a tape manufacturer charges $p(x)$ dollars per tape, where

$$p(x) = 6 - \frac{x}{8},$$

then $x$ thousand tapes will be sold.

(a) Find an expression for the total revenue from the sale of $x$ thousand tapes.

(b) Find the value of $x$ that leads to maximum revenue.

(c) Find the maximum revenue.

**Cost** 15. A truck burns fuel at the rate of

$$G(x) = \frac{1}{32}\left(\frac{64}{x} + \frac{x}{50}\right)$$

gallons per mile while traveling at $x$ mph.

(a) If fuel costs $1.60 per gallon, find the speed that will produce minimum total cost for a 400-mile trip. (See Example 3.)

(b) Find the minimum total cost.

**Cost** 16. A rock-and-roll band travels from engagement to engagement in a large bus. This bus burns fuel at the rate of

$$G(x) = \frac{1}{50}\left(\frac{200}{x} + \frac{x}{15}\right),$$

gallons per mile while traveling at $x$ mph.

(a) If fuel costs $2 per gallon, find the speed that will produce minimum total cost for a 250-mile trip.

(b) Find the minimum total cost.

**Area with Fixed Cost** 17. A rectangular field is to be enclosed on all four sides with a fence. Fencing material costs $3 per foot for two opposite sides, and $6 per foot for the other two sides. Find the maximum area that can be enclosed for $2400.

**Area with Fixed Cost** 18. A rectangular field is to be enclosed with a fence. One side of the field is against an existing fence, so that no fence is needed on that side. If material for the fence costs $2 per foot for the two ends and $4 per foot for the side parallel to the existing fence, find the dimensions of the field of largest area that can be enclosed for $1000.

**Revenue** 19. The manager of a peach orchard is trying to decide when to arrange for picking the peaches. If they are picked now, the average yield per tree will be 100 lb, which can be sold for 40¢ per pound. Past experience shows that the yield per tree will increase about 5 lb per week, while the price will decrease about 2¢ per pound per week.

(a) Let $x$ represent the number of weeks that the manager should wait. Find the income per pound.

(b) Find the number of pounds per tree.

(c) Find the total revenue from a tree.

(d) When should the peaches be picked in order to produce maximum revenue?

(e) What is the maximum revenue?

**Income** 20. The manager of an 80-unit apartment complex is trying to decide what rent to charge. Experience has shown that at a rent of $200, all the units will be full. On the average, one additional unit will remain vacant for each $20 increase in rent.

(a) Let $x$ represent the number of $20 increases. Find an expression for the rent for each apartment. (See Example 2.)

(b) Find an expression for the number of apartments rented.

**(c)** Find an expression for the total revenue from all rented apartments.

**(d)** What value of $x$ leads to maximum revenue?

**(e)** What is the maximum revenue?

**Profit** **21.** In planning a small restaurant, it is estimated that a profit of $5 per seat will be made if the number of seats is between 60 and 80, inclusive. On the other hand, the profit on each seat will decrease by 5 cents for each seat above 80.

**(a)** Find the number of seats that will produce the maximum profit.

**(b)** What is the maximum profit?

**Timing Income** **22.** A local group of scouts has been collecting old aluminum cans for recycling. The group has already collected 12,000 lb of cans, for which they could currently receive $4 per hundred pounds. The group can continue to collect cans at the rate of 400 lb per day. However, a glut in the old-can market has caused the recycling company to announce that it will lower its price, starting immediately, by $.10 per hundred pounds per day. The scouts can make only one trip to the recycling center. Find the best time for the trip. What total income will be received?

**Packaging Design** **23.** A television manufacturing firm needs to design an open-topped box with a square base. The box must hold 32 cubic inches. Find the dimensions of the box that can be built with the minimum amount of materials. (See the figure.)

**Revenue** **24.** A local club is arranging a charter flight to Hawaii. The cost of the trip is $425 each for 75 passengers, with a refund of $5 per passenger for each passenger in excess of 75.

**(a)** Find the number of passengers that will maximize the revenue received from the flight.

**(b)** Find the maximum revenue.

**Packaging Design** **25.** A company wishes to manufacture a box with a volume of 36 cubic feet that is open on top and is twice as long as it is wide. Find the dimensions of the box produced from the minimum amount of material.

**Packaging Cost** **26.** A closed box with a square base is to have a volume of 16,000 cubic centimeters. The material for the top and bottom of the box costs $3 per square centimeter, while the material for the sides costs $1.50 per square centimeter. Find the dimensions of the box that will lead to minimum total cost. What is the minimum total cost?

**Fuel Cost** **27.** In Example 3, we found the speed in miles per hour that minimized cost when we considered only the cost of the fuel. Rework the problem taking into account the driver's salary of $12 per hour. (*Hint:* If the trip is 1000 mi at $x$ mph, the driver will be paid for $1000/x$ hours.)

**Use of Materials**   **28.** A mathematics book is to contain 36 square inches of printed matter per page, with margins of 1 inch along the sides and $1\frac{1}{2}$ inches along the top and bottom. Find the dimensions of the page that will require the minimum amount of paper. (See the figure.)

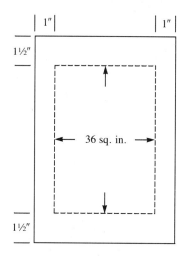

**Cost**   **29.** A company wishes to run a utility cable from point $A$ on the shore (see the figure) to an installation at point $B$ on the island. The island is 6 mi from the shore. It costs $400 per mile to run the cable on land and $500 per mile underwater. Assume that the cable starts at $A$ and runs along the shoreline, then angles and runs underwater to the island. Find the point at which the line should begin to angle in order to yield the minimum total cost. (*Hint:* The length of the line underwater is $\sqrt{x^2 + 36}$.)

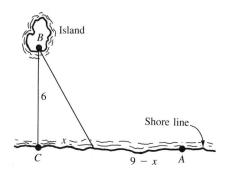

**Pricing**   **30.** Decide what you would do if your assistant presented the following contract for your signature:

> Your firm offers to deliver 300 tables to a dealer, at $90 per table, and to reduce the price per table on the entire order by 25¢ for each additional table over 300.

Find the dollar total involved in the largest possible transaction between the manufacturer and the dealer; then find the smallest possible dollar amount.

### LIFE SCIENCES

**Maximum Sustainable Harvest**     *Find the maximum sustainable harvest in Exercises 31–36. See Example 5.*

**31.** $f(S) = -.1 S^2 + 11 S$          **32.** $f(S) = -S^2 + 2.2 S$          **33.** $f(S) = 15\sqrt{S}$

**34.** $f(S) = 12 S^{.25}$          **35.** $f(S) = \dfrac{25 S}{S + 2}$          **36.** $f(S) = .999 S$

**Pigeon Flight**     **37.** Homing pigeons avoid flying over large bodies of water, preferring to fly around them instead. (One possible explanation is the fact that extra energy is required to fly over water because air pressure drops over water in the daytime.) Assume that a pigeon released from a boat 1 mi from the shore of a lake (point $B$ in the figure) flies first to point $P$ on the shore and then along the straight edge of the lake to reach its home at $L$. If $L$ is 2 mi from point $A$, the point on the shore closest to the boat, and if a pigeon needs 4/3 as much energy to fly over water as over land, find the location of point $P$.

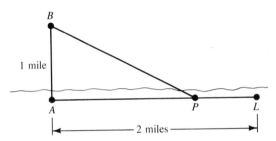

**Salmon Spawning**     **38.** When salmon struggle upstream to their spawning grounds, it is essential that they conserve energy, for they no longer feed once they have left the ocean. Let $v_0 =$ speed (in miles per hour) of the current, $d =$ distance the salmon must travel, and $v =$ speed of the salmon. Hence, $(v - v_0)$ is the net velocity of the salmon upstream. (See the figure.)*

**(a)** Show that if the journey takes $t$ hr, then

$$t = \frac{d}{v - v_0}.$$

We next will assume that the amount of energy expended per hour when the speed of the salmon is $v$ is directly proportional to $v^\alpha$ for some $\alpha > 1$. (Empirical data suggest this.)

**(b)** Show that the total energy $T$ expended over the journey is given by

$$T = k\frac{v^\alpha}{v - v_0} \quad \text{for } v > v_0.$$

*Exercises 38–39 from *Mathematics for the Biosciences* by Michael R. Cullen. Copyright © 1983 PWS Publishers. Reprinted by permission.

(c) Show that $T$ is minimized by selecting velocity

$$v = \frac{\alpha v_0}{\alpha - 1}.$$

(d) If $v_0 = 2$ mph and the salmon make the 20-mi journey by swimming for 40 hr, estimate $\alpha$. What must you assume in order to do the computations?

**Lung Function**    **39.** It is well known that during coughing, the diameters of the trachea and bronchi decrease. Let $r_0$ be the normal radius of an airway at atmospheric pressure $P_0$. For the trachea, $r_0 \approx .5$ inch. The glottis is the opening at the entrance of the trachea through which air must pass as it either enters or leaves the lungs, as depicted in the illustration.† Assume that, after a deep inspiration of air, the glottis is closed. Then pressure develops in the airways and the radius of the airway decreases. We will assume a simple linear relation between $r$ and $P$:

$$r - r_0 = a(P - P_0)$$
$$\text{or} \quad \Delta r = a\Delta P$$

where $a < 0$ and $r_0/2 \leq r \leq r_0$.

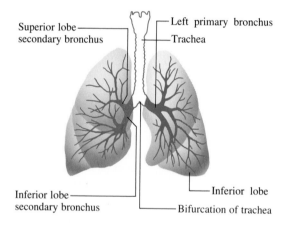

Superior lobe— secondary bronchus
Left primary bronchus
Trachea
Inferior lobe— secondary bronchus
Inferior lobe
Bifurcation of trachea

**EXERCISE 39**

When the glottis is opened, how does the air flow through these passages? We will assume that the flow is governed by **Poiseuille's laws:**

$$\textbf{(i)} \ v = \frac{P - P_0}{k}(r^2 - x^2), \quad 0 \leq x \leq r$$

$$\textbf{(ii)} \ F = \frac{dV}{dt} = \frac{\pi(P - P_0)}{2k}r^4$$

Here, $v$ is the velocity at a distance $x$ from the center of the airway (in cm/sec, e.g.) and $F$ is the flow rate (in cm³/sec, e.g.). The average velocity $\bar{v}$ over the circular cross section is given by

$$\textbf{(iii)} \ \bar{v} = \frac{F}{(\pi r^2)} = \frac{P - P_0}{2k}r^2.$$

†The figure for Exercise 39 was adapted from Barbara R. Landau, *Essential Anatomy and Physiology*, 2nd ed. Glenview, Ill.: Scott, Foresman and Co., Copyright 1980.

(a) Write the flow rate as a function of $r$ only. Find the value of $r$ that maximizes the rate of flow.

(b) Write both the average velocity $\bar{v}$ and $v(0)$, the velocity in the center of the airway, as functions of $r$ alone. Find the value of $r$ that maximizes these two velocities.

## 3.6  FURTHER BUSINESS APPLICATIONS: ECONOMIC LOT SIZE; ECONOMIC ORDER QUANTITY; ELASTICITY OF DEMAND

In this section we introduce three common business applications of calculus. The first two, *economic lot size* and *economic order quantity,* are related. A manufacturer must determine the production lot (or batch) size that will result in minimum production and storage costs, while a purchaser must decide what quantity of an item to order in an effort to minimize reordering and storage costs. The third application, *elasticity of demand,* deals with the sensitivity of demand for a product to changes in the price of the product.

**Economic Lot Size**   Suppose that a company manufactures a constant number of units of a product per year and that the product can be manufactured in several batches of equal size during the year. If the company were to manufacture the item only once per year, it would minimize setup costs but incur high warehouse costs. On the other hand, if it were to make many small batches, this would increase setup costs. Calculus can be used to find the number of batches per year that should be manufactured in order to minimize total cost. This number is called the **economic lot size.**

Figure 52 shows several of the possibilities for a product having an annual demand of 12,000 units. The top graph shows the results if only one batch of the product is made annually: in this case an average of 6000 items will be held in a warehouse. If four batches (of 3000 each) are made at equal time intervals during a year, the average number of units in the warehouse falls to only 1500. If twelve batches are made, an average of 500 items will be in the warehouse.

The following variables will be used in our discussion of economic lot size.

$x$ = number of batches to be manufactured annually

$k$ = cost of storing one unit of the product for one year

$f$ = fixed setup cost to manufacture the product

$g$ = variable cost of manufacturing a single unit of the product

$M$ = total number of units produced annually

The company has two types of costs associated with the production of its product: a cost associated with manufacturing the item and a cost associated with storing the finished product.

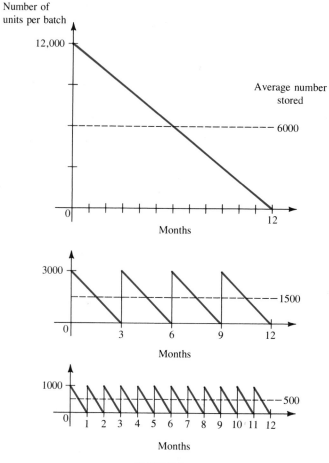

**FIGURE 52**

During a year the company will produce $x$ batches of the product, with $M/x$ units of the product produced per batch. Each batch has a fixed cost $f$ and a variable cost $g$ per unit, so that the manufacturing cost per batch is

$$f + \frac{gM}{x}.$$

There are $x$ batches per year, so the total annual manufacturing cost is

$$\left(f + \frac{gM}{x}\right)x. \tag{1}$$

Since each batch consists of $M/x$ units, and since demand is constant, it is common to assume an average inventory of

$$\frac{1}{2}\left(\frac{M}{x}\right) = \frac{M}{2x}$$

units per year. The cost of storing one unit of the product for a year is $k$, so the total storage cost is

$$k\left(\frac{M}{2x}\right) = \frac{kM}{2x}. \tag{2}$$

The total production cost is the sum of the manufacturing and storage costs, or the sum of expressions (1) and (2). If $T(x)$ is the total cost of producing $x$ batches,

$$T(x) = \left(f + \frac{gM}{x}\right)x + \frac{kM}{2x} = fx + gM + \frac{kM}{2x}.$$

To find the value of $x$ that will minimize $T(x)$, remember that $f$, $g$, $k$, and $M$ are constants and find $T'(x)$.

$$T'(x) = f - \frac{kM}{2}x^{-2}.$$

Set this derivative equal to 0.

$$f - \frac{kM}{2}x^{-2} = 0$$

$$f = \frac{kM}{2x^2}$$

$$2fx^2 = kM$$

$$x^2 = \frac{kM}{2f}$$

$$x = \sqrt{\frac{kM}{2f}} \tag{3}$$

The second derivative test can be used to show that $\sqrt{kM/(2f)}$ is the annual number of batches that gives minimum total production costs. (See Exercise 1.)

This application is referred to as the *inventory problem* and is treated in more detail in management science courses.

### ■■■ EXAMPLE 1

A paint company has a steady annual demand for 24,500 cans of automobile primer. The cost accountant for the company says that it costs $2 to hold one can of paint for one year and $500 to set up the plant for the production of the primer. Find the number of batches of primer that should be produced in order to minimize total production costs.

Use equation (3) above.

$$x = \sqrt{\frac{kM}{2f}} = \sqrt{\frac{2(24,500)}{2(500)}} = \sqrt{49} = 7$$

Seven batches of primer per year will lead to minimum production costs.

***Economic Order Quantity***    We can extend our previous discussion to the problem of reordering an item that is used at a constant rate throughout the year. Here, the company using a product must decide how many units to request each time an order is placed; that is, it must identify the **economic order quantity.** In this case, the variables are as follows:

$$x = \text{number of units per order;}$$
$$k = \text{cost of storing one unit for one year;}$$
$$f = \text{fixed cost to place an order;}$$
$$M = \text{total units needed per year.}$$

The goal is to minimize the total cost of ordering over a year's time, where

$$\text{Total Cost} = \text{Storage Cost} + \text{Reorder Cost.}$$

Again assume an average inventory of $x/2$. Since it costs $k$ to store one unit for a year, the yearly storage cost is

$$k\left(\frac{x}{2}\right) = \frac{kx}{2}.$$

The reorder cost is the product of the number of batches ordered each year, $M/x$, and the cost per order, $f$. (As shown above, variable costs per order do not affect the final result.) Thus, the reorder cost is

$$f\left(\frac{M}{x}\right) = \frac{fM}{x},$$

and the total cost is

$$T(x) = \frac{kx}{2} + \frac{fM}{x}.$$

The total cost is minimized when $T'(x) = 0$.

$$T'(x) = \frac{k}{2} - \frac{fM}{x^2} = 0$$

$$\frac{k}{2} = \frac{fM}{x^2}$$

$$kx^2 = 2fM$$

$$x^2 = \frac{2fM}{k}$$

$$x = \sqrt{\frac{2fM}{k}}$$

This result shows that $\sqrt{(2fM)/k}$ items should be ordered each time. Verify that this order quantity minimizes total cost.

### ▬ EXAMPLE 2

A large pharmacy has an annual need for 200 units of a certain anti iotic. It costs $10 to store one unit for one year. The fixed cost of placing an order (clerical time, mailing, and so on) amounts to $40. Find the number of units to order each time.

Here $k = 10$, $M = 200$, and $f = 40$. We have

$$x = \sqrt{\frac{2(40)(200)}{10}} = \sqrt{1600} = 40.$$

The drug should be ordered in lots of 40 units.  ▬

***Elasticity of Demand***   Anyone who sells a product or service is concerned with how a change in price affects demand. The sensitivity of demand to price changes varies with different items. For items such as soft drinks, pepper, and light bulbs, relatively small percentage changes in price will not change the demand for the item much. For cars, home loans, furniture, and computer equipment, however, small percentage changes in price have significant effects on demand.

One way to measure the sensitivity of demand to changes in price is by the ratio of percent change in demand to percent change in price. If $x$ represents demand and $p$ price, then this ratio can be written as

$$\frac{\dfrac{\Delta x}{x}}{\dfrac{\Delta p}{p}},$$

where $\Delta x$ gives the change in $x$ and $\Delta p$ gives the change in $p$. This ratio is always negative since $x$ and $p$ are positive, while $\Delta x$ and $\Delta p$ have opposite signs. (An *increase* in price causes a *decrease* in demand.) If the absolute value of this ratio is large, it suggests that a relatively large drop (decrease) in demand results from a relatively small increase in price.

This ratio can be rewritten as

$$\frac{\dfrac{\Delta x}{x}}{\dfrac{\Delta p}{p}} = \frac{\Delta x}{x} \cdot \frac{p}{\Delta p} = \frac{p}{x} \cdot \frac{\Delta x}{\Delta p}.$$

If $x = f(p)$, then $\Delta x = f(p + \Delta p) - f(p)$, and

$$\frac{\Delta x}{\Delta p} = \frac{f(p + \Delta p) - f(p)}{\Delta p}.$$

As $\Delta p \to 0$, this quotient becomes

$$\lim_{\Delta p \to 0} \frac{\Delta x}{\Delta p} = \lim_{\Delta p \to 0} \frac{f(p + \Delta p) - f(p)}{\Delta p} = \frac{dx}{dp},$$

and

$$\lim_{\Delta p \to 0} \frac{p}{x} \cdot \frac{\Delta x}{\Delta p} = \frac{p}{x} \cdot \frac{dx}{dp}.$$

The *positive* quantity

$$E = -\frac{p}{x} \cdot \frac{dx}{dp}$$

is called the **elasticity of demand.** It measures the instantaneous responsiveness of demand to price.

For example, $E = .2$ for medical services, but $E = 1.2$ for purchasing a new home. The demand for essential medical services is *much less* responsive to price changes than is the demand for nonessential new home purchases.

If $E < 1$, the relative change in demand is less than the relative change in price, and the demand is called **inelastic.** If $E > 1$, the relative change in demand is greater than the relative change in price, and the demand is called **elastic.** When $E = 1$, the percentage changes in price and demand are relatively equal and the demand is said to have **unit elasticity.**

### ■■■ EXAMPLE 3

Given the demand function $x = 300 - 3p$, $0 \le p \le 100$, find the following.

**(a)** Calculate and interpret the elasticity of demand when $p = 25$ and when $p = 75$.

Since $x = 300 - 3p$, $dx/dp = -3$, and

$$E = -\frac{p}{x} \cdot \frac{dx}{dp}$$

$$= -\frac{p}{300 - 3p}(-3)$$

$$= \frac{3p}{300 - 3p}.$$

Let $p = 25$ to get

$$E = \frac{3(25)}{300 - 3(25)} = \frac{75}{225} = \frac{1}{3} \approx .33.$$

Since $.33 < 1$, the demand is inelastic, and a percentage change in price will result in a smaller percentage change in demand. For example, a 10% increase in price will cause a 3.3% decrease in demand.

If $p = 75$, then

$$E = \frac{3(75)}{300 - 3(75)} = \frac{225}{75} = 3,$$

and since $3 > 1$, demand is elastic. At this point a percentage increase in price will result in a *greater* percentage decrease in demand. Here, a 10% increase in price will cause a 30% decrease in demand.

**(b)** Determine the price where demand has unit elasticity. What is the significance of this price?

Demand will have unit elasticity at the price $p$ that makes $E = 1$. Here

$$E = \frac{3p}{300 - 3p} = 1$$

$$3p = 300 - 3p$$

$$6p = 300$$

$$p = 50.$$

Demand has unit elasticity at a price of 50. This means that at this price the percentage changes in price and demand are about the same. ▰

The definitions from this discussion can be summarized as follows.

**ELASTICITY OF DEMAND**

> Let $x = f(p)$, where $x$ is demand at a price $p$. The elasticity of demand is $E = -\dfrac{p}{x} \cdot \dfrac{dx}{dp}$.
>
> Demand is inelastic if $E < 1$.
>
> Demand is elastic if $E > 1$.
>
> Demand has unit elasticity if $E = 1$.

Elasticity can be related to the total revenue, $R$, by considering the derivative of $R$. Since revenue is given by price times sales (demand),

$$R = px.$$

Differentiate with respect to $p$ using the product rule.

$$\frac{dR}{dp} = p \cdot \frac{dx}{dp} + x \cdot 1$$

$$= \frac{x}{x} \cdot p \cdot \frac{dx}{dp} + x \qquad \text{Multiply by } \frac{x}{x} \text{ (or 1)}$$

$$= x\left(\frac{p}{x} \cdot \frac{dx}{dp}\right) + x$$

$$= x(-E) + x$$

$$= x(-E + 1)$$

$$= x(1 - E)$$

Total revenue $R$ is increasing, optimized, or decreasing depending on whether $dR/dp > 0$, $dR/dp = 0$, or $dR/dp < 0$. These three situations correspond to $E < 1$, $E = 1$, or $E > 1$. See Figure 53.

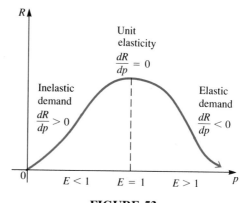

**FIGURE 53**

In summary, total revenue is related to elasticity as follows.

**REVENUE AND ELASTICITY**

1. If the demand is inelastic, total revenue increases as price increases.
2. If the demand is elastic, total revenue decreases as price increases.
3. Total revenue is maximized at the price where demand has unit elasticity.

 EXAMPLE 4

If the demand for a product is $x = 216 - 2p^2$, where $p$ is the price, find the following.

(a) The price intervals where demand is elastic and where demand is inelastic

Since $x = 216 - 2p^2$, $dx/dp = -4p$, and

$$E = -\frac{p}{x} \cdot \frac{dx}{dp}$$

$$= -\frac{p}{216 - 2p^2}(-4p)$$

$$= \frac{4p^2}{216 - 2p^2}.$$

To decide where $E < 1$ or $E > 1$, solve the corresponding *equation*.

$$E = 1$$

$$\frac{4p^2}{216 - 2p^2} = 1$$

$$4p^2 = 216 - 2p^2$$

$$6p^2 = 216$$

$$p^2 = 36$$

$$p = 6$$

Substitute a test number on either side of 6 in the expression for $E$ to see which values make $E < 1$ and which make $E > 1$.

$$\text{Let } p = 1: E = \frac{4(1)^2}{216 - 2(1)^2} = \frac{4}{214} < 1.$$

$$\text{Let } p = 10: E = \frac{4(10)^2}{216 - 2(10)^2} = \frac{400}{216 - 200} > 1.$$

Demand is inelastic when $E < 1$. This occurs when $p < 6$. Demand is elastic when $E > 1$—that is, when $p > 6$.

(b) What prices will cause revenue to increase or decrease?

Revenue will increase when demand is inelastic, so keeping prices below $6 per item will keep demand high enough to continue to increase revenue. When demand is elastic, revenue is decreasing, so any price over $6 per item will cause demand to decrease to the point where revenue also decreases. ▰

## 3.6 EXERCISES

1. In the discussion of economic lot size, use the second derivative to show that the value of $x$ obtained in the text [equation (3)] really leads to the minimum cost.

2. Why do you think that the variable cost $g$ does not appear in the equation for $x$ [equation (3)]?

▀▀▀ APPLICATIONS

### BUSINESS AND ECONOMICS

**Lot Size**     **3.** Find the approximate number of batches of an item that should be produced anually if 100,000 units are to be manufactured, it costs $1 to store a unit for one year, and it costs $500 to set up the factory to produce each batch.

**Lot Size**     **4.** A manufacturer has a steady annual demand for 16,800 cases of sugar. It costs $3 to store one case for one year and it costs $7 to produce each batch. Find the number of batches of sugar that should be produced each year.

**Lot Size**     **5.** Find the number of units per batch that will be manufactured in Exercise 3.

**Lot Size**     **6.** Find the number of cases per batch in Exercise 4.

**Order Quantity**     **7.** A bookstore has an annual demand for 100,000 copies of a best-selling book. It costs $.50 to store one copy for one year, and it costs $60 to place an order. Find the optimum number of copies per order.

**Order Quantity**     **8.** A restaurant has an annual demand for 810 bottles of a California wine. It costs $1 to store one bottle for one year, and it costs $5 to place a reorder. Find the optimum number of bottles per order.

**Elasticity**     *For each of the following demand functions, find* **(a)** *E, and* **(b)** *values of x (if any) at which total revenue is maximized.*

**9.** $x = 25,000 - 50p$      **10.** $x = 50 - \dfrac{p}{4}$

**11.** $x = \dfrac{3000}{p}$      **12.** $x = \sqrt[3]{\dfrac{2000}{p}}$

**Elasticity**     *In Exercises 13–14, find the elasticity of demand (E) for the given demand function at the indicated values of p. Is the demand elastic, inelastic, or neither at the indicated values? Interpret your results.*

**13.** $x = 300 - 2p$
   **(a)** $p = \$100$      **(b)** $p = \$50$

**14.** $x = 400 - .2p^2$
   **(a)** $p = \$20$      **(b)** $p = \$40$

**Elasticity**     **15.** The demand function for $x$ units of a commodity is given by
$$p = (25 - x)^{1/2}, \qquad 0 < x < 25.$$
   **(a)** Find $E$ when $x = 16$.
   **(b)** Find the values of $x$ and $p$ that maximize total revenue.
   **(c)** Find the value of $E$ for the $x$-value found in part (b).

**Elasticity**     **16.** Suppose that a demand function is linear—that is, $x = m - np$ for $0 \le p \le m/n$, where $m$ and $n$ are positive constants. Show that $E = 1$ at the midpoint of the demand curve on the interval $0 \le p \le m/n$ (that is, at $p = m/(2n)$).

# ▤ 3.7 IMPLICIT DIFFERENTIATION

In almost all of the examples and applications so far, any necessary functions have been defined as

$$y = f(x),$$

with $y$ given **explicitly** in terms of $x$, or as an **explicit function** of $x$. For example,

$$y = 3x - 2, \quad y = x^2 + x + 6, \quad \text{and} \quad y = -x^3 + 2$$

are all explicit functions of $x$. The equation $4xy - 3x = 6$ can be expressed as an explicit function of $x$ by solving for $y$. This gives

$$4xy - 3x = 6$$
$$4xy = 3x + 6$$
$$y = \frac{3x + 6}{4x}.$$

On the other hand, some equations in $x$ and $y$ cannot be readily solved for $y$, and some equations cannot be solved for $y$ at all. For example, while it would be possible (but tedious) to use the quadratic formula to solve for $y$ in the equation $y^2 + 2yx + 4x^2 = 0$, it is not possible to solve for $y$ in the equation $y^5 + 8y^3 + 6y^2x^2 + 2yx^3 + 6 = 0$. In equations such as these last two, $y$ is said to be given **implicitly** in terms of $x$.

In such cases, it may still be possible to find the derivative $dy/dx$ by a process called **implicit differentiation.** In doing so, we assume that there exists some function or functions $f$, which we may or may not be able to find, such that $y = f(x)$ and $dy/dx$ exists. It is useful to use $dy/dx$ here rather than $f'(x)$ to make it clear which variable is independent and which is dependent.

## ▤ EXAMPLE 1

Find $\dfrac{dy}{dx}$ if $3xy + 4y^2 = 10$.

Differentiate with respect to $x$ on both sides of the equation.

$$3xy + 4y^2 = 10$$
$$\frac{d}{dx}(3xy + 4y^2) = \frac{d}{dx}(10). \tag{1}$$

Now differentiate each term on the left side of the equation. Think of $3xy$ as the product $(3x)(y)$ and use the product rule. Since

$$\frac{d}{dx}(3x) = 3 \quad \text{and} \quad \frac{d}{dx}(y) = \frac{dy}{dx},$$

the derivative of $(3x)(y)$ is

$$(3x)\frac{dy}{dx} + (y)3 = 3x\frac{dy}{dx} + 3y.$$

To differentiate the second term, $4y^2$, use the generalized power rule, since $y$ is assumed to be some function of $x$.

Derivative
of $y^2$

$$\frac{d}{dx}(4y^2) = 4(2y^1)\overbrace{\frac{dy}{dx}}^{} = 8y\frac{dy}{dx}$$

On the right side of equation (1), the derivative of 10 is 0. Taking the indicated derivatives in equation (1) term by term gives

$$3x\frac{dy}{dx} + 3y + 8y\frac{dy}{dx} = 0.$$

Now solve this result for $dy/dx$.

$$(3x + 8y)\frac{dy}{dx} = -3y$$

$$\frac{dy}{dx} = \frac{-3y}{3x + 8y} \quad ∎$$

## EXAMPLE 2

Find $dy/dx$ for $x^2 + 2xy^2 + 3x^2y = 0$.

Again, differentiate on each side with respect to $x$.

$$\frac{d}{dx}(x^2 + 2xy^2 + 3x^2y) = \frac{d}{dx}(0)$$

Use the product rule to find the derivatives of $2xy^2$ and $3x^2y$.

| Derivative of $x^2$ | Derivative of $2xy^2$ | Derivative of $3x^2y$ | Derivative of 0 |
|:---:|:---:|:---:|:---:|
| ↓ | | | ↓ |

$$2x \quad + \quad 2x\left(2y\frac{dy}{dx}\right) + 2y^2 \quad + \quad 3x^2\left(\frac{dy}{dx}\right) + 6xy = \quad 0$$

Simplify to get

$$2x + 4xy\left(\frac{dy}{dx}\right) + 2y^2 + 3x^2\left(\frac{dy}{dx}\right) + 6xy = 0,$$

from which

$$(4xy + 3x^2)\frac{dy}{dx} = -2x - 2y^2 - 6xy,$$

or, finally,

$$\frac{dy}{dx} = \frac{-2x - 2y^2 - 6xy}{4xy + 3x^2}. \quad \blacksquare$$

## ■ EXAMPLE 3

Find $dy/dx$ for $x + \sqrt{xy} = y^2$.

Take the derivative on each side.

$$\frac{d}{dx}(x + \sqrt{xy}) = \frac{d}{dx}(y^2)$$

Since $\sqrt{xy} = \sqrt{x} \cdot \sqrt{y} = x^{1/2} \cdot y^{1/2}$ for nonnegative values of $x$ and $y$, taking derivatives gives

$$1 + x^{1/2}\left(\frac{1}{2}y^{-1/2} \cdot \frac{dy}{dx}\right) + y^{1/2}\left(\frac{1}{2}x^{-1/2}\right) = 2y\frac{dy}{dx}$$

$$1 + \frac{x^{1/2}}{2y^{1/2}} \cdot \frac{dy}{dx} + \frac{y^{1/2}}{2x^{1/2}} = 2y\frac{dy}{dx}.$$

Multiply both sides by $2x^{1/2} \cdot y^{1/2}$.

$$2x^{1/2} \cdot y^{1/2} + x\frac{dy}{dx} + y = 4x^{1/2} \cdot y^{3/2} \cdot \frac{dy}{dx}$$

Combine terms and solve for $dy/dx$.

$$2x^{1/2} \cdot y^{1/2} + y = (4x^{1/2} \cdot y^{3/2} - x)\frac{dy}{dx}$$

$$\frac{dy}{dx} = \frac{2x^{1/2} \cdot y^{1/2} + y}{4x^{1/2} \cdot y^{3/2} - x} \quad \blacksquare$$

## ■ EXAMPLE 4

The graph of $x^2 + 5y^2 = 36$ is the ellipse shown in Figure 54. Find the equation of the tangent line at the point $(4, 2)$.

A vertical line can cut the graph of the ellipse of Figure 54 in more than one point, so the graph is not the graph of a function. This means that $y$ *cannot* be written as a function of $x$, so $dy/dx$ would not exist. It seems likely, however, that a tangent line could be drawn to the graph at the point $(4, 2)$. To get around this difficulty, we try to decide if a function $f$ does exist whose graph is exactly the same as our graph in the "vicinity" of the point $(4, 2)$. Then the tangent line to $f$ at $(4, 2)$ is also the tangent line to the ellipse. See Figure 55.

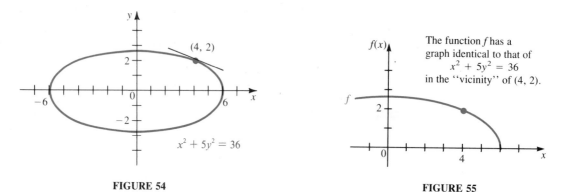

**FIGURE 54**          **FIGURE 55**

Assuming that such a function $f$ exists, $dy/dx$ can be found by implicit differentiation.

$$2x + 10y \cdot \frac{dy}{dx} = 0$$

$$\frac{dy}{dx} = \frac{-2x}{10y} = -\frac{x}{5y}$$

To find the slope of the tangent line at the point $(4, 2)$, let $x = 4$ and $y = 2$. The slope, $m$, is

$$m = -\frac{x}{5y} = -\frac{4}{5 \cdot 2} = -\frac{2}{5}.$$

The equation of the tangent line is then found by using the point-slope form of the equation of a line.

$$y - y_1 = m(x - x_1)$$

$$y - 2 = -\frac{2}{5}(x - 4)$$

$$5y - 10 = -2x + 8$$

$$2x + 5y = 18$$

This tangent line is graphed in Figure 54. ▰

The steps used in implicit differentiation can be summarized as follows.

**IMPLICIT**
**DIFFERENTIATION**

> To find $dy/dx$ for an equation containing $x$ and $y$:
>
> **1.** Differentiate on both sides with respect to $x$.
> **2.** Place all terms with $dy/dx$ on one side of the equals sign, and all terms without $dy/dx$ on the other side.
> **3.** Factor out $dy/dx$, and then solve for $dy/dx$.

When an applied problem involves an equation that is not given in explicit form, implicit differentiation can be used to locate maxima and minima or to find rates of change.

### ▰ EXAMPLE 5

The demand function for a certain commodity is given by

$$p = \frac{500,000}{2x^3 + 400x + 5000},$$

where $p$ is the price in dollars and $x$ is the demand in hundreds of units. Find the rate of change of demand with respect to price when $x = 100$ (that is, find $dx/dp$).

Rewrite the equation as

$$2x^3 + 400x + 5000 = 500,000p^{-1}.$$

Use implicit differentiation to find $dx/dp$.

$$6x^2\frac{dx}{dp} + 400\frac{dx}{dp} = -500,000\,p^{-2}$$

$$(6x^2 + 400)\frac{dx}{dp} = \frac{-500,000}{p^2}$$

$$\frac{dx}{dp} = \frac{-500,000}{p^2(6x^2 + 400)}$$

When $x = 100$,

$$p = \frac{500,000}{2(100)^3 + 400(100) + 5000} = .244,$$

and

$$\frac{dx}{dp} = \frac{-500,000}{(.244)^2\,[6(100)^2 + 400]} \approx -139.$$

This means that when demand ($x$) is 100 hundreds, or 10,000, demand is decreasing at the rate of 139 hundred, or 13,900, units per dollar change in price. ▰

## 3.7 EXERCISES

*Find dy/dx by implicit differentiation for each of the following.*

1. $4x^2 + 3y^2 = 6$

2. $2x^2 - 5y^2 = 4$

3. $2xy + y^2 = 8$

4. $-3xy - 4y^2 = 2$

5. $y^2 = 4x + 1$

6. $y^2 - 2x = 6$

7. $6xy^2 - 8y + 1 = 0$

8. $-4y^2x^2 - 3x + 2 = 0$

9. $x^2 + 2xy = 6$

10. $2x^2 - 3xy = 10$

11. $6x^2 + 8xy + y^2 = 6$

12. $8x^2 = 6y^2 + 2xy$

13. $x^3 = y^2 + 4$

14. $x^3 - 6y^2 = 10$

15. $\dfrac{1}{x} - \dfrac{1}{y} = 2$

16. $\dfrac{3}{2x} + \dfrac{1}{y} = y$

17. $3x^2 = \dfrac{2 - y}{2 + y}$

18. $2y^2 = \dfrac{5 + x}{5 - x}$

19. $x^2y + y^3 = 4$

20. $2xy^2 + 2y^3 + 5x = 0$

21. $\sqrt{x} + \sqrt{y} = 4$

22. $2\sqrt{x} - \sqrt{y} = 1$

23. $\sqrt{xy} + y = 1$

24. $\sqrt{2xy} - 1 = 3y^2$

*Find the equation of the tangent line at the given point on each curve in Exercises 25–32.*

25. $x^2 + y^2 = 25;$  $(-3, 4)$

26. $x^2 + y^2 = 100;$  $(8, -6)$

27. $x^2y^2 = 1;$  $(-1, 1)$

28. $x^2y^3 = 8;$  $(-1, 2)$

29. $x^2 + \sqrt{y} = 7;$  $(2, 9)$

30. $2y^2 - \sqrt{x} = 4;$  $(16, 2)$

31. $y + \dfrac{\sqrt{x}}{y} = 3;$  $(4, 2)$

32. $x + \dfrac{\sqrt{y}}{3x} = 2;$  $(1, 9)$

33. The graph of $x^2 + y^2 = 100$ is a circle having center at the origin and radius 10.
    (a) Write the equations of the tangent lines at the points where $x = 6$.
    (b) Graph the circle and the tangent lines.

34. The graph of $xy = 1$, shown in the figure, is a hyperbola.
    (a) Write the equations of the tangent lines at $x = -1$ and $x = 1$.
    (b) Graph the hyperbola and the tangent lines.

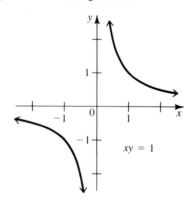

**EXERCISE 34**

**35.** The graph of $y^5 - y - x^2 = -1$ is shown in the figure. Find the equation of the tangent line to the graph at the point $(1, 1)$.

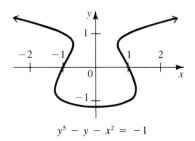

$y^5 - y - x^2 = -1$

**EXERCISE 35**

$3(x^2 + y^2)^2 = 25(x^2 - y^2)$

**EXERCISE 36**

**36.** The graph of $3(x^2 + y^2)^2 = 25(x^2 - y^2)$ is shown in the figure. Find the equation of the tangent line at the point $(2, 1)$.

**37.** Suppose $x^2 + y^2 + 1 = 0$. Use implicit differentiation to find $dy/dx$. Then explain why the result you got is meaningless. (*Hint:* Can $x^2 + y^2 + 1$ equal 0?)

*Let $\sqrt{u} + \sqrt{2v + 1} = 5$. Find the derivatives in Exercises 38–39.*

**38.** $du/dv$                                        **39.** $dv/du$

---

## ▤ APPLICATIONS

### BUSINESS AND ECONOMICS

**Demand**   **40.** The demand equation for a certain product is $2p^2 + x^2 = 1600$, where $p$ is the price per unit in dollars and $x$ is the number of units demanded.
   **(a)** Find and interpret $dx/dp$.
   **(b)** Find and interpret $dp/dx$.

**Cost and Revenue**   **41.** For a certain product, cost $C$ and revenue $R$ are given as follows, where $x$ is the number of units sold (in hundreds).

$$\text{Cost: } C^2 = x^2 + 100\sqrt{x} + 50$$
$$\text{Revenue: } 900(x - 5)^2 + 25R^2 = 22{,}500$$

   **(a)** Find and interpret the marginal cost $dC/dx$ at $x = 5$.
   **(b)** Find and interpret the marginal revenue $dR/dx$ at $x = 5$.

### PHYSICAL SCIENCES

**Velocity**   *The position of a particle at time $t$ is given by $s$. Find the velocity $ds/dt$.*

**42.** $s^3 - 4st + 2t^3 - 5t = 0$
**43.** $2s^2 + \sqrt{st} - 4 = 3t$

# ▰▰ 3.8 RELATED RATES

It is common for variables to be functions of time—for example, sales of an item may depend on the season of the year, or a population of animals may be increasing at a certain rate several months after being introduced into an area. Time is often present implicitly in a mathematical model, meaning that derivatives with respect to time must be found by the method of implicit differentiation discussed in the previous section. For example, if a particular mathematical model leads to the equation

$$x^2 + 5y - 3x + 1 = 0,$$

differentiating on both sides with respect to $t$ gives

$$\frac{d}{dt}(x^2 + 5y - 3x + 1) = \frac{d}{dt}(0)$$

$$2x \cdot \frac{dx}{dt} + 5 \cdot \frac{dy}{dt} - 3 \cdot \frac{dx}{dt} = 0$$

$$(2x - 3)\frac{dx}{dt} + 5\frac{dy}{dt} = 0.$$

The derivatives (or rates of change) $dx/dt$ and $dy/dt$ are related by this last equation; for this reason they are called **related rates.**

## ▰▰ EXAMPLE 1

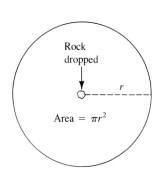

Rock dropped

$r$

Area = $\pi r^2$

**FIGURE 56**

A small rock is dropped into a lake. Circular ripples spread over the surface of the water, with the radius of each circle increasing at the rate of 3/2 feet per second. Find the rate of change of the area inside the circle formed by a ripple at the instant that the radius is 4 feet.

As shown in Figure 56, the area $A$ and the radius $r$ are related by

$$A = \pi r^2.$$

Take the derivative of each side with respect to time.

$$\frac{d}{dt}(A) = \frac{d}{dt}(\pi r^2)$$

$$\frac{dA}{dt} = 2\pi r \cdot \frac{dr}{dt} \qquad \textbf{(1)}$$

Since the radius is increasing at the rate of 3/2 feet per second,

$$\frac{dr}{dt} = \frac{3}{2}.$$

The rate of change of area at the instant $r = 4$ is given by $dA/dt$ evaluated at $r = 4$. Substituting into equation (1) gives

$$\frac{dA}{dt} = 2\pi \cdot 4 \cdot \frac{3}{2}$$

$$\frac{dA}{dt} = 12\pi \approx 37.7 \text{ square feet per second.} \quad \blacksquare$$

As suggested by Example 1, four basic steps are involved in solving problems about related rates.

**SOLVING RELATED RATE PROBLEMS**

1. Identify all given quantities, as well as the quantities to be found. Draw a sketch when possible.
2. Write an equation relating the variables of the problem.
3. Use implicit differentiation to find the derivative of both sides of the equation in Step 2 with respect to time.
4. Solve for the derivative giving the unknown rate of change and substitute the given values.

**Note:** Differentiate *first*, and *then* substitute values for the variables. If the substitutions were performed first, differentiating would not lead to useful results.

**▬▬ EXAMPLE 2**

A 50-foot ladder is placed against a large building. The base of the ladder is resting on an oil spill, and it slips (to the right in Figure 57) at the rate of 3 feet per minute. Find the rate of change of the height of the top of the ladder above the ground at the instant when the base of the ladder is 30 feet from the base of the building.

Let $y$ be the height of the top of the ladder above the ground, and let $x$ be the distance of the base of the ladder from the base of the building. By the Pythagorean theorem,

$$x^2 + y^2 = 50^2 \qquad (2)$$

Both $x$ and $y$ are functions of time $t$ in minutes after the moment that the ladder starts slipping. Take the derivative of both sides of equation (2) with respect to time, getting

$$\frac{d}{dt}(x^2 + y^2) = \frac{d}{dt}(50^2)$$

$$2x\frac{dx}{dt} + 2y\frac{dy}{dt} = 0. \qquad (3)$$

Since the base is sliding at the rate of 3 feet per minute,

$$\frac{dx}{dt} = 3.$$

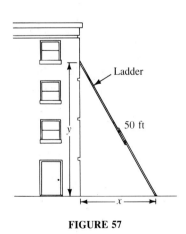

**FIGURE 57**

Ladder

50 ft

$y$

$x$

Also, the base of the ladder is 30 feet from the base of the building. Use this to find $y$.

$$50^2 = 30^2 + y^2$$
$$2500 = 900 + y^2$$
$$1600 = y^2$$
$$y = 40$$

In summary, $y = 40$, $x = 30$, and the rate of change of $x$ over time $t$ is $dx/dt = 3$. Substituting these values into equation (3) to find the rate of change of $y$ over time gives

$$2(30)(3) + 2(40)\frac{dy}{dt} = 0$$

$$180 + 80\frac{dy}{dt} = 0$$

$$80\frac{dy}{dt} = -180$$

$$\frac{dy}{dt} = \frac{-180}{80} = \frac{-9}{4} = -2.25.$$

At the instant when the base of the ladder is 30 feet from the base of the building, the top of the ladder is sliding down the building at the rate of 2.25 feet per minute. (The minus sign shows that the ladder is sliding *down*.) ▬

▬ EXAMPLE 3

A Florida real estate agent handles sales for a new housing development. She estimates that her gross monthly sales are given by

$$S = 32y - xy - \frac{y^2}{3200},$$

where $y$ is the average cost (in dollars) of a new house in the development and $x$ percent is the current interest rate for home mortgages. If the current interest rate is 12% and is rising $\frac{1}{2}$% each month, and if the current average price of a new house is $80,000 and is increasing by $1000 per month, how fast are her expected gross sales changing? Are her sales increasing or decreasing at this time?

From the statement of the problem, $S$, $x$, and $y$ are related by $S = 32y - xy - y^2/3200$. Also, $dx/dt = \frac{1}{2}$, $dy/dt = 1000$, $x = 12$, $y = 80,000$, and $dS/dt$ must be found. The first step is to differentiate with respect to $t$ on both sides of the given equation relating $S$, $x$, and $y$.

$$S = 32y - xy - \frac{y^2}{3200}$$

$$\frac{dS}{dt} = 32\frac{dy}{dt} - \left(x\frac{dy}{dt} + y\frac{dx}{dt}\right) - \frac{2y}{3200}\cdot\frac{dy}{dt}$$

Substituting for $dx/dt$, $dy/dt$, $x$, and $y$ gives

$$\frac{dS}{dt} = 32(1000) - 12(1000) - (80{,}000)\left(\frac{1}{2}\right) - \frac{(2)(80{,}000)(1000)}{3200}$$

$$\frac{dS}{dt} = -70{,}000$$

The agent's sales are decreasing at the rate of \$70,000 per month under the stated conditions. ▬

### ▬▬ EXAMPLE 4

A company is increasing production of an item at the rate of 50 units per day. All units produced can be sold. The daily demand function is given by

$$p = 50 - \frac{x}{200},$$

where $x$ is the number of units produced (and sold) and $p$ is price in dollars. Find the rate of change of revenue with respect to time (in days) when the daily production is 200 units.

   The revenue function,

$$R = xp = x\left(50 - \frac{x}{200}\right) = 50x - \frac{x^2}{200},$$

relates $R$ and $x$. The rate of change of $x$ over time (in days) is $dx/dt = 50$. The rate of change of revenue over time, $dR/dt$, is to be found when $x = 200$. Differentiate both sides of the equation

$$R = 50x - \frac{x^2}{200}$$

with respect to $t$.

$$\frac{dR}{dt} = 50\frac{dx}{dt} - \frac{1}{100}x\frac{dx}{dt} = \left(50 - \frac{1}{100}x\right)\frac{dx}{dt}$$

Now substitute the known values for $x$ and $dx/dt$.

$$\frac{dR}{dt} = \left[50 - \frac{1}{100}(200)\right](50) = 2400$$

Thus revenue is increasing at the rate of \$2400 per day. ▬

### ■ EXAMPLE 5

Blood flows faster the closer it is to the center of a blood vessel. According to Poiseuille's laws (see Exercise 39 in Section 5), the velocity $V$ of blood is given by

$$V = k(R^2 - r^2),$$

**FIGURE 58**

where $R$ is the radius of the blood vessel, $r$ is the distance of a layer of blood flow from the center of the vessel, and $k$ is a constant. See Figure 58.

**(a)** Find $dV/dt$ when $k = 375$. Treat $r$ as a constant. Assume the given units are compatible.

$$V = 375(R^2 - r^2)$$

$$\frac{dV}{dt} = 375\left(2R\frac{dR}{dt} - 0\right) \qquad r \text{ is a constant}$$

$$\frac{dV}{dt} = 750R\frac{dR}{dt}$$

**(b)** Suppose a skier's blood vessel has radius $R = .008$ millimeters and that cold weather is causing the vessel to contract at a rate of $dR/dt = -.001$ millimeters per minute. How fast is the velocity of blood changing? Use the constants given in part (a).

Find $dV/dt$. From part (a),

$$\frac{dV}{dt} = 750R\frac{dR}{dt}.$$

Here $R = .008$ and $dR/dt = -.001$, so

$$\frac{dV}{dt} = 750(.008)(-.001) = -.006.$$

That is, the velocity of the blood is decreasing at a rate of $-.006$ millimeters per minute. The units indicate that this is a deceleration (negative acceleration), since it gives the rate of change of velocity.  ■

## ▬ 3.8 EXERCISES

*Evaluate dy/dt for each of the following.*

**1.** $y = 9x^2 + 2x$;  $dx/dt = 4, x = 6$

**2.** $y = \dfrac{3x + 2}{1 - x}$;  $dx/dt = -1, x = 3$

**3.** $y^2 - 5x^2 = -1$;  $dx/dt = -3, x = 1, y = 2$

**4.** $8y^3 + x^2 = 1$;  $dx/dt = 2, x = 3, y = -1$

**5.** $xy - 5x + 2y^3 = -70$;  $dx/dt = -5, x = 2, y = -3$

**6.** $4x^3 - 9xy^2 + y = -80$;  $dx/dt = 4, x = -3, y = 1$

**7.** $\dfrac{x^2 + y}{x - y} = 9$;  $dx/dt = 2, x = 4, y = 2$

**8.** $\dfrac{y^3 - x^2}{x + 2y} = \dfrac{17}{7}$;  $dx/dt = 1, x = -3, y = -2$

### ■ APPLICATIONS

#### BUSINESS AND ECONOMICS

**Cost**  **9.** A manufacturer of handcrafted wine racks has determined that the cost to produce $x$ units per month is given by $C = .1x^2 + 10,000$. How fast is cost per month changing when production is changing at the rate of 10 units per month and the production level is 100 units?

**Cost/Revenue**  **10.** The manufacturer in Exercise 9 has found that the cost $C$ and revenue $R$ (in dollars) from the production and sale of $x$ units are related by the equation

$$C = \frac{R^2}{400,000} + 10,000.$$

Find the rate of change of revenue per unit when the cost per unit is changing by $10 and the revenue is $20,000.

**Revenue/Cost/Profit**  **11.** Given the revenue and cost functions $R = 50x - .4x^2$ and $C = 5x + 15$, where $x$ is the daily production (and sales), find the following when 40 units are produced daily and the rate of change of production is 10 units per day.

  **(a)** The rate of change of revenue with respect to time

  **(b)** The rate of change of cost with respect to time

  **(c)** The rate of change of profit with respect to time

**Production**  **12.** Repeat Exercise 11, given that 200 units are produced daily and the rate of change of production is 50 units per day.

**Demand**  **13.** A product sells for $3.50 currently and has a demand function of

$$p = \frac{8000}{x}.$$

Suppose manufacturing costs are increasing over time at a rate of 15% and the company plans to increase the price $p$ at this rate as well. Find the rate of change of demand over time.

**Revenue**  **14.** A company is increasing production at the rate of 25 units per day. The daily demand function is given by

$$p = 70 - \frac{x^2}{120}.$$

Find the rate of change of revenue with respect to time (in days) when the daily production (and sales) is 20.

#### LIFE SCIENCES

**Blood Velocity**  *Exercises 15–17 refer to Example 5 in this section.*

  **15.** Find $dV/dt$ if $r = 2$ mm, $k = 3$, $dr/dt = .02$ mm per minute, and $R$ is fixed.

  **16.** Find $dV/dt$ if $r = 1$ mm, $k = 4$, $dr/dt = .004$ mm per minute, and $R$ is fixed.

  **17.** A cross-country skier has a history of heart problems. She takes nitroglycerin to dilate blood vessels, thus avoiding angina (chest pain) due to blood vessel contraction. Use Poiseuille's law with $K = 555.6$ to find the rate of change of the blood velocity when $R = .02$ mm and $R$ is changing at .003 mm per minute. Assume $r$ is constant.

## SOCIAL SCIENCES

**Crime Rate**  18. Sociologists have found that crime rates are influenced by temperature. In a midwestern town of 100,000 people, the crime rate has been approximated as

$$C = \frac{1}{10}(T - 60)^2 + 100,$$

where $C$ is the number of crimes per month and $T$ is the average monthly temperature. The average temperature for May was 76°, and by the end of May the temperature was rising at the rate of 8° per month. How fast is the crime rate rising at the end of May?

**Learning Skills**  19. It is estimated that a person learning a certain assembly-line task takes

$$T(x) = \frac{2 + x}{2 + x^2}$$

minutes to perform the task after $x$ repetitions. Find $dT/dt$ if $dx/dt$ is 4, and 4 repetitions of the task have been performed.

**Memorization Skills**  20. Under certain conditions, a person can memorize $W$ words in $t$ minutes, when

$$W(t) = \frac{-.02t^2 + t}{t + 1}.$$

Find $dW/dt$ when $t = 5$.

## PHYSICAL SCIENCES

**Sliding Ladder**  21. A 25-ft ladder is placed against a building. The base of the ladder is slipping away from the building at the rate of 4 ft per minute. Find the rate at which the top of the ladder is sliding down the building at the instant when the bottom of the ladder is 7 ft from the base of the building.

**Distance**  22. One car leaves a given point and travels north at 30 mph. Another car leaves the same point at the same time and travels west at 40 mph. At what rate is the distance between the two cars changing at the instant when the cars have traveled 2 hr?

**Area**  23. A rock is thrown into a still pond. The circular ripples move outward from the point of impact of the rock so that the radius of the circle formed by a ripple increases at the rate of 2 ft per minute. Find the rate at which the area is changing at the instant the radius is 4 ft.

**Volume**  24. A spherical snowball is placed in the sun. The sun melts the snowball so that its radius decreases 1/4 inch per hour. Find the rate of change of the volume with respect to time at the instant the radius is 4 inches.

**Volume**  25. A sand storage tank used by the highway department for winter storms is leaking. As the sand leaks out, it forms a conical pile. The radius of the base of the pile increases at the rate of 1 inch per minute. The height of the pile is always twice the radius of the base. Find the rate at which the volume of the pile is increasing at the instant the radius of the base is 5 inches.

**Shadow Length**  26. A man 6 ft tall is walking away from a lamp post at the rate of 50 ft per minute. When the man is 8 ft from the lamp post, his shadow is 10 ft long. Find the rate at which the length of the shadow is increasing when he is 25 ft from the lamp post. (See the figure on the next page.)

**Water Level** 27. A trough has a triangular cross section. The trough is 6 ft across the top, 6 ft deep, and 16 ft long. Water is being pumped into the trough at the rate of 4 cubic feet per minute. Find the rate at which the height of water is increasing at the instant that the height is 4 ft.

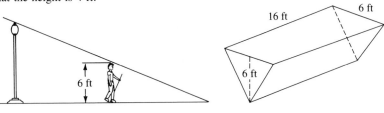

**EXERCISE 26**                               **EXERCISE 27**

**Velocity** 28. A pulley is on the edge of a dock, 8 ft above the water level. A rope is being used to pull in a boat. The rope is attached to the boat at water level. The rope is being pulled in at the rate of 1 ft per second. Find the rate at which the boat is approaching the dock at the instant the boat is 8 ft from the dock.

**EXERCISE 28**

## ▤ 3.9  DIFFERENTIALS

As mentioned earlier, the symbol $\Delta x$ represents a change in the variable $x$. Similarly, $\Delta y$ represents a change in $y$. An important problem that arises in many applications is to determine $\Delta y$ given specific values of $x$ and $\Delta x$. This quantity is often difficult to evaluate. In this section we show a method of approximating $\Delta y$ that uses the derivative $dy/dx$.

For values $x_1$ and $x_2$,

$$\Delta x = x_2 - x_1.$$

Solving for $x_2$ gives

$$x_2 = x_1 + \Delta x.$$

For a function $y = f(x)$, the symbol $\Delta y$ represents a change in $y$:

$$\Delta y = f(x_2) - f(x_1).$$

Replacing $x_2$ with $x_1 + \Delta x$ gives

$$\Delta y = f(x_1 + \Delta x) - f(x_1).$$

If $\Delta x$ is used instead of $h$, the derivative of a function $f$ at $x_1$ could be defined as

$$\frac{dy}{dx} = \lim_{\Delta x \to 0} \frac{\Delta y}{\Delta x}.$$

If the derivative exists, then

$$\frac{dy}{dx} \approx \frac{\Delta y}{\Delta x}$$

as long as $\Delta x$ is close to 0. Multiplying both sides by $\Delta x$ (assume $\Delta x \neq 0$) gives

$$\Delta y \approx \frac{dy}{dx} \cdot \Delta x.$$

Until now, $dy/dx$ has been used as a single symbol representing the derivative of $y$ with respect to $x$. In this section, separate meanings for $dy$ and $dx$ are introduced in such a way that their quotient, when $dx \neq 0$, is the derivative of $y$ with respect to $x$. These meanings of $dy$ and of $dx$ are then used to find an approximate value of $\Delta y$.

To define $dy$ and $dx$, look at Figure 59, which shows the graph of a function $y = f(x)$. The tangent line to the graph has been drawn at the point $P$. Let $\Delta x$ be any nonzero real number (in practical problems, $\Delta x$ is a small number) and locate the point $x + \Delta x$ on the $x$-axis. Draw a vertical line through $x + \Delta x$. Let this vertical line cut the tangent line at $M$ and the graph of the function at $Q$.

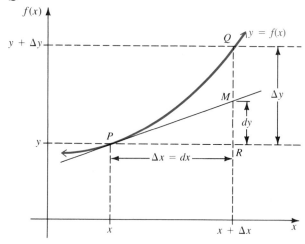

**FIGURE 59**

Define the new symbol $dx$ to be the same as $\Delta x$. Define the new symbol $dy$ to equal the length $MR$. The slope of $PM$ is $f'(x)$. By the definition of slope, the slope of $PM$ is also $dy/dx$, so that

$$f'(x) = \frac{dy}{dx},$$

or

$$dy = f'(x)dx.$$

In summary, the definitions of the symbols $dy$ and $dx$ are as follows.

**DIFFERENTIALS**

> For a function $y = f(x)$ whose derivative exists, the **differential** of $x$, written $dx$, is an arbitrary real number (usually small); the **differential** of $y$, written $dy$, is the product of $f'(x)$ and $dx$, or
>
> $$dy = f'(x)dx.$$

The usefulness of the differential is suggested by Figure 59. As $dx$ approaches 0, the value of $dy$ gets closer and closer to that $\Delta y$, so that for small nonzero values of $dx$

$$dy \approx \Delta y,$$

or

$$\Delta y \approx f'(x)dx.$$

**■ EXAMPLE 1**

Find $dy$ for the following functions.

**(a)** $y = 6x^2$

The derivative is $y' = 12x$, so that

$$dy = 12x\,dx.$$

**(b)** If $y = 8x^{-3/4}$, then $dy = -6x^{-7/4}\,dx$.

**(c)** If $f(x) = \dfrac{2 + x}{2 - x}$, then $dy = \dfrac{4}{(2 - x)^2}\,dx.$  ■

**■ EXAMPLE 2**

Suppose $y = 5x - x^2 + 4$. Find $dy$ for the following values of $x$ and $dx$.

**(a)** $x = -3,\ dx = .05$

First, by definition

$$dy = (5 - 2x)dx.$$

If $x = -3$ and $dx = .05$, then

$$dy = [5 - 2(-3)](.05) = 11(.05) = .55.$$

**(b)** $x = 0$, $dx = -.004$

$$dy = (5 - 2 \cdot 0)(-.004)$$
$$= 5(-.004)$$
$$= -.02 \quad \blacksquare$$

As discussed above,

$$\Delta y = f(x + \Delta x) - f(x).$$

For small nonzero values of $\Delta x$, $\Delta y \approx dy$, *so that*

$$dy \approx f(x + \Delta x) - f(x),$$

or
$$f(x) + dy \approx f(x + \Delta x). \tag{1}$$

Replacing $dy$ with $f'(x)dx$ gives the following result.

**DIFFERENTIAL APPROXIMATION**

> Let $f$ be a function whose derivative exists. For small nonzero values of $\Delta x$,
> $$dy \approx \Delta y,$$
> and
> $$f(x + \Delta x) \approx f(x) + dy = f(x) + f'(x)dx.$$

Differentials are used to find an approximate value of the change in a value of the dependent variable corresponding to a given change in the independent variable. When the concept of marginal cost (or profit or revenue) was used to approximate the change in cost for nonlinear functions, the same idea was developed. Thus the differential $dy$ approximates $\Delta y$ in much the same way as the marginal quantities approximate changes in functions.

For example, for a cost function $C(x)$,

$$dC = C'(x)dx = C'(x)\Delta x.$$

Since $\Delta C \approx dC$,

$$\Delta C \approx C'(x)\Delta x.$$

If the change in production, $\Delta x$, is equal to 1, then

$$C(x + 1) - C(x) = \Delta C$$
$$\approx C'(x)\Delta x$$
$$= C'(x),$$

which shows that marginal cost $C'(x)$ approximates the cost of the next unit produced, as mentioned earlier.

## ▬ EXAMPLE 3
Let $C(x) = 2x^3 + 300$.

(a) Find $\Delta C$ and $C'(x)$ when $\Delta x = 1$ and $x = 3$.

$$\Delta C = C(4) - C(3) = 428 - 354 = 74$$
$$C'(x) = 6x^2$$
$$C'(3) = 54$$

Here, the approximation of $C'(3)$ for $\Delta C$ is poor, since $\Delta x = 1$ is large relative to $x = 3$.

(b) Find $\Delta C$ and $C'(x)$ when $\Delta x = 1$ and $x = 50$.

$$\Delta C = C(51) - C(50) = 265,602 - 250,300 = 15,302$$

$$C'(50) = 6(2500) = 15,000$$

This approximation is quite good since $\Delta x = 1$ is small compared to $x = 50$.  ▬

## ▬ EXAMPLE 4
An analyst for a manufacturer of small appliances estimates that the total profit in hundreds of dollars from the sale of $x$ hundred microwave ovens is given by

$$P(x) = -2x^3 + 51x^2 + 88x.$$

In a report to management, the analyst projected sales for the coming month to be 1000 units for a total profit of \$389,000. He now realizes that his sales estimate may have been as much as 100 units too high. Approximately how far off is his profit estimate?

Differentials can be used to find the approximate change in $P$ resulting from decreasing $x$ by $100 = 1$ hundred units. This change can be approximated by $dP = P'(x)dx$ where $x = 1000 = 10$ hundred units and $dx = -1$ (hundred) units. Since $P'(x) = -6x^2 + 102x + 88$,

$$\Delta P \approx dP = (-6x^2 + 102x + 88)\, dx$$
$$= [-6(100) + 102(10) + 88]\,(-1)$$
$$= (-600 + 1020 + 88)\,(-1)$$
$$= (508)\,(-1)$$
$$= -508$$

hundred dollars. Thus, the profit estimate may have been as much as \$50,800 too high.  ▬

### ■ EXAMPLE 5

The mathematical model

$$y = A(x) = \frac{-7}{480}x^3 + \frac{127}{120}x, \qquad 0 \le x < 9,$$

gives the approximate alcohol concentration (in tenths of a percent) in an average person's bloodstream $x$ hours after drinking 8 ounces of 100-proof whiskey.

**(a)** Approximate the change in $y$ as $x$ changes from 3 to 3.5. Use $dy$ as an approximation of $\Delta y$.

Here,

$$dy = \left(\frac{-7}{160}x^2 + \frac{127}{120}\right)dx.$$

In this example, $x = 3$ and $dx = \Delta x = 3.5 - 3 = .5$. Substitution gives

$$\Delta y \approx dy = \left(\frac{-7}{160} \cdot 3^2 + \frac{127}{120}\right)(.5) \approx .33,$$

so $\Delta y \approx .33$. This means that the alcohol concentration increases by about .33 tenths of a percent as $x$ changes from 3 hours to 3.5 hours. (The *exact* increase, found by calculating $A(3.5) - A(3)$, is .30 tenths of a percent.)

**(b)** Approximate the change in $y$ as $x$ changes from 6 to 6.25 hours.

Let $x = 6$ and $dx = 6.25 - 6 = .25$.

$$\Delta y \approx dy = \left(\frac{-7}{160} \cdot 6^2 + \frac{127}{120}\right)(.25) \approx -.13$$

The minus sign shows that alcohol concentration decreases by .13 tenths of a percent as $x$ changes from 6 to 6.25 hours.  ■

The final example in this section shows how differentials are used to estimate errors that might enter into measurements of a physical quantity.

### ■ EXAMPLE 6

In a precision manufacturing process, ball bearings must be made with a radius of .6 mm, with a maximum error in the radius of $\pm$ .015 mm. Estimate the maximum error in the volume of the ball bearing.

The formula for the volume of a sphere is

$$V = \frac{4}{3}\pi r^3.$$

If an error of $\Delta r$ is made in measuring the radius of the sphere, the maximum error in the volume is

$$\Delta V = \frac{4}{3}\pi(r + \Delta r)^3 - \frac{4}{3}\pi r^3.$$

Approximate $\Delta V$ with $dV$, where

$$dV = 4\pi r^2 dr.$$

Replacing $r$ with .6 and $dr = \Delta r$ with $\pm$ .015 gives

$$dV = 4\pi(.6)^2(\pm .015) \approx \pm .0679.$$

The maximum error in the volume is about .07 cubic millimeters.

## 3.9 EXERCISES

*Find dy for each of the following.*

**1.** $y = 6x^2$

**2.** $y = -8x^4$

**3.** $y = 7x^2 - 9x + 6$

**4.** $y = -3x^3 + 2x^2$

**5.** $y = 2\sqrt{x}$

**6.** $y = 8\sqrt{2x - 1}$

**7.** $y = \dfrac{8x - 2}{x - 3}$

**8.** $y = \dfrac{-4x + 7}{3x - 1}$

**9.** $y = x^2\left(x - \dfrac{1}{x} + 2\right)$

**10.** $y = -x^3\left(2 + \dfrac{3}{x^2} - \dfrac{5}{x}\right)$

**11.** $y = \left(2 - \dfrac{3}{x}\right)\left(1 + \dfrac{1}{x}\right)$

**12.** $y = \left(9 - \dfrac{2}{x^2}\right)\left(3 + \dfrac{1}{x}\right)$

*For Exercises 13–24, find dy for the given values of x and $\Delta x$.*

**13.** $y = 2x^2 - 5x; \quad x = -2, \Delta x = .2$

**14.** $y = x^2 - 3x; \quad x = 3, \Delta x = .1$

**15.** $y = x^3 - 2x^2 + 3; \quad x = 1, \Delta x = -.1$

**16.** $y = 2x^3 + x^2 - 4x; \quad x = 2, \Delta x = -.2$

**17.** $y = \sqrt{3x}; \quad x = 1, \Delta x = .15$

**18.** $y = \sqrt{4x - 1}; \quad x = 5, \Delta x = .08$

**19.** $y = \dfrac{2x - 5}{x + 1}; \quad x = 2, \Delta x = -.03$

**20.** $y = \dfrac{6x - 3}{2x + 1}; \quad x = 3, \Delta x = -.04$

**21.** $y = -6\left(2 - \dfrac{1}{x^2}\right); \quad x = -1, \Delta x = .02$

**22.** $y = 9\left(3 + \dfrac{1}{x^4}\right); \quad x = -2, \Delta x = -.015$

**23.** $y = \dfrac{1 + x}{\sqrt{x}}; \quad x = 9, \Delta x = -.03$

**24.** $y = \dfrac{2 - 5x}{\sqrt{x + 1}}; \quad x = 3, \Delta x = .02$

## ▰▰ APPLICATIONS

### BUSINESS AND ECONOMICS

**Demand**   25. The demand for grass seed (in thousands of pounds) at a price of $x$ dollars is
$$d(x) = -5x^3 - 2x^2 + 1500.$$
Use the differential to approximate the changes in demand for the following changes in $x$.
(a) \$2 to \$2.50    (b) \$6 to \$6.30

**Average Cost**   26. The average cost to manufacture $x$ dozen marking pencils is
$$A(x) = .04x^3 + .1x^2 + .5x + 6.$$
Use the differential to approximate the changes in the average cost for the following changes in $x$.
(a) 3 to 4    (b) 5 to 6

**Mortgage Rates**   27. A county realty group estimates that the number of housing starts per year is
$$H(r) = \frac{300}{1 + .03r^2},$$
where $r$ is the mortgage rate in percent ($r = 8$ means 8%). Use the differential to approximate the change in housing starts when $r = 10$, if mortgage rates change by .5%.

**Revenue**   28. A company estimates that the revenue (in dollars) from the sale of $x$ units of doghouses is given by
$$R(x) = 625 + .03x + .0001x^2.$$
Use the differential to approximate the change in revenue from the sale of one more doghouse when 1000 doghouses are sold.

**Profit**   29. The profit function for the company in Exercise 28 is
$$P(x) = -390 + 24x + 5x^2 - \frac{1}{3}x^3,$$
where $x$ represents the demand for the product. Find the approximate change in profit for a one-unit change in demand when demand is at a level of 1000 doghouses. Use the differential.

**Material Requirement**   30. A cube 4 inches on an edge is given a protective coating .1 inch thick. About how much coating should a production manager order for 1000 such cubes?

**Material Requirement**   31. Beach balls 1 ft in diameter have a thickness of .03 inch. How much material would be needed to make 5000 beach balls?

### LIFE SCIENCES

**Drug Concentration**   32. The concentration of a certain drug in the bloodstream $x$ hr after being administered is approximately
$$C(x) = \frac{5x}{9 + x^2}.$$
Use the differential to approximate the changes in concentration for the following changes in $x$.
(a) 1 to 1.5    (b) 2 to 2.25

**Bacteria Population** **33.** The population of bacteria (in millions) in a certain culture $x$ hr after an experimental nutrient is introduced into the culture is

$$P(x) = \frac{25x}{8 + x^2}.$$

Use the differential to approximate the changes in population for the following changes in $x$.

**(a)** 2 to 2.5  **(b)** 3 to 3.25

**Area of a Blood Vessel** **34.** The radius of a blood vessel is 17 mm. A drug causes the radius to change to 16 mm. Find the approximate change in the area of a cross section of the vessel.

**Volume of a Tumor** **35.** A tumor is approximately spherical in shape. If the radius of the tumor changes from 14 mm to 16 mm, find the approximate change in volume.

**Area of an Oil Slick** **36.** An oil slick is in the shape of a circle. Find the approximate increase in the area of the slick if its radius increases from 1.2 mi to 1.4 mi.

**Area of a Bacteria Colony** **37.** The shape of a colony of bacteria on a Petri dish is circular. Find the approximate increase in its area if the radius increases from 20 mm to 22 mm.

## PHYSICAL SCIENCES

**Volume** **38.** A spherical beach ball is being inflated. Find the approximate change in volume if the radius increases from 4 cm to 4.2 cm.

**Volume** **39.** A spherical snowball is melting. Find the approximate change in volume if the radius decreases from 3 cm to 2.8 cm.

## GENERAL

**Measurement Error** **40.** The edge of a square is measured as 3.45 inches, with a possible error of $\pm.002$ inches. Estimate the maximum error in the area of the square.

**Measurement Error** **41.** The radius of a circle is measured as 4.87 inches, with a possible error of $\pm.040$ inches. Estimate the maximum error in the area of the circle.

**Measurement Error** **42.** A sphere has a radius of 5.81 inches, with a possible error of $\pm.003$ inches. Estimate the maximum error in the volume of the sphere.

**Measurement Error** **43.** A cone has a known height of 7.284 inches. The radius of the base is measured as 1.09 inches, with a possible error of $\pm.007$ inch. Estimate the maximum error in the volume of the cone. (The volume of a cone is given by $V = \frac{1}{3}\pi r^2 h$.)

| ▇▇ **KEY WORDS** ▇▇ | | | |
|---|---|---|---|
| **3.1** | increasing function | **3.4** | second derivative |

**KEY WORDS**

**3.1** increasing function
decreasing function
**3.2** relative maximum
relative minimum
critical number
first derivative test
**3.3** absolute maximum
absolute minimum
extreme value
theorem

**3.4** second derivative
acceleration
concavity
concave upward
and downward
point of inflection
second derivative
test
**3.6** economic lot size
economic order
quantity

elasticity of
demand
**3.7** explicit function
implicit
differentiation
explicit
differentiation
**3.8** related rates
**3.9** differential
differential
approximation

## CHAPTER 3 REVIEW EXERCISES

*Find the largest open intervals where f is increasing or decreasing.*

**1.** $f(x) = x^2 - 5x + 3$

**2.** $f(x) = -2x^2 - 3x + 4$

**3.** $f(x) = -x^3 - 5x^2 + 8x - 6$

**4.** $f(x) = 4x^3 + 3x^2 - 18x + 1$

**5.** $f(x) = \dfrac{6}{x - 4}$

**6.** $f(x) = \dfrac{5}{2x + 1}$

*Find the locations and values of all relative maxima and minima.*

**7.** $f(x) = -x^2 + 4x - 8$

**8.** $f(x) = x^2 - 6x + 4$

**9.** $f(x) = 2x^2 - 8x + 1$

**10.** $f(x) = -3x^2 + 2x - 5$

**11.** $f(x) = 2x^3 + 3x^2 - 36x + 20$

**12.** $f(x) = 2x^3 + 3x^2 - 12x + 5$

**13.** $f(x) = -2x^3 - \dfrac{1}{2}x^2 + x - 3$

**14.** $f(x) = -\dfrac{4}{3}x^3 + x^2 + 30x - 7$

**15.** $f(x) = x^4 - \dfrac{4}{3}x^3 - 4x^2 + 1$

**16.** $f(x) = -\dfrac{2}{3}x^3 + \dfrac{9}{2}x^2 + 5x + 1$

**17.** $f(x) = \dfrac{x - 1}{2x + 1}$

**18.** $f(x) = \dfrac{2x - 5}{x + 3}$

*Find the largest open intervals where the graph of f is concave upward or concave downward. Find the locations of any points of inflection. Graph each function.*

**19.** $f(x) = -4x^3 - x^2 + 4x + 5$

**20.** $f(x) = x^3 + \dfrac{5}{2}x^2 - 2x - 3$

**21.** $f(x) = x^4 + 2x^2$

**22.** $f(x) = 6x^3 - x^4$

**23.** $f(x) = \dfrac{x^2 + 4}{x}$

**24.** $f(x) = x + \dfrac{8}{x}$

**25.** $f(x) = \dfrac{2x}{3 - x}$

**26.** $f(x) = \dfrac{-4x}{1 + 2x}$

*Find the second derivative of each of the following, and then find f″(1) and f″(−3).*

**27.** $f(x) = 3x^4 - 5x^2 - 11x$

**28.** $f(x) = 9x^3 + \dfrac{1}{x}$

**29.** $f(x) = \dfrac{5x - 1}{2x + 3}$

**30.** $f(x) = \dfrac{4 - 3x}{x + 1}$

**31.** $f(t) = \sqrt{t^2 + 1}$

**32.** $f(t) = -\sqrt{5 - t^2}$

*Find the locations of all absolute maxima and minima on the given intervals.*

**33.** $f(x) = -x^2 + 5x + 1;\quad [1, 4]$

**34.** $f(x) = 4x^2 - 8x - 3;\quad [-1, 2]$

**35.** $f(x) = x^3 + 2x^2 - 15x + 3;\quad [-4, 2]$

**36.** $f(x) = -2x^3 - 2x^2 + 2x - 1;\quad [-3, 1]$

*Find dy/dx in Exercises 37–44.*

**37.** $x^2y^3 + 4xy = 2$

**38.** $\dfrac{x}{y} - 4y = 3x$

**39.** $9\sqrt{x} + 4y^3 = \dfrac{2}{x}$

**40.** $2\sqrt{y - 1} = 8x^{2/3}$

**41.** $\dfrac{x + 2y}{x - 3y} = y^{1/2}$

**42.** $\dfrac{6 + 5x}{2 - 3y} = \dfrac{1}{5x}$

**43.** $(4y^2 - 3x)^{2/3} = 6x$

**44.** $(8x + y^{1/2})^3 = 9y^2$

**45.** Find the equation of the tangent to the graph of $\sqrt{2x} - 4yx = -22$ at the point (2, 3).

**46.** The graph of $x^2 + y^2 = 25$ is a circle of radius 5 with center at the origin. Find the equation of the tangent lines when $x = -3$. Graph the circle and the tangent lines.

*Find dy/dt.*

**47.** $y = 8x^3 - 7x^2$;   $dx/dt = 4, x = 2$

**48.** $y = \dfrac{9 - 4x}{3 + 2x}$;   $dx/dt = -1, x = -3$

**49.** $y = \dfrac{1 + \sqrt{x}}{1 - \sqrt{x}}$;   $dx/dt = -4, x = 4$

**50.** $\dfrac{x^2 + 5y}{x - 2y} = 2$;   $dx/dt = 1, x = 2, y = 0$

*Find dy.*

**51.** $y = 8x^3 - 2x^2$

**52.** $y = 4(x^2 - 1)^3$

**53.** $y = \dfrac{6 - 5x}{2 + x}$

**54.** $y = \sqrt{9 + x^3}$

*Evaluate dy in Exercises 55–56.*

**55.** $y = 8 - x^2 + x^3$;   $x = -1, \Delta x = .02$

**56.** $y = \dfrac{3x - 7}{2x + 1}$;   $x = 2, \Delta x = .003$

**57.** Find two numbers whose sum is 25 and whose product is a maximum.

**58.** Find $x$ and $y$ such that $x = 2 + y$, and $xy^2$ is minimized.

# ▦ APPLICATIONS

## BUSINESS AND ECONOMICS

**Profit**   **59.** Suppose the profit from a product is $P(x) = 40x - x^2$, where $x$ is the price in hundreds of dollars.

 (a) At what price will the maximum profit occur?

 (b) What is the maximum profit?

**Profit**   **60.** The total profit in tens of dollars from the sale of $x$ hundred boxes of candy is given by

$$P(x) = -x^3 + 10x^2 - 12x.$$

 (a) Find the number of boxes of candy that should be sold in order to produce maximum profit.

 (b) Find the maximum profit.

**Packaging Design**   **61.** The packaging department of a corporation is designing a box with a square base and no top. The volume is to be 32 cubic meters. To reduce cost, the box is to have minimum surface area. What dimensions (height, length, and width) should the box have?

**Packaging Design**     **62.** Another product (see Exercise 61) will be packaged in a closed cylindrical tin can with a volume of $54\pi$ cubic inches. Find the radius and height of the can if it is to have minimum surface area.

**Packaging Design**     **63.** A company plans to package its product in a cylinder that is open at one end. The cylinder is to have a volume of $27\pi$ cubic inches. What radius should the circular bottom of the cylinder have to minimize the cost of the material? (*Hint:* The volume of a circular cylinder is $\pi r^2 h$, where $r$ is the radius of the circular base and $h$ is the height; the surface area of the cylinder open at one end is $2\pi rh + \pi r^2$.)

**Order Quantity**     **64.** A store sells 980,000 cases of a product annually. It costs $2 to store one case for a year and $20 to place a reorder. Find the number of cases that should be ordered each time.

**Order Quantity**     **65.** A very large camera store sells 320,000 rolls of film annually. It costs 10 cents to store one roll for one year and $10 to place a reorder. Find the number of rolls that should be ordered each time.

**Lot Size**     **66.** In one year, a health food manufacturer produces and sells 240,000 cases of vitamins. It costs $2 to store a case for one year and $15 to produce each batch. Find the number of batches that should be produced annually.

**Lot Size**     **67.** A company produces 128,000 cases of soft drink annually. It costs $1 to store a case for one year and $10 to produce one lot. Find the number of lots that should be produced annually.

## LIFE SCIENCES

**Pollution**     **68.** A circle of pollution is spreading from a broken underwater waste disposal pipe, with the radius increasing at the rate of 4 ft/min. Find the rate of change of the area of the circle when the radius is 7 ft.

## SOCIAL SCIENCES

**Playground Area**     **69.** The city park department is planning an enclosed play area in a new park. One side of the area will be against an existing building, with no fence needed there. Find the dimensions of the maximum rectangular area that can be made with 900 m of fence.

## PHYSICAL SCIENCES

**Sliding Ladder**     **70.** A 50-ft ladder is placed against a building. The top of the ladder is sliding down the building at the rate of 2 ft/min. Find the rate at which the base of the ladder is slipping away from the building at the instant that the base is 30 ft from the building.

**Spherical Radius**     **71.** A large weather balloon is being inflated at the rate of 1.2 cu ft of air per minute. Find the rate of change of the radius when the radius is 1.2 ft.

**Water Level**     **72.** A water trough 2 ft across, 4 ft long, and 1 ft deep has ends in the shape of isosceles triangles. (See the figure.) it is being filled with 3.5 cu ft of water per minute. Find the rate at which the depth of water in the tank is changing when the water is 1/3 ft deep.

GENERAL

**Volume** **73.** Approximate the volume of coating on a sphere of radius 4 inches if the coating is .02 inch thick.

**Area** **74.** A square has an edge of 9.2 inches, with a possible error in the measurement of $\pm.04$ inch. Estimate the possible error in the area of the square.

## EXTENDED APPLICATION

## A TOTAL COST MODEL FOR A TRAINING PROGRAM*

In this application, we set up a mathematical model for determining the total costs in setting up a training program. Then we use calculus to find the time between training programs that produces the minimum total cost. The model assumes that the demand for trainees is constant and that the fixed cost of training a batch of trainees is known. Also, it is assumed that people who are trained, but for whom no job is readily available, will be paid a fixed amount per month while waiting for a job to open up.

The model uses the following variables.

$D$ = demand for trainees per month
$N$ = number of trainees per batch
$C_1$ = fixed cost of training a batch of trainees
$C_2$ = variable cost of training per trainee per month
$C_3$ = salary paid monthly to a trainee who has not yet been given a job after training
$m$ = time interval in months between successive batches of trainees
$t$ = length of training program in months
$Z(m)$ = total monthly cost of program

The total cost of training a batch of trainees is given by $C_1 + NtC_2$. However $N = mD$, so that the total cost per batch is $C_1 + mDtC_2$.

After training, personnel are given jobs at the rates of $D$ per month. Thus, $N - D$ of the trainees will not get a job the first month, $N - 2D$ will not get a job the second month, and so on. The $N - D$ trainees who do not get a job the first month produce total costs of $(N - D)C_3$, those not getting jobs during the second month produce costs of $(N - 2D)C_3$, and so on. Since $N = mD$, the costs during the first month can be written as

$$(N - D)C_3 = (mD - D)C_3 = (m - 1)DC_3,$$

while the costs during the second month are $(m - 2)DC_3$, and so on. The total cost for keeping the trainees without a job is thus

$$(m - 1)DC_3 + (m - 2)DC_3 + (m - 3)DC_3 + \cdots + 2DC_3 + DC_3,$$

which can be factored to give

$$DC_3[(m - 1) + (m - 2) + (m - 3) + \cdots + 2 + 1].$$

The expression in brackets is the sum of the terms of an arithmetic sequence, discussed in most algebra texts. Using formulas for arithmetic sequences, the expression in brackets can be shown to equal $m(m - 1)/2$, so that we have

$$DC_3\left[\frac{m(m - 1)}{2}\right] \tag{1}$$

---

*Based on "A Total Cost Model for a Training Program" by P. L. Goyal and S. K. Goyal, Faculty of Commerce and Administration, Concordia University. Used with permission.

as the total cost for keeping jobless trainees.

The total cost per batch is the sum of the training cost per batch, $C_1 + mDtC_2$, and the cost of keeping trainees without a proper job, given by (1). Since we assume that a batch of trainees is trained every $m$ months, the total cost per month, $Z(m)$, is given by

$$Z(m) = \frac{C_1 + mDtC_2}{m} + \frac{DC_3\left[\dfrac{m(m-1)}{2}\right]}{m} = \frac{C_1}{m} + DtC_2 + DC_3\left(\frac{m-1}{2}\right).$$

## EXERCISES

1. Find $Z'(m)$.

2. Solve the equation $Z'(m) = 0$.
   As a practical matter, it is usually required that $m$ be a whole number. If $m$ does not come out to be a whole number in Exercise 2, then $m^+$ and $m^-$, the two whole numbers closest to $m$, must be chosen. Calculate both $Z(m^+)$ and $Z(m^-)$; the smaller of the two provides the optimum value of $Z$.

3. Suppose a company finds that its demand for trainees is 3 per month, that a training program requires 12 months, that the fixed cost of training a batch of trainees is $15,000, that the variable cost per trainee per month is $100, and that trainees are paid $900 per month after training but before going to work. Use your result from Exercise 2 and find $m$.

4. Since $m$ is not a whole number, find $m^+$ and $m^-$.

5. Calculate $Z(m^+)$ and $Z(m^-)$.

6. What is the optimum time interval between successive batches of trainees? How many trainees should be in a batch?

# 4

# Exponential and Logarithmic Functions

Exponential functions may be the single most important category of functions used in practical applications. These functions, and the closely related logarithmic functions, are used to describe growth and decay, which are important ideas in management, social science, and biology.

The following problem illustrates the behavior of exponential functions. A person contracts to work for a period of thirty days and is given the option of receiving a lump sum of $1000 or receiving $.01 on the first day of work, $.02 on the second, $.04 on the third, $.08 on the fourth, and so on, with each day's pay double that of the previous day. While the first choice may appear more lucrative (after all, the *total* amount earned by the second method after seven days is only $1.27), the second method actually will earn the worker over five million dollars on the thirtieth day alone, and over ten million dollars for the month!

A function that defines the amount of payment on the $n$th day in the example above is

$$f(n) = .01(2^n).$$

The base 2 appears as a result of the doubling that occurs. A function in which the variable appears in the exponent is an *exponential function*. Exponential functions tend to increase or decrease rapidly. The function in the example above is an increasing function. An example of a decreasing function is one that gives the amount of a radioactive substance present as time passes, since the amount gradually diminishes. In a decreasing exponential function, the base is a number between 0 and 1.

For many years *logarithms* were used primarily to assist in involved calculations. Current technology has made this use of logarithms obsolete, but *logarithmic functions* play an important role in many applications of mathematics. We shall see in this chapter that exponential and logarithmic functions are intimately related; in fact, they are inverses of each other.

## 4.1    EXPONENTIAL FUNCTIONS

In earlier chapters we discussed functions involving expressions such as $x^2$, $(2x + 1)^3$, or $x^{3/4}$, where the variable or variable expression is the base of an exponential expression, and the exponent is a constant. In an exponential function, the variable is in the exponent and the base is a constant.

**EXPONENTIAL FUNCTION**

> An **exponential function** with base $a$ is defined as
> $$f(x) = a^x, \qquad \text{where } a > 0 \text{ and } a \neq 1.$$

(If $a = 1$, the function is the constant function $f(x) = 1$.)

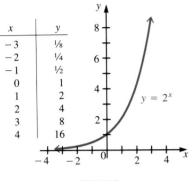

**FIGURE 1**

The exponential function $y = 2^x$ can be graphed by making a table of values of $x$ and $y$, as shown in Figure 1. Plotting these points and drawing a smooth curve through them gives the graph shown in Figure 1. The graph approaches the negative $x$-axis but will never actually touch it, since $y$ cannot be 0, so the $x$-axis is a horizontal asymptote. The graph suggests that the domain is the set of all real numbers and the range is the set of all positive real numbers. The function $f$, where $f(x) = 2^x$, has a base greater than 1, and the function is increasing over its domain. This graph is typical of the graphs of exponential functions of the form $y = a^x$, where $a > 1$.

## ■■ EXAMPLE 1

Graph $y = 2^{-x}$.

By the properties of exponents,

$$2^{-x} = \frac{1}{2^x} = \left(\frac{1}{2}\right)^x.$$

Construct a table of values and draw a smooth curve through the resulting points (see Figure 2). This graph is typical of the graphs of exponential functions of the form $y = a^x$ where $0 < a < 1$. The domain includes all real numbers and the range includes all positive numbers. Notice that this function, with $f(x) = 2^{-x} = (1/2)^x$, is decreasing over its domain. ■■

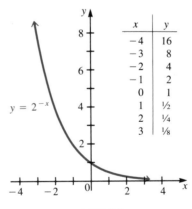

**FIGURE 2**

In the definition of an exponential function, notice that the base $a$ is restricted to positive values, with negative or zero bases not allowed. For example, the function $y = (-4)^x$ could not include such numbers as $x = 1/2$ or $x = 1/4$ in the domain. The resulting graph would be at best a series of separate points having little practical use.

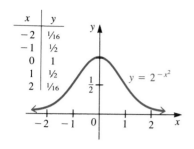

| x | y |
|---|---|
| -2 | 1/16 |
| -1 | 1/2 |
| 0 | 1 |
| 1 | 1/2 |
| 2 | 1/16 |

**FIGURE 3**

## ■ EXAMPLE 2

Graph $y = 2^{-x^2}$.

Preparing a table of values of $x$ and $y$ and plotting the corresponding points gives the graph in Figure 3. Both the negative sign and the fact that $x$ is squared affect the shape of the graph, with the final result looking quite different from the typical graphs of exponential functions in Figures 1 and 2. Graphs like this are important in probability, where the normal curve has an equation similar to the one in this example. ■

In Figures 1 and 2, which are typical graphs of exponential functions, a given value of $x$ leads to exactly one value of $a^x$. Because of this, an equation with a variable in the exponent, called an **exponential equation,** often can be solved using the following property.

> If $a \neq 1$ and $a^x = a^y$, then $x = y$. Also, if $x = y$, then $a^x = a^y$.

(Both bases must be the same.) The value $a = 1$ is excluded since $1^2 = 1^3$, for example, even though $2 \neq 3$. To solve $2^{3x} = 2^7$ using this property, work as follows.

$$2^{3x} = 2^7$$
$$3x = 7$$
$$x = \frac{7}{3}$$

## ■ EXAMPLE 3

**(a)** Solve $9^x = 27$.

First rewrite both sides of the equation so the bases are the same. Since $9 = 3^2$ and $27 = 3^3$,

$$(3^2)^x = 3^3$$
$$3^{2x} = 3^3$$
$$2x = 3$$
$$x = \frac{3}{2}.$$

**(b)** Solve $32^{2x-1} = 128^{x+3}$.

Since the bases must be the same, write 32 as $2^5$ and 128 as $2^7$, giving

$$32^{2x-1} = 128^{x+3}$$
$$(2^5)^{2x-1} = (2^7)^{x+3}$$
$$2^{10x-5} = 2^{7x+21}.$$

Now use the property above to get

$$10x - 5 = 7x + 21$$
$$3x = 26$$
$$x = \frac{26}{3}.$$

Verify this solution in the original equation.   ▬

***Compound Interest***   The calculation of compound interest is an important application of exponential functions. The cost of borrowing money or the return on an investment is called **interest.** The amount borrowed or invested is the **principal,** $P$. The **rate of interest** $r$ is given as a percent per year, and $t$ is the **time,** measured in years.

**SIMPLE INTEREST**

> ▬ The product of the principal $P$, rate $r$, and time $t$ gives **simple interest,** $I$:
> $$I = Prt.$$

With **compound interest,** interest is charged (or paid) on interest as well as on principal. To find a formula for compound interest, first suppose that $P$ dollars, the principal, is deposited at a rate of interest $i$ per year. The interest earned during the first year is found by using the formula for simple interest:

$$\text{First-Year Interest} = P \cdot i \cdot 1 = Pi.$$

At the end of one year, the amount on deposit will be the sum of the original principal and the interest earned, or

$$P + Pi = P(1 + i). \qquad (1)$$

If the deposit earns compound interest, the interest earned during the second year is found from the total amount on deposit at the end of the first year. Thus, the interest earned during the second year (again found by the formula for simple interest), is given by

$$[P(1 + i)](i)(1) = P(1 + i)i, \qquad (2)$$

so that the total amount on deposit at the end of the second year is given by the sum of amounts from (1) and (2) above, or

$$P(1 + i) + P(1 + i)i = P(1 + i)(1 + i) = P(1 + i)^2.$$

In the same way, the total amount on deposit at the end of three years is

$$P(1 + i)^3.$$

After $j$ years, the total amount on deposit, called the **compound amount,** is $P(1 + i)^j$.

When interest is compounded more than once a year, the compound interest formula is adjusted. For example, if interest is to be paid quarterly (four times a year), 1/4 of the interest rate is used each time interest is calculated, so the rate becomes $i/4$, and the number of compounding periods in $n$ years becomes $4n$. Generalizing from this idea gives the following formula.

**COMPOUND INTEREST**

> ▬▬ If a deposit of $P$ dollars is invested at a yearly rate of interest $i$ compounded $m$ times per year for $n$ years, the compound amount is
>
> $$A = P\left(1 + \frac{i}{m}\right)^{mn} \qquad \text{dollars.}$$

▬▬ **EXAMPLE 4**

Find the amount of interest paid on an investment of \$9000 at 12% compounded quarterly for 8 years.

Use the formula for compound interest with $P = 9000$, $i = .12$, $m = 4$, and $n = 8$.

$$A = P\left(1 + \frac{i}{m}\right)^{mn}$$

$$= 9000\left(1 + \frac{.12}{4}\right)^{4(8)} = 9000(1 + .03)^{32} = 9000(1.03)^{32}$$

To find $(1.03)^{32}$, use a calculator or the compound interest table given in the Appendix. Look for 3% across the top and 32 down the side. You should find 2.57508, which gives

$$A = 9000(2.57508) = 23,175.72.$$

The investment plus the interest is \$23,175.72. The interest amounts to \$23,175.72 − \$9000 = \$14,175.72. ▬▬

***The Number e***    Perhaps the single most useful base for an exponential function is the number $e$, an irrational number that occurs often in practical applications. The letter $e$ was chosen to represent this number in honor of the Swiss mathematician Leonhard Euler (pronounced ''oiler'') (1707–1783). To see how the number $e$ occurs in an application, begin with the formula for compound interest,

$$P\left(1 + \frac{i}{m}\right)^{nm}.$$

Suppose that a lucky investment produces annual interest of 100%, so that $i = 1.00 = 1$. Suppose also that you can deposit only \$1 at this rate, and for only one year. Then $P = 1$ and $n = 1$. Substituting these values into the formula for compound interest gives

$$P\left(1 + \frac{i}{m}\right)^{nm} = 1\left(1 + \frac{1}{m}\right)^{1(m)} = \left(1 + \frac{1}{m}\right)^{m}.$$

As interest is compounded more and more often, $m$ gets larger and the value of this expression will increase. For example, if $m = 1$ (interest is compounded annually),

$$\left(1 + \frac{1}{m}\right)^m = \left(1 + \frac{1}{1}\right)^1 = 2^1 = 2,$$

so that your $1 becomes $2 in one year. Using a calculator with a $y^x$ key gives the results shown in the following table for larger and larger values of $m$. These results have been rounded to five decimal places.

| $m$ | $\left(1 + \dfrac{1}{m}\right)^m$ |
|---|---|
| 1 | 2 |
| 2 | 2.25 |
| 5 | 2.48832 |
| 10 | 2.59374 |
| 25 | 2.66584 |
| 50 | 2.69159 |
| 100 | 2.70481 |
| 500 | 2.71557 |
| 1000 | 2.71692 |
| 10,000 | 2.71815 |
| 1,000,000 | 2.71828 |

The table suggests that as $m$ increases, the value of $(1 + 1/m)^m$ gets closer and closer to some fixed number. It turns out that this is indeed the case. This fixed number is called $e$.

**DEFINITION OF** $e$

$$e = \lim_{m \to +\infty} \left(1 + \frac{1}{m}\right)^m \approx 2.718281828$$

(This last approximation gives the value of $e$ to nine decimal places.) A table in the appendix to this book gives various powers of $e$. Also, some calculators give values of $e^x$. In Figure 4, the functions $y = 2^x$, $y = e^x$, and $y = 3^x$ are graphed for comparison.

The number $e$ is often used as the base in an exponential equation because it provides a good model for many natural, as well as economic, phenomena.

Euler investigated the exponential function $y = e^x$. He approximated $e$ to 23 decimal places using the definition $\lim\limits_{m \to +\infty} (1 + 1/m)^m$. Euler also established a relationship among the five most famous constants in mathematics: $e^{\pi i} + 1 = 0$.

Any exponential function $y = a^x$ can be written as an exponential function with base $e$. For example, there exists a real number $k$ such that

$$2 = e^k.$$

Raising both sides to the power $x$ gives

$$2^x = e^{kx},$$

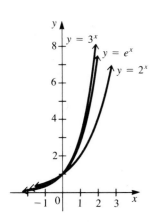

**FIGURE 4**

so that powers of 2 can be found by evaluating appropriate powers of $e$. To find the necessary number $k$, use the table in the Appendix. Look down the $e^x$ column for the number closest to 2. You should find that

$$e^{.70} \approx 2.01375,$$

so that $k \approx .70$ and

$$2^x = e^{kx} \approx e^{.70x}.$$

Better accuracy could be obtained with a calculator or a more complete table. In Section 4 of this chapter we will see why this change of base may be useful. A general statement can be drawn from this example.

---

For every positive real number $a$, there is a real number $k$ such that
$$a^x = e^{kx}.$$

---

## ▅▅ EXAMPLE 5

Use the table for powers of $e$ to approximate the value of each exponential.

**(a)** $e^{1.5}$

Use the first two columns of the table to find

$$e^{1.5} \approx 4.48168.$$

**(b)** $e^{-3} \approx 0.04978$

Look up 3 in the first column and read across to find $e^{-3}$ in the third column.

**(c)** $2^{3.57}$

Use the result found above,

$$2^x \approx e^{.70x},$$

with $x = 3.57$. Then

$$2^{3.57} \approx e^{(.70)(3.57)} = e^{2.499} \approx e^{2.5} \approx 12.$$

**(d)** $5^{-1.38}$

Some calculators have a $y^x$ key. With such a calculator, enter 5, press the $y^x$ key, enter 1.38, press the $\pm$ key to get a negative, and then press $=$. The result should be .108497915.

To find this value using the table let $5^x = e^{kx}$, so that $5 = e^k$. From the table, $k \approx 1.60$ and $5^x \approx e^{1.60x}$. Then

$$5^{-1.38} \approx e^{1.60(-1.38)} \approx e^{-2.208}.$$

Using the table, approximate $e^{-2.208}$ as $e^{-2.2} \approx .11$.  ▅▅

As mentioned earlier, exponential functions have many practical applications. In situations that involve growth or decay of a population, the size of the population at a given time $t$ often is determined by an exponential function of $t$.

### ▰ EXAMPLE 6

Biologists studying salmon have found that the oxygen consumption of yearling salmon (in appropriate units) increases exponentially with speed of swimming according to the function

$$f(x) = 100(3)^{.6x},$$

where $x$ is the speed in feet per second. Find each of the following.

**(a)** The oxygen consumption when the fish are still

When the fish are still, their speed is 0. Substitute 0 for $x$:

$$f(0) = 100(3)^{(.6)(0)} = 100(3)^0$$
$$= 100 \cdot 1 = 100.$$

When the fish are still, their oxygen consumption is 100 units.

**(b)** The oxygen consumption at a speed of 5 feet per second

Replace $x$ with 5.

$$f(5) = 100(3)^{(.6)(5)} = 100(3)^3 = 2700$$

A speed of 5 feet per second increases the oxygen consumption to 2700 units.

**(c)** The oxygen consumption at a speed of 2 feet per second

Find $f(2)$ as follows.

$$f(2) = 100(3)^{(.6)(2)} = 100(3)^{1.2}$$

Use a calculator or use the table to find $3^x \approx e^{1.1x}$, so that $3^{1.2} \approx e^{1.32}$. The closest entry in the table is for $e^{1.30}$. Using that entry,

$$3^{1.2} \approx e^{1.3} \approx 3.66929.$$

Now find $f(2)$.

$$f(2) = 100(3)^{1.2} \approx 100(3.7) = 370$$

At a speed of 2 feet per second, oxygen consumption is about 370 units. ▰

---

### ▰ 4.1 EXERCISES

*Graph each of the following.*

**1.** $y = 3^x$

**2.** $y = 4^x$

**3.** $y = 3^{-x}$

**4.** $y = 4^{-x}$

**5.** $y = \left(\dfrac{1}{4}\right)^x$

**6.** $y = \left(\dfrac{1}{3}\right)^x$

**7.** $y = 3^{2x}$

**8.** $y = 3^{x^2}$

**9.** $y = e^{-x^2}$

**10.** $y = e^{x+1}$

**11.** $y = 10 - 5e^{-x}$

**12.** $y = 100 - 80e^{-x}$

*Graph the following equations by plotting points in the interval* $[-3, 2]$.

**13.** $y = x \cdot 2^x$    **14.** $y = x^2 \cdot 2^x$    **15.** $y = \dfrac{e^x - e^{-x}}{2}$    **16.** $y = \dfrac{e^x + e^{-x}}{2}$

*Solve each equation.*

**17.** $2^x = \dfrac{1}{8}$    **18.** $4^x = 64$    **19.** $e^x = e^2$    **20.** $e^x = \dfrac{1}{e^2}$

**21.** $4^x = 8$    **22.** $25^x = 125$    **23.** $16^x = 64$    **24.** $\left(\dfrac{1}{8}\right)^x = 8$

**25.** $\left(\dfrac{3}{4}\right)^x = \dfrac{16}{9}$    **26.** $16^{-x+1} = 8$    **27.** $25^{-2x} = 3125$    **28.** $16^{x+2} = 64^{2x-1}$

**29.** $(e^4)^{-2x} = e^{-x+1}$    **30.** $e^{-x} = (e^2)^{x+3}$    **31.** $2^{|x|} = 16$    **32.** $5^{-|x|} = \dfrac{1}{25}$

**33.** $2^{x^2-4x} = \dfrac{1}{16}$    **34.** $5^{x^2+x} = 1$    **35.** $8^{x^2} = 2^{5x}$    **36.** $9^x = 3^{x^2}$

*Approximate the value of the each expression.*

**37.** $e^{2.5}$    **38.** $e^{-.04}$    **39.** $e^{-.13}$    **40.** $e^{-2.8}$

**41.** $6^{.5}$    **42.** $4^{3.2}$    **43.** $1.1^{-2.4}$    **44.** $4.5^{-.8}$

**45.** Suppose the domain of $y = 2^x$ is restricted to rational values of $x$ (which are the only values discussed prior to this section.) Describe in words the resulting graph.

**46.** In our definition of exponential function, we ruled out negative values of $a$. The author of a textbook on mathematical economics, however, obtained a "graph" of $y = (-2)^x$ by plotting the following points and drawing a smooth curve through them.

| $x$ | $-4$ | $-3$ | $-2$ | $-1$ | $0$ | $1$ | $2$ | $3$ |
|---|---|---|---|---|---|---|---|---|
| $y$ | $1/16$ | $-1/8$ | $1/4$ | $-1/2$ | $1$ | $-2$ | $4$ | $-8$ |

The graph oscillates very neatly from positive to negative values of $y$. Comment on this approach. (This example shows the dangers of point plotting when drawing graphs.)

## ▬ APPLICATIONS

### BUSINESS AND ECONOMICS

**Compound Interest**    **47.** Find the interest earned on \$10,000 invested for 5 years at 8% interest compounded as follows.

(a) Annually    (b) Semiannually (twice a year)    (c) Quarterly    (d) Monthly

**Compound Interest**    **48.** Suppose \$26,000 is borrowed for 3 years at 12% interest. Find the interest paid over this period if the interest is compounded as follows.

(a) Annually    (b) Semiannually    (c) Quarterly    (d) Monthly

**Compound Interest**   **49.** Terry Wong deposits $9430 commission earned on a real estate sale in the bank at 6% interest compounded quarterly. How much will be on deposit in 4 years?

**Compound Interest**   **50.** Julie Grey borrows $17,500 to open a new restaurant. The money is to be repaid at the end of 6 years, with 10% interest compounded semiannually. How much will she owe in 6 years?

**Compound Interest**   **51.** Michael Anderson needs to choose between two investments: one pays 8% compounded semiannually, and the other pays 7½% compounded monthly. If he plans to invest $18,000 for 1½ years, which investment should he choose? How much extra interest will he earn by making the better choice?

**Compound Interest**   **52.** Find the interest rate required for an investment of $5000 to grow to $8000 in 4 years if interest is compounded as follows.
   **(a)** Annually      **(b)** Quarterly

**Compound Interest**   **53.** If $1 is deposited into an account paying 12% per year compounded annually, then after $t$ years the account will contain

$$y = (1 + .12)^t = (1.12)^t$$

dollars.

   **(a)** Use a calculator to help you complete the following table.

| $t$ | 0 | 1 | 2 | 3 | 4 | 5 | 6 | 7 | 8 | 9 | 10 |
|---|---|---|---|---|---|---|---|---|---|---|---|
| $y$ | 1 | | | | | 1.76 | | | | | 3.11 |

   **(b)** Graph $y = (1.12)^t$.

**Inflation**   **54.** If money loses value at the rate of 8% per year, the value of $1 in $t$ years is given by

$$y = (1 - .08)^t = (.92)^t.$$

   **(a)** Use a calculator to help complete the following table.

| $t$ | 0 | 1 | 2 | 3 | 4 | 5 | 6 | 7 | 8 | 9 | 10 |
|---|---|---|---|---|---|---|---|---|---|---|---|
| $y$ | 1 | | | | | .66 | | | | | .43 |

   **(b)** Graph $y = (.92)^t$.

**Employee Training**   **55.** A person learning certain skills involving repetition tends to learn quickly at first. Then learning tapers off and approaches some upper limit. Suppose the number of symbols per minute that a keypunch operator can produce is given by

$$p(t) = 250 - 120(2.8)^{-.5t},$$

   where $t$ is the number of months the operator has been in training. Find each of the following.
   **(a)** $p(2)$      **(b)** $p(4)$      **(c)** $p(10)$      **(d)** Graph $p(t)$.

**Cost Estimation**   **56.** Use the results of Exercise 54(a) to answer the following questions.
   **(a)** Suppose a house costs $65,000 today. Estimate the cost of a similar house in 10 years. (*Hint:* Solve the equation $.43x = \$65,000$.)
   **(b)** Find the cost of a $20 textbook in 8 years.

## LIFE SCIENCES

**Number of Cells in an Organism**    *Generally speaking, the larger an organism, the greater its complexity (as measured by counting the number of different types of cells present). The figure shows an estimate of the maximum number of cells (or the largest volume) in various organisms plotted against the number of types of cells found in those organisms.\**

57. Use the graph to estimate the maximum number of cells and the corresponding volume for each of the following organisms.
    **(a)** Whale    **(b)** Sponge    **(c)** Green alga

58. From the graph, estimate the number of cell types in each of the following organisms.
    **(a)** Mushroom    **(b)** Kelp    **(c)** Sequoia

**Growth of Bacteria**    59. *Escherichia coli* is a strain of bacteria that occurs naturally in many different situations. Under certain conditions, the number of these bacteria present in a colony is given by

$$E(t) = E_0 \cdot 2^{t/30},$$

where $E(t)$ is the number of bacteria present $t$ minutes after the beginning of an experiment, and $E_0$ is the number present when $t = 0$. Let $E_0 = 1,000,000$, and use a calculator with a $y^x$ key to find the number of bacteria at the following times.

**(a)** $t = 5$    **(b)** $t = 10$    **(c)** $t = 15$    **(d)** $t = 20$
**(e)** $t = 30$    **(f)** $t = 60$    **(g)** $t = 90$    **(h)** $t = 120$

**Growth of Species**    60. Under certain conditions, the number of individuals of a species that is newly introduced into an area can double every year. That is, if $t$ represents the number of years since the species was introduced into the area, and $y$ represents the number of individuals, then

$$y = 6 \cdot 2^t$$

if 6 animals were introduced into the area originally.

---

*\*From *On Size and Life* by Thomas A. McMahon and John Tyler Bonner. Copyright © 1983 by Thomas A. McMahon and John Tyler Bonner. Reprinted by permission of W. H. Freeman and Company.*

**(a)** Complete the following table.

| t | 0 | 1 | 2 | 3 | 4 | 5 | 6 | 7 | 8 | 9 | 10 |
|---|---|---|---|---|---|---|---|---|---|---|----|
| y | 6 | | | | | 192 | | | | | 6144 |

**(b)** Graph $y = 6 \cdot 2^t$.

## SOCIAL SCIENCES

**City Planning**   **61.** City planners have determined that the population of a city is

$$P(t) = 1,000,000(2^{.2t}),$$

where $t$ represents time measured in years.
Find each of the following values.

**(a)** $P(0)$   **(b)** $P\left(\dfrac{5}{2}\right)$   **(c)** $P(5)$   **(d)** $P(10)$

**(e)** Graph $P(t)$.

## PHYSICAL SCIENCES

**Radioactive Decay**   **62.** Suppose the quantity in grams of a radioactive substance present at time $t$ is

$$Q(t) = 500(3^{-.5t}),$$

where $t$ is measured in months.
Find the quantity present at each of the following times.

**(a)** $t = 0$   **(b)** $t = 2$   **(c)** $t = 4$   **(d)** $t = 10$

**(e)** Graph $Q(t)$.

## 4.2   LOGARITHMIC FUNCTIONS

Suppose the general level of inflation in the economy averages 5% per year. The number of years it will take for prices to double under these conditions is called the **years to double.** For $1 to double (become $2) in $n$ years, assuming annual compounding means that

$$A = P\left(1 + \frac{i}{m}\right)^{mn}$$

becomes

$$2 = 1\left(1 + \frac{.05}{1}\right)^{1(n)}$$

or

$$2 = (1.05)^n.$$

This equation would be easier to solve if the variable were not in the exponent. *Logarithms* are defined for just this purpose.

**DEFINITION OF LOGARITHM**

For $a > 0$ and $a \neq 1$,
$$y = \log_a x \quad \text{means} \quad a^y = x.$$

(Read $y = \log_a x$ as "$y$ is the logarithm of $x$ to the base $a$.") For example, the exponential statement $2^4 = 16$ can be translated into the logarithmic statement $4 = \log_2 16$. Also, in the problem discussed above $(1.05)^n = 2$ can be rewritten with this definition as $n = \log_{1.05} 2$. Think of a logarithm as an exponent: $\log_a x$ is the exponent used with the base $a$ to get $x$.

### ■■ EXAMPLE 1

This example shows the same statements written in both exponential and logarithmic forms.

| *Exponential Form* | *Logarithmic Form* |
|---|---|
| **(a)** $3^2 = 9$ | $\log_3 9 = 2$ |
| **(b)** $(1/5)^{-2} = 25$ | $\log_{1/5} 25 = -2$ |
| **(c)** $10^5 = 100{,}000$ | $\log_{10} 100{,}000 = 5$ |
| **(d)** $4^{-3} = 1/64$ | $\log_4 1/64 = -3$ |
| **(e)** $2^{-4} = 1/16$ | $\log_2 1/16 = -4$ |
| **(f)** $e^0 = 1$ | $\log_e 1 = 0$ ■■ |

For a given positive value of $x$, the definition of logarithm leads to exactly one value of $y$, making $y = \log_a x$ the *logarithmic function* of base $a$ (the base $a$ must be positive, with $a \neq 1$).

**LOGARITHMIC FUNCTION**

If $a > 0$ and $a \neq 1$, then the logarithmic function of base $a$ is
$$f(x) = \log_a x.$$

The graphs of the exponential function $f(x) = 2^x$ and the logarithmic function $g(x) = \log_2 x$ are shown in Figure 5. The graphs show that $f(3) = 2^3 = 8$, while $g(8) = \log_2 8 = 3$. Thus, $f(3) = 8$ and $g(8) = 3$. Also, $f(2) = 4$ and $g(4) = 2$. In fact, for any number $m$, if $f(m) = p$, then $g(p) = m$. Functions related in this way are called *inverses* of each other. The graphs also show that the domain of the exponential function (the set of real numbers) is the range of the logarithmic function. Also, the range of the exponential function (the set of positive real numbers) is the domain of the logarithmic function. Every logarithmic function is the inverse of some exponential function. This means that we can graph logarithmic functions by rewriting them as exponential functions using the definition of logarithm. The graphs in Figure 5 show a characteristic of inverse functions: their graphs are mirror images about the line $y = x$. A more complete discussion of inverse functions is given in most standard intermediate algebra and college algebra books.

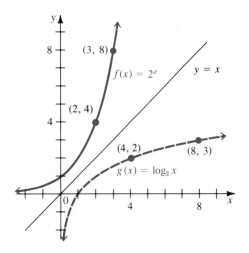

**FIGURE 5**

Plotting points and connecting them with a smooth curve gives the graph of $y = \log_{1/2} x$ shown in Figure 6, which also includes the graph of $y = (1/2)^x$. Again, the logarithmic and exponential graphs are mirror images with respect to the line $y = x$.

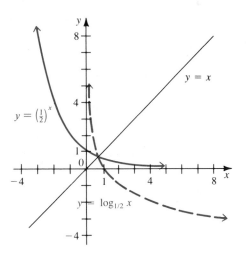

**FIGURE 6**

The graph of $y = \log_2 x$ shown in Figure 5 is typical of the graphs of logarithmic functions $y = \log_a x$, where $a > 1$. The graph of $y = \log_{1/2} x$, shown in Figure 6, is typical of logarithmic functions of the form $y = \log_a x$, where $0 < a < 1$. For both logarithmic graphs, the y-axis is a vertical asymptote.

***Properties of Logarithms*** The usefulness of logarithmic functions depends in large part on the following **properties of logarithms.**

PROPERTIES OF
LOGARITHMS

> Let $x$ and $y$ be any positive real numbers and $r$ be any real number. Let $a$ be a positive real number, $a \neq 1$. Then
>
> (a) $\log_a xy = \log_a x + \log_a y;$     Product rule
>
> (b) $\log_a \dfrac{x}{y} = \log_a x - \log_a y;$     Quotient rule
>
> (c) $\log_a x^r = r \log_a x;$     Power rule
>
> (d) $\log_a a = 1;$
>
> (e) $\log_a 1 = 0.$

To prove property (a), let $m = \log_a x$ and $n = \log_a y$. Then, by the definition of logarithm,

$$a^m = x \quad \text{and} \quad a^n = y.$$

Hence,

$$a^m a^n = xy.$$

By a property of exponents, $a^m a^n = a^{m+n}$, so

$$a^{m+n} = xy.$$

Now use the definition of logarithm to write

$$\log_a xy = m + n.$$

Since $m = \log_a x$ and $n = \log_a y$,

$$\log_a xy = \log_a x + \log_a y.$$

Proofs of parts (b) and (c) are left for the exercises. Parts (d) and (e) depend on the definition of a logarithm.

### ■ EXAMPLE 2

If all the following variable expressions represent positive numbers, then for $a > 0$, $a \neq 1$, the statements in (a)–(c) are true.

(a) $\log_a x + \log_a (x - 1) = \log_a x(x - 1)$

(b) $\log_a \dfrac{x^2 - 4x}{x + 6} = \log_a (x^2 - 4x) - \log_a (x + 6)$

(c) $\log_a (9x^5) = \log_a 9 + \log_a(x^5) = \log_a 9 + 5 \cdot \log_a x$     ■

The invention of logarithms is credited to John Napier (1550–1617), who first called logarithms "artificial numbers." Later he joined the Greek words *logos* (ratio) and *arithmos* (number) to form the word used today. The development of logarithms was motivated by a need for faster computation. Tables of logarithms and slide rule devices were developed by Napier, Henry Briggs (1561–1631), Edmund Gunter (1581–1626), and others.

Historically, one of the main applications of logarithms has been as an aid to numerical calculation. The properties of logarithms and a table of logarithms were used to simplify many numerical problems. With today's widespread use of calculators, this application of logarithms has declined. Since our number system has base 10, logarithms to base 10 were most convenient for numerical calculations, so base 10 logarithms were called **common logarithms.** Common logarithms are still useful in other applications. For simplicity,

$$\log_{10} x \quad \text{is abbreviated} \quad \log x.$$

With this notation,

$$\log 1000 = \log 10^3 \ = 3,$$
$$\log 100 = \log 10^2 \ = 2,$$
$$\log 1 = \log 10^0 \ = 0,$$
$$\log .01 = \log 10^{-2} = -2, \qquad \text{and so on.}$$

British mathematician Henry Briggs (mentioned above) is credited with the development of common logarithms. In his honor they are sometimes called Briggsian logarithms.

One application of common logarithms is their use in chemistry to define the pH of a solution as

$$\text{pH} = -\log [H_3O^+],$$

where $[H_3O^+]$ is the hydronium ion concentration in moles per liter. The number pH is a measure of the acidity or alkalinity of solutions. Pure water has a pH of 7.0, with pH values greater than 7.0 indicating alkalinity and values less than 7.0 indicating acidity.

Most practical applications of logarithms use the number $e$ as base. (Recall that to seven decimal places, $e = 2.7182818$.) Logarithms to base $e$ are called **natural logarithms,** and

$$\log_e x \quad \text{is abbreviated} \quad \ln x.$$

Although common logarithms may seem more "natural" than logarithms to base $e$, there are several good reasons for using natural logarithms instead. The most important reason is discussed in Section 4 of this chapter.

A table of natural logarithms is given in the back of the book. From this table, for example,

$$\ln 55 = 4.0073$$
$$\ln 1.9 = 0.6419$$

and

$$\ln .4 = -.9163.$$

■■■ EXAMPLE 3

Use a calculator or the table of natural logarithms to find the following logarithms.

**(a)** ln 85

With a calculator, simply enter 85, then press the ln key, and read the result, 4.4427.

The table of natural logarithms gives ln 85 = 4.4427, the same as the answer found when using a calculator. The results from using a table or a calculator may differ sometimes due to rounding error.

**(b)** ln 36

A calculator gives ln 36 = 3.5835. To use the table, first use properties of logarithms, since 36 is not listed.

$$\ln 36 = \ln 6^2 = 2 \ln 6 \approx 2(1.7918) = 3.5836$$

Alternatively, find ln 36 as follows:

$$\ln 36 = \ln 9 \cdot 4 = \ln 9 + \ln 4 = 2.1972 + 1.3863 = 3.5835. \quad ■■■$$

Just as any exponential function can be expressed with base $e$, a logarithmic function to any base can be expressed as a natural (base $e$) logarithmic function. The following theorem is used to change logarithms from one base to another.

**CHANGE-OF-BASE THEOREM**

If $x$ is any positive number and if $a$ and $b$ are positive real numbers, $a \neq 1, b \neq 1$, then

$$\log_a x = \frac{\log_b x}{\log_b a}.$$

To prove this result, use the definition of logarithm to write $y = \log_a x$ as $x = a^y$ or $x = a^{\log_a x}$ (for positive $x$ and positive $a$, $a \neq 1$). Now take base $b$ logarithms of both sides of this last equation. This gives

$$\log_b x = \log_b a^{\log_a x}$$

or

$$\log_b x = (\log_a x)(\log_b a),$$

from which

$$\log_a x = \frac{\log_b x}{\log_b a}.$$

If the base $b$ is equal to $e$, then by the change-of-base theorem,

$$\log_a x = \frac{\log_e x}{\log_e a}.$$

Using ln $x$ for $\log_e x$ gives the special case of the theorem using natural logarithms.

For any positive numbers $a$ and $x$, $a \neq 1$,

$$\log_a x = \frac{\ln x}{\ln a}.$$

## ■ EXAMPLE 4

Use natural logarithms to find each of the following. Round to the nearest hundredth.

**(a)** $\log_5 27$

Let $x = 27$ and $a = 5$. Using the second form of the theorem gives

$$\log_5 27 = \frac{\ln 27}{\ln 5}.$$

Now use a calculator or the table of natural logarithms.

$$\log_5 27 \approx \frac{3.2958}{1.6094} \approx 2.05$$

To check, use a calculator with a $y^x$ key, along with the definition of logarithm, to verify that $5^{2.05} \approx 27$.

**(b)** $\log_2 .1$

$$\log_2 .1 = \frac{\ln .1}{\ln 2} \approx \frac{-2.3026}{.6931} = -3.32 \quad \blacksquare$$

Equations involving logarithms are often solved by using the fact that exponential functions and logarithmic functions are inverses, so a logarithmic equation can be rewritten (with the definition of logarithm) as an exponential equation. In other cases, the properties of logarithms may be useful in simplifying an equation involving logarithms.

## ■ EXAMPLE 5

Solve each of the following equations for $x$.

**(a)** $\log_x \dfrac{8}{27} = 3$

First, use the definition of logarithm and write the expression in exponential form. To solve for $x$, take the cube root on both sides.

$$x^3 = \frac{8}{27}$$

$$x = \frac{2}{3}$$

**(b)** $\log_4 x = \dfrac{5}{2}$

In exponential form, the given statement becomes

$$4^{5/2} = x$$
$$(4^{1/2})^5 = x$$
$$2^5 = x$$
$$32 = x.$$

**(c)** $\log_2 x - \log_2 (x-1) = 1$

By a property of logarithms,

$$\log_2 x - \log_2 (x - 1) = \log_2 \frac{x}{x - 1},$$

so the original equation becomes

$$\log_2 \frac{x}{x - 1} = 1.$$

Use the definition of logarithms to write this result in exponential form:

$$\frac{x}{x - 1} = 2^1 = 2.$$

Solve this equation:

$$\frac{x}{x - 1} (x - 1) = 2(x - 1)$$
$$x = 2(x - 1)$$
$$x = 2x - 2$$
$$-x = -2$$
$$x = 2.$$

It is important to check solutions when solving equations involving logarithms because $\log_a u$, where $u$ is an expression in $x$, has domain $u > 0$.  ■

In the previous section exponential equations like $(1/3)^x = 81$ were solved by writing each side of the equation as a power of 3. That method cannot be used to solve an equation such as $3^x = 5$, however, since 5 cannot easily be written as a power of 3. A more general method for solving these equations depends on the following property of logarithms, which is supported by the graphs of logarithmic functions (Figures 5 and 6).

For $x > 0$, $y > 0$, $b > 0$, and $b \neq 1$,

$$\text{if} \qquad x = y, \qquad \text{then} \quad \log_b x = \log_b y,$$
$$\text{and} \qquad \text{if} \quad \log_b x = \log_b y, \qquad \text{then} \qquad x = y.$$

## EXAMPLE 6

Solve $3^{2x} = 4^{x+1}$.

Taking natural logarithms (logarithms to any base could be used) on both sides gives

$$\ln 3^{2x} = \ln 4^{x+1}.$$

Now use the power rule for logarithms.

$$2x \ln 3 = (x + 1)\ln 4$$
$$(2 \ln 3)x = (\ln 4)x + \ln 4$$
$$(2 \ln 3)x - (\ln 4)x = \ln 4$$
$$(2 \ln 3 - \ln 4)x = \ln 4$$
$$x = \frac{\ln 4}{2 \ln 3 - \ln 4}$$

Use the table or a calculator to evaluate the logarithms, then divide, to get

$$x \approx \frac{1.3863}{2(1.0986) - 1.3863} \approx 1.710. \quad \blacksquare$$

The final examples in this section illustrate the use of logarithms in practical problems.

## EXAMPLE 7

Complete the solution of the problem posed at the beginning of this section.

Recall that if prices will double after $n$ years at an inflation level of 5%, compounded annually, $n$ is given by the equation

$$2 = (1.05)^n.$$

We can solve this equation by first taking natural logarithms on both sides.

$$\ln 2 = \ln (1.05)^n$$
$$\ln 2 = n \ln 1.05 \qquad \text{Property of logarithms}$$
$$n = \frac{\ln 2}{\ln 1.05} \approx 14.2$$

It will take about 14 years for prices to double.   $\blacksquare$

The problem solved in Example 7 can be generalized for the compound interest equation

$$A = P(1 + i)^n.$$

Solving for $n$ as in Example 7 gives the doubling time in years as

$$n = \frac{\ln 2}{\ln (1 + i)}.$$

It can be shown that

$$n = \frac{\ln 2}{\ln (1 + i)} \approx \frac{.693}{i},$$

and that

$$\frac{70}{100i} \leq \frac{\ln 2}{\ln (1 + i)} \leq \frac{72}{100i}.$$

The **rule of 70** says that for $.001 \leq i \leq .05$, $70/100i$ gives a good approximation of $n$. The **rule of 72** says that for $.05 \leq i \leq .12$, $72/100i$ approximates $n$ quite well.

### ▆▆ EXAMPLE 8

Approximate the years to double at an interest rate of 6% using first the rule of 70, then the rule of 72.

By the rule of 70, money will double at 6% interest after

$$\frac{70}{100i} = \frac{70}{100(.06)} = \frac{70}{6} = 11.67 \left( \text{or } 11\frac{2}{3} \right)$$

years.

Using the rule of 72 gives

$$\frac{72}{100i} = \frac{72}{6} = 12$$

years doubling time. Since a more precise answer is given by

$$\frac{\ln 2}{\ln(1 + i)} = \frac{\ln 2}{\ln(1.06)} = \frac{.693}{.058} \approx 11.9,$$

the rule of 72 gives a better approximation than the rule of 70. This agrees with the statement that the rule of 72 works well for values of $i$ where $.05 \leq i \leq .12$, since $i = .06$ falls into this category.  ▆▆

### ▆▆ EXAMPLE 9

One measure of the diversity of the species in an ecological community is given by the *index of diversity H*, where

$$H = \frac{-1}{\ln 2} [P_1 \ln P_1 + P_2 \ln P_2 + \ldots + P_n \ln P_n],$$

and $P_1, P_2, \ldots, P_n$ are the proportions of a sample belonging to each of $n$ species found in the sample. For example, in a community with two species, where there are 90 of one species and 10 of the other, $P_1 = 90/100 = .9$ and $P_2 = 10/100 = .1$, with

$$H = \frac{-1}{\ln 2} [.9 \ln .9 + .1 \ln .1].$$

From the table or a calculator,

$$\ln 2 = .6931, \quad \ln .9 = -.1054, \quad \text{and} \quad \ln .1 = -2.3026.$$

Therefore,

$$H \approx \frac{-1}{.6931} [(.9)(-.1054) + (.1)(-2.3026)] \approx .469.$$

Verify that $H \approx .971$ if there are 60 of one species and 40 of the other and that $H = 1$ if there are 50 of each species. As the index of diversity gets close to 1, there is more diversity, with "perfect" diversity equal to 1.  ■

## 4.2 EXERCISES

*Write each exponential equation in logarithmic form.*

**1.** $2^3 = 8$

**2.** $5^2 = 25$

**3.** $3^4 = 81$

**4.** $6^3 = 216$

**5.** $\left(\frac{1}{3}\right)^{-2} = 9$

**6.** $\left(\frac{3}{4}\right)^{-2} = \frac{16}{9}$

*Write each logarithmic equation in exponential form.*

**7.** $\log_2 128 = 7$

**8.** $\log_3 81 = 4$

**9.** $\log_{25} \frac{1}{25} = -1$

**10.** $\log_2 \frac{1}{8} = -3$

**11.** $\log 10{,}000 = 4$

**12.** $\log .00001 = -5$

*Evaluate each logarithm.*

**13.** $\log_5 25$

**14.** $\log_9 81$

**15.** $\log_4 64$

**16.** $\log_6 216$

**17.** $\log_2 \frac{1}{4}$

**18.** $\log_3 \frac{1}{27}$

**19.** $\log_2 \sqrt[3]{\frac{1}{4}}$

**20.** $\log_8 \sqrt[4]{\frac{1}{2}}$

**21.** Complete the following table of values for $y = \log_3 x$.

| $x$ | 1/27 | | | | | | 9 |
|---|---|---|---|---|---|---|---|
| $y$ | $-3$ | $-2$ | $-1$ | 0 | 1 | 2 | 3 |

Graph the function using the same scale on both axes.

**22.** Complete the following table of values for $y = 3^x$.

| $x$ | $-3$ | $-2$ | $-1$ | 0 | 1 | 2 | 3 |
|---|---|---|---|---|---|---|---|
| $y$ | | 1/9 | | | | | 27 |

Graph the function on the same axes you used in Exercise 21. Compare the two graphs. How are they related?

*Graph each of the following.*

**23.** $y = \log_4 x$

**24.** $y = \log x$

**25.** $y = \log_{1/3} (x - 1)$

**26.** $y = \log_{1/2} (1 + x)$

**27.** $y = \log_2 x^2$

**28.** $y = \log_2 |x|$

*Use the properties of logarithms to write each expression as a sum, difference, or product.*

**29.** $\log_9 (7m)$

**30.** $\log_5 (8p)$

**31.** $\log_3 \dfrac{3p}{5k}$

**32.** $\log_7 \dfrac{11p}{13y}$

**33.** $\log_3 \dfrac{5\sqrt{2}}{\sqrt[4]{7}}$

**34.** $\log_2 \dfrac{9\sqrt[3]{5}}{\sqrt[4]{3}}$

*Suppose $\log_b 2 = a$ and $\log_b 3 = c$. Use the properties of logarithms to find each of the following.*

**35.** $\log_b 8$   **36.** $\log_b 24$   **37.** $\log_b 54$   **38.** $\log_b 144$   **39.** $\log_b (72b)$   **40.** $\log_b (4b^2)$

*Find the following natural logarithms.*

**41.** $\ln 20$   **42.** $\ln 35$   **43.** $\ln 800$   **44.** $\ln 920$   **45.** $\ln 1800$

**46.** $\ln 250$   **47.** $\ln 55{,}000$   **48.** $\ln 12{,}000$   **49.** $\ln .45$   **50.** $\ln .39$

*Use natural logarithms to evaluate each logarithm to the nearest hundredth.*

**51.** $\log_5 20$

**52.** $\log_{12} 170$

**53.** $\log_{1.2} 5.5$

**54.** $\log_{2.8} .12$

**55.** $\log_{10} 420$

**56.** $\log_{10} .008$

*Solve each equation. Round decimal answers to the nearest hundredth.*

**57.** $\log_x 25 = -2$

**58.** $\log_x \dfrac{1}{16} = -2$

**59.** $\log_9 27 = m$

**60.** $\log_8 4 = z$

**61.** $\log_y 8 = \dfrac{3}{4}$

**62.** $\log_r 7 = \dfrac{1}{2}$

**63.** $\log_3 (5x + 1) = 2$

**64.** $\log_5 (9x - 4) = 1$

**65.** $\log_4 x - \log_4 (x + 3) = -1$

**66.** $\log_9 m - \log_9 (m - 4) = -2$

**67.** $3^x = 5$

**68.** $4^x = 12$

**69.** $e^{k-1} = 4$

**70.** $e^{2y} = 12$

**71.** $2e^{5a+2} = 8$

**72.** $10e^{3z-7} = 5$

**73.** $\left(1 + \dfrac{k}{2}\right)^4 = 8$

**74.** $\left(1 + \dfrac{a}{5}\right)^3 = 9$

**75.** Prove: $\log_a \dfrac{x}{y} = \log_a x - \log_a y$

**76.** Prove: $\log_a x^r = r \log_a x$

---

### ▤ APPLICATIONS

#### BUSINESS AND ECONOMICS

**Inflation**   **77.** Assuming annual compounding, find the time it would take for the general level of prices in the economy to double at the following average inflation rates.

(a) 3%   (b) 6%   (c) 8%

(d) Check your answers using either the rule of 70 or the rule of 72, whichever applies.

**Compound Interest** **78.** Elena Jorgensen invests $15,000 in an account paying 6% per year compounded semiannually.

(a) In how many years will the compounded amount double?

(b) In how many years will the amount triple?

(c) Check your answer to part (a) using the rule of 72.

## LIFE SCIENCES

**Index of Diversity** *For Exercises 79 and 80 refer to Example 9.*

**79.** Suppose a sample of a small community shows two species with 50 individuals each. Find the index of diversity $H$.

**80.** A virgin forest in northwestern Pennsylvania has 4 species of large trees with the following proportions of each: hemlock, .521; beech, .324; birch, .081; maple, .074. Find the index of diversity $H$.

**Number of Species in a Sample** **81.** The number of species in a sample is given by

$$S(n) = a \ln \left( 1 + \frac{n}{a} \right).$$

Here $n$ is the number of individuals in the sample and $a$ is a constant that indicates the diversity of species in the community. If $a = .36$, find $S(n)$ for the following values of $n$.

(a) 100 (b) 200 (c) 150 (d) 10

**82.** In Exercise 81, find $n$ if $S(n) = 9$ and $a = .36$.

**Growth of a Species** **83.** The population of an animal species that is introduced into a certain area may grow rapidly at first but then grow more slowly as time goes on. A logarithmic function can provide an excellent model of such growth. Suppose that the population of foxes $F(t)$ in an area $t$ months after the foxes were introduced there is

$$F(t) = 500 \log (2t + 3).$$

Find the population of foxes at the following times.

(a) When they are first released into the area (that is, when $t = 0$)

(b) After 3 months

(c) After 15 months

(d) Graph $y = F(t)$.

*The graphs for Exercises 84–85\* are plotted on a* logarithmic scale *where differences between successive measurements are not always the same. Data that do not plot in a linear pattern on the usual Cartesian axes often form a linear pattern when plotted on a logarithmic scale. Notice that on the vertical scale, the distance from 1 to 2 is not the same as the distance from 2 to 3, and so on. This is characteristic of a graph drawn with logarithmic scales.*

**Weight Lifting** **84.** The accompanying graph gives the world weight-lifting records as $\log W_T$ plotted against the logarithm of body weight. Here $W_T$ is the total weight lifted in three lifts: the press, the snatch, and the clean-and-jerk. The numbers beside each point indicate the body weight class, in pounds.

---

\*From *On Size and Life* by Thomas A. McMahon and John Tyler Bonner. Copyright © 1983 by Thomas A. McMahon and John Tyler Bonner. Reprinted by permission of W. H. Freeman and Company.

(a) Find the records for a weight of 150 pounds and for a weight of 165 pounds. (Use base 10 logarithms.)

(b) Find the body weight that corresponds to a record of 750 pounds.

**Oxygen Consumption**  **85.** The accompanying graph gives the rate of oxygen consumption for resting guinea pigs of various sizes. This rate is proportional to body mass raised to the power .67. Estimate the oxygen consumption for a guinea pig with body mass of .3 kg. Do the same for one with body mass of .7.

**EXERCISE 84**

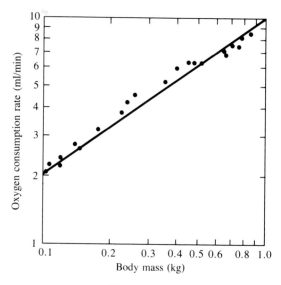

**EXERCISE 85**

## SOCIAL SCIENCES

**Evolution of Languages**  **86.** The number of years $N(r)$ since two independently evolving languages split off from a common ancestral language is approximated by

$$N(r) = -5000 \ln r,$$

where $r$ is the proportion of the words from the ancestral language that are common to both languages now.

Find each of the following.

(a) $N(.9)$    (b) $N(.5)$    (c) $N(.3)$

(d) How many years have elapsed since the split if 70% of the words of the ancestral language are common to both languages today?

(e) If two languages split off from a common ancestral language about 1000 years ago, find $r$.

## PHYSICAL SCIENCES

*For Exercises 87-90, recall that log x represents the common (base 10) logarithm of x.*

**Intensity of Sound**  **87.** The loudness of sounds is measured in a unit called a *decibel*. To do this, a very faint sound, called the *threshold sound*, is assigned an intensity $I_0$. If a particular sound has intensity $I$, then the decibel rating of this louder sound is

$$10 \cdot \log \frac{I}{I_0}.$$

Find the decibel ratings of sounds having the following intensities.

(a) $100I_0$    (b) $1000I_0$    (c) $100,000I_0$    (d) $1,000,000I_0$

**Intensity of Sound** 88. Find the decibel ratings of the following sounds having intensities as given. Round answers to the nearest whole number.

(a) Whisper, $115I_0$

(b) Busy street, $9,500,000I_0$

(c) Heavy truck, 20 m away, $1,200,000,000I_0$

(d) Rock music, $895,000,000,000I_0$

(e) Jetliner at takeoff, $109,000,000,000,000I_0$

**Earthquake Intensity** 89. The intensity of an earthquake, measured on the Richter scale, is given by

$$\log \frac{I}{I_0},$$

where $I_0$ is the intensity of an earthquake of a certain (small) size. Find the Richter scale ratings of earthquakes with the following intensities.

(a) $1000I_0$    (b) $1,000,000I_0$    (c) $100,000,000I_0$

**Earthquake Intensity** 90. The San Francisco earthquake of 1906 had a Richter scale rating of 8.6. Use a calculator with a $y^x$ key to express the intensity of this earthquake as a multiple of $I_0$. (See Exercise 89.)

## 4.3 DERIVATIVES OF LOGARITHMIC FUNCTIONS

In this section formulas will be developed for the derivatives of $y = \ln x$ and other logarithmic functions. To find the derivative of the function $y = \ln x$, assuming $x > 0$, go back to the definition of the derivative given earlier: if $y = f(x)$, then

$$y' = f'(x) = \lim_{h \to 0} \frac{f(x + h) - f(x)}{h},$$

provided this limit exists. (Remember that $h$ represents the variable in this expression, while $x$ is constant.)

Here, $f(x) = \ln x$, with

$$y' = f'(x) = \lim_{h \to 0} \frac{\ln (x + h) - \ln x}{h}.$$

This limit can be found using properties of logarithms. By the quotient rule for logarithms,

$$y' = \lim_{h \to 0} \frac{\ln (x + h) - \ln x}{h} = \lim_{h \to 0} \frac{\ln \left( \dfrac{x + h}{x} \right)}{h} = \lim_{h \to 0} \left[ \frac{1}{h} \ln \left( \frac{x + h}{x} \right) \right].$$

The power rule for logarithms is used to get

$$y' = \lim_{h \to 0} \ln \left( \frac{x + h}{x} \right)^{1/h} = \lim_{h \to 0} \ln \left( 1 + \frac{h}{x} \right)^{1/h}.$$

Now make a substitution: let $m = x/h$, so that $h/x = 1/m$. As $h \to 0$, with $x$ fixed, $m \to +\infty$ (since $x > 0$), so

$$\lim_{h \to 0} \ln \left( 1 + \frac{h}{x} \right)^{1/h} = \lim_{m \to +\infty} \ln \left( 1 + \frac{1}{m} \right)^{m/x}$$

$$= \lim_{m \to +\infty} \ln \left[ \left( 1 + \frac{1}{m} \right)^m \right]^{1/x}.$$

By the power rule for logarithms, this becomes

$$= \lim_{m \to +\infty} \left[ \frac{1}{x} \cdot \ln \left( 1 + \frac{1}{m} \right)^m \right].$$

Since $x$ is fixed here (a constant), by the product rule for limits, this last limit becomes

$$\lim_{m \to +\infty} \left[ \frac{1}{x} \cdot \ln \left( 1 + \frac{1}{m} \right)^m \right] = \left( \lim_{m \to +\infty} \frac{1}{x} \right) \lim_{m \to +\infty} \left[ \ln \left( 1 + \frac{1}{m} \right)^m \right]$$

$$= \frac{1}{x} \cdot \lim_{m \to +\infty} \left[ \ln \left( 1 + \frac{1}{m} \right)^m \right].$$

Since the logarithmic function is continuous,

$$\frac{1}{x} \cdot \lim_{m \to +\infty} \left[ \ln \left( 1 + \frac{1}{m} \right)^m \right] = \frac{1}{x} \cdot \ln \left[ \lim_{m \to +\infty} \left( 1 + \frac{1}{m} \right)^m \right].$$

In Section 1 of this chapter we saw that the limit on the right equals $e$, so that

$$\frac{1}{x} \cdot \ln \left[ \lim_{m \to +\infty} \left( 1 + \frac{1}{m} \right)^m \right] = \frac{1}{x} \cdot \ln e = \frac{1}{x} \cdot 1 = \frac{1}{x}.$$

Summarizing, if $y = \ln x$, then $y' = 1/x$.

## ■■■ EXAMPLE 1

Find the derivative of $y = \ln 6x$. Assume $x > 0$.

Use the properties of logarithms and the rules for derivatives.

$$y' = \frac{d}{dx} (\ln 6x)$$

$$= \frac{d}{dx} (\ln 6 + \ln x)$$

$$= \frac{d}{dx} (\ln 6) + \frac{d}{dx} (\ln x) = 0 + \frac{1}{x} = \frac{1}{x} \quad ■■$$

The chain rule can be used to find the derivative of the more general logarithmic function $y = \ln g(x)$. Recall the chain rule:

If $y = f[g(x)]$, then $y' = f'[g(x)] \cdot g'(x)$.

Let $y = f(u) = \ln u$ and $u = g(x)$, so that $f[g(x)] = \ln g(x)$. Then

$$f'[g(x)] = f'(u) = \frac{1}{u} = \frac{1}{g(x)},$$

and by the chain rule,

$$y' = f'[g(x)] \cdot g'(x)$$

$$= \frac{1}{g(x)} \cdot g'(x) = \frac{g'(x)}{g(x)}.$$

## ■ EXAMPLE 2

Find the derivative of $y = \ln (x^2 + 1)$.

Here $g(x) = x^2 + 1$ and $g'(x) = 2x$. Thus,

$$y' = \frac{g'(x)}{g(x)} = \frac{2x}{x^2 + 1}. \quad ■$$

If $y = \ln (-x)$, where $x < 0$, the chain rule with $g(x) = -x$ and $g'(x) = -1$ gives

$$y' = \frac{g'(x)}{g(x)} = \frac{-1}{-x} = \frac{1}{x}.$$

The derivative of $y = \ln (-x)$ is the same as the derivative of $y = \ln x$. For this reason, these two results can be combined into one rule using the absolute value of $x$. A similar situation holds true for $y = \ln [g(x)]$ and $y = \ln [-g(x)]$. These results are summarized below.

**DERIVATIVES OF $\ln |x|$ AND $\ln |g(x)|$**

$$\frac{d}{dx} \ln |x| = \frac{1}{x}$$

$$\frac{d}{dx} \ln |g(x)| = \frac{g'(x)}{g(x)}$$

## ■ EXAMPLE 3

Find the derivative of each function.

**(a)** $y = \ln |5x|$

Let $g(x) = 5x$, so that $g'(x) = 5$. From the formula above,

$$y' = \frac{g'(x)}{g(x)} = \frac{5}{5x} = \frac{1}{x}.$$

Notice that the derivative of $\ln |5x|$ is the same as the derivative of $\ln |x|$. Also, in Example 1, the derivative of $\ln 6x$ was the same as that for $\ln x$. This suggests that for any constant $a$,

$$\frac{d}{dx} \ln |ax| = \frac{d}{dx} \ln |x|$$

$$= \frac{1}{x}.$$

For a proof of this result, see Exercise 43.

**(b)** $y = \ln |3x^2 - 4x|$

$$y' = \frac{6x - 4}{3x^2 - 4x}$$

**(c)** $y = 3x \ln x^2$

This function is the product of the two functions $3x$ and $\ln x^2$, so use the product rule.

$$y' = (3x) \left[ \frac{d}{dx} \ln x^2 \right] + (\ln x^2) \left[ \frac{d}{dx} 3x \right]$$

$$= 3x \left( \frac{2x}{x^2} \right) + (\ln x^2)(3)$$

$$= 6 + 3 \ln x^2$$

By the power rule for logarithms,

$$y' = 6 + \ln (x^2)^3$$
$$= 6 + \ln x^6.$$

Alternatively, write the answer as $y' = 6 + 6 \ln x$. ■

## ▄▄ EXAMPLE 4

Find the derivative of $y = \log |3x + 2|$.

This is a base 10 logarithm, while the derivative rule developed above applies only to natural logarithms. To find the derivative, first use the change-of-base theorem to convert the function to one involving natural logarithms.

$$y = \log_{10} |3x + 2|$$

$$= \frac{\ln |3x + 2|}{\ln 10}$$

$$= \frac{1}{\ln 10} \ln |3x + 2|$$

Now find the derivative. (Remember: 1/ln 10 is a constant.)

$$y' = \frac{1}{\ln 10} \cdot \frac{d}{dx} [\ln |3x + 2|]$$

$$= \frac{1}{\ln 10} \cdot \frac{3}{3x + 2}$$

$$= \frac{3}{\ln 10 \, (3x + 2)} \quad \blacksquare$$

The procedure in Example 4 can be used to derive the following results.

**DERIVATIVES WITH OTHER BASES**

For any suitable value of $a$,

$$D_x(\log_a |x|) = \frac{1}{\ln a} \cdot \frac{1}{x}.$$

When $y = \log_a |g(x)|$, the chain rule gives

$$D_x(\log_a |g(x)|) = \frac{1}{\ln a} \cdot \frac{g'(x)}{g(x)}.$$

## ▰ EXAMPLE 5

Find any extrema or inflection points and sketch the graph of $y = (\ln x)/x^2$, $x > 0$.

To locate any extrema, begin by finding the first derivative.

$$y' = \frac{x^2(1/x) - 2x \ln x}{x^4}$$

$$= \frac{x - 2x \ln x}{x^4}$$

$$= \frac{x(1 - 2 \ln x)}{x^4}$$

$$= \frac{1 - 2 \ln x}{x^3}$$

The derivative function exists everywhere on the domain $(x > 0)$, so any maxima or minima will occur only at points where the derivative equals 0.

$$\frac{1 - 2 \ln x}{x^3} = 0$$

$$1 - 2 \ln x = 0$$

$$2 \ln x = 1$$

$$\ln x = \frac{1}{2}$$

Use the definition of logarithm to get the equivalent exponential statement

$$x = e^{1/2} \approx 1.65.$$

To locate inflection points and determine whether $x = e^{1/2}$ represents a maximum or a minimum, find the second derivative.

$$y'' = \frac{x^3(-2/x) - 3x^2(1 - 2 \ln x)}{x^6}$$

$$= \frac{-2x^2 - 3x^2 + 6x^2 \ln x}{x^6}$$

$$= \frac{-5x^2 + 6x^2 \ln x}{x^6}$$

$$= \frac{x^2(-5 + 6 \ln x)}{x^6}$$

$$= \frac{-5 + 6 \ln x}{x^4}$$

Use a calculator to show that the second derivative is negative for $x = e^{1/2}$ with a maximum where $x = e^{1/2}$. Set the second derivative equal to 0 and solve the resulting equation to identify any inflection points.

$$\frac{-5 + 6 \ln x}{x^4} = 0$$

$$-5 + 6 \ln x = 0$$

$$6 \ln x = 5$$

$$\ln x = 5/6$$

$$x = e^{5/6} \approx 2.30$$

Verify that $f''(1) < 0$ and $f''(3) > 0$, with an inflection point when $x = e^{5/6} \approx 2.3$. Also, verify that the graph is concave downward for $x < e^{5/6}$ and concave upward for $x > e^{5/6}$. Use this information and plot a few points as necessary to get the graph in Figure 7. ▬

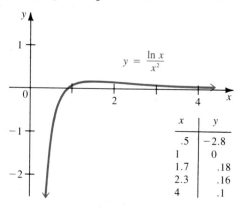

| x | y |
|-----|------|
| .5 | -2.8 |
| 1 | 0 |
| 1.7 | .18 |
| 2.3 | .16 |
| 4 | .1 |

**FIGURE 7**

## ▰ 4.3 EXERCISES

*Find the derivative of each of the functions defined as follows.*

**1.** $y = \ln |8x|$

**2.** $y = \ln |-4x|$

**3.** $y = \ln |3 - x|$

**4.** $y = \ln |1 + x^2|$

**5.** $y = \ln |2x^2 - 7x|$

**6.** $y = \ln |-8x^2 + 6x|$

**7.** $y = \ln \sqrt{x + 5}$

**8.** $y = \ln \sqrt{2x + 1}$

**9.** $y = \ln (x^4 + 5x^2)^{3/2}$

**10.** $y = \ln |(5x^3 - 2x)^{3/2}|$

**11.** $y = -3x \ln |x + 2|$

**12.** $y = (3x + 1) \ln |x - 1|$

**13.** $y = x^2 \ln |x|$

**14.** $y = x \ln |2 - x^2|$

**15.** $y = \dfrac{2 \ln |x + 3|}{x^2}$

**16.** $y = \dfrac{\ln |x|}{x^3}$

**17.** $y = \dfrac{\ln |x|}{4x + 7}$

**18.** $y = \dfrac{-2 \ln |x|}{3x - 1}$

**19.** $y = \dfrac{3x^2}{\ln |x|}$

**20.** $y = \dfrac{x^3 - 1}{2 \ln |x|}$

**21.** $y = (\ln |x + 1|)^4$

**22.** $y = \sqrt{\ln |x - 3|}$

**23.** $y = \ln |\ln |x||$

**24.** $y = (\ln 4)(\ln |3x|)$

**25.** $y = \log |6x|$

**26.** $y = \log |2x - 3|$

**27.** $y = \log |1 - x|$

**28.** $y = \log |-3x|$

**29.** $y = \log_5 \sqrt{5x + 2}$

**30.** $y = \log_7 \sqrt{2x - 3}$

**31.** $y = \log_3 |(x^2 + 2x)^{3/2}|$

**32.** $y = \log_2 |(2x^2 - x)^{5/2}|$

*Find all relative maxima or minima for the functions defined as follows. Sketch the graphs.*

**33.** $y = x \ln x, \quad x > 0$

**34.** $y = x - \ln x, \quad x > 0$

**35.** $y = x \ln |x|$

**36.** $y = x - \ln |x|$

**37.** $y = (\ln |x|)^2$

**38.** $y = (\ln |x|)^3$

**39.** $y = \dfrac{\ln x}{x}, \quad x > 0$

**40.** $y = \dfrac{\ln x^2}{x^2}$

**41.** What is true about the slope of the tangent line to the graph of $f(x) = \ln x$ as $x \to +\infty$? as $x \to 0$?

**42.** Let $f(x) = \ln x$.
   **(a)** Compute $f'(x), f''(x), f'''(x), f^{(4)}(x)$, and $f^{(5)}(x)$.
   **(b)** Guess a formula for $f^{(n)}(x)$, where $n$ is any positive integer.

**43.** Prove $\dfrac{d}{dx} \ln |ax| = \dfrac{d}{dx} \ln |x|$ for any constant $a$.

## ▰ APPLICATIONS

### BUSINESS AND ECONOMICS

**Profit**  **44.** Assume that the total revenue received from the sale of $x$ items is given by

$$R(x) = 30 \ln (2x + 1),$$

while the total cost to produce $x$ items is $C(x) = x/2$. Find the number of items that should be manufactured so that profit, $R(x) - C(x)$, is a maximum.

**Revenue**  **45.** Suppose the demand function for $x$ thousand of a certain item is

$$p = 100 + \frac{50}{\ln x}, \quad x > 1,$$

where $p$ is in dollars.
   **(a)** Find the marginal revenue.
   **(b)** Find the revenue from the 8001st item ($x = 8$).

**Profit**    **46.** If the cost function in dollars for $x$ thousand of the item in Exercise 45 is $C(x) = 100x + 100$, find the following.

(a) The marginal cost

(b) The profit function $P(x)$

(c) The profit from the 8001st item ($x = 8$).

**Marginal Revenue Product**    **47.** The demand function for $x$ units of a product is

$$p = 100 - 10 \ln x, \qquad 1 < x < 20,000,$$

where $x = 6n$ and $n$ is the number of employees producing the product.

(a) Find the revenue function $R(x)$.

(b) Find the marginal revenue product function. (See Example 11 in Section 6 of the chapter on the derivative.)

(c) Evaluate and interpret the marginal revenue product when $x = 20$.

## LIFE SCIENCES

**Insect Mating**    **48.** Consider an experiment in which equal numbers of male and female insects of a certain species are permitted to intermingle. Assume that

$$M(t) = (.1t + 1) \ln \sqrt{t}$$

represents the number of matings observed among the insects in an hour, where $t$ is the temperature in degrees Celsius. (Note: The formula is an approximation at best and holds only for specific temperature intervals.)

(a) Find $M(15)$.    (b) Find $M(25)$.

(c) Find the rate of change of $M(t)$ when $t = 15$.

**Population Growth**    **49.** Suppose that the population of a certain collection of rare Brazilian ants is given by

$$P(t) = (t + 100) \ln (t + 2),$$

where $t$ represents the time in days. Find the rates of change of the population when $t = 2$ and when $t = 8$.

## GENERAL

**Information Content**    **50.** Suppose dots and dashes are transmitted over a telegraph line so that dots occur a fraction $p$ of the time (where $0 \le p \le 1$) and dashes occur a fraction $1 - p$ of the time. The **information content** of the telegraph line is given by $I(p)$, where

$$I(p) = -p \ln p - (1 - p) \ln (1 - p).$$

(a) Show that $I'(p) = -\ln p + \ln (1 - p)$.

(b) Let $I'(p) = 0$ and find the value of $p$ that maximizes $I(p)$.

## FOR THE COMPUTER

*Graph the functions in Exercises 51 and 52 by plotting points in the given intervals.*

**51.** $y = x + \ln x$,    $(0, 9]$

**52.** $y = \ln x - \dfrac{1}{x}$,    $(0, 9]$

# 4.4 DERIVATIVES OF EXPONENTIAL FUNCTIONS

The derivatives of exponential functions can be found using the formula for the derivative of the natural logarithm function. To find the derivative of the function $y = e^x$, first take the natural logarithm of each side so that the power rule for logarithms can be used to get $x$ out of the exponent.

$$y = e^x$$
$$\ln y = \ln e^x$$

By the power rule for logarithms and the fact that $\ln e = 1$,

$$\ln y = x \cdot \ln e$$
$$\ln y = x.$$

Now, using implicit differentiation, take the derivative of each side. (Since $y$ is a function of $x$, we can use the chain rule.)

$$\frac{1}{y} \cdot \frac{dy}{dx} = 1$$
$$\frac{dy}{dx} = y$$

Since $y = e^x$,

$$\frac{d}{dx}(e^x) = e^x \quad \text{or} \quad y' = e^x.$$

This result shows one of the main reasons for the widespread use of $e$ as a base—the function $y = e^x$ is its own derivative. Furthermore, it is the *only* useful function with this property. For a more general result, let $y = e^u$ and $u = g(x)$ so that $y = e^{g(x)}$. Then

$$\frac{dy}{du} = e^u \quad \text{and} \quad \frac{du}{dx} = g'(x).$$

By the chain rule,

$$\frac{dy}{dx} = \frac{dy}{du} \cdot \frac{du}{dx}$$
$$= e^u \cdot g'(x)$$
$$= e^{g(x)} \cdot g'(x).$$

Now replace $y$ with $e^{g(x)}$ to get

$$\frac{d}{dx}e^{g(x)} = e^{g(x)} \cdot g'(x).$$

These results are summarized on the next page.

**DERIVATIVES OF $e^x$**
**AND $e^{g(x)}$**

$$\frac{d}{dx}e^x = e^x$$

$$\frac{d}{dx}e^{g(x)} = g'(x)e^{g(x)}$$

■ **EXAMPLE 1**

Find derivatives of the following functions.

**(a)** $y = e^{5x}$

Let $g(x) = 5x$, with $g'(x) = 5$. Then

$$y' = 5e^{5x}.$$

**(b)** $y = 3e^{-4x}$

$$y' = 3 \cdot e^{-4x}(-4) = -12e^{-4x}$$

**(c)** $y = 10e^{3x^2}$

$$y' = 10(e^{3x^2})(6x) = 60xe^{3x^2} \quad ■$$

■ **EXAMPLE 2**

Let $y = e^x(\ln |x|)$. Find $y'$.

Use the product rule.

$$y' = e^x\left(\frac{1}{x}\right) + (\ln |x|)e^x$$

$$y' = e^x\left(\frac{1}{x} + \ln |x|\right) \quad ■$$

■ **EXAMPLE 3**

Let $y = \dfrac{100,000}{1 + 100e^{-.3x}}$. Find $y'$.

Use the quotient rule.

$$y' = \frac{(1 + 100e^{-.3x})(0) - 100,000(-30e^{-.3x})}{(1 + 100e^{-.3x})^2}$$

$$= \frac{3,000,000e^{-.3x}}{(1 + 100e^{-.3x})^2} \quad ■$$

■ **EXAMPLE 4**

Find $D_x(3^x)$.

Change $3^x$ to $e^{kx}$. From the table of powers of $e$, $k = 1.10$ and $3^x \approx e^{1.10x}$,
so

$$D_x(3^x) \approx D_x(e^{1.10x}) = 1.10e^{1.10x}. \quad ■$$

In Example 4, we found $e^{1.10} \approx 3$. Taking natural logarithms on both sides gives

$$e^{1.10} \approx 3$$
$$\ln e^{1.10} \approx \ln 3$$
$$1.10 \ln e \approx \ln 3$$
$$1.10 \approx \ln 3.$$

By this result, the derivative of $f(x) = 3^x$ can also be written as

$$D_x(3^x) = (\ln 3)e^{x \ln 3}.$$

Extending this idea gives a general rule for the derivative of an exponential function with any appropriate base $a$.

$$D_x a^x = (\ln a)e^{x \ln a}$$
$$= (\ln a)e^{\ln a^x}$$
$$D_x a^x = (\ln a)a^x \qquad e^{\ln a^x} = a^x$$

Use the chain rule to get the following result.

**DERIVATIVE OF $a^{g(x)}$**

$$D_x[a^{g(x)}] = (\ln a)e^{(\ln a)g(x)}g'(x) = (\ln a)a^{g(x)}g'(x)$$

This formula and the formula

$$D_x \log_a|x| = \frac{1}{\ln a} \cdot \frac{1}{x}$$

are simplest when $a = e$, since $\ln e = 1$. This is a major reason for using base $e$ exponential and logarithmic functions.

■ **EXAMPLE 5**

The amount in grams of a radioactive substance present after $t$ years is given by

$$A(t) = 100e^{-.12t}.$$

**(a)** Find the rate of change of the amount present after 3 years.

The rate of change is given by the derivative $dy/dt$.

$$\frac{dy}{dt} = 100 (e^{-.12t})(-.12) = -12e^{-.12t}$$

After 3 years ($t = 3$), the rate of change is

$$\frac{dy}{dt} = -12e^{-.12(3)} = -12e^{-.36} \approx -8.4$$

grams per year.

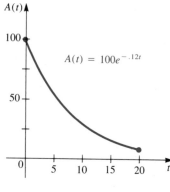

$A(t) = 100e^{-.12t}$

**FIGURE 8**

**(b)** In part (a) is the amount present increasing or decreasing when $t = 3$?

Since the rate of change is negative, the amount is decreasing.

**(c)** Graph $A(t)$ on $[0, 20]$.

From part (a), $dy/dt = -12e^{-.12t}$. Since $e^{-.12t}$ is never equal to 0, there is no maximum or minimum function value. Find the second derivative to determine any points of inflection and the concavity.

$$\frac{d^2y}{dt^2} = -12(e^{-.12t})(-.12) = 1.44e^{-.12t}$$

The second derivative is always positive, so the graph of $A$ is always concave upward. Plot a few points to get the graph shown in Figure 8. ■

### ■ EXAMPLE 6

Use the information given by the first and second derivatives to graph $f(x) = 2^{-x^2}$.

The first derivative is

$$f'(x) = \ln 2(2^{-x^2})(-2x)$$

$$= (-2x \ln 2)(2^{-x^2}) \qquad \text{or} \qquad \frac{-2x \ln 2}{2^{x^2}}.$$

Set $f'(x)$ equal to 0 and solve for $x$.

$$(-2x \ln 2)(2^{-x^2}) = 0$$
$$-2x \ln 2 = 0$$
$$x = 0$$

Note that $2^{x^2}$ can never be zero, so $x = 0$ is the only critical number. We can find the second derivative by using the product rule since $f'(x)$ is the product of two factors.

$$f''(x) = (-2x \ln 2)[\ln 2(2^{-x^2})(-2x)] + (2^{-x^2})(-2 \ln 2)$$
$$= 4x^2 (\ln 2)^2(2^{-x^2}) - 2 \ln 2(2^{-x^2})$$

Factoring out the common factors gives

$$f''(x) = (2 \ln 2)(2^{-x^2})(2x^2 \ln 2 - 1).$$

Set $f''(x)$ equal to 0 and solve for $x$ to find any inflection points.

$$2 \ln 2(2^{-x^2})(2x^2 \ln 2 - 1) = 0$$
$$2x^2 \ln 2 - 1 = 0$$
$$2x^2 \ln 2 = 1$$
$$x^2 = \frac{1}{2 \ln 2}$$
$$x = \pm \sqrt{\frac{1}{2 \ln 2}} \approx \pm .85$$

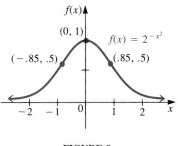

**FIGURE 9**

Test a value from each of the intervals

$$\left(-\infty, -\sqrt{\frac{1}{2\ln 2}}\right), \quad \left(-\sqrt{\frac{1}{2\ln 2}}, \sqrt{\frac{1}{2\ln 2}}\right), \quad \text{and} \quad \left(\sqrt{\frac{1}{2\ln 2}}, +\infty\right)$$

to verify that there are inflection points at $x \approx -.85$ and $x \approx .85$. Use the second derivative test to verify that there is a maximum of 1 at $x = 0$. Plot the points where $x \approx -.85$, $x = 1$, and $x \approx .85$, and use information about concavity from testing the intervals given above to get the graph in Figure 9. ∎

Frequently a population, or the sales of a certain product, will start growing slowly, then grow more rapidly, and then gradually level off. Such growth can often be approximated by a mathematical model of the form.

$$f(x) = \frac{b}{1 + ae^{kx}}$$

for appropriate constants $a$, $b$, and $k$.

### ▰▰ EXAMPLE 7

Suppose that the sales of a new product can be approximated for its first few years on the market by the function

$$S(x) = \frac{100,000}{1 + 100e^{-.3x}},$$

where $x$ is time in years since the introduction of the product.

**(a)** Find the rate of change of sales when $x = 4$.

The derivative of this sales function, which gives the rate of change of sales, was found in Example 3. Using that derivative,

$$S'(4) = \frac{3,000,000e^{-.3(4)}}{[1 + 100e^{-.3(4)}]^2} = \frac{3,000,000e^{-1.2}}{(1 + 100e^{-1.2})^2}.$$

By using a calculator or the table of powers of $e$, $e^{-1.2} \approx .301$, and

$$S'(4) \approx \frac{3,000,000(.301)}{[1 + 100(.301)]^2}$$

$$\approx \frac{903,000}{(1 + 30.1)^2}$$

$$= \frac{903,000}{967.21} \approx 934.$$

The rate of change of sales at time $x = 4$ is about 934 units per year. The positive number indicates that sales are increasing at this time.

**(b)** Graph $S(x)$.

Use the first derivative to check for extrema. Set $S'(x) = 0$ and solve for $x$.

$$S'(x) = \frac{3{,}000{,}000e^{-.3x}}{(1 + 100e^{-.3x})^2} = 0$$

$$3{,}000{,}000e^{-.3x} = 0$$

$$\frac{3{,}000{,}000}{e^{.3x}} = 0$$

The expression on the left can never equal 0, so $S'(x) = 0$ leads to no extrema. Next, find any critical values for which $S'(x)$ does not exist, that is, values that make the denominator equal 0.

$$1 + 100e^{-.3x} = 0$$

$$100e^{-.3x} = -1$$

$$\frac{100}{e^{.3x}} = -1$$

Since the expression on the left is always positive, it cannot equal $-1$. There are no critical values and no extrema. Further, $e^{-.3x}$ is always positive, so $S'(x)$ is always positive. This means the graph is always increasing over the domain of the function. Use the second derivative to verify that the graph has an inflection point at approximately 15.4 and is concave upward on $(-\infty, 15.4)$ and concave downward on $(15.4, +\infty)$. Plotting a few points leads to the graph in Figure 10.   ▰

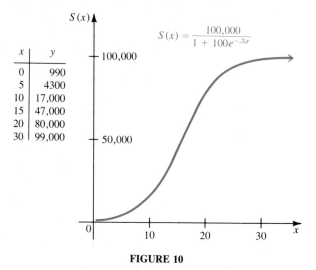

| $x$ | $y$ |
|----|--------|
| 0  | 990    |
| 5  | 4300   |
| 10 | 17,000 |
| 15 | 47,000 |
| 20 | 80,000 |
| 30 | 99,000 |

$$S(x) = \frac{100{,}000}{1 + 100e^{-.3x}}$$

**FIGURE 10**

Environmental situations place effective limits on the population growth of an organism in an area. Many such limited growth situations are described by the **logistic function,**

$$G(t) = \frac{mG_0}{G_0 + (m - G_0)e^{-kmt}},$$

where $t$ represents time in appropriate units, $G_0$ is the initial number present, $m$ is the maximum possible size of the population, and $k$ is a positive constant.

### ■ EXAMPLE 8

The growth of a population of rare South American insects is given by the logistic function with $k = .00001$ and $t$ in months. Assume that there are 200 insects initially and that the maximum population size is 10,000.

**(a)** Find the growth function $G(t)$ for these insects.

Substitute the given values for $k$, $G_0$, and $m$ into the logistic function to get

$$G(t) = \frac{mG_0}{G_0 + (m - G_0)e^{-kmt}}$$

$$= \frac{(10,000)(200)}{200 + (10,000 - 200)e^{-(.00001)(10,000)t}}$$

$$= \frac{2,000,000}{200 + 9800e^{-.1t}}.$$

**(b)** Find $G(5)$ and $G(10)$.

$$G(5) = \frac{2,000,000}{200 + 9800e^{-.5}} = 325.5$$

There are about 326 insects after 5 months.

$$G(10) = \frac{2,000,000}{200 + 9800e^{-1}} = 525.6$$

There are about 526 insects after 10 months.

**(c)** Find any maxima or minima for the function and interpret the results.

For

$$G(t) = \frac{2,000,000}{200 + 9800e^{-.1t}},$$

use the quotient rule as follows.

$$G'(t) = \frac{(200 + 9800e^{-.1t})0 - 2,000,000(-980e^{-.1t})}{(200 + 9800e^{-.1t})^2}$$

$$= \frac{1,960,000,000e^{-.1t}}{(200 + 9800e^{-.1t})^2}$$

Find $t$ such that $G'(t) = 0$ by setting the numerator equal to 0 and solving the equation for $t$. The equation

$$1,960,000,000e^{-.1t} = 0$$

or

$$e^{-.1t} = 0$$

has no solution, so there are no maxima nor minima. Since both the numerator and denominator of $G'(t)$ are always positive, $G'(t)$ is always positive and therefore $G(t)$ is an increasing function.

**(d)** Graph $G(t)$ for $t \geq 0$. There is an inflection point at $t = 38.9$, with $G''(t) > 0$ on the interval $(0, 38.9)$ and $G''(t) < 0$ on the interval $(38.9, +\infty)$. What happens to the population at the inflection point?

The graph is shown in Figure 11. From the given information, the graph is concave upward on the interval $(0, 38.9)$ and concave downward for $t$ in $(38.9, +\infty)$. Compare this with the graph in Figure 10. At the inflection point the growth rate (given by $G'(t)$) changes from an increasing rate to a decreasing rate. This means the population grows quite rapidly for about 39 months (3¼ years) and then grows at a slower and slower pace as the population approaches 10,000. ■

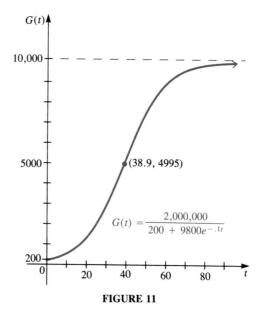

**FIGURE 11**

*Find derivatives of the functions defined as follows.*

**1.** $y = e^{4x}$
**2.** $y = e^{-2x}$
**3.** $y = -6e^{-2x}$
**4.** $y = 8e^{4x}$
**5.** $y = -8e^{2x}$

**6.** $y = .2e^{5x}$
**7.** $y = -16e^{x+1}$
**8.** $y = -4e^{-.1x}$
**9.** $y = e^{x^2}$
**10.** $y = e^{-x^2}$

**11.** $y = 3e^{2x^2}$
**12.** $y = -5e^{4x^3}$
**13.** $y = 4e^{2x^2-4}$
**14.** $y = -3e^{3x^2+5}$
**15.** $y = xe^x$

**16.** $y = x^2e^{-2x}$
**17.** $y = (x - 3)^2e^{2x}$
**18.** $y = (3x^2 - 4x)e^{-3x}$

**19.** $y = e^{x^2} \ln x, \quad x > 0$
**20.** $y = e^{2x-1} \ln (2x - 1), \quad x > \dfrac{1}{2}$
**21.** $y = \dfrac{e^x}{\ln x}, \quad x > 0$

**22.** $y = \dfrac{\ln x}{e^x}, \quad x > 0$
**23.** $y = \dfrac{x^2}{e^x}$
**24.** $y = \dfrac{e^x}{2x + 1}$

**25.** $y = \dfrac{e^x + e^{-x}}{x}$

**26.** $y = \dfrac{e^x - e^{-x}}{x}$

**27.** $y = \dfrac{5000}{1 + 10e^{.4x}}$

**28.** $y = \dfrac{600}{1 - 50e^{.2x}}$

**29.** $y = \dfrac{10,000}{9 + 4e^{-.2x}}$

**30.** $y = \dfrac{500}{12 + 5e^{-.5x}}$

**31.** $y = (2x + e^{-x^2})^2$

**32.** $y = (e^{2x} + \ln x)^3, \quad x > 0$

*Find the derivative of each of the following.*

**33.** $y = 8^{5x}$

**34.** $y = 2^{-x}$

**35.** $y = 3 \cdot 4^{x^2 + 2}$

**36.** $y = -10^{3x^2 - 4}$

**37.** $y = 2 \cdot 3^{\sqrt{x}}$

**38.** $y = 5 \cdot 7^{\sqrt{x - 2}}$

*Find all relative maxima or minima for the functions defined as follows. Sketch the graphs.*

**39.** $y = -xe^x$

**40.** $y = xe^{-x}$

**41.** $y = x^2 e^{-x}$

**42.** $y = (x - 1)e^{-x}$

**43.** $y = e^x + e^{-x}$

**44.** $y = -x^2 e^x$

**45.** Describe the slope of the tangent line to the graph of $f(x) = e^x$ for each of the following.

(a) $x \to -\infty$     (b) $x \to 0$

**46.** Prove: $\dfrac{d}{dx} e^{ax} = ae^{ax}$ for any constant $a$.

---

### ▓ APPLICATIONS

#### BUSINESS AND ECONOMICS

**Sales**   **47.** The sales of a new personal computer, in thousands, are given by

$$S(t) = 100 - 90e^{-.3t}$$

where $t$ represents time in years. Find the rate of change of sales at each of the following times.

(a) $t = 1$     (b) $t = 5$

**Product Durability**   **48.** Suppose $P(x) = e^{-.02x}$ represents the proportion of shoes manufactured by a given company that are still wearable after $x$ days of use.
Find the proportions of shoes wearable after the following periods of time.

(a) 1 day     (b) 10 days     (c) 100 days

(d) Calculate and interpret $P'(100)$.

**Product Durability**   **49.** Using data in a car magazine, we constructed the mathematical model

$$y = 100e^{-.03045t}$$

for the percent of cars of a certain type still on the road after $t$ years.
Find the percent of cars on the road after the following numbers of years.

(a) 0     (b) 2     (c) 4     (d) 6

Find $y'$ for the following values of $t$.

(e) 0     (f) 2

(g) Interpret your answers to (e) and (f).

**Elasticity**    **50.** Refer to the discussion of elasticity in the previous chapter. For the demand function $p = 400\, e^{-.2x}$, find each of the following.

    **(a)** The elasticity $E$

    **(b)** The value of $x$ for which total revenue is maximum

## LIFE SCIENCES

**Pollution Concentration**    **51.** The concentration of pollutants, in grams per liter, in the east fork of the Big Weasel River is approximated by

$$P(x) = .04e^{-4x},$$

where $x$ is the number of miles downstream from a paper mill that the measurement is taken.

Find each of the following values.

    **(a)** $P(.5)$     **(b)** $P(1)$     **(c)** $P(2)$

Find the rate of change of concentration with respect to distance for each of the following distances.

    **(d)** $x = .5$     **(e)** $x = 1$     **(f)** $x = 2$

**Radioactive Decay**    **52.** Assume that the amount in grams of a radioactive substance present at time $t$ is given by

$$A(t) = 500e^{-.25t}.$$

Find the rate of change of the quantity present at each of the following times.

    **(a)** $t = 0$     **(b)** $t = 4$     **(c)** $t = 6$     **(d)** $t = 10$

## SOCIAL SCIENCES

**Habit Strength**    **53.** According to work by the psychologist C. L. Hull, the strength of a habit is a function of the number of times the habit is repeated. If $N$ is the number of repetitions and $H(N)$ is the strength of the habit, then

$$H(N) = 1000(1 - e^{-kN})$$

where $k$ is a constant.

Find $H'(N)$ if $k = .1$ and $N$ is as follows.

    **(a)** $N = 10$     **(b)** $N = 100$     **(c)** $N = 1000$

    **(d)** Show that $H'(N)$ is always positive. What does this mean?

## FOR THE COMPUTER

*Graph the functions in Exercises 54 and 55 by plotting points in the given intervals.*

**54.** $y = e^x \ln x$,   $(0, 5]$     **55.** $y = \dfrac{5 \ln 10x}{e^{.5x}}$,   $(0, 10]$

---

## ■ 4.5 APPLICATIONS: GROWTH AND DECAY

In many situations that occur in biology, economics, and the social sciences, a quantity changes at a rate proportional to the amount of the quantity present. In such cases the amount present at time $t$ is a function of $t$ called the **exponential growth function.***

---

*The derivation of this function is presented in a later chapter.

**EXPONENTIAL GROWTH FUNCTION**

> Let $y_0$ be the amount or number of something present at time $t = 0$. Then, under certain conditions, the amount present at any time $t$ is given by
>
> $$y = y_0 e^{kt},$$
>
> where $k$ is a constant.

If $k > 0$, then $k$ is called the **growth constant;** if $k < 0$, then $k$ is called the **decay constant.** A common example is the growth of bacteria in a culture. The more bacteria present, the faster the population increases.

### ▬ EXAMPLE 1

Yeast in a sugar solution is growing at a rate such that 1 gram becomes 1.5 grams after 20 hours. Find the growth function, assuming exponential growth.

The values for $y_0$ and $k$ in the exponential growth function $y = y_0 e^{kt}$ must be found. Since $y_0$ is the amount present at time $t = 0$, $y_0 = 1$. To find $k$, substitute $y = 1.5$, $t = 20$, and $y_0 = 1$ into the equation.

$$y = y_0 e^{kt}$$
$$1.5 = 1 \; e^{k(20)}$$

Now take natural logarithms on both sides and use the power rule for logarithms and the fact that $\ln e = 1$.

$$1.5 = e^{20k}$$
$$\ln 1.5 = \ln e^{20k}$$
$$\ln 1.5 = 20k \ln e$$
$$\ln 1.5 = 20k$$
$$\frac{\ln 1.5}{20} = k$$
$$k \approx .02 \text{ (to the nearest hundredth)}$$

The exponential growth function is $y = e^{.02t}$, where $y$ is the number of grams of yeast present after $t$ hours.   ▬

### ▬ EXAMPLE 2

A population of fruit flies is contained in a large glass jar that can hold 1000 flies. The population $y$ is increasing according to the function

$$y = 100e^{.03t},$$

where $t$ is time in days.

**(a)** What was the initial population?

In the equation $y = y_0 e^{kt}$, the initial population is given by $y_0$. In the given function, $y_0 = 100$, so there were 100 fruit flies in the jar initially.

**(b)** How long will it take for the fruit fly population to fill the jar to capacity?

The maximum population is 1000. To find the value of $t$ that corresponds to $y = 1000$, solve the equation

$$1000 = 100e^{.03t}$$

for $t$. First divide both sides by 100:

$$10 = e^{.03t}.$$

Take natural logarithms on both sides.

$$\ln 10 = \ln e^{.03t}$$

By the power rule for logarithms, and using the fact that $\ln e = 1$,

$$\ln 10 = .03t \ln e$$
$$\ln 10 = .03t$$
$$t = \frac{\ln 10}{.03}.$$

Use a calculator or the table of natural logarithms to find $\ln 10 \approx 2.3026$, so that

$$t \approx 76.8$$

It will take about 76.8 days for the jar to be filled to capacity. ■

The decline of a population or decay of a substance may also be described by the exponential growth function. In this case the decay constant $k$ is negative, since an increase in time leads to a decrease in the quantity present. Radioactive substances provide a good example of exponential decay. By definition, the **half-life** of a radioactive substance is the time it takes for exactly half of the initial quantity to decay.

### ■ EXAMPLE 3

Suppose that $A(t)$, the amount of a certain radioactive substance present at time $t$, is given by

$$A(t) = 1000e^{-.1t},$$

where $t$ is measured in days and $A(t)$ in grams. Find the half-life of this substance.

To find the half-life, we must find a value of $t$ such that $A(t) = \frac{1}{2}(1000) = 500$ grams. That is, we must solve the equation

$$500 = 1000e^{-.1t}$$

for $t$. First divide both sides by 1000, obtaining

$$\frac{1}{2} = e^{-.1t}.$$

Now take natural logarithms of both sides:

$$\ln \frac{1}{2} = \ln e^{-.1t}.$$

Using the power rule of logarithms and the fact that $\ln e = 1$,

$$\ln \frac{1}{2} = -.1t,$$

and
$$t = \frac{\ln \frac{1}{2}}{-.1}.$$

Since $\ln \frac{1}{2} = \ln .5$, use the table of natural logarithms or a calculator to get

$$t \approx \frac{-.6931}{-.1} \approx 6.9.$$

It will take about 6.9 days for half the sample to decay. ■

## ■ EXAMPLE 4

Carbon 14 is a radioactive isotope of carbon that has a half-life of about 5600 years. The earth's atmosphere contains much carbon, mostly in the form of carbon dioxide gas, with small traces of carbon 14. Most atmospheric carbon is the nonradioactive isotope, carbon 12. The ratio of carbon 14 to carbon 12 is virtually constant in the atmosphere. As a plant absorbs carbon dioxide from the air in the process of photosynthesis, however, the carbon 12 stays in the plant while the carbon 14 is converted into nitrogen. Thus, in a plant, the ratio of carbon 14 to carbon 12 is smaller than the ratio in the atmosphere. Even when the plant dies or is consumed by an animal or a person, this ratio will continue to decrease. Based on these facts, a method of dating objects called carbon-14 dating has been developed.

**(a)** Suppose a mummy has been discovered in which the ratio of carbon 14 to carbon 12 is only about half the ratio found in the atmosphere. How long ago did the individual who became the mummy die?

As mentioned above, in 5600 years half the carbon 14 in a specimen will decay. This means the individual who became the mummy died about 5600 years ago.

**(b)** Let $R$ be the (nearly constant) ratio of carbon 14 to carbon 12 in the atmosphere, and let $r$ be the ratio in an observed specimen. What is the relationship between $R$ and $r$?

Assuming that $R$ and $r$ are related by an exponential growth function, the value of $r$, where $R$ is the ratio at time $t = 0$, is

$$r = Re^{kt}.$$

To find $k$, use the fact that $r = R/2$ if $t = 5600$. Substitution gives

$$\frac{1}{2}R = Re^{5600k} \qquad \text{or} \qquad \frac{1}{2} = e^{5600k}.$$

Taking natural logarithms on both sides and using the power rule on the right,

$$\ln \frac{1}{2} = 5600\, k \ln e$$

or $\qquad\qquad\qquad \ln 1 - \ln 2 = 5600\, k \ln e.$

Since $\ln 1 = 0$ and $\ln e = 1$,

$$-\ln 2 = 5600k$$

$$k = -\frac{\ln 2}{5600}.$$

The relationship between $R$ and $r$ is given by

$$r = Re^{(-t \ln 2)/5600}. \qquad\qquad (*)$$

**(c)** Verify equation (*) for $t = 0$.

Substitute 0 for $t$ in the equation (*). This gives

$$r = Re^{(-0\, \cdot\, \ln 2)/5600}$$

$$= R \cdot e^0 = R \cdot 1 = R$$

This result is correct; when $t = 0$, the specimen has just died, so that $R$ and $r$ should be the same.

**(d)** Suppose a specimen is found in which $r = (2/3)R$. Estimate the age of the specimen.

Use equation (*) and substitute $(2/3)R$ for $r$.

$$r = Re^{(-t \ln 2)/5600}$$

$$\frac{2}{3}R = Re^{(-t \ln 2)/5600}$$

Dividing through by $R$ gives

$$\frac{2}{3} = e^{(-t \ln 2)/5600}.$$

Take natural logarithms on both sides:

$$\ln \frac{2}{3} = \ln e^{(-t \ln 2)/5600}.$$

Using properties of logarithms gives

$$\ln 2 - \ln 3 = \frac{-t \ln 2}{5600}.$$

To solve this equation for $t$, the age of the specimen, multiply both sides by $-5600/\ln 2$. This gives

$$\frac{-5600(\ln 2 - \ln 3)}{\ln 2} = t.$$

Using the table or a calculator, $\ln 3 \approx 1.0986$ and $\ln 2 \approx .6931$. These values lead to

$$t \approx \frac{-5600(.6931 - 1.0986)}{.6931} \approx 3280,$$

so the specimen is about 3280 years old. ■

In economics, the formula for *continuous compounding* is a good example of an exponential growth function. As interest is compounded more and more often, the compound amount $A$ approaches

$$\lim_{m \to +\infty} P\left(1 + \frac{i}{m}\right)^{mn}.$$

For example, if $P = 1$ and $n = 1$,

$$A = \left(1 + \frac{i}{m}\right)^{m}.$$

Let $i/m = 1/s$ so that $m = si$; then

$$\lim_{m \to +\infty} \left(1 + \frac{i}{m}\right)^{m} = \lim_{s \to +\infty} \left(1 + \frac{1}{s}\right)^{si}$$

$$= \lim_{s \to +\infty} \left[\left(1 + \frac{1}{s}\right)^{s}\right]^{i}$$

$$= \left[\lim_{s \to +\infty} \left(1 + \frac{1}{s}\right)^{s}\right]^{i}.$$

In Section 1 of this chapter we saw that

$$\lim_{m \to +\infty} \left(1 + \frac{1}{m}\right)^{m} = e.$$

Using $s$ as the variable instead of $m$ gives

$$\lim_{s \to +\infty} \left(1 + \frac{1}{s}\right)^{s} = e.$$

Thus,

$$\left[\lim_{s \to +\infty} \left(1 + \frac{1}{s}\right)^{s}\right]^{i} = e^{i},$$

and so

$$\lim_{m \to +\infty} \left(1 + \frac{i}{m}\right)^{m} = e^{i}.$$

Generalizing from this example gives the formula below for **continuous compounding.**

CONTINUOUS
COMPOUNDING

> If a deposit of $P$ dollars is invested at a rate of interest $i$ compounded continuously for $n$ years, the compound amount is
> $$A = Pe^{ni} \quad \text{dollars.}$$

### ■ EXAMPLE 5

Suppose $1000 is deposited in an account paying 12% compounded continuously for 10 years. Find the compound amount.

Let $P = 1000$, $n = 10$, and $i = .12$. Then

$$A = 1000e^{10(.12)} = 1000e^{1.2}.$$

From the table of powers of $e$, or a calculator, $e^{1.2} \approx 3.32012$, and

$$A = 1000(3.32012) = 3320.12,$$

or $3320.12. Verify that daily compounding produces only 66 cents less.  ■

### ■ EXAMPLE 6

Assuming continuous compounding, if the inflation rate averaged 6% a year for 5 years, how much would a $1 item cost at the end of the 5 years?

In the formula for continuous compounding, let $P = 1$, $n = 5$, and $i = .06$ to get

$$A = 1 \, e^{5(.06)} = e^{.3} \approx 1.34986.$$

An item that cost $1 at the beginning of the 5-year period would cost $1.35 at the end of the period, an increase of 35% or about 1/3.  ■

The exponential growth functions discussed so far all continued to grow without bound. More realistically, many populations grow exponentially for a while, but then the growth is slowed by some external constraint that eventually limits the growth. For example, an animal population may grow to the point where its habitat can no longer support the population and the growth rate begins to dwindle until a stable population size is reached. Models that reflect this pattern are called **limited growth functions.** The next two examples discuss functions of this type that occur in industry and social science.

### ■ EXAMPLE 7

Suppose the sales, $S(x)$, in some appropriate unit, of a new model of typewriter are approximated by the function.

$$S(x) = 1000 - 800e^{-x},$$

where $x$ represents the number of years the typewriter has been on the market.

**(a)** In how many years will sales reach 500 units?

Replace $S(x)$ with 500 and solve the equation for $x$.

$$500 = 1000 - 800e^{-x}$$
$$-500 = -800e^{-x}$$
$$.625 = e^{-x}$$

Take natural logarithms on both sides.

$$\ln .625 = \ln e^{-x}$$
$$\ln .625 = -x \ln e$$
$$\ln .625 = -x$$
$$x = -\ln .625 \approx .47$$

Sales will reach 500 units in about 1/2 year or 6 months.

**(b)** Will sales ever reach 1000 units?

Replacing $S(x)$ with 1000 gives

$$1000 = 1000 - 800e^{-x}$$
$$0 = -800e^{-x}$$
$$e^{-x} = 0.$$

Since $e^{-x} > 0$ for all $x$, sales will never reach 1000 units. As $x$ gets larger and larger, $800e^{-x} = 800/e^x$ will get smaller and smaller. This means sales will tend to level off with time and gradually approach a limit of 1000. The horizontal line $y = 1000$ is an asymptote to the graph of the function, as shown in Figure 12.  ▰

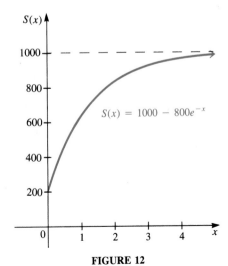

**FIGURE 12**

### ▬ EXAMPLE 8

Assembly-line operations tend to have a high turnover of employees, forcing companies to spend much time and effort in training new workers. It has been found that a worker new to a task on the line will produce items according to the function

$$P(x) = 25 - 25e^{-.3x},$$

where $P(x)$ items are produced by the worker on day $x$.

**(a)** What is the limit on the number of items a worker on this assembly line can produce?

Since the exponent on $e$ is negative, write the function as

$$P(x) = 25 - \frac{25}{e^{.3x}}.$$

As $x \to +\infty$, the term $(25/e^{.3x}) \to 0$, so $P(x) \to 25$. The limit is 25 items.

**(b)** How many days will it take for a new worker to produce 20 items?

Let $P(x) = 20$ and solve for $x$.

$$P(x) = 25 - 25e^{-.3x}$$
$$20 = 25 - 25e^{-.3x}$$
$$-5 = -25e^{-.3x}$$
$$.2 = e^{-.3x}$$

Now take natural logarithms of both sides and use properties of logarithms.

$$\ln .2 = -.3x \ln e$$
$$\ln .2 = -.3x$$
$$x = \frac{\ln .2}{-.3} \approx 5.4$$

In about 5 1/2 days on the job a new worker will be producing 20 items. A graph of the function $P$ is shown in Figure 13.  ▬

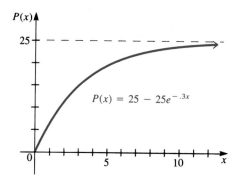

**FIGURE 13**

Graphs such as the one in Figure 13 are called **learning curves.** According to such a graph, a new worker tends to learn quickly at first; then learning tapers off and approaches some upper limit. This is characteristic of the learning of certain types of skills involving the repetitive performance of the same task.

The next example discusses still another type of exponential function.

### ■■■ EXAMPLE 9

Under certain conditions the total number of facts of a certain kind that are remembered is approximated by

$$N(t) = y_0\left(\frac{1 + e}{1 + e^{t+1}}\right),$$

where $N(t)$ is the number of facts remembered at time $t$, measured in days, and $y_0$ is the number of facts remembered initially. Graph the function.

The graph of $N(t)$ has no vertical asymptotes, since $1 + e^{t+1} > 0$ for all $t$. The numerator is constant and the denominator increases as $t$ increases, so $y = 0$ is a horizontal asymptote. The first derivative is

$$N'(t) = \frac{-y_0(1 + e)e^{t+1}}{(1 + e^{t+1})^2}.$$

Neither the numerator nor denominator of $N'(t)$ can equal 0, so there are no extrema. Since $N'(t) < 0$ for all values of $t$, the graph is always decreasing. Plotting a few points will complete the graph. For example, if $t = 0$,

$$N(0) = y_0\left(\frac{1 + e}{1 + e^1}\right) = y_0(1) = y_0.$$

If $t = 1$,

$$N(1) = y_0\left(\frac{1 + e}{1 + e^2}\right) \approx y_0\left(\frac{3.718}{8.389}\right) = .44y_0.$$

The graph, shown in Figure 14, is called a *forgetting curve*.  ■■■

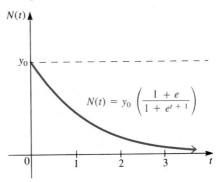

**FIGURE 14**

## 4.5 EXERCISES

## APPLICATIONS

### BUSINESS AND ECONOMICS

**Inflation**
**1.** Assuming continuous compounding, what will it cost to buy a $10 item in 3 years at the following inflation rates?

(a) 3%     (b) 4%     (c) 5%

**Compound Interest**
**2.** Bert Bezzone invests a $25,000 inheritance in a fund paying 9% per year compounded continuously. What will be the amount on deposit after each of the following time periods?

(a) 1 year     (b) 5 years     (c) 10 years

**Compound Interest**
**3.** Linda Youngman, who is self-employed, wants to invest $60,000 in a pension plan. One investment offers 10% compounded quarterly. Another offers 9.75% compounded continuously. Which investment will earn the most interest in 5 years? How much more will the better plan earn?

**Compound Interest**
**4.** If Ms. Youngman (see Exercise 3) chooses the plan with continuous compounding, how long will it take for her $60,000 to grow to $80,000?

**Sales**
**5.** Sales of a new model of can opener are approximated by

$$S(x) = 5000 - 4000e^{-x},$$

where $x$ represents the number of years that the can opener has been on the market, and $S(x)$ represents sales in thousands.
Find each of the following.

(a) $S_0 = S(0)$     (b) $S(2)$     (c) $S(10)$

(d) When will sales reach 4,500,000?

(e) Find the limit on sales.

(f) Graph $S(x)$.

**Sales**
**6.** Experiments have shown that sales of a product, under relatively stable market conditions, but in the absence of promotional acitivties such as advertising, tend to decline at a constant yearly rate. This sales decline, which varies considerably from product to product, can often be expressed by an exponential function of the form

$$S(t) = S_0e^{-at},$$

where $S(t)$ is the sales at time $t$ measured in years, $S_0$ is sales at time $t = 0$, and $a$ is the sales decay constant. Suppose a certain product had sales of 50,000 at the beginning of the year and 45,000 at the end of the year.

(a) Write an equation for the sales decline.

(b) Find $S(2)$.

(c) How long will it take for sales to decrease to 40,000?

### LIFE SCIENCES

**Growth of Lice**
**7.** A population of 100 lice is growing exponentially. After 2 months the population has increased to 125.

(a) Write an exponential equation to express the exponential growth function $y$ in terms of time $t$ in months.

(b) How long will it take for the population to reach 500?

**Growth of Bacteria**   **8.** A culture contains 25,000 bacteria, with the population increasing exponentially. The culture contains 40,000 bacteria after 10 hours.

(a) Write an exponential equation to express the growth function $y$ in terms of time $t$ in hours.

(b) How long will it be until there are 60,000 bacteria?

**Decrease in Bacteria**   **9.** When a bactericide is introduced into a culture of 50,000 bacteria, the number of bacteria decreases exponentially. After 9 hours, there are only 20,000 bacteria.

(a) Write an exponential equation to express the decay function $y$ in terms of time $t$ in hours.

(b) In how many hours will half the number of bacteria remain?

**Decrease in Bacteria**   **10.** The number of bacteria in a certain culture, $B(t)$, is approximated by

$$B(t) = 250,000e^{-.04t}$$

where $t$ is time measured in hours. Find each of the following.

(a) $B_0$    (b) $B(5)$    (c) $B(20)$

(d) The time until only 125,000 bacteria are present

(e) The time until only 25,000 bacteria are present

**Growth of Bacteria**   **11.** Under certain conditions, the number of a type of bacteria present in a colony is given by

$$E(t) = E_0 e^{kt},$$

where $E(t)$ is the number of bacteria present $t$ minutes after the beginning of an experiment, and $E_0$ is the number present when $t = 0$. Suppose there were $10^6$ bacteria present initially and $1.8 \times 10^6$ bacteria present after 60 minutes.

(a) Write an equation for the number of bacteria present at time $t$.

(b) Find $E(10)$.

(c) How long will it take for the population to reach 1,500,000?

**Limited Growth**   **12.** In the previous section we discussed the *logistic function:*

$$G(t) = \frac{mG_0}{G_0 + (m - G_0)e^{-kmt}},$$

where $G_0$ is the initial number present, $m$ is the maximum possible size of the population, and $k$ is a positive constant. Assume $G_0 = 1000$, $m = 2500$, $k = .0004$, and $t$ is time in decades (10-year periods).

(a) Find $G(.2)$.    (b) Find $G(1)$.    (c) Find $G(3)$.

(d) At what time $t$ will the population reach 2000?

(e) Graph $G(t)$. (*Hint:* There is an inflection point at $t \approx .4$. What does this mean in terms of population?)

## SOCIAL SCIENCES

**Population Growth**   **13.** The population of a boomtown in Alaska is increasing exponentially. There were 10,000 people in town when the boom began. Two years later the population had reached 12,000.

(a) Write an equation to express the growth function $y$ in terms of time $t$ in years.

(b) How long will it be until the population doubles?

**Medical School Applications**    **14.** The higher a student's grade-point average, the fewer applications the student must send to medical schools (other things being equal). Using information given in a guidebook for prospective medical school students, we constructed the following mathematical model of the number of applications that a student should send out:

$$y = 540e^{-1.3x},$$

where $y$ is the number of applications that should be sent out by a person whose grade-point average is $x$. Here $2.0 \le x \le 4.0$.

Use a calculator with a $y^x$ key to find the number of applications that should be sent out by students having the following grade-point averages.

**(a)** 2.0    **(b)** 2.5    **(c)** 3.0

What grade-point averages require the following numbers of applications?

**(d)** 15    **(e)** 3

**Skills Training**    **15.** Suppose the number of symbols per minute a keypunch operator can produce is given by

$$p(t) = 350 - 80e^{-.3t},$$

where $t$ is the number of months the operator has been training. Find each of the following.

**(a)** $p(2)$    **(b)** $p(4)$    **(c)** $p(10)$

**(d)** When will an operator produce 300 symbols per minute?

**(e)** What is the limit on the number of symbols produced per minute?

**(f)** Graph $p(t)$.

**Skills Training**    **16.** The number of words per minute that an average typist can type is given by

$$W(t) = 60 - 30e^{-.5t},$$

where $t$ is time in months after the beginning of a typing class. Find each of the following.

**(a)** $W_0 = W(0)$    **(b)** $W(1)$    **(c)** $W(4)$

**(d)** When will the average typist type 45 words per minute?

**(e)** What is the upper limit on the number of words that can be typed according to this model?

**(f)** Graph $W(t)$.

**Skills Training**    **17.** Assuming that a person new to an assembly line will produce

$$P(x) = 500 - 500e^{-x}$$

items per day, where $x$ is time measured in days. Find each of the following.

**(a)** $P_0 = P(0)$    **(b)** $P(2)$    **(c)** $P(5)$

**(d)** When will a new person produce 400 items?

**(e)** Find the limit on the number of items produced per day.

**(f)** Graph $P(x)$.

**Spread of a Rumor**    *A sociologist has shown that the fraction $y(t)$ of people who have heard a rumor after t days is approximated by*

$$y(t) = \frac{y_0 e^{kt}}{1 - y_0(1 - e^{kt})},$$

*where $y_0$ is the fraction of people who have heard the rumor at time $t = 0$, and k is a constant. A graph of $y(t)$ for a particular value of k is shown in the figure that follows.*

**18.** If $k = .1$ and $y_0 = .05$, find $y(10)$.

**19.** If $k = .2$ and $y_0 = .10$, find $y(5)$.

**20.** If $k = .1$ and $y_0 = .02$, find the number of days until half the people have heard the rumor.

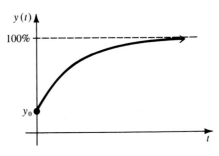

### PHYSICAL SCIENCES

**Half-Life of a Sample**

**21.** The half-life of plutonium 241 is approximately 13 years. How much of a 2-gram sample will remain after 100 years?

**Half-Life of a Sample**

**22.** The half-life of radium 226 is approximately 1620 years. How much of a 2-gram sample will remain after 500 years?

**Half-Life of a Sample**

**23.** The half-life of iodine 131 is approximately 8 days. What fractional part of an initial amount of iodine 131 remains after one month has elapsed? (Assume a 30-day month.)

**Radioactive Decay**

**24.** Five hundred grams of a radioactive substance is decaying exponentially. The substance is reduced to 400 grams after 4 days.

    **(a)** Write an exponential equation to express the decay function $y$ in terms of time $t$ in days.

    **(b)** Find the half-life of the substance.

**Chemical Dissolution**

**25.** The amount of chemical that will dissolve in a solution increases exponentially as the temperature is increased. At $0°$ C 10 grams of the chemical dissolves and at $10°$ C 11 grams dissolves.

    **(a)** Write an equation to express the amount of chemical dissolved, $y$, in terms of temperature, $t$, in degrees Celsius.

    **(b)** At what temperature will 15 grams dissolve?

**Carbon-14 Dating**

*Using Example 4, find the age of a specimen for each of the following.*

**26.** $r = .8R$       **27.** $r = .4R$       **28.** $r = .1R$       **29.** $r = .01R$

**Decay of Radioactivity**

**30.** A large cloud of radioactive debris from a nuclear explosion has floated over the Pacific Northwest, contaminating much of the hay supply. Consequently, farmers in the area are concerned that the cows who eat this hay will give contaminated milk. (The tolerance level for radioactive iodine in milk is 0.) The percent of the initial amount of radioactive iodine still present in the hay after $t$ days is approximated by $P(t)$, which is given by the mathematical model

$$P(t) = 100e^{-.1t}.$$

    **(a)** Find the percent remaining after 4 days.

    **(b)** Find the percent remaining after 10 days.

    **(c)** Some scientists feel that the hay is safe after the percent of radioactive iodine has declined to 10% of the original amount. Solve the equation $10 = 100e^{-.1t}$ to find the number of days before the hay may be used.

    **(d)** Other scientists believe that the hay is not safe until the level of radioactive iodine has declined to only 1% of the original level. Find the number of days that this would take.

**Newton's Law of Cooling**    *Newton's law of cooling says that the rate at which a body cools is proportional to the difference in temperature between the body and an environment into which it is introduced. This leads to an equation where the temperature $f(t)$ of the body at time $t$ after being introduced into an environment having constant temperature $T_0$ is*

$$f(t) = T_0 + Ce^{-kt},$$

*where $C$ and $k$ are constants. Use this result in Exercises 31–34.*

**31.** Find the temperature of an object when $t = 4$ if $T_0 = 125$, $C = 8$, and $k = .2$.

**32.** Find the temperature of an object when $t = 9$ if $T_0 = 18$, $C = 5$, and $k = .6$.

**33.** If $C = 100$, $k = .1$, and $t$ is in minutes, how long will it take a hot cup of coffee to cool to a temperature of 25° Celsius in a room at 20° Celsius?

**34.** If $C = -14.6$ and $k = .6$ and $t$ is in hours, how long will it take a frozen pizza to thaw to 10° Celsius in a room at 18° Celsius?

---

## ▤ 4.6 APPLICATIONS: MATHEMATICS OF FINANCE

In this section we extend the concept of compound interest introduced in the first section of this chapter. The material on interest rates that we have studied up to this point is summarized below.

**SUMMARY OF INTEREST FORMULAS**

> **Simple Interest**    The simple interest $I$ on $P$ dollars at a yearly rate of interest $i$ for $t$ years is
>
> $$I = Prt \quad \text{dollars.}$$
>
> **Compound Interest**    If a deposit of $P$ dollars is invested at a yearly rate of interest $i$ compounded $m$ times per year for $n$ years, the compound amount is
>
> $$A = P\left(1 + \frac{i}{m}\right)^{mn} \quad \text{dollars.}$$
>
> **Continuous Compounding**    If a deposit of $P$ dollars is invested at a rate of interest $i$ compounded continuously for $n$ years, the compound amount is
>
> $$A = Pe^{ni} \quad \text{dollars.}$$

*Effective Rate*    We could use a compound interest table or a calculator to see that \$1 at 8% interest (per year) compounded semiannually is $1(1.04)^2 = 1.0816$ or \$1.0816. The actual increase of \$.0816 is 8.16% rather than the 8% that would be earned with interest compounded annually. To distinguish between these two amounts, 8% (the annual interest rate) is called the **nominal** or **stated** interest rate, and 8.16% is called the **effective** interest rate.

**EFFECTIVE RATE**

> If $i$ is the annual stated rate of interest and $m$ is the number of compounding periods per year, the effective rate of interest is
>
> $$\left(1 + \frac{i}{m}\right)^m - 1.$$

### EXAMPLE 1

Find the effective rate corresponding to a nominal interest rate of 8% compounded quarterly.

Use the formula given above or look in the compound interest table in the Appendix for $8\%/4 = 2\%$ for 4 periods. Either way, the effective rate is $1.08243 - 1 = .08243$ or 8.243%. ▅

The formula for interest compounded $m$ times a year, $A = P(1 + i/m)^{mn}$, has five variables, $A$, $P$, $i$, $m$, and $n$. If the values of any four are known, then the value of the fifth can be found. In particular, if $A$, the amount of money we wish to end up with, is given as well as $i$, $m$, and $n$, then $P$ can be found. Here $P$ is the amount that should be deposited today to produce $A$ dollars in $n$ years. The amount $P$ is called the **present value of** $A$ dollars.

### EXAMPLE 2

Ed Calvin has a balloon payment of $100,000 due in 3 years. What is the present value of that amount if the money earns interest at 12% annually?

Here $P$ in the compound interest formula is unknown, with $A = 100{,}000$, $i = .12$, $n = 3$, and $m = 1$. Substitute the known values into the formula to get $100{,}000 = P(1.12)^3$. Solve for $P$ using a calculator to find $(1.12)^3$.

$$P = \frac{100{,}000}{(1.12)^3} = 71{,}178.03$$

The present value of $100,000 in 3 years at 12% a year is $71,178.03. ▅

The equation in Example 2 could have been solved as follows.

$$100{,}000 = P(1.12)^3$$

Multiply both sides by $(1.12)^{-3}$ to solve for $P$, so

$$P = 100{,}000(1.12)^{-3}.$$

This suggests a general formula for present value.

**PRESENT VALUE**

> The present value of $A$ dollars at a rate of interest $i$ compounded $m$ times per year for $n$ years is
>
> $$P = A\left(1 + \frac{i}{m}\right)^{-mn}.$$

The equation $A = Pe^{ni}$ also can be solved for any of the variables $A$, $P$, $n$, or $i$, as the following example shows.

### ■■■ EXAMPLE 3

Suppose the rate of inflation in the economy averages 8% per year, assuming continuous compounding. Find the number of years it would take for the general level of prices to double.

To find the number of years it will take for $1 worth of goods or services to cost $2, find $n$ in the equation

$$A = Pe^{ni}$$

or

$$2 = 1 \, e^{.08n}.$$

Taking natural logarithms on both sides,

$$\ln 2 = \ln e^{.08n}.$$

Using the power rule for logarithms and the fact that $\ln e = 1$ gives

$$\ln 2 = .08n$$

$$n = \frac{\ln 2}{.08} \approx 8.7.$$

The general level of prices will double in about 9 years.    ■■

***Annuities***    Each of the formulas discussed above involved a *single payment* at the end of a specified time period. Sometimes loans are repaid or savings are accumulated with a series of *equal payments* made at equal periods of time, called an **annuity.** The time between payments is the **payment period,** and the time from the beginning of the first payment period to the end of the last is the **term** of the annuity. The **future value** of the annuity, the final sum on deposit, is defined to be the sum of the compound amounts of all the payments, compounded to the end of the term.

### ■■■ EXAMPLE 4

Find the value of an annuity in which a payment of $900 is made at the end of each year for 5 years into an account paying 6% interest.

The first payment produces a compound amount of

$$900(1 + .06)^4 = 900(1.06)^4.$$

Use 4 as the exponent since the money is deposited at the *end* of the first year. The second payment produces a compound amount of $900(1.06)^3$, and so on. The last payment will earn no interest at all. The future value of the annuity is

$$900(1.06)^4 + 900(1.06)^3 + 900(1.06)^2 + 900(1.06) + 900$$
$$= 1136.23 + 1071.91 + 1011.24 + 954.00 + 900.00$$
$$= 5073.38,$$

or $5073.38    ■■

To generalize the result in Example 4, suppose that $R$ dollars is paid into an account at the end of each period for $n$ periods, at a rate of interest $i$ per period. If $A$ represents the future value of the annuity, then

$$A = R(1 + i)^{n-1} + R(1 + i)^{n-2} + \cdots + R(1 + i) + R,$$

or, in reverse order,

$$A = R + R(1 + i) + \cdots + R(1 + i)^{n-1}.$$

This sum is the sum of the first $n$ terms of the geometric sequence with first term $R$ and common ratio $1 + i$. The sum of the first $n$ terms of such a sequence is

$$A = \frac{R[(1 + i)^n - 1]}{(1 + i) - 1} = \frac{R[(1 + i)^n - 1]}{i}.$$

This discussion is summarized below.

**FUTURE VALUE OF AN ANNUITY**

> The future value of an annuity of $n$ payments of $R$ dollars each at the end of consecutive interest periods, with interest compounded at a rate of interest $i$ per period, is given by
>
> $$A = \frac{R[(1 + i)^n - 1)]}{i}.$$

### EXAMPLE 5

Verify the formula given above for the annuity in Example 4.

From Example 4, $R = 900$, $i = .06$, and $n = 5$. Then

$$A = \frac{R[(1 + i)^n - 1)]}{i} = \frac{900[(1.06)^5 - 1]}{.06} = 5073.38,$$

which agrees with the future value found in Example 4. ■

### EXAMPLE 6

To save money for a trip, Adam Bryer plans to deposit \$1000 twice each year for three years in an account paying 5% interest compounded semiannually. How much will he accumulate with this plan?

Let $R = 1000$, $i = .05/2 = .025$, and $n = 3 \cdot 2 = 6$ in the future value formula.

$$A = \frac{R[(1 + i)^n - 1]}{i} = \frac{1000[(1.025)^6 - 1]}{.025} = 6387.74$$

The annuity will amount to \$6387.74 at the end of three years. ■

### ▰ EXAMPLE 7

Suppose Bryer (from Example 6) has determined that he needs $5300 for the trip. What equal payments should he deposit to accumulate the necessary amount?

Here, $A = 5300$, $R$ is to be found, and $i = .025$ and $n = 6$ as before. Substitute these values into the future value formula and solve for $R$.

$$A = \frac{R[(1 + i)^n - 1]}{i}$$

$$5300 = \frac{R[(1.025)^6 - 1]}{.025}$$

$$(5300)(.025) = R[(1.025)^6 - 1]$$

$$R = \frac{5300(.025)}{(1.025)^6 - 1} = 829.71$$

Payments of $829.71 twice each year for three years will produce $5300.    ▰

***Individual Retirement Accounts****   Money deposited in an individual retirement account (IRA) produces earnings that are not taxed until they are withdrawn from the account. Thus, an IRA effectively defers the tax on earnings, but there is a corresponding loss of liquidity since there are penalties for withdrawals from an IRA before the age of $59\frac{1}{2}$ years. Suppose that you have $P$ dollars to invest for $n$ years in either an IRA or a regular account, both of which pay $i$ percent compounded annually. Earnings from the regular account are immediately taxable, in contrast to the IRA earnings, which are tax sheltered until withdrawal. To compare these two investment options mathematically, assume for simplicity that the same interest rate, $i$ percent, and the same income tax rate, $t$ percent, apply throughout the $n$ years of the investment.

By the compound interest formula, the principal $P$ grows to $P(1 + i)^n$ after $n$ years in the IRA. If the entire amount is then withdrawn (and the investor is past $59\frac{1}{2}$ years of age), the tax due on the *earnings* is $[P(1 + i)^n - P]t$. Thus, the amount left after taxes is

$$A = P(1 + i)^n - [P(1 + i)^n - P]t$$
$$= P[(1 + i)^n(1 - t) + t].$$

If $P = 1$, then

$$M = (1 + i)^n(1 - t) + t$$

is called the **IRA multiplier,** because one dollar will grow to $M$ dollars (after taxes) after $n$ years in the IRA.

---

*The discussion and examples about IRAs are courtesy of Thomas B. Muenzenberger, Kansas State University, Manhattan, Kansas.

### ■ EXAMPLE 8

Suppose that you deposit $2000 in an IRA that pays 10% interest compounded annually. Suppose further that the money is left in the account for 30 years, then withdrawn, and the earnings are taxed at 37%. Find the tax due. What is the IRA multiplier?

Here the principal grows to $2000(1 + .1)^{30} = 34,899$ dollars (to the nearest dollar), and the tax due on the *earnings* is

$$(\$34,899 - \$2000)(.37) = \$12,173.$$

Thus, the amount left after taxes is $A = 34,899 - 12,173 = 22,726$ dollars. The IRA multiplier is $M = 11.4$. ■

Each year in the regular account, $t$ percent of the $i$ percent earned is withdrawn to pay the tax on the earnings, leaving $(1 - t)i$ percent. Thus, by the compound interest formula, the principal $P$ grows to

$$A = P[1 + (1 - t)i]^n$$

after $n$ years in the regular account. If $P = 1$, then

$$m = [1 + (1 - t)i]^n$$

is the multiplier, because one dollar will grow to $m$ dollars (after taxes) after $n$ years in the regular account.

### ■ EXAMPLE 9

Suppose that you deposit $2000 in a regular account that pays 10% compounded annually. Suppose further that the income tax rate is 37% and that the money is left in the account for 30 years and then withdrawn. Find the multiplier, $m$.

Here, the principal grows to $A = 2,000[1 + (1 - .37)(.10)]^{30} = 12,503$ dollars. The multiplier is 6.3. ■

From the previous two examples, $M/m = 11.4/6.3 = 1.8$, indicating that the IRA yields almost twice as much as the regular account over a thirty-year period. The **multiplier function**

$$\frac{M}{m} = \frac{(1 + i)^n (1 - t) + t}{[1 + (1 - t) i]^n}$$

can be used in the same way to compare the two investment options for other values of $n$, $i$, and $t$. Note that the multiplier function is a function in three variables.* The following table contains values of $M/m$ for $n = 30$ and various values of $i$ and $t$.

---

*Functions of more than one independent variable are discussed in the chapter entitled "Functions of Several Variables."

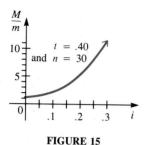

**FIGURE 15**

| $t$ \ $i$ | .05 | .10 | .15 | .20 | .25 | .30 |
|---|---|---|---|---|---|---|
| .10 | 1.07 | 1.19 | 1.34 | 1.49 | 1.65 | 1.81 |
| .20 | 1.13 | 1.41 | 1.77 | 2.21 | 2.72 | 3.30 |
| .30 | 1.18 | 1.64 | 2.33 | 3.27 | 4.48 | 6.02 |
| .40 | 1.23 | 1.89 | 3.02 | 4.77 | 7.33 | 11.0 |
| .50 | 1.27 | 2.13 | 3.83 | 6.83 | 11.8 | 19.8 |
| .60 | 1.29 | 2.34 | 4.72 | 9.50 | 18.6 | 35.0 |
| .70 | 1.28 | 2.45 | 5.49 | 12.5 | 27.8 | 59.3 |
| .80 | 1.23 | 2.37 | 5.79 | 14.9 | 37.6 | 91.4 |
| .90 | 1.15 | 1.96 | 4.81 | 13.6 | 38.9 | 108 |
| 1 | 1 | 1 | 1 | 1 | 1 | 1 |

It is apparent from the table that for $n = 30$ and a fixed $t$, the multiplier function $M/m$ is an increasing function of $i$ (see Figure 15 for $t = .40$). In fact, for a fixed $t$ and for *any* $n > 1$, the multiplier function is an increasing function of $i$. Thus, the advantage of the IRA over the regular account widens as the interest rate $i$ increases. This effect is particularly dramatic for high income tax rates (see the table).

## 4.6 EXERCISES

*Find the effective rate corresponding to each of the following nominal rates of interest.*

**1.** 5% compounded monthly

**2.** 15% compounded quarterly

**3.** 18% compounded semiannually

**4.** 10% compounded monthly

**5.** 11% compounded quarterly

**6.** 13% compounded semiannually

*Find the present value of each deposit.*

**7.** $2000 at 6% compounded semiannually for 11 years

**8.** $5000 at 8% compounded annually for 12 years

**9.** $10,000 at 10% compounded quarterly for 8 years

**10.** $7300 at 11% compounded semiannually for 15 years

**11.** $10,000, if interest is 12% compounded semiannually for 5 years

**12.** $25,000, if interest is 16% compounded quarterly for 11 quarters

**13.** $45,678.93 if interest is 12.6% compounded monthly for 11 months

**14.** $123,788 if interest is 14.7% compounded daily for 195 days (assume 365 days in a year)

*Find the future value of each of the following annuities.*

**15.** $R = \$7800$, 8% interest compounded semiannually for 7 years

**16.** $R = \$2600$, 10% interest compounded quarterly for 4 years

**17.** $R = \$500$, 6% interest compounded quarterly for 3 years

**18.** $R = \$13,000$, 9% interest compounded monthly for 2 years

**19.** $R = \$42,000$, 6% interest compounded semiannually for 5 years

**20.** $R = \$220$, 12% interest compounded monthly for 4 years

### ■ APPLICATIONS

#### BUSINESS AND ECONOMICS

**Effective Rate**   **21.** Janet Tilden bought a television set with money borrowed from the bank at 9% interest compounded semiannually. What effective interest rate did she pay?

**Effective Rate**   **22.** A firm deposits some funds in a special account at 7.2% compounded quarterly. What effective rate will they earn?

**Effective Rate**   **23.** Virginia Nicolai deposits $7500 of lottery winnings in an account paying 6% interest compounded monthly. What effective rate does the account earn?

**Present Value**   **24.** Scott Whitten must make a balloon payment of $20,000 in 4 years. Find the present value of the payment if it includes annual interest of 8%.

**Present Value**   **25.** A company must pay a $307,000 settlement in 3 years.
  **(a)** What amount must be deposited now at 6% semiannually to have enough money for the settlement?
  **(b)** How much interest will be earned?
  **(c)** Suppose the company can deposit only $200,000 now. How much more will be needed in 3 years?

**Present Value**   **26.** A couple wants to have $20,000 in 5 years for a down payment on a new house.
  **(a)** How much could they deposit today, at 8% compounded quarterly, so that they will have the required amount in 5 years?
  **(b)** How much interest will be earned?
  **(c)** If they can deposit only $10,000 now, how much more will they need to complete the $20,000 after 5 years?

**Value of an Annuity**   **27.** To settle a debt, Kevin Sharp has agreed to pay $100 each quarter into an account that earns 5% interest compounded quarterly for 4 years. How much will he have at the end of this period?

**Value of an Annuity**   **28.** A land developer purchases some land in Arizona, agreeing to make monthly payments of $1000 into an account paying 6% monthly for 6 years. Find the future value of the account.

**Value of an Annuity**   **29.** An accountant wants to buy a new computer in 3 years. The computer will cost $12,000. To accumulate the money, the accountant plans to deposit a sum of money at the end of each six-month period in an account paying 7.8% interest compounded semiannually. How much should each payment be?

**Value of an Annuity**   **30.** Michael Branson wants to buy a $15,000 car in 4 years. How much money must he deposit at the end of each month in an account paying 6% compounded monthly in order to have enough money to pay for the car?

**IRA Accounts**   **31. (a)** Suppose that you deposit $2000 in an IRA that pays 10% compounded annually. Suppose further that the money is left in the account for 20 years, then withdrawn, and the earnings are taxed at 37%. Find the amount left after taxes, then find the IRA multiplier $M$.
  **(b)** Suppose instead that the money was deposited in a regular account for the same period of time at the same rates. Find the amount left after taxes; then find the multiplier $m$.
  **(c)** Find $M/m$. What can you conclude about investment options (a) and (b)?

**IRA Accounts**    **32.** Redo Exercise 31 for $n = 10$.

**IRA Accounts**    **33.** Draw a conclusion from Example 8 and the results of Exercises 31(a) and 32(a). (*Hint:* It is called the *time value of money*.)

**IRA Accounts**    **34.** Draw a conclusion from Example 9 and Exercises 31(b) and 32(b).

## FOR THE COMPUTER

*Find the effective rate for a nominal rate of 7½% compounded as follows.*

**35.** 360 times per year

**36.** 365 times per year

*Find the present value of each of the following amounts.*

**37.** $5270, interest at 4¾% compounded annually for 18 years

**38.** $36,950, interest at 7.2% compounded semiannually for 15.5 years

**39.** $12,650, interest at 8.15% compounded quarterly for 25 years

**40.** $7516.28, interest at 6.3% compounded monthly for 8 years

**41.** Construct tables for $n = 10$ and $n = 20$ and various values of $i$ and $t$ similar to the table for $n = 30$ given after Example 9. Draw conclusions from your data.

### KEY WORDS

**4.1**  **exponential function**
**exponential equation**
**simple interest**
**compound interest**
**compound amount**
*e*

**4.2**  **years to double**
**logarithm**
**logarithmic function**
**properties of logarithms**
**common logarithms**
**natural logarithms**

**4.4**  **logistic function**

**4.5**  **exponential growth function**
**growth constant**
**decay constant**
**half-life**
**continuous compounding**
**limited growth function**
**learning curve**
**forgetting curve**

**4.6**  **effective rate**
**nominal (stated) rate**
**present value**
**future value**
**annuity**

## CHAPTER 4        REVIEW EXERCISES

*Solve each equation.*

**1.** $2^{3x} = \dfrac{1}{8}$

**2.** $\left(\dfrac{9}{16}\right)^x = \dfrac{3}{4}$

**3.** $9^{2y-1} = 27^y$

**4.** $\dfrac{1}{2} = \left(\dfrac{b}{4}\right)^{1/4}$

*Graph the functions defined as follows.*

**5.** $y = 5^x$

**6.** $y = 5^{-x}$

**7.** $y = \left(\dfrac{1}{5}\right)^x$

**8.** $y = \left(\dfrac{1}{2}\right)^{x-1}$

*Write the following equations using logarithms.*

**9.** $2^6 = 64$

**10.** $3^{1/2} = \sqrt{3}$

**11.** $e^{.09} = 1.09417$

**12.** $10^{1.07918} = 12$

*Write the following equations using exponents.*

**13.** $\log_2 32 = 5$

**14.** $\log_{10} 100 = 2$

**15.** $\ln 82.9 = 4.41763$

**16.** $\log 15.46 = 1.18921$

*Evaluate each expression.*

**17.** $\log_3 81$

**18.** $\log_{1/3} 81$

**19.** $\log_{32} 16$

**20.** $\log_{25} 5$

**21.** $\log_{100} 1000$

**22.** $\log_{1/2} 4$

*Find each of the following natural logarithms.*

**23.** $\ln 6.2$

**24.** $\ln 700$

**25.** $\ln 483$

**26.** $\ln .504$

*Simplify each expression using the properties of logarithms.*

**27.** $\log_5 3k + \log_5 7k^3$

**28.** $\log_3 2y^3 - \log_3 8y^2$

**29.** $2 \log_2 x - 3 \log_2 m$

**30.** $5 \log_4 r - 3 \log_4 r^2$

*Solve each equation. Round each answer to the nearest thousandth.*

**31.** $8^p = 19$

**32.** $3^z = 11$

**33.** $2^{-m} = 7$

**34.** $15^{-k} = 9$

**35.** $e^{-5-2x} = 5$

**36.** $e^{3x-1} = 12$

**37.** $\left(1 + \dfrac{m}{3}\right)^5 = 10$

**38.** $\left(1 + \dfrac{2p}{5}\right)^2 = 3$

*Find the derivative of each function defined as follows.*

**39.** $y = -6e^{2x}$

**40.** $y = 8e^{.5x}$

**41.** $y = e^{-2x^3}$

**42.** $y = -4e^{x^2}$

**43.** $y = 5xe^{2x}$

**44.** $y = -7x^2e^{-3x}$

**45.** $y = \ln |2 + x^2|$

**46.** $y = \ln |5x + 3|$

**47.** $y = \dfrac{\ln |3x|}{x - 3}$

**48.** $y = \dfrac{\ln |2x - 1|}{x + 3}$

**49.** $y = \dfrac{x\,e^x}{\ln |x^2 - 1|}$

**50.** $y = \dfrac{(x^2 + 1)e^{2x}}{\ln |x|}$

**51.** $y = (x^2 + e^x)^2$

**52.** $y = (e^{2x+1} - 2)^4$

*Find all relative maxima or minima for the functions defined as follows. Graph each function.*

**53.** $y = xe^x$

**54.** $y = 3xe^{-x}$

**55.** $y = \dfrac{e^x}{x - 1}$

**56.** $y = \dfrac{\ln |5x|}{2x}$

*Find the amount of interest earned by each of the following deposits.*

**57.** $6902 at 12% compounded semiannually for 8 years

**58.** $2781.36 at 8% compounded quarterly for 6 years

*Find the compound amount if $12,104 is invested at 8% compounded continuously for each of the following periods.*

**59.** 2 yr

**60.** 4 yr

**61.** 7 yr

**62.** 9 yr

**63.** 12 yr

*Find the compound amounts for the following deposits if interest is compounded continuously.*

**64.** $1500 at 10% for 9 years

**65.** $12,000 at 5% for 8 years

**66.** $68,100 at 9% for 2 years

**67.** $7590 at 15% for 4 years

*Find the present value of each of the following amounts.*

**68.** $2000 at 6% interest compounded annually for 5 years

**69.** $10,000 at 8% interest compounded semiannually for 6 years

**70.** $43,200 at 8% interest compounded quarterly for 4 years

**71.** $9760 at 12% interest compounded monthly for 3 years

*Find the future value of each of the following annuities.*

**72.** $5000 each year at 7% compounded annually for 6 years

**73.** $800 each quarter for 8 years at 6% interest compounded quarterly

**74.** $1500 twice a year at 8% compounded semiannually for 3 years

**75.** $360 each month for 4 years at 12% compounded monthly

## ■ APPLICATIONS

### BUSINESS AND ECONOMICS

**Employee Training**  **76.** A company finds that its new workers produce

$$P(x) = 100 - 100e^{-.8x}$$

items per day after $x$ days on the job.
Find each of the following.

(a) $P_0 = P(0)$     (b) $P(1)$     (c) $P(5)$

(d) How many items per day should an experienced worker produce?

(e) How long will it be before a new employee will produce 50 items per day?

(f) Graph $P(x)$.

**Compound Interest**  **77.** To help pay for college expenses, Ginny Guerrant borrowed $10,000 at 10% interest compounded semiannually for 8 years. How much will she owe at the end of the 8-year period?

**Continuous Compounding**  **78.** How long will it take for $1 to triple at an average annual inflation rate of 8% compounded continuously?

**Continuous Compounding**  **79.** Find the interest rate needed for $6000 to grow to $8000 in 3 years with continuous compounding.

**Present Value**  **80.** Bill Poole wants to open a camera shop. How much must he deposit now at 6% interest compounded monthly to have $25,000 at the end of 3 years?

**Value of an Annuity**  **81.** Suppose Poole (see Exercise 80) deposits $635 each month in an account paying 6% compounded monthly for 3 years.

(a) Will he have $25,000 at the end of the period?

(b) If not, how much will he lack?

### LIFE SCIENCES

**Population Growth**  **82.** A population of 15,000 small deer in a specific region has grown exponentially to 17,000 in 4 years.

(a) Write an exponential equation to express the population growth $y$ in terms of time $t$ in years.

(b) At this rate, how long will it take for the population to reach 45,000?

**Intensity of Light**   83. The intensity of light (in appropriate units) passing through water decreases exponentially with the depth it penetrates beneath the surface according to the function

$$I(x) = 10e^{-.3x},$$

where $x$ is the depth in meters. A certain water plant requires light of an intensity of 1 unit. What is the greatest depth of water in which it will grow?

**Drug Concentration**   84. The concentration of a certain drug in the bloodstream at time $t$ in minutes is given by

$$c(t) = e^{-t} - e^{-2t}.$$

Find the concentrations at the following times.

(a) $t = 0$   (b) $t = 1$   (c) $t = 5$

(d) Find the maximum concentration and the value of $t$ where it occurs.

(e) Graph $c(t)$.

**Glucose Concentration**   85. When glucose is infused into a person's bloodstream at a constant rate of $c$ grams per minute, the glucose is converted and removed from the bloodstream at a rate proportional to the amount present. The amount of glucose in the bloodstream at time $t$ (in minutes) is given by

$$g(t) = \frac{c}{a} + \left(g_0 - \frac{c}{a}\right)e^{-at},$$

where $a$ is a positive constant. Assume $g_0 = .08$, $c = .1$, and $a = 1.3$. Find each of the following.

(a) $g(5)$   (b) $g(10)$   (c) $g(20)$

(d) At what time is the amount of glucose in the bloodstream .1 gram?

## SOCIAL SCIENCES

**Legislative Turnover**   86. The turnover of legislators is a problem of interest to political scientists. One model of legislative turnover in the U.S. House of Representatives is given by the exponential function

$$M = 434e^{-.08t},$$

where $M$ is the number of continuously serving members at time $t$.* This model is based on the 1965 membership of the House. Thus, 1965 corresponds to $t = 0$, 1966 to $t = 1$, and so on. Find the number of continuously serving members in each of the following years.

(a) 1969   (b) 1973   (c) 1979

**Average Height**   87. The average height, in meters, of the members of a certain tribe is approximated by

$$h = .5 + \log t,$$

where $t$ is the tribe member's age in years, and $1 < t < 20$. Find the heights of tribe members of the following ages.

(a) 2 years   (b) 5 years   (c) 10 years   (d) 20 years

(e) Graph $h$.

---

*From UMAP Unit 296, "Exponential Models of Legislative Turnover" by Thomas W. Casstevens, published by Birkhauser Boston, Inc. Reprinted by permission of COMAP, Inc.

## PHYSICAL SCIENCES

**Oil Production**    **88.** The production of an oil well has decreased exponentially from 128,000 barrels per year five years ago to 100,000 barrels per year at present.

   **(a)** Letting $t = 0$ represent the present time, write an exponential equation for production $y$ in terms of time $t$ in years.

   **(b)** Find the time it will take for production to fall to 70,000 barrels per year.

**Radioactive Decay**    **89.** Radium 228 decays exponentially. A sample that contained 100 grams 6 years ago ($t = 0$) has decreased to 25 grams at present.

   **(a)** Write an exponential equation to express the amount $y$ present after $t$ years.

   **(b)** What is the half-life of radium 228?

**EXTENDED APPLICATION**

## THE VAN MEEGEREN ART FORGERIES*

After the liberation of Belgium at the end of World War II, officials began a search for Nazi collaborators. One person arrested as a collaborator was a minor painter, H. A. Van Meegeren; he was charged with selling a valuable painting by the Dutch artist Vermeer (1632–1675) to the Nazi Hermann Goering. He defended himself from the very serious charge of collaboration by claiming that the painting was a fake—he had forged it himself.

He also claimed that the beautiful and famous painting "Disciples at Emmaus," as well as several other supposed Vermeers, was his own work. To prove this, he did another "Vermeer" in his prison cell. An international panel of experts was assembled, which pronounced as forgeries all the "Vermeers" in question.

Many people would not accept the verdict of this panel for the painting "Disciples at Emmaus"; it was felt to be too beautiful to be the work of a minor talent such as Van Meegeren. In fact, the painting was declared genuine by a noted art scholar and sold for $170,000. The question of the authenticity of this painting continued to trouble art historians, who began to insist on conclusive proof one way or the other. This proof was given by a group of scientists at Carnegie-Mellon University, using the idea of radioactive decay.

The dating of objects is based on radioactivity; the atoms of certain radioactive elements are unstable, and within a given time period a fixed fraction of such atoms will spontaneously disintegrate, forming atoms of a new element. If $t_0$ represents some initial time, $N_0$ represents the number of atoms present at time $t_0$, and $N$ represents the number present at some later time $t$, then it can be shown (using physics and calculus) that

$$t - t_0 = \frac{1}{\lambda} \cdot \ln \frac{N_0}{N},$$

where $\lambda$ is a "decay constant" that depends on the radioactive substance under consideration.

---

*From "The Van Meegeren Art Forgeries" by Martin Braum from *Applied Mathematical Sciences*, Vol. 15. Copyright © 1975. Published by Springer-Verlag New York, Inc. Reprinted by permission.

If $t_0$ is the time that the substance was formed or made, then $t - t_0$ is the age of the item. Thus, the age of an item is given by

$$\frac{1}{\lambda} \cdot \ln \frac{N_0}{N}.$$

The decay constant $\lambda$ can be readily found, as can $N$, the number of atoms present now. The problem is $N_0$—we can't find a value for this variable. It is possible to get reasonable ranges for the values of $N_0$, however, by studying the white lead in the painting. This pigment has been used by artists for over 2000 years. It contains a small amount of radioactive lead 210 and an even smaller amount of radium 226.

Radium 226 disintegrates through a series of intermediate steps to produce lead 210. The lead 210, in turn, decays to form polonium 210. This last process, lead 210 to polonium 210, has a half-life of 22 years. That is, in 22 years half the initial quantity of lead 210 will decay to polonium 210.

When lead ore is processed to form white lead, most of the radium is removed with other waste products. Thus, most of the supply of lead 210 is cut off, with the remainder beginning to decay very rapidly. This process continues until the lead 210 in the white lead is once more in equilibrium with the small amount of radium then present. Let $y(t)$ be the amount of lead 210 per gram of white lead present at time of manufacture of the pigment, $t_0$. Let $r$ represent the number of disintegrations of radium 226 per minute per gram of white lead. (Actually, $r$ is a function of time, but the half-life of radium 226 is so long in comparison to the time interval in question that we assume it to be a constant.) If $\lambda$ is the decay constant for lead 210, then it can be shown that

$$y(t) = \frac{r}{\lambda}[1 - e^{-\lambda(t - t_0)}] + y_0 e^{-\lambda(t - t_0)}. \tag{1}$$

All variables in this result can be evaluated except $y_0$. To get around this problem, we use the fact that the original amount of lead 210 was in radioactive equilibrium with the larger amount of radium 226 in the ore from which the metal was extracted. We therefore take samples of different ores and compute the rate of disintegration of radium 226. The results are as shown in the table.

| Location of Ore | Disintegrations of Radium 226 per Minute per Gram of White Lead |
|---|---|
| Oklahoma | 4.5 |
| S.E. Missouri | .7 |
| Idaho | .18 |
| Idaho | 2.2 |
| Washington | 140 |
| British Columbia | .4 |

The numbers in the table vary from .18 to 140—quite a range. Since the number of disintegrations is proportional to the amount of lead 210 present originally, we must conclude that $y_0$ also varies over a tremendous range. Thus, equation (1) cannot be used to obtain even a crude estimate of the age of a painting. However, we want to distinguish only between a modern forgery and a genuine painting 300 years old.

To do this, we observe that if the painting is very old compared to the 22-year half-life of lead 210, then the amount of radioactivity from the lead 210 will almost equal the amount of radioactivity from the radium 226. On the other hand, if the painting is modern, then the amount of radioactivity from the lead 210 will be much greater than the amount from the radium 226.

We want to know if the painting is modern or 300 years old. To find out, let $t - t_0 = 300$ in equation (1), getting

$$\lambda y_0 = \lambda \cdot y(t) \cdot e^{300\lambda} + r(e^{300\lambda} - 1) \qquad \text{(2)}$$

after some rearrangement of terms. If the painting is modern, then $\lambda y_0$ should be a very large number; $\lambda y_0$ represents the number of disintegrations of the lead 210 per minute per gram of white lead at the time of manufacture. By studying samples of white lead, we can conclude that $\lambda y_0$ should never be anywhere near as large as 30,000. Thus, we use equation (2) to calculate $\lambda y_0$; if our result is greater than 30,000 we conclude that the painting is a modern forgery. The details of this calculation are left for the exercises.

## EXERCISES

1. To calculate $\lambda$, use the formula

$$\lambda = \frac{\ln 2}{\text{half-life}}.$$

Find $\lambda$ for lead 210, whose half-life is 22 years.

2. For the painting "Disciples at Emmaus," the current rate of disintegration of the lead 210 was measured and found to be $\lambda \cdot y(t) = 8.5$. Also, $r$ was found to be .8. Use this information and equation (2) to calculate $\lambda y_0$. Based on your results, what do you conclude about the age of the painting?

Several other possible forgeries are listed below. Decide which of them must be modern forgeries.

| Title | $\lambda \cdot y(t)$ | $r$ |
|---|---|---|
| 3. "Washing of Feet" | 12.6 | .26 |
| 4. "Lace Maker" | 1.5 | .4 |
| 5. "Laughing Girl" | 5.2 | 6 |
| 6. "Woman Reading Music" | 10.3 | .3 |

# 5 Integration

In the previous three chapters, we studied the mathematical process of finding the derivative of a function, and we considered various applications of derivatives. The material that was covered belongs to the branch of calculus called *differential calculus*. In this chapter and the next we will study another branch of calculus, called *integral calculus*. Like the derivative of a function, the indefinite integral of a function is a special limit with many diverse applications. Geometrically, the derivative is related to the slope of the tangent line to a curve, while the definite integral is related to the area under a curve. As this chapter will show, differential and integral calculus are connected by the fundamental theorem of calculus.

## 5.1 ANTIDERIVATIVES

Functions used in applications in previous chapters have provided information about a *total amount* of a quantity, such as cost, revenue, profit, temperature, gallons of oil, or distance. Working with these functions provided information about the rate of change of these quantities and allowed us to answer important questions about the extrema of the functions. It is not always possible to find ready-made functions that provide information about the total amount of a quantity, but it is often possible to collect enough data to come up with a function that gives the *rate* of *change* of a quantity. Since we know that derivatives give the rate of change when the total amount is known, is it possible to reverse the process and use a known rate of change to get a function that gives the total amount of a quantity? The answer is yes; this reverse process, called *antidifferentiation*, is the topic of this section. The *antiderivative* of a function is defined as follows.

**ANTIDERIVATIVE**

> If $F'(x) = f(x)$, then $F(x)$ is an **antiderivative** of $f(x)$.

### EXAMPLE 1

**(a)** If $F(x) = 10x$, then $F'(x) = 10$, so $F(x) = 10x$ is an antiderivative of $f(x) = 10$.

**(b)** For $F(x) = x^2$, $F'(x) = 2x$, making $F(x) = x^2$ an antiderivative of $f(x) = 2x$. ■

### EXAMPLE 2

Find an antiderivative of $f(x) = 5x^4$.

To find a function $F(x)$ whose derivative is $5x^4$, work backwards. Recall that the derivative of $x^n$ is $nx^{n-1}$. If

$$nx^{n-1} \text{ is } 5x^4,$$

then $n - 1 = 4$ and $n = 5$, so $x^5$ is an antiderivative of $5x^4$. ■

### ▬ EXAMPLE 3

Suppose a population is growing at a rate given by $f(x) = e^x$, where $x$ is time in years from some initial date. Find a function giving the population at time $x$.

Let the population function be $F(x)$. Then

$$f(x) = F'(x) = e^x.$$

The derivative of the function defined by $F(x) = e^x$ is $F'(x) = e^x$, so a population function with the given growth rate is $F(x) = e^x$. ▬

The function from Example 1(b), defined by $F(x) = x^2$, is not the only function whose derivative is $f(x) = 2x$; for example, both $G(x) = x^2 + 2$ and $H(x) = x^2 - 7$ have $f(x) = 2x$ as a derivative. In fact, for any real number $C$, the function $F(x) = x^2 + C$ has $f(x) = 2x$ as its derivative. This means that there is a *family* or *class* of functions having $2x$ as an antiderivative. As the next theorem states, if two functions $F(x)$ and $G(x)$ are antiderivatives of $f(x)$, then $F(x)$ and $G(x)$ can differ only by a constant.

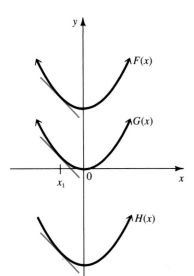

Slopes of the tangent lines at $x = x_1$ are the same.

**FIGURE 1**

> ▬ If $F(x)$ and $G(x)$ are both antiderivatives of a function $f(x)$, then there is a constant $C$ such that
>
> $$F(x) - G(x) = C.$$
>
> (Two antiderivatives of a function can differ only by a constant.)

For example,

$$F(x) = x^2 + 2, \qquad G(x) = x^2, \qquad \text{and} \qquad H(x) = x^2 - 4$$

are all antiderivatives of $f(x) = 2x$, and any two of them differ only by a constant. The derivative of a function gives the slope of the tangent line at any $x$-value. The fact that these three functions have the same derivative, $f(x) = 2x$, means that their slopes at any particular value of $x$ are the same, as shown in Figure 1.

The family of all antiderivatives of the function $f$ is indicated by

$$\int f(x)\ dx.$$

The symbol $\int$ is the **integral sign,** $f(x)$ is the **integrand,** and $\int f(x)\ dx$ is called an **indefinite integral,** the most general antiderivative of $f$.

**INDEFINITE INTEGRAL**

> ▬ If $F'(x) = f(x)$, then
>
> $$\int f(x)\ dx = F(x) + C,$$
>
> for any real number $C$.

For example, using this notation,

$$\int (2x)\, dx = x^2 + C.$$

The *dx* in the indefinite integral indicates that ∫*f*(*x*) *dx* is the "integral of *f*(*x*) *with respect to x*" just as the symbol *dy/dx* denotes the "derivative of *y* with respect to *x*." For example, in the indefinite integral ∫*2axdx*, *dx* indicates that *a* is to be treated as a constant and *x* as the variable, so that

$$\int 2axdx = 2a \cdot \frac{x^2}{2} = ax^2 + C.$$

On the other hand,

$$\int 2axda = 2x \cdot \frac{a^2}{2} = xa^2 + C.$$

A more complete interpretation of *dx* will be discussed later.

The symbol ∫*f*(*x*)*dx* was created by G.W. Leibniz (1646–1716) in the latter part of the seventeenth century. The ∫ is an elongated S from *summa*, the Latin word for *sum*. The word *integral* as a term in the calculus was coined by Jakob Bernoulli (1654–1705), a Swiss mathematician who corresponded frequently with Leibniz. The relationship between sums and integrals will be clarified in the next two sections.

The method of working backwards, as above, to find an antiderivative is not very satisfactory. Some rules for finding antiderivatives are needed. Since the process of finding an indefinite integral is the inverse of the process of finding a derivative, each formula for derivatives leads to a rule for indefinite integrals.

As mentioned above, the derivative of $x^n$ is found by multiplying *x* by *n* and reducing the exponent on *x* by 1. To find an indefinite integral, that is, to undo what was done, *increase* the exponent by 1 and *divide* by the new exponent, $n + 1$.

**POWER RULE**

For any real number $n \neq -1$,

$$\int x^n\, dx = \frac{1}{n + 1} x^{n+1} + C.$$

This rule can be verified by differentiating the expression on the right above:

$$\frac{d}{dx}\left(\frac{1}{n + 1} x^{n+1} + C\right) = \frac{n + 1}{n + 1} x^{(n+1)-1} + 0 = x^n.$$

(If $n = -1$, the expression in the denominator is 0, and the above rule cannot be used. Finding an antiderivative for this case is discussed later.)

### ■ EXAMPLE 4

Find each indefinite integral.

**(a)** $\int t^3 \, dt$

Use the power rule with $n = 3$.

$$\int t^3 \, dt = \frac{1}{3+1} t^{3+1} + C = \frac{1}{4} t^4 + C$$

**(b)** $\int \frac{1}{t^2} \, dt$

First, write $1/t^2$ as $t^{-2}$. Then

$$\int \frac{1}{t^2} \, dt = \int t^{-2} \, dt = \frac{t^{-1}}{-1} + C = \frac{-1}{t} + C.$$

**(c)** $\int \sqrt{u} \, du$

Since $\sqrt{u} = u^{1/2}$,

$$\int \sqrt{u} \, du = \int u^{1/2} \, du = \frac{1}{1/2 + 1} u^{3/2} + C = \frac{2}{3} u^{3/2} + C.$$

To check this, differentiate $(2/3)u^{3/2} + C$; the derivative is $u^{1/2}$, the original function.

**(d)** $\int dx$

Write $dx$ as $1 \cdot dx$ and use the fact that $x^0 = 1$ for any nonzero number $x$ to get

$$\int dx = \int 1 dx = \int x^0 dx = \frac{1}{1} x^1 + C = x + C. \quad ■$$

As shown earlier, the derivative of the product of a constant and a function is the product of the constant and the derivative of the function. A similar rule applies to indefinite integrals. Also, since derivatives of sums or differences are found term by term, indefinite integrals also can be found term by term.

**CONSTANT MULTIPLE RULE AND SUM OR DIFFERENCE RULE**

If all indicated integrals exist,

$$\int k \cdot f(x)dx = k \int f(x)dx \qquad \text{for any real number } k,$$

and

$$\int [f(x) \pm g(x)]dx = \int f(x) \, dx \pm \int g(x)dx.$$

■ EXAMPLE 5

Find each of the following.

**(a)** $\int 2v^3 \, dv$

By the constant multiple rule and the power rule,

$$\int 2v^3 \, dv = 2\int v^3 \, dv = 2\left(\frac{1}{4}v^4\right) + C = \frac{1}{2}v^4 + C.$$

**(b)** $\int \frac{12}{z^5} \, dz$

Use negative exponents.

$$\int \frac{12}{z^5} \, dz = \int 12z^{-5} \, dz = 12\int z^{-5} \, dz = 12\left(\frac{z^{-4}}{-4}\right) + C$$

$$= -3z^{-4} + C = \frac{-3}{z^4} + C$$

**(c)** $\int (3z^2 - 4z + 5) dz$

Using the rules in this section,

$$\int (3z^2 - 4z + 5) dz = 3\int z^2 \, dz - 4\int z \, dz + 5\int dz$$

$$= 3\left(\frac{1}{3}z^3\right) - 4\left(\frac{1}{2}z^2\right) + 5z + C$$

$$= z^3 - 2z^2 + 5z + C.$$

Only one constant $C$ is needed in the answer: the three constants from integrating term by term are combined. In Example 5(a), $C$ represents any real number, so it is not necessary to multiply it by 2 in the next-to-last step. ■

■ EXAMPLE 6

Find each of the following.

**(a)** $\int \frac{x^2 + 1}{\sqrt{x}} \, dx$

First rewrite the integrand as follows.

$$\int \frac{x^2 + 1}{\sqrt{x}} dx = \int \left(\frac{x^2}{\sqrt{x}} + \frac{1}{\sqrt{x}}\right) dx$$

$$= \int \left(\frac{x^2}{x^{1/2}} + \frac{1}{x^{1/2}}\right) dx$$

$$= \int (x^{3/2} + x^{-1/2}) dx$$

Now find the antiderivative.

$$\int(x^{3/2} + x^{-1/2})dx = \frac{x^{5/2}}{5/2} + \frac{x^{1/2}}{1/2} + C$$

$$= \frac{2}{5}x^{5/2} + 2x^{1/2} + C$$

**(b)** $\int(x^2 - 1)^2dx$

Square the binomial first, and then find the antiderivative.

$$\int(x^2 - 1)^2dx = \int(x^4 - 2x^2 + 1)dx$$

$$= \frac{x^5}{5} - \frac{2x^3}{3} + x + C \quad \blacksquare$$

To check integration, take the derivative of the result. For instance, in Example 5(c) check that $z^3 - 2z^2 + 5z + C$ is the required indefinite integral by taking the derivative

$$\frac{d}{dz}(z^3 - 2z^2 + 5z + C) = 3z^2 - 4z + 5,$$

which agrees with the original information.

It was shown earlier that the derivative of $f(x) = e^x$ is $f'(x) = e^x$. Also, the derivative of $f(x) = e^{kx}$ is $f'(x) = k \cdot e^{kx}$. These results lead to the following formulas for indefinite integrals of exponential functions.

**INDEFINITE INTEGRALS OF EXPONENTIAL FUNCTIONS**

$$\int e^x dx = e^x + C$$

$$\int e^{kx}dx = \frac{1}{k} \cdot e^{kx} + C, \qquad k \neq 0$$

■ **EXAMPLE 7**

**(a)** $\int 9e^t \, dt = 9\int e^t dt = 9e^t + C$

**(b)** $\int e^{9t} \, dt = \frac{1}{9} e^{9t} + C$

**(c)** $\int 3e^{(5/4)u}du = 3\left(\frac{1}{5/4} e^{(5/4)u}\right) + C$

$$= 3\left(\frac{4}{5}\right) e^{(5/4)u} + C$$

$$= \frac{12}{5} e^{(5/4)u} + C \quad \blacksquare$$

The restriction $n \neq -1$ was necessary in the formula for $\int x^n dx$ since $n = -1$ made the denominator of $1/(n + 1)$ equal to 0. To find $\int x^n dx$ when $n = -1$, that is, to find $\int x^{-1} dx$, recall the differentiation formula for the logarithmic function: the derivative of $f(x) = \ln |x|$, where $x \neq 0$, is $f'(x) = 1/x = x^{-1}$. This formula for the derivative of $f(x) = \ln |x|$ gives a formula for $\int x^{-1} dx$.

**INDEFINITE INTEGRAL OF $x^{-1}$**

$$\int x^{-1} dx = \int \frac{1}{x} dx = \ln |x| + C$$

### EXAMPLE 8

**(a)** $\displaystyle\int \frac{4}{x} dx = 4 \int \frac{1}{x} dx = 4 \ln |x| + C$

**(b)** $\displaystyle\int \left( -\frac{5}{x} + e^{-2x} \right) dx = -5 \ln |x| - \frac{1}{2} e^{-2x} + C$

In all the examples above, the antiderivative family of functions was found. In many applications, however, the given information allows us to determine the value of the integration constant $C$. The next examples illustrate this idea.

### EXAMPLE 9

Suppose a publishing company has found that the marginal cost at a level of production of $x$ thousand books is given by

$$C'(x) = \frac{50}{\sqrt{x}}$$

and that the fixed cost (the costs before the first book can be produced) is $25,000. Find the cost function $C(x)$.

Write $50/\sqrt{x}$ as $50/x^{1/2}$ or $50x^{-1/2}$, and then use the indefinite integral rules to integrate the function.

$$\int \frac{50}{\sqrt{x}} dx = \int 50x^{-1/2} dx = 50(2x^{1/2}) + k = 100x^{1/2} + k$$

(Here $k$ is used instead of $C$ to avoid confusion with the cost function $C(x)$.) To find the value of $k$, use the fact that $C(0)$ is 25,000.

$$C(x) = 100x^{1/2} + k$$
$$25,000 = 100 \cdot 0 + k$$
$$k = 25,000$$

With this result, the cost function is $C(x) = 100x^{1/2} + 25,000.$

### ■ EXAMPLE 10

Suppose the marginal revenue from a product is given by $50 - 3x - x^2$. Find the demand function for the product.

The marginal revenue is the derivative of the revenue function

$$\frac{dR}{dx} = 50 - 3x - x^2,$$

so
$$R = \int (50 - 3x - x^2)dx$$
$$= 50x - \frac{3x^2}{2} - \frac{x^3}{3} + C.$$

If $x = 0$, then $R = 0$ (no items sold means no revenue), and

$$0 = 50(0) - \frac{3(0)^2}{2} - \frac{(0)^3}{3} + C$$

$$0 = C.$$

Thus,

$$R = 50x - \frac{3x^2}{2} - \frac{x^3}{3}$$

gives the revenue function. Now, recall that $R = xp$, where $p$ is the demand function. Then

$$50x - \frac{3}{2}x^2 - \frac{1}{3}x^3 = xp$$

$$50 - \frac{3}{2}x - \frac{1}{3}x^2 = p,$$

which gives the demand function. ■

In the next example integrals are used to find the position of a particle when the acceleration of the particle is given.

### ■ EXAMPLE 11

Recall that if the function $s(t)$ gives the position of a particle at time $t$, then its velocity $v(t)$ and its acceleration $a(t)$ are given by

$$v(t) = s'(t) \quad \text{and} \quad a(t) = v'(t) = s''(t).$$

**(a)** Suppose the velocity of an object is $v(t) = 6t^2 - 8t$ and that the object is at $-5$ when time is 0. Find $s(t)$.

Since $v(t) = s'(t)$, the function $s(t)$ is an antiderivative of $v(t)$:

$$s(t) = \int v(t)dt = \int (6t^2 - 8t)\, dt$$
$$= 2t^3 - 4t^2 + C$$

for some constant $C$. Find $C$ from the given information that $s = -5$ when $t = 0$.

$$s(t) = 2t^3 - 4t^2 + C$$
$$-5 = 2(0)^3 - 4(0)^2 + C$$
$$-5 = C$$
$$s(t) = 2t^3 - 4t^2 - 5$$

**(b)** Many experiments have shown that when an object is dropped, its acceleration (ignoring air resistance) is constant. This constant has been found to be approximately 32 feet per second every second; that is,

$$a(t) = -32.$$

The negative sign is used because the object is falling. Suppose an object is thrown down from the top of the 1100-foot-tall Sears Tower in Chicago. If the initial velocity of the object is $-20$ feet per second, find $s(t)$.

First find $v(t)$ by integrating $a(t)$:

$$v(t) = \int (-32) \, dt = -32t + k.$$

When $t = 0$, $v(t) = -20$:

$$-20 = -32(0) + k$$
$$-20 = k$$

and

$$v(t) = -32t - 20.$$

Now integrate $v(t)$ to find $s(t)$.

$$s(t) = \int (-32t - 20) dt = -16t^2 - 20t + C$$

Since $s(t) = 1100$ when $t = 0$, we can substitute these values into the equation for $s(t)$ to get $C = 1100$ and

$$s(t) = -16t^2 - 20t + 1100$$

as the distance of the object from the ground after $t$ seconds.

**(c)** Use the equations derived in (b) to find out how fast the object was falling when it hit the ground and how long it took to strike the ground.

When the object strikes the ground, $s = 0$, so

$$0 = -16t^2 - 20t + 1100.$$

To solve this equation for $t$, factor out the common factor of $-4$ and then use the quadratic formula.

$$0 = -4(4t^2 + 5t - 275)$$
$$t = \frac{-5 \pm \sqrt{25 + 4400}}{8} \approx \frac{-5 \pm 66.5}{8}$$

Only the positive value of $t$ is meaningful here: $t \approx 7.69$. It takes the object about 7.69 seconds to strike the ground. From the velocity equation, with $t = 7.69$, we find

$$v(t) = -32t - 20$$
$$v(7.69) = -32(7.69) - 20 \approx -266,$$

so the object was falling (as indicated by the negative sign) at about 266 feet per second when it hit the ground. ▬

▬ **EXAMPLE 12**

Find a function $f$ whose graph has slope $f'(x) = 6x^2 + 4$ and goes through the point $(1, 1)$.

Since $f'(x) = 6x^2 + 4$,

$$f(x) = \int (6x^2 + 4)dx = 2x^3 + 4x + C.$$

The graph of $f$ goes through $(1, 1)$, so $C$ can be found by substituting 1 for $x$ and 1 for $f(x)$.

$$1 = 2(1)^3 + 4(1) + C$$
$$1 = 6 + C$$
$$C = -5$$

Finally, $f(x) = 2x^3 + 4x - 5$. ▬

▬ **5.1 EXERCISES**

*Find each of the following.*

**1.** $\int 4x \, dx$

**2.** $\int 8x \, dx$

**3.** $\int 5t^2 dt$

**4.** $\int 6x^3 dx$

**5.** $\int 6 \, dk$

**6.** $\int 2 \, dy$

**7.** $\int (2z + 3)dz$

**8.** $\int (3x - 5)dx$

**9.** $\int (x^2 + 6x)dx$

**10.** $\int (t^2 - 2t)dt$

**11.** $\int (t^2 - 4t + 5)dt$

**12.** $\int (5x^2 - 6x + 3)dx$

**13.** $\int (4z^3 + 3z^2 + 2z - 6)dz$

**14.** $\int (12y^3 + 6y^2 - 8y + 5)dy$

**15.** $\int 5\sqrt{z} \, dz$

**16.** $\int t^{1/4} dt$

**17.** $\int (u^{1/2} + u^{3/2})du$

**18.** $\int (4\sqrt{v} - 3v^{3/2})dv$

**19.** $\int (15x\sqrt{x} + 2\sqrt{x})dx$

**20.** $\int (x^{1/2} - x^{-1/2})dx$

**21.** $\int (10u^{3/2} - 14u^{5/2})du$

**22.** $\int (56t^{5/2} + 18t^{7/2})dt$

**23.** $\int \left(\frac{1}{z^2}\right) dz$

**24.** $\int \left(\frac{4}{x^3}\right) dx$

**25.** $\int \left(\frac{1}{y^3} - \frac{1}{\sqrt{y}}\right) dy$

**26.** $\int \left(\sqrt{u} + \frac{1}{u^2}\right) du$

**27.** $\int (-9t^{-2} - 2t^{-1})dt$

**28.** $\int (8x^{-3} + 4x^{-1})dx$

**29.** $\int e^{2t}dt$

**30.** $\int e^{-3y}dy$

**31.** $\int 3e^{-.2x}dx$

**32.** $\int -4e^{.2v}dv$

**33.** $\int \left(\frac{3}{x} + 4e^{-.5x}\right)dx$

**34.** $\int \left(\frac{9}{x} - 3e^{-.4x}\right)dx$

**35.** $\int \frac{1 + 2t^3}{t} dt$

**36.** $\int \frac{2y^{1/2} - 3y^2}{y} dy$

**37.** $\int (e^{2u} + 4u)du$

**38.** $\int (v^2 - e^{3v})dv$

**39.** $\int (x + 1)^2 dx$

**40.** $\int (2y - 1)^2 dy$

**41.** $\int \frac{\sqrt{x} + 1}{\sqrt[3]{x}} dx$

**42.** $\int \frac{1 - 2\sqrt[3]{z}}{\sqrt[3]{z}} dz$

**43.** Find an equation of the curve whose tangent line has a slope of
$$f'(x) = x^{2/3},$$
given that the point (1, 3/5) is on the curve.

**44.** The slope of the tangent line to a curve is given by
$$f'(x) = 6x^2 - 4x + 3.$$
If the point (0, 1) is on the curve, find an equation of the curve.

---

### ▦ APPLICATIONS

#### BUSINESS AND ECONOMICS

**Cost**    *Find the cost function for each of the following marginal cost functions.*

**45.** $C'(x) = 4x - 5$,   fixed cost is $8

**46.** $C'(x) = 2x + 3x^2$,   fixed cost is $15

**47.** $C'(x) = .2x^2 + 5x$,   fixed cost is $10

**48.** $C'(x) = .8x^2 - x$,   fixed cost is $5

**49.** $C'(x) = x^{1/2}$,   16 units cost $40

**50.** $C'(x) = x^{2/3} + 2$,   8 units cost $58

**51.** $C'(x) = x^2 - 2x + 3$,   3 units cost $15

**52.** $C'(x) = x + 1/x^2$,   2 units cost $5.50

**53.** $C'(x) = 1/x + 2x$,   7 units cost $58.40

**54.** $C'(x) = 5x - 1/x$,   10 units cost $94.20

**Profit**    **55.** The marginal profit of a small fast-food stand is given by
$$P'(x) = 2x + 20,$$
where $x$ is the sales volume in thousands of hamburgers. The "profit" is $-50$ dollars when no hamburgers are sold. Find the profit function.

**Profit**    **56.** Suppose the marginal profit from the sale of $x$ hundred items is
$$P'(x) = 4 - 6x + 3x^2,$$
and the profit on 0 items is $-$40. Find the profit function.

### LIFE SCIENCES

**Biochemical Excretion**    **57.** If the rate of excretion of a biochemical compound is given by

$$f'(t) = .01e^{-.01t},$$

the total amount excreted by time $t$ (in minutes) is $f(t)$.

(a) Find an expression for $f(t)$.

(b) If 0 units are excreted at time $t = 0$, how many units are excreted in 10 minutes?

**Concentration of a Solute**    **58.** According to Fick's law, the diffusion of a solute across a cell membrane is given by

$$c'(t) = \frac{kA}{V}[C - c(t)], \tag{1}$$

where $A$ is the area of the cell membrane, $V$ is the volume of the cell, $c(t)$ is the concentration inside the cell at time $t$, $C$ is the concentration outside the cell, and $k$ is a constant. If $c_0$ represents the concentration of the solute inside the cell when $t = 0$, then it can be shown that

$$c(t) = (c_0 - C)e^{-kAt/V} + M. \tag{2}$$

(a) Use the last result to find $c'(t)$.

(b) Substitute back into equation (1) to show that (2) is indeed the correct antiderivative of (1).

### PHYSICAL SCIENCES

*Exercises 59–61 refer to Example 8 in this section.*

**Velocity**    **59.** For a particular object, $a(t) = t^2 + 1$ and $v(0) = 6$. Find $v(t)$.

**Distance**    **60.** Suppose $v(t) = 6t^2 - 2/t^2$ and $s(1) = 8$. Find $s(t)$.

**Time**    **61.** An object is dropped from a small plane flying at 6400 feet. Assume that $a(t) = -32$ feet per second per second and $v(0)$ is 0, and find $s(t)$. How long will it take the object to hit the ground?

---

### ≡ 5.2 SUBSTITUTION

We saw how to integrate a few simple functions in the previous section. More complicated functions can sometimes be integrated by *substitution*. The substitution technique depends on the idea of a differential, discussed in an earlier chapter. If $u = f(x)$, the *differential* of $u$, written $du$, is defined as

$$du = f'(x)dx.$$

For example, if $u = 6x^4$, then $du = 24x^3\, dx$.

Recall the chain rule for derivatives as used in the following example:

$$\frac{d}{dx}(x^2 - 1)^5 = 5(x^2 - 1)^4(2x) = 10x(x^2 - 1)^4.$$

As in this example, the result of using the chain rule is often a product of two functions. Because of this, functions formed by the product of two functions can sometimes be integrated by using the chain rule in reverse. In the example above, working backwards from the derivative gives

$$\int 10x(x^2 - 1)^4 \, dx = (x^2 - 1)^5 + C.$$

To find an antiderivative involving products, it often helps to make a substitution: let $u = x^2 - 1$, so that $du = 2x \, dx$. Now substitute $u$ for $x^2 - 1$ and $du$ for $2x \, dx$ in the indefinite integral above.

$$\int 10x(x^2 - 1)^4 \, dx = \int 5 \cdot 2x(x^2 - 1)^4 \, dx$$
$$= 5\int (x^2 - 1)^4 \, (2x \, dx)$$
$$= 5\int u^4 \, du$$

This last integral can now be found by the power rule.

$$5\int u^4 \, du = 5 \cdot \frac{1}{5}u^5 + C = u^5 + C$$

Finally, substitute $x^2 - 1$ for $u$.

$$\int 10x(x^2 - 1)^4 \, dx = (x^2 - 1)^5 + C$$

This method of integration is called **integration by substitution.** As shown above, it is simply the chain rule for derivatives in reverse. The results can always be verified by differentiation.

This discussion leads to the following integration formula, which is sometimes called the general power rule for integrals.

**GENERAL POWER RULE**

For $u = f(x)$ and $du = f'(x)dx$,

$$\int u^n du = \frac{u^{n+1}}{n + 1} + C.$$

### ▬ EXAMPLE 1

Find $\int 6x(3x^2 + 4)^4 \, dx$.

A certain amount of trial and error may be needed to decide on the expression to set equal to $u$. The integrand must be written as two factors, one of which is the derivative of the other. In this example, if $u = 3x^2 + 4$, then $du = 6x dx$. Now substitute.

$$\int 6x(3x^2 + 4)^4 \, dx = \int (3x^2 + 4)^4 (6x \, dx) = \int u^4 \, du$$

Find this last indefinite integral.

$$\int u^4 \, du = \frac{u^5}{5} + C$$

Now replace $u$ with $3x^2 + 4$.

$$\int 6x(3x^2 + 4)^4 \, dx = \frac{u^5}{5} + C = \frac{(3x^2 + 4)^5}{5} + C$$

To verify this result, find the derivative:

$$\frac{d}{dx}\left[ \frac{(3x^2 + 4)^5}{5} + C \right] = \frac{5}{5}(3x^2 + 4)^4(6x) + 0 = (3x^2 + 4)^4(6x).$$

The derivative is the original function, as required. ▬

▬ **EXAMPLE 2**

Find $\int x^2 \sqrt{x^3 + 1} \, dx$.

An expression raised to a power is usually a good choice for $u$, so because of the square root or 1/2 power, let $u = x^3 + 1$; then $du = 3x^2 \, dx$. The integrand does not contain the constant 3, which is needed for $du$. To take care of this, multiply by 3/3, placing 3 inside the integral sign and 1/3 outside.

$$\int x^2 \sqrt{x^3 + 1} \, dx = \frac{1}{3} \int 3x^2 \sqrt{x^3 + 1} \, dx = \frac{1}{3} \int \sqrt{x^3 + 1} \, (3x^2 \, dx)$$

Now substitute $u$ for $x^3 + 1$ and $du$ for $3x^2 \, dx$, and then integrate.

$$\frac{1}{3} \int \sqrt{x^3 + 1} \, 3x^2 \, dx = \frac{1}{3} \int \sqrt{u} \, du = \frac{1}{3} \int u^{1/2} \, du$$

$$= \frac{1}{3} \cdot \frac{u^{3/2}}{3/2} + C = \frac{2}{9} u^{3/2} + C$$

Since $u = x^3 + 1$,

$$\int x^2 \sqrt{x^3 + 1} \, dx = \frac{2}{9}(x^3 + 1)^{3/2} + C. \quad ▬$$

The substitution method given in the examples above *will not always work*. For example, you might try to find

$$\int x^3 \sqrt{x^3 + 1} \, dx$$

by substituting $u = x^3 + 1$, so that $du = 3x^2 \, dx$. However, there is no *constant* that can be inserted inside the integral sign to give $3x^2$. This integral, and a great many others, cannot be evaluated by substitution.

### EXAMPLE 3

Find $\int \dfrac{2x + 5}{(x^2 + 5x)^2}\, dx$.

Let $u = x^2 + 5x$, so that $du = (2x + 5)dx$. This gives

$$\int \frac{2x + 5}{(x^2 + 5x)^2}\, dx = \int \frac{du}{u^2} = \int u^{-2}\, du$$

$$= \frac{u^{-1}}{-1} + C = \frac{-1}{u} + C.$$

Substituting $x^2 + 5x$ for $u$ gives

$$\int \frac{2x + 5}{(x^2 + 5x)^2}\, dx = \frac{-1}{x^2 + 5x} + C. \quad \blacksquare$$

Recall the formula for $\dfrac{d}{dx}(e^u)$, where $u = f(x)$:

$$\frac{d}{dx}(e^u) = e^u \frac{d}{dx}u.$$

For example, if $u = x^2$ then $\dfrac{d}{dx}u = \dfrac{d}{dx}(x^2) = 2x$, and

$$\frac{d}{dx}(e^{x^2}) = e^{x^2} \cdot 2x.$$

Working backwards, if $u = x^2$, then $du = 2x\, dx$, so

$$\int e^{x^2} \cdot 2x\, dx = \int e^u\, du = e^u + C = e^{x^2} + C.$$

The work above suggests the following rule for the indefinite integral of $e^u$, where $u = f(x)$.

**INDEFINITE INTEGRAL**
**OF $e^u$**

> If $u = f(x)$, then $du = f'(x)dx$ and
>
> $$\int e^u\, du = e^u + C.$$

### EXAMPLE 4

Find $\int x^2 \cdot e^{x^3}\, dx$.

Let $u = x^3$, the exponent on $e$. Then $du = 3x^2\, dx$. Multiplying by $3/3$ gives

$$\int x^2 \cdot e^{x^3}\, dx = \frac{1}{3} \int e^{x^3}\, (3x^2\, dx)$$

$$= \frac{1}{3} \int e^u\, du = \frac{1}{3}e^u + C = \frac{1}{3}e^{x^3} + C. \quad \blacksquare$$

Recall that the antiderivative of $f(x) = 1/x$ is $\ln |x|$. The next example uses that fact that $\int x^{-1}\, dx = \ln |x| + C$, together with the method of substitution.

### ■ EXAMPLE 5

Find $\displaystyle\int \frac{(2x - 3)dx}{x^2 - 3x}$.

Let $u = x^2 - 3x$, so that $du = (2x - 3)dx$. Then

$$\int \frac{(2x - 3)dx}{x^2 - 3x} = \int \frac{du}{u} = \ln |u| + C = \ln |x^2 - 3x| + C. \quad ■$$

Generalizing from the results of Example 5 suggests the following rule for finding the indefinite integral of $u^{-1}$, where $u = f(x)$.

**INDEFINITE INTEGRAL OF $u^{-1}$**

> If $u = f(x)$, then $du = f'(x)dx$ and
> $$\int u^{-1}\, du = \int \frac{du}{u} = \ln |u| + C.$$

The next example shows a more complicated integral evaluated by the method of substitution.

### ■ EXAMPLE 6

Find $\int x\sqrt{1 - x}\, dx$.

Let $u = 1 - x$. Then $x = 1 - u$ and $dx = -du$. Now substitute:

$$\int x\sqrt{1 - x}\, dx = \int (1 - u)\sqrt{u}(-du) = \int (u - 1)u^{1/2}\, du$$

$$= \int (u^{3/2} - u^{1/2})\, du = \frac{2}{5}u^{5/2} - \frac{2}{3}u^{3/2} + C$$

$$= \frac{2}{5}(1 - x)^{5/2} - \frac{2}{3}(1 - x)^{3/2} + C. \quad ■$$

The substitution method is useful if the integral can be written in one of the following forms, where $u(x)$ is some function of $x$.

**SUBSTITUTION METHOD**

| *Form of the Integral* | *Form of the Antiderivative* |
|---|---|
| **1.** $\displaystyle\int [u(x)]^n \cdot u'(x)\, dx,\ n \neq -1$ | $\dfrac{[u(x)]^{n+1}}{n + 1} + C$ |
| **2.** $\displaystyle\int e^{u(x)} \cdot u'(x)\, dx$ | $e^{u(x)} + C$ |
| **3.** $\displaystyle\int \frac{u'(x)\, dx}{u(x)}$ | $\ln |u(x)| + C$ |

## EXAMPLE 7

The research department for a hardware chain has determined that at one store the marginal price of $x$ boxes per week of a particular type of nails is

$$p'(x) = \frac{-4000}{(2x + 15)^3}.$$

Find the demand equation if the weekly demand for this type of nails is 10 boxes when the price of a box of nails is \$4.

To find the demand function, first integrate $p'(x)$ as follows.

$$p(x) = \int p'(x)dx$$

$$= \int \frac{-4000}{(2x + 15)^3} dx$$

Let $u = 2x + 15$. Then $du = 2dx$, and

$$p(x) = -2000 \int (2x + 15)^{-3} 2\, dx$$

$$= -2000 \int u^{-3} du$$

$$= (-2000)\frac{u^{-2}}{-2} + C$$

$$= \frac{1000}{u^2} + C \qquad \qquad \textbf{(1)}$$

$$p(x) = \frac{1000}{(2x + 15)^2} + C$$

Find the value of $C$ by using the given information that $p = 4$ when $x = 10$.

$$4 = \frac{1000}{(2 \cdot 10 + 15)^2} + C$$

$$4 = \frac{1000}{35^2} + C$$

$$4 = .82 + C$$

$$3.18 = C$$

Replacing $C$ with 3.18 in equation (1) gives the demand function,

$$p(x) = \frac{1000}{(2x + 15)^2} + 3.18. \quad \blacksquare$$

## EXAMPLE 8

To determine the top 100 popular songs of each year since 1956, Jim Quirin and Barry Cohen developed a function that represents the rate of change on the charts of *Billboard* magazine required for a song to earn a "star" on the *Billboard* "Hot 100" survey.* They developed the function

$$f(x) = \frac{A}{B + x},$$

where $f(x)$ represents the rate of change in position on the charts, $x$ is the position on the "Hot 100" survey, and $A$ and $B$ are appropriate constants. The function

$$F(x) = \int f(x)dx$$

is defined as the "Popularity Index." Find $F(x)$.

Integrating $f(x)$ gives

$$F(x) = \int f(x)dx$$

$$= \int \frac{A}{B + x}dx$$

$$= A\int \frac{1}{B + x}dx$$

Let $u = B + x$, so that $du = dx$. Then

$$F(x) = A\int \frac{1}{u} du = A \ln u + C$$

$$= A \ln (B + x) + C.$$

(The absolute value bars are not necessary, since $B + x$ is always positive here.) ▰

## 5.2 EXERCISES

*Use substitution to find the following indefinite integrals.*

**1.** $\int 4(2x + 3)^4 \, dx$

**2.** $\int (-4t + 1)^3 \, dt$

**3.** $\int \frac{4}{(y - 2)^3} \, dy$

**4.** $\int \frac{-3}{(x + 1)^4} \, dx$

**5.** $\int \frac{2 \, dm}{(2m + 1)^3}$

**6.** $\int \frac{3 \, du}{\sqrt{3u - 5}}$

**7.** $\int \frac{2x + 2}{(x^2 + 2x - 4)^4} \, dx$

**8.** $\int \frac{6x^2 \, dx}{(2x^3 + 7)^{3/2}}$

**9.** $\int z\sqrt{z^2 - 5} \, dz$

**10.** $\int r\sqrt{r^2 + 2} \, dr$

**11.** $\int (-4e^{2p}) \, dp$

**12.** $\int 5e^{-.3g} \, dg$

**13.** $\int 3x^2 e^{2x^3} \, dx$

**14.** $\int re^{-r^2} \, dr$

**15.** $\int (1 - t)e^{2t - t^2} \, dt$

**16.** $\int (x^2 - 1)e^{x^3 - 3x} \, dx$

**17.** $\int \frac{e^{1/z}}{z^2} \, dz$

**18.** $\int \frac{e^{\sqrt{y}}}{2\sqrt{y}} \, dy$

**19.** $\int \frac{-8}{1 + 3x} \, dx$

**20.** $\int \frac{9}{2 + 5t} \, dt$

*Formula for the "Popularity Index" from *Chartmasters' Rock 100,* Fourth Edition, by Jim Quirin and Barry Cohen. Copyright © 1987 by Chartmasters. Reprinted by permission.

**21.** $\displaystyle\int \frac{dt}{2t + 1}$

**22.** $\displaystyle\int \frac{dw}{5w - 2}$

**23.** $\displaystyle\int \frac{v\,dv}{(3v^2 + 2)^4}$

**24.** $\displaystyle\int \frac{x\,dx}{(2x^2 - 5)^3}$

**25.** $\displaystyle\int \frac{x - 1}{(2x^2 - 4x)^2}\,dx$

**26.** $\displaystyle\int \frac{2x + 1}{(x^2 + x)^3}\,dx$

**27.** $\displaystyle\int \left(\frac{1}{r} + r\right)\left(1 - \frac{1}{r^2}\right)dr$

**28.** $\displaystyle\int \left(\frac{2}{A} - A\right)\left(\frac{-2}{A^2} - 1\right)dA$

**29.** $\displaystyle\int \frac{x^2 + 1}{(x^3 + 3x)^{2/3}}\,dx$

**30.** $\displaystyle\int \frac{B^3 - 1}{(2B^4 - 8B)^{3/2}}\,dB$

**31.** $\displaystyle\int p(p + 1)^5\,dp$

**32.** $\displaystyle\int x^3(1 + x^2)^{1/4}\,dx$

**33.** $\displaystyle\int t\sqrt{5t - 1}\,dt$

**34.** $\displaystyle\int 4r\sqrt{8 - r}\,dr$

**35.** $\displaystyle\int \frac{u}{\sqrt{u - 1}}\,du$

**36.** $\displaystyle\int \frac{2x}{(x + 5)^6}\,dx$

**37.** $\displaystyle\int (\sqrt{x^2 + 12x})(x + 6)\,dx$

**38.** $\displaystyle\int (\sqrt{x^2 - 6x})(x - 3)\,dx$

**39.** $\displaystyle\int \frac{t}{t^2 + 2}\,dt$

**40.** $\displaystyle\int \frac{-4x}{x^2 + 3}\,dx$

**41.** $\displaystyle\int ze^{2z^2}\,dz$

**42.** $\displaystyle\int x^2 e^{-x^3}\,dx$

---

## ▤ APPLICATIONS

### BUSINESS AND ECONOMICS

**Revenue**    **43.** Suppose the marginal revenue in hundreds of thousands of dollars from the sale of $x$ jet planes is
$$R'(x) = 2x(x^2 + 50)^2.$$
Find the total revenue function if the revenue from 3 planes is \$206,379.

**Maintenance**    **44.** The rate of expenditure for maintenance for a particular machine is given by
$$M'(x) = \sqrt{x^2 + 12x}\,(2x + 12),$$
where $x$ is time measured in years. Maintenance costs for the fourth year are \$612. Find the total maintenance function.

**Profit**    **45.** The rate of growth of the profit (in millions of dollars) from a new technology is approximated by
$$p'(x) = xe^{-x^2},$$
where $x$ represents time measured in years. The total profit in the third year that the new technology is in operation is \$10,000. Find the total profit function.

### LIFE SCIENCES

**Drug Sensitivity**    **46.** If $x$ milligrams of a certain drug are administered to a person, the rate of change in the person's temperature, in degrees Celsius, with respect to the dosage (the person's *sensitivity* to the drug) is given by
$$D'(x) = \frac{2}{x + 9}.$$
One milligram raises the person's temperature 2.5° C. Find the function giving the total change in body temperature.

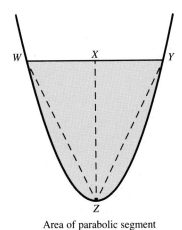

Area of parabolic segment

$$= \frac{4}{3} \text{ (area of triangle } WYZ)$$

**FIGURE 2**

## 5.3 AREA AND THE DEFINITE INTEGRAL

To calculate the areas of geometric figures such as rectangles, squares, triangles, and circles, we use specific formulas. In this section we consider the problem of finding the area of a figure or region that is bounded by curves, such as the shaded region in Figure 2.

The brilliant Greek mathematician Archimedes (about 287 B.C.–212 B.C.) is considered one of the greatest mathematicians of all time. His development of a rigorous method known as "exhaustion" to derive results was a forerunner of the ideas of integral calculus. Archimedes used a method that would later be verified by the more rigorous theory of integration. His method involved viewing a geometric figure as a sum of other figures. For example, he thought of a plane surface area as a figure consisting of infinitely many parallel line segments. Among the results established by Archimedes' method was the fact that the area of a segment of a parabola (shown in color in Figure 2) is equal to 4/3 the area of a triangle with the same base and the same height.

Under certain conditions the area of a region can be thought of as a sum of parts. Figure 3 shows the region bounded by the $y$-axis, the $x$-axis, and the graph of $f(x) = \sqrt{4 - x^2}$. A very rough approximation of the area of this region can be found by using two rectangles as in Figure 4. The height of the rectangle on the left is $f(0) = 2$ and the height of the rectangle on the right is $f(1) = \sqrt{3}$. The width of each rectangle is 1, making the total area of the two rectangles

$$1 \cdot f(0) + 1 \cdot f(1) = 2 + \sqrt{3} \approx 3.7321 \text{ square units.}$$

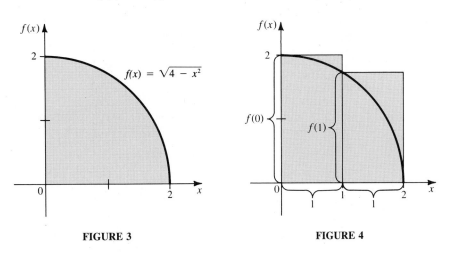

**FIGURE 3**                    **FIGURE 4**

As shown in Figure 4, this approximation is greater than the actual area. To improve the accuracy of the approximation, we could divide the interval from $x = 0$ to $x = 2$ into four equal parts, each of width 1/2, as shown in

Figure 5. As before, the height of each rectangle is given by the value of $f$ at the left side of the rectangle, and its area is the width, 1/2, multiplied by the height. The total area of the four rectangles is

$$\frac{1}{2} \cdot f(0) + \frac{1}{2} \cdot f\left(\frac{1}{2}\right) + \frac{1}{2} \cdot f(1) + \frac{1}{2} \cdot f\left(1\frac{1}{2}\right)$$

$$= \frac{1}{2}(2) + \frac{1}{2}\left(\frac{\sqrt{15}}{2}\right) + \frac{1}{2}(\sqrt{3}) + \frac{1}{2}\left(\frac{\sqrt{7}}{2}\right)$$

$$= 1 + \frac{\sqrt{15}}{4} + \frac{\sqrt{3}}{2} + \frac{\sqrt{7}}{4} \approx 3.4957 \text{ square units.}$$

This approximation looks better, but it is still greater than the actual area desired. To improve the approximation, divide the interval from $x = 0$ to $x = 2$ into 8 parts with equal widths of 1/4. (See Figure 6.) The total area of all these rectangles is

$$\frac{1}{4} \cdot f(0) + \frac{1}{4} \cdot f\left(\frac{1}{4}\right) + \frac{1}{4} \cdot f\left(\frac{1}{2}\right) + \frac{1}{4} \cdot f\left(\frac{3}{4}\right) + \frac{1}{4} \cdot f(1) + \frac{1}{4} \cdot f\left(\frac{5}{4}\right)$$

$$+ \frac{1}{4} \cdot f\left(\frac{3}{2}\right) + \frac{1}{4} \cdot f\left(\frac{7}{4}\right)$$

$$\approx 3.3398 \text{ square units.}$$

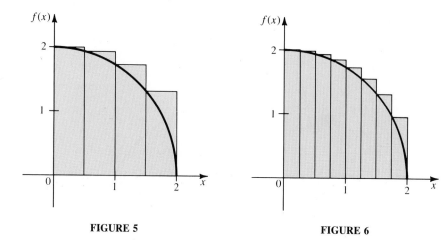

**FIGURE 5**          **FIGURE 6**

This process of approximating the area under a curve by using more and more rectangles to get a better and better approximation can be generalized. To do this, divide the interval from $x = 0$ to $x = 2$ into $n$ equal parts. Each of these $n$ intervals has width

$$\frac{2 - 0}{n} = \frac{2}{n},$$

so each rectangle has width $2/n$ and height determined by the function-value at the left side of the rectangle. Using a computer to find approximations to the area for several values of $n$ gives the results in the following table.

| $n$ | Area |
|---|---|
| 2 | 3.7321 |
| 4 | 3.4957 |
| 8 | 3.3398 |
| 10 | 3.3045 |
| 20 | 3.2285 |
| 50 | 3.1783 |
| 100 | 3.1512 |
| 500 | 3.1455 |

The areas in the last column in the table are approximations of the area under the curve, above the $x$-axis, and between the lines $x = 0$ and $x = 2$. As $n$ becomes larger and larger, the approximation is better and better, getting closer to the actual area. In this example, the exact area can be found by a formula from plane geometry. Write the given function as

$$y = \sqrt{4 - x^2},$$

then square both sides to get

$$y^2 = 4 - x^2$$
$$x^2 + y^2 = 4,$$

the equation of a circle centered at the origin with radius 2. The region in Figure 3 is the quarter of this circle that lies in the first quadrant. The actual area of this region is one-quarter of the area of the entire circle, or

$$\frac{1}{4}\pi(2)^2 = \pi \approx 3.1416.$$

As the number of rectangles increases without bound, the sum of the areas of these rectangles gets closer and closer to the actual area of the region, $\pi$. This can be written as

$$\lim_{n \to +\infty} (\text{sum of areas of } n \text{ rectangles}) = \pi.$$

(The value of $\pi$ was originally found by a process similar to this.)*

---

*The number $\pi$ is the ratio of the circumference of a circle to its diameter. It is an example of an *irrational number* and as such it cannot be expressed as a terminating or repeating decimal. Many approximations have been used for $\pi$ over the years. A passage in the Bible (I Kings 7:23) indicates a value of 3. The Egyptians used the value 3.16, and Archimedes showed that its value must be between 22/7 and 223/71. A Hindu writer, Brahmagupta, used $\sqrt{10}$ as its value in the seventh century. The search for the digits of $\pi$ has continued throughout the years. Two Japanese mathematicians recently computed the value to over 10 million decimal places!

This approach could be used to find the area of any region bounded by the $x$-axis, the lines $x = a$, and $x = b$, and the graph of a function $y = f(x)$. At this point, we need some new notation and terminology to write sums concisely. We will indicate addition (or summation) by using the Greek letter sigma, $\Sigma$, as shown in the next example.

### ■ EXAMPLE 1

Find the following sums.

**(a)** $\displaystyle\sum_{i=1}^{5} i$

Replace $i$ in turn with the integers 1 through 5 and add the resulting terms.

$$\sum_{i=1}^{5} i = 1 + 2 + 3 + 4 + 5 = 15$$

**(b)** $\displaystyle\sum_{i=1}^{4} a_i = a_1 + a_2 + a_3 + a_4$

**(c)** $\displaystyle\sum_{i=1}^{3} (6x_i - 2)$ if $x_1 = 2$, $x_2 = 4$, $x_3 = 6$

Let $i = 1$, 2, and 3, respectively, to get

$$\sum_{i=1}^{3} (6x_i - 2) = (6x_1 - 2) + (6x_2 - 2) + (6x_3 - 2).$$

Now substitute the given values for $x_1$, $x_2$, and $x_3$.

$$\sum_{i=1}^{3} (6x_i - 2) = (6 \cdot 2 - 2) + (6 \cdot 4 - 2) + (6 \cdot 6 - 2)$$
$$= 10 + 22 + 34$$
$$= 66$$

**(d)** $\displaystyle\sum_{i=1}^{4} f(x_i)\Delta x$ if $f(x) = x^2$, $x_1 = 0$, $x_2 = 2$, $x_3 = 4$, $x_4 = 6$, and $\Delta x = 2$

$$\sum_{i=1}^{4} f(x_i)\Delta x = f(x_1)\Delta x + f(x_2)\Delta x + f(x_3)\Delta x + f(x_4)\Delta x$$
$$= x_1^2\Delta x + x_2^2\Delta x + x_3^2\Delta x + x_4^2\Delta x$$
$$= 0^2(2) + 2^2(2) + 4^2(2) + 6^2(2)$$
$$= 0 + 8 + 32 + 72 = 112 \quad ■$$

Now we can generalize to get a method of finding the area bounded by the curve $y = f(x)$, the $x$-axis, and the vertical lines $x = a$ and $x = b$, as shown in Figure 7. To approximate this area, we could divide the region under the curve first into 10 rectangles (Figure 7(a)) and then into 20 rectangles (Figure 7(b)). The sums of the areas of the rectangles give approximations to the area under the curve.

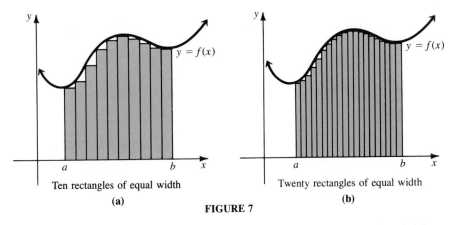

Ten rectangles of equal width

**(a)**

Twenty rectangles of equal width

**(b)**

**FIGURE 7**

To get a number that can be defined as the *exact* area, begin by dividing the interval from $a$ to $b$ into $n$ pieces of equal width, using each of these $n$ pieces as the base of a rectangle. (See Figure 8.) The left endpoints of the $n$ intervals are labeled $x_1, x_2, x_3, \ldots, x_{n+1}$, where $a = x_1$ and $b = x_{n+1}$. In the graph of Figure 8, the symbol $\Delta x$ is used to represent the width of each of the intervals. The darker rectangle is an arbitrary rectangle called the *i*th rectangle. Its area is the product of its length and width. Since the width of the *i*th rectangle is $\Delta x$ and the length of the *i*th rectangle is given by the height $f(x_i)$,

$$\text{Area of } i\text{th Rectangle} = f(x_i) \cdot \Delta x.$$

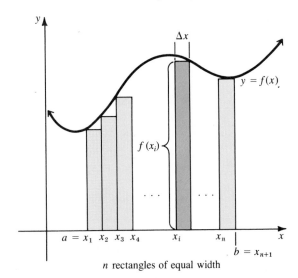

$n$ rectangles of equal width

**FIGURE 8**

The total area under the curve is approximated by the sum of the areas of all *n* of the rectangles. With sigma notation, the approximation to the total area becomes

$$\text{Area of All } n \text{ Rectangles } = \sum_{i=1}^{n} f(x_i) \cdot \Delta x.$$

The exact area is defined to be the limit of this sum (if the limit exists) as the number of rectangles increases without bound:

$$\text{Exact Area } = \lim_{n \to +\infty} \sum_{i=1}^{n} f(x_i)\Delta x.$$

This limit is called the *definite integral* of *f(x)* from *a* to *b* and is written as follows.

**THE DEFINITE INTEGRAL**

> If *f* is defined on the interval [*a*, *b*], the **definite integral** of *f* from *a* to *b* is given by
>
> $$\int_{a}^{b} f(x)dx = \lim_{n \to +\infty} \sum_{i=1}^{n} f(x_i)\Delta x,$$
>
> provided the limit exists, where $\Delta x = (b - a)/n$ and $x_i$ is *any* value of *x* in the *i*th interval.

As indicated in this definition, although the left endpoint of the *i*th interval has been used to find the height of the *i*th rectangle, any number in the *i*th interval can be used. (A more general definition is possible in which the rectangles do not necessarily all have the same width.) The *b* above the integral sign is called the **upper limit** of integration, and the *a* is the **lower limit** of integration. This use of the word "limit" has nothing to do with the limit of the sum; it refers to the limits, or boundaries, on *x*.

The sum in the definition of the definite integral is an example of a Riemann sum, named for the German mathematician Georg Riemann (1826–1866) who was responsible for the early development of integrability of functions. The concepts of "Riemann sum" and "Riemann integral" are still studied in rigorous calculus textbooks.

In the example at the beginning of this section, the area bounded by the *x*-axis, the curve $y = \sqrt{4 - x^2}$, the lines $x = 0$ and $x = 2$ could be written as the definite integral

$$\int_{0}^{2} \sqrt{4 - x^2} \, dx = \pi.$$

Notice that unlike the indefinite integral, which is a set of *functions,* the definite integral represents a *number.* The next section will show how antiderivatives are used in finding the definite integral and thus the area under a curve.

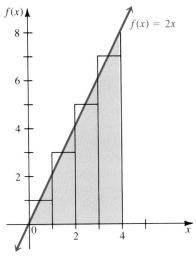

**FIGURE 9**

Keep in mind that finding the definite integral of a function can be thought of as a mathematical process that gives the sum of an infinite number of individual parts (within certain limits). The definite integral represents area only if the function involved is *nonnegative* ($f(x) \geq 0$) at every *x*-value in the interval [*a*, *b*]. There are many other interpretations of the definite integral, and all of them involve this idea of approximation by appropriate sums.

## ▬ EXAMPLE 2

Approximate $\int_0^4 2x\ dx$, the area of the region under the graph of $f(x) = 2x$, above the *x*-axis, and between $x = 0$ and $x = 4$, by using four rectangles of equal width whose heights are the values of the function at the midpoint of each rectangle.

We want to find the area of the shaded region in Figure 9. The heights of the four rectangles given by $f(x_i)$ for $i = 1, 2, 3,$ and 4 are as follows.

| $i$ | $x_i$ | $f(x_i)$ |
|---|---|---|
| 1 | $x_1 = .5$ | $f(.5) = 1.0$ |
| 2 | $x_2 = 1.5$ | $f(1.5) = 3.0$ |
| 3 | $x_3 = 2.5$ | $f(2.5) = 5.0$ |
| 4 | $x_4 = 3.5$ | $f(3.5) = 7.0$ |

The width of each rectangle is $\Delta x = \dfrac{4 - 0}{4} = 1$. The sum of the areas of the four rectangles is

$$\sum_{i=1}^{4} f(x_i)\Delta x = f(x_1)\Delta x + f(x_2)\Delta x + f(x_3)\Delta x + f(x_4)\Delta x$$

$$= f(.5)\Delta x + f(1.5)\Delta x + f(2.5)\Delta x + f(3.5)\Delta x$$

$$= (1)(1) + (3)(1) + 5(1) + 7(1)$$

$$= 16.$$

Using the formula for the area of a triangle, $A = (1/2)bh$, with *b*, the length of the base, equal to 4 and *h*, the height, equal to 8, gives

$$A = \frac{1}{2}\,bh = \frac{1}{2}\,(4)(8) = 16,$$

the exact value of the area. The approximation equals the exact area in this case because our use of the midpoints of each subinterval distributed the error evenly above and below the graph. ▬

***Total Change*** Suppose the function $f(x) = x^2 + 20$ gives the marginal cost of some item at a particular $x$-value. Then $f(2) = 24$ gives the rate of change of cost at $x = 2$. That is, a unit change in $x$ (at this point) will produce a change of 24 units in the cost function. Also, $f(3) = 29$ means that each unit of change in $x$ (when $x = 3$) will produce a change of 29 units in the cost function.

To find the *total* change in the cost function as $x$ changes from 2 to 3, we could divide the interval from 2 to 3 into $n$ equal parts, using each part as the base of a rectangle as we did above. The area of each rectangle would approximate the change in cost at the $x$-value that is the left endpoint of the base of the rectangle. Then the sum of the areas of these rectangles would approximate the net total change in cost from $x = 2$ to $x = 3$. The limit of this sum as $n \to +\infty$ would give the exact total change.

This result produces another application of the definite integral: the area of the region under the graph of the marginal cost function $f(x)$ that is above the $x$-axis and between $x = a$ and $x = b$ gives the *net total change in the cost* as $x$ goes from $a$ to $b$.

**TOTAL CHANGE IN $F(x)$**

> If $f(x)$ gives the rate of change of $F(x)$ for $x$ in $[a, b]$, then the **total change** in $F(x)$ as $x$ goes from $a$ to $b$ is given by
>
> $$\lim_{n \to +\infty} \sum_{i=1}^{n} f(x_i)\Delta x = \int_a^b f(x)\,dx.$$

In other words, the total change in a quantity can be found from the function that gives the rate of change of the quantity, using the same methods used to approximate the area under a curve.

### ■ EXAMPLE 3

Figure 10 shows the rate of change of the annual maintenance charges for a certain machine. To approximate the total maintenance charges over the 10-year life of the machine, use approximating rectangles, dividing the interval from 0 to 10 into ten equal subdivisions. Each rectangle has width 1; using the left endpoint of each rectangle to determine the height of the rectangle, the approximation becomes

$$1 \cdot 0 + 1 \cdot 500 + 1 \cdot 750 + 1 \cdot 1800 + 1 \cdot 1800 + 1 \cdot 3000 + 1 \cdot 3000$$
$$+ 1 \cdot 3400 + 1 \cdot 4200 + 1 \cdot 5100 = 23{,}550.$$

About \$23,550 will be spent on maintenance over the 10-year life of the machine. ■

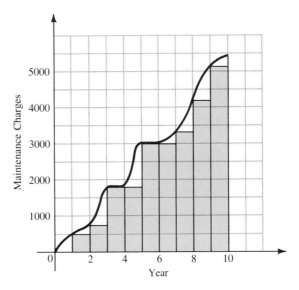

**FIGURE 10**

Before discussing further applications of the definite integral, we need a more efficient method for evaluating it. This method will be developed in the next section.

## 5.3 EXERCISES

*Evaluate each sum.*

**1.** $\displaystyle\sum_{i=1}^{3} 3i$

**2.** $\displaystyle\sum_{i=1}^{6} (-5i)$

**3.** $\displaystyle\sum_{i=1}^{5} (2i + 7)$

**4.** $\displaystyle\sum_{i=1}^{10} (5i - 8)$

**5.** Let $x_1 = -5$, $x_2 = 8$, $x_3 = 7$, and $x_4 = 10$. Find $\displaystyle\sum_{i=1}^{4} x_i$.

**6.** Let $x_1 = 10$, $x_2 = 15$, $x_3 = -8$, $x_4 = -12$, and $x_5 = 0$. Find $\displaystyle\sum_{i=1}^{5} x_i$.

**7.** Let $f(x) = x - 3$ and $x_1 = 4$, $x_2 = 6$, $x_3 = 7$. Find $\displaystyle\sum_{i=1}^{3} f(x_i)$.

**8.** Let $f(x) = x^2 + 1$ and $x_1 = -2$, $x_2 = 0$, $x_3 = 2$, $x_4 = 4$. Find $\displaystyle\sum_{i=1}^{4} f(x_i)$.

**9.** Let $f(x) = 2x + 1$, $x_1 = 0$, $x_2 = 2$, $x_3 = 4$, $x_4 = 6$, and $\Delta x = 2$. Find $\displaystyle\sum_{i=1}^{4} f(x_i)\Delta x$.

**10.** Let $f(x) = 1/x$, $x_1 = 1/2$, $x_2 = 1$, $x_3 = 3/2$, $x_4 = 2$, and $\Delta x = 1/2$. Find $\displaystyle\sum_{i=1}^{4} f(x_i)\Delta x$.

*In Exercises 11–22, first approximate the area under the given curve and above the x-axis by using two rectangles. Let the height of the rectangle be given by the value of the function at the left side of the rectangle. Then repeat the process and approximate the area using four rectangles.*

**11.** $f(x) = 3x + 2$ from $x = 1$ to $x = 5$

**12.** $f(x) = -2x + 1$ from $x = -4$ to $x = 0$

**13.** $f(x) = x + 5$ from $x = 2$ to $x = 4$

**14.** $f(x) = 3 + x$ from $x = 1$ to $x = 3$

**15.** $f(x) = x^2$ from $x = 1$ to $x = 5$

**16.** $f(x) = x^2$ from $x = 0$ to $x = 4$

**17.** $f(x) = x^2 + 2$ from $x = -2$ to $x = 2$

**18.** $f(x) = -x^2 + 4$ from $x = -2$ to $x = 2$

**19.** $f(x) = e^x - 1$ from $x = 0$ to $x = 4$

**20.** $f(x) = e^x + 1$ from $x = -2$ to $x = 2$

**21.** $f(x) = \dfrac{1}{x}$ from $x = 1$ to $x = 5$

**22.** $f(x) = \dfrac{2}{x}$ from $x = 1$ to $x = 9$

**23.** Consider the region below $f(x) = x/2$, above the x-axis, between $x = 0$ and $x = 4$. Let $x_i$ be the left endpoint of the $i$th subinterval.

  **(a)** Approximate the area of the region using four rectangles.

  **(b)** Approximate the area of the region using eight rectangles.

  **(c)** Find $\int_0^4 f(x)\,dx$ by using the formula for the area of a triangle.

**24.** Find $\int_0^5 (5 - x)\,dx$ by using the formula for the area of a triangle.

## ■ APPLICATIONS

*In Exercises 25–28, estimate the area under each curve by summing the area of rectangles. Let the function value at the left side of each rectangle give the height of the rectangle.*

### BUSINESS AND ECONOMICS

**Sales** **25.** The graph shows the rate of sales of new cars in a recent year. Estimate the total sales during that year. Use rectangles with a width of 1 month.

**Electricity Consumption**   **26.** The graph shows the rate of use of electrical energy (in kilowatt hours) in a certain city on a very hot day. Estimate the total usage of electricity on that day. Let the width of each rectangle be 2 hr.

## LIFE SCIENCES

**Alcohol Concentration**   **27.** The graph shows the approximate concentration of alcohol in a person's bloodstream $t$ hr after drinking 2 oz of alcohol. Estimate the total amount of alcohol in the bloodstream by estimating the area under the curve. Use rectangles of width 1 hr.

**Oxygen Inhalation**   **28.** The graph shows the rate of inhalation of oxygen by a person riding a bicycle very rapidly for 10 min. Estimate the total volume of oxygen inhaled in the first 20 min after the beginning of the ride. Use rectangles of width 1 min.

**Snow Depth**  *When doing research on the wolf-moose relationships in Michigan's Isle Royale National Park, biologist Rolf Peterson needed a snow-depth index to help correlate populations with snow levels. The graph shown at the right is given a snow depth index of 1.0. The vertical scale gives snow depth in inches, and the horizontal scale gives months.†*

*Find the snow depth index for each of the years represented by the following graphs by forming the ratio of the areas under the given curves to the area under the curve graphed above.*

**29.**

**30.**

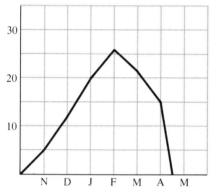

## PHYSICAL SCIENCES

**Distance**  *The next two graphs are from* Road & Track *magazine.\* The curve shows the velocity at t sec after the car accelerates from a dead stop. To find the total distance traveled by the car in reaching 100 mph, we must estimate the definite integral*

$$\int_0^T v(t)\,dt,$$

*where T represents the number of seconds it takes for the car to reach 100 mph.*

---

†From *Wolf Ecology and Prey Relationships on Isle Royale* by Rolf Olin Peterson. Copyright © 1977 by the U.S. Government Printing Office, p. 187.

*From *Road & Track,* April and May, 1978. Reprinted with permission of Road & Track.

*Use the graphs to estimate this distance by adding the areas of rectangles with widths of 3 seconds. The last rectangle will have a width of 1 second. To adjust your answer to miles per hour, divide by 3600 (the number of seconds in an hour). You then have the number of miles that the car traveled in reaching 100 mph. Finally, multiply by 5280 feet per mile to convert the answer to feet.*

**31.** Estimate the distance traveled by the Porsche 928, using the graph on the left.

**32.** Estimate the distance traveled by the BMW 733i using the graph on the right.

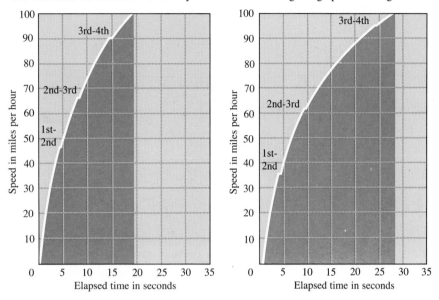

### FOR THE COMPUTER

*In Exercises 33–36, use the method given in the text to approximate the area between the x-axis and the graph of each function on the given interval.*

**33.** $f(x) = x \ln x$;   $[1, 5]$

**34.** $f(x) = x^2 e^{-x}$;   $[-1, 3]$

**35.** $f(x) = \dfrac{\ln x}{x}$;   $[1, 5]$

**36.** $f(x) = \dfrac{e^x - e^{-x}}{2}$;   $[0, 4]$

## 5.4   THE FUNDAMENTAL THEOREM OF CALCULUS

The work from the last two sections can now be put together. We have seen that, if $f(x) > 0$,

$$\int_a^b f(x)\, dx$$

gives the area between the graph of $f(x)$ and the *x*-axis, from $x = a$ to $x = b$. We can find this definite integral by using the antiderivatives discussed earlier.

The definite integral was defined and evaluated in the previous section using the limit of a sum. In that section, we also saw that if $f(x)$ gives the rate of change of $F(x)$, the definite integral $\int_a^b f(x)\,dx$ gives the total change of $F(x)$ as $x$ changes from $a$ to $b$. If $f(x)$ gives the rate of change of $F(x)$, then $F(x)$ is an antiderivative of $f(x)$. Writing the total change in $F(x)$ from $x = a$ to $x = b$ as $F(b) - F(a)$ shows the connection between antiderivatives and definite integrals. This relationship is called the **fundamental theorem of calculus.**

**FUNDAMENTAL
THEOREM OF CALCULUS**

Let $f$ be continuous on the interval $[a, b]$, and let $F$ be *any* antiderivative of $f$. Then

$$\int_a^b f(x)\,dx = F(b) - F(a) = F(x)\Big|_a^b.$$

The symbol $F(x)\Big|_a^b$ is used to represent $F(b) - F(a)$. It is important to note that the fundamental theorem does not require $f(x) > 0$. The condition $f(x) > 0$ is necessary only when using the fundamental theorem to find area. Also, note that the fundamental theorem does not *define* the definite integral; it just provides a method for evaluating it.

**EXAMPLE 1**

First find $\int 4t^3\,dt$ and then find $\int_1^2 4t^3\,dt$.

By the rules given earlier,

$$\int 4t^3\,dt = t^4 + C.$$

By the fundamental theorem, the value of $\int_1^2 4t^3\,dt$ is found by evaluating $t^4\Big|_1^2$, with no constant $C$ required.

$$\int_1^2 4t^3\,dt = t^4\Big|_1^2 = 2^4 - 1^4 = 15 \quad\blacksquare$$

Example 1 illustrates the difference between the definite integral and the indefinite integral. A definite integral is a real number; an indefinite integral is a family of functions in which all the functions are antiderivatives of a function $f$.

To see why the fundamental theorem of calculus is true for $f(x) > 0$, look at Figure 11. Define the function $A(x)$ as the area between the $x$-axis and the graph of $y = f(x)$ from $a$ to $x$. We first show that $A$ is an antiderivative of $f$, that is, $A' = f$.

To do this, let $h$ be a small positive number. Then $A(x + h) - A(x)$ is the shaded area in Figure 11. This area can be approximated with a rectangle having width $h$ and height $f(x)$. The area of the rectangle is $h \cdot f(x)$, and

$$A(x + h) - A(x) \approx h \cdot f(x).$$

Dividing both sides by $h$ gives

$$\frac{A(x + h) - A(x)}{h} \approx f(x).$$

This approximation improves as $h$ gets smaller and smaller. Take the limit on the left as $h$ approaches 0;

$$\lim_{h \to 0} \frac{A(x + h) - A(x)}{h} = f(x).$$

This limit is simply $A'(x)$, so

$$A'(x) = f(x).$$

This result means that $A$ is an antiderivative of $f$, as we set out to show.

Since $A(x)$ is the area under the curve $y = f(x)$ from $a$ to $x$, $A(a) = 0$. The expression $A(b)$ is the area from $a$ to $b$, the desired area. Since $A$ is an antiderivative of $f$,

$$\int_a^b f(x)\, dx = A(x)\Big|_a^b = A(b) - A(a) = A(b) - 0 = A(b).$$

This suggests the proof of the fundamental theorem—for $f(x) > 0$ the area under the curve $y = f(x)$ from $a$ to $b$ is given by $\int_a^b f(x)\, dx$.

The fundamental theorem of calculus certainly deserves its name, which sets it apart as the most important theorem of calculus. It is the key connection between differential calculus and integral calculus, which originally were developed separately without knowledge of this connection between them.

The variable used in the integrand does not matter; each of the following definite integrals represents the number $F(b) - F(a)$.

$$\int_a^b f(x)\, dx = \int_a^b f(t)\, dt = \int_a^b f(u)\, du$$

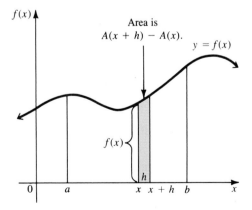

**FIGURE 11**

Key properties of definite integrals are listed below. Some of them are just restatements of properties from Section 1.

**PROPERTIES OF DEFINITE INTEGRALS**

If all indicated definite integrals exist,

1. $\displaystyle\int_a^a f(x)\ dx = 0;$

2. $\displaystyle\int_a^b k \cdot f(x)\ dx = k \cdot \int_a^b f(x)\ dx$    for any real constant $k$
   (constant multiple of a function);

3. $\displaystyle\int_a^b [f(x) \pm g(x)]dx = \int_a^b f(x)\ dx \pm \int_a^b g(x)\ dx$
   (sum or difference of functions);

4. $\displaystyle\int_a^b f(x)\ dx = \int_a^c f(x)\ dx + \int_c^b f(x)\ dx$    for any real number $c$.

For $f(x) > 0$, since the distance from $a$ to $a$ is 0, the first property says that the "area" under the graph of $f$ bounded by $x = a$ and $x = a$ is 0. Also, since $\int_a^c f(x)\ dx$ represents the darker region in Figure 12, and $\int_c^b f(x)\ dx$ represents the lighter region,

$$\int_a^b f(x)\ dx = \int_a^c f(x)\ dx + \int_c^b f(x)\ dx,$$

as stated in the fourth property. While the figure shows $a < c < b$, the property is true for any value of $c$ where both $f(x)$ and $F(x)$ are defined.

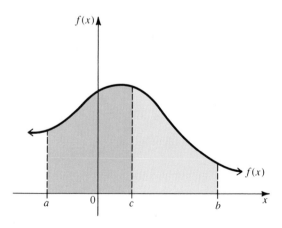

**FIGURE 12**

An algebraic proof is given here for the third property; proofs of the other properties are left for the exercises. If $F(x)$ and $G(x)$ are antiderivatives of $f(x)$ and $g(x)$ respectively,

$$\int_a^b [f(x) + g(x)]dx = [F(x) + G(x)]\Big|_a^b$$

$$= [F(b) + G(b)] - [F(a) + G(a)]$$

$$= [F(b) - F(a)] + [G(b) - G(a)]$$

$$= \int_a^b f(x)\, dx + \int_a^b g(x)\, dx.$$

### ▅▅ EXAMPLE 2
Find $\int_2^5 (6x^2 - 3x + 5)\, dx$.

Use the properties above, and the fundamental theorem, along with properties from Section 1.

$$\int_2^5 (6x^2 - 3x + 5)\, dx = 6\int_2^5 x^2 dx - 3\int_2^5 x\, dx + 5\int_2^5 dx$$

$$= 2x^3\Big|_2^5 - \frac{3}{2}x^2\Big|_2^5 + 5x\Big|_2^5$$

$$= 2(5^3 - 2^3) - \frac{3}{2}(5^2 - 2^2) + 5(5 - 2)$$

$$= 2(125 - 8) - \frac{3}{2}(25 - 4) + 5(3)$$

$$= 234 - \frac{63}{2} + 15 = \frac{435}{2} \ \ \blacksquare$$

### ▅▅ EXAMPLE 3

$$\int_1^2 \frac{dy}{y} = \ln|y|\Big|_1^2 = \ln|2| - \ln|1|$$

$$= \ln 2 - \ln 1 \approx .6931 - 0 = .6931 \ \ \blacksquare$$

### ▅▅ EXAMPLE 4

Evaluate $\int_0^5 x\sqrt{25 - x^2}\, dx$.

Use substitution. Let $u = 25 - x^2$, so that $du = -2x\, dx$. With a definite integral, the limits should be changed, too. The new limits on $u$ are found as follows.

$$\text{If } x = 5, \text{ then } u = 25 - 5^2 = 0;$$
$$\text{if } x = 0, \text{ then } u = 25 - 0^2 = 25.$$

Then

$$\int_0^5 x\sqrt{25 - x^2}\, dx = -\frac{1}{2}\int_0^5 \sqrt{25 - x^2}(-2x\, dx)$$

$$= -\frac{1}{2}\int_{25}^0 \sqrt{u}\, du$$

$$= -\frac{1}{2}\int_{25}^0 u^{1/2}\, du$$

$$= -\frac{1}{2}\cdot\frac{u^{3/2}}{3/2}\Big|_{25}^0$$

$$= -\frac{1}{2}\cdot\frac{2}{3}[0^{3/2} - 25^{3/2}]$$

$$= -\frac{1}{3}(-125)$$

$$= \frac{125}{3}. \quad \blacksquare$$

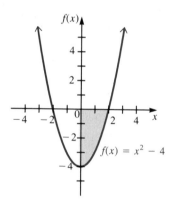

**FIGURE 13**

***Area*** In the previous section we saw that, if $f(x) > 0$ in $[a, b]$, the definite integral $\int_a^b f(x)\, dx$ gives the area below the graph of the function $y = f(x)$, above the $x$-axis, and between the lines $x = a$ and $x = b$.

To see how to work around the requirement that $f(x) > 0$, look at the graph of $f(x) = x^2 - 4$ in Figure 13. The area bounded by the graph of $f$, the $x$-axis, and the vertical lines $x = 0$ and $x = 2$ lies below the $x$-axis. Using the fundamental theorem to find this area gives

$$\int_0^2 (x^2 - 4)\, dx = \left(\frac{x^3}{3} - 4x\right)\Big|_0^2 = \left(\frac{8}{3} - 8\right) - (0 - 0) = \frac{-16}{3}.$$

The result is a negative number because $f(x)$ is negative for values of $x$ in the interval $[0, 2]$. Since $\Delta x$ is always positive, if $f(x) < 0$ the product $f(x) \cdot \Delta x$ is negative, so $\int_0^2 f(x)\, dx$ is negative. Since area is nonnegative, the required area is given by $|-16/3|$ or $16/3$. Using a definite integral, the area could be written as

$$\left|\int_0^2 (x^2 - 4)dx\right| = \left|\frac{-16}{3}\right| = \frac{16}{3}.$$

## ▬ EXAMPLE 5

Find the area of the region between the $x$-axis and the graph of $f(x) = x^2 - 3x$ from $x = 1$ to $x = 3$.

**FIGURE 14**

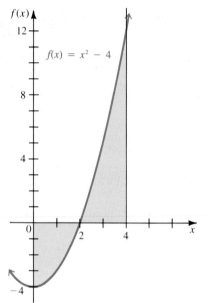

**FIGURE 15**

The region is shown in Figure 14. Since the region lies below the $x$-axis, the area is given by

$$\left| \int_1^3 (x^2 - 3x)dx \right|.$$

By the fundamental theorem,

$$\int_1^3 (x^2 - 3x)dx = \left( \frac{x^3}{3} - \frac{3x^2}{2} \right)\Big|_1^3 = \left( \frac{27}{3} - \frac{27}{2} \right) - \left( \frac{1}{3} - \frac{3}{2} \right) = -\frac{10}{3}.$$

The required area is $\left| -10/3 \right| = 10/3.$ ▬

▬ EXAMPLE 6

Find the area between the $x$-axis and the graph of $f(x) = x^2 - 4$ from $x = 0$ to $x = 4$.

Figure 15 shows the required region. Part of the region is below the $x$-axis. The definite integral over that interval will have a negative value. To find the area, integrate the negative and positive portions separately and take the absolute value of the first result before combining the two results to get the total area. Start by finding the point where the graph crosses the $x$-axis. This is done by solving the equation

$$x^2 - 4 = 0.$$

The solutions of this equation are 2 and $-2$. The only solution in the interval $[0, 4]$ is 2. The total area of the region in Figure 15 is

$$\left| \int_0^2 (x^2 - 4)dx \right| + \int_2^4 (x^2 - 4)dx = \left| \left( \frac{1}{3}x^3 - 4x \right)\Big|_0^2 \right| + \left( \frac{1}{3}x^3 - 4x \right)\Big|_2^4$$

$$= \left| \frac{8}{3} - 8 \right| + \left( \frac{64}{3} - 16 \right) - \left( \frac{8}{3} - 8 \right)$$

$$= 16. ▬$$

Incorrectly using one integral over the entire interval to find the area in Example 6 would have given

$$\int_0^4 (x^2 - 4)dx = \left( \frac{x^3}{3} - 4x \right)\Big|_0^4 = \left( \frac{64}{3} - 16 \right) - 0 = \frac{16}{3},$$

which is not the correct area. This definite integral represents no area, but is just a real number.

**FINDING AREA**

In summary, to find the area bounded by $f(x)$, $x = a$, $x = b$, and the $x$-axis, use the following steps.

1. Sketch a graph.
2. Find any $x$-intercepts of $f(x)$ in $[a, b]$. These divide the total region into subregions.
3. The definite integral will be *positive* for subregions above the $x$-axis and *negative* for subregions below the $x$-axis. Use separate integrals to find the areas of the subregions.
4. The total area is the sum of the areas of all of the subregions.

In the last section, we saw that the area under a rate of change function $f'(x)$ from $x = a$ to $x = b$ gives the total value of $f(x)$ on $[a, b]$. Now we can use the definite integral to solve these problems.

**EXAMPLE 7**

The yearly rate of consumption of natural gas in trillions of cubic feet for a certain city is

$$C'(t) = t + e^{.01t},$$

where $t$ is in years and $t = 0$ corresponds to 1990. At this consumption rate, what is the total amount the city will use in the 10-year period of the nineties?

To find the consumption over the 10-year period from 1990 through 1999, use the definite integral.

$$\int_0^{10} (t + e^{.01t})dt = \left( \frac{t^2}{2} + \frac{e^{.01t}}{.01} \right) \Big|_0^{10}$$
$$\approx (50 + 100e^{.1}) - (0 + 100)$$
$$\approx -50 + 100(1.10517) \approx 60.5$$

Therefore, a total of 60.5 trillion cubic feet of natural gas will be used during the nineties if the consumption rate remains the same. ■

## 5.4 EXERCISES

*Evaluate each definite integral.*

**1.** $\int_{-2}^{4} (-1)dp$

**2.** $\int_{-4}^{1} 6x \, dx$

**3.** $\int_{-1}^{2} (3t - 1)dt$

**4.** $\int_{-2}^{2} (4z + 3)dz$

**5.** $\int_0^2 (5x^2 - 4x + 2)dx$

**6.** $\int_{-2}^{3} (-x^2 - 3x + 5)dx$

**7.** $\int_0^2 3\sqrt{4u + 1} \, du$

**8.** $\int_3^9 \sqrt{2r - 2} \, dr$

**9.** $\int_0^1 2(t^{1/2} - t)dt$

**10.** $\int_0^4 -(3x^{3/2} + x^{1/2})dx$

**11.** $\int_1^4 (5y\sqrt{y} + 3\sqrt{y})dy$

**12.** $\int_4^9 (4\sqrt{r} - 3r\sqrt{r})dr$

**13.** $\int_4^6 \dfrac{2}{(x-3)^2} dx$

**14.** $\int_1^4 \dfrac{-3}{(2p+1)^2} dp$

**15.** $\int_1^5 (5n^{-2} + n^{-3})dn$

**16.** $\int_2^3 (3x^{-2} - x^{-4})dx$

**17.** $\int_2^3 \left(2e^{-.1A} + \dfrac{3}{A}\right)dA$

**18.** $\int_1^2 \left(\dfrac{-1}{B} + 3e^{.2B}\right)dB$

**19.** $\int_1^2 \left(e^{5u} - \dfrac{1}{u^2}\right)du$

**20.** $\int_{.5}^1 (p^3 - e^{4p})dp$

**21.** $\int_{-1}^0 (2y - 3)^5 dy$

**22.** $\int_0^3 (4m + 2)^3 dm$

**23.** $\int_1^{64} \dfrac{\sqrt{z} - 2}{\sqrt[3]{z}} dz$

**24.** $\int_1^8 \dfrac{3 - y^{1/3}}{y^{2/3}} dy$

*In each of the following, use the definite integral to find the area between the x-axis and f(x) over the indicated interval. Check first to see if the graph crosses the x-axis in the given interval.*

**25.** $f(x) = 2x + 3;$   $[8, 10]$

**26.** $f(x) = 4x - 7;$   $[5, 10]$

**27.** $f(x) = 2 - 2x^2;$   $[0, 5]$

**28.** $f(x) = 9 - x^2;$   $[0, 6]$

**29.** $f(x) = x^2 + 4x - 5;$   $[-1, 3]$

**30.** $f(x) = x^2 - 6x + 5;$   $[-1, 4]$

**31.** $f(x) = x^3;$   $[-1, 3]$

**32.** $f(x) = x^3 - 2x;$   $[-2, 4]$

**33.** $f(x) = e^x - 1;$   $[-1, 2]$

**34.** $f(x) = 1 - e^{-x};$   $[-1, 2]$

**35.** $f(x) = \dfrac{1}{x};$   $[1, e]$

**36.** $f(x) = \dfrac{1}{x};$   $[e, e^2]$

*Find the area of each shaded region.*

**37.**

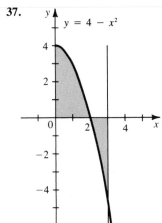

$y = 4 - x^2$

**38.**

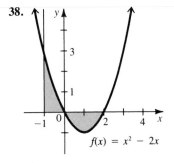

$f(x) = x^2 - 2x$

**39.**

**40.**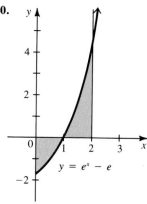

*Show that each of the following is true.*

**41.** $\int_a^a f(x)dx = 0$

**42.** $\int_a^b kf(x)dx = k\int_a^b f(x)dx$

**43.** $\int_a^b f(x)dx = \int_a^c f(x)dx + \int_c^b f(x)dx$

**44.** Use Exercise 43 to find $\int_{-1}^4 f(x)dx$, given

$$f(x) = \begin{cases} 2x + 3 & \text{if } x \le 0 \\ -\dfrac{x}{4} - 3 & \text{if } x > 0. \end{cases}$$

---

### ▤ APPLICATIONS

#### BUSINESS AND ECONOMICS

**Expenditures**  **45.** De Win Enterprises has found that its expenditure rate per day (in hundreds of dollars) on a certain type of job is given by

$$E'(x) = 4x + 2,$$

where $x$ is the number of days since the start of the job.

(a) Find the total expenditure if the job takes 10 days.

(b) How much will be spent on the job from the tenth to the twentieth day?

(c) If the company wants to spend no more than $5000 on the job, in how many days must they complete it?

**Income**  **46.** De Win Enterprises (see Exercise 45) also knows that the rate of income per day (in hundreds of dollars) for the same job is

$$I'(x) = 100 - x,$$

where $x$ is the number of days since the job was started.

(a) Find the total income for the first 10 days.

(b) Find the income from the tenth to the twentieth day.

(c) How many days must the job last for the total income to be at least $5000?

**Profit**    **47.** After a new firm starts in business, it finds that its rate of profits (in hundreds of dollars) after $t$ years of operation is given by

$$P'(t) = 6t^2 + 4t + 5.$$

(a) Find the total profits in the first three years.

(b) Find the profit in the fourth year of operation.

**Worker Efficiency**    **48.** A worker new to a job will improve his efficiency with time so that it takes him fewer hours to produce an item with each day on the job, up to a certain point. Suppose the rate of change of the number of hours it takes a worker in a certain factory to produce the $x$th item is given by

$$H'(x) = 20 - 2x.$$

(a) What is the total number of hours required to produce the first 5 items?

(b) What is the total number of hours required to produce the first 10 items?

### LIFE SCIENCES

**Pollution**    **49.** Pollution from a factory is entering a lake. The rate of concentration of the pollutant at time $t$ is given by

$$P'(t) = 140t^{5/2},$$

where $t$ is the number of years since the factory started introducing pollutants into the lake. Ecologists estimate that the lake can accept a total level of pollution of 4850 units before all the fish life in the lake ends. Can the factory operate for 4 years without killing all the fish in the lake?

**Spread of an Oil Leak**    **50.** An oil tanker is leaking oil at the rate of $20t + 50$ barrels per hour, where $t$ is time in hours after the tanker hits a hidden rock (when $t = 0$).

(a) Find the total number of barrels that the ship will leak on the first day.

(b) Find the number of barrels that the ship will leak on the second day.

**Tree Growth**    **51.** After long study, tree scientists conclude that a eucalyptus tree will grow at the rate of $.2 + 4t^{-4}$ feet per year, where $t$ is time in years.

(a) Find the number of feet that the tree will grow in the second year.

(b) Find the number of feet the tree will grow in the third year.

**Growth of a Substance**    **52.** The rate at which a substance grows is given by

$$R'(x) = 200e^x,$$

where $x$ is the time in days. What is the total accumulated growth after 2.5 days?

**Drug Reaction**    **53.** For a certain drug, the rate of reaction in appropriate units is given by

$$R'(t) = \frac{5}{t} + \frac{2}{t^2},$$

where $t$ is measured in hours after the drug is administered. Find the total reaction to the drug over the following time periods.

(a) From $t = 1$ to $t = 12$

(b) From $t = 12$ to $t = 24$

**Blood Flow**    **54.** In the exercises for an earlier chapter, the speed $s$ of the blood in a blood vessel was given as

$$s = k(R^2 - r^2),$$

where $R$ is the (constant) radius of the blood vessel, $r$ is the distance of the flowing blood from the center of the blood vessel, and $k$ is a constant. Total blood flow in millimeters per minute is given by.

$$Q(R) = \int_0^R 2\pi sr\, dr.$$

**(a)** Find the general formula for $Q$ in terms of $R$ by evaluating the definite integral given above.

**(b)** Evaluate $Q(.4)$.

## PHYSICAL SCIENCES

**Oil Consumption**    **55.** Suppose that the rate of consumption of a natural resource is $c(t)$, where

$$c'(t) = ke^{rt}.$$

Here $t$ is time in years, $r$ is a constant, and $k$ is the consumption in the year when $t = 0$. In 1989, an oil company sold 1.2 billion barrels of oil. Assume that $r = .04$.

**(a)** Write $c(t)$ for the oil company, letting $t = 0$ represent 1989.

**(b)** Set up a definite integral for the amount of oil that the company will sell in the next ten years.

**(c)** Evaluate the definite integral of part (b).

**(d)** The company has about 20 billion barrels of oil in reserve. To find the number of years that this amount will last, solve the equation

$$\int_0^T 1.2e^{.04t}dt = 20.$$

**(e)** Rework part (d), assuming that $r = .02$.

**Mine Production**    **56.** A mine begins producing at time $t = 0$. After $t$ years, the mine is producing at the rate of

$$P'(t) = 10t - \frac{15}{\sqrt{t}}$$

tons per year. Write an expression for the total output of the mine from year 1 to year $T$.

**Consumption of a Natural Resource**    **57.** The rate of consumption of one natural resource in one country is

$$C'(t) = 72e^{.014t},$$

where $t = 0$ corresponds to 1989. How much of the resource will be used, altogether, from 1989 to year $T$?

**Oil Consumption**    **58.** The rate of consumption of oil (in billions of barrels) by the company in Exercise 55 was given as

$$1.2e^{.04t},$$

where $t = 0$ corresponds to 1989. Find the total amount of oil used by the company from 1989 to year $T$. At this rate, how much will be used in 5 years?

## 5.5   THE AREA BETWEEN TWO CURVES

Many important applications of integrals require finding the area between two graphs. The method used in previous sections to find the area between the graph of a function and the $x$-axis from $x = a$ to $x = b$ can be generalized to find such an area. For example, the area between the graphs of $f(x)$ and $g(x)$ from $x = a$ to $x = b$ in Figure 16(a) is the same as the difference of the area from $a$ to $b$ between $f(x)$ and the $x$-axis, shown in Figure 16(b), and the area from $a$ to $b$ between $g(x)$ and the $x$-axis (see Figure 16(c)). That is, the area between the graphs is given by

$$\int_a^b f(x) \; dx \; - \; \int_a^b g(x) \; dx,$$

which can be written as

$$\int_a^b [f(x) \; - \; g(x)] \; dx.$$

**AREA BETWEEN TWO CURVES**

If $f$ and $g$ are continuous functions and $f(x) \geq g(x)$ on $[a, b]$, then the area between the curves $f(x)$ and $g(x)$ from $x = a$ to $x = b$ is given by

$$\int_a^b [f(x) \; - \; g(x)] \; dx.$$

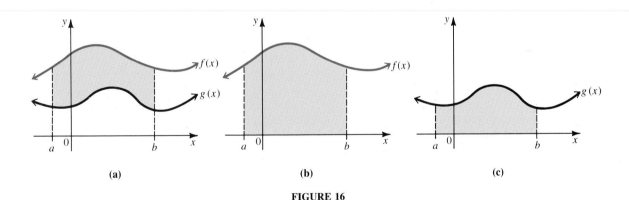

|     |     |     |
| :-: | :-: | :-: |
| (a) | (b) | (c) |

**FIGURE 16**

## EXAMPLE 1

Find the area bounded by $f(x) = -2x^2 + 1$, $g(x) = (1/5)x + 2$, $x = -1$, and $x = 2$.

Sketch the graphs of the four equations to determine the enclosed region, which is the shaded area shown in Figure 17. Since $g(x) \geq f(x)$ for $x$ in the interval $[-1, 2]$, the area is given by

$$\int_{-1}^{2} [g(x) - f(x)]dx = \int_{-1}^{2} [(\frac{1}{5}x + 2) - (-2x^2 + 1)]dx$$

$$= \int_{-1}^{2} (\frac{1}{5}x + 2 + 2x^2 - 1)dx$$

$$= \int_{-1}^{2} (\frac{1}{5}x + 1 + 2x^2)dx$$

$$= (\frac{x^2}{10} + x + \frac{2x^3}{3}) \Big|_{-1}^{2}$$

$$= (\frac{4}{10} + 2 + \frac{16}{3}) - (\frac{1}{10} - 1 - \frac{2}{3})$$

$$= \frac{4}{10} + 2 + \frac{16}{3} - \frac{1}{10} + 1 + \frac{2}{3}$$

$$= \frac{3}{10} + 3 + \frac{18}{3}$$

$$= \frac{93}{10} \quad \text{or} \quad 9.3. \quad \blacksquare$$

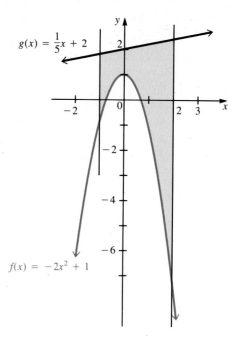

$g(x) = \frac{1}{5}x + 2$

$f(x) = -2x^2 + 1$

**FIGURE 17**

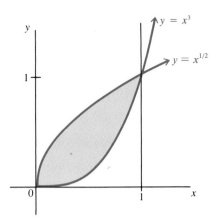

**FIGURE 18**

### ■ EXAMPLE 2

Find the area between the curves $y = x^{1/2}$ and $y = x^3$ from $x = 0$ to $x = 1$.

As shown in Figure 18, $f(x) = x^{1/2}$, $g(x) = x^3$, and the area between these two curves is given by

$$\int_a^b [f(x) - g(x)]\, dx = \int_0^1 (x^{1/2} - x^3)\, dx.$$

Using the fundamental theorem,

$$\int_0^1 (x^{1/2} - x^3)\, dx = \left(\frac{x^{3/2}}{3/2} - \frac{x^4}{4}\right)\Bigg|_0^1$$

$$= \left(\frac{2}{3}x^{3/2} - \frac{x^4}{4}\right)\Bigg|_0^1$$

$$= \frac{2}{3}\cdot 1 - \frac{1}{4} - 0$$

$$= \frac{5}{12}. \quad ■$$

The difference between two integrals can be used to find the area between the graphs of two functions even if one graph lies below the *x*-axis or both graphs lie below the *x*-axis. In fact, if $f(x) \geq g(x)$ for all values of *x* in the interval $[a, b]$, then the area between the two graphs is always given by

$$\int_a^b [f(x) - g(x)]\, dx.$$

To see this, look at the graphs in Figure 19(a), where $f(x) \geq g(x)$ for *x* in $[a, b]$. Suppose a constant *C* is added to both functions, with *C* large enough so that both graphs lie above the *x*-axis, as in Figure 19(b). The region between the graphs is not changed; by the work done above, the area is given by $\int_a^b [f(x) - g(x)]dx$ regardless of where the graphs of $f(x)$ and $g(x)$ are located. As long as $f(x) \geq g(x)$ on $[a, b]$, then the area between the graphs from $x = a$ to $x = b$ will equal $\int_a^b [f(x) - g(x)]dx$.

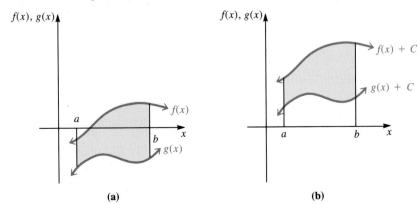

(a)          (b)

**FIGURE 19**

■ EXAMPLE 3

Find the area of the region enclosed by $y = x^2 - 2x$ and $y = x$ on [0, 4].

The required region, shown in Figure 20, is composed of two separate regions. Let $f(x) = x^2 - 2x$ and $g(x) = x$. Then, in the interval from 0 to 3, $g(x) \geq f(x)$, but from 3 to 4, $f(x) \geq g(x)$. Because of this switch, the area is found by taking the sum of two integrals, as follows.

$$\text{Area} = \int_0^3 [x - (x^2 - 2x)] \, dx + \int_3^4 [(x^2 - 2x) - x] \, dx$$

$$= \int_0^3 (-x^2 + 3x) \, dx + \int_3^4 (x^2 - 3x) \, dx$$

$$= \left( \frac{-x^3}{3} + \frac{3x^2}{2} \right) \Big|_0^3 + \left( \frac{x^3}{3} - \frac{3x^2}{2} \right) \Big|_3^4$$

$$= \left( -9 + \frac{27}{2} - 0 \right) + \left( \frac{64}{3} - 24 - 9 + \frac{27}{2} \right)$$

$$= \frac{19}{3}$$

This example illustrates the importance of a sketch when using the definite integral to find the area between two curves. ■

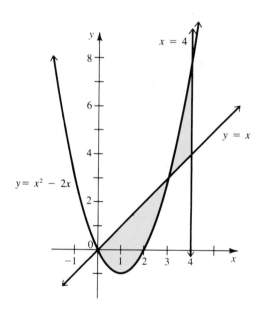

**FIGURE 20**

In the rest of this section we will look at some typical applications that require finding the area between two curves.

### ■ EXAMPLE 4

A company is considering a new manufacturing process in one of its plants. The new process provides substantial initial savings, with the savings declining with time $x$ according to the rate-of-savings function

$$S(x) = 100 - x^2,$$

where $S(x)$ is in thousands of dollars. At the same time, the cost of operating the new process increases with time $x$, according to the rate-of-cost function (in thousands of dollars)

$$C(x) = x^2 + \frac{14}{3}x.$$

**(a)** For how many years will the company realize savings?

Figure 21 shows the graphs of the rate-of-savings and rate-of-cost functions. The rate-of-cost (marginal cost) is increasing, while the rate-of-savings (marginal savings) is decreasing. The company should use this new process until the difference between these quantities is zero; that is, until the time at which these graphs intersect. The graphs intersect when

$$C(x) = S(x),$$

or

$$100 - x^2 = x^2 + \frac{14}{3}x.$$

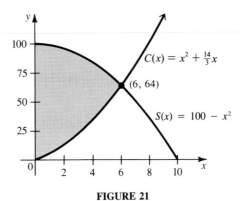

**FIGURE 21**

Solve this equation as follows.

$$0 = 2x^2 + \frac{14}{3}x - 100$$

$$= 3x^2 + 7x - 150 \qquad \text{Multiply by 3/2}$$

$$= (x - 6)(3x + 25) \qquad \text{Factor}$$

Set each factor equal to 0 separately to get

$$x = 6 \qquad \text{or} \qquad x = -25/3.$$

Only 6 is a meaningful solution here. The company should use the new process for 6 years.

**(b)** What will be the net total savings during this period?

Since the total savings over the 6-year period is given by the area under the rate-of-savings curve and the total additional cost by the area under the rate-of-cost curve, the net total savings over the 6-year period is given by the area between the rate of cost and the rate of savings curves and the lines $x = 0$ and $x = 6$. This area can be evaluated with a definite integral as follows.

$$\begin{aligned}
\text{Total Savings} &= \int_0^6 \left[ (100 - x^2) - \left( x^2 + \frac{14}{3}x \right) \right] dx \\
&= \int_0^6 \left( 100 - \frac{14}{3}x - 2x^2 \right) dx \\
&= 100x - \frac{7}{3}x^2 - \frac{2}{3}x^3 \Big|_0^6 \\
&= 100(6) - \frac{7}{3}(36) - \frac{2}{3}(216) = 372.
\end{aligned}$$

The company will save a total of $372,000 over the 6-year period. ■

## ■ EXAMPLE 5

A farmer has been using a new fertilizer that gives him a better yield, but because it exhausts the soil of other nutrients he must use other fertilizers in greater and greater amounts, so that his costs increase each year. The new fertilizer produces a rate of increase in revenue (in hundreds of dollars) that is given by

$$R(t) = -.4t^2 + 8t + 10,$$

where $t$ is measured in years. The rate of increase in yearly costs (also in hundreds of dollars) caused by using the fertilizer is given by

$$C(t) = 2t + 5.$$

How long can the farmer profitably use the fertilizer? What will be his net increase in revenue over this period?

The farmer should use the new fertilizer until the marginal costs equal the marginal revenue. Find this point by solving the equation $R(t) = C(t)$ as follows.

$$\begin{aligned}
-.4t^2 + 8t + 10 &= 2t + 5 \\
-4t^2 + 80t + 100 &= 20t + 50 \\
-4t^2 + 60t + 50 &= 0
\end{aligned}$$

Use the quadratic formula to get

$$t = 15.8.$$

The new fertilizer will be profitable for about 15.8 years.

 To find the total amount of additional revenue over the 15.8-year period, find the area between the graphs of the rate-of-revenue and the rate-of-cost functions, as shown in Figure 22.

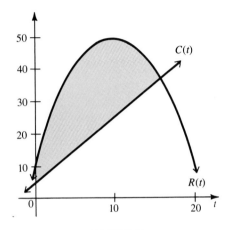

**FIGURE 22**

$$\text{Total Savings} = \int_0^{15.8} [R(t) - C(t)] \, dt$$

$$= \int_0^{15.8} [(-.4t^2 + 8t + 10) - (2t + 5)] \, dt$$

$$= \int_0^{15.8} (-.4t^2 + 6t + 5) dt$$

$$= \left( \frac{-.4t^3}{3} + \frac{6t^2}{2} + 5t \right) \Big|_0^{15.8}$$

$$= 302.01$$

The total savings will amount to about $30,000 over the 15.8-year period.

 It is not realistic to say that the farmer will need to use the new process for 15.8 years. He will probably have to use it for 15 years or 16 years. In this case, when the mathematical model produces results that are not in the domain of the function, it is necessary to find the total savings after 15 years and after 16 years and then select the best result. ▬

*Consumers' Surplus*   The market determines the price at which a product is sold. As indicated earlier, the point of intersection of the demand curve and the supply curve for a product gives the equilibrium price. At the equilibrium price consumers will purchase the same amount of the product that the manu-

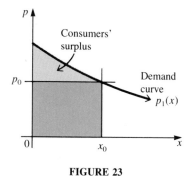

**FIGURE 23**

## CONSUMERS' SURPLUS

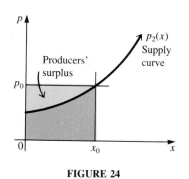

**FIGURE 24**

## PRODUCERS' SURPLUS

facturers want to sell. Sometimes, however, consumers are willing to spend more for an item than the equilibrium price. The total of the differences between the equilibrium price of the item and the higher prices individuals are willing to pay is thought of as savings realized by those individuals and is called the **consumers' surplus.**

In Figure 23, the area under the demand curve is the total amount consumers are willing to spend for $x_0$ items. The heavily shaded area under the line $y = p_0$ shows the total amount consumers actually will spend at the equilibrium price of $p_0$. The lightly shaded area represents the consumers' surplus. As the figure suggests, the consumers' surplus is given by an area between the two curves $p = p_1(x)$ and $p = p_0$, so its value can be found with a definite integral as follows.

> If $p_1(x)$ is a demand function with equilibrium price $p_0$ and equilibrium demand $x_0$, then
>
> $$\text{Consumers' Surplus} = \int_0^{x_0} \left[ p_1(x) - p_0 \right] dx.$$

Similarly, if manufacturers are willing to supply a product at a price *lower* than the equilibrium price $p_0$, the total of the differences between the equilibrium price and the lower prices at which the manufacturers will sell the product is considered added income for the manufacturers and is called the **producers' surplus.** Figure 24 shows the (heavily shaded) total area under the supply curve from $x = 0$ to $x = x_0$, which is the minimum total amount the manufacturers are willing to realize from the sale of $x_0$ items. The total area under the line $p = p_0$ is the amount actually realized. The difference in these two areas, the producers' surplus, also is given by a definite integral.

> If $p_2(x)$ is a supply function with equilibrium price $p_0$ and equilibrium supply $x_0$, then
>
> $$\text{Producers' Surplus} = \int_0^{x_0} \left[ p_0 - p_2(x) \right] dx.$$

### ■■■ EXAMPLE 6

Suppose the price, in cents, for a certain product is

$$p(x) = 900 - 20x - x^2,$$

when the demand for the product is $x$ units. Also, suppose the function

$$p(x) = x^2 + 10x$$

gives the price, in cents, when the supply is $x$ units. Find the consumers' surplus and the producers' surplus.

Begin by finding the equilibrium supply or demand. The graphs of both functions are shown in Figure 25 along with the equilibrium point at which supply and demand are equal. To find the equilibrium supply or demand $x$, solve the following equation.

$$900 - 20x - x^2 = x^2 + 10x$$

$$0 = 2x^2 + 30x - 900$$

$$0 = x^2 + 15x - 450$$

The only positive solution of the equation is $x = 15$. At the equilibrium point where supply and demand are both 15 units, the price is

$$p(15) = 900 - 20(15) - 15^2 = 375,$$

or \$3.75. The consumers' surplus, represented by the area shown in Figure 25, is

$$\int_0^{15} \left[ (900 - 20x - x^2) - 375 \right] dx.$$

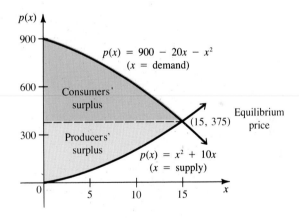

**FIGURE 25**

Evaluating the definite integral gives

$$\left( 900x - 10x^2 - \frac{1}{3}x^3 - 375x \right) \Big|_0^{15}$$

$$= \left[ 900(15) - 10(15)^2 - \frac{1}{3}(15)^3 - (375)(15) \right] - 0$$

$$= 4500.$$

Here the consumers' surplus is 4500 cents, or $45. The producers' surplus, also shown in Figure 25, is given by

$$\int_0^{15}\left[375 - (x^2 + 10x)\right]dx = \left[375x - \left(\frac{1}{3}x^3 + 5x^2\right)\right]\Big|_0^{15}$$

$$= 5625 - \left[\frac{1}{3}(15)^3 + 5(15)^2\right]$$

$$= 3375.$$

The producers' surplus is $33.75.  ▬▬

## 5.5 EXERCISES

*Find the area between the curves in Exercises 1–16.*

**1.** $x = -2,$  $x = 1,$  $y = x^2 + 4,$  $y = 0$

**2.** $x = 1,$  $x = 2,$  $y = x^3,$  $y = 0$

**3.** $x = -3,$  $x = 1,$  $y = x + 1,$  $y = 0$

**4.** $x = -2,$  $x = 0,$  $y = 1 - x^2,$  $y = 0$

**5.** $x = 0,$  $x = 6,$  $y = 5x,$  $y = 3x + 10$

**6.** $x = -2,$  $x = 1,$  $y = 2x,$  $y = x^2 - 3$

**7.** $y = x^2 - 30,$  $y = 10 - 3x$

**8.** $y = x^2 - 18,$  $y = x - 6$

**9.** $y = x^2,$  $y = 2x$

**10.** $y = x^2,$  $y = x^3$

**11.** $x = 1,$  $x = 6,$  $y = \dfrac{1}{x},$  $y = -1$

**12.** $x = 0,$  $x = 4,$  $y = \dfrac{1}{x + 1},$  $y = \dfrac{x - 1}{2}$

**13.** $x = -1,$  $x = 1,$  $y = e^x - 2,$  $y = x + 1$

**14.** $x = 0,$  $x = 2,$  $y = e^{-x},$  $y = e^x$

**15.** $x = 1,$  $x = 2,$  $y = e^x,$  $y = \dfrac{1}{x}$

**16.** $x = 2,$  $x = 4,$  $y = \dfrac{x}{2} + 3,$  $y = \dfrac{1}{x - 1}$

## ▤ APPLICATIONS

### BUSINESS AND ECONOMICS

**Net Savings**    **17.** Suppose a company wants to introduce a new machine that will produce a rate of annual savings given by

$$S(x) = 150 - x^2,$$

where $x$ is the number of years of operation of the machine, while producing a rate of annual costs of

$$C(x) = x^2 + \frac{11}{4}x.$$

**(a)** For how many years will it be profitable to use this new machine?

**(b)** What are the net total savings during the first year of use of the machine?

**(c)** What are the net total savings over the entire period of use of the machine?

**Net Savings** 18. A new smog-control device will reduce the output of sulfur oxides from automobile exhausts. It is estimated that the rate of savings to the community from the use of this device will be approximated by

$$S(x) = -x^2 + 4x + 8,$$

where $S(x)$ is the rate of savings in millions of dollars after $x$ years of use of the device. The new device cuts down on the production of sulfur oxides but it causes an increase in the production of nitrous oxides. The rate of additional costs in millions to the community after $x$ years is approximated by

$$C(x) = \frac{3}{25}x^2.$$

(a) For how many years will it pay to use the new device?

(b) What will be the net total savings over this period of time?

**Profit** 19. De Win Enterprises had an expenditure rate (in hundreds of dollars) of $E(x) = 4x + 2$, and an income rate (in hundreds of dollars) of $I(x) = 100 - x$ on a particular job, where $x$ was the number of days from the start of the job. Their profit on that job will equal total income less total expenditure. Profit will be maximized if the job ends at the optimum time, which is the point where the two curves meet. Find the following.

(a) The optimum number of days for the job to last.

(b) The total income for the optimum number of days.

(c) The total expenditure for the optimum number of days.

(d) The maximum profit for the job.

**Net Savings** 20. A factory has installed a new process that will produce an increased rate of revenue (in thousands of dollars) of

$$R(t) = -t^2 + 15.5t + 16,$$

where $t$ is measured in years. The new process produces additional costs (in thousands of dollars) at the rate of

$$C(t) = .25t + 4.$$

(a) When will it no longer be profitable to use this new process?

(b) Find the total net savings.

**Consumers' and Producers' Surplus** *For Exercises 21–26 refer to Example 6.*

21. Find the producers' surplus if the supply function of some item is given by

$$p(x) = x^2 + 2x + 50.$$

Assume supply and demand are in equilibrium at $x = 20$.

22. Suppose the supply function for a certain commodity is given by

$$p(x) = 100 + 3x + x^2,$$

and that supply and demand are in equilibrium at $x = 3$. Find the producers' surplus.

23. Find the consumers' surplus if the demand function for an item is given by

$$p(x) = 50 - x^2,$$

assuming supply and demand are in equilibrium at $x = 5$.

24. Find the consumers' surplus if the demand for an item is given by

$$p(x) = -(x + 4)^2 + 66,$$

and if supply and demand are in equilibrium at $x = 3$.

**25.** Suppose the supply function of a certain item is given by

$$p(x) = \frac{7}{5}x,$$

and the demand function is

$$p(x) = -\frac{3}{5}x + 10.$$

**(a)** Graph the supply and demand curves.
**(b)** Find the point at which supply and demand are in equilibrium.
**(c)** Find the consumers' surplus.
**(d)** Find the producers' surplus.

**26.** Repeat the four steps in Exercise 25 for the supply function

$$p(x) = x^2 + \frac{11}{4}x$$

and the demand function

$$p(x) = 150 - x^2.$$

## SOCIAL SCIENCES

**Distribution of Income**    **27.** Suppose that all the people in a country are ranked according to their incomes, starting at the bottom. Let $x$ represent the fraction of the community making the lowest income ($0 \le x \le 1$); $x = .4$, therefore, represents the lower 40% of all income producers. Let $I(x)$ represent the proportion of the total income earned by the lowest $x$ of all people. Thus, $I(.4)$ represents the fraction of total income earned by the lowest 40% of the population. Suppose

$$I(x) = .9x^2 + .1x.$$

Find and interpret the following.

**(a)** $I(.1)$    **(b)** $I(.4)$    **(c)** $I(.6)$    **(d)** $I(.9)$

If income were distributed uniformly, we would have $I(x) = x$. The area under this line of complete equality is 1/2. As $I(x)$ dips farther below $y = x$, there is less equality of income distribution. This inequality can be quantified by the ratio of the area between $I(x)$ and $y = x$ to 1/2. This ratio is called the *coefficient of inequality* and equals $2\int_0^1(x - I(x))dx$.

**(e)** Graph $I(x) = x$ and $I(x) = .9x^2 + .1x$ for $0 \le x \le 1$ on the same axes.
**(f)** Find the area between the curves.

## FOR THE COMPUTER

*Approximate the area between the graphs of each pair of functions on the given interval.*

**28.** $y = \ln x$ and $y = xe^x$;    $[1, 4]$

**29.** $y = \ln x$ and $y = 4 - x^2$;    $[2, 4]$

**30.** $y = \sqrt{9 - x^2}$ and $y = \sqrt{x + 1}$;    $[-1, 3]$

**31.** $y = \sqrt{4 - 4x^2}$ and $y = \sqrt{\frac{9 - x^2}{3}}$;    $[-1, 1]$

| ═ KEY WORDS ═ | 5.1 | antiderivative |  | total change |
|---|---|---|---|---|

═ CHAPTER 5   REVIEW EXERCISES

*Find each indefinite integral.*

**1.** $\int 6\,dx$

**2.** $\int (-4)dx$

**3.** $\int (2x + 3)dx$

**4.** $\int (5x - 1)dx$

**5.** $\int (x^2 - 3x + 2)dx$

**6.** $\int (6 - x^2)dx$

**7.** $\int 3\sqrt{x}\,dx$

**8.** $\int \frac{\sqrt{x}}{2}\,dx$

**9.** $\int (x^{1/2} + 3x^{-2/3})dx$

**10.** $\int (2x^{4/3} + x^{-1/2})dx$

**11.** $\int \frac{-4}{x^3}\,dx$

**12.** $\int \frac{5}{x^4}\,dx$

**13.** $\int -3e^{2x}\,dx$

**14.** $\int 5e^{-x}\,dx$

**15.** $\int \frac{2}{x - 1}\,dx$

**16.** $\int \frac{-4}{x + 2}\,dx$

*Use substitution to find each indefinite integral.*

**17.** $\int xe^{3x^2}\,dx$

**18.** $\int 2xe^{x^2}\,dx$

**19.** $\int \frac{3x}{x^2 - 1}\,dx$

**20.** $\int \frac{-x}{2 - x^2}\,dx$

**21.** $\int 2x\sqrt{x^2 - 3}\,dx$

**22.** $\int x\sqrt{5x^2 + 6}\,dx$

**23.** $\int \frac{x^2\,dx}{(x^3 + 5)^4}$

**24.** $\int (x^2 - 5x)^4(2x - 5)dx$

**25.** $\int \frac{4x - 5}{2x^2 - 5x}\,dx$

**26.** $\int \frac{12(2x + 9)}{x^2 + 9x + 1}\,dx$

**27.** $\int \frac{x^3}{e^{3x^4}}\,dx$

**28.** $\int e^{3x^2 + 4}\,x\,dx$

**29.** $\int -2e^{-5x}\,dx$

**30.** $\int e^{-4x}\,dx$

*Find the area of the region enclosed by each group of curves.*

**31.** $f(x) = 5 - x^2,\ g(x) = x^2 - 3$

**32.** $f(x) = x^2 - 4x,\ g(x) = x + 1,\ x = 2,\ x = 4$

**33.** $f(x) = x^2 - 4x,\ g(x) = x - 6$

**34.** $f(x) = 5 - x^2,\ g(x) = x^2 - 3,\ x = 0,\ x = 4$

**35.** Evaluate $\sum_{i=1}^{4}(i^2 - i)$.

**36.** Let $f(x) = 3x + 1,\ x_1 = -1,\ x_2 = 0,\ x_3 = 1,\ x_4 = 2,$ and $x_5 = 3$. Find
$\sum_{i=1}^{5} f(x_i)$.

**37.** Approximate the area under the graph of $f(x) = 2x + 3$ and above the x-axis from $x = 0$ to $x = 4$ using four rectangles. Let the height of each rectangle be the function value on the left side.

**38.** Find $\int_0^4 (2x + 3)\, dx$ by using the formula for the area of a trapezoid: $A = \frac{1}{2}(B + b)h$, where $B$ and $b$ are the lengths of the parallel sides and $h$ is the distance between them. (See Exercise 37.)

*Find each definite integral.*

**39.** $\displaystyle\int_1^2 (3x^2 + 5)dx$    **40.** $\displaystyle\int_1^6 (2x^2 + x)dx$    **41.** $\displaystyle\int_1^5 (3x^{-2} + x^{-3})dx$    **42.** $\displaystyle\int_2^3 (5x^{-2} + x^{-4})dx$

**43.** $\displaystyle\int_1^3 2x^{-1}\, dx$    **44.** $\displaystyle\int_1^6 8x^{-1}\, dx$    **45.** $\displaystyle\int_0^4 2e^x\, dx$    **46.** $\displaystyle\int_1^6 \frac{5}{2}e^{4x}\, dx$

*Find the area between the x-axis and f(x) over each of the given intervals.*

**47.** $f(x) = \sqrt{x - 1};\quad [1, 10]$

**48.** $f(x) = 9 - x^2;\quad [-2, 2]$

**49.** $f(x) = e^x;\quad [0, 2]$

**50.** $f(x) = 1 + e^{-x};\quad [0, 4]$

---

### ☰ APPLICATIONS

#### BUSINESS AND ECONOMICS

**Cost**    *Find the cost function for each of the marginal cost functions in Exercises 51–54.*

**51.** $C'(x) = 10 - 2x;$   fixed cost is \$4.

**52.** $C'(x) = 2x + 3x^2;$   2 units cost \$12.

**53.** $C'(x) = 3\sqrt{2x - 1};$   13 units cost \$270.

**54.** $C'(x) = \dfrac{1}{x + 1};$   fixed cost is \$18.

**Loan Principal**    **55.** The two curves in the figure give the rate of change of principal in each year of the loan for two different types of 30-year mortgages of \$50,000 at 15% interest.

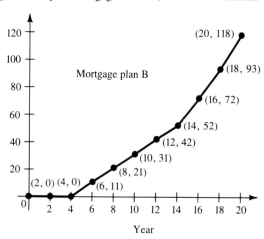

Use rectangles of width 2 units and height determined by the function value on the left side to find the following:

(a) The total amount of principal accumulated in the first 20 years under mortgage plan A.

(b) The total amount of principal accumulated in the first 20 years of mortgage plan B.

**Utilization of Reserves** **56.** A manufacturer of electronic equipment requires a certain rare metal. He has a reserve supply of 4,000,000 units that he will not be able to replace. If the rate at which the metal is used is given by

$$f(t) = 100,000e^{.03t},$$

where $t$ is the time in years, how long will it be before he uses up the supply? (*Hint:* Find an expression for the total amount used in $t$ years and set it equal to the known reserve supply.)

**Sales** **57.** The rate of change of sales of a new brand of tomato soup, in thousands, is given by

$$S'(x) = \sqrt{x} + 2,$$

where $x$ is the time in months that the new product has been on the market. Find the total sales after 9 months.

**Producers' and Consumers' Surplus** **58.** Suppose that the supply function of some commodity is

$$p(x) = x^2 + 5x + 100$$

and the demand function for the commodity is

$$p(x) = 350 - x^2.$$

(a) Find the producers' surplus.      (b) Find the consumers' surplus.

**Net Savings** **59.** A company has installed new machinery that will produce a savings rate (in thousands of dollars) of

$$S'(x) = 225 - x^2,$$

where $x$ is the number of years the machinery is to be used. The rate of additional costs to the company due to the new machinery is expected to be

$$C'(x) = x^2 + 25x + 150.$$

For how many years should the company use the new machinery? Find the net savings in thousands of dollars over this period.

## LIFE SCIENCES

**Population Growth** **60.** The rate of change of a population of prairie dogs, in terms of the number of coyotes, $x,$ that prey on them, is given by

$$P(x) = 25 - .1x.$$

Find the total number of prairie dogs as the coyote population grows from 100 to 200.

**Population Growth** **61.** The rate of change of the population of a rare species of Australian spider is given by

$$y' = 100 - .2x,$$

where $y$ is the number of spiders present at time $x$, measured in months. Find the total number of spiders in the first 10 months.

### PHYSICAL SCIENCES

**Linear Motion**  **62.** A particle is moving along a straight line with velocity $v(t) = t^2 - 2t$. Its distance from the starting point after 3 sec is 8 cm. Find $s(t)$, the distance of the particle from the starting point after $t$ sec.

**EXTENDED APPLICATION**

## ESTIMATING DEPLETION DATES FOR MINERALS

It is becoming more and more obvious that the earth contains only a finite quantity of minerals. The "easy and cheap" sources of minerals are being used up, forcing an ever more expensive search for new sources. For example, oil from the North Slope of Alaska would never have been used in the United States during the 1930s because a great deal of Texas and California oil was readily available.

We said in an earlier chapter that population tends to follow an exponential growth curve. Mineral usage also follows such a curve. Thus, if $q$ represents the rate of consumption of a certain mineral at time $t$, while $q_0$ represents consumption when $t = 0$, then

$$q = q_0 e^{kt},$$

where $k$ is the growth constant. For example, the world consumption of petroleum in a recent year was about 19,600 million barrels, with the value of $k$ about 6%. If we let $t = 0$ correspond to this base year, then $q_0 = 19,600$, $k = .06$, and

$$q = 19,600 e^{.06t}$$

is the rate of consumption at time $t$, assuming that all present trends continue.

Based on estimates of the National Academy of Science, 2,000,000 millions of barrels of oil are now in provable reserves or are likely to be discovered in the future. At the present rate of consumption, in how many years will these reserves be depleted? We can use the integral calculus of this chapter to find out.

To begin, we need to know the total quantity of petroleum that would be used between time $t = 0$ and some future time $t = t_1$. Figure 26 shows a typical graph of the function $q = q_0 e^{kt}$.

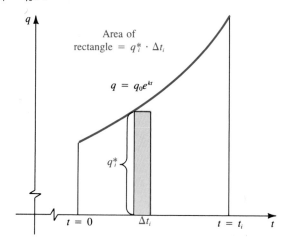

**FIGURE 26**

Following the work we did in Section 3, divide the time interval from $t = 0$ to $t = t_1$ into $n$ subintervals. Let the $i$th subinterval have width $\Delta t_i$. Let the rate of consumption for the $i$th subinterval be approximated by $q_i^*$. Thus, the approximate total consumption for the subinterval is given by

$$q_i^* \cdot \Delta t_i,$$

and the total consumption over the interval from time $t = 0$ to $t = t_1$ is approximated by

$$\sum_{i=1}^{n} q_i^* \cdot \Delta t_i.$$

The limit of this sum as each of the $\Delta t_i$'s approaches 0 gives the total consumption from time $t = 0$ to $t = t_1$. That is,

$$\text{Total Consumption} = \lim_{\Delta t_i \to 0} \sum q_i^* \cdot \Delta t_i.$$

We have seen, however, that this limit is the definite integral of the function $q = q_0 e^{kt}$ from $t = 0$ to $t = t_1$, or

$$\text{Total Consumption} = \int_0^{t_1} q_0 e^{kt}\, dt.$$

We can now evaluate this definite integral.

$$\int_0^{t_1} q_0 e^{kt}\, dt = q_0 \int_0^{t_1} e^{kt}\, dt = q_0 \left(\frac{1}{k} e^{kt}\right)\Big|_0^{t_1}$$

$$= \frac{q_0}{k} e^{kt}\Big|_0^{t_1} = \frac{q_0}{k} e^{kt_1} - \frac{q_0}{k} e^0$$

$$= \frac{q_0}{k} e^{kt_1} - \frac{q_0}{k}(1) \qquad\qquad \textbf{(*)}$$

$$= \frac{q_0}{k}(e^{kt_1} - 1)$$

Now let us return to the numbers we gave for petroleum. We said that $q_0 = 19{,}600$ million barrels, where $q_0$ represents consumption in the base year. We have $k = .06$, with total petroleum reserves estimated as 2,000,000 million barrels. Thus, using equation (*) we have

$$2{,}000{,}000 = \frac{19{,}600}{.06}(e^{.06t_1} - 1).$$

Multiply both sides of the equation by .06:

$$120{,}000 = 19{,}600(e^{.06t_1} - 1).$$

Divide both sides of the equation by 19,600.

$$6.1 = e^{.06t_1} - 1$$

Add 1 to both sides.

$$7.1 = e^{.06t_1}$$

Take natural logarithms of both sides:

$$\ln 7.1 = \ln e^{.06t_1}$$
$$= .06t_1 \ln e$$
$$= .06t_1 \qquad (\text{since } \ln e = 1).$$

Finally,

$$t_1 = \frac{\ln 7.1}{.06}.$$

From the table of natural logarithms or a calculator, estimate ln 7.1 as about 1.96. Thus,

$$t_1 = \frac{1.96}{.06} = 33.$$

By this result, petroleum reserves will last the world for thirty-three years.

    The results of mathematical analyses such as this must be used with great caution. By the analysis above, the world would use all the petroleum that it wants in the thirty-second year after the base year, but there would be none at all in thirty-four years. This is not at all realistic. As petroleum reserves decline, the price will increase, causing demand to decline and supplies to increase.

**EXERCISES**

1. Find the number of years that the estimated petroleum reserves would last if used at the same rate as in the base year.

2. How long would the estimated petroleum reserves last if the growth constant was only 2% instead of 6%?

Estimate the length of time until depletion for each of the following minerals.

3. Bauxite (the ore from which aluminum is obtained), estimated reserves in base year 15,000,000 thousand tons, rate of consumption in base year 63,000 thousand tons, growth constant 6%

4. Bituminous coal, estimated world reserves 2,000,000 million tons, rate of consumption in base year 2200 million tons, growth constant 4%

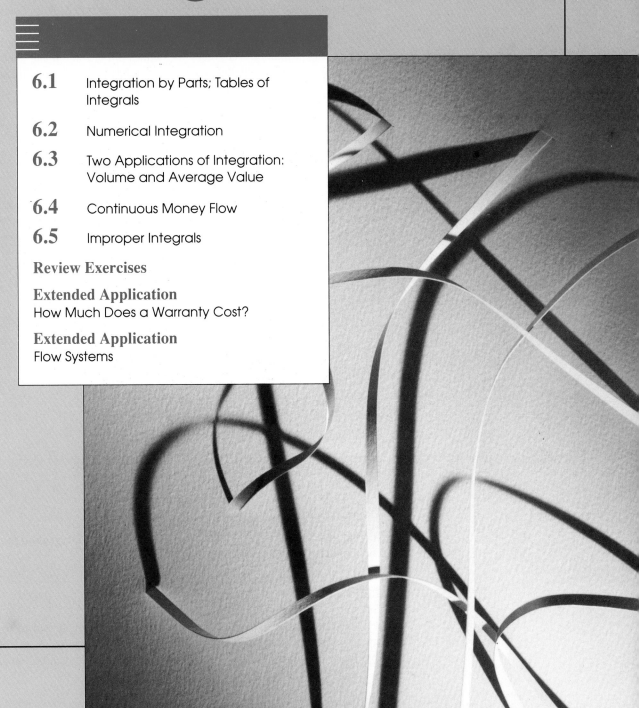

# 6 Further Techniques and Applications of Integration

In the previous chapter we discussed indefinite and definite integrals and presented rules for finding the antiderivatives of several types of functions. We saw how these techniques could be used in various applications. In this chapter we develop additional methods of integrating functions, including numerical methods of integration. Numerical methods often are used in applications involving experimental data or with functions that cannot be integrated by other methods. In this chapter we also show how to evaluate an integral that has one or both limits at infinity. These new techniques provide additional applications of integration, such as finding volumes of solids of revolution, the average value of a function, and continuous money flow.

## 6.1 INTEGRATION BY PARTS; TABLES OF INTEGRALS

The technique of *integration by parts* often makes it possible to reduce a complicated integral to a simpler integral. If $u$ and $v$ are both differentiable functions, then $uv$ is also differentiable and, by the product rule for derivatives,

$$\frac{d(uv)}{dx} = u\frac{dv}{dx} + v\frac{du}{dx}.$$

This expression can be rewritten, using differentials, as

$$d(uv) = u\,dv + v\,du.$$

Integrating both sides of this last equation gives

$$\int d(uv) = \int u\,dv + \int v\,du,$$

or

$$uv = \int u\,dv + \int v\,du.$$

Rearranging terms gives the following formula.

**INTEGRATION BY PARTS**

> If $u$ and $v$ are differentiable functions, then
>
> $$\int u\,dv = uv - \int v\,du.$$

The process of finding integrals by this formula is called **integration by parts.** The method is shown in the following examples.

### EXAMPLE 1

Find $\int xe^{5x}\,dx$.

While this integral cannot be found by using any method studied so far, it can be found with integration by parts. First write the expression $xe^{5x}\,dx$ as a product of two functions $u$ and $dv$ in such a way that $\int dv$ can be found. One way to do this is to choose the two functions $x$ and $e^{5x}$. Both $x$ and $e^{5x}$ can be integrated, but $\int x\,dx$, which is $x^2/2$, is more complicated than $x$ itself, while the derivative of $x$ is 1, which is simpler than $x$. Since $e^{5x}$ remains the same (except for the coefficient) whether it is integrated or differentiated, it is best here to choose

$$dv = e^{5x}\,dx \quad \text{and} \quad u = x.$$

Then

$$du = dx,$$

and $v$ is found by integrating $dv$:

$$v = \int dv = \int e^{5x}\,dx = \frac{1}{5}e^{5x} + C.$$

For simplicity, ignore the constant $C$ and add it at the end of the process of integration by parts. Now substitute into the formula for integration by parts and complete the integration.

$$\int u\,dv = uv - \int v\,du$$

$$\int \underbrace{x}_{u}\underbrace{e^{5x}\,dx}_{dv} = \underbrace{x}_{u}\underbrace{\left(\frac{1}{5}e^{5x}\right)}_{v} - \int \underbrace{\frac{1}{5}e^{5x}}_{v}\underbrace{dx}_{du}$$

$$= \frac{1}{5}xe^{5x} - \frac{1}{25}e^{5x} + C$$

The constant $C$ was added in the last step. As before, check the answer by taking its derivative. ▬

## ▬ EXAMPLE 2
Find $\int \ln x\,dx$ for $x > 0$.

No rule has been given for integrating $\ln x$, so choose

$$dv = dx \quad \text{and} \quad u = \ln x.$$

Then

$$v = x \quad \text{and} \quad du = \frac{1}{x}dx,$$

and

$$\int u \cdot dv = v \cdot u - \int v \cdot du$$

$$\int \overbrace{\ln x\,dx}^{} = \overbrace{x \ln x}^{} - \int x \cdot \frac{1}{x}dx$$

$$= x \ln x - \int dx$$

$$= x \ln x - x + C. \;\; ▬$$

■■■ EXAMPLE 3

Find $\int_0^3 x\sqrt{1 + x}\, dx$.

First find the indefinite integral $\int x\sqrt{1 + x}\, dx$. The objective is to simplify $v\,du$. Since the derivative of $x$ is 1, try

$$dv = \sqrt{1 + x}\, dx = (1 + x)^{1/2}\, dx \qquad \text{and} \qquad u = x.$$

Then

$$v = \frac{2}{3}(1 + x)^{3/2} \qquad \text{and} \qquad du = dx.$$

Substitute these values into the formula for integration by parts and integrate the second term on the right.

$$\int u\, dv = uv - \int v\, du$$

$$\int x\sqrt{1 + x}\, dx = \frac{2}{3}x(1 + x)^{3/2} - \int \frac{2}{3}(1 + x)^{3/2}\, dx$$

$$= \frac{2}{3}x(1 + x)^{3/2} - \frac{2}{3} \cdot \frac{2}{5}(1 + x)^{5/2} + C$$

Now find the definite integral.

$$\int_0^3 x\sqrt{1 + x}\, dx = \left[ \frac{2}{3}x(1 + x)^{3/2} - \frac{4}{15}(1 + x)^{5/2} \right]_0^3$$

$$= \left[ \frac{2}{3}(3)(1 + 3)^{3/2} - \frac{4}{15}(1 + 3)^{5/2} \right]$$

$$- \left[ \frac{2}{3}(0)(1 + 0)^{3/2} - \frac{4}{15}(1 + 0)^{5/2} \right]$$

$$= \left[ 2(4)^{3/2} - \frac{4}{15}(4)^{5/2} \right] - \left[ 0 - \frac{4}{15} \right]$$

$$= 16 - \frac{128}{15} + \frac{4}{15} = \frac{116}{15} \quad ■■$$

The preceding examples illustrate the following general principles for identifying integrals that can be found with integration by parts.

**CONDITIONS FOR INTEGRATION BY PARTS**

■■■ Integration by parts can be used only if the integrand satisfies the following conditions.

**1.** The integrand can be written as the product of two factors, $u$ and $dv$.
**2.** It is possible to integrate $dv$ to get $v$ and to differentiate $u$ to get $du$.
**3.** The integral $\int v\, du$ can be found.

Sometimes it is necessary to use the technique of integrating by parts more than once, as in the following example.

## ■■ EXAMPLE 4

Find $\int 2x^2 e^x\, dx$.

Since $\int e^x\, dx$ can be found and since the derivative of $x^2$ is $2x$, a simpler expression, choose

$$dv = e^x dx \qquad \text{and} \qquad u = 2x^2$$

Then

$$v = e^x \qquad \text{and} \qquad du = 4x\, dx.$$

Now substitute into the formula for integration by parts.

$$\int u\, dv = uv - \int v\, du$$

$$\int \underbrace{2x^2}_{u}\, \underbrace{e^x\, dx}_{dv} = \underbrace{2x^2 e^x}_{u\ \ v} - \int e^x \underbrace{(4x\, dx)}_{v\ \ \ du} \tag{1}$$

Find $\int e^x(4x\, dx)$ by parts. Start by choosing $dv = e^x\, dx$, which gives $v = e^x$ and $u = 4x$, so that $du = 4\, dx$. Substituting again into the formula for integration by parts gives

$$\int \underbrace{4x}_{u}\, \underbrace{e^x\, dx}_{dv} = \underbrace{4x\, e^x}_{u\ \ v} - \int e^x \underbrace{(4\, dx)}_{v\ \ \ du}$$

$$= 4x\, e^x - 4e^x.$$

Substitute this antiderivative back into equation (1) to get the desired integral.

$$\int 2x^2 e^x dx = 2x^2 e^x - \int e^x(4x dx) \tag{1}$$

$$= 2x^2 e^x - (4xe^x - 4e^x)$$

$$= 2x^2 e^x - 4xe^x + 4e^x + C$$

(A constant $C$ was added in the last step.) ■■

***Integration Using Tables of Integrals*** The method of integration by parts requires choosing the factor $dv$ so that $\int dv$ can be found. If this is not possible, or if the remaining factor, which becomes $u$, does not have a differential $du$ such that $\int v\, du$ can be found, the technique cannot be used. For example, to integrate

$$\int \frac{1}{4 - x^2}\, dx,$$

we might choose $dv = dx$ and $u = (4 - x^2)^{-1}$. Integration gives $v = x$ and differentiation gives $du = 2x\, dx/(4 - x^2)^2$, with

$$\int \frac{1}{4 - x^2}\, dx = \frac{x}{4 - x^2} - \int \frac{2x^2\, dx}{(4 - x^2)^2}.$$

The integral on the right is more complicated than the original integral, however. A second use of integration by parts on the new integral would make matters even worse. Since we cannot choose $dv = (4 - x^2)^{-1} \, dx$ because it cannot be integrated by the methods studied so far, integration by parts is not possible for this problem. In fact, there are many functions whose integrals cannot be found by any of the methods described in this text. Many of these can be found by more advanced methods and are listed in **tables of integrals.** Such a table is given at the back of this book. The next examples show how to use this table.

### ■ EXAMPLE 5

Find $\displaystyle\int \frac{1}{\sqrt{x^2 + 16}} \, dx.$

If $a = 4$, this antiderivative is the same as entry 5 in the table,

$$\int \frac{1}{\sqrt{x^2 + a^2}} \, dx = \ln \left| \frac{x + \sqrt{x^2 + a^2}}{a} \right| + C.$$

Substituting 4 for $a$ gives

$$\int \frac{1}{\sqrt{x^2 + 16}} \, dx = \ln \left| \frac{x + \sqrt{x^2 + 16}}{4} \right| + C.$$

This result could be verified by taking the derivative of the right side of this last equation.  ■

### ■ EXAMPLE 6

Find $\displaystyle\int \frac{8}{16 - x^2} \, dx.$

Convert this antiderivative into the one given in entry 7 of the table by writing the 8 in front of the integral sign (permissible only with constants) and by letting $a = 4$. Doing this gives

$$\begin{aligned} 8 \int \frac{1}{16 - x^2} \, dx &= 8 \left[ \frac{1}{2 \cdot 4} \ln \left( \frac{4 + x}{4 - x} \right) \right] + C \\ &= \ln \left( \frac{4 + x}{4 - x} \right) + C. \end{aligned}$$

In entry 7 of the table, the condition $x^2 < a^2$ is given. Since $a = 4$ in this example, the result given above is valid only for $x^2 < 16$, so that the final answer should be written

$$\int \frac{8}{16 - x^2} \, dx = \ln \left( \frac{4 + x}{4 - x} \right) + C \qquad \text{for } x^2 < 16.$$

Because of the condition $x^2 < 16$, the expression in parentheses is always positive, so that absolute value bars are not needed.  ■

### ▆ EXAMPLE 7

Find $\int \sqrt{9x^2 + 1}\ dx$.

This antiderivative seems most similar to entry 15 of the table. However, entry 15 requires that the coefficient of the $x^2$ term be 1. That requirement can be satisfied here by factoring out the 9.

$$\int \sqrt{9x^2 + 1}\ dx = \int \sqrt{9\left(x^2 + \frac{1}{9}\right)}\,dx$$

$$= \int 3\sqrt{x^2 + \frac{1}{9}}dx$$

$$= 3\int \sqrt{x^2 + \frac{1}{9}}dx$$

Now use entry 15 with $a = 1/3$.

$$\int \sqrt{9x^2 + 1}\ dx = 3\left[\frac{x}{2}\sqrt{x^2 + \frac{1}{9}} + \frac{(1/3)^2}{2} \ln\left|x + \sqrt{x^2 + \frac{1}{9}}\right|\right] + C$$

$$= \frac{3x}{2}\sqrt{x^2 + \frac{1}{9}} + \frac{1}{6}\ln\left|x + \sqrt{x^2 + \frac{1}{9}}\right| + C \quad ▆$$

### ▆ 6.1 EXERCISES

*Use integration by parts to find the integrals in Exercises 1–10.*

**1.** $\int xe^x\ dx$

**2.** $\int (x + 1)e^x\ dx$

**3.** $\int (5x - 9)e^{-3x}\ dx$

**4.** $\int (6x + 3)e^{-2x}\ dx$

**5.** $\int_0^1 \frac{2x + 1}{e^x}\ dx$

**6.** $\int_0^1 \frac{1 - x}{3e^x}\ dx$

**7.** $\int_1^4 \ln 2x\ dx$

**8.** $\int_1^2 \ln 5x\ dx$

**9.** $\int x \ln x\ dx, \quad x > 0$

**10.** $\int x^2 \ln x\ dx, \quad x > 0$

**11.** Find the area between $y = (x - 2)e^x$ and the *x*-axis from $x = 2$ to $x = 4$.

**12.** Find the area between $y = xe^x$ and the *x*-axis from $x = 0$ to $x = 1$.

*Exercises 13–22 are mixed—some require integration by parts, while others can be integrated by using techniques discussed earlier. Some problems may require using integration by parts more than once.*

**13.** $\int x^2 e^{2x}\ dx$

**14.** $\int_1^2 (1 - x^2)e^{2x}\ dx$

**15.** $\int_0^5 x\sqrt[3]{x^2 + 2}\ dx$

**16.** $\int (2x - 1)\ln (3x)dx, \quad x > 0$

**17.** $\int (8x + 7)\ln(5x)dx, \quad x > 0$

**18.** $\int xe^{x^2}\ dx$

**19.** $\int x^2\sqrt{x + 2}\, dx$     **20.** $\int_0^1 \dfrac{x^2\, dx}{2x^3 + 1}$     **21.** $\int_0^1 \dfrac{x^3\, dx}{\sqrt{3 + x^2}}$     **22.** $\int \dfrac{x^2\, dx}{2x^3 + 1}$

*Use the table of integrals to find each indefinite integral.*

**23.** $\int \dfrac{-4}{\sqrt{x^2 + 36}}\, dx$     **24.** $\int \dfrac{9}{\sqrt{x^2 + 9}}\, dx$     **25.** $\int \dfrac{6}{x^2 - 9}\, dx$

**26.** $\int \dfrac{-12}{x^2 - 16}\, dx$     **27.** $\int \dfrac{-4}{x\sqrt{9 - x^2}}\, dx$     **28.** $\int \dfrac{3}{x\sqrt{121 - x^2}}\, dx$

**29.** $\int \dfrac{-2x}{3x + 1}\, dx$     **30.** $\int \dfrac{6x}{4x - 5}\, dx$     **31.** $\int \dfrac{2}{3x(3x - 5)}\, dx$

**32.** $\int \dfrac{-4}{3x(2x + 7)}\, dx$     **33.** $\int \dfrac{4}{4x^2 - 1}\, dx$     **34.** $\int \dfrac{-6}{9x^2 - 1}\, dx$

**35.** $\int \dfrac{3}{x\sqrt{1 - 9x^2}}\, dx$     **36.** $\int \dfrac{-2}{x\sqrt{1 - 16x^2}}\, dx$     **37.** $\int \dfrac{4x}{2x + 3}\, dx$

**38.** $\int \dfrac{4x}{6 - x}\, dx$     **39.** $\int \dfrac{-x}{(5x - 1)^2}\, dx$     **40.** $\int \dfrac{-3}{x(4x + 3)^2}\, dx$

---

## ▤ APPLICATIONS

*Use integration by parts to solve the following problems.*

### BUSINESS AND ECONOMICS

**Rate of Change of Revenue**   **41.** The rate of change of revenue in dollars from the sale of $x$ units of small desk calculators is

$$R'(x) = 20x(2x + 3)^{1/2}.$$

Find the total revenue from the sale of the first 39 calculators.

### LIFE SCIENCES

**Reaction to a Drug**   **42.** The rate of reaction to a drug is given by

$$r'(x) = 2x^2 e^{-x},$$

where $x$ is the number of hours since the drug was administered. Find the total reaction to the drug from $x = 1$ to $x = 6$.

**Growth of a Population**   **43.** The rate of growth of a microbe population is given by

$$m'(x) = 30xe^{2x},$$

where $x$ is time in days. What is the total accumulated growth after 3 days?

### SOCIAL SCIENCES

**Production Rate**   **44.** The rate (in hours per item) at which a worker in a certain job produces the $x$th item is

$$h'(x) = \sqrt[3]{1 + x}.$$

What is the total number of hours it will take this worker to produce the first 7 items?

## 6.2   NUMERICAL INTEGRATION

As mentioned in the previous section, some integrals cannot be evaluated by any technique or found in any table. Since $\int_a^b f(x)dx$ represents an area when $f(x) > 0$, any approximation of that area also approximates the definite integral. In Section 3 of the previous chapter, the region under a curve was divided into rectangles and the sum of the areas of the rectangles was used as an approximation of the area under the curve. Many methods of approximating definite integrals by areas have been made feasible by the availability of pocket calculators and high-speed computers. These methods are referred to as **numerical integration** methods. We shall discuss two methods of numerical integration, the trapezoidal rule and Simpson's rule.

***Trapezoidal Rule***   To illustrate the use of the trapezoidal rule, consider the integral

$$\int_1^5 \frac{1}{x}\,dx.$$

The shaded region in Figure 1 shows the area representing that integral, the area under the graph $f(x) = 1/x$, above the x-axis, and between the lines $x = 1$ and $x = 5$. As shown in the figure, if the area under the curve is approximated with trapezoids rather than rectangles, the approximation should be improved.

Since $\int (1/x)dx = \ln |x| + C$,

$$\int_1^5 \frac{1}{x}\,dx = \ln |x| \Big|_1^5 = \ln 5 - \ln 1 = \ln 5 - 0 = \ln 5.$$

From the table of natural logarithms or a calculator, $\ln 5 \approx 1.609438$.

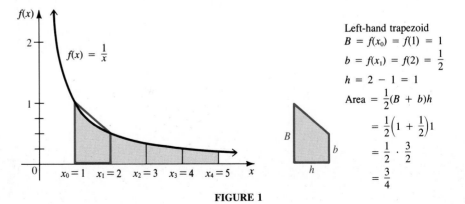

**FIGURE 1**

As in earlier work, to approximate this area we divide the interval [1, 5] into subintervals of equal width. To get a first approximation to ln 5 by the trapezoidal rule, find the sum of the areas of the four trapezoids shown in Figure 1. From geometry, the area of a trapezoid is half the product of the sum of the bases and the altitude. Each of the trapezoids in Figure 1 has altitude 1. (In this case, the bases of the trapezoid are vertical and the altitudes are horizontal.) Adding the areas gives

$$\ln 5 = \int_1^5 \frac{1}{x}\, dx \approx \frac{1}{2}\left(\frac{1}{1} + \frac{1}{2}\right)(1) + \frac{1}{2}\left(\frac{1}{2} + \frac{1}{3}\right)(1) + \frac{1}{2}\left(\frac{1}{3} + \frac{1}{4}\right)(1) + \frac{1}{2}\left(\frac{1}{4} + \frac{1}{5}\right)(1)$$

$$= \frac{1}{2}\left(\frac{3}{2} + \frac{5}{6} + \frac{7}{12} + \frac{9}{20}\right) \approx 1.68333.$$

To get a better approximation, divide the interval [1, 5] into more subintervals. Generally speaking, the larger the number of subintervals, the better the approximation. The results for selected values of *n* are shown below to 5 decimal places.

| $n$ | $\int_1^5 \frac{1}{x}\, dx = \ln 5$ |
|---|---|
| 6 | 1.64360 |
| 8 | 1.62897 |
| 10 | 1.62204 |
| 20 | 1.61263 |
| 100 | 1.60957 |
| 1000 | 1.60944 |

When $n = 1000$, the approximation agrees with the table value to 5 decimal places.

Generalizing from this example, let *f* be a continuous function on an interval [*a*, *b*]. Divide the interval from *a* to *b* into *n* equal subintervals by the points $a = x_0, x_1, x_2, \ldots, x_n = b$, as shown in Figure 2. Use the subintervals to make trapezoids that approximately fill in the region under the curve. The approximate value of the definite integral $\int_a^b f(x)\, dx$ is given by the sum of the areas of trapezoids, or

$$\int_a^b f(x)\, dx \approx \frac{1}{2}\left[f(x_0) + f(x_1)\right]\left(\frac{b-a}{n}\right) + \frac{1}{2}\left[f(x_1) + f(x_2)\right]\left(\frac{b-a}{n}\right) + \cdots + \frac{1}{2}\left[f(x_{n-1}) + f(x_n)\right]\left(\frac{b-a}{n}\right)$$

$$= \left(\frac{b-a}{n}\right)\left[\frac{1}{2}f(x_0) + \frac{1}{2}f(x_1) + \frac{1}{2}f(x_1) + \frac{1}{2}f(x_2) + \frac{1}{2}f(x_2) + \cdots + \frac{1}{2}f(x_{n-1}) + \frac{1}{2}f(x_n)\right]$$

$$= \left(\frac{b-a}{n}\right)\left[\frac{1}{2}f(x_0) + f(x_1) + f(x_2) + \cdots + f(x_{n-1}) + \frac{1}{2}f(x_n)\right].$$

This result gives the following rule.

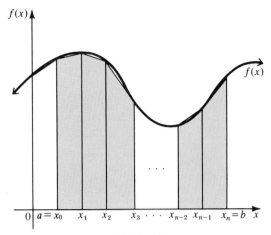

**FIGURE 2**

**TRAPEZOIDAL RULE**

Let $f$ be a continuous function on $[a, b]$ and let $[a, b]$ be divided into $n$ equal subintervals by the points $a = x_0, x_1, x_2, \ldots, x_n = b$. Then, by the **trapezoidal rule,**

$$\int_a^b f(x) \, dx \approx \left(\frac{b - a}{n}\right)\left[\frac{1}{2} f(x_0) + f(x_1) + \cdots + f(x_{n-1}) + \frac{1}{2} f(x_n)\right].$$

### EXAMPLE 1

Use the trapezoidal rule with $n = 4$ to approximate

$$\int_3^7 \frac{1}{x - 2} \, dx.$$

Here $a = 3$, $b = 7$, and $n = 4$, with $(b - a)/n = (7 - 3)/4 = 1$ as the altitude of each trapezoid. Then $x_0 = 3$, $x_1 = 4$, $x_2 = 5$, $x_3 = 6$, and $x_4 = 7$. Now find the corresponding function values. The work can be organized into a table, as follows.

| $i$ | $x_i$ | $f(x_i)$ |
|-----|-------|----------|
| 0 | 3 | $\dfrac{1}{3 - 2} = 1$ |
| 1 | 4 | $\dfrac{1}{4 - 2} = \dfrac{1}{2}$ |
| 2 | 5 | $\dfrac{1}{5 - 2} = \dfrac{1}{3}$ |
| 3 | 6 | $\dfrac{1}{6 - 2} = \dfrac{1}{4}$ |
| 4 | 7 | $\dfrac{1}{7 - 2} = \dfrac{1}{5}$ |

Substitution into the trapezoidal rule gives

$$\int_3^7 \frac{1}{x-2} \, dx \approx \frac{7-3}{4}\left[\frac{1}{2}(1) + \frac{1}{2} + \frac{1}{3} + \frac{1}{4} + \frac{1}{2}\left(\frac{1}{5}\right)\right]$$

$$= 1\left(\frac{1}{2} + \frac{1}{2} + \frac{1}{3} + \frac{1}{4} + \frac{1}{10}\right) \approx 1.6833.$$

A calculator was used in the last step. In this case, the integral can be found and then evaluated using a table or calculator.

$$\int_3^7 \frac{1}{x-2} \, dx = \ln(x-2)\Big|_3^7 = \ln 5 - \ln 1$$

$$\approx 1.6094 - 0 = 1.6094$$

The approximation 1.6833 found above using the trapezoidal rule with $n = 4$ differs from the value in the table by .0739. As mentioned above, this error would be reduced if larger values were used for $n$. Techniques for estimating such errors are considered in more advanced courses. ■

***Simpson's Rule***    Another numerical method, *Simpson's rule*, approximates consecutive portions of the curve with portions of parabolas rather than the line segments of the trapezoidal rule. Simpson's rule usually gives a better approximation than the trapezoidal rule for the same number of subintervals. As shown in Figure 3, one parabola is fitted through points *A*, *B*, and *C*, another through *C*, *D*, and *E*, and so on. Then the sum of the areas under these parabolas will approximate the area under the graph of the function. Because of the way the parabolas overlap, it is necessary to have an even number of intervals, and therefore an odd number of points, to apply Simpson's rule.

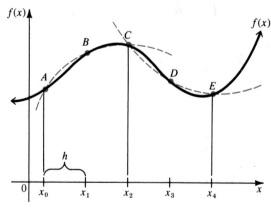

**FIGURE 3**

If $h$, the length of each subinterval, is $(b - a)/n$, the area under the parabola through points $A$, $B$, and $C$ can be found by a definite integral. The details are omitted; the result is

$$\frac{h}{3}\left[f(x_0) + 4f(x_1) + f(x_2)\right].$$

Similarly, the area under the parabola through points $C$, $D$, and $E$ is

$$\frac{h}{3}\left[f(x_2) + 4f(x_3) + f(x_4)\right].$$

When these expressions are added, the last term of one expression equals the first term of the next. For example, the sum of the two areas given above is

$$\frac{h}{3}\left[f(x_0) + 4f(x_1) + 2f(x_2) + 4f(x_3) + f(x_4)\right].$$

This illustrates the origin of the pattern of the terms in the following rule.

**SIMPSON'S RULE**

Let $f$ be a continuous function on $[a, b]$ and let $[a, b]$ be divided into an even number $n$ of equal subintervals by the points $a = x_0, x_1, x_2,$ $\ldots , x_n = b$. Then by **Simpson's rule,**

$$\int_a^b f(x)\,dx \approx \frac{b - a}{3n}\,[f(x_0) + 4f(x_1) + 2f(x_2) + 4f(x_3) + \cdots$$
$$+ 2f(x_{n-2}) + 4f(x_{n-1}) + f(x_n)].$$

Thomas Simpson (1710–1761), a British mathematician, wrote texts on many branches of mathematics. Some of these texts went through as many as ten editions. His name became attached to this numerical method of approximating definite integrals even though the method preceded his work.

**■ EXAMPLE 2**

Use Simpson's rule with $n = 4$ to evaulate

$$\int_0^4 \sqrt{x^2 + 1}\,dx.$$

Here $a = 0$, $b = 4$, and $n = 4$. The endpoints of the four intervals are $x_0 = 0$, $x_1 = 1$, $x_2 = 2$, $x_3 = 3$, and $x_4 = 4$. Set up a table and find the function values for each endpoint.

| $i$ | $x_i$ | $f(x_i)$ |
|---|---|---|
| 0 | 0 | 1 |
| 1 | 1 | 1.4142 |
| 2 | 2 | 2.2361 |
| 3 | 3 | 3.1623 |
| 4 | 4 | 4.1231 |

Since $(b - a)/(3n) = 4/12 = 1/3$, substituting into Simpson's rule gives

$$\int_0^4 \sqrt{x^2 + 1} \, dx \approx \frac{1}{3}[1 + 4(1.4142) + 2(2.2361) + 4(3.1623) + 4.1231]$$

$$= \frac{1}{3}(27.9013)$$

$$= 9.3004. \quad \blacksquare$$

Numerical methods make it possible to approximate

$$\int_a^b f(x) \, dx$$

even when $f(x)$ is not known. The next example shows how this is done.

▰▰▰ **EXAMPLE 3**

As mentioned earlier, the velocity $v(t)$ gives the rate of change of distance $s(t)$ with respect to time $t$. Suppose a vehicle travels an unknown distance. The passengers keep track of the velocity at 10-minute intervals (every 1/6 of an hour) with the following results.

| Time in Hours, $t$ | 1/6 | 2/6 | 3/6 | 4/6 | 5/6 | 1 | 7/6 |
|---|---|---|---|---|---|---|---|
| Velocity in Miles per Hour, $v(t)$ | 45 | 55 | 52 | 60 | 64 | 58 | 47 |

What is the total distance traveled in the 60-minute period from $t = 1/6$ to $t = 7/6$?

The distance traveled in $t$ hours is $s(t)$, with $s'(t) = v(t)$. The total distance traveled between $t = 1/6$ and $t = 7/6$ is given by

$$\int_{1/6}^{7/6} v(t) \, dt.$$

Even though this integral cannot be evaluated since we do not have an expression for $v(t)$, either the trapezoid rule or Simpson's rule can be used to approximate its value and give the total distance traveled. In either case, let $n = 6$, $a = t_0 = 1/6$, and $b = t_6 = 7/6$. By the trapezoidal rule,

$$\int_{1/6}^{7/6} v(t) \, dt \approx \frac{7/6 - 1/6}{6}\left[\frac{1}{2}(45) + 55 + 52 + 60 + 64 + 58 + \frac{1}{2}(47)\right]$$

$$\approx 55.83.$$

By Simpson's rule,

$$\int_{1/6}^{7/6} v(t) \, dt \approx \frac{7/6 - 1/6}{3(6)}[45 + 4(55) + 2(52) + 4(60) + 2(64) + 4(58) + 47]$$

$$= \frac{1}{18}(45 + 220 + 104 + 240 + 128 + 232 + 47)$$

$$\approx 56.44.$$

The distance traveled in the 1-hour period was about 56 miles.

As mentioned above, Simpson's rule generally gives a better approximation than the trapezoidal rule. As $n$ increases, the two approximations get closer and closer. For the same accuracy, however, a smaller value of $n$ generally can be used with Simpson's rule so that less computation is necessary. ▬

The branch of mathematics that studies methods of approximating definite integrals (as well as many other topics) is called *numerical analysis*. Some calculators give such approximations by using Simpson's rule, as shown in the following excerpt from the manual for model T1–55 II, a calculator produced by Texas Instruments.

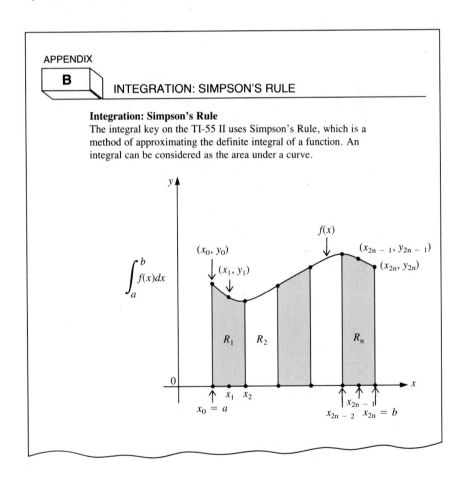

APPENDIX

**B**      INTEGRATION: SIMPSON'S RULE

**Integration: Simpson's Rule**
The integral key on the TI-55 II uses Simpson's Rule, which is a method of approximating the definite integral of a function. An integral can be considered as the area under a curve.

## 6.2 EXERCISES

In Exercises 1–10, use $n = 4$ to approximate the value of each of the given integrals by the following methods: **(a)** the trapezoidal rule, and **(b)** Simpson's rule. **(c)** In Exercises 1–8 only, find the exact value by integration.

**1.** $\int_0^2 x^2 \, dx$    **2.** $\int_0^2 (2x + 1) \, dx$    **3.** $\int_{-1}^3 \frac{1}{4 - x} \, dx$    **4.** $\int_1^5 \frac{1}{x + 1} \, dx$    **5.** $\int_{-2}^2 (2x^2 + 1) \, dx$

**6.** $\int_0^3 (2x^2 + 1) \, dx$    **7.** $\int_1^5 \frac{1}{x^2} \, dx$    **8.** $\int_2^4 \frac{1}{x^3} \, dx$    **9.** $\int_0^4 \sqrt{x^2 + 1} \, dx$    **10.** $\int_1^4 x\sqrt{2x - 1} \, dx$

**11.** Find the area under the semicircle $y = \sqrt{4 - x^2}$ and above the x-axis by using $n = 8$ with the following methods: **(a)** the trapezoidal rule, and **(b)** Simpson's rule. **(c)** Compare the results with the area found by the formula for the area of a circle. Which of the two approximation techniques was more accurate?

**12.** Find the area between the x-axis and the ellipse $4x^2 + 9y^2 = 36$ by using $n = 12$ with the following methods: **(a)** the trapezoidal rule, and **(b)** Simpson's rule. (*Hint:* Solve the equation for y and find the area of the semiellipse.) **(c)** Compare the results with the actual area, $6\pi \approx 18.8496$ (which can be found by methods not considered in this text). Which approximation technique was more accurate?

## APPLICATIONS

### BUSINESS AND ECONOMICS

**Total Sales**    **13.** A sales manager presented the following results at a sales meeting.

| Year, $x$ | 1 | 2 | 3 | 4 | 5 | 6 | 7 |
|---|---|---|---|---|---|---|---|
| Rate of Sales, $f(x)$ | .4 | .6 | .9 | 1.1 | 1.3 | 1.4 | 1.6 |

Find the total sales over the given period as follows.

**(a)** Plot these points. Connect the points with line segments.

**(b)** Use the trapezoidal rule to find the area bounded by the broken line of part (a), the x-axis, the line $x = 1$, and the line $x = 7$.

**(c)** Find the same area using Simpson's rule.

**Total Sales**    **14.** A company's marginal costs in hundreds of dollars were as follows over a certain period.

| Year, $x$ | 1 | 2 | 3 | 4 | 5 | 6 | 7 |
|---|---|---|---|---|---|---|---|
| Marginal Cost, $f(x)$ | 9.0 | 9.2 | 9.5 | 9.4 | 9.8 | 10.1 | 10.5 |

Repeat steps (a)–(c) of Exercise 13 for this data to find the total sales over the given period.

### LIFE SCIENCES

**Drug Reaction Rate**   **15.** The reaction rate to a new drug is given by

$$y = e^{-t^2} + \frac{1}{t},$$

where $t$ is time measured in hours after the drug is administered. Find the total reaction to the drug from $t = 1$ to $t = 9$ by letting $n = 8$ and using the following methods: **(a)** the trapezoidal rule, and **(b)** Simpson's rule.

**Growth Rate**   **16.** The growth rate of a certain tree in feet is given by

$$y = \frac{2}{t} + e^{-t^2/2},$$

where $t$ is time in years. Find the total growth through the end of the sixth year by using $n = 12$ with the following methods: **(a)** the trapezoidal rule, and **(b)** Simpson's rule.

**Blood Level Curves**   *In the study of bioavailability in pharmacy, a drug is given to a patient. The level of concentration of the drug is then measured periodically, producing blood level curves such as the ones shown in the figure. The areas under the curves give the total amount of the drug available to the patient.\* Use the trapezoidal rule with $n = 10$ to find the following areas.*

**17.** Find the total area under the curve for Formulation A. What does this area represent?

**18.** Find the total area under the curve for Formulation B. What does this area represent?

**19.** Find the area between the curve for Formulation A and the minimum effective concentration line. What does your answer represent?

**20.** Find the area between the curve for Formulation B and the minimum effective concentration line. What does this area represent?

---

\*These graphs are from *Basics of Bioavailability* by D. J. Chodos and A. R. DeSantos, copyright © 1978 by the Upjohn Company.

## SOCIAL SCIENCES

**Educational Psychology**  **21.** The results from a research study in psychology were as follows.

| Number of Hours of Study, $x$ | 1 | 2 | 3 | 4 | 5 | 6 | 7 |
|---|---|---|---|---|---|---|---|
| Rate of Extra Points Earned on a Test, $f(x)$ | 4 | 7 | 11 | 9 | 15 | 16 | 23 |

Repeat steps (a)–(c) of Exercise 13 for this data.

## PHYSICAL SCIENCES

**Chemical Formation**  **22.** The table below shows the results from a chemical experiment.

| Concentration of Chemical A, $x$ | 1 | 2 | 3 | 4 | 5 | 6 | 7 |
|---|---|---|---|---|---|---|---|
| Rate of Formation of Chemical B, $f(x)$ | 12 | 16 | 18 | 21 | 24 | 27 | 32 |

Repeat steps (a)–(c) of Exercise 13 for this data.

## FOR THE COMPUTER

*Use the trapezoidal rule and then use Simpson's rule to approximate each of the following integrals. Use n = 100.*

**23.** $\displaystyle\int_4^8 \ln{(x^2 - 10)}\, dx$

**24.** $\displaystyle\int_{-2}^2 e^{-x^2}\, dx$

**25.** $\displaystyle\int_{-2}^2 \sqrt{9 - 2x^2}\, dx$

**26.** $\displaystyle\int_{-1}^1 \sqrt{16 + 5x^2}\, dx$

**27.** $\displaystyle\int_1^5 (2x^2 + 3x - 1)^{2/5}\, dx$

**28.** $\displaystyle\int_1^4 (x^3 + 4)^{5/4}\, dx$

*Use either the trapezoidal rule or Simpson's rule with n = 100 for Exercises 29 and 30.*

**Total Revenue**  **29.** An electronics company analyst has determined that the rate per month at which revenue comes in from the calculator division is given by

$$R(x) = 105e^{.01x} + 32,$$

where $x$ is the number of months the division has been in operation. Find the total revenue between the twelfth and thirty-sixth months.

**Blood Pressure**  **30.** Blood pressure in an artery changes rapidly over a very short time for a healthy young adult, from a high of about 120 to a low of about 80. Suppose the blood pressure function over an interval of 1.5 sec is given by

$$f(x) = .2x^5 - .68x^4 + .8x^3 - .39x^2 + .055x + 100,$$

where $x$ is the time in seconds after a peak reading. The area under the curve for one cycle is important in some blood pressure studies. Find the area under $f(x)$ from .1 sec to 1.1 sec.

**FIGURE 4**

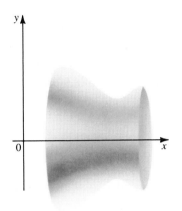

**FIGURE 5**

### 6.3  TWO APPLICATIONS OF INTEGRATION: VOLUME AND AVERAGE VALUE

In this section we discuss two very useful topics—finding the volume of a solid by integration, and finding the average value of a function.

*Volume*   Figure 4 shows the region below the graph of some function $y = f(x)$, above the *x*-axis, and between $x = a$ and $x = b$. We have seen how to use integrals to find the area of such a region. Now, suppose this region is revolved about the *x*-axis as shown in Figure 5. The resulting figure is called a **solid of revolution.** In many cases, the volume of a solid of revolution can be found by integration.

To begin, divide the interval $[a, b]$ into *n* subintervals of equal width $\Delta x$ by the points $a = x_0, x_1, x_2, \ldots, x_i, \ldots, x_n = b$. Then think of slicing the solid into *n* slices of equal thickness $\Delta x$, as shown in Figure 6. If the slices are thin enough, each slice is very close to being a right circular cylinder. The formula for the volume of a right circular cylinder is $\pi r^2 h$, where *r* is the radius of the circular base and *h* is the height of the cylinder. As shown in Figure 7, the height of each slice is $\Delta x$. (The height is horizontal here, since the cylinder is on its side.) The radius of the circular base of each slice is $f(x_i)$. Thus, the volume of the slice is closely approximated by $\pi [f(x_i)]^2 \Delta x$. The volume of the solid of revolution will be approximated by the sum of the volumes of the slices:

$$V \approx \sum_{i=1}^{n} \pi [f(x_i)]^2 \Delta x.$$

**(a)**          **FIGURE 6**          **(b)**

**FIGURE 7**

By definition, the volume of the solid of revolution is the limit of this sum as the thickness of the slices approaches 0, or

$$V = \lim_{\Delta x \to 0} \sum_{i=1}^{n} \pi [f(x_i)]^2 \, \Delta x.$$

This limit, like the one discussed earlier for area, is a definite integral.

**VOLUME OF A SOLID OF REVOLUTION**

If $f(x)$ is nonnegative and $R$ is the region between $f(x)$ and the $x$-axis from $x = a$ to $x = b$, the volume of the solid formed by rotating $R$ about the $x$-axis is given by

$$V = \lim_{\Delta x \to 0} \sum_{i=1}^{n} \pi [f(x_i)]^2 \Delta x = \int_a^b \pi [f(x)]^2 \, dx.$$

The technique of summing disks to approximate volumes was originated by Johannes Kepler (1571–1630), a famous German astronomer who discovered three laws of planetary motion. He estimated volumes of wine casks used at his wedding by means of solids of revolution.

### EXAMPLE 1

Find the volume of the solid of revolution formed by rotating about the $x$-axis the region bounded by $y = x + 1$, $y = 0$, $x = 1$, and $x = 4$. (See Figure 8(a).)

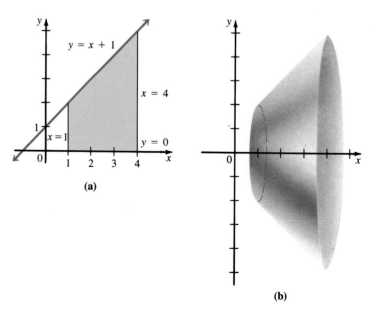

(a)

(b)

**FIGURE 8**

The solid is shown in Figure 8(b). Use the formula given above for the volume, with $a = 1$, $b = 4$, and $f(x) = x + 1$.

$$V = \int_1^4 \pi(x + 1)^2 \, dx = \pi \left( \frac{(x + 1)^3}{3} \right) \bigg|_1^4$$

$$= \frac{\pi}{3}(5^3 - 2^3) = \frac{117\pi}{3} = 39\pi \quad \blacksquare$$

### ▰ EXAMPLE 2

Find the volume of the solid of revolution formed by rotating about the *x*-axis the area bounded by $f(x) = 4 - x^2$ and the *x*-axis. (See Figure 9(a).)

The solid is shown in Figure 9(b). Find *a* and *b* from the *x*-intercepts. If $y = 0$, then $x = 2$ or $x = -2$, so that $a = -2$ and $b = 2$. The volume is

$$V = \int_{-2}^2 \pi(4 - x^2)^2 \, dx = \int_{-2}^2 \pi(16 - 8x^2 + x^4) \, dx$$

$$= \pi \left( 16x - \frac{8x^3}{3} + \frac{x^5}{5} \right) \bigg|_{-2}^2 = \frac{512\pi}{15}. \quad \blacksquare$$

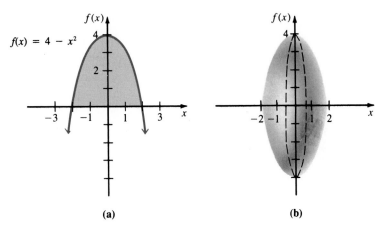

**(a)**

**(b)**

**FIGURE 9**

### ▰ EXAMPLE 3

Find the volume of a right circular cone with height *h* and base radius *r*.

Figure 10(a) shows the required cone, while Figure 10(b) shows an area that could be rotated about the *x*-axis to get such a cone (See Figure 10(c).) Here $y = f(x)$ is the equation (in slope-intercept form) of the line through $(0, r)$ and $(h, 0)$. The slope of this line is

$$\frac{0 - r}{h - 0} = -\frac{r}{h}.$$

(a)

(b)

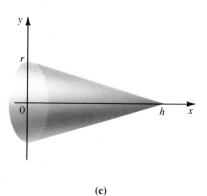

(c)

**FIGURE 10**

Using the point-slope formula with the point $(h, 0)$ gives

$$y - y_1 = m(x - x_1)$$

$$y = -\frac{r}{h}(x - h)$$

as the equation of the line. Then the volume is

$$V = \int_0^h \pi \left[ -\frac{r}{h}(x - h) \right]^2 dx = \pi \int_0^h \frac{r^2}{h^2}(x - h)^2 \, dx.$$

Since $r$ and $h$ are constants,

$$V = \frac{\pi r^2}{h^2} \int_0^h (x - h)^2 \, dx.$$

Using substitution with $u = x - h$ and $du = dx$ gives

$$V = \frac{\pi r^2}{h^2} \left[ \frac{(x - h)^3}{3} \right] \Bigg|_0^h = \frac{\pi r^2}{3h^2}[0 - (-h)^3] = \frac{\pi r^2 h}{3}.$$

This is the familiar formula for the volume of a right circular cone.  ■

***Average Value of a Function***    The average of the $n$ numbers $v_1, v_2, v_3,$ . . . , $v_i,$ . . . , $v_n$ is given by

$$\frac{v_1 + v_2 + v_3 + \cdots + v_n}{n} = \frac{\displaystyle\sum_{i=1}^n v_i}{n}.$$

The average value of a function $f$ on $[a, b]$ can be defined in a similar manner; divide the interval $[a, b]$ into $n$ subintervals, each of width $\Delta x$. Then choose an $x$-value, $x_i$, in each interval, and find $f(x_i)$. The average function value for the $n$ subintervals and the given choices of $x_i$ is

$$\frac{f(x_1) + f(x_2) + f(x_3) + \cdots + f(x_n)}{n} = \frac{\displaystyle\sum_{i=1}^n f(x_i)}{n}.$$

Since $(b - a)/n = \Delta x$, multiply on the left by $(b - a)/(b - a)$ and rearrange the expression to get

$$\frac{b - a}{b - a} \cdot \frac{\displaystyle\sum_{i=1}^{n} f(x_i)}{n} = \frac{b - a}{n} \cdot \frac{\displaystyle\sum_{i=1}^{n} f(x_i)}{b - a}$$

$$= \Delta x \cdot \frac{\displaystyle\sum_{i=1}^{n} f(x_i)}{b - a}$$

$$= \frac{1}{b - a} \sum_{i=1}^{n} f(x_i) \, \Delta x.$$

Now, take the limit as $n \to +\infty$. If the limit exists, then

$$\lim_{n \to +\infty} \frac{1}{b - a} \sum_{i=1}^{n} f(x_i) \, \Delta x = \frac{1}{b - a} \lim_{n \to +\infty} \sum_{i=1}^{n} f(x_i) \, \Delta x$$

$$= \frac{1}{b - a} \int_a^b f(x) \, dx.$$

The following definition summarizes this discussion.

**AVERAGE VALUE OF A FUNCTION**

> The **average value** of a function $f$ on the interval $[a, b]$ is
>
> $$\frac{1}{b - a} \int_a^b f(x) \, dx,$$
>
> provided the indicated definite integral exists.

The average value, sometimes denoted $\bar{y}$, can be thought of as the height of the rectangle with base $b - a$. See Figure 11. For $f(x) \geq 0$, this rectangle has area $\bar{y}(b - a)$, which equals the area under the graph of $f(x)$ from $x = a$ to $x = b$, so that

$$\bar{y}(b - a) = \int_a^b f(x) dx.$$

**FIGURE 11**

#### ■ EXAMPLE 4

An analyst determines that the cost (in dollars) of the $x$th item of a new product is given by

$$f(x) = 3\sqrt{x} + 8.$$

Find the average cost per unit of the first 100 units produced.

With $a = 0$ and $b = 100$, the average cost is

$$\frac{1}{100} \int_0^{100} (3\sqrt{x} + 8)\, dx = \frac{1}{100}(2x^{3/2} + 8x)\Big|_0^{100}$$

$$= \frac{1}{100}(2000 + 800 - 0)$$

$$= 28,$$

for an average cost of \$28 per unit.  ■

#### ■ EXAMPLE 5

A magazine has a monthly circulation of ~~1500~~ 2,000 copies. An advertising campaign is started to increase the circulation. Suppose the monthly circulation (in hundreds of copies) $t$ months after the beginning of the campaign is given by

$$C(t) = 25 - 5\, e^{-.01t}.$$

What is the average monthly circulation during the first 6 months of the campaign?

Use the formula for average value with $a = 0$ and $b = 6$. The average monthly circulation in hundreds is

$$\frac{1}{6-0} \int_0^6 (25 - 5\, e^{-.01t})\, dt = \frac{1}{6}\left[25t - \frac{5}{-.01} e^{-.01t}\right]\Big|_0^6$$

$$= \frac{1}{6}\left[25t + 500\, e^{-.01t}\right]\Big|_0^6$$

$$= \frac{1}{6}\left[150 + 500\, e^{-.06} - 500\right]$$

$$= 20.147,$$

or approximately 2015 copies.  ■

## 6.3 EXERCISES

*Find the volume of the solid of revolution formed by rotating about the x-axis each region bounded by the given curves.*

**1.** $f(x) = x$, $y = 0$, $x = 0$, $x = 2$

**2.** $f(x) = 2x$, $y = 0$, $x = 0$, $x = 3$

**3.** $f(x) = 2x + 1$, $y = 0$, $x = 0$, $x = 4$

**4.** $f(x) = x - 4$, $y = 0$, $x = 4$, $x = 10$

**5.** $f(x) = \dfrac{1}{3}x + 2$, $y = 0$, $x = 1$, $x = 3$

**6.** $f(x) = \dfrac{1}{2}x + 4$, $y = 0$, $x = 0$, $x = 5$

**7.** $f(x) = \sqrt{x}$, $y = 0$, $x = 1$, $x = 2$

**8.** $f(x) = \sqrt{x + 1}$, $y = 0$, $x = 0$, $x = 3$

**9.** $f(x) = \sqrt{2x + 1}$, $y = 0$, $x = 1$, $x = 4$

**10.** $f(x) = \sqrt{3x + 2}$, $y = 0$, $x = 1$, $x = 2$

**11.** $f(x) = e^x$, $y = 0$, $x = 0$, $x = 2$

**12.** $f(x) = 2e^x$, $y = 0$, $x = -2$, $x = 1$

**13.** $f(x) = \dfrac{1}{\sqrt{x}}$, $y = 0$, $x = 1$, $x = 4$

**14.** $f(x) = \dfrac{1}{\sqrt{x + 1}}$, $y = 0$, $x = 0$, $x = 2$

**15.** $f(x) = x^2$, $y = 0$, $x = 1$, $x = 5$

**16.** $f(x) = \dfrac{x^2}{2}$, $y = 0$, $x = 0$, $x = 4$

**17.** $f(x) = 1 - x^2$, $y = 0$

**18.** $f(x) = 2 - x^2$, $y = 0$

*The function defined by $y = \sqrt{r^2 - x^2}$ has as its graph a semicircle of radius r with center at (0, 0). (See the figure.) In Exercises 19–21, find the volume that results when each semicircle is rotated about the x-axis. (The result of Exercise 21 gives a formula for the volume of a sphere with radius r.)*

**19.** $f(x) = \sqrt{1 - x^2}$

**20.** $f(x) = \sqrt{16 - x^2}$

**21.** $f(x) = \sqrt{r^2 - x^2}$

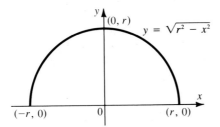

**22.** Find a formula for the volume of an ellipsoid. See Exercises 19–21 and the following figures.

Ellipsoid

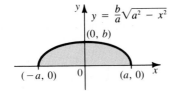

*Find the average value of each function on the given interval.*

**23.** $f(x) = 3 - 2x^2$;  [1, 9]

**24.** $f(x) = x^2 - 2$;  [0, 5]

**25.** $f(x) = (2x - 1)^{1/2}$;  [1, 13]

**26.** $f(x) = \sqrt{x + 1}$;  [3, 8]

**27.** $f(x) = e^{.1x}$;  [0, 10]

**28.** $f(x) = e^{x/5}$;  [0, 5]

# ▤ APPLICATIONS

## LIFE SCIENCES

**Blood Flow**    **29.** The figure shows the blood flow in a small artery of the body. The flow of blood is *laminar* (in layers), with the velocity very low near the artery walls and highest in the center of the artery. To calculate the total flow in the artery we think of the flow as being made up of many layers of concentric tubes sliding one on the other.

Suppose $R$ is the radius of an artery and $r$ is the distance from a given layer to the center. Then the velocity of blood in a given layer can be shown to equal

$$v(r) = k(R^2 - r^2),$$

where $k$ is a numerical constant.

Since the area of a circle is $A = \pi r^2$, the change in the area of the cross section of one of the layers, corresponding to a small change in the radius, $\Delta r$, can be approximated by differentials. For $dr = \Delta r$, the differential of the area $A$ is

$$dA = 2\pi r \, dr = 2\pi r \Delta r,$$

where $\Delta r$ is the thickness of the layer. The total flow in the layer is defined to be the product of volume and cross-section area, or

$$F(r) = 2\pi r k(R^2 - r^2)\Delta r.$$

**(a)** Set up a definite integral to find the total flow in the artery.

**(b)** Evaluate this definite integral.

**Drug Reaction**    **30.** The intensity of the reaction to a certain drug, in appropriate units, is given by

$$R(t) = \frac{1}{t} + \frac{2}{t^2},$$

where $t$ is time in hours after the drug is administered. Find the average intensity during each of the following hours.

**(a)** Second hour    **(b)** Twelfth hour    **(c)** Twenty-fourth hour

## SOCIAL SCIENCES

**Production Rate**    **31.** Suppose the number of items a new worker on an assembly line produces daily after $t$ days on the job is given by

$$I(t) = 35 + 2t.$$

Find the average number of items produced daily by this employee after the following numbers of days.

**(a)** 5    **(b)** 10    **(c)** 15

**Typing Speed**    **32.** The function $W(t) = -6t^2 + 10t + 80$ describes a typist's speed (in words per minute) over a time interval $[0, 5]$.

**(a)** Find $W(0)$.

**(b)** Find the maximum $W$ value and the time $t$ when it occurs.

**(c)** Find the average speed over $[0, 5]$.

# ▬ 6.4 CONTINUOUS MONEY FLOW

In an earlier chapter we looked at the concepts of present value and future value, and we discussed interest payments, loan payments, mortgage payments, and other payments made periodically at scheduled times. In some situations, however, money flows into and out of an account almost continuously over a period of time. Examples include income in a store, bank receipts and dispersals, and highway tolls. Although the flow of money in such cases is not exactly continuous, it can be treated, with useful results, as though it *were* continuous.

## ▬ EXAMPLE 1

The rate of change of income produced by a soda machine (in dollars per year) is given by

$$f(t) = 500 \, e^{.02t},$$

where $t$ is time in years since the installation of the machine. Find the total income produced by the machine during its first 3 years of operation.

Since the rate of change of income is given, the total income can be determined by using the definite integral.

$$
\begin{aligned}
\text{Total Income} &= \int_0^3 500 \, e^{.02t} dt \\
&= \frac{500}{.02} e^{.02t} \Big|_0^3 \\
&= 25,000 \, e^{.02t} \Big|_0^3 \\
&= 25,000[e^{.06} - 1] \\
&= 1545.91
\end{aligned}
$$

Thus, the soda machine will produce \$1545.91 total income in its first 3 years of operation. ▬

The money in Example 1 is not received as a one-time lump sum payment of \$1545.91. Instead, it comes in on a regular basis, perhaps daily, weekly, or monthly. In discussions of such problems it is usually assumed that the income is received continuously over a period of time, and so we speak of a "flow of money into income" and the rate of that flow.

***Total Money Flow***    Let the continuous function $f(x)$ represent the rate of flow of money per unit of time. If $x$ is in years and $f(x)$ is in dollars per year, the area under $f(x)$ between two points in time gives the total dollar flow over the given time interval.

The function $f(x) = 2000$, shown in Figure 12, represents a uniform rate of money flow of \$2000 per year. The graph of this money flow is a horizontal line and the total money flow over a specified time $t$ is given by the rectangular area below the graph of $f(x)$ and above the $x$-axis between $x = 0$ and $x = t$. For example, the total money flow over $t = 5$ years would be 2000(5) = 10,000, or \$10,000.

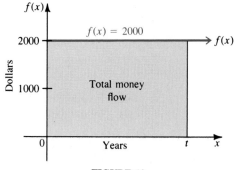

**FIGURE 12**

The area in the uniform rate example could be found by using an area formula from geometry. For a variable function like the function in Example 1, however, a definite integral is needed to find the total money flow over a specific time interval. For the function $f(x) = 2000e^{.08x}$, for example, the total money flow over a 5-year period would be given by

$$\int_0^5 2000e^{.08x} \, dx \approx 12{,}295.62,$$

or \$12,295.62. See Figure 13.

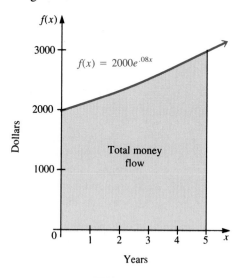

**FIGURE 13**

**TOTAL MONEY FLOW**

> If $f(x)$ is the rate of money flow, then the total money flow over the time interval from $x = 0$ to $x = t$ is given by
>
> $$\int_0^t f(x)\ dx.$$

It should be noted that this "total money flow" does not take into account the interest the money could earn after it is received. It is simply the sum of the periodic (continuous) income.

***Present Value of Money Flow***   As mentioned earlier, an amount of money that can be deposited today to yield a given sum in the future is called the *present value* of this future sum. The future sum may be called the *future value* or *final amount*. To find the present value of a continuous money flow with interest compounded continuously, let $f(x)$ represent the rate of the continuous flow. In Figure 14, the time axis from 0 to $x$ is divided into $n$ subintervals, each of width $\Delta x_i$. The amount of money that flows during any interval of time is given by the area between the $x$-axis and the graph of $f(x)$ over the specified time interval. The area of each subinterval is approximated by the area of a rectangle with height $f(x_i)$, where $x_i$ is the left endpoint of the $i$th subinterval. The area of each rectangle is $f(x_i)\Delta x_i$ which (approximately) gives the amount of money flow over that subinterval.

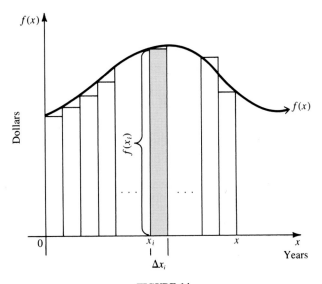

**FIGURE 14**

Earlier, we saw that the present value $P$ of an amount $A$ compounded continuously for $n$ years at a rate of interest $i$ is $P = Ae^{-ni}$. Using $j$ for the interest rate instead of $i$, letting $x_i$ represent the time instead of $n$, and replacing $A$ with $f(x_i)\Delta x_i$, the present value of the money flow over the $i$th subinterval is approximately equal to

$$P_i = f(x_i)\Delta x_i e^{-jx_i}.$$

The total present value is approximately equal to the sum

$$\sum_{i=1}^{n} f(x_i)\Delta x_i e^{-jx_i}.$$

This approximation is improved as $n$ is increased, and if we take the limit of the sum as $n$ increases without bound, we have the present value

$$P = \lim_{n\to\infty} \sum_{i=1}^{n} f(x_i)\Delta x_i e^{-jx_i}.$$

This limit of a summation is given by the definite integral below.

**PRESENT VALUE OF MONEY FLOW**

> If $f(x)$ is the rate of continuous money flow at an interest rate $j$ at time $x$, then the present value is
>
> $$P = \int_0^t f(x)\, e^{-jx}\, dx.$$

This present value of money flow is really the amount of money that would have to be deposited now in an account paying an annual interest rate $j$ compounded continuously so that the total sum in the account at time $t$ would equal the continuous money flow with interest up to time $t$.

### ▬ EXAMPLE 2

A company expects its annual income during the next 3 years to be given by

$$f(x) = 75,000x, \qquad 0 \le x \le 3.$$

What is the present value of this income over the 3-year period, assuming an annual interest rate of 8%?

Use the formula for present value given above, with $f(x) = 75,000x$ and $j = .08$.

$$P = \int_0^3 75,000x\, e^{-.08x} dx$$

$$= 75,000 \int_0^3 x\, e^{-.08x} dx$$

Using the table of integrals with $n = 1$ and $a = -.08$ gives

$$75,000 \int_0^3 xe^{-.08x} \, dx = 75,000 \left[ \frac{xe^{-.08x}}{-.08} \Bigg|_0^3 - \frac{1}{-.08} \int_0^3 e^{-.08x} \, dx \right]$$

$$= 75,000 \left[ -12.5x \, e^{-.08x} \Bigg|_0^3 - \int_0^3 (-12.5 \, e^{-.08x}) \, dx \right]$$

$$= 75,000 \left[ -12.5x \, e^{-.08x} \Bigg|_0^3 + \frac{12.5}{-.08} (e^{-.08x}) \Bigg|_0^3 \right]$$

$$= 75,000[-12.5(3) \, e^{-.24} - 156.25(e^{-.24} - 1)]$$

$$= 75,000[-29.4985 + 33.3393]$$

$$= 288,060,$$

or \$288,060. Notice that the actual income over the 3-year period is given by

$$\text{Total Money Flow} = \int_0^3 75,000x \, dx$$

$$= \frac{75,000x^2}{2} \Bigg|_0^3$$

$$= 337,500,$$

or \$337,500. This means that it would take a lump-sum deposit of \$288,060 today paying a continuously compounded interest rate of 8% over a 3-year period to equal the total cash flow of \$337,500 with interest. This approach is used as a basis for determining insurance claims involving income considerations. ■

***Accumulated Amount of Money Flow at Time t*** To find the amount of the money flow with interest at any time $t$, use the formula $A = Pe^{ni}$ with $n = t$, $i = j$, and $P = \int_0^t f(x) \, e^{-jx} \, dx$, as follows.

**ACCUMULATED AMOUNT OF MONEY FLOW AT TIME $t$**

■ If $f(x)$ is the rate of money flow at an interest rate $j$ at time $x$, the amount of flow at time $t$ is

$$A = e^{jt} \int_0^t f(x) \, e^{-jx} \, dx.$$

Here, the amount of money $A$ represents the accumulated value or final amount of the money flow *including* interest received on the money after it comes in.

It turns out that most money flows can be expressed as exponential or polynomial functions. When these are multiplied by $e^{-jx}$, the result is a function that can be integrated. The next example illustrates uniform flow, where $f(x)$ is a constant function. (This is a special case of the polynomial function.)

■ EXAMPLE 3

If money is flowing continuously at a constant rate of $2000 per year over 5 years at 12% interest compounded continuously, find each of the following.

**(a)** The total amount of the flow over the 5-year period

The total amount is given by $\int_0^t f(x)\, dx$. Here $f(x) = 2000$ and $t = 5$.

$$\int_0^5 2000\, dx = 2000x \Big|_0^5 = 2000(5) = 10,000$$

The total money flow over the 5-year period is $10,000.

**(b)** The accumulated amount, compounded continuously, at time $t = 5$

At $t = 5$ with $j = .12$, the amount is

$$A = e^{jt} \int_0^t f(x)\, e^{-jx}\, dx = e^{(.12)5} \int_0^5 (2000)e^{-.12x}\, dx$$

$$= (e^{.6})(2000) \int_0^5 e^{-.12x}\, dx = (e^{.6})(2000)\left(\frac{1}{-.12}\right)\left(e^{-.12x}\Big|_0^5\right)$$

$$= \frac{2000e^{.6}}{-.12}(e^{-.6} - 1) = \frac{2000}{-12}(1 - e^{.6}) \qquad (e^{.6})(e^{-.6}) = 1$$

$$= 13,701.98,$$

or $13,701.98. The answer to part (a), $10,000.00, was the amount of money flow over the 5-year period. The $13,701.98 gives that amount with interest compounded continuously over the 5-year period.

**(c)** The present value of the amount with interest

Use $P = \int_0^t f(x)\, e^{-jx}\, dx$ with $f(x) = 2000$, $j = .12$, and $t = 5$.

$$P = \int_0^5 2000e^{-.12x}\, dx = 2000\left(\frac{e^{-.12x}}{-.12}\right)\Big|_0^5$$

$$= \frac{2000}{-.12}(e^{-.6} - 1) = 7519.81$$

The present value of the amount with interest in 5 years is $7519.81, which can be checked by substituting $13,701.98 for $A$ in $A = Pe^{jt}$. The present value, $P$, could have been found by dividing the amount found in (b) by $e^{jt} = e^{.6}$. Check that this would give the same result. ■

If the money flow is increasing or decreasing exponentially, then $f(x) = Ce^{kx}$, where $C$ is a constant that represents the initial amount and $k$ is the (nominal) continuous rate of change, which may be positive or negative.

### ■ EXAMPLE 4

A continuous money flow starts at $1000 and increases exponentially at 2% a year.

(a) Find the accumulated amount at the end of 5 years at 10% interest compounded continuously.

Here $C = 1000$ and $k = .02$, so that $f(x) = 1000e^{.02x}$. Using $j = .10$ and $t = 5$.

$$A = e^{(.10)5} \int_0^5 1000e^{.02x} e^{-.10x} \, dx$$

$$= (e^{.5})(1000) \int_0^5 e^{-.08x} \, dx \qquad e^{.02x} \cdot e^{-.10x} = e^{-.08x}$$

$$= 1000e^{.5} \left( \frac{1}{-.08} e^{-.08x} \right) \Big|_0^5$$

$$= \frac{1000e^{.5}}{-.08} (e^{-.4} - 1)$$

$$= \frac{1000}{-.08} (e^{.1} - e^{.5}) = 6794.38,$$

or $6794.38.

(b) Find the present value at 15% interest compounded continuously.

Using $f(x) = 1000e^{.02x}$ with $j = .15$ and $t = 5$ in the present value expression,

$$P = \int_0^5 1000e^{.02x} e^{-.15x} \, dx$$

$$= 1000 \int_0^5 e^{-.13x} \, dx$$

$$= 1000 \left( \frac{1}{-.13} e^{-.13x} \Big|_0^5 \right)$$

$$= \frac{1000}{-.13} (e^{-.65} - 1)$$

$$= 3676.57,$$

or $3676.57. ■

If the rate of change of the continuous money flow is given by the polynomial function $f(x) = a_n x^n + a_{n-1} x^{n-1} + \cdots + a_0$, the expressions for present value and accumulated amount can be integrated term by term using integration by parts or a table of integrals.

## EXAMPLE 5

The rate of change of a continuous flow of money is given by

$$f(x) = 1000x^2 + 100x.$$

Find the present value of this money flow at the end of 10 years at 10% compounded continuously.

Evaluate

$$P = \int_0^{10} (1000x^2 + 100x)e^{-.10x}\, dx$$

$$= \int_0^{10} 1000x^2 e^{-.10x}\, dx + \int_0^{10} 100x e^{-.10x}\, dx.$$

First find each indefinite integral using the table of integrals. For the first integral, $n = 2$, $a = -1$, and

$$1000 \int x^2 e^{-.1x}\, dx = 1000 \left[ \frac{x^2 e^{-.1x}}{-.1} - \frac{2}{-.1} \int x e^{-.1x}\, dx \right].$$

Use the table again with $n = 1$ and $a = -.1$ to get

$$1000 \int x^2 e^{-.1x}\, dx = 1000 \left[ \frac{1}{-.1} x^2 e^{-.1x} - \frac{2}{-.1}\left( \frac{x e^{-.1x}}{-.1} - \frac{1}{-.1} \int e^{-.1x}\, dx \right) \right]. \quad (*)$$

Use this result to find the definite integral.

$$1000 \int_0^{10} x^2 e^{-.1x}\, dx = 1000 \left[ \frac{1}{-.1} x^2 e^{-.1x} - \frac{2}{(-.1)^2}\left( x e^{-.1x} - \frac{1}{-.1} e^{-.1x} \right) \right]\Bigg|_0^{10}$$

$$= 1000[(-1000e^{-1} - 4000e^{-1}) - (0 - 2000)]$$

$$= 160{,}602.79$$

For the second integral use the result in the parentheses in equation (*).

$$100 \int_0^{10} x e^{-.1x}\, dx = 100 \left[ \frac{x e^{-.1x}}{-.1} - \frac{1}{-.1} \int e^{-.1x}\, dx \right]\Bigg|_0^{10}$$

$$= 100 \left[ \frac{x e^{-.1x}}{-.1} + \frac{1}{.1}\left( \frac{e^{-.1x}}{-.1} \right) \right]\Bigg|_0^{10}$$

$$= 100(-10x e^{-.1x} - 100 e^{-.1x})\Bigg|_0^{10}$$

$$= 100(-10 \cdot 10 e^{-1} - 100 e^{-1} + 0 + 100)$$

$$= 100(-200 e^{-1} + 100)$$

$$= 2642.41$$

The present value is

$$P = 160,602.79 + 2642.41 = 163,245.20,$$

or $163,245.20.

## 6.4 EXERCISES

*Each of the functions in Exercises 1–14 represents the rate of flow of money per unit of time. Assume a 10-year period at 12% compounded continuously and find each of the following:* **(a)** *the present value;* **(b)** *the amount at t = 10.*

**1.** $f(x) = 1000$

**2.** $f(x) = 300$

**3.** $f(x) = 500$

**4.** $f(x) = 2000$

**5.** $f(x) = 400e^{.03x}$

**6.** $f(x) = 800e^{.05x}$

**7.** $f(x) = 5000e^{-.01x}$

**8.** $f(x) = 1000e^{-.02x}$

**9.** $f(x) = .1x$

**10.** $f(x) = .5x$

**11.** $f(x) = .01x + 100$

**12.** $f(x) = .05x + 500$

**13.** $f(x) = 1000 - 100x$

**14.** $f(x) = 2000 - 150x$

## APPLICATIONS

### BUSINESS AND ECONOMICS

**15.** An investment is expected to yield a uniform continuous rate of change of income flow of $20,000 per year for 3 years. Find the final amount at an interest rate of 14% compounded continuously.

**16.** A real estate investment is expected to produce a uniform continuous rate of change of income flow of $8000 per year for 6 years. Find the present value at each of the following rates, compounded continuously.

 **(a)** 12%   **(b)** 10%   **(c)** 15%

**17.** The rate of change of a continuous flow of money starts at $5000 and decreases exponentially at 1% per year for 8 years. Find the present value and final amount at an interest rate of 18% compounded continuously.

**18.** The rate of change of a continuous money flow starts at $1000 and increases exponentially at 5% per year for 4 years. Find the present value and final amount if interest earned is 11% compounded continuously.

**19.** A money market fund has a continuous flow of money that changes at a rate of $f(x) = 1500 - 60x^2$, reaching 0 in 5 years. Find the present value of this flow if interest is 10% compounded continuously.

**20.** Find the amount of a continuous money flow in 3 years if the rate of change is given by $f(x) = 1000 - x^2$ and if interest is 10% compounded continuously.

## 6.5 IMPROPER INTEGRALS

Sometimes it is useful to be able to integrate a function over an infinite period of time. For example, we might want to find the total amount of income generated by an apartment building into the indefinite future, or the total amount of pollution into a bay from a source that is continuing indefinitely. In this section we define integrals with one or more infinite limits of integration that can be used to solve such problems.

The graph in Figure 15(a) shows the area bounded by the curve $f(x) = x^{-3/2}$, the $x$-axis, and the vertical line $x = 1$. Think of the shaded region below the curve as extending indefinitely to the right. Does this shaded region have an area?

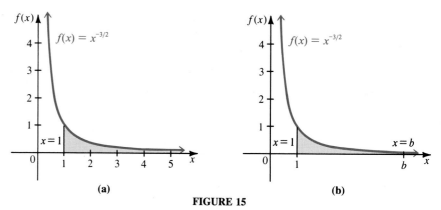

(a)  (b)

**FIGURE 15**

To see if the area of this region can be defined, introduce a vertical line at $x = b$, as shown in Figure 15(b). This vertical line gives a region with both upper and lower limits of integration. The area of this new region is given by the definite integral

$$\int_1^b x^{-3/2} \, dx.$$

By the fundamental theorem of calculus,

$$\int_1^b x^{-3/2} \, dx = (-2x^{-1/2}) \Big|_1^b$$

$$= -2b^{-1/2} - (-2 \cdot 1^{-1/2})$$

$$= -2b^{-1/2} + 2 = 2 - \frac{2}{b^{1/2}}.$$

Suppose we now let the vertical line $x = b$ in Figure 15(b) move farther to the right. That is, suppose $b \to +\infty$. The expression $-2/b^{1/2}$ would then approach 0, and

$$\lim_{b \to +\infty} \left( 2 - \frac{2}{b^{1/2}} \right) = 2 - 0 = 2.$$

This limit is defined to be the *area* of the region shown in Figure 15(a), so that

$$\int_1^{+\infty} x^{-3/2}\, dx = 2.$$

An integral of the form

$$\int_a^{+\infty} f(x)\, dx, \qquad \int_{-\infty}^b f(x)\, dx, \qquad \text{or} \qquad \int_{-\infty}^{+\infty} f(x)\, dx$$

is called an improper integral. These **improper integrals** are defined as follows.

**IMPROPER INTEGRALS**

> If $f$ is continuous on the indicated interval and if the indicated limits exist, then
>
> $$\int_a^{+\infty} f(x)\, dx = \lim_{b \to +\infty} \int_a^b f(x)\, dx,$$
>
> $$\int_{-\infty}^b f(x)\, dx = \lim_{a \to -\infty} \int_a^b f(x)\, dx,$$
>
> $$\int_{-\infty}^{+\infty} f(x)\, dx = \int_{-\infty}^c f(x)\, dx + \int_c^{+\infty} f(x)\, dx,$$
>
> for real numbers $a$, $b$, and $c$.

If the expressions on the right side exist, the integrals are **convergent;** otherwise, they are **divergent.**

### ■ EXAMPLE 1

Find the following integrals.

**(a)** $\displaystyle\int_1^{+\infty} \frac{dx}{x}$

A graph of this region is shown in Figure 16. By the definition of an improper integral,

$$\int_1^{+\infty} \frac{dx}{x} = \lim_{b \to +\infty} \int_1^b \frac{dx}{x}.$$

Find $\displaystyle\int_1^b \frac{dx}{x}$ by the fundamental theorem of calculus.

$$\int_1^b \frac{dx}{x} = \ln |x| \Big|_1^b = \ln |b| - \ln |1| = \ln |b| - 0 = \ln |b|$$

As $b \to +\infty$, $\ln |b| \to +\infty$, so

$$\lim_{b \to +\infty} \ln |b| \text{ does not exist.}$$

Since the limit does not exist, $\displaystyle\int_1^{+\infty} \frac{dx}{x}$ is divergent.

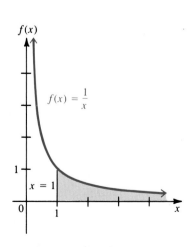

$f(x)$

$f(x) = \dfrac{1}{x}$

$x = 1$

**FIGURE 16**

**FIGURE 17**

**(b)** $\int_{-\infty}^{-2} \frac{1}{x^2} \, dx = \lim_{a \to -\infty} \int_{a}^{-2} \frac{1}{x^2} \, dx = \lim_{a \to -\infty} \left( \frac{-1}{x} \right) \Bigg]_{a}^{-2} = \lim_{a \to -\infty} \left( \frac{1}{2} + \frac{1}{a} \right) = \frac{1}{2}$

A graph of this region is shown in Figure 17. Since the limit exists, this integral converges.  ▬

▬▬ **EXAMPLE 2**

Find $\int_{0}^{+\infty} 4e^{-3x} \, dx$.

By definition,

$$\int_{0}^{+\infty} 4e^{-3x} \, dx = \lim_{b \to +\infty} \int_{0}^{b} 4e^{-3x} \, dx$$

$$= \lim_{b \to +\infty} \left( -\frac{4}{3} e^{-3x} \right) \Bigg|_{0}^{b}$$

$$= \lim_{b \to +\infty} \left[ -\frac{4}{3} e^{-3b} - \left( -\frac{4}{3} e^{0} \right) \right]$$

$$= \lim_{b \to +\infty} \left( \frac{-4}{3e^{3b}} + \frac{4}{3} \right) = 0 + \frac{4}{3} = \frac{4}{3}.$$

This integral converges.  ▬

The following examples describe applications of improper integrals.*

▬▬ **EXAMPLE 3**

The rate at which a chemical is being released into a small stream at time $t$ is given by $P_0 e^{-kt}$, where $P_0$ is the amount of chemical released into the stream initially. Suppose $P_0 = 1000$ and $k = .06$. Find the total amount of the chemical that will be released into the stream into the indefinite future.

Find

$$\int_{0}^{+\infty} P_0 e^{-kt} \, dt = \int_{0}^{+\infty} 1000 e^{-.06t} \, dt.$$

Work as above.

$$\int_{0}^{+\infty} 1000 e^{-.06t} \, dt = \lim_{b \to +\infty} \int_{0}^{b} 1000 e^{-.06t} \, dt$$

$$= \lim_{b \to +\infty} \left( \frac{1000}{-.06} e^{-.06t} \right) \Bigg|_{0}^{b}$$

$$= \lim_{b \to +\infty} \left( \frac{1000}{-.06 e^{.06b}} - \frac{1000}{-.06} e^{0} \right) = \frac{-1000}{-.06} = 16{,}667$$

A total of 16,667 units of the chemical eventually will be released.  ▬

---

*Probability density functions, discussed in a later chapter, are an important application of improper integrals.

The *capital value* of an asset is sometimes defined as the present value of all future net earnings of the asset. If $R(t)$ gives the rate of change of the annual earnings produced by an asset at time $t$, as shown in Section 4, the **capital value** is

$$\int_0^{+\infty} R(t) \, e^{-kt} \, dt,$$

where $k$ is the annual rate of interest, compounded continuously, and the first payment is assumed to be made in one year.

### ■■ EXAMPLE 4

Suppose income from a rental property is generated at the annual rate of $4000 per year. Find the capital value of this property at an interest rate of 10% compounded continuously.

This is a continuous income stream with a rate of flow of 4000, so $R(t) = 4000$. Also, $k = .10$ or $.1$. The capital value is given by

$$\int_0^{+\infty} 4000 \, e^{-.1t} dt = \lim_{b \to +\infty} \int_0^b 4000 \, e^{-.1t} dt$$

$$= \lim_{b \to +\infty} \left[ \frac{4000}{-.1} e^{-.1t} \right]\Bigg|_0^b$$

$$= \lim_{b \to +\infty} [-40,000 \, e^{-.1b} + 40,000] = 40,000,$$

or $40,000. ■■

### ■■ EXAMPLE 5

Find the capital value of a property when rent is paid *perpetually* if $k = .04$ and $R(x) = 5000$.†

The capital value will be

$$\int_0^{+\infty} 5000 e^{-.04t} \, dt.$$

Find this improper integral as follows.

$$\int_0^{+\infty} 5000 e^{-.04t} \, dt = \lim_{b \to +\infty} \int_0^b 5000 e^{-.04t} \, dt = \lim_{b \to +\infty} \left( \frac{5000}{-.04} e^{-.04t} \right)\Bigg|_0^b$$

$$= \lim_{b \to +\infty} (-125,000 e^{-.04b} + 125,000) = 125,000$$

The capital value is $125,000. ■■

---

†There are historical situations where governments have agreed to pay a fixed rent perpetually.

## 6.5 EXERCISES

Determine whether the improper integrals in Exercises 1–22 converge or diverge, and find the value of each that converges.

**1.** $\int_{2}^{+\infty} \frac{1}{x^2} dx$

**2.** $\int_{5}^{+\infty} \frac{1}{x^2} dx$

**3.** $\int_{1}^{+\infty} \frac{1}{\sqrt{x}} dx$

**4.** $\int_{16}^{+\infty} \frac{-3}{\sqrt{x}} dx$

**5.** $\int_{-\infty}^{-1} \frac{2}{x^3} dx$

**6.** $\int_{-\infty}^{-4} \frac{3}{x^4} dx$

**7.** $\int_{1}^{+\infty} \frac{1}{x^{1.001}} dx$

**8.** $\int_{1}^{+\infty} \frac{1}{x^{.999}} dx$

**9.** $\int_{-\infty}^{-1} x^{-2} dx$

**10.** $\int_{-\infty}^{-4} x^{-2} dx$

**11.** $\int_{-\infty}^{-1} x^{-8/3} dx$

**12.** $\int_{-\infty}^{-27} x^{-5/3} dx$

**13.** $\int_{0}^{+\infty} 4e^{-4x} dx$

**14.** $\int_{0}^{+\infty} 10e^{-10x} dx$

**15.** $\int_{-\infty}^{0} 4e^{x} dx$

**16.** $\int_{-\infty}^{0} 3e^{4x} dx$

**17.** $\int_{-\infty}^{-1} \ln |x| \, dx$

**18.** $\int_{1}^{+\infty} \ln |x| \, dx$

**19.** $\int_{0}^{+\infty} \frac{dx}{(x+1)^2}$

**20.** $\int_{0}^{+\infty} \frac{dx}{(2x+1)^3}$

**21.** $\int_{-\infty}^{-1} \frac{2x-1}{x^2-x} dx$

**22.** $\int_{1}^{+\infty} \frac{2x+3}{x^2+3x} dx$

Use the table of integrals as necessary to evaluate the following integrals.

**23.** $\int_{-\infty}^{-1} \frac{2}{3x(2x-7)} dx$

**24.** $\int_{1}^{+\infty} \frac{7}{2x(5x+1)} dx$

**25.** $\int_{1}^{+\infty} \frac{4}{9x(x+1)^2} dx$

**26.** $\int_{-\infty}^{-5} \frac{5}{4x(x+2)^2} dx$

**27.** $\int_{0}^{+\infty} xe^{2x} dx$

**28.** $\int_{-\infty}^{0} xe^{3x} dx$

For Exercises 29–32, find the area between the graph of the given function and the x-axis over the given interval, if possible.

**29.** $f(x) = \dfrac{1}{x-1}$  for $(-\infty, 0]$

**30.** $f(x) = e^{-x}$  for $(-\infty, e]$

**31.** $f(x) = \dfrac{1}{(x-1^2)}$  for $(-\infty, 0]$

**32.** $f(x) = \dfrac{1}{(x-1)^3}$  for $(-\infty, 0]$.

**33.** Find $\int_{-\infty}^{+\infty} xe^{-x^2} dx$.

**34.** Find $\int_{-\infty}^{+\infty} \dfrac{x}{(1+x^2)^2} dx$.

## APPLICATIONS

### BUSINESS AND ECONOMICS

**Capital Value**    Find the capital values of the properties in Exercises 35–38.

**35.** A castle for which annual rent of $60,000 will be paid in perpetuity; the interest rate is 8% compounded continuously

**36.** A fort on a strategic peninsula in the North Sea, annual rent $500,000, paid in perpetuity; the interest rate is 6% compounded continuously

**37.** Find the capital value of an asset that generates $6000 yearly income if the interest rate is as follows.

(a) 8% compounded continuously

(b) 10% compounded continuously

**38.** An investment produces a perpetual stream of income with a flow rate of

$$K(t) = 1000 \, e^{.02t}.$$

Find the capital value at an interest rate of 7% compounded continuously.

**39.** The Drucker family wants to establish an ongoing scholarship award at a college. Each year in June $3000 will be awarded, starting one year from now. What amount must the Druckers provide the college, assuming funds will be invested at 10% compounded continuously?

## PHYSICAL SCIENCES

**Radioactive Waste**     *Radioactive waste is entering the atmosphere over an area at a decreasing rate. Use the improper integral*

$$\int_0^{+\infty} Pe^{-kt} \, dt$$

*with P = 50 to find the total amount of the waste that will enter the atmosphere for each of the following values of k.*

**40.** $k = .06$

**41.** $k = .04$

---

**KEY WORDS**

| | |
|---|---|
| 6.1 **integration by parts** | **average value of a function** |
| **tables of integrals** | 6.4 **total money flow** |
| 6.2 **numerical integration** | **present value of money flow** |
| **trapezoidal rule** | 6.5 **improper integral** |
| **Simpson's rule** | **convergent integral** |
| 6.3 **solid of revolution** | **divergent integral** |

---

## CHAPTER 6     REVIEW EXERCISES

*Use integration by parts to find the following integrals.*

**1.** $\int x(8 - x)^{3/2} \, dx$

**2.** $\int \dfrac{3x}{\sqrt{x - 2}} dx$

**3.** $\int xe^x \, dx$

**4.** $\int (x + 2)e^{-3x} \, dx$

**5.** $\int \ln |2x + 3| \, dx$

**6.** $\int (x - 1) \ln |x| \, dx$

**7.** Find the area between $y = (3 + x^2)e^{2x}$ and the *x*-axis from $x = 0$ to $x = 1$.

**8.** Find the area between $y = x^3(x^2 - 1)^{1/3}$ and the *x*-axis from $x = 1$ to $x = 3$.

*In Exercises 9–11 use the trapezoidal rule with n = 4 to approximate the value of each integral.*

**9.** $\int_2^6 \dfrac{dx}{x^2 - 1}$

**10.** $\int_2^{10} \dfrac{x \, dx}{x - 1}$

**11.** $\int_1^5 \ln x \, dx$

*In Exercises 12–14 use Simpson's rule with n = 4 to approximate the value of each integral. Compare your answers with the answers to Exercises 9–11.*

**12.** $\displaystyle\int_{2}^{6} \frac{dx}{x^2 - 1}$

**13.** $\displaystyle\int_{2}^{10} \frac{x\, dx}{x - 1}$

**14.** $\displaystyle\int_{1}^{5} \ln x\, dx$

**15.** Find the area under the semicircle $y = \sqrt{1 - x^2}$ and above the $x$-axis by the trapezoidal rule, using $n = 6$.

**16.** Repeat Exercise 15 using Simpson's rule.

*Find the volume of the solid of revolution formed by rotating each of the following bounded regions about the x-axis.*

**17.** $f(x) = 2x - 1$, $y = 0$, $x = 3$

**18.** $f(x) = \sqrt{x - 2}$, $y = 0$, $x = 11$

**19.** $f(x) = e^{-x}$, $y = 0$, $x = -2$, $x = 1$

**20.** $f(x) = \dfrac{1}{\sqrt{x - 1}}$, $y = 0$, $x = 2$, $x = 4$

**21.** $f(x) = 4 - x^2$, $y = 0$, $x = -1$, $x = 1$

**22.** $f(x) = \dfrac{x^2}{4}$, $y = 0$, $x = 4$

**23.** Use Simpson's rule with $n = 8$ to approximate the volume of the *ellipsoid* formed by rotating about the $x$-axis the region between the graph of $f(x) = (3/2)\sqrt{4 - x^2}$ and the $x$-axis.

**24.** Find a formula for the volume of the right circular cylinder formed by rotating about the $x$-axis the region below the line $y = r$ from $x = 0$ to $x = h$ and above the $x$-axis.

**25.** Find the average value of $f(x) = \sqrt{x + 1}$ over the interval $[0, 8]$.

*Find the value of each integral that converges.*

**26.** $\displaystyle\int_{1}^{+\infty} x^{-1}\, dx$

**27.** $\displaystyle\int_{-\infty}^{-2} x^{-2}\, dx$

**28.** $\displaystyle\int_{0}^{+\infty} \frac{dv}{(5x + 2)^2}$

**29.** $\displaystyle\int_{1}^{+\infty} 6e^{-x}\, dx$

**30.** $\displaystyle\int_{-\infty}^{0} \frac{x}{x^2 + 3}\, dx$

**31.** $\displaystyle\int_{10}^{+\infty} \ln (2x)\, dx$

*Find the area between the graph of each function and the x-axis over the given interval, if possible.*

**32.** $f(x) = \dfrac{3}{(x - 2)^2}$  for $(-\infty, 1]$

**33.** $f(x) = 3e^{-x}$  for $[0, +\infty)$

## APPLICATIONS

### BUSINESS AND ECONOMICS

**Total Revenue**  **34.** The rate of change of revenue from the sale of $x$ units of toaster ovens is

$$R'(x) = x(x - 50)^{1/2}.$$

Find the total revenue from the sale of the 50th to the 75th ovens.

**Total Sales**  **35.** Use the values of $f(x)$ given in the table to find total sales in millions for the given period by the trapezoidal rule, using $n = 6$.

| Year, $x$ | 1 | 2 | 3 | 4 | 5 | 6 | 7 |
|---|---|---|---|---|---|---|---|
| Rate of Growth of Sales, $f(x)$ | .7 | 1.2 | 1.5 | 1.9 | 2.2 | 2.4 | 2.0 |

**Total Sales**  **36.** Repeat Exercise 35 using Simpson's rule.

**Present Value**
**of Money Flow**  *Each of the functions in Exercises 37–40 represents the rate of flow of money per unit of time over the given time period, compounded continuously at the given annual interest rate. Find the present value in each case.*

**37.** $f(x) = 5000$; 8 years; 9%  **38.** $f(x) = 25{,}000$; 12 years; 10%

**39.** $f(x) = 100e^{.02x}$; 5 years; 11%  **40.** $f(x) = 30x$; 18 months; 15%

**Amount of Money at Time $t$**  *Assume that each of the following functions gives the rate of flow of money per unit of time over the given period, with continuous compounding at the given rate. Find the amount at the end of the time period.*

**41.** $f(x) = 2000$; 5 months; 1% per month

**42.** $f(x) = 500e^{-.03x}$; 8 years; 10% per year

**43.** $f(x) = 20x$; 6 years; 12% per year

**44.** $f(x) = 1000 + 200x$; 10 years; 9% per year

**Money Flow**  **45.** An investment scheme is expected to produce a continuous flow of money, starting at $1000 and increasing exponentially at 5% a year for 7 years. Find the present value at an interest rate of 11% compounded continuously.

**Money Flow**  **46.** The proceeds from the sale of a building will yield a uniform continuous flow of $10,000 a year for 10 years. Find the final amount at an interest rate of 10.5% compounded continuously.

**Capital Value**  **47.** Find the capital value of an office building for which annual rent of $50,000 will be paid in perpetuity, if the interest rate is 9%.

## LIFE SCIENCES

**Drug Reaction**  **48.** The reaction rate to a new drug $x$ hours after the drug is administered is

$$r'(x) = .5xe^{-x}.$$

Find the total reaction over the first 5 hours.

**Oil Leak Pollution**  **49.** An oil leak from an uncapped well is polluting a bay at a rate of $f(x) = 100e^{-.05x}$ gallons per year. Use an improper integral to find the total amount of oil that will enter the bay, assuming the well is never capped.

## PHYSICAL SCIENCES

**Average Temperatures**  **50.** Suppose the temperature in a river at a point $x$ meters downstream from a factory that is discharging hot water into the river is given by

$$T(x) = 400 - .25x^2.$$

Find the average temperature over each of the following intervals.

**(a)** [0, 10]  **(b)** [10, 40]  **(c)** [0, 40]

## EXTENDED APPLICATION

## HOW MUCH DOES A WARRANTY COST?*

This application uses some of the ideas of probability. The probability of an event is a number $p$, where $0 \le p \le 1$, such that $p$ is equal to the number of ways that the event can happen divided by the total number of possible outcomes. For example, the probability of drawing a red card from a deck of 52 cards (of which 26 are red) is given by

$$P\text{(red card)} = \frac{26}{52} = \frac{1}{2}.$$

In the same way, the probability of drawing a black queen from a deck of 52 cards is

$$P\text{(black queen)} = \frac{2}{52} = \frac{1}{26}.$$

In this application we find the cost of a warranty program to a manufacturer. This cost depends on the quality of the products made, as we might expect. We use the following variables.

$\quad c$ = constant product price, per unit, including cost of warranty (We assume the
$\qquad$ price charged per unit is constant, since this price is likely to be fixed by
$\qquad$ competition.)
$\quad m$ = expected lifetime of the product
$\quad w$ = length of the warranty period
$\quad N$ = size of a production lot, such as a year's production
$\quad r$ = warranty cost per unit
$C(t)$ = pro rata customer rebate at time $t$
$P(t)$ = probability of product failure at any time $t$
$F(t)$ = number of failures occurring at time $t$

We assume that the warranty is of the pro rata customer rebate type, in which the customer is paid for the proportion of the warranty left. Hence, if the product has a warranty period of $w$, and fails at time $t$, then the product worked for the fraction $t/w$ of the warranty period. The customer is reimbursed for the unused portion of the warranty, or the fraction

$$1 - \frac{t}{w}.$$

If we assume the product cost $c$ originally, and if we use $C(t)$ to represent the customer rebate at time $t$, we have

$$C(t) = c\left(1 - \frac{t}{w}\right).$$

For many different types of products, it has been shown by experience that

$$P(t) = 1 - e^{-t/m}$$

provides a good estimate of the probability of product failure at time $t$. The total number of failures at time $t$ is given by the product $P(t)$ and $N$, the total number of items per batch. If we use $F(t)$ to represent this total, we have

$$F(t) = N \cdot P(t) = N(1 - e^{-t/m}).$$

---

*Reprinted by permission of Warren W. Menke, "Determination of Warranty Reserves," *Management Sciences,* Volume 15, Number 10, June 1969. Copyright © 1969 The Institute of Management Sciences.

The total number of failures in some "tiny time interval" of width $dt$ can be shown to be the derivative of $F(t)$,

$$F'(t) = \left(\frac{N}{m}\right) e^{-t/m},$$

while the cost for the failures in this "tiny time interval" is

$$C(t) \cdot F'(t) = c\left(1 - \frac{t}{w}\right)\left(\frac{N}{m}\right)e^{-t/m}$$

The total cost for all failures during the warranty period is thus given by the definite integral

$$\int_0^w c\left(1 - \frac{t}{w}\right)\left(\frac{N}{m}\right)e^{-t/m} \, dt.$$

Using integration by parts, this definite integral can be shown to equal

$$Nc\left[ -e^{-t/m} + \frac{t}{w} \cdot e^{-t/m} + \frac{m}{w}(e^{-t/m}) \right]\Bigg|_0^w \quad \text{or} \quad Nc\left(1 - \frac{m}{w} + \frac{m}{w} e^{-w/m}\right).$$

This last quantity is the total warranty cost for all the units manufactured. Since there are $N$ units per batch, the warranty cost per item is

$$r = \frac{1}{N}\left[ Nc\left(1 - \frac{m}{w} + \frac{m}{w} e^{-w/m}\right) \right] = c\left(1 - \frac{m}{w} + \frac{m}{w} e^{-w/m}\right).$$

For example, suppose a product that costs $100 has an expected life of 24 months, with a 12-month warranty. Then we have $c = \$100$, $m = 24$, $w = 12$, with $r$, the warranty cost per unit, given by

$$r = 100(1 - 2 + 2e^{-.5}) = 100[-1 + 2(.6065)] = 100(.2130) = 21.30.$$

where we found $e^{-.5}$ from the table of powers of $e$.

 **EXERCISES**

*Find r for each of the following.*

**1.** $c = \$50$, $m = 48$ months, $w = 24$ months
**2.** $c = \$1000$, $m = 60$ months, $w = 24$ months
**3.** $c = \$1200$, $m = 30$ months, $w = 30$ months

**EXTENDED APPLICATION**

## FLOW SYSTEMS*

One of the most important biological applications of integration is the determination of properties of flow systems. Among the significant characteristics of a flow system are its flow rate (a heart's output in liters per minute, for example), its volume (for blood, the vascular volume), and the quantity of a substance flowing through it (a pollutant in a stream, perhaps).

A flow system can be anything from a river to an oil pipeline to an artery. We assume that the system has some well-defined volume between two points, and a more or less constant flow rate between them. In arteries the blood pulses, but we shall measure the flow over long enough time intervals for the flow rate to be regarded as constant.

*From "Flow System Integrals," by Arthur C. Segal, *The UMAP Journal*, Vol. III, No. 1, 1982. Copyright © 1982 Educational Development Center, Inc. Reprinted by permission of COMAP, Inc.

The standard method for analyzing flow system behavior is the so-called indicator dilution method, but we shall also discuss the more recent thermodilution technique for determining cardiac output. In the indicator dilution method, a known or unknown quantity of a substance such as a dye, a pollutant, or a radioactive tracer is injected or seeps into a flow system at some entry point. The substance is assumed to mix well with the fluid, and at some downstream measuring point its concentration is sampled either serially or continuously. The concentration function $c(t)$ is then integrated over an appropriate time interval, either by numerical approximation, or a planimeter, or electronically, or by curve-fitting and the fundamental theorem of calculus. As we shall see, the integral

$$\int_0^T c(t)\, dt$$

(whose units are mass $\times$ time/volume) is extremely useful for finding unknown flow system properties.

The basis for the analysis of flow systems is the principle of mass conservation, or "what goes in must come out." We are going to assume that the indicator does not leak from the stream (for example, by diffusion into tissue), and that the time interval is over before any of the indicator recirculates (less than a minute in the case of blood). What goes *into* the system is a (known or unknown) quantity $Q_{in}$ of indicator. What comes *out* of the system, or more precisely, what passes by the downstream measuring point at a flow rate $F$, is a volume of fluid with a variable concentration $c(t)$ of the indicator. The product of $F$ and $c(t)$ is called the **mass rate** of the indicator flowing past the measuring point. The total quantity $Q_{out}$ of indicator flowing by during the time interval from 0 to $T$ is the integral of the mass rate from 0 to $T$:

$$Q_{out} = \int_0^T Fc(t)\, dt.$$

By our assumption of mass conservation (no leakage and no recirculation), we must have $Q_{in} = Q_{out}$, or

$$Q_{in} = \int_0^T Fc(t)\, dt = F\int_0^T c(t)\, dt, \tag{1}$$

provided of course that $F$ is essentially constant.

Equation (1) is the basic relationship of indicator dilution. We shall illustrate its use with an example.

### EXAMPLE 1

Suppose that an industrial plant is discharging toxic Kryptonite into the Cahaba River at an unknown variable rate. Downriver, a field station of the Environmental Protection Agency measures the following concentrations:

| Time | Concentration |
|------|---------------|
| 2 A.M. | 3.0 mg/m$^3$ |
| 6 A.M. | 3.5 mg/m$^3$ |
| 10 A.M. | 2.5 mg/m$^3$ |
| 2 P.M. | 1.0 mg/m$^3$ |
| 6 P.M. | .5 mg/m$^3$ |
| 10 P.M. | 1.0 mg/m$^3$ |

The EPA has found the flow rate of the Cahaba in the vicinity to be 1000 ft$^3$/sec $= 1.02 \times 10^5$ m$^3$/hr. How much Kryptonite is being discharged?

We use equation (1) to estimate the total amount of Kryptonite discharged during a 24-hour day by approximating the concentration integral. Dividing the day into six 4-

hour subintervals (so that $\Delta t = 4$) and using the EPA data as our function values $c(t)_i$, we compute as follows:

$$\int_0^{24} c(t)\, dt \approx \sum_{i=1}^{6} c(t_i)\Delta t$$

$$= \Delta t \sum_{i=1}^{6} c(t_i) = 4(3.0 + 3.5 + 2.5 + 1.0 + .5 + 1.0) = 4(11.5) = 46,$$

or 46 mg $\times$ hrs/m$^3$. Hence, by equation (1), the total daily discharge is

$$Q_{in} = F \int_0^{24} c(t)\, dt$$

$$\approx \left(1.02 \times 10^5 \frac{m^3}{hour}\right)\left(46 \frac{mg \times hour}{m^3}\right)$$

$$= 4.7 \times 10^6 \text{ mg} = 4.7 \text{ kilograms of Kryptonite.} \quad \blacksquare$$

*DETERMINING CARDIAC OUTPUT BY THERMODILUTION* Knowledge of a patient's cardiac output is a valuable aid to diagnosis and treatment of heart damage. By cardiac output we mean the volume flow rate of venous blood pumped by the right ventricle through the pulmonary artery to the lungs (to be oxygenated). Normal resting cardiac output is 4 to 5 liters per minute, but in critically ill patients it may fall well below 3 liters per minute.

Unfortunately, cardiac output is difficult to determine by dye dilution because of the need to calibrate the dye and to sample the blood repeatedly. There are also problems with dye instability, recirculation, and slow dissipation. With the development in 1970 of a suitable pulmonary artery catheter, it became possible to make rapid and routine bedside measurements of cardiac output by thermodilution, and to do so with very few patient complications.

In a coronary intensive care unit (typically), a balloon-tipped catheter is inserted in an arm vein and guided to the superior vena cava or right atrium (upper chamber of the heart). A temperature-sensing device called a thermistor is moved through the right side of the heart and then positioned a few centimeters into the pulmonary artery. Then, 10 ml of cold D5W (5% dextrose in water) is injected into the vena cava or atrium. As the cold D5W mixes with blood in the right heart, temperature variations in the flow are detected by the thermistor. Figure 18 shows a typical record of temperature change as a function of time. With computerized equipment, cardiac output can be displayed within a minute after injection. Since the cold D5W is warmed by the body before it recirculates, repeated measurements can be made reliably.

This curve shows the temperature variation in the pulmonary artery that resulted from injecting a few milliliters of cold dextrose solution into the right atrium of the heart.

**FIGURE 18**

We now derive an equation for thermodilution comparable to equation (1). First, we need to define the thermal equivalent of mass. We define the "quantity of cold" injected to be

$$Q_{in} = V_i(T_b - T_i)$$

where $V_i$ is the volume of injectate, $T_i$ is its temperature, and $T_b$ is body temperature. Thus, "quantity of cold" injected is jointly proportional to how cold the injectate is (below body) and to how much is injected.

As with indicator dilution, we shall assume "conservation of cold," that is, no loss and no recirculation. (Actually, there is some loss of cold as the injectate travels up through the catheter to the injection point, but that can be corrected for.) The quantity of cold passing by the thermistor in a short time interval $dt$ can be calculated by multiplying the flow volume $F\,dt$ and the temperature variation $\Delta T$ during the interval. The total quantity of cold flowing by is

$$Q_{out} = \int_0^{+\infty} F\Delta T dt$$
$$= F\int_0^{+\infty} \Delta T dt.$$

Equating $Q_{in}$ to $Q_{out}$ and solving for the flow rate $F$ produces the analog of equation (1):

$$Q_{out} = Q_{in} = V_i(T_b - T_i)$$
$$F\int_0^{+\infty} \Delta T dt = V_i(T_b - T_i)$$
$$F = \frac{V_i(T_b - T_i)}{\int_0^{+\infty} \Delta T dt}. \tag{2}$$

If time is measured in seconds and temperature in degrees Celsius, then the denominator of equation (2) has units of (°C)(sec) and the numerator has units of (ml)(°C). Hence, $F$ has units ml/sec, which is converted to liters/minute by multiplying by 60/1000 = .06.

Although thermistor signals are usually calibrated and integrated electronically, we shall illustrate equation (2) with an example.

### EXAMPLE 2

Suppose $T_b = 37°C$, $T_i = 0°C$, and $V_i = 10$ ml. Suppose that the change in temperature with respect to time is represented by the curve $\Delta T = .1t^2e^{-.31t}$, shown in Figure 19. Finally, suppose that the net effect of all the correction factors is to multiply the right side of equation (2) by .891. Then integration by parts gives

$$F = \frac{(10)(37 - 0)(.891)}{\int_0^{+\infty} (.1)t^2e^{-.31t}\, dt}$$

$$= \frac{330}{(.1)\dfrac{2}{(.31)^3}}$$

$$= 49.16 \text{ ml/sec}$$

$$= 2.95 \text{ liters/minute.} \quad \blacksquare$$

$\Delta T$ (°C)

.4

.2

$\Delta T = .1t^2 e^{-.31t}$

0   10   20   30   40   $t$

(seconds)

**FIGURE 19**

The illustration of equation (2) uses this smooth curve in place of the thermodilution curve in Figure 18.

**EXERCISES**

**1.** Approximate the integral in Example 1 using **(a)** the trapezoidal rule, and **(b)** Simpson's rule. Let $t$ represent the number of hours after 2 A.M., so that 2 A.M. = 0, 6 A.M. = 4, and so on. Assume the reading after 24 hours is 3.0 mg/m$^3$ again.

**2.** A dye dilution technique for measuring cardiac output involves injection of a dye into a main vein near the heart and measurement by a catheter in the aorta at regular intervals. Suppose the results from such measurements are as given in the table.

| Time (seconds) | 0 | 4 | 8 | 12 | 16 | 20 | 24 |
|---|---|---|---|---|---|---|---|
| Concentration (ml/liter) | 0 | .6 | 2.7 | 4.1 | 2.9 | .9 | 0 |

If 8 ml of dye is injected, approximate the cardiac output $F$ in the equation

$$Q_{in} = F \int_0^t c(t)\, dt$$

by **(a)** the trapezoidal rule, and **(b)** Simpson's rule.

**3.** In Example 2 let $T_b = 37.1$°C, $T_i = 5$°C, and $v_i = 8$ ml. Find the cardiac output $F$.

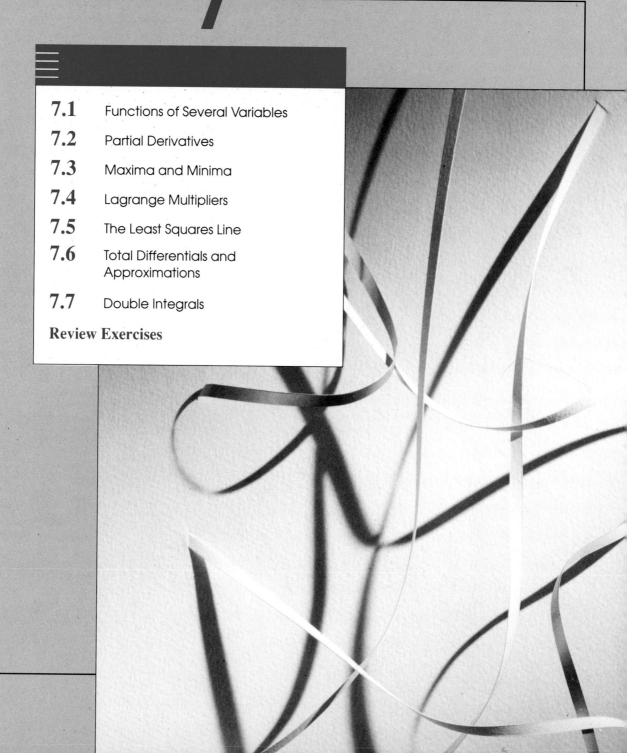

# 7 Functions of Several Variables

If a company produces $x$ items at a cost of $10 per item, then the total cost $C(x)$ of producing the items is given by

$$C(x) = 10x.$$

The cost is a function of one independent variable, the number of items produced. If the company produces two products, with $x$ of one product at a cost of $10 each, and $y$ of another product at a cost of $15 each, then the total cost to the firm is a function of *two* independent variables, $x$ and $y$. By generalizing $f(x)$ notation, the total cost can be written as $C(x, y)$, where

$$C(x, y) = 10x + 15y.$$

In this chapter we discuss such functions of more than one independent variable. Many of the ideas developed for functions of one variable also apply to functions of more than one variable. In particular, the fundamental idea of derivative generalizes in a very natural way to functions of more than one variable.

## ▬ 7.1  FUNCTIONS OF SEVERAL VARIABLES

The total cost for the firm in the example above is given by

$$C(x, y) = 10x + 15y.$$

When $x = 5$ and $y = 12$ the total cost is written $C(5, 12)$, with

$$C(5, 12) = 10 \cdot 5 + 15 \cdot 12 = 230.$$

### ▬ EXAMPLE 1 .
Let $f(x, y) = 4x^2 + 2xy + 3/y$ and find each of the following.

**(a)** $f(-1, 3)$.

Replace $x$ with $-1$ and $y$ with 3.

$$f(-1, 3) = 4(-1)^2 + 2(-1)(3) + \frac{3}{3} = 4 - 6 + 1 = -1$$

**(b)** $f(2, 0)$

Because of the quotient $3/y$, it is not possible to replace $y$ with 0, so $f(2, 0)$ does not exist.  ▬

**FUNCTION OF TWO VARIABLES**

> ▬ $z = f(x, y)$ is a **function of two variables** if a unique value of $z$ is obtained from each ordered pair of real numbers $(x, y)$. The variables $x$ and $y$ are **independent variables,** and $z$ is the **dependent variable.** The set of all ordered pairs of real numbers $(x, y)$ such that $f(x, y)$ exists is the **domain** of $f$; the set of all values of $f(x, y)$ is the **range.**

Unless otherwise stated, the domain of the functions we will study is assumed to be the largest set of ordered pairs of real numbers $(x, y)$ for which $f(x, y)$ exists.

### ▰▰ EXAMPLE 2

Let $x$ represent the number of milliliters (ml) of carbon dioxide released by the lungs in one minute. Let $y$ be the change in the carbon dioxide content of the blood as it leaves the lungs ($y$ is measured in ml of carbon dioxide per 100 ml of blood). The total output of blood from the heart in one minute (measured in ml) is given by $C$, where $C$ is a function of $x$ and $y$ such that

$$C = C(x, y) = \frac{100x}{y}.$$

Find $C(320, 6)$.

Replace $x$ with 320 and $y$ with 6 to get

$$C(320, 6) = \frac{100(320)}{6}$$

$$\approx 5333 \text{ ml of blood per minute.} \quad ▰▰$$

The definition given after Example 1 was for a function of two independent variables, but similar definitions could be given for functions of three, four, or more independent variables.

### ▰▰ EXAMPLE 3

Let $f(x, y, z) = 4xz - 3x^2y + 2z^2$. Find each of the following.

**(a)** $f(2, -3, 1)$

Replace $x$ with 2, $y$ with $-3$, and $z$ with 1.

$$f(2, -3, 1) = 4(2)(1) - 3(2)^2(-3) + 2(1)^2 = 8 + 36 + 2 = 46$$

**(b)** $f(-4, 3, -2) = 4(-4)(-2) - 3(-4)^2(3) + 2(-2)^2$

$$= 32 - 144 + 8 = -104. \quad ▰▰$$

*Graphing Functions of Two Independent Variables*  The best way to describe a new concept is often by referring to a picture. In earlier sections, we have frequently used graphs of functions to illustrate ideas. We can learn about the behavior of functions of two independent variables by looking at the graphs of equations with three variables, such as $z = f(x, y)$. Functions of one independent variable are graphed by using an $x$-axis and a $y$-axis to locate points in a plane. The plane determined by the $x$- and $y$-axes is called the *xy-plane*. A third axis is needed to graph functions of two independent variables— the $z$-axis, which goes through the origin in the *xy*-plane and is perpendicular to both the $x$-axis and the $y$-axis.

Figure 1 shows one possible way to draw the three axes. In Figure 1, the *yz*-plane is in the plane of the page, with the *x*-axis perpendicular to the plane of the page.

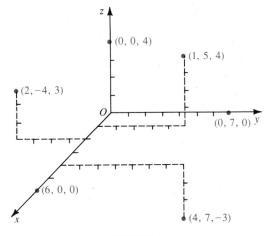

**FIGURE 1**

Just as we graphed ordered pairs earlier we can now graph **ordered triples** of the form $(x, y, z)$. For example, to locate the point corresponding to the ordered triple $(2, -4, 3)$, start at the origin and go 2 units along the positive *x*-axis. Then go 4 units in a negative direction (to the left) parallel to the *y*-axis. Finally, go up 3 units parallel to the *z*-axis. The point representing $(2, -4, 3)$ is shown in Figure 1, together with several other points. The region of three-dimensional space where all coordinates are positive is called the **first octant.**

Many of the results from two dimensions suggest similar results for three dimensions. For example, the distance formula from previous work generalizes in a natural way. By the Pythagorean theorem, the diagonal of the base of the box in Figure 2(a) has length $\sqrt{a^2 + b^2}$. The diagonal of the box itself is the hypotenuse of a right triangle having the diagonal of the base as one side and a vertical edge as the other side. Again by the Pythagorean theorem, the length of the diagonal of the box is the square root of $(\sqrt{a^2 + b^2})^2 + c^2$, or $\sqrt{a^2 + b^2 + c^2}$, as indicated in Figure 2(b).

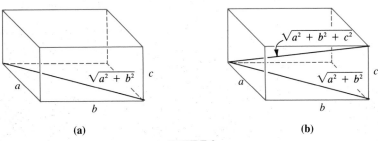

**(a)**                                                      **(b)**

**FIGURE 2**

Generalizing this approach leads to the *distance formula*.

**DISTANCE FORMULA**

> The distance $d$ between $(x_1, y_1, z_1)$ and $(x_2, y_2, z_2)$ is
>
> $$d = \sqrt{(x_2 - x_1)^2 + (y_2 - y_1)^2 + (z_2 - z_1)^2}.$$

A **sphere** is the set of all points in space a fixed distance from a fixed point. The equation of a sphere is found from the distance formula.

**EQUATION OF A SPHERE**

> An equation of the sphere with center at $(h, k, j)$ and radius $r$ is
>
> $$(x - h)^2 + (y - k)^2 + (z - j)^2 = r^2.$$

### EXAMPLE 4

Graph $x^2 - 4x + y^2 + 6y + z^2 - 8z = -25$.

To get the equation in the form of the equation of a sphere, complete the square on $x$, $y$, and $z$, as shown in Section 1.4. Add the necessary constants on both sides to get three perfect square trinomials. In each case, add the square of half the coefficient of the first-degree term.

$$x^2 - 4x \quad\quad + y^2 + 6y \quad\quad + z^2 - 8z \quad\quad = -25$$
$$x^2 - 4x + 4 + y^2 + 6y + 9 + z^2 - 8z + 16 = -25 + 4 + 9 + 16$$
$$(x - 2)^2 + (y + 3)^2 + (z - 4)^2 = 4$$

The graph of this equation is a sphere of radius 2 with center at $(2, -3, 4)$. See Figure 3.

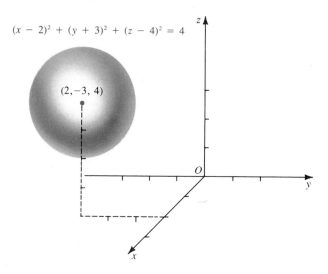

$$(x - 2)^2 + (y + 3)^2 + (z - 4)^2 = 4$$

$(2, -3, 4)$

**FIGURE 3**

As shown in Chapter 1, the graph of $ax + by = c$ (where $a$ and $b$ are not both 0) is a straight line. This result generalizes to three dimensions.

**PLANE**

> The graph of
> $$ax + by + cz = d$$
> is a **plane** if $a$, $b$, and $c$ are not all 0.

### ■ EXAMPLE 5

Graph $2x + y + z = 6$.

By the result above the graph of this equation is a plane. Earlier, we graphed straight lines by finding $x$- and $y$-intercepts. A similar idea helps in graphing a plane. To find the $x$-intercept, which is the point where the graph crosses the $x$-axis, let $y = 0$ and $z = 0$.

$$2x + 0 + 0 = 6$$
$$x = 3$$

The point $(3, 0, 0)$ is on the graph. Letting $x = 0$ and $z = 0$ gives the point $(0, 6, 0)$, while $x = 0$ and $y = 0$ lead to $(0, 0, 6)$. The plane through these three points includes the triangular surface shown in Figure 4. This region is the first-octant part of the plane that is the graph of $2x + y + z = 6$. ■

### ■ EXAMPLE 6

Graph $x + z = 6$.

To find the $x$-intercept, let $z = 0$, giving $(6, 0, 0)$. If $x = 0$, we get the point $(0, 0, 6)$. Because there is no $y$ in the equation $x + z = 6$, there can be no $y$-intercept. A plane that has no $y$-intercept is parallel to the $y$-axis. The first-octant portion of the graph of $x + z = 6$ is shown in Figure 5. ■

**FIGURE 4**

**FIGURE 5**

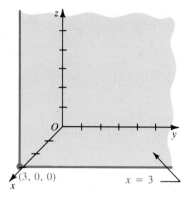

**FIGURE 6**

## EXAMPLE 7

Graph $x = 3$.

This graph, which goes through (3, 0, 0), can have no $y$-intercept and no $z$-intercept. It is therefore a plane parallel to the $y$-axis and the $z$-axis and, therefore, to the $xz$-plane. The first-octant portion of the graph is shown in Figure 6. ▬

The graph of a function $z = f(x, y)$ is called a **surface.** It may be difficult to draw the surface resulting from a function of two variables. One useful way of picturing these graphs is by finding various **traces**—the curves that result when a surface is cut by a plane. For example, the sphere $x^2 + y^2 + z^2 = 25$ is graphed in Figure 7. The **xy-trace,** the intersection of the sphere and the $xy$-plane, is a circle of radius 5. The **yz-trace** and the **xz-trace,** defined in a similar manner, are also circles of radius 5.

The $xy$-trace is the function of $x$ and $y$ that results from setting $z = 0$. More generally, choosing a value $k$ for $z$ and substituting it into the equation $z = f(x, y)$ gives an equation in $x$ and $y$ that describes a plane figure parallel to the $xy$-plane and located at a height $k$. Letting $k$ vary gives a family of such curves, called **level curves.** See Figure 8.

One application of level curves in economics occurs with production functions. A **production function** $z = f(x, y)$ is a function that gives the quantity $z$ of an item produced as a function of $x$ and $y$, where $x$ is the amount of labor and $y$ is the amount of capital (in appropriate units) needed to produce $z$ units. If the production function has the special form $z = P(x, y) = Ax^a y^{1-a}$, where $A$ is a constant and $0 < a < 1$, the function is called a **Cobb-Douglas production function.**

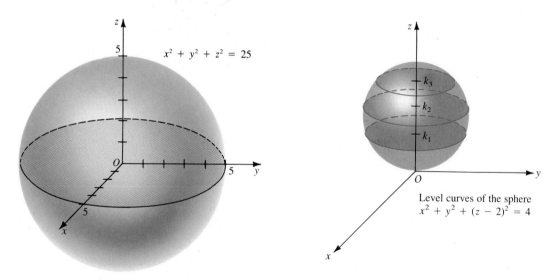

**FIGURE 7**

Level curves of the sphere
$x^2 + y^2 + (z - 2)^2 = 4$

**FIGURE 8**

### ■ EXAMPLE 8

Find the level curve at a production of 100 items for the Cobb-Douglas production function $z = x^{2/3}y^{1/3}$.

Let $z = 100$ to get

$$100 = x^{2/3}y^{1/3}$$

$$\frac{100}{x^{2/3}} = y^{1/3}.$$

Now cube both sides to express $y$ as a function of $x$.

$$y = \frac{100^3}{x^2} = \frac{1{,}000{,}000}{x^2}. \quad ■$$

The level curve of height 100 from Example 8 is shown graphed in three dimensions in Figure 9(a) and on the familiar $xy$-plane in Figure 9(b). The points of the graph correspond to those values of $x$ and $y$ that lead to production of 100 items. The curve is called an **isoquant,** from *iso* (equal) and *quant* (amount). In Example 8, the "amounts" all "equal" 100.

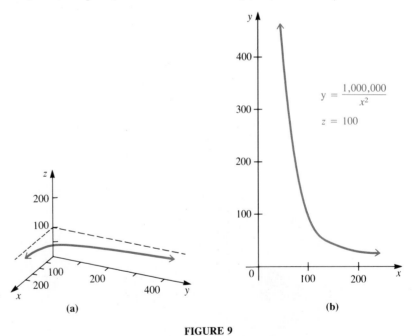

**(a)**  **(b)**

**FIGURE 9**

Because of the difficulty of drawing the graphs of more complicated functions, we now merely list some common equations and their graphs. These graphs were drawn by computer, a very useful method of depicting three-dimensional surfaces.

**Paraboloid,** $z = x^2 + y^2$

$xy$-trace: point
$yz$-trace: parabola
$xz$-trace: parabola

**Ellipsoid,** $\dfrac{x^2}{a^2} + \dfrac{y^2}{b^2} + \dfrac{z^2}{c^2} = 1$

$xy$-trace: ellipse
$yz$-trace: ellipse
$xz$-trace: ellipse

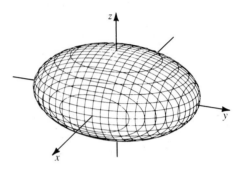

**Hyperbolic Paraboloid,** $x^2 - y^2 = z$
(sometimes called a **saddle**)

$xy$-trace: two intersecting lines
$yz$-trace: parabola
$xz$-trace: parabola

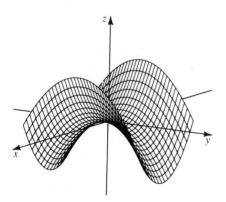

**Hyperboloid of Two Sheets,**
$-x^2 - y^2 + z^2 = 1$

$xy$-trace: none
$yz$-trace: hyperbola
$xz$-trace: hyperbola

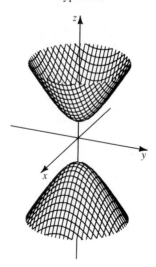

## 7.1 EXERCISES

1. Let $f(x, y) = 4x + 5y + 3$. Find the following.
   (a) $f(2, -1)$          (b) $f(-4, 1)$
   (c) $f(-2, -3)$        (d) $f(0, 8)$

2. Let $g(x, y) = -x^2 - 4xy + y^3$. Find the following.
   (a) $g(-2, 4)$        (b) $g(-1, -2)$
   (c) $g(-2, 3)$        (d) $g(5, 1)$

3. Let $h(x, y) = \sqrt{x^2 + 2y^2}$. Find the following.
   (a) $h(5, 3)$        (b) $h(2, 4)$
   (c) $h(-1, -3)$      (d) $h(-3, -1)$

4. Let $f(x, y) = \dfrac{\sqrt{9x + 5y}}{\log x}$. Find the following.
   (a) $f(10, 2)$        (b) $f(100, 1)$
   (c) $f(1000, 0)$      (d) $f\left(\dfrac{1}{10}, 5\right)$

*Graph the first-octant portion of each plane.*

5. $x + y + z = 6$
6. $x + y + z = 12$
7. $2x + 3y + 4z = 12$
8. $4x + 2y + 3z = 24$
9. $3x + 2y + z = 18$
10. $x + 3y + 2z = 9$
11. $x + y = 4$
12. $y + z = 5$
13. $x = 2$
14. $z = 3$

*Find the distance between the points in each pair.*

15. $(1, 4, 2)$ and $(0, 3, 1)$
16. $(3, 7, 9)$ and $(1, 5, 3)$
17. $(-2, 1, 7)$ and $(3, -4, 0)$
18. $(-5, -3, -8)$ and $(-2, -4, -6)$
19. $(1, 0, -4)$ and $(-2, 1, 3)$
20. $(5, 8, -6)$ and $(-3, 0, 4)$

*Find an equation for each sphere.*

21. Center $(-1, 4, 2)$, radius 3
22. Center $(2, -5, 8)$, radius 1
23. Center $(2, 0, -3)$, radius 4
24. Center $(0, 0, 3)$, radius 5

*Find the center and radius of each sphere.*

25. $x^2 + 10x + y^2 - 6y + z^2 + 8z = -41$
26. $x^2 - 4x + y^2 - 10y + z^2 + 6z = -13$
27. $x^2 + 6x + y^2 + z^2 - 2z = 2$
28. $x^2 + 2x + y^2 - 6y + z^2 - 4z = -13$

*Match each equation in Exercises 29–34 with its graph in (a)–(f) below.*

**29.** $z = x^2 + y^2$

**30.** $z^2 - y^2 - x^2 = 1$

**31.** $x^2 - y^2 = z$

**32.** $z = y^2 - x^2$

**33.** $\dfrac{x^2}{16} + \dfrac{y^2}{25} + \dfrac{z^2}{4} = 1$

**34.** $z = 5(x^2 + y^2)^{-1/2}$

**(a)**

**(b)**

**(c)**

**(d)**

**(e)**

**(f)**

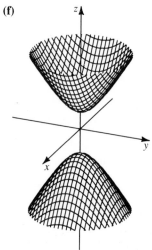

**35.** Let $f(x, y) = 9x^2 - 3y^2$, and find each of the following.

    **(a)** $\dfrac{f(x + h, y) - f(x, y)}{h}$    **(b)** $\dfrac{f(x, y + h) - f(x, y)}{h}$

**36.** Let $f(x, y) = 7x^3 + 8y^2$, and find each of the following.

    **(a)** $\dfrac{f(x + h, y) - f(x, y)}{h}$    **(b)** $\dfrac{f(x, y + h) - f(x, y)}{h}$

### ▰ APPLICATIONS

#### BUSINESS AND ECONOMICS

**Production Functions**  **37.** Production of a precision camera is given by

$$P(x, y) = 100\left[\frac{3}{5}x^{-2/5} + \frac{2}{5}y^{-2/5}\right]^{-5},$$

where $x$ is the amount of labor in work hours and $y$ is the amount of capital. Find the following.

(a) $P(32, 1)$    (b) $P(1, 32)$

(c) If 32 work hours and 243 units of capital are used, what is the production output?

**Individual Retirement Accounts**  The multiplier function discussed in an earlier chapter,

$$M = \frac{(1 + i)^n(1 - t) + t}{[1 + (1 - t)i]^n},$$

compares the growth of an Individual Retirement Account (IRA) with the growth of the same deposit in a regular savings account. The function M depends on the three variables n, i, and t, where n represents the number of years left at interest, i represents the interest rate in both types of account, and t represents the income tax rate. Let M = f(n, i, t) and find the following.

**38.** $f(25, .05, .28)$

**39.** $f(25, .06, .33)$

**Production Functions**  Find the level curve at a production of 500 for each of the production functions in Exercises 40–41. Graph each function on the xy-plane.

**40.** The production function $z$ for the United States was once estimated as $z = x^{.7}y^{.3}$, where $x$ stands for the amount of labor and $y$ stands for the amount of capital.

**41.** If $x$ represents the amount of labor and $y$ the amount of capital, a production function for Canada is approximately $z = x^{.4}y^{.6}$.

#### LIFE SCIENCES

**Cattle Ranching**  **42.** The number of cows that can graze on a certain ranch without causing overgrazing is approximated by

$$C(x, y) = 9x + 5y - 4,$$

where $x$ is the number of acres of grass and $y$ the number of acres of alfalfa. Find the following.

(a) $C(50, 0)$    (b) $C(30, 4)$

(c) How many cows may graze if there are 80 acres of grass and 20 acres of alfalfa?

(d) If the ranch has 10 acres of alfalfa and 5 acres of grass, how many cows may graze?

Use a calculator with a $y^x$ key to work the following problems.*

**Oxygen Consumption**    **43.** The oxygen consumption of a well-insulated mammal that is not sweating is approximated by

$$m = \frac{2.5(T - F)}{w^{.67}},$$

where $T$ is the internal body temperature of the animal (in °C), $F$ is the temperature of the outside of the animal's fur (in °C), and $w$ is the animal's weight in kilograms. Find $m$ for the following data.

(a) $T = 38°$, $F = 6°$, $w = 32$ kg     (b) $T = 40°$, $F = 20°$, $w = 43$ kg

**Body Surface Area**    **44.** The surface area of a human (in square meters) is approximated by

$$A = .202W^{.425}H^{.725}$$

where $W$ is the weight of the person in kilograms and $H$ is the height in meters. Find $A$ for the following data.

(a) $W = 72$, $H = 1.78$     (b) $W = 65$, $H = 1.40$     (c) $W = 70$, $H = 1.60$

(d) Find your own surface area.

## 7.2  PARTIAL DERIVATIVES

A small firm makes only two products, radios and audiocassette recorders, with the profits of the firm given by

$$P(x, y) = 40x^2 - 10xy + 5y^2 - 80,$$

where $x$ is the number of units of radios sold and $y$ is the number of units of recorders sold. How will a change in $x$ or $y$ affect $P$?

Suppose that sales of radios have been steady at 10 units; only the sales of recorders vary. The management would like to find the marginal profit with respect to $y$, the number of recorders sold. Recall that marginal profit is given by the derivative of the profit function. Here, $x$ is fixed at 10. Using this information, we begin by finding a new function, $f(y) = P(10, y)$. Let $x = 10$ to get

$$f(y) = P(10, y) = 40(10)^2 - 10(10)y + 5y^2 - 80$$
$$= 3920 - 100y + 5y^2.$$

The function $f(y)$ shows the profit from the sale of $y$ recorders, assuming that $x$ is fixed at 10 units. Find the derivative $df/dy$ to get the marginal profit with respect to $y$.

$$\frac{df}{dy} = -100 + 10y$$

*From *Mathematics in Biology* by Duane J. Clow and N. Scott Urquhart. Copyright © 1974 by W. W. Norton & Company, Inc. Used by permission.

In this example, the derivative of the function $f(y)$ was taken with respect to $y$ only; we assumed that $x$ was fixed. To generalize, let $z = f(x, y)$. An intuitive definition of the *partial derivatives* of $f$ with respect to $x$ and $y$ follows.

**PARTIAL DERIVATIVES (INFORMAL DEFINITION)**

> The **partial derivative of $f$ with respect to $x$** is the derivative of $f$ obtained by treating $x$ as a variable and $y$ as a constant.
>
> The **partial derivative of $f$ with respect to $y$** is the derivative of $f$ obtained by treating $y$ as a variable and $x$ as a constant.

The symbols $f_x(x, y)$ (no prime is used), $\partial z/\partial x$, and $\partial f/\partial x$ are used to represent the partial derivative of $z = f(x, y)$ with respect to $x$, with similar symbols used for the partial derivative with respect to $y$. The symbol $\partial z/\partial x$ is used to suggest the notation $dy/dx$ used earlier for derivatives. The symbol $f_x(x, y)$ is often abbreviated as just $f_x$, with $f_y(x, y)$ abbreviated $f_y$.

Generalizing from the definition of derivative given earlier, partial derivatives of a function $z = f(x, y)$ are formally defined as follows.

**PARTIAL DERIVATIVES (FORMAL DEFINITION)**

> Let $z = f(x, y)$ be a function of two independent variables. Let all indicated limits exist. Then the partial derivative of $f$ with respect to $x$ is
>
> $$f_x = \frac{\partial f}{\partial x} = \lim_{h \to 0} \frac{f(x + h, y) - f(x, y)}{h},$$
>
> and the partial derivative of $f$ with respect to $y$ is
>
> $$f_y = \frac{\partial f}{\partial y} = \lim_{h \to 0} \frac{f(x, y + h) - f(x, y)}{h}.$$
>
> If the indicated limits do not exist, then the partial derivatives do not exist.

Similar definitions could be given for functions of more than two independent variables.

Figure 10 shows a surface $z = f(x, y)$ and a plane that is parallel to the $xz$-plane. The equation of the plane is $y = a$. (This corresponds to holding $y$ fixed). Since $y = a$ for points on the plane, any point on the curve that represents the intersection of the plane and the surface must have the form $(x, a, f(x, a))$. Thus, this curve can be described as $z = f(x, a)$. Since $a$ is constant, $z = f(x, a)$ is a function of one variable. When the derivative of $z = f(x, a)$ is evaluated at $x = b$, it gives the slope of the line tangent to this curve at the point $(b, a, f(b, a))$, as shown in Figure 10. Thus, the partial derivative of $f$ with respect to $x$, $f_x(b, a)$, gives the rate of change of the surface $z = f(x, y)$ in the $x$-direction at the point $(b, a, f(b, a))$. In the same way, the partial derivative with respect to $y$ will give the slope of the line tangent to the surface in the $y$-direction at the point $(b, a, f(b, a))$.

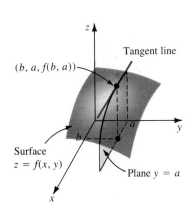

**FIGURE 10**

## EXAMPLE 1

Let $f(x, y) = 4x^2 - 9xy + 6y^3$. Find $f_x$ and $f_y$.

To find $f_x$, treat $y$ as a constant and $x$ as a variable. The derivative of the first term, $4x^2$, is $8x$. In the second term, $-9xy$, the constant coefficient of $x$ is $-9y$, so the derivative with $x$ as the variable is $-9y$. The derivative of $6y^3$ is zero, since we are treating $y$ as a constant. Thus,

$$f_x = 8x - 9y.$$

Now, to find $f_y$, treat $y$ as a variable and $x$ as a constant. Since $x$ is a constant, the derivative of $4x^2$ is zero. In the second term, the coefficient of $y$ is $-9x$ and the derivative of $-9xy$ is $-9x$. The derivative of the third term is $18y^2$. Thus,

$$f_y = -9x + 18y^2. \quad \blacksquare$$

The next example shows how the chain rule can be used to find partial derivatives.

## EXAMPLE 2

Let $f(x, y) = \ln|x^2 + y|$. Find $f_x$ and $f_y$.

Recall the formula for the derivative of a natural logarithm function. If $g(x) = \ln|x|$, then $g'(x) = 1/x$. Using this formula and the chain rule,

$$f_x = \frac{1}{x^2 + y} \cdot D_x(x^2 + y) = \frac{1}{x^2 + y} \cdot 2x = \frac{2x}{x^2 + y},$$

and

$$f_y = \frac{1}{x^2 + y} \cdot D_y(x^2 + y) = \frac{1}{x^2 + y} \cdot 1 = \frac{1}{x^2 + y}. \quad \blacksquare$$

The notation

$$f_x(a, b) \qquad \text{or} \qquad \frac{\partial f}{\partial x}(a, b)$$

represents the value of a partial derivative when $x = a$ and $y = b$, as shown in the next example.

## EXAMPLE 3

Let $f(x, y) = 2x^2 + 3xy^3 + 2y + 5$. Find the following.

(a) $f_x(-1, 2)$

First, find $f_x$ by holding $y$ constant.

$$f_x = 4x + 3y^3$$

Now let $x = -1$ and $y = 2$.

$$f_x(-1, 2) = 4(-1) + 3(2)^3 = -4 + 24 = 20$$

**(b)** $\dfrac{\partial f}{\partial y}(-4, -3)$

Since $\partial f/\partial y = 9xy^2 + 2,$

$$\frac{\partial f}{\partial y}(-4, -3) = 9(-4)(-3)^2 + 2 = 9(-36) + 2 = -322. \quad \blacksquare$$

### ▬ EXAMPLE 4

Suppose that the temperature of the water at the point on a river where a nuclear power plant discharges its hot waste water is approximated by

$$T(x, y) = 2x + 5y + xy - 40,$$

where $x$ represents the temperature of the river water in degrees Celsius before it reaches the power plant and $y$ is the number of megawatts (in hundreds) of electricity being produced by the plant.

**(a)** Find the temperature of the discharged water if the water reaching the plant has a temperature of 8°C and if 300 megawatts of electricity are being produced.

Here $x = 8$ and $y = 3$ (since 300 megawatts of electricity are being produced), with

$$T(8, 3) = 2(8) + 5(3) + 8(3) - 40 = 15.$$

The water at the outlet of the plant has a temperature of 15°C.

**(b)** Find and interpret $T_x(9, 5)$.

First, find the partial derivative $T_x$.

$$T_x = 2 + y$$

This partial derivative gives the rate of change of $T$ with respect to $x$. Replacing $x$ with 9 and $y$ with 5 gives

$$T_x(9, 5) = 2 + 5 = 7.$$

Just as marginal cost is the approximate cost of one more item, this result, 7, is the approximate change in temperature of the output water if input water temperature changes by 1 degree, from $x = 9$ to $x = 9 + 1 = 10$, while $y$ remains constant at 5 (500 megawatts) of electricity produced.

**(c)** Find and interpret $T_y(9, 5)$.

The partial derivative $T_y$ is

$$T_y = 5 + x.$$

This partial derivative gives the rate of change of $T$ with respect to $y$ as

$$T_y(9, 5) = 5 + 9 = 14.$$

This result, 14, is the approximate change in temperature resulting from a 1-unit increase in production of electricity from $y = 5$ to $y = 5 + 1 = 6$ (from 500 to 600 megawatts), while the input water temperature $x$ remains constant at 9°C. ■

As mentioned in the previous section, if $P(x, y)$ gives the output $P$ produced by $x$ units of labor and $y$ units of capital, $P(x, y)$ is a production function. The partial derivatives of this production function have practical implications. For example, $\partial P/\partial x$ gives the marginal productivity of labor. This represents the rate at which the output is changing with respect to a one-unit change in labor for a fixed capital investment. That is, if the capital investment is held constant and labor is increased by 1 work hour, $\partial P/\partial x$ will yield the approximate change in the production level. Likewise, $\partial P/\partial y$ gives the marginal productivity of capital, which represents the rate at which the output is changing with respect to a one-unit change in capital for a fixed labor value. So if the labor force is held constant and the capital investment is increased by 1 unit, $\partial P/\partial y$ will approximate the corresponding change in the production level.

## ■ EXAMPLE 5

A company that manufactures computers has determined that its production function is given by

$$P(x, y) = 500x + 800y + 3x^2y - x^3 - \frac{y^4}{4},$$

where $x$ is the size of the labor force (measured in work hours per week) and $y$ is the amount of capital (measured in units of $1000) invested. Find the marginal productivity of labor and capital when $x = 50$ and $y = 20$, and interpret the results.

The marginal productivity of labor is found by taking the derivative of $P$ with respect to $x$.

$$\frac{\partial P}{\partial x} = 500 + 6xy - 3x^2$$

$$\frac{\partial P}{\partial x}(50, 20) = 500 + 6(50)(20) - 2(50)^2$$

$$= 1500$$

Thus, if the capital investment is held constant at $20,000 and labor is increased from 50 to 51 work hours per week, production will increase by about 1500 units. In the same way, the marginal productivity of capital is $\partial P/\partial y$.

$$\frac{\partial P}{\partial y} = 800 + 3x^2 - y^3$$

$$\frac{\partial P}{\partial y}(50, 20) = 800 + 3(50)^2 - (20)^3$$

$$= 300$$

If work hours are held constant at 50 hours per week and the capital investment is increased from $20,000 to $21,000, production will increase by about 300 units. ▬

▬ EXAMPLE 6

If an isosceles triangle has two equal sides of length $a$ and a third side of length $b$, then its area is given by

$$f(a, b) = \frac{1}{4}b\sqrt{4a^2 - b^2}.$$

**(a)** Find and interpret $f_a$ and $f_b$.

$$
\begin{aligned}
f_a &= \frac{1}{4}b \cdot \frac{1}{2}(4a^2 - b^2)^{-1/2}(8a) \\
&= \frac{ab}{\sqrt{4a^2 - b^2}}
\end{aligned}
$$

$$
\begin{aligned}
f_b &= \frac{1}{4}b \cdot \frac{1}{2}(4a^2 - b^2)^{-1/2}(-2b) + \frac{1}{4}(4a^2 - b^2)^{1/2} \\
&= \frac{1}{4}(4a^2 - b^2)^{-1/2}(-b^2 + 4a^2 - b^2) \\
&= \frac{-2b^2 + 4a^2}{4(4a^2 - b^2)^{1/2}} \\
&= \frac{2a^2 - b^2}{2\sqrt{4a^2 - b^2}}
\end{aligned}
$$

The partial derivative $f_a$ gives the rate at which the area is changing per unit of change in the length of the equal sides if the third side is held constant. That is, if $a$ changes by 1 unit, while $b$ remains constant, then the area will change by approximately $ab/\sqrt{4a^2 - b^2}$. The partial derivative $f_b$ is interpreted similarly as the approximate change in area corresponding to a one-unit change in the side with measure $b$, while $a$ is held constant.

**(b)** If the bottom of a planter is to be made in the shape of an isosceles triangle with the two equal sides 3 feet long and the third side 2 feet long, what will the area be? What would be the approximate effect on the area if the third side were increased to 3 feet?

The area is

$$
\begin{aligned}
f(3, 2) &= \frac{1}{4}(2)\sqrt{4(3)^2 - (2)^2} \\
&= \frac{1}{2}\sqrt{32} \\
&\approx 2.8,
\end{aligned}
$$

or 2.8 square feet. If the third side, with measure $b$, is increased from 2 to 3 feet, an increase of 1 foot, the area will be increased by approximately

$$f_b(3, 2) = \frac{2(3)^2 - (2)^2}{2\sqrt{4(3)^2 - (2)^2}} = \frac{7}{\sqrt{32}} \approx 1.2,$$

or 1.2 square feet.  ■

***Second-Order Partial Derivatives***    Earlier, we saw with functions of the form $y = f(x)$ that the second derivative could be used to decide if a critical number gave a relative maximum or a relative minimum. Second-order partial derivatives are used for the same purpose with functions of the form $z = f(x,y)$. We saw that a second derivative is the derivative of the first derivative. A similar method can be used to find **second-order partial derivatives,** defined as follows.

**SECOND-ORDER PARTIAL DERIVATIVES**

■ For a function $z = f(x, y)$, if the indicated partial derivative exists, then

$$\frac{\partial}{\partial x}\left(\frac{\partial z}{\partial x}\right) = \frac{\partial^2 z}{\partial x^2} = f_{xx}; \qquad \frac{\partial}{\partial y}\left(\frac{\partial z}{\partial y}\right) = \frac{\partial^2 z}{\partial y^2} = f_{yy};$$

$$\frac{\partial}{\partial y}\left(\frac{\partial z}{\partial x}\right) = \frac{\partial^2 z}{\partial y \partial x} = f_{xy}; \qquad \frac{\partial}{\partial x}\left(\frac{\partial z}{\partial y}\right) = \frac{\partial^2 z}{\partial x \partial y} = f_{yx}.$$

As seen above, $f_{xx}$ is used as an abbreviation for $f_{xx}(x, y)$, with $f_{yy}$, $f_{xy}$, and $f_{yx}$ used in a similar way. The symbol $f_{xx}$ is read "the partial derivative of $f_x$ with respect to $x$." Also, the symbol $\partial^2 z/\partial y^2$ is read "the partial derivative of $\partial z/\partial y$ with respect to $y$."

Be careful with a mixed partial derivative such as

$$\frac{\partial^2 z}{\partial y \partial x}.$$

This symbol says to start with a function $z = f(x, y)$ and find $f_x$ or $\partial z/\partial x$, the partial derivative with respect to $x$. Then find the derivative of $f_x$ (or $\partial z/\partial x$) with respect to $y$, producing $f_{xy}$ (or $\partial^2 z/\partial y \partial x$). Based on this order of finding partial derivatives,

$$\frac{\partial^2 z}{\partial y \partial x} = f_{xy}.$$

Notice that the order of $x$ and $y$ in the two symbols is reversed.

#### ▰▰ EXAMPLE 7

Find all second-order partial derivatives for

$$f(x, y) = -4x^3 - 3x^2y^3 + 2y^2.$$

First find $f_x$ and $f_y$.

$$f_x = -12x^2 - 6xy^3 \quad \text{and} \quad f_y = -9x^2y^2 + 4y$$

To find $f_{xx}$, take the partial derivative of $f_x$ with respect to $x$.

$$f_{xx} = -24x - 6y^3$$

Take the partial derivative of $f_y$, with respect to $y$; this gives $f_{yy}$.

$$f_{yy} = -18x^2y + 4$$

Find $f_{xy}$ by starting with $f_x$, then taking the partial derivative of $f_x$ with respect to $y$.

$$f_{xy} = -18xy^2$$

Finally, find $f_{yx}$ by starting with $f_y$; take its partial derivative with respect to $x$.

$$f_{yx} = -18xy^2 \quad ▰▰$$

#### ▰▰ EXAMPLE 8

Let $f(x, y) = 2e^x - 8x^3y^2$. Find all second-order partial derivatives.

Here $f_x = 2e^x - 24x^2y^2$ and $f_y = -16x^3y$. [Recall: If $g(x) = e^x$, then $g'(x) = e^x$.] Now find the second-order partial derivatives.

$$f_{xx} = 2e^x - 48xy^2 \qquad f_{xy} = -48x^2y$$
$$f_{yy} = -16x^3 \qquad f_{yx} = -48x^2y \quad ▰▰$$

In both examples of second-order partial derivatives, $f_{xy} = f_{yx}$. It can be proved that this happens for many functions, including most functions found in applications, as well as all the functions in this book.

### ▰▰ 7.2 EXERCISES

**1.** Let $z = f(x, y) = 12x^2 - 8xy + 3y^2$. Find each of the following.

(a) $\dfrac{\partial z}{\partial x}$  (b) $\dfrac{\partial z}{\partial y}$  (c) $\dfrac{\partial f}{\partial x}(2, 3)$  (d) $f_y(1, -2)$

**2.** Let $z = g(x, y) = 5x + 9x^2y + y^2$. Find each of the following.

(a) $\dfrac{\partial g}{\partial x}$  (b) $\dfrac{\partial g}{\partial y}$  (c) $\dfrac{\partial z}{\partial y}(-3, 0)$  (d) $g_x(2, 1)$

*In Exercises 3–16, find $f_x$ and $f_y$. Then find $f_x(2, -1)$ and $f_y(-4, 3)$. Leave the answers in terms of e in Exercises 7–10 and 15–16.*

**3.** $f(x, y) = -2xy + 6y^3 + 2$

**4.** $f(x, y) = 4x^2y - 9y^2$

**5.** $f(x, y) = 3x^3y^2$

**6.** $f(x, y) = -2x^2y^4$

**7.** $f(x, y) = e^{x+y}$

**8.** $f(x, y) = 3e^{2x+y}$

**9.** $f(x, y) = -5e^{3x-4y}$

**10.** $f(x, y) = 8e^{7x-y}$

**11.** $f(x, y) = \dfrac{x^2 + y^3}{x^3 - y^2}$

**12.** $f(x, y) = \dfrac{3x^2y^3}{x^2 + y^2}$

**13.** $f(x, y) = \ln|1 + 3x^2y^3|$

**14.** $f(x, y) = \ln|2x^5 - xy^4|$

**15.** $f(x, y) = xe^{x^2y}$

**16.** $f(x, y) = y^2e^{(x+3y)}$

*Find all second-order partial derivatives for the following.*

**17.** $f(x, y) = 6x^3y - 9y^2 + 2x$

**18.** $g(x, y) = 5xy^4 + 8x^3 - 3y$

**19.** $R(x, y) = 4x^2 - 5xy^3 + 12y^2x^2$

**20.** $h(x, y) = 30y + 5x^2y + 12xy^2$

**21.** $r(x, y) = \dfrac{4x}{x + y}$

**22.** $k(x, y) = \dfrac{-5y}{x + 2y}$

**23.** $z = 4xe^y$

**24.** $z = -3ye^x$

**25.** $r = \ln|x + y|$

**26.** $k = \ln|5x - 7y|$

**27.** $z = x \ln|xy|$

**28.** $z = (y + 1) \ln|x^3y|$

*For the functions defined as follows, find values of x and y such that both $f_x(x, y) = 0$ and $f_y(x, y) = 0$.*

**29.** $f(x, y) = 6x^2 + 6y^2 + 6xy + 36x - 5$

**30.** $f(x, y) = 50 + 4x - 5y + x^2 + y^2 + xy$

**31.** $f(x, y) = 9xy - x^3 - y^3 - 6$

**32.** $f(x, y) = 2200 + 27x^3 + 72xy + 8y^2$

*Find $f_x$, $f_y$, $f_z$, and $f_{yz}$ for the following.*

**33.** $f(x, y, z) = x^2 + yz + z^4$

**34.** $f(x, y, z) = 3x^5 - x^2 + y^5$

**35.** $f(x, y, z) = \dfrac{6x - 5y}{4z + 5}$

**36.** $f(x, y, z) = \dfrac{2x^2 + xy}{yz - 2}$

**37.** $f(x, y, z) = \ln|x^2 - 5xz^2 + y^4|$

**38.** $f(x, y, z) = \ln|8xy + 5yz - x^3|$

---

### ▉ APPLICATIONS

#### BUSINESS AND ECONOMICS

**Manufacturing Cost**   **39.** Suppose that the manufacturing cost of a precision electronic calculator is approximated by

$$M(x, y) = 40x^2 + 30y^2 - 10xy + 30,$$

where $x$ is the cost of electronic chips and $y$ is the cost of labor. Find the following.

**(a)** $M_y(4, 2)$     **(b)** $M_x(3, 6)$     **(c)** $(\partial M/\partial x)_{(2,5)}$     **(d)** $(\partial M/\partial y)_{(6,7)}$

**Revenue**   **40.** The revenue from the sale of $x$ units of a tranquilizer and $y$ units of an antibiotic is given by

$$R(x, y) = 5x^2 + 9y^2 - 4xy.$$

(a) Suppose $x = 9$ and $y = 5$. What is the approximate effect on revenue if $x$ is increased to 10, while $y$ is fixed?

(b) Suppose $x = 9$ and $y = 5$. What is the approximate effect on revenue if $y$ is increased to 6, while $x$ is fixed?

**Sales**   **41.** A car dealership estimates that the total weekly sales of its most popular model is a function of the car's list price $p$, and the interest rate in percent, $i$, offered by the manufacturer. The approximate weekly sales are given by

$$f(p, i) = 132p - 2pi - .01p^2.$$

(a) Find the weekly sales if the average list price is $9400 and the manufacturer is offering an 8% interest rate.

(b) Find and interpret $f_p$ and $f_i$.

(c) What would be the effect on weekly sales if the price is $9400 and interest rates rise from 8% to 9%?

**Marginal Productivity**   **42.** Suppose the production function of a company is given by

$$P(x, y) = 100\sqrt{x^2 + y^2},$$

where $x$ represents units of labor and $y$ represents units of capital. Find the following when $x = 4$ and $y = 3$.

(a) The marginal productivity of labor

(b) The marginal productivity of capital

**Marginal Productivity**   **43.** A manufacturer estimates that production (in hundreds of units) is a function of the amounts $x$ and $y$ of labor and capital used, as follows.

$$f(x, y) = \left[ \frac{1}{3}x^{-1/3} + \frac{2}{3}y^{-1/3} \right]^{-3}$$

(a) Find the number of units produced when 27 units of labor and 64 units of capital are utilized.

(b) Find and interpret $f_x(27, 64)$ and $f_y(27, 64)$.

(c) What would be the approximate effect on production of increasing labor by 1 unit?

**Marginal Productivity**   **44.** The production function $z$ for the United States was once estimated as

$$z = x^{.7}y^{.3},$$

where $x$ stands for the amount of labor and $y$ the amount of capital. Find the marginal productivity of labor and of capital.

**Marginal Productivity**   **45.** A similar production function for Canada is

$$z = x^{.4}y^{.6},$$

with $x$, $y$, and $z$ as in Exercise 44. Find the marginal productivity of labor and of capital.

**Marginal Productivity**   **46.** A manufacturer of automobile batteries estimates that his total production in thousands of units is given by

$$f(x, y) = 3x^{1/3}y^{2/3},$$

where $x$ is the number of units of labor and $y$ is the number of units of capital utilized.

(a) Find and interpret $f_x(64, 125)$ and $f_y(64, 125)$, if the current level of production uses 64 units of labor and 125 units of capital.

**(b)** What would be the approximate effect on production of increasing labor to 65 units while holding capital at the current level?

**(c)** Suppose that sales have been good and management wants to increase either capital or labor by 1 unit. Which option would result in a larger increase in production?

## LIFE SCIENCES

**Grasshopper Matings**    **47.** The total number of matings per day between individuals of a certain species of grasshoppers is approximated by

$$M(x, y) = 2xy + 10xy^2 + 30y^2 + 20,$$

where $x$ represents the temperature in °C and $y$ represents the number of days since the last rainfall. Find the following.

**(a)** $(\partial M/\partial x)_{(20,4)}$    **(b)** $(\partial M/\partial y)_{(24,10)}$    **(c)** $M_x(17, 3)$    **(d)** $M_y(21, 8)$

*The next two exercises continue the discussions in Exercise 43 and 44 from Section 1.*

**Oxygen Consumption**    **48.** The oxygen consumption of a well-insulated mammal that is not sweating is approximated by

$$m = m(T, F, w) = \frac{2.5(T - F)}{w^{.67}} = 2.5(T - F)w^{-.67},$$

where $T$ is the internal body temperature of the animal (in °C), $F$ is the temperature of the outside of the animal's fur (in °C), and $w$ is the animal's weight in kilograms.

**(a)** Find $m_T$.    **(b)** Find $m_T(38, 12, 30)$.
**(c)** Find $m_w$.    **(d)** Find $m_w(40, 20, 40)$.
**(e)** Find $m_F$.    **(f)** Find $m_F(36, 14, 25)$.

**Body Surface Area**    **49.** The surface area of a human, in square meters, is approximated by

$$A(W, H) = .202W^{.425}H^{.725},$$

where $W$ is the weight of the person in kilograms and $H$ is the height in meters.

**(a)** Find $\partial A/\partial W$.    **(b)** Find $(\partial A/\partial W)_{(72,1.8)}$.
**(c)** Find $\partial A/\partial H$.    **(d)** Find $(\partial A/\partial H)_{(70,1.6)}$.

**Blood Flow**    **50.** In one method of computing the quantity of blood pumped through the lungs in one minute, a researcher first finds each of the following (in milliliters).

$b$ = quantity of oxygen used by body in one minute
$a$ = quantity of oxygen per liter of blood that has just gone through the lungs
$v$ = quantity of oxygen per liter of blood that is about to enter the lungs

In one minute,

> Amount of Oxygen Used = Amount of Oxygen per Liter
> × Liters of Blood Pumped.

If $C$ is the number of liters of blood pumped through the lungs in one minute, then

$$b = (a - v) \cdot C \quad \text{or} \quad C = \frac{b}{a - v}.$$

**(a)** Find $C$ if $a = 160$, $b = 200$, and $v = 125$.

**(b)** Find $C$ if $a = 180$, $b = 260$, and $v = 142$.

Find the following partial derivatives.

**(c)** $\partial C/\partial b$    **(d)** $\partial C/\partial v$

**Health**     **51.** A weight-loss counselor has prepared a program of diet and exercise for a client. If the client sticks to the program, the weight loss that can be expected in pounds per week is given by

$$\text{Weight Loss} = f(n, c) = \frac{1}{8}n^2 - \frac{1}{5}c + \frac{1937}{8},$$

where $c$ is the average daily calorie intake for the week and $n$ is the number of 40-min aerobic workouts per week.

(a) How many pounds can the client expect to lose by eating an average of 1200 cal per day and participating in four 40-min workouts in a week?

(b) Find and interpret $\partial f / \partial n$.

(c) The client currently averages 1100 cal per day and does three 40-min workouts each week. What would be the approximate impact on weekly weight loss of adding a fourth workout per week?

**Drug Reaction**     **52.** The reaction to $x$ units of a drug $t$ hours after it was administered is given by

$$R(x, t) = x^2(a - x)t^2 e^{-t},$$

for $0 \le x \le a$ (where $a$ is a constant). Find the following.

(a) $\dfrac{\partial R}{\partial x}$     (b) $\dfrac{\partial R}{\partial t}$     (c) $\dfrac{\partial^2 R}{\partial x^2}$     (d) $\dfrac{\partial^2 R}{\partial t^2}$     (e) $\dfrac{\partial^2 R}{\partial x \partial t}$     (f) $\dfrac{\partial^2 R}{\partial t \partial x}$

## SOCIAL SCIENCES

*The following problem involves a function of three independent variables. Partial derivatives for such functions are defined in the same way as for functions of two independent variables. For example, given $f(x, y, z)$, we can find $f_x$ by holding $y$ and $z$ constant and differentiating with respect to $x$.*

**Education**     **53.** A developmental mathematics instructor at a large university has determined that a student's probability of success in the university's pass/fail remedial algebra course is a function of $s$, $n$, and $a$, where $s$ is the student's score on the departmental placement exam, $n$ is the number of semesters of mathematics passed in high school, and $a$ is the student's mathematics SAT score. She estimates that $p$, the probability of passing the course, will be

$$p = f(s, n, a) = .003a + .1(sn)^{1/2}$$

for $200 \le a \le 800$, $0 \le s \le 10$, and $0 \le n \le 8$. Assuming that the above model has some merit, find the following.

(a) If a student scores 8 on the placement exam, has taken 6 semesters of high-school math, and has an SAT score of 450, what is the probability of passing the course?

(b) Find $p$ for a student with 3 semesters of high school mathematics, a placement score of 3, and an SAT score of 320.

(c) Find $f_n(3, 3, 320)$ and $f_a(3, 3, 320)$.

## PHYSICAL SCIENCES

**Gravitational Attraction**     **54.** The gravitational attraction $F$ on a body a distance $r$ from the center of the earth, where $r$ is greater than the radius of the earth, is a function of its mass $m$ and the distance $r$ as follows:

$$F = \frac{mgR^2}{r^2},$$

where $R$ is the radius of the earth and $g$ is the force of gravity—about 32 ft/sec$^2$.

(a) Find and interpret $F_m$ and $F_r$.

(b) Show that $F_m > 0$ and $F_r < 0$. Why is this reasonable?

**Light Intensity**    **55.** The light intensity in water at a depth of $x$ ft below the surface $t$ sec after dawn is given by

$$i(x,\ t) = (e^{-ax})\left(\frac{N}{1 + be^{-kt}}\right),$$

for constants $a$. $N$, $b$, and $k$. Find the following.

(a) $\dfrac{\partial i}{\partial x}$    (b) $\dfrac{\partial i}{\partial t}$    (c) $\dfrac{\partial^2 i}{\partial x^2}$    (d) $\dfrac{\partial^2 i}{\partial t^2}$    (e) $\dfrac{\partial^2 i}{\partial x \partial t}$    (f) $\dfrac{\partial^2 i}{\partial t \partial x}$

# 7.3 MAXIMA AND MINIMA

One of the most important applications of calculus is in finding maxima and minima for functions. Earlier, we studied this idea extensively for functions of a single independent variable; now we will see that extrema can be found for functions of two variables. In particular, an extension of the second derivative test can be defined and used to identify maxima or minima. We begin with the definitions of relative maxima and minima.

**RELATIVE MAXIMA AND MINIMA**

> Let $(a,\ b)$ be the center of a circular region contained in the $xy$-plane. Then, for a function $z = f(x,\ y)$ defined for every $(x,\ y)$ in the region, $f(a,\ b)$ is a **relative maximum** if
>
> $$f(a,\ b) \geq f(x,\ y)$$
>
> for all points $(x,\ y)$ in the circular region, and $f(a,\ b)$ is a **relative minimum** if
>
> $$f(a,\ b) \leq f(x,\ y)$$
>
> for all points $(x,\ y)$ in the circular region.

As before, the word *extremum* is used for either a relative maximum or a relative minimum. Examples of a relative maximum and a relative minimum are given in Figures 11 and 12.

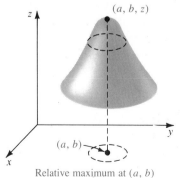

Relative maximum at $(a, b)$

**FIGURE 11**

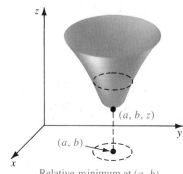

Relative minimum at $(a, b)$

**FIGURE 12**

When functions of a single variable were discussed, a distinction was made between relative extrema and absolute extrema. The methods for finding absolute extrema are quite involved for functions of two variables, so we will discuss only relative extrema here. In most practical applications the relative extrema coincide with the absolute extrema.

As suggested by Figure 13, at a relative maximum the tangent line parallel to the x-axis has a slope of 0, as does the tangent line parallel to the y-axis. (Notice the similarity to functions of one variable.) That is, if the function $z = f(x, y)$ has a relative extremum at $(a, b)$, then $f_x(a, b) = 0$ and $f_y(a, b) = 0$, as stated in the next theorem.

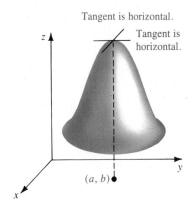

Tangent is horizontal.

Tangent is horizontal.

**FIGURE 13**

**LOCATION OF EXTREMA**

Let a function $z = f(x, y)$ have a relative maximum or relative minimum at the point $(a, b)$. Let $f_x(a, b)$ and $f_y(a, b)$ both exist. Then

$$f_x(a, b) = 0 \quad \text{and} \quad f_y(a, b) = 0.$$

Just as with functions of one variable, the fact that the slopes of the tangent lines are 0 is no guarantee that a relative extremum has been located. For example, Figure 14 shows the graph of $z = f(x, y) = x^2 - y^2$. Both $f_x(0, 0) = 0$ and $f_y(0, 0) = 0$, and yet $(0, 0)$ leads to neither a relative maximum nor a relative minimum for the function. The point $(0, 0, 0)$ on the graph of this function is called a **saddle point**; it is a minimum when approached from one direction but a maximum when approached from another direction. A saddle point is neither a maximum nor a minimum.

The theorem on location of extrema suggests a useful strategy for finding extrema. First, locate all points $(a, b)$ where $f_x(a, b) = 0$ and $f_y(a, b) = 0$. Then test each of these points separately, using the test given after the next example. For a function $f(x, y)$, the points $(a, b)$ such that $f_x(a, b) = 0$ and $f_y(a, b) = 0$ are called **critical points.**

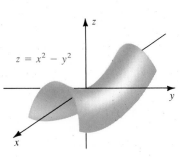

$z = x^2 - y^2$

**FIGURE 14**

### ▬▬ EXAMPLE 1

Find all critical points for

$$f(x, y) = 6x^2 + 6y^2 + 6xy + 36x - 5.$$

Find all points $(a, b)$ such that $f_x(a, b) = 0$ and $f_y(a, b) = 0$. Here

$$f_x = 12x + 6y + 36 \quad \text{and} \quad f_y = 12y + 6x.$$

Place each of these two partial derivatives equal to 0.

$$12x + 6y + 36 = 0 \quad \text{and} \quad 12y + 6x = 0$$

These two equations make up a system of linear equations. We can use the substitution method to solve this system. First, rewrite $12y + 6x = 0$ as follows:

$$12y + 6x = 0$$
$$6x = -12y$$
$$x = -2y.$$

Now substitute $-2y$ for $x$ in the other equation.

$$12x + 6y + 36 = 0$$
$$12(-2y) + 6y + 36 = 0$$
$$-24y + 6y + 36 = 0$$
$$-18y + 36 = 0$$
$$-18y = -36$$
$$y = 2$$

From the equation $x = -2y$, $x = -2(2) = -4$, and the solution of the system of equations is $(-4, 2)$. Since this is the only solution of the system, $(-4, 2)$ is the only critial point for the given function. By the theorem above, if the function has a relative extremum, it will occur at $(-4, 2)$. ▬▬

The results of the next theorem can be used to decide whether $(-4, 2)$ in Example 1 leads to a relative maximum, a relative minimum, or neither.

**TEST FOR RELATIVE EXTREMA**

▬▬ For a function $z = f(x, y)$, let $f_{xx}$, $f_{yy}$, and $f_{xy}$ all exist. Let $(a, b)$ be a point for which

$$f_x(a, b) = 0 \quad \text{and} \quad f_y(a, b) = 0.$$

Define the number $D$ by

$$D = f_{xx}(a, b) \cdot f_{yy}(a, b) - [f_{xy}(a, b)]^2.$$

Then

(a) $f(a, b)$ is a relative maximum if $D > 0$ and $f_{xx}(a, b) < 0$;
(b) $f(a, b)$ is a relative minimum if $D > 0$ and $f_{xx}(a, b) > 0$;
(c) $f(a, b)$ is a saddle point (neither a maximum nor a minimum) if $D < 0$;
(d) if $D = 0$, the test gives no information.

The chart below summarizes the conclusions of the theorem.

|  | $f_{xx}(a, b) < 0$ | $f_{xx}(a, b) > 0$ |
|---|---|---|
| $D > 0$ | Relative Maximum | Relative Minimum |
| $D = 0$ | No Information | |
| $D < 0$ | Saddle Point | |

### ■ EXAMPLE 2

The previous example showed that the only critical point for the function

$$f(x, y) = 6x^2 + 6y^2 + 6xy + 36x - 5$$

is $(-4, 2)$. Does $(-4, 2)$ lead to a relative maximum, a relative minimum, or neither?

Find out by using the test above. From Example 1,

$$f_x(-4, 2) = 0 \quad \text{and} \quad f_y(-4, 2) = 0.$$

Now find the various second partial derivatives used in finding $D$. From $f_x = 12x + 6y + 36$ and $f_y = 12y + 6x$,

$$f_{xx} = 12, \quad f_{yy} = 12, \quad \text{and} \quad f_{xy} = 6.$$

(If these second-order partial derivatives had not all been constants, they would have had to be evaluated at the point $(-4, 2)$.) Now

$$D = f_{xx}(-4, 2) \cdot f_{yy}(-4, 2) - [f_{xy}(-4, 2)]^2 = 12 \cdot 12 - 6^2 = 108.$$

Since $D > 0$ and $f_{xx}(-4, 2) = 12 > 0$, part (b) of the theorem applies, showing that $f(x, y) = 6x^2 + 6y^2 + 6xy + 36x - 5$ has a relative minimum at $(-4, 2)$. This relative minimum is $f(-4, 2) = -77$. ■

### ■ EXAMPLE 3

Find all points where the function

$$f(x, y) = 9xy - x^3 - y^3 - 6$$

has any relative maxima or relative minima.

First find any critical points. Here

$$f_x = 9y - 3x^2 \quad \text{and} \quad f_y = 9x - 3y^2.$$

Place each of these partial derivatives equal to 0.

$$
\begin{array}{cc}
f_x = 0 & f_y = 0 \\
9y - 3x^2 = 0 & 9x - 3y^2 = 0 \\
9y = 3x^2 & 9x = 3y^2 \\
3y = x^2 & 3x = y^2
\end{array}
$$

In the first equation ($3y = x^2$), notice that since $x^2 \geq 0$, $y \geq 0$. Also, in the second equation ($3x = y^2$), $y^2 \geq 0$, so $x \geq 0$.

The substitution method can be used again to solve the system of equations

$$3y = x^2$$
$$3x = y^2.$$

The first equation, $3y = x^2$, can be rewritten as $y = x^2/3$. Substitute this into the second equation to get

$$3x = y^2 = \left(\frac{x^2}{3}\right)^2 = \frac{x^4}{9}.$$

Solve this equation as follows.

$$27x = x^4$$
$$x^4 - 27x = 0$$
$$x(x^3 - 27) = 0 \qquad \text{Factor}$$
$$x = 0 \quad \text{or} \quad x^3 - 27 = 0 \qquad \text{Set each factor equal to 0}$$
$$x^3 = 27$$
$$x = 3 \qquad \text{Take the cube root on each side}$$

Use these values of $x$, along with the equation $3x = y^2$, to find $y$.

If $x = 0$, 

$$3x = y^2$$
$$3(0) = y^2$$
$$0 = y^2$$
$$0 = y$$

If $x = 3$,

$$3x = y^2$$
$$3(3) = y^2$$
$$9 = y^2$$
$$3 = y \quad \text{or} \quad -3 = y$$

The points $(0, 0)$, $(3, 3)$, and $(3, -3)$ appear to be critical points; however, $(3, -3)$ does not satisfy $y \geq 0$. The only possible relative extrema for $f(x, y) = 9xy - x^3 - y^3 - 6$ occur at the critical points $(0, 0)$ or $(3, 3)$. To identify any extrema, use the test. Here

$$f_{xx} = -6x, \qquad f_{yy} = -6y, \qquad \text{and} \qquad f_{xy} = 9.$$

Test each of the possible critical points.

For $(0, 0)$:

$f_{xx}(0, 0) = -6(0) = 0$

$f_{yy}(0, 0) = -6(0) = 0$

$f_{xy}(0, 0) = 9$

$D = 0 \cdot 0 - 9^2 = -81$

Since $D < 0$, there is a saddle point at $(0, 0)$.

For $(3, 3)$:

$f_{xx}(3, 3) = -6(3) = -18$

$f_{yy}(3, 3) = -6(3) = -18$

$f_{xy}(3, 3) = 9$

$D = -18(-18) - 9^2 = 243$

Here $D > 0$ and $f_{xx}(3, 3) = -18 < 0$; there is a relative maximum at $(3, 3)$. ▬

## ▬ EXAMPLE 4

A company is developing a new soft drink. The cost in dollars to produce a batch of the drink is approximated by

$$C(x, y) = 2200 + 27x^3 - 72xy + 8y^2,$$

where $x$ is the number of kilograms of sugar per batch and $y$ is the number of grams of flavoring per batch.

**(a)** Find the amounts of sugar and flavoring that result in minimum cost per batch.

Start with the following partial derivatives.

$$C_x = 81x^2 - 72y \qquad \text{and} \qquad C_y = -72x + 16y$$

Set each of these equal to 0 and solve for $y$.

$$81x^2 - 72y = 0 \qquad\qquad -72x + 16y = 0$$
$$-72y = -81x^2 \qquad\qquad 16y = 72x$$
$$y = \frac{9}{8}x^2 \qquad\qquad y = \frac{9}{2}x$$

From the equation on the left, $y \geq 0$. Since $(9/8)x^2$ and $(9/2)x$ both are equal to $y$, they are equal to each other. Place $(9/8)x^2$ and $(9/2)x$ equal, and solve the resulting equation for $x$.

$$\frac{9}{8}x^2 = \frac{9}{2}x$$
$$9x^2 = 36x$$
$$9x^2 - 36x = 0$$
$$9x(x - 4) = 0$$
$$9x = 0 \qquad \text{or} \qquad x - 4 = 0$$

The equation $9x = 0$ leads to $x = 0$, which is not a useful answer for the problem. Substitute $x = 4$ into $y = (9/2)x$ to find $y$.

$$y = \frac{9}{2}x = \frac{9}{2}(4) = 18$$

Now check to see if the critical point $(4, 18)$ leads to a relative minimum. For $(4, 18)$,

$$C_{xx} = 162x = 162(4) = 648, \qquad C_{yy} = 16, \qquad \text{and} \qquad C_{xy} = -72.$$

Also,

$$D = (648)(16) - (-72)^2 = 5184.$$

Since $D > 0$ and $C_{xx}(4, 18) > 0$, the cost at $(4, 18)$ is a minimum.

**(b)** What is the minimum cost?

To find the minimum cost, go back to the cost function and evaluate $C(4, 18)$.

$$C(x, y) = 2200 + 27x^3 - 72xy + 8y^2$$
$$C(4, 18) = 2200 + 27(4)^3 - 72(4)(18) + 8(18)^2 = 1336$$

The minimum cost for a batch of soft drink is $1336.00. ▬

## ▬ 7.3 EXERCISES

*Find all points where the functions defined as follows have any relative extrema.*
*Identify any saddle points.*

**1.** $f(x, y) = xy + x - y$

**2.** $f(x, y) = 4xy + 8x - 9y$

**3.** $f(x, y) = x^2 - 2xy + 2y^2 + x - 5$

**4.** $f(x, y) = x^2 + xy + y^2 - 6x - 3$

**5.** $f(x, y) = x^2 - xy + y^2 + 2x + 2y + 6$

**6.** $f(x, y) = x^2 + xy + y^2 + 3x - 3y$

**7.** $f(x, y) = x^2 + 3xy + 3y^2 - 6x + 3y$

**8.** $f(x, y) = 5xy - 7x^2 - y^2 + 3x - 6y - 4$

**9.** $f(x, y) = 4xy - 10x^2 - 4y^2 + 8x + 8y + 9$

**10.** $f(x, y) = x^2 + xy + 3x + 2y - 6$

**11.** $f(x, y) = x^2 + xy - 2x - 2y + 2$

**12.** $f(x, y) = x^2 + xy + y^2 - 3x - 5$

**13.** $f(x, y) = x^2 - y^2 - 2x + 4y - 7$

**14.** $f(x, y) = 4x + 2y - x^2 + xy - y^2 + 3$

**15.** $f(x, y) = 2x^3 + 3y^2 - 12xy + 4$

**16.** $f(x, y) = 5x^3 + 2y^2 - 60xy - 3$

**17.** $f(x, y) = x^2 + 4y^3 - 6xy - 1$

**18.** $f(x, y) = 3x^2 + 7y^3 - 42xy + 5$

**19.** $f(x, y) = e^{xy}$

**20.** $f(x, y) = x^2 + e^y$

*Figures (a)–(f) show the graphs of the functions defined in Exercises 21–26. Find all relative extrema for each function, and then match the equation to its graph.*

**21.** $z = -3xy + x^3 - y^3 + \dfrac{1}{8}$

**22.** $z = \dfrac{3}{2}y - \dfrac{1}{2}y^3 - x^2y + \dfrac{1}{16}$

**23.** $z = y^4 - 2y^2 + x^2 - \dfrac{17}{16}$

**24.** $z = -2x^3 - 3y^4 + 6xy^2 + \dfrac{1}{16}$

**25.** $z = -x^4 + y^4 + 2x^2 - 2y^2 + \dfrac{1}{16}$

**26.** $z = -y^4 + 4xy - 2x^2 + \dfrac{1}{16}$

**(a)**

**(b)**

**(c)**

**(d)**

**(e)**

**(f)**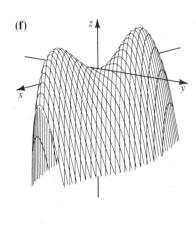

**27.** Show that $f(x, y) = 1 - x^4 - y^4$ has a relative maximum, even though $D$ in the theorem is 0.

**28.** Show that $D = 0$ for $f(x, y) = x^3 + (x - y)^2$ and that the function has no relative extrema.

## ▰ APPLICATIONS

### BUSINESS AND ECONOMICS

**Profit**  **29.** Suppose that the profit of a certain firm in hundreds of dollars is approximated by
$$P(x, y) = 1000 + 24x - x^2 + 80y - y^2,$$
where $x$ is the cost of a unit of labor and $y$ is the cost of a unit of goods. Find values of $x$ and $y$ that maximize profit. Find the maximum profit.

**Labor Costs**  **30.** Suppose the labor cost in dollars for manufacturing a precision camera can be approximated by
$$L(x, y) = \frac{3}{2}x^2 + y^2 - 2x - 2y - 2xy + 68,$$
where $x$ is the number of hours required by a skilled craftsperson and $y$ is the number of hours required by a semiskilled person. Find values of $x$ and $y$ that minimize the labor cost. Find the minimum labor cost.

**Chicken Production**  **31.** The number of chickens that can be fed from $x$ lb of Super chicken feed and $y$ lb of Brand A feed is given by
$$R(x, y) = 800 - 2x^3 + 12xy - y^2.$$
Find the number of pounds of each kind of feed that produces the maximum number of chickens.

**Profit**  **32.** The total profit from one acre of a certain crop depends on the amount spent on fertilizer, $x$, and hybrid seed, $y$, according to the model
$$P(x, y) = -x^2 + 3xy + 160x - 5y^2 + 200y + 2,600.$$
Find values of $x$ and $y$ that lead to maximum profit. Find the maximum profit.

**Cost** **33.** The total cost in dollars to produce $x$ units of electrical tape and $y$ units of packing tape is given by

$$C(x, y) = 2x^2 + 3y^2 - 2xy + 2x - 126y + 3800.$$

Find the number of units of each kind of tape that should be produced so that the total cost is a minimum. Find the minimum total cost.

**Revenue** **34.** The total revenue in hundreds of dollars from the sale of $x$ spas and $y$ solar heaters is approximated by

$$R(x, y) = 12 + 74x + 85y - 3x^2 - 5y^2 - 5xy.$$

Find the number of each that should be sold to produce maximum revenue. Find the maximum revenue.

## 7.4 LAGRANGE MULTIPLIERS

A builder wants to maximize the floor space in a new building while keeping the costs fixed at \$500,000. The building will be 40 feet high, with a rectangular floor plan and three stories. The costs, which depend on the dimensions of the rectangular floor plan, are given by

$$\text{Costs} = xy + 20y + 20x + 474{,}000,$$

where $x$ is the width and $y$ the length of the rectangle. Thus, the builder wishes to maximize the area $A(x) = xy$ and satisfy the condition

$$xy + 20y + 20x + 474{,}000 = 500{,}000.$$

In the previous section, we showed how to find relative extrema using partial derivatives. The builder's problem here is different, since a secondary condition, called a **constraint,** must also be satisfied.

A typical problem of this type might require the smallest possible value of the function $z = x^2 + y^2$, subject to the constraint $x + y = 4$. To see how to find this minimum value, we might first graph both the surface $z = x^2 + y^2$ and the plane $x + y = 4$, as in Figure 15. The necessary minimum value would be found on the curve formed by the intersection of the two graphs.

Problems with constraints often are solved by the method of *Lagrange multipliers,* named for the French mathematician Joseph Louis Lagrange (1736–1813). The proof for the method is complicated and is not given here. The method of Lagrange multipliers is used for problems of the form

Find the relative extrema for $\quad z = f(x, y)$,
subject to $\quad g(x, y) = 0$.

We state the method only for functions of two independent variables, but it is valid for any number of variables.

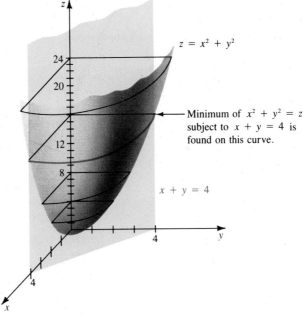

$z = x^2 + y^2$

Minimum of $x^2 + y^2 = z$ subject to $x + y = 4$ is found on this curve.

$x + y = 4$

**FIGURE 15**

**LAGRANGE MULTIPLIERS**

All relative extrema of the function $z = f(x, y)$, subject to a constraint $g(x, y) = 0$, will be found among those points $(x, y)$ for which there exists a value of $\lambda$ such that

$$F_x(x, y, \lambda) = 0, \qquad F_y(x, y, \lambda) = 0, \qquad F_\lambda(x, y, \lambda) = 0,$$

where $$F(x, y, \lambda) = f(x, y) + \lambda \cdot g(x, y),$$

and all indicated partial derivatives exist.

In the theorem, the function $F(x, y, \lambda) = f(x, y) + \lambda \cdot g(x, y)$ is called the Lagrange function; $\lambda$, the Greek letter *lambda*, is called a **Lagrange multiplier.**

### ■ EXAMPLE 1

Find the minimum value of

$$f(x, y) = 5x^2 + 6y^2 - xy,$$

subject to the constraint $x + 2y = 24$.

Go through the following steps.

**1.** Rewrite the constraint in the form $g(x, y) = 0$.

In this example the constraint $x + 2y = 24$ becomes

$$x + 2y - 24 = 0,$$

with $$g(x, y) = x + 2y - 24.$$

2. Form the Lagrange function $F(x, y, \lambda)$, the sum of the function $f(x, y)$ and the product of $\lambda$ and $g(x, y)$.
   Here,

$$
\begin{aligned}
F(x, y, \lambda) &= f(x, y) + \lambda \cdot g(x, y) \\
&= 5x^2 + 6y^2 - xy + \lambda(x + 2y - 24) \\
&= 5x^2 + 6y^2 - xy + \lambda x + 2\lambda y - 24\lambda.
\end{aligned}
$$

3. Find $F_x$, $F_y$, and $F_\lambda$.

$$
\begin{aligned}
F_x &= 10x - y + \lambda \\
F_y &= 12y - x + 2\lambda \\
F_\lambda &= x + 2y - 24
\end{aligned}
$$

4. Form the system of equations $F_x = 0$, $F_y = 0$, and $F_\lambda = 0$.

$$
\begin{aligned}
10x - y + \lambda &= 0 & \text{(1)} \\
12y - x + 2\lambda &= 0 & \text{(2)} \\
x + 2y - 24 &= 0 & \text{(3)}
\end{aligned}
$$

5. Solve the system of equations from Step 4 for $x$, $y$, and $\lambda$.

   A good general approach to solving these systems of equations is to solve each of the first two equations for $\lambda$, then set the two results equal and simplify, as follows.

$$
10x - y + \lambda = 0 \quad \text{becomes} \quad \lambda = -10x + y
$$

$$
12y - x + 2\lambda = 0 \quad \text{becomes} \quad \lambda = \frac{x - 12y}{2}
$$

$$
\begin{aligned}
-10x + y &= \frac{x - 12y}{2} \\
-20x + 2y &= x - 12y \\
-21x &= -14y \\
x &= \frac{2y}{3}
\end{aligned}
$$

Now substitute $(2y)/3$ for $x$ in equation (3).

$$
\begin{aligned}
x + 2y - 24 &= 0 \\
\frac{2y}{3} + 2y - 24 &= 0 \\
2y + 6y - 72 &= 0 \\
8y &= 72 \\
y &= 9
\end{aligned}
$$

Since $x = (2y)/3$ and $y = 9$, $x = 6$. It is not necessary to find the value of $\lambda$.

Thus, the minimum value for $f(x, y) = 5x^2 + 6y^2 - xy$, subject to the constraint $x + 2y = 24$, is at the point $(6, 9)$. The minimum value is $f(6, 9) = 612$. The second derivative test for relative extrema from the previous section can be used to verify that 612 is indeed a minimum: since $f_{xx} = 10$, $f_{yy} = 12$, and $f_{xy} = -1$, $D = 10 \cdot 12 - (-1)^2 > 0$, so part (b) applies and indicates a minimum. ▀

Before looking at applications of Lagrange multipliers, let us summarize the steps involved in solving a problem by this method.

**USING LAGRANGE MULTIPLIERS**

1. Write the constraint in the form $g(x, y) = 0$.
2. Form the Lagrange function
$$F(x, y, \lambda) = f(x, y) + \lambda \cdot g(x, y).$$
3. Find $F_x$, $F_y$, and $F_\lambda$.
4. Form the system of equations
$$F_x = 0, \qquad F_y = 0, \qquad F_\lambda = 0.$$
5. Solve the system in Step 4; the relative extrema for $f$ are among the solutions of the system.

▀ **EXAMPLE 2**

Complete the solution of the problem given in the introduction to this section. Maximize the area, $A(x) = xy$, subject to the cost constraint

$$xy + 20y + 20x + 474{,}000 = 500{,}000.$$

Go through the four steps presented above.

1. $g(x, y) = xy + 20y + 20x - 26{,}000$
2. $F(x, y) = xy + \lambda(xy + 20y + 20x - 26{,}000)$
3. $F_x = y + \lambda y + 20\lambda$
   $F_y = x + \lambda x + 20\lambda$
   $F_\lambda = xy + 20y + 20x - 26{,}000$
4. $y + \lambda y + 20\lambda = 0$                                                    (6)
   $x + \lambda x + 20\lambda = 0$                                                    (7)
   $xy + 20y + 20x - 26{,}000 = 0$                                             (8)
5. Solving equations (6) and (7) for $\lambda$ gives

$$\lambda = \frac{-y}{(y + 20)} \quad \text{and} \quad \lambda = \frac{-x}{(x + 20)}$$

$$\frac{-y}{(y + 20)} = \frac{-x}{(x + 20)}$$
$$y(x + 20) = x(y + 20)$$
$$xy + 20y = xy + 20x$$
$$x = y.$$

Now substitute $y$ for $x$ in equation (8) to get

$$y^2 + 20y + 20y - 26,000 = 0$$
$$y^2 + 40y - 26,000 = 0$$

Use the quadratic formula to find $y \approx 142.5$. Since $x = y$, $x \approx 142.5$. The maximum area of $(142.5)^2 \approx 20,306$ square feet will be achieved if the floor plan is a square with a side of 142.5 feet. ▰

As mentioned earlier, the method of Lagrange multipliers works for more than two independent variables. The next example shows how to find extrema for a function of three independent variables.

**FIGURE 16**

### ▰ EXAMPLE 3

Find the dimensions of the closed rectangular box of maximum volume that can be produced from 6 square feet of material.

Let $x$, $y$, and $z$ represent the dimensions of the box, as shown in Figure 16. The volume of the box is given by

$$f(x, y, z) = xyz.$$

As shown in Figure 16, the total amount of material required for the two ends of the box is $2xy$, the total needed for the sides is $2xz$, and the total needed for the top and bottom is $2yz$. Since 6 square feet of material is available,

$$2xy + 2xz + 2yz = 6 \quad \text{or} \quad xy + xz + yz = 3.$$

In summary, $f(x, y, z) = xyz$ is to be maximized subject to the constraint $xy + xz + yz = 3$. Go through the steps that were given above.

**1.** $g(x, y, z) = xy + xz + yz - 3$

**2.** $F(x, y, z, \lambda) = xyz + \lambda(xy + xz + yz - 3)$

**3.** $F_x = yz + \lambda y + \lambda z$

$\phantom{3.}\; F_y = xz + \lambda x + \lambda z$

$\phantom{3.}\; F_z = xy + \lambda x + \lambda y$

$\phantom{3.}\; F_\lambda = xy + xz + yz - 3$

**4.** $yz + \lambda y + \lambda z = 0$

$\phantom{4.}\; xz + \lambda x + \lambda z = 0$

$\phantom{4.}\; xy + \lambda x + \lambda y = 0$

$\phantom{4.}\; xy + xz + yz - 3 = 0$

**5.** Solve each of the first three equations for $\lambda$. You should get

$$\lambda = \frac{-yz}{y + z}, \qquad \lambda = \frac{-xz}{x + z}, \qquad \text{and} \qquad \lambda = \frac{-xy}{x + y}.$$

Set these expressions for $\lambda$ equal, and simplify as follows.

$$\frac{-yz}{y+z} = \frac{-xz}{x+z} \quad \text{and} \quad \frac{-xz}{x+z} = \frac{-xy}{x+y}$$

$$\frac{y}{y+z} = \frac{x}{x+z} \qquad\qquad \frac{z}{x+z} = \frac{y}{x+y}$$

$$xy + yz = xy + xz \qquad\qquad zx + zy = yx + yz$$

$$yz = xz \qquad\qquad\qquad zx = yx$$

$$y = x \qquad\qquad\qquad\quad z = y$$

Thus $x = y = z$. From the fourth equation in Step 4, with $x = y$ and $z = y$,

$$xy + xz + yz - 3 = 0$$
$$y^2 + y^2 + y^2 - 3 = 0$$
$$3y^2 = 3$$
$$y^2 = 1$$
$$y = \pm 1.$$

The negative solution is not applicable, so the solution of the system of equations is $x = 1$, $y = 1$, $z = 1$. In other words, the box is a cube that measures 1 foot on a side. ■

## ■ EXAMPLE 4

Suppose the Cobb-Douglas production function for a particular manufacturer is given by

$$f(x, y) = 50x^{2/5}y^{3/5},$$

where $x$ represents the units of labor and $y$ represents the units of capital. Suppose units of labor and capital cost \$100 and \$200 each, respectively. If the total expense for labor and capital is limited to \$30,000, find the maximum production level for this manufacturer.

The maximum value of $f(x, y) = 50x^{2/5}y^{3/5}$ is to be found, subject to the constraint $100x + 200y = 30,000$. First, let

$$F(x, y, \lambda) = 50x^{2/5}y^{3/5} - \lambda(100x + 200y - 30,000).$$

Then
$$F_x = 20x^{-3/5}y^{3/5} - 100\lambda = 0 \qquad\qquad \textbf{(1)}$$
$$F_y = 30x^{2/5}y^{-2/5} - 200\lambda = 0 \qquad\qquad \textbf{(2)}$$
$$F_\lambda = 100x + 200y - 30,000 = 0. \qquad\qquad \textbf{(3)}$$

Now solve equations (1) and (2) for $\lambda$.

$$20x^{-3/5}y^{3/5} - 100\lambda = 0 \qquad\qquad 30x^{2/5}y^{-2/5} - 200\lambda = 0$$

$$\lambda = \frac{20x^{-3/5}y^{3/5}}{100} \qquad\qquad \lambda = \frac{30x^{2/5}y^{-2/5}}{200}$$

$$= \frac{x^{-3/5}y^{3/5}}{5} \qquad\qquad = \frac{3x^{2/5}y^{-2/5}}{20}$$

Set these expressions for $\lambda$ equal and solve for $x$.

$$\frac{x^{-3/5}y^{3/5}}{5} = \frac{3x^{2/5}y^{-2/5}}{20}$$

$$20x^{-3/5}y^{3/5} = 15x^{2/5}y^{-2/5}$$

$$20y = 15x \qquad \text{Multiply both sides by } x^{3/5}y^{2/5}$$

$$\frac{4}{3}y = x$$

Now substitute this value into equation (3).

$$100\left(\frac{4}{3}y\right) + 200y = 30{,}000$$

$$\frac{400}{3}y + 200y = 30{,}000$$

$$400y + 600y = 90{,}000$$

$$1000y = 90{,}000$$

$$y = 90$$

Since $x = (4/3)y$, $x = (4/3)(90) = 120$. The manufacturer should use 120 units of labor and 90 units of capital. The maximum production will be

$$f(120, 90) = 50(120)^{2/5}(90)^{3/5} \approx 5049 \text{ units.} \quad \blacksquare$$

## 7.4 EXERCISES

*Find the relative maxima or minima in Exercises 1–10.*

1. Maximum of $f(x, y) = 2xy$, subject to $x + y = 12$

2. Maximum of $f(x, y) = 4xy + 2$, subject to $x + y = 24$

3. Maximum of $f(x, y) = x^2y$, subject to $2x + y = 4$

4. Maximum of $f(x, y) = 4xy^2$, subject to $3x - 2y = 5$

5. Minimum of $f(x, y) = x^2 + 2y^2 - xy$, subject to $x + y = 8$

6. Minimum of $f(x, y) = 3x^2 + 4y^2 - xy - 2$, subject to $2x + y = 21$

7. Maximum of $f(x, y) = x^2 - 10y^2$, subject to $x - y = 18$

8. Maximum of $f(x, y) = 12xy - x^2 - 3y^2$, subject to $x + y = 16$

9. Maximum of $f(x, y, z) = xyz^2$, subject to $x + y + z = 6$

10. Maximum of $f(x, y, z) = xy + 2xz + 2yz$ subject to $xyz = 32$

11. Find two numbers whose sum is 20 and whose product is a maximum.

12. Find two numbers whose sum is 100 and whose product is a maximum.

13. Find two numbers $x$ and $y$ such that $x + y = 18$ and $xy^2$ is maximized.

14. Find two numbers $x$ and $y$ such that $x + y = 36$ and $x^2y$ is maximized.

15. Find three numbers whose sum is 90 and whose product is a maximum.

16. Find three numbers whose sum is 240 and whose product is a maximum.

## ▤ APPLICATIONS

### BUSINESS AND ECONOMICS

**Maximum Area for Fixed Expenditure** **17.** Because of terrain difficulties, two sides of a fence can be built for $6 per foot, while the other two sides cost $4 per foot. (See the sketch.) Find the field of maximum area that can be enclosed for $1200.

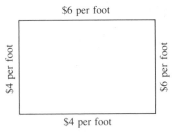

$6 per foot

$4 per foot        $6 per foot

$4 per foot

**Maximum Area for Fixed Expenditure** **18.** To enclose a yard, a fence is built against a large building, so that fencing material is used only on three sides. Material for the ends costs $8 a foot; material for the side opposite the building costs $6 per foot. Find the dimensions of the yard of maximum area that can be enclosed for $1200.

**Cost** **19.** The total cost to produce $x$ large needlepoint kits and $y$ small ones is given by

$$C(x, y) = 2x^2 + 6y^2 + 4xy + 10.$$

If a total of ten kits must be made, how should production be allocated so that total cost is minimized?

**Profit** **20.** The profit from the sale of $x$ units of radiators for automobiles and $y$ units of radiators for generators is given by

$$P(x, y) = -x^2 - y^2 + 4x + 8y.$$

Find values of $x$ and $y$ that lead to a maximum profit if the firm must produce a total of 6 units of radiators.

**Production** **21.** A manufacturing firm estimates that its total production of automobile batteries in thousands of units is

$$f(x, y) = 3x^{1/3}y^{2/3},$$

where $x$ is the number of units of labor and $y$ is the number of units of capital utilized. Labor costs are $80 per unit, and capital costs are $150 per unit. How many units each of labor and capital will maximize production, if the firm can spend $40,000 for these costs?

**Production** **22.** For another product, the manufacturing firm in Exercise 21 estimates that production is a function of labor $x$ and capital $y$ as follows:

$$f(x, y) = 12x^{3/4}y^{1/4}.$$

If $25,200 is available for labor and capital, and if the firm's costs are $100 and $180 per unit, respectively, how many units of labor and capital will give maximum production?

GENERAL

**Area**    **23.** A farmer has 200 m of fencing. Find the dimensions of the rectangular field of maximum area that can be enclosed by this amount of fencing.

**Area**    **24.** Find the area of the largest rectangular field that can be enclosed with 600 m of fencing. Assume that no fencing is needed along one side of the field.

**Surface Area**    **25.** A cylindrical can is to be made that will hold $250\pi$ cubic inches of candy. Find the dimensions of the can with minimum surface area.

**Surface Area**    **26.** An ordinary 12-oz beer or soda pop can holds about 25 cubic inches. Use a calculator and find the dimensions of a can with minimum surface area. Measure a can and see how close its dimensions are to the results you found.

**Volume**    **27.** A rectangular box with no top is to be built from 500 square meters of material. Find the dimensions of such a box that will enclose the maximum volume.

**Surface Area**    **28.** A 1-lb soda cracker box has a volume of 185 cubic inches. The end of the box is square. Find the dimensions of such a box that has minimum surface area.

## 7.5  THE LEAST SQUARES LINE—A MINIMIZATION APPLICATION

In trying to predict the future sales of a product or the total number of matings between two species of insects, it is common to gather as much data as possible and then draw a graph showing the data. This graph, called a **scatter diagram,** can then be inspected to see if a reasonably simple mathematical curve will fit fairly well through all the given points.

As an example, suppose a firm gathers data showing the relationship between the price $y$ of an item in dollars and the number $x$ of units sold, with results as follows.

| Units Sold, $x$ | 10 | 15 | 20 |
|---|---|---|---|
| Price, $y$ | 80 | 68 | 62 |

A graph of this data is shown in Figure 17.

The graph suggests that a straight line fits reasonably well through these data points. If all the data points were to lie on the straight line, the point-slope form of the equation of a line (Chapter 1) could be used to find the equation of the line through the points.

In practice, the points on a scatter diagram almost never fit a straight line exactly. Usually we must decide on the "best" straight line through the points. One way to define the "best" line is as follows. Figure 18 shows the same scattered points from Figure 17. This time, however, vertical line segments indicate the distances of these points from a possible line. The "best" straight line is often defined as the one for which the sum of the squares of the distances, $d_1^2 + d_2^2 + d_3^2$ here, is minimized.

**FIGURE 17**

**FIGURE 18**

To find the equation of the "best" line, let $y' = mx + b$ be the equation of the line in Figure 18. (The purpose of using $y'$ here is to distinguish the calculated $y$-values from the given $y$-values. This use of $y'$ has nothing to do with the derivative.) Then for the point with distance $d_1$, $x = 10$ and $y = 80$, so that $y' = mx + b$ becomes $y_1' = 10m + b$, and

$$d_1^2 = [80 - (10m + b)]^2.$$

Also,

$$d_2^2 = [68 - (15m + b)]^2,$$

and

$$d_3^2 = [62 - (20m + b)]^2.$$

Values of $m$ and $b$ must be found that will minimize $d_1^2 + d_2^2 + d_3^2$. If $d_1^2 + d_2^2 + d_3^2 = S$, then

$$S = [80 - (10m + b)]^2 + [68 - (15m + b)]^2 + [62 - (20m + b)]^2.$$

To minimize $S$ using the method shown earlier in this chapter, the partial derivatives $\partial S/\partial m$ and $\partial S/\partial b$ are needed.

$$\frac{\partial S}{\partial m} = -20[80 - (10m + b)] - 30[68 - (15m + b)] - 40[62 - (20m + b)]$$

$$\frac{\partial S}{\partial b} = -2[80 - (10m + b)] - 2[68 - (15m + b)] - 2[62 - (20m + b)]$$

Simplify each of these, obtaining

$$\frac{\partial S}{\partial m} = -6120 + 1450m + 90b$$

and

$$\frac{\partial S}{\partial b} = -420 + 90m + 6b.$$

Set $\partial S/\partial m = 0$ and $\partial S/\partial b = 0$ and simplify.

$$-6120 + 1450m + 90b = 0 \quad \text{or} \quad 145m + 9b = 612$$
$$-420 + 90m + 6b = 0 \quad \text{or} \quad 15m + b = 70$$

Solving the system on the right leads to the solution $m = -1.8$, $b = 97$. The equation that best fits through the points in Figure 17 is $y' = mx + b$, or

$$y' = -1.8x + 97.$$

The second derivative test can be used to show that this line *minimizes* the sum of the squares of the distances. Thus, $y' = -1.8x + 97$ is the "best" straight line that will fit through the data points. This line is called the **least squares line,** or sometimes the **regression line.**

Once a least squares line is obtained from a set of data, the equation can be used to predict a value of one variable, given a value of the other. In the example above, management might want an estimate of the number of units that would be sold at a price of $75. To get this estimate, replace $y'$ with 75.

$$y' = -1.8x + 97$$
$$75 = -1.8x + 97$$
$$-22 = -1.8x$$
$$12.2 \approx x$$

About 12 units would be sold at a price of $75.

The example above used three data points on the scatter diagram. A practical problem, however, might well involve a very large number of points. The method used above can be extended to find the least squares line for any finite number of data points. To simplify the notation for this more complicated case, *summation notation,* or *sigma notation,* is used as explained in the chapter on integration.

$$\sum_{i=1}^{n} x_i = x_1 + x_2 + x_3 + \cdots + x_n$$

Let $y' = mx + b$ be the least squares line for the set of known data points $(x_1, y_1)$, $(x_2, y_2)$, . . . , $(x_n, y_n)$. (See Figure 19). As above, the sum of squares,

$$d_1^2 + d_2^2 + \cdots + d_n^2,$$

is to be minimized. The square of the distance of the first point $(x_1, y_1)$ from the line $y' = mx + b$ is given by

$$d_1^2 = [y_1 - (mx_1 + b)]^2.$$

The square of the distance of the point $(x_i, y_i)$ from the line is

$$d_i^2 = [y_i - (mx_i + b)]^2.$$

The sum

$$\sum_{i=1}^{n} [y_i - (mx_i + b)]^2.$$

is to be minimized.

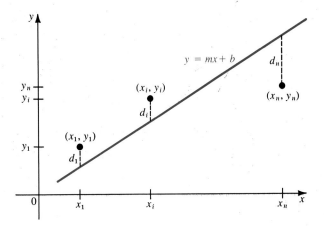

**FIGURE 19**

Since the $x_i$ and $y_i$ values represent known data points, the unknowns in this sum are the numbers $m$ and $b$. To emphasize this fact, write the sum as a function of $m$ and $b$:

$$f(m, b) = \sum_{i=1}^{n}[y_i - (mx_i + b)]^2$$

$$= \sum_{i=1}^{n}[y_i - mx_i - b]^2$$

$$= (y_1 - mx_1 - b)^2 + (y_2 - mx_2 - b)^2$$

$$+ \cdots + (y_n - mx_n - b)^2.$$

To find the minimum value of this function, find the partial derivatives with respect to $m$ and to $b$ and place each equal to 0. (Recall that all the $x$'s and $y$'s are constants here.)

$$\frac{\partial f}{\partial m} = -2x_1(y_1 - mx_1 - b) - 2x_2(y_2 - mx_2 - b) - \cdots$$

$$- 2x_n(y_n - mx_n - b) = 0$$

$$\frac{\partial f}{\partial b} = -2(y_1 - mx_1 - b) - 2(y_2 - mx_2 - b) - \cdots$$

$$- 2(y_n - mx_n - b) = 0.$$

Using some algebra and rearranging terms, these two equations become

$$(x_1^2 + x_2^2 + \cdots + x_n^2)m + (x_1 + x_2 + \cdots + x_n)b$$

$$= x_1y_1 + x_2y_2 + \cdots + x_ny_n,$$

and   $(x_1 + x_2 + \cdots + x_n)m + nb = y_1 + y_2 + \cdots + y_n.$

These last two equations can be rewritten using abbreviated sigma notation as follows. (Remember that $n$ terms are being added.)

$$\left(\sum x^2\right)m + \left(\sum x\right)b = \sum xy \tag{1}$$

$$\left(\sum x\right)m + nb = \sum y \tag{2}$$

To solve this system of equations, multiply the first equation on both sides by $-n$ and the second on both sides by $\sum x$, and then add to eliminate the term with $b$.

$$-n\left(\sum x^2\right)m - n\left(\sum x\right)b = -n\left(\sum xy\right)$$

$$\left(\sum x\right)\left(\sum x\right)m + \left(\sum x\right)nb = \left(\sum x\right)\left(\sum y\right)$$

$$\overline{\left(\sum x\right)\left(\sum x\right)m - n\left(\sum x^2\right)m = \left(\sum x\right)\left(\sum y\right) - n\left(\sum xy\right)}$$

Write the product $(\sum x)(\sum x)$ as $(\sum x)^2$. Using this notation and solving the last equation for $m$ gives

$$m = \frac{\left(\sum x\right)\left(\sum y\right) - n\left(\sum xy\right)}{\left(\sum x\right)^2 - n\left(\sum x^2\right)}. \tag{3}$$

The easiest way to find $b$ is to solve equation (2) for $b$:

$$\left(\sum x\right)m + nb = \sum y \tag{2}$$

$$nb = \sum y - \left(\sum x\right)m$$

$$b = \frac{\sum y - m\left(\sum x\right)}{n}.$$

A summary of the formulas for $m$ and $b$ in the least squares equation follows.

**LEAST SQUARES EQUATION**

The **least squares equation** for the $n$ points $(x_1, y_1)$, $(x_2, y_2)$, . . . , $(x_n, y_n)$ is given by

$$y' = mx + b,$$

where

$$m = \frac{(\sum x)(\sum y) - n(\sum xy)}{(\sum x)^2 - n(\sum x^2)} \quad \text{and} \quad b = \frac{\sum y - m(\sum x)}{n}.$$

**FIGURE 20**

As the formulas suggest, find $m$ first. Next, use the value of $m$ and the second formula to find $b$.

## ▆▆ EXAMPLE 1

Find the least squares line for the data in the following chart.

| $x$ | 1 | 2 | 3 | 5 | 6 |
|-----|-----|-----|-----|-----|-----|
| $y$ | $-2$ | 2 | 5 | 12 | 14 |

Start by drawing the scatter diagram of Figure 20. Since the points in the scatter diagram lie approximately in a straight line, it is appropriate to find the least squares line for the data.

The formulas for $m$ and $b$ require $\Sigma x$, $\Sigma y$, $\Sigma x^2$, and $\Sigma xy$. Organize the work as follows.

| $x$ | $y$ | $x^2$ | $xy$ |
|-----|-----|-----|-----|
| 1 | $-2$ | 1 | $-2$ |
| 2 | 2 | 4 | 4 |
| 3 | 5 | 9 | 15 |
| 5 | 12 | 25 | 60 |
| 6 | 14 | 36 | 84 |
| Totals  17 | 31 | 75 | 161 |

The chart shows that $\Sigma x = 17$, $\Sigma y = 31$, $\Sigma x^2 = 75$, $\Sigma xy = 161$, and $n = 5$. Using the equation for $m$ given above,

$$m = \frac{17(31) - 5(161)}{17^2 - 5(75)} = \frac{527 - 805}{289 - 375} = \frac{-278}{-86} \approx 3.2.$$

Now find $b$:

$$b = \frac{\Sigma y - m(\Sigma x)}{n} = \frac{31 - 3.2(17)}{5} \approx -4.7.$$

The least squares equation for the given data is $y' = mx + b$, or

$$y' = 3.2x - 4.7.$$

This line is graphed in Figure 21.  ▆

The equation $y' = 3.2x - 4.7$ can be used to estimate values of $y$ for given values of $x$. For example, if $x = 4$,

$$y' = 3.2x - 4.7 = 3.2(4) - 4.7 = 12.8 - 4.7 = 8.1.$$

Also, if $x = 7$, then $y' = 3.2(7) - 4.7 = 17.7$.

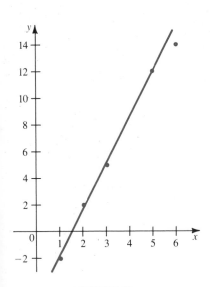

**FIGURE 21**

Sir Francis Galton (1822–1911), a British scientist, is credited with setting forth the ideas of regression and correlation in his statistical work *Natural Inheritance*. His interest in statistics developed as a result of his research on weather prediction.

## 7.5 EXERCISES

*In Exercises 1–4, draw a scatter diagram for the given set of data points and then find the least squares equation.*

**1.**

| $x$ | 1 | 2 | 3 | 5 | 9 |
|---|---|---|---|---|---|
| $y$ | 9 | 13 | 18 | 25 | 41 |

Estimate $y$ when $x$ is 4; 7.

**2.**

| $x$ | 2 | 3 | 5 | 6 | 8 |
|---|---|---|---|---|---|
| $y$ | 8 | 13 | 23 | 28 | 38 |

Estimate $y$ when $x$ is 7; 9.

**3.**

| $x$ | 4 | 5 | 8 | 12 | 14 |
|---|---|---|---|---|---|
| $y$ | 3 | 7 | 17 | 28 | 35 |

Estimate $x$ when $y$ is 12; 32.

**4.**

| $x$ | 3 | 4 | 5 | 6 | 8 |
|---|---|---|---|---|---|
| $y$ | 8 | 12 | 16 | 18 | 28 |

Estimate $x$ when $y$ is 17; 26.

## APPLICATIONS

### BUSINESS AND ECONOMICS

**Insurance Sales**    **5.** Records show that the annual sales for the Sweet Palms Life Insurance Company in 5-year periods for the last 20 years were as follows.

| Year | Year in Coded Form, $x$ | Sales in Millions, $y$ |
|---|---|---|
| 1968 | 1 | 1.0 |
| 1973 | 2 | 1.3 |
| 1978 | 3 | 1.7 |
| 1983 | 4 | 1.9 |
| 1988 | 5 | 2.1 |

(a) Plot the five points on a scatter diagram.

(b) Find the least squares equation and graph it on the scatter diagram from part (a).

(c) Predict the company's sales for 1993.

**Sales and Profit**    **6.** A fast-food chain wishes to find the relationship between annual store sales (in thousands of dollars) and percent of pretax profit in order to estimate increases in profit due to increased sales volume. The data shown below were obtained from a sample of stores across the country.

| Annual Sales, $x$ | 250 | 300 | 375 | 425 | 450 | 475 | 500 | 575 | 600 | 650 |
|---|---|---|---|---|---|---|---|---|---|---|
| Pretax Profit, $y$ | 9.3 | 10.8 | 14.8 | 14.0 | 14.2 | 15.3 | 15.9 | 19.1 | 19.2 | 21.0 |

(a) Plot the ten pairs of values on a scatter diagram.

(b) Find the equation of the least squares line and graph it on the scatter diagram of part (a).

(c) Using the equation of part (b), predict the percent of pretax profit for annual sales of $700,000; of $750,000.

**Sales Analysis**      **7.** Sales (in thousands of dollars) of a certain company are shown here.

| Year, $x$ | 0 | 1 | 2 | 3 | 4 | 5 |
|-----------|-----|-----|-----|-----|-----|-----|
| Sales, $y$ | 48 | 59 | 66 | 75 | 80 | 90 |

Find the equation of the least squares line.

**Repair Costs and Production**   **8.** The following data were used to determine whether there is a relationship between repair costs and barrels of beer produced by a brewery. The data (in thousands) are given for a 10-month period.

| Month | Barrels of Beer, $x$ | Repairs, $y$ |
|-------|------|------|
| Jan | 369 | 299 |
| Feb | 379 | 280 |
| Mar | 482 | 393 |
| April | 493 | 388 |
| May | 496 | 385 |
| June | 567 | 423 |
| July | 521 | 374 |
| Aug | 482 | 357 |
| Sept | 391 | 106 |
| Oct | 417 | 332 |

(a) Find the equation of the least squares line.

(b) If 500,000 barrels of beer are produced, what will the equation from part (a) give as the predicted repair costs?

*It is sometimes possible to get a better prediction for a variable by considering its relationship with more than one other variable. For example, one should be able to predict college GPAs more precisely if both high school GPAs and scores on the ACT are considered. To do this, we alter the equation used to find a least squares line by adding a term for the new variable as follows. If y represents college GPAs, $x_1$ high school GPAs, and $x_2$ ACT scores, then y', the predicted GPA, is given by*

$$y' = ax_1 + bx_2 + c.$$

*This equation represents a **least squares plane**. The equations for the constants a, b, and c are more complicated than those given in the text for m and b, so calculating a least squares equation for three variables is more likely to require the aid of a computer.*

**Revenue and Price**

**9.** Alcoa* used a least squares line with two independent variables, $x_1$ and $x_2$, to predict the effect on revenue of the price of aluminum forged truck wheels, as follows.

$$x_1 = \text{the average price per wheel}$$
$$x_2 = \text{large truck production in thousands}$$
$$y = \text{sales of aluminum forged truck wheels in thousands}$$

Using data for the past eleven years, the company found the equation of the least squares plane (see the discussion on the preceding page) to be

$$y' = 49.2755 - 1.1924x_1 + .1631x_2.$$

The following figures were then forecast for truck production.

| 1982 | 1983 | 1984 | 1985 | 1986 | 1987 |
|------|------|------|------|------|------|
| 160.0 | 165.0 | 170.0 | 175.0 | 180.0 | 185.0 |

Three possible price levels per wheel were considered: $42, $45, and $48.

**(a)** Use the least squares plane equation given above to find the estimated sales of wheels ($y$) for 1984 at each of the three price levels.

**(b)** Repeat part (a) for 1987.

**(c)** For which price level, on the basis of the 1984 and 1987 figures, are total estimated sales greatest?

(By comparing total estimated sales for the years 1982 through 1987 at each of the three price levels, the company found that the selling price of $42 per wheel would generate the greatest sales volume over the six-year period.)

## LIFE SCIENCES

**Size of Corn**

**10.** In a study to determine the linear relationship between $y$, the length (in decimeters) of an ear of corn, and $x$, the amount (in tons per acre) of fertilizer used, the following data were collected.

$$n = 10 \qquad \Sigma x = 30 \qquad \Sigma y = 24$$
$$\Sigma xy = 75 \qquad \Sigma x^2 = 100$$

**(a)** Find an equation for the least squares line.

**(b)** If 3 tons per acre of fertilizer are used, what length (in decimeters) would the equation in (a) predict for an ear of corn?

**Height and Weight**

**11.** A sample of 10 adult men gave the following data on their heights (in inches) and weights (in pounds).

| Height, $x$ | 62 | 62 | 63 | 65 | 66 | 67 | 68 | 68 | 70 | 72 |
|-------------|----|----|----|----|----|----|----|----|----|----|
| Weight, $y$ | 120 | 140 | 130 | 150 | 142 | 130 | 135 | 175 | 149 | 168 |

**(a)** Find the equation of the least squares line.

Using the results above, predict the weight of a man whose height is as follows.

**(b)** 60 inches    **(c)** 70 inches

---

*This example supplied by John H. Van Denender, Public Relations Department, Aluminum Company of America.

*Exercise 12 is for students who have studied logarithms.*

**Bacterial Growth**    **12.** Sometimes the scatter diagram of the data does not have a linear pattern. This is particularly true in some biological and chemical applications. In these applications, however, often the scatter diagram of the *logarithms* of the data has a linear pattern. A least squares line can then be used to predict the logarithm of any desired value from which the value itself can be found. Suppose that a certain kind of bacterium grows in number as shown in Table A. The actual number of bacteria present at each time period is replaced with the common logarithm of that number in Table B.

| Table A | | Table B | |
|---|---|---|---|
| *Time in Hours* | *Number of Bacteria* | *Time* $x$ | *Log,* $y$ |
| 0 | 1000 | 0 | 3.0000 |
| 1 | 1649 | 1 | 3.2172 |
| 2 | 2718 | 2 | 3.4343 |
| 3 | 4482 | 3 | 3.6515 |
| 4 | 7389 | 4 | 3.8686 |
| 5 | 12182 | 5 | 4.0857 |

We can now find a least squares line that will predict $y$, given $x$.

**(a)** Plot the original pairs of numbers. The pattern should be nonlinear.

**(b)** Plot the log values against the time values. The pattern should be almost linear.

**(c)** Find the equation of the least squares line. (First round off the log values to the nearest hundredth.)

**(d)** Predict the log value for a time of 7 hr. Find the number whose logarithm is your answer. This will be the predicted number of bacteria.

## SOCIAL SCIENCES

**Test Scores**    **13.** The ACT scores of eight students were compared to their GPAs after one year in college, with the following results.

| ACT Score, $x$ | 19 | 20 | 22 | 24 | 25 | 26 | 27 | 29 |
|---|---|---|---|---|---|---|---|---|
| GPA, $y$ | 2.2 | 2.4 | 2.7 | 2.6 | 3.0 | 3.5 | 3.4 | 3.8 |

**(a)** Plot the eight points on a scatter diagram.

**(b)** Find the least squares equation and graph it on the scatter diagram of part (a).

**(c)** Using the results of (b), predict the GPA of a student with an ACT score of 28.

## PHYSICAL SCIENCES

**Temperature**    **14.** In an experiment to determine the linear relationship between temperatures on the Celsius scale ($y$) and on the Fahrenheit scale ($x$), a student got the following results.

$$n = 5 \qquad \Sigma x = 376 \qquad \Sigma y = 120$$
$$\Sigma xy = 28,050 \qquad \Sigma x^2 = 62,522$$

**(a)** Find an equation for the least squares line.

**(b)** Find the reading on the Celsius scale that corresponds to a reading of 120° Fahrenheit, using the equation from part (a).

## ≣ 7.6 TOTAL DIFFERENTIALS AND APPROXIMATIONS

In the second section of this chapter we used partial derivatives to find the marginal productivity of labor and of capital for a production function. The marginal productivity gives the rate of change of production for a 1-unit change in labor or for a 1-unit change in capital. To find the change in productivity for any (usually small) change in labor and/or in capital, we need the concept of a *differential*. Recall from an earlier chapter that the differential of a function $y = f(x)$, written $dy$, is defined as

$$dy = f'(x) \cdot dx,$$

where $dx$, the differential of $x$, is any real number (usually small). This definition shows that $dy$ is a function of *two* variables, $x$ and $dx$. We can extend this idea to three dimensions by defining a *total differential*.

**TOTAL DIFFERENTIAL FOR TWO VARIABLES**

Let $z = f(x, y)$ be a function of $x$ and $y$. Let $dx$ and $dy$ be real numbers. Then the **total differential** of $f$ is

$$df = f_x(x, y) \cdot dx + f_y(x, y) \cdot dy.$$

(Sometimes $df$ is written $dz$.)

By this definition, $df$ is a function of *four* variables, $x$, $dx$, $y$, and $dy$.

### ≣ EXAMPLE 1

Find $df$ for each function.

**(a)** $f(x, y) = 9x^3 - 8x^2y + 4y^3$

First find $f_x$ and $f_y$.

$$f_x = 27x^2 - 16xy \qquad \text{and} \qquad f_y = -8x^2 + 12y^2$$

By the definition,

$$df = (27x^2 - 16xy)dx + (-8x^2 + 12y^2)dy.$$

**(b)** $z = f(x, y) = \ln(x^3 + y^2)$

Since

$$f_x = \frac{3x^2}{x^3 + y^2} \qquad \text{and} \qquad f_y = \frac{2y}{x^3 + y^2},$$

the total differential is

$$dz = df = \left(\frac{3x^2}{x^3 + y^2}\right)dx + \left(\frac{2y}{x^3 + y^2}\right)dy. \quad \blacksquare$$

■ EXAMPLE 2

Let $f(x, y) = 9x^3 - 8x^2y + 4y^3$. Find $df$ when $x = 1$, $y = 3$, $dx = .01$, and $dy = -.02$.

Example 1 (a) gave the total differential

$$df = (27x^2 - 16xy)dx + (-8x^2 + 12y^2)dy.$$

Replace $x$ with 1, $y$ with 3, $dx$ with .01, and $dy$ with $-.02$.

$$df = [27 \cdot 1^2 - 16(1)(3)](.01) + (-8 \cdot 1^2 + 12 \cdot 3^2)(-.02)$$
$$= (27 - 48)(.01) + (-8 + 108)(-.02)$$
$$= (-21)(.01) + (100)(-.02)$$
$$= -2.21 \quad ■$$

The idea of a total differential can be extended to include functions of three independent variables.

**TOTAL DIFFERENTIAL**
**FOR THREE VARIABLES**

■ If $w = f(x, y, z)$, then the total differential $dw$ is

$$dw = f_x(x, y, z)dx + f_y(x, y, z)dy + f_z(x, y, z)dz,$$

provided all indicated partial derivatives exist.

As this definition shows, $dw$ is a function of six variables, $x$, $y$, $z$, $dx$, $dy$, and $dz$. Similar definitions could be given for functions of more than three independent variables.

■ EXAMPLE 3

Let $w = f(x, y, z) = x^2 + yz^4$.

**(a)** Find $dw$.

Find the necessary three partial derivatives.

$$f_x = 2x \qquad f_y = z^4 \qquad f_z = 4yz^3$$

By the definition of the total differential,

$$dw = 2x\,dx + z^4\,dy + 4yz^3\,dz.$$

**(b)** Evaluate $dw$ for $x = -1$, $y = 2$, $z = -3$, $dx = .02$, $dy = -.03$, and $dz = .05$.

Substitute these values into the total differential $dw$.

$$dw = 2(-1)(.02) + (-3)^4(-.03) + 4(2)(-3)^3(.05) = -13.27 \quad ■$$

***Approximations***   Recall that with a function of one variable, $y = f(x)$, the differential $dy$ can be used to approximate the change in $y$, $\Delta y$, corresponding to a change in $x$, $\Delta x$. A similar approximation can be made for a function of two variables.

> For small values of $\Delta x$ and $\Delta y$,
> $$dz \approx \Delta z,$$
> where $\Delta z = f(x + \Delta x, y + \Delta y) - f(x, y)$.

## ▄▄ EXAMPLE 4

Let $f(x, y) = 6x^2 + xy + y^3$. Find $\Delta z$ and $dz$ when $x = 2$, $y = -1$, $\Delta x = -.03$, and $\Delta y = .02$.

Here $x + \Delta x = 2 + (-.03) = 1.97$ and $y + \Delta y = -1 + .02 = -.98$. Find $\Delta z$ with the definition above.

$$
\begin{aligned}
\Delta z &= f(x + \Delta x, y + \Delta y) - f(x, y) \\
&= f(1.97, -.98) - f(2, -1) \\
&= [6(1.97)^2 + (1.97)(-.98) + (-.98)^3] - [6(2)^2 + 2(-1) + (-1)^3] \\
&= [23.2854 - 1.9306 - .941192] - [24 - 2 - 1] \\
&= 20.413608 - 21 \\
&= -.586392
\end{aligned}
$$

To find $dz$, the total differential, first find

$$f_x = 12x + y \qquad \text{and} \qquad f_y = x + 3y^2.$$

Then $\qquad\qquad\qquad dz = (12x + y)dx + (x + 3y^2)dy.$

Since $dx = \Delta x$ and $dy = \Delta y$, substitution gives

$$
\begin{aligned}
dz &= [12 \cdot 2 + (-1)](-.03) + [2 + 3 \cdot (-1)^2](.02) \\
&= (23)(-.03) + (5)(.02) \\
&= -.59.
\end{aligned}
$$

In this example the values of $\Delta z$ and $dz$ are very close. (Compare the amount of work needed to find $\Delta z$ with that needed for $dz$.) ▄▄

For small values of $dx$ and $dy$ the values of $\Delta z$ and $dz$ are approximately equal. Since $\Delta z = f(x + dx, y + dy) - f(x, y)$,

$$f(x + dx, y + dy) = f(x, y) + \Delta z$$

or $\qquad\qquad f(x + dx, y + dy) \approx f(x, y) + dz.$

Replacing $dz$ with the expression for the total differential gives the following result.

**APPROXIMATIONS BY DIFFERENTIALS**

> For a function $f$ having all indicated partial derivatives, and for small values of $dx$ and $dy$,
> $$f(x + dx, y + dy) \approx f(x, y) + dz,$$
> or $\quad f(x + dx, y + dy) \approx f(x, y) + f_x(x, y) \cdot dx + f_y(x, y) \cdot dy.$

## EXAMPLE 5

The total profit in hundreds of dollars from the sale of $x$ tons of fertilizer for wheat and $y$ tons of fertilizer for corn is given by

$$P(x, y) = 6x^{1/2} + 9y^{2/3} + \frac{5000}{xy}.$$

Suppose a firm is producing 64 tons of fertilizer for wheat and 125 tons of fertilizer for corn. Find the approximate change in profit if production is changed to 63 tons and 127 tons, respectively. Use this result to estimate the total profit from the sale of 63 tons and 127 tons.

The approximate change in profit, $dP$, is

$$dP = P_x(x, y) \cdot dx + P_y(x, y) \cdot dy.$$

Finding the necessary partial derivatives gives

$$dP = \left(3x^{-1/2} - \frac{5000}{x^2y}\right)dx + \left(6y^{-1/3} - \frac{5000}{xy^2}\right)dy.$$

Substitute 64 for $x$, 125 for $y$, $-1$ for $dx$, and 2 for $dy$.

$$dP = \left(3 \cdot 64^{-1/2} - \frac{5000}{64^2 \cdot 125}\right)(-1) + \left(6 \cdot 125^{-1/3} - \frac{5000}{64 \cdot 125^2}\right)(2)$$

$$= \left(3 \cdot \frac{1}{8} - \frac{5000}{512,000}\right)(-1) + \left(6 \cdot \frac{1}{5} - \frac{5000}{1,000,000}\right)(2)$$

$$= (.375 - .0098)(-1) + (1.2 - .005)(2)$$

$$\approx 2.0248$$

A change in production from 64 tons of fertilizer for wheat and 125 tons for corn, to 63 and 127 tons respectively, will increase profits by about 2.02 hundred dollars, or $202.

To estimate the total profit from the production of 63 tons and 127 tons, use the approximation formula given above.

$$P(63, 127) \approx P(64, 125) + dP$$

$$= \left(6 \cdot 64^{1/2} + 9 \cdot 125^{2/3} + \frac{5000}{64 \cdot 125}\right) + 2.0248$$

$$= \left(6 \cdot 8 + 9 \cdot 25 + \frac{5000}{8000}\right) + 2.0248 = 275.65$$

hundred dollars, or a total profit of $27,565. ▪

## EXAMPLE 6

A short length of blood vessel is in the shape of a right circular cylinder. (See Figure 22.) The length of the vessel is measured as 42 millimeters, and the radius is measured as 2.5 millimeters. Suppose the maximum error in the measurement of the length is .9 millimeters, with an error of no more than .2 millimeters in the measurement of the radius. Find the maximum possible error in calculating the volume of the blood vessel.

**FIGURE 22**

The volume of a right circular cylinder is given by $V = \pi r^2 h$. To approximate the error in the volume, find the total differential, $dV$.

$$dV = (2\pi rh) \cdot dr + (\pi r^2) \cdot dh$$

Here, $r = 2.5$, $h = 42$, $dr = .2$, and $dh = .9$. Substitution gives

$$dV = [(2\pi)(2.5)(42)(.2) + [\pi(2.5)^2](.9) \approx 149.6$$

The maximum possible error in calculating the volume is 149.6 cubic millimeters.

## 7.6 EXERCISES

*Find dz or dw, as appropriate, for each of the following.*

**1.** $z = 9x^4 - 5y^3$

**2.** $z = x^2 + 7y^4$

**3.** $z = x^2y^3 + y$

**4.** $z = 8x^2 - xy^2$

**5.** $z = \dfrac{x + y}{x - y}$

**6.** $z = \dfrac{x + y^2}{y - 2}$

**7.** $z = 2\sqrt{xy} - \sqrt{x + y}$

**8.** $z = \sqrt{x^2 + y^2} + \sqrt{xy}$

**9.** $z = (3x + 2)\sqrt{1 - 2y}$

**10.** $z = (5x^2 + 6)\sqrt{4 + 3y}$

**11.** $z = \ln(x^2 + 2y^4)$

**12.** $z = \ln\left(\dfrac{8 + x}{8 - y}\right)$

**13.** $z = xy^2e^{x+y}$

**14.** $z = (x + y)e^{-x^2}$

**15.** $z = x^2 - y \ln x$

**16.** $z = x^2 + 3y - x \ln y$

**17.** $w = x^4yz^3$

**18.** $w = 6x^3y^2z^5$

**19.** $w = 6\left(1 - \dfrac{2}{y} + \dfrac{1}{x} - \dfrac{3}{z}\right)$

**20.** $w = 8\left(\dfrac{4 + x}{y + z}\right)$

*Evaluate dz using the given information.*

**21.** $z = x^2 + 3xy + y^2$;  $x = 4, y = -2, dx = .02, dy = -.03$

**22.** $z = 8x^3 + 2x^2y - y$;  $x = 1, y = 3, dx = .01, dy = .02$

**23.** $z = \dfrac{x - 4y}{x + 2y};\quad x = 0,\ y = 5,\ dx = -.03,\ dy = .05$

**24.** $z = \dfrac{y^2 + 3x}{y^2 - x};\quad x = 4,\ y = -4,\ dx = .01,\ dy = .03$

**25.** $z = \sqrt{x + y} + \sqrt{x(y + 1)};\quad x = 2,\ y = 1,\ dx = -.03,\ dy = -.02$

**26.** $z = (6x + y) + \sqrt{x^2 + y^2};\quad x = 1,\ y = 2,\ dx = -.01,\ dy = .04$

**27.** $z = \ln(x^2 + y^2);\quad x = 2,\ y = 3,\ dx = .02,\ dy = -.03$

**28.** $z = \ln\left(\dfrac{x + y}{x - y}\right);\quad x = 4,\ y = -2,\ dx = .03,\ dy = .02$

*Evalute dw using the given information.*

**29.** $w = x^4y + z^2x;\quad x = 1,\ y = -1,\ z = 2,\ dx = .01,\ dy = .03,\ dz = .02$

**30.** $w = x^2y^3 - x^3y^4z;\quad x = 2,\ y = 3,\ z = -1,\ dx = -.03,\ dy = -.01,\ dz = .01$

**31.** $w = \dfrac{5x^2 + y^2}{z + 1};\quad x = -2,\ y = 1,\ z = 1,\ dx = .02,\ dy = -.03,\ dz = .02$

**32.** $w = x\ln(yz) - y\ln\dfrac{x}{z};\quad x = 2,\ y = 1,\ z = 4,\ dx = .03,\ dy = .02,\ dz = -.01$

---

## ▤ APPLICATIONS

### BUSINESS AND ECONOMICS

**Manufacturing**  **33.** Approximate the amount of aluminum needed for a beverage can of radius 2.5 cm and height 14 cm. Assume the walls of the can are .08 cm thick.

**Manufacturing**  **34.** Approximate the amount of material needed to make a water tumbler of diameter 3 cm and height 9 cm. Assume the walls of the tumbler are .2 cm thick.

**Volume of a Coating**  **35.** An industrial coating .2 inches thick is applied to all sides of a box of dimensions 10 inches by 9 inches by 14 inches. Estimate the volume of the coating used.

**Manufacturing Cost**  **36.** The manufacturing cost of a precision electronic calculator is approximated by

$$M(x, y) = 40x^2 + 30y^2 - 10xy + 30,$$

where $x$ is the cost of the chips and $y$ is the cost of labor. Right now, the company spends $4 on chips and $7 on labor. Use differentials to approximate the change in cost if the company spends $5 on chips and $6.50 on labor.

**Cost**  **37.** The cost in dollars to produce $x$ satellite receiving dishes and $y$ transmitters is given by

$$C(x, y) = \ln(x^2 + y) + e^{xy/20}.$$

Production schedules now call for 15 receiving dishes and 9 transmitters. Use differentials to approximate the change in costs if 1 more dish and 1 fewer transmitter are made.

**Profit**  **38.** The profit from the sale of $x$ small computers and $y$ electronic games is given by

$$P(x, y) = \frac{x}{x + 5y} + \frac{y + x}{y}.$$

Right now, 75 small computers and 50 games are being sold. Use differentials to approximate the change in profit if 3 fewer computers and 2 more games were sold.

**Production**    **39.** The production function for one country is

$$z = x^{.65}y^{.35},$$

where $x$ stands for units of labor and $y$ for units of capital. At present, $x$ is 50 and $y$ is 29. Use differentials to estimate the change in $z$ if $x$ becomes 52 and $y$ becomes 27.

**Production**    **40.** The production function for another country is

$$z = x^{.8}y^{.2},$$

where $x$ stands for units of labor and $y$ for units of capital. At present, $x$ is 20 and $y$ is 18. Use differentials to estimate the change in $z$ if $x$ becomes 21 and $y$ becomes 16.

## LIFE SCIENCES

**Bone Preservative Volume**    **41.** A piece of bone in the shape of a right circular cylinder is 7 cm long and has a radius of 1.4 cm. It is coated with a layer of preservative .09 cm thick. Estimate the volume of preservative used.

**Blood Vessel Volume**    **42.** A portion of a blood vessel is measured to have length 7.9 cm and radius .8 cm. If each measurement could be off by as much as .15 cm, estimate the maximum possible error in calculating the volume of the vessel.

**Grasshopper Matings**    **43.** The total number of matings per day between individuals of a certain species of grasshoppers is approximated by

$$M(x, y) = 2xy + 10xy^2 + 30y^2 + 20,$$

where $x$ represents the temperature in °C and $y$ represents the number of days since the last rainfall. Currently, the temperature is 20°C, and it has been 7 days since the last rainfall. Use differentials to approximate the change in $M$ if the temperature changes to 18°C and it has been 8 days since the last rainfall.

**Blood Volume**    **44.** In Exercise 50 of Section 2 of this chapter, we found that the number of liters of blood pumped through the lungs in one minute is given by

$$C = \frac{b}{a - v}.$$

Suppose $a = 160$, $b = 200$, $v = 125$. Estimate the change in $C$ if $a$ becomes 145, $b$ becomes 190, and $v$ changes to 130.

**Oxygen Consumption**    **45.** In Exercise 48 of Section 2 of this chapter, we found that the oxygen consumption of a mammal is

$$m = \frac{2.5(T - F)}{w^{.67}}.$$

Suppose $T$ is 38°, $F$ is 12°, and $w$ is 30 kg. Approximate the change in $m$ if $T$ changes to 36°, $F$ changes to 13°, and $w$ becomes 31 kg.

## GENERAL

**Estimating Area**    **46.** The width of a rectangle is measured as 15.8 cm, while the length is measured as 29.6 cm. The width measurement could be off by .8 cm, and the length could be off by 1.1 cm. Estimate the maximum possible error in calculating the area of the rectangle.

**Estimating Area**    **47.** The height of a triangle is measured as 42.6 cm, with the base measured as 23.4 cm. The measurement of the height can be off as much as 1.2 cm, and that of the base off by no more than .9 cm. Estimate the maximum possible error in calculating the area of the triangle.

**Estimating Volume**    **48.** The height of a cone is measured to be 8.4 cm, with a radius of 2.9 cm. Each measurement could be off by as much as .1 cm. Estimate the maximum possible error in calculating the volume of the cone.

## 7.7  DOUBLE INTEGRALS

Earlier in this chapter we found partial derivatives of functions of two or more variables by holding constant all variables except one. In this section a similar process is considered for antiderivatives of functions of two or more variables. For example, in

$$\int (5x^3y^4 - 6x^2y + 2)dy$$

the notation $dy$ indicates integration with respect to $y$, so we treat $y$ as the variable and $x$ as a constant. Using the rules for antiderivatives gives

$$\int (5x^3y^4 - 6x^2y + 2)dy = x^3y^5 - 3x^2y^2 + 2y + C(x).$$

The constant $C$ used earlier must be replaced with $C(x)$ to show that the "constant of integration" here can be any function involving only the variable $x$. Just as before, check this work by taking the derivative (actually the partial derivative) of the answer:

$$\frac{\partial}{\partial y}\left[x^3y^5 - 3x^2y^2 + 2y + C(x)\right] = 5x^3y^4 - 6x^2y + 2 + 0,$$

which shows that the antiderivative is correct.

### EXAMPLE 1
Find $\int (5x^3y^4 - 6x^2y + 2)dx$.

Because of the $dx$, treat $y$ as a constant and $x$ as the variable.

$$\int (5x^3y^4 - 6x^2y + 2)dx = \frac{5}{4}x^4y^4 - 2x^3y + 2x + C(y)$$

### EXAMPLE 2
**(a)** Find $\int x(x^2 + y)dx$.

Multiply $x$ and $x^2 + y$. Then integrate each term with $x$ as the variable and $y$ as a constant.

$$\int x(x^2 + y)dx = \int (x^3 + xy)dx$$
$$= \frac{x^4}{4} + \frac{x^2}{2} \cdot y + f(y) = \frac{1}{4}x^4 + \frac{1}{2}x^2y + f(y)$$

**(b)** Find $\int x(x^2 + y)dy$.

Since $y$ is the variable and $x$ is held constant,

$$\int x(x^2 + y)dy = \int (x^3 + xy)dy = x^3 y + \frac{1}{2}xy^2 + g(x). \quad \blacksquare$$

The analogy to integration of functions of one variable can be continued for evaluating definite integrals. We do this by holding one variable constant and using the fundamental theorem of calculus with the other variable.

■■ **EXAMPLE 3**

**(a)** Evaluate $\int_3^5 (6xy^2 + 12x^2y + 4y)dx$.

First, find an antiderivative:

$$\int (6xy^2 + 12x^2y + 4y)dx = 3x^2y^2 + 4x^3y + 4xy + h(y).$$

Now replace each $x$ with 5, and then with 3, and subtract the results.

$$(3x^2y^2 + 4x^3y + 4xy + h(y))\Big|_3^5 = [3 \cdot 5^2 \cdot y^2 + 4 \cdot 5^3 \cdot y + 4 \cdot 5 \cdot y + h(y)]$$

$$- [3 \cdot 3^2 \cdot y^2 + 4 \cdot 3^3 \cdot y + 4 \cdot 3 \cdot y + h(y)]$$
$$= 75y^2 + 500y + 20y + h(y)$$
$$- (27y^2 + 108y + 12y + h(y))$$
$$= 48y^2 + 400y$$

The "function of integration," $h(y)$, drops out, just as the constant of integration does with definite integrals of functions of one variable.

**(b)** Evaluate $\int_1^2 (6xy^2 + 12x^2y + 4y)dy$.

Integrate with respect to $y$; then substitute 2 and 1 for $y$ and subtract.

$$\int_1^2 (6xy^2 + 12x^2y + 4y)dy = (2xy^3 + 6x^2y^2 + 2y^2)\Big|_1^2$$

$$= (2x \cdot 2^3 + 6x^2 \cdot 2^2 + 2 \cdot 2^2)$$
$$- (2x \cdot 1^3 + 6x^2 \cdot 1^2 + 2 \cdot 1^2)$$
$$= 16x + 24x^2 + 8 - (2x + 6x^2 + 2)$$
$$= 14x + 18x^2 + 6 \quad \blacksquare$$

As Example 3 suggests, an integral of the form

$$\int_a^b f(x, y)dy$$

produces a result that is a function of $x$, while

$$\int_a^b f(x, y)dx$$

produces a function of $y$. These resulting functions of one variable can themselves be integrated, as in the next example.

## ■ EXAMPLE 4

**(a)** Find $\int_1^2 \left[ \int_3^5 (6xy^2 + 12x^2y + 4y)dx \right] dy.$

In Example 3(a), we found the quantity in brackets to be $48y^2 + 400y$. Thus,

$$\int_1^2 \left[ \int_3^5 (6xy^2 + 12x^2y + 4y)dx \right] dy = \int_1^2 (48y^2 + 400y)dy$$

$$= (16y^3 + 200y^2) \Big|_1^2$$

$$= 16 \cdot 2^3 + 200 \cdot 2^2 - (16 \cdot 1^3 + 200 \cdot 1^2)$$

$$= 128 + 800 - (16 + 200)$$

$$= 712.$$

**(b)** Evaluate $\int_3^5 \left[ \int_1^2 (6xy^2 + 12x^2y + 4y)dy \right] dx.$ (This is the same integrand, with the same limits of integration as in part (a), but the order of integration is reversed.)

Use the result from Example 3(b).

$$\int_3^5 \left[ \int_1^2 (6xy^2 + 12x^2y + 4y)dy \right] dx$$

$$= \int_3^5 (14x + 18x^2 + 6)dx$$

$$= (7x^2 + 6x^3 + 6x) \Big|_3^5$$

$$= 7 \cdot 5^2 + 6 \cdot 5^3 + 6 \cdot 5 - (7 \cdot 3^2 + 6 \cdot 3^3 + 6 \cdot 3)$$

$$= 175 + 750 + 30 - (63 + 162 + 18) = 712 \quad ■$$

The answers in the two parts of Example 4 are equal. It can be proved that for a large class of functions, including most functions that occur in applications, the following equation holds true.

$$\int_a^b \left[ \int_c^d f(x, y)dx \right] dy = \int_c^d \left[ \int_a^b f(x, y)dy \right] dx$$

Because these two integrals are equal, the brackets are not needed, and either of the integrals is given by

$$\int_a^b \int_c^d f(x, y)dx\, dy.$$

This integral is called an **iterated integral** since it is evaluated by integrating twice, first using one variable and then using the other. (The order in which $dx$ and $dy$ are written tells the order of integration with the innermost differential used first.)

The fact that the iterated integrals above are equal makes it possible to define a *double integral*. First, the set of points $(x, y)$ with $c \leq x \leq d$ and $a \leq y \leq b$ defines a rectangular region $R$ in the plane, as shown in Figure 23. Then, the *double integral over R* is defined as follows.

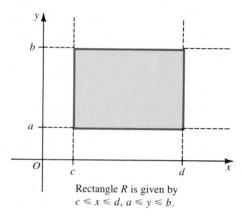

Rectangle $R$ is given by
$c \leq x \leq d, a \leq y \leq b.$

**FIGURE 23**

**DOUBLE INTEGRAL**

The **double integral** of $f(x, y)$ over a rectangular region $R$ is written

$$\iint_R f(x, y)dx\, dy \qquad \text{or} \qquad \iint_R f(x, y)dy\, dx,$$

and equals either

$$\int_a^b \int_c^d f(x, y)dx\, dy \qquad \text{or} \qquad \int_c^d \int_a^b f(x, y)dy\, dx.$$

Extending earlier definitions, $f(x, y)$ is the **integrand** and $R$ is the **region of integration**.

■■■ **EXAMPLE 5**

Find $\displaystyle\iint_R \sqrt{x} \cdot \sqrt{y-2}\ dx\ dy$ over the rectangular region $R$ defined by $0 \le x$

$\le 4,\ 3 \le y \le 11$.

Integrate first with respect to $x$; then, as a check, use $y$ first. Values of $x$ for $(x, y)$ in $R$ go from 0 to 4; values of $y$ from 3 to 11.

$$\iint_R \sqrt{x} \cdot \sqrt{y-2}\ dx\ dy = \int_3^{11}\left[\int_0^4 \sqrt{x}\cdot\sqrt{y-2}\ dx\right]dy$$

$$= \int_3^{11}\left(\frac{2}{3}x^{3/2}\sqrt{y-2}\right)\Bigg|_0^4\ dy$$

$$= \int_3^{11}\left[\frac{2}{3}(4^{3/2}\sqrt{y-2}) - \frac{2}{3}(0^{3/2})\sqrt{y-2}\right]dy$$

$$= \int_3^{11}\left(\frac{16}{3}\sqrt{y-2} - 0\right)dy = \int_3^{11}\left(\frac{16}{3}\sqrt{y-2}\right)dy$$

$$= \frac{32}{9}(y-2)^{3/2}\Bigg|_3^{11} = \frac{32}{9}(9)^{3/2} - \frac{32}{9}(1)^{3/2}$$

$$= 96 - \frac{32}{9} = \frac{832}{9}$$

Now rework the problem by integrating with respect to $y$ first.

$$\iint_R \sqrt{x} \cdot \sqrt{y-2}\ dx\ dy = \int_0^4\left[\int_3^{11} \sqrt{x}\cdot\sqrt{y-2}\ dy\right]dx$$

$$= \int_0^4\left(\frac{2}{3}\sqrt{x}(y-2)^{3/2}\right)\Bigg|_3^{11}\ dx$$

$$= \int_0^4\left[\frac{2}{3}\sqrt{x}(9)^{3/2} - \frac{2}{3}\sqrt{x}(1)^{3/2}\right]dx$$

$$= \int_0^4\left(18\sqrt{x} - \frac{2}{3}\sqrt{x}\right)dx$$

$$= \int_0^4 \frac{52}{3}\sqrt{x}\ dx = \frac{104}{9}x^{3/2}\Bigg|_0^4$$

$$= \frac{104}{9}(4)^{3/2} - \frac{104}{9}(0)^{3/2} = \frac{832}{9}$$

As expected, both answers are the same. ■■

***Volume*** As shown earlier, the definite integral $\int_a^b f(x)dx$ can be used to find the area under a curve. In a similar manner, double integrals are used to find the *volume under a surface*. Figure 24 shows that portion of a surface $f(x, y)$ directly over a rectangle $R$ in the $xy$-plane. Just as areas were approximated by a large number of small rectangles, volume could be approximated by adding the volumes of a large number of properly drawn small boxes. The height of a typical box would be $f(x, y)$ with the length and width given by $dx$ and $dy$. The formula for the volume of a box would then suggest the following result.

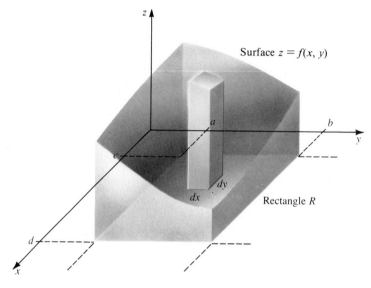

**FIGURE 24**

**VOLUME**

Let $z = f(x, y)$ be a function that is never negative on the rectangular region $R$ defined by $c \le x \le d$, $a \le y \le b$. The volume of the solid under the graph of $f$ and over the region $R$ is

$$\iint_R f(x, y)dx \, dy.$$

### EXAMPLE 6

Find the volume under the surface $z = x^2 + y^2$ shown in Figure 25.

By the results given above, the volume is

$$\iint_R f(x, y)dx \, dy,$$

where $f(x, y) = x^2 + y^2$ and $R$ is the region $0 \le x \le 4$, $0 \le y \le 4$.

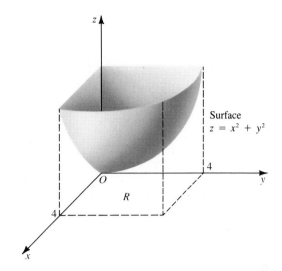

**FIGURE 25**

By definition,

$$\iint\limits_{R} f(x, y)dx\, dy = \int_0^4 \left[ \int_0^4 (x^2 + y^2)dx \right] dy$$

$$= \int_0^4 \left( \frac{1}{3}x^3 + xy^2 \right) \Big|_0^4 dy$$

$$= \int_0^4 \left( \frac{64}{3} + 4y^2 \right) dy = \left( \frac{64}{3}y + \frac{4}{3}y^3 \right) \Big|_0^4$$

$$= \frac{64}{3} \cdot 4 + \frac{4}{3} \cdot 4^3 - 0 = \frac{512}{3}. \quad \blacksquare$$

3 cm

4 cm

**FIGURE 26**

## ▇ EXAMPLE 7

A product design consultant for a cosmetics company has been asked to design a bottle for the company's newest perfume. The thickness of the glass is to vary so that the outside of the bottle has straight sides and the inside has curved sides, as shown in Figure 26. Before presenting the design to management, the consultant needs to make a reasonably accurate estimate of the amount each bottle will hold. If the base of the bottle is to be 4 cm by 3 cm, and if a cross section of its interior is to be a parabola of the form $z = -y^2 + 4y$, what is its internal volume?

The interior of the bottle can be graphed in three-dimensional space as shown in Figure 27. Its volume is simply the volume above the region $R$ in the

$xy$ plane and below the graph of $f(x, y) = -y^2 + 4y$. This volume is given by the double integral

$$\int_0^3 \int_0^4 (-y^2 + 4y) \, dy \, dx = \int_0^3 \left( \frac{-y^3}{3} + \frac{4y^2}{2} \right) \Big|_0^4 \, dx$$

$$= \int_0^3 \left( \frac{-64}{3} + 32 - 0 \right) dx$$

$$= \frac{32}{3} x \Big|_0^3$$

$$= 32 - 0 = 32.$$

The bottle holds 32 cubic centimeters. ▬▬

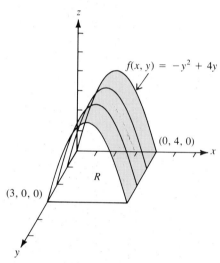

FIGURE 27

The next example illustrates the process of building a mathematical model to solve a practical problem.

▬▬ EXAMPLE 8

Consider a right cylinder $C$ of ocean water with its top on the surface of the ocean, as shown in Figure 28. Suppose that $C$ has depth $D$. Light incident on top of $C$ certainly fluctuates in intensity during the day, but at any time during the day, the light also diminishes in intensity as it passes down through $C$. Thus the intensity of light at a point in $C$ depends on the depth and the time. So let

FIGURE 28

$$x = \text{the depth below the surface in feet,}$$
$$t = \text{the time after dawn in seconds,}$$
$$\text{and} \quad i(x, t) = \text{the light intensity at depth } x \text{ and time } t.$$

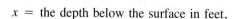

The photosynthesis rate of the phytoplankton in $C$ depends on several variables, but the light intensity is the primary factor. So let

$r(i(x, t)) = $ the instantaneous rate of photosynthesis production
per (vertical) foot per second at depth $x$ and time $t$.

Let $T > 0$ and divide the rectangular region

$$R = \{(x, t) | 0 \leq x \leq D \quad \text{and} \quad 0 \leq t \leq T\}$$

into smaller rectangles, as shown in Figure 29(a). Now let $\Delta x_i = x_i - x_{i-1}$ (see Figure 29(b)) and $\Delta t_j = t_j - t_{j-1}$ for each $i$ and each $j$. Then the product

$$r(i(x_i, t_j)) \cdot \Delta x_i \cdot \Delta t_j$$

approximates the photosynthesis production that occurs in the layer $x_{i-1} \leq x \leq x_i$ during the time interval $t_{j-1} \leq t \leq t_j$. The double sum

$$\sum_{j=1}^{n} \sum_{i=1}^{m} r(i(x_i, t_j)) \cdot \Delta x_i \cdot \Delta t_j$$

approximates the total photosynthesis production that occurs in $C$ during the time interval $0 \leq t \leq T$. Taking the limit as $n$ and $m$ become infinitely large, the total photosynthesis production that occurs in $C$ during the time interval $0 \leq t \leq T$ is

$$\int_0^T \int_0^D r(i(x, t))dx \, dt. \quad \blacksquare$$

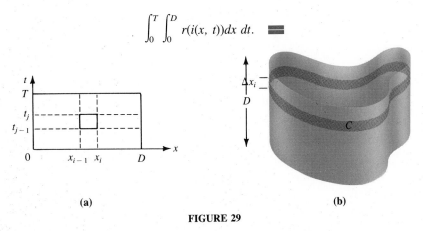

(a)                                (b)

**FIGURE 29**

***Double Integrals Over Other Regions***   In the work in this section, we found double integrals over rectangular regions, with constant limits of integration. Now this work can be extended to include *variable* limits of integration. (Notice in the following examples that the variable limits always go on the *inner* integral sign.)

### ▬ EXAMPLE 9

Evaluate $\displaystyle\int_1^2 \int_y^{y^2} xy \, dx \, dy$.

Integrate first with respect to $x$, then with respect to $y$.

$$\int_1^2 \int_y^{y^2} xy \, dx \, dy = \int_1^2 \left[ \int_y^{y^2} xy \, dx \right] dy = \int_1^2 \left( \frac{1}{2} x^2 y \right) \Big|_y^{y^2} dy$$

Replace $x$ first with $y^2$ and then with $y$, and subtract.

$$\int_1^2 \int_y^{y^2} xy \, dx \, dy = \int_1^2 \left[ \frac{1}{2}(y^2)^2 y - \frac{1}{2}(y)^2 y \right] dy$$

$$= \int_1^2 \left( \frac{1}{2} y^5 - \frac{1}{2} y^3 \right) dy = \left( \frac{1}{12} y^6 - \frac{1}{8} y^4 \right) \Big|_1^2$$

$$= \left( \frac{1}{12} \cdot 2^6 - \frac{1}{8} \cdot 2^4 \right) - \left( \frac{1}{12} \cdot 1^6 - \frac{1}{8} \cdot 1^4 \right)$$

$$= \frac{64}{12} - \frac{16}{8} - \frac{1}{12} + \frac{1}{8} = \frac{27}{8} \quad ▬$$

The use of variable limits of integration permits evaluation of double integrals over the types of regions shown in Figure 30. Double integrals over more complicated regions are discussed in more advanced books. Integration over regions such as those of Figure 30 is done with the results of the following theorem.

**DOUBLE INTEGRALS OVER VARIABLE REGIONS**

▬ Let $z = f(x, y)$ be a function of two variables.
If $R$ is the region (in Figure 30(a)) defined by $c \leq x \leq d$ and $g(x) \leq y \leq h(x)$, then

$$\iint_R f(x, y) \, dy \, dx = \int_c^d \left[ \int_{g(x)}^{h(x)} f(x, y) \, dy \right] dx.$$

If $R$ is the region (in Figure 30(b)) defined by $g(y) \leq x \leq h(y)$ and $a \leq y \leq b$, then

$$\iint_R f(x, y) \, dx \, dy = \int_a^b \left[ \int_{g(y)}^{h(y)} f(x, y) \, dx \right] dy.$$

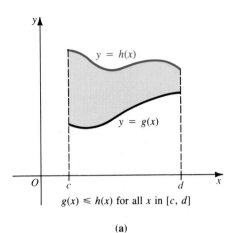

$g(x) \le h(x)$ for all $x$ in $[c, d]$

**(a)**

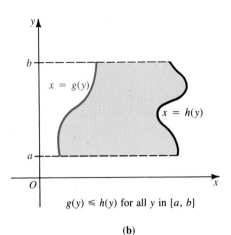

$g(y) \le h(y)$ for all $y$ in $[a, b]$

**(b)**

**FIGURE 30**

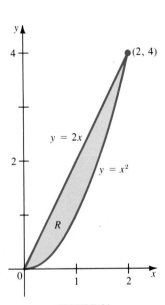

**FIGURE 31**

### ▬ EXAMPLE 10

Let $R$ be the shaded region in Figure 31, and evaluate

$$\iint_R (x + 2y) \, dy \, dx.$$

Region $R$ is bounded by $h(x) = 2x$ and $g(x) = x^2$, with $0 \le x \le 2$. By the first result in the theorem above,

$$\iint_R (x + 2y) \, dydx = \int_0^2 \int_{x^2}^{2x} (x + 2y) \, dy \, dx$$

$$= \int_0^2 (xy + y^2) \Big|_{x^2}^{2x} \, dx$$

$$= \int_0^2 [x(2x) + (2x)^2 - (x \cdot x^2 + (x^2)^2)] \, dx$$

$$= \int_0^2 [2x^2 + 4x^2 - (x^3 + x^4)] \, dx$$

$$= \int_0^2 (6x^2 - x^3 - x^4) \, dx$$

$$= \left( 2x^3 - \frac{1}{4}x^4 - \frac{1}{5}x^5 \right) \Big|_0^2$$

$$= 2 \cdot 2^3 - \frac{1}{4} \cdot 2^4 - \frac{1}{5} \cdot 2^5 - 0$$

$$= 16 - 4 - \frac{32}{5} = \frac{28}{5}. \quad ▬$$

### ▬ EXAMPLE 11

Let $R$ be the shaded region in Figure 32, and evaluate

$$\iint_R (x + 2y)\, dx\, dy.$$

This is the same region shown in Figure 31, but with the equations of the boundaries given in terms of $x$ rather than $y$. That is, $R$ is defined by $y/2 \le x \le \sqrt{y}$, $0 \le y \le 4$, and

$$\iint_R (x + 2y)\, dxdy = \int_0^4 \int_{y/2}^{\sqrt{y}} (x + 2y)\, dx\, dy$$

$$= \int_0^4 \left( \frac{1}{2}x^2 + 2xy \right) \Bigg|_{y/2}^{\sqrt{y}} dy$$

$$= \int_0^4 \left[ \left( \frac{1}{2}(\sqrt{y})^2 + 2(\sqrt{y})y \right) - \left( \frac{1}{2}\left(\frac{1}{2}y\right)^2 + 2\left(\frac{1}{2}y\right)y \right) \right] dy$$

$$= \int_0^4 \left[ \frac{1}{2}y + 2y^{3/2} - \left( \frac{1}{8}y^2 + y^2 \right) \right] dy$$

$$= \int_0^4 \left[ \frac{1}{2}y + 2y^{3/2} - \frac{9}{8}y^2 \right] dy$$

$$= \left( \frac{1}{4}y^2 + \frac{4}{5}y^{5/2} - \frac{3}{8}y^3 \right) \Bigg|_0^4$$

$$= \frac{1}{4} \cdot 4^2 + \frac{4}{5} \cdot 4^{5/2} - \frac{3}{8} \cdot 4^3 - 0$$

$$= 4 + \frac{128}{5} - 24 = \frac{28}{5}. \quad ▬$$

The answers in Examples 10 and 11 are the same, as we would expect them to be.

***Interchanging Limits of Integration***    Sometimes it is easier to integrate first with respect to $x$, and then $y$, while with other integrals the reverse process is easier. The limits of integration can be reversed whenever the region $R$ is like one of the regions in Figure 32. The next example shows how this process works.

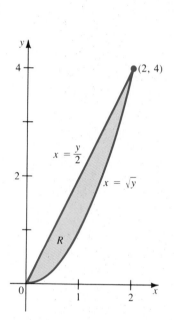

**FIGURE 32**

### ■■ EXAMPLE 12

Interchange the limits of integration in

$$\int_0^{16} \int_{\sqrt{y}}^{4} f(x, y) \, dx \, dy.$$

For this integral, region $R$ is given by $\sqrt{y} \le x \le 4$, $0 \le y \le 16$. A graph of $R$ is shown in Figure 33.

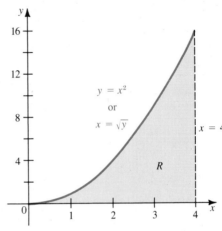

**FIGURE 33**

The same region $R$ can be written in an alternate way. As Figure 33 shows, one boundary of $R$ is $x = \sqrt{y}$. Solving for $y$ gives $y = x^2$. Also, Figure 33 shows that $0 \le x \le 4$. Since $R$ can be written as $0 \le y \le x^2$, $0 \le x \le 4$, the double integral above can be written

$$\int_0^4 \int_0^{x^2} f(x, y) \, dy \, dx. \quad ■$$

### ■ 7.7 EXERCISES

*Evaluate the following integrals.*

**1.** $\displaystyle\int_0^3 (x^3 y + y) \, dx$

**2.** $\displaystyle\int_1^4 (xy^2 - x) \, dy$

**3.** $\displaystyle\int_2^5 (x + y^3 x^2 - 2) \, dy$

**4.** $\displaystyle\int_1^3 (x^3 - 2x^2 y^2 + 5) \, dx$

**5.** $\displaystyle\int_4^8 \sqrt{6x + y} \, dx$

**6.** $\displaystyle\int_3^7 \sqrt{x + 5y} \, dy$

**7.** $\displaystyle\int_3^6 x\sqrt{x^2 + 3y} \, dx$

**8.** $\displaystyle\int_4^5 x\sqrt{x^2 + 3y} \, dy$

**9.** $\displaystyle\int_4^9 \frac{3 + 5y}{\sqrt{x}} \, dx$

**10.** $\displaystyle\int_2^7 \frac{3 + 5y}{\sqrt{x}} \, dy$

**11.** $\displaystyle\int_3^5 \frac{6x + 2y}{3x^2 + 2xy} \, dx$

**12.** $\displaystyle\int_1^4 \frac{30y^2 + 16yx^2}{10y^3 + 8y^2x^2} \, dy$

**13.** $\displaystyle\int_{-1}^1 e^{x + 4y} \, dy$

**14.** $\displaystyle\int_2^6 e^{x + 4y} \, dx$

**15.** $\displaystyle\int_0^5 xe^{x^2 + 9y} \, dx$

**16.** $\displaystyle\int_1^6 xe^{x^2 + 9y} \, dy$

**17.** $\displaystyle\int_0^4 y\sqrt{y^2 + 3x} \, dy$

**18.** $\displaystyle\int_1^5 y\sqrt{y^2 + 3x} \, dy$

**19.** $\displaystyle\int_3^6 \frac{10y}{\sqrt{3x + 5y^2}} \, dx$

**20.** $\displaystyle\int_0^2 \frac{10y}{\sqrt{3x + 5y^2}} \, dy$

Evaluate the following iterated integrals. (Many of these use results from Exercises 1–20).

**21.** $\int_1^2 \left[ \int_0^3 (x^3y + y) \, dx \right] dy$

**22.** $\int_0^3 \left[ \int_1^4 (xy^2 - x) \, dy \right] dx$

**23.** $\int_{-1}^1 \left[ \int_2^5 (x + y^3x^2 - 2) \, dy \right] dx$

**24.** $\int_{-2}^1 \left[ \int_1^3 (x^3 - 2x^2y^2 + 5) \, dx \right] dy$

**25.** $\int_0^1 \left[ \int_3^6 x\sqrt{x^2 + 3y} \, dx \right] dy$

**26.** $\int_0^3 \left[ \int_4^5 x\sqrt{x^2 + 3y} \, dy \right] dx$

**27.** $\int_1^2 \left[ \int_4^9 \frac{3 + 5y}{\sqrt{x}} \, dx \right] dy$

**28.** $\int_{16}^{25} \left[ \int_2^7 \frac{3 + 5y}{\sqrt{x}} \, dy \right] dx$

**29.** $\int_0^3 \left[ \int_0^4 y\sqrt{y^2 + 3x} \, dy \right] dx$

**30.** $\int_0^2 \left[ \int_2^6 e^{x+4y} \, dx \right] dy$

**31.** $\int_1^2 \int_1^2 \frac{dx \, dy}{xy}$

**32.** $\int_1^4 \int_2^5 \frac{dy \, dx}{x}$

**33.** $\int_2^4 \int_3^5 \left( \frac{x}{y} + \frac{y}{3} \right) dx \, dy$

**34.** $\int_3^4 \int_1^2 \left( \frac{6x}{5} + \frac{y}{x} \right) dx \, dy$

Find each double integral over the rectangular region R with the given boundaries.

**35.** $\iint_R (x + 3y^2) \, dx \, dy; \quad 0 \le x \le 2, \, 1 \le y \le 5$

**36.** $\iint_R (4x^3 + y^2) \, dx \, dy; \quad 1 \le x \le 4, \, 0 \le y \le 2$

**37.** $\iint_R \sqrt{x + y} \, dy \, dx; \quad 1 \le x \le 3, \, 0 \le y \le 1$

**38.** $\iint_R x^2\sqrt{x^3 + 2y} \, dx \, dy; \quad 0 \le x \le 2, \, 0 \le y \le 3$

**39.** $\iint_R \frac{2}{(x + y)^2} \, dy \, dx; \quad 2 \le x \le 3, \, 1 \le y \le 5$

**40.** $\iint_R \frac{y}{\sqrt{6x + 5y^2}} \, dx \, dy; \quad 0 \le x \le 3, \, 1 \le y \le 2$

**41.** $\iint_R ye^{(x+y^2)} \, dx \, dy; \quad 2 \le x \le 3, \, 0 \le y \le 2$

**42.** $\iint_R x^2 e^{(x^3 + 2y)} \, dx \, dy; \quad 1 \le x \le 2, \, 1 \le y \le 3$

Find the volume under the given surface $z = f(x, y)$ and above the rectangle with the given boundaries.

**43.** $z = 6x + 2y + 5; \quad -1 \le x \le 1, \, 0 \le y \le 3$

**44.** $z = 9x + 5y + 12; \quad 0 \le x \le 3, \, -2 \le y \le 1$

**45.** $z = x^2; \quad 0 \le x \le 1, \, 0 \le y \le 4$

**46.** $z = \sqrt{y}; \quad 0 \le x \le 4, \, 0 \le y \le 9$

**47.** $z = x\sqrt{x^2 + y}; \quad 0 \le x \le 1, \, 0 \le y \le 1$

**48.** $z = yx\sqrt{x^2 + y^2}; \quad 0 \le x \le 4, \, 0 \le y \le 1$

**49.** $z = \frac{xy}{(x^2 + y^2)^2}; \quad 1 \le x \le 2, \, 1 \le y \le 4$

**50.** $z = e^{x+y}; \quad 0 \le x \le 1, \, 0 \le y \le 1$

While it is true that a double integral can be evaluated by using either dx or dy first, sometimes one choice over the other makes the work easier. Evaluate the double integrals in Exercise 51 and 52 in the easiest way possible.

**51.** $\iint_R xe^{xy} \, dx \, dy; \quad 0 \le x \le 2, \, 0 \le y \le 1$

**52.** $\iint_R x^2 e^{2x^3 + 6y} \, dx \, dy; \quad 0 \le x \le 1, \, 0 \le y \le 1$

*Evaluate each double integral.*

**53.** $\int_1^4 \int_0^y (x + 4y) \, dx \, dy$

**54.** $\int_0^2 \int_1^x (3x + 5y) \, dy \, dx$

**55.** $\int_2^4 \int_2^{x^2} (x^2 + y^2) \, dy \, dx$

**56.** $\int_0^5 \int_0^{2y} (x^2 + y) \, dx \, dy$

**57.** $\int_0^4 \int_0^x \sqrt{xy} \, dy \, dx$

**58.** $\int_1^4 \int_0^x \sqrt{x + y} \, dy \, dx$

**59.** $\int_1^2 \int_0^{x^2 - 1} xy^2 \, dy \, dx$

**60.** $\int_2^4 \int_{1+y}^{2+3y} (x - y^2) \, dx \, dy$

**61.** $\int_1^2 \int_y^{3y} \frac{1}{x} \, dx \, dy$

**62.** $\int_1^4 \int_x^{x^2} \frac{1}{y} \, dy \, dx$

**63.** $\int_0^4 \int_1^{e^x} \frac{x}{y} \, dy \, dx$

**64.** $\int_0^1 \int_{2x}^{4x} e^{x+y} \, dy \, dx$

*Use the region R with the indicated boundaries to evaluate each of the following double integrals.*

**65.** $\iint_R (4x + 7y) \, dy \, dx; \quad 1 \le x \le 3, \, 0 \le y \le x + 1$

**66.** $\iint_R (3x + 9y) \, dy \, dx; \quad 2 \le x \le 4, \, 2 \le y \le 3x$

**67.** $\iint_R (4 - 4x^2) \, dy \, dx; \quad 0 \le x \le 1, \, 0 \le y \le 2 - 2x$

**68.** $\iint_R \frac{dy \, dx}{x}; \quad 1 \le x \le 2, \, 0 \le y \le x - 1$

**69.** $\iint_R e^{x/y^2} \, dx \, dy; \quad 1 \le y \le 2, \, 0 \le x \le y^2$

**70.** $\iint_R (x^2 - y) \, dy \, dx; \quad -1 \le x \le 1, \, -x^2 \le y \le x^2$

**71.** $\iint_R x^3 y \, dx \, dy; \quad R \text{ bounded by } y = x^2, \, y = 2x$

**72.** $\iint_R x^2 y^2 \, dx \, dy; \quad R \text{ bounded by } y = x, \, y = 2x, \, x = 1$

**73.** $\iint_R (x + y) \, dy \, dx; \quad R \text{ bounded by } 4y = x^2, \, x = 2y - 4$

**74.** $\iint_R xy^2 \, dy \, dx; \quad R \text{ bounded by } y = 2x, \, y = 3 - x, \, y = 0$

**75.** $\iint_R \frac{dy \, dx}{y}; \quad R \text{ bounded by } y = x, \, y = \frac{1}{x}, \, x = 2$

The idea of the average value of a function, discussed earlier for functions of the form $y = f(x)$, can be extended to functions of more than one independent variable. For a function $z = f(x, y)$, the **average value** of $f$ over a region $R$ is defined as

$$\frac{1}{A} \iint_R f(x, y) \, dx \, dy,$$

where $A$ is the area of the region $R$. Find the average value for each of the following functions over the regions $R$ having the given boundaries.

**76.** $f(x, y) = 5xy + 2y; \, 1 \le x \le 4, \, 1 \le y \le 2$

**77.** $f(x, y) = x^2 + y^2; \, 0 \le x \le 2, \, 0 \le y \le 3$

**78.** $f(x, y) = e^{-5y+3x}; \, 0 \le x \le 2, \, 0 \le y \le 2$

**79.** $f(x, y) = e^{2x+y}; \, 1 \le x \le 2, \, 2 \le y \le 3$

# ≡ APPLICATIONS

## BUSINESS AND ECONOMICS

**Packaging**   **80.** The manufacturer of fruit juice drink has decided to try innovative packaging in order to revitalize sagging sales. The fruit juice drink is to be packaged in containers in the shape of tetrahedra in which three edges are perpendicular, as shown in the figure. Two of the perpendicular edges will be 3 inches long, and the third edge will be 6 inches long. Find the volume of the container. [*Hint:* The equation of the plane shown in the figure is $z = f(x, y) = 6 - 2x - 2y$.)

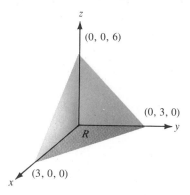

**Average Cost**   **81.** A company's total cost for operating its two warehouses is

$$C(x, y) = \frac{1}{9}x^2 + 2x + y^2 + 5y + 100$$

dollars, where $x$ represents the number of units stored at the first warehouse and $y$ represents the number of units stored at the second. Find the average cost to store a unit if the first warehouse has between 48 and 75 units, and the second has between 20 and 60 units.

**Average Production**   **82.** A production function is given by

$$P(x, y) = 500x^{.2}y^{.8},$$

where $x$ is the number of units of labor and $y$ is the number of units of capital. Find the average production level if $x$ varies from 10 to 50 and $y$ from 20 to 40.

**Average Profit**   **83.** The profit in dollars from selling $x$ units of one product and $y$ units of a second product is

$$P = -(x - 100)^2 - (y - 50)^2 + 2000.$$

The weekly sales for the first product vary from 100 to 150 units, and the weekly sales for the second product vary from 40 to 80 units. Estimate average weekly profit for these two products.

**Average Revenue**   **84.** A company sells two products. The demand functions of the products are given by

$$x_1 = 300 - 2p_1 \quad \text{and} \quad x_2 = 500 - 1.2p_2,$$

where $x_1$ units of the first product are demanded at price $p_1$ and $x_2$ units of the second product are demanded at price $p_2$. The total revenue will be given by

$$R = x_1p_1 + x_2p_2.$$

Find the average revenue if the price $p_1$ varies from \$25 to \$50 and the price $p_2$ varies from \$50 to \$75.

## PHYSICAL SCIENCES

**Effects of Nuclear Weapons**   **85.** *The Effects of Nuclear Weapons,* prepared by the U.S. Department of Defense, contains these remarks on the computation of the radioactive dose after a 1-kiloton atomic explosion.*

"If all the residues from 1-kiloton fission yield were deposited on a smooth surface in varying concentrations typical of an early fallout pattern, instead of uniformly, the product of the dose rate at 1 hour and the area would be replaced by the 'area integral' of the 1-hour dose rate defined by

$$\text{Area Integral} = \int_A R_1 \, dA,$$

where $R_1$ is the 1-hour dose rate over an element of area $dA$, and $A$ square miles is the total area covered by the residues."

Explain why an integral is involved. (Note that in this case $R_1$ denotes the function and $A$ the region. In the diverse applications of integrals many notations are employed.)

---

**KEY WORDS**

**7.1**   **functions of several variables**
**ordered triple**
**first octant**
**distance formula**
**sphere**
**plane**
**surface**
**trace**
**level curves**
**production function**
**isoquant**
**7.2**   **partial derivative**

**second-order partial derivative**
**7.3**   **saddle point**
**critical point**
**7.4**   **constraint**
**Lagrange multiplier**
**7.5**   **scatter diagram**
**least squares line**
**regression line**
**7.6**   **total differential**
**7.7**   **iterated integral**
**double integral**
**region of integration**

---

## CHAPTER 7   REVIEW EXERCISES

*Find f(−1, 2) and f(6, −3) for each of the following.*

**1.** $f(x, y) = -4x^2 + 6xy - 3$

**2.** $f(x, y) = 3x^2y^2 - 5x + 2y$

**3.** $f(x, y) = \dfrac{x - 3y}{x + 4y}$

**4.** $f(x, y) = \dfrac{\sqrt{x^2 + y^2}}{x - y}$

*Find the distance between the points in each pair.*

**5.** $(-1, 4, 0)$ and $(2, -1, 3)$

**6.** $(-2, 1, 5)$ and $(1, 1, 4)$

**7.** $(0, -2, 5)$ and $(3, 2, -1)$

**8.** $(5, 0, 3)$ and $(-2, 4, 3)$

*Graph the first-octant portion of each plane.*

**9.** $x + y + z = 4$

**10.** $x + y + 4z = 8$

**11.** $5x + 2y = 10$

**12.** $3x + 5z = 15$

**13.** $x = 3$

**14.** $y = 2$

---

*Find the center and radius of each sphere in Exercises 15–16.*

**15.** $x^2 - 4x + y^2 + 6y + z^2 - 8z + 20 = 0$

**16.** $x^2 + 6x + y^2 - 10y + z^2 + 12z + 54 = 0$

**17.** Let $z = f(x, y) = -5x^2 + 7xy - y^2$. Find each of the following.

**(a)** $\dfrac{\partial z}{\partial x}$   **(b)** $\left(\dfrac{\partial z}{\partial y}\right)(-1, 4)$   **(c)** $f_{xy}(2, -1)$

**18.** Let $z = f(x, y) = \dfrac{x + y^2}{x - y^2}$. Find each of the following.

**(a)** $\dfrac{\partial z}{\partial y}$   **(b)** $\left(\dfrac{\partial z}{\partial x}\right)(0, 2)$   **(c)** $f_{xx}(-1, 0)$

*Find $f_x$ and $f_y$.*

**19.** $f(x, y) = 9x^3y^2 - 5x$

**20.** $f(x, y) = 6x^5y - 8xy^9$

**21.** $f(x, y) = \sqrt{4x^2 + y^2}$

**22.** $f(x, y) = \dfrac{2x + 5y^2}{3x^2 + y^2}$

**23.** $f(x, y) = x^2 \cdot e^{2y}$

**24.** $f(x, y) = (y - 2)^2 \cdot e^{(x + 2y)}$

**25.** $f(x, y) = \ln|2x^2 + y^2|$

**26.** $f(x, y) = \ln|2 - x^2y^3|$

*Find $f_{xx}$ and $f_{xy}$.*

**27.** $f(x, y) = 4x^3y^2 - 8xy$

**28.** $f(x, y) = -6xy^4 + x^2y$

**29.** $f(x, y) = \dfrac{2x}{x - 2y}$

**30.** $f(x, y) = \dfrac{3x + y}{x - 1}$

**31.** $f(x, y) = x^2e^y$

**32.** $f(x, y) = ye^{x^2}$

**33.** $f(x, y) = \ln|2 - x^2y|$

**34.** $f(x, y) = \ln|1 + 3xy^2|$

*Find all points where the functions defined below have any relative extrema. Find any saddle points.*

**35.** $z = x^2 + 2y^2 - 4y$

**36.** $z = x^2 + y^2 + 9x - 8y + 1$

**37.** $f(x, y) = x^2 + 5xy - 10x + 3y^2 - 12y$

**38.** $z = x^3 - 8y^2 + 6xy + 4$

**39.** $z = \dfrac{1}{2}x^2 + \dfrac{1}{2}y^2 + 2xy - 5x - 7y + 10$

**40.** $f(x, y) = 3x^2 + 2xy + 2y^2 - 3x + 2y - 9$

**41.** $z = x^3 + y^2 + 2xy - 4x - 3y - 2$

**42.** $f(x, y) = 7x^2 + y^2 - 3x + 6y - 5xy$

*Use Lagrange multipliers to find the extrema of the functions defined as follows, subject to the given constraints.*

**43.** $f(x, y) = x^2y$;   $x + y = 4$

**44.** $f(x, y) = x^2 + y^2$;   $x = y + 2$

**45.** $z = 3x^2 - 2y^2$;   $y - 5x = 0$

**46.** $z = x^2 + 2xy + 2y^2$;   $x + y = 1$

**47.** Find two numbers $x$ and $y$, whose sum is 80, such that $x^2y$ is maximized.

**48.** Find two numbers $x$ and $y$, whose sum is 50, such that $xy^2$ is maximized.

**49.** A closed box with square ends must have a volume of 125 cubic inches. Find the dimensions of such a box that has minimum surface area.

**50.** Find the maximum rectangular area that can be enclosed with 400 feet of fencing, if no fencing is needed along one side.

*Find dz or dw, as appropriate, for each of the following.*

**51.** $z = 6x^3 - 10y^2$

**52.** $z = 7x^3y - 4y^3$

**53.** $z = \dfrac{x - 2y}{x + 2y}$

**54.** $z = 3x^2 + \sqrt{x + y}$    **55.** $z = x^2 y e^{x-y}$    **56.** $z = \ln|x + 4y| + y^2 \ln x$

**57.** $w = x^5 + y^4 - z^3$    **58.** $w = \dfrac{3 + 5xy}{2 - z}$

*Evaluate dz using the given information.*

**59.** $z = 2x^2 - 4y^2 + 6xy;\quad x = 2,\ y = -3,\ dx = .01,\ dy = .05$

**60.** $z = \dfrac{x + 5y}{x - 2y};\quad x = 1,\ y = -2,\ dx = -.04,\ dy = .02$

*Evaluate each of the following.*

**61.** $\displaystyle\int_0^4 (x^2 y^2 + 5x)dx$    **62.** $\displaystyle\int_0^3 (x + 5y + y^2)dy$    **63.** $\displaystyle\int_2^5 \sqrt{6x + 3y}\ dx$

**64.** $\displaystyle\int_1^3 6y^4 \sqrt{8x + 3y}\ dx$    **65.** $\displaystyle\int_4^9 \dfrac{6y - 8}{\sqrt{x}}dx$    **66.** $\displaystyle\int_3^5 e^{2x - 7y}dx$

**67.** $\displaystyle\int_0^5 \dfrac{6x}{\sqrt{4x^2 + 2y^2}}dx$    **68.** $\displaystyle\int_1^3 \dfrac{y^2}{\sqrt{7x + 11y^3}}dy$

*Evaluate each iterated integral.*

**69.** $\displaystyle\int_0^2 \left[\int_0^4 (x^2 y^2 + 5x)dx\right]dy$    **70.** $\displaystyle\int_0^2 \left[\int_0^3 (x + 5y + y^2)dy\right]dx$    **71.** $\displaystyle\int_3^4 \left[\int_2^5 \sqrt{6x + 3y}\ dx\right]dy$

**72.** $\displaystyle\int_1^2 \left[\int_3^5 e^{2x - 7y}dx\right]dy$    **73.** $\displaystyle\int_2^4 \int_2^4 \dfrac{dx\ dy}{y}$    **74.** $\displaystyle\int_1^2 \int_1^2 \dfrac{dx\ dy}{x}$

*Find each double integral over the region R with boundaries as indicated.*

**75.** $\displaystyle\iint_R (x^2 + y^2)dx\ dy;\quad 0 \le x \le 2,\ 0 \le y \le 3$    **76.** $\displaystyle\iint_R \sqrt{2x + y}\ dx\ dy;\quad 1 \le x \le 3,\ 2 \le y \le 5$

**77.** $\displaystyle\iint_R \sqrt{y + x}\ dx\ dy;\quad 0 \le x \le 7,\ 1 \le y \le 9$    **78.** $\displaystyle\iint_R y e^{y^2 + x}dx\ dy;\quad 0 \le x \le 1,\ 0 \le y \le 1$

*Find the volume under the given surface z = f(x, y) and above the given rectangle.*

**79.** $z = x + 9y + 8;\quad 1 \le x \le 6,\ 0 \le y \le 8$    **80.** $z = x^2 + y^2;\quad 3 \le x \le 5,\ 2 \le y \le 4$

*Evaluate each double integral.*

**81.** $\displaystyle\int_0^1 \int_0^{2x} xy\ dy\ dx$    **82.** $\displaystyle\int_0^1 \int_0^{x^3} y\ dy\ dx$    **83.** $\displaystyle\int_0^1 \int_{x^2}^x x^3 y\ dy\ dx$    **84.** $\displaystyle\int_0^1 \int_y^{\sqrt{y}} x\ dx\ dy$

*Use the region R, with boundaries as indicated, to evaluate the given double integral.*

**85.** $\displaystyle\iint_R (2x + 3y)dx\ dy;\quad 0 \le y \le 1,\ y \le x \le 2 - y$

**86.** $\displaystyle\iint_R (2 - x^2 - y^2)dy\ dx;\quad 0 \le x \le 1,\ x^2 \le y \le x$

## ▦ APPLICATIONS

### BUSINESS AND ECONOMICS

**Charge for Auto Painting**   **87.** The charge in dollars for painting a sports car is given by

$$C(x, y) = 2x^2 + 4y^2 - 3xy + \sqrt{x},$$

where $x$ is the number of hours of labor needed and $y$ is the number of gallons of paint and sealer used. Find each of the following.
(a) $C(10, 5)$   (b) $C(15, 10)$   (c) $C(20, 20)$

**Manufacturing Costs**   **88.** The manufacturing cost in dollars for a medium-sized business computer is given by

$$c(x, y) = 2x + y^2 + 4xy + 25,$$

where $x$ is the memory capacity of the computer in kilobytes and $y$ is the number of hours of labor required. Find each of the following.

(a) $\dfrac{\partial c}{\partial x}(64, 6)$   (b) $\dfrac{\partial c}{\partial y}(128, 12)$

**Productivity**   **89.** The production function $z$ for one country is

$$z = x^{.6}y^{.4},$$

where $x$ represents the amount of labor and $y$ the amount of capital. Find the marginal productivity of each of the following.
(a) Labor   (b) Capital

**Total Cost**   *The total cost in dollars to manufacture $x$ solar cells and $y$ solar collectors is*

$$c(x, y) = x^2 + 5y^2 + 4xy - 70x - 164y + 1800.$$

**90.** Find values of $x$ and $y$ that produce minimum total cost.

**91.** Find the minimum total cost.

**Earnings**   **92.** (a) Find the least squares line for the following data, which give the earnings in ten-thousands for a certain company after $x$ yr.

| $x$ | 3 | 5 | 7 | 8 |
|---|---|---|---|---|
| $y$ | 4 | 11 | 20 | 23 |

(b) Predict the earnings after 6 yr.

**Production Materials**   **93.** Approximate the amount of material needed to manufacture a cone of radius 2 cm, height 8 cm, and wall thickness .21 cm.

**Production Materials**   **94.** A sphere of radius 2 ft is to receive an insulating coating 1 inch thick. Approximate the volume of the coating needed.

**Production Error**   **95.** The height of a sample cone from a production line is measured as 11.4 cm, while the radius is measured as 2.9 cm. Each of these measurements could be off by .2 cm. Approximate the maximum possible error in the volume of the cone.

### LIFE SCIENCES

**Animal Interaction**   **96.** The number of foxes and birds that can coexist in an area is approximated by

$$N(x, y) = 5x^2 - 3x + y^2 + xy,$$

where $x$ is the average temperature in degrees Celsius and $y$ is the annual rainfall in centimeters. Find each of the following.

(a) $N(20, 30)$   (b) $N(24, 50)$   (c) $N(14, 70)$

**Blood Sugar and Cholesterol Levels**

97. The following data show the connection between blood sugar levels $x$ and cholesterol levels $y$ for 8 different diabetic patients.

| Patient | 1 | 2 | 3 | 4 | 5 | 6 | 7 | 8 |
|---|---|---|---|---|---|---|---|---|
| Blood Sugar Level, $x$ | 130 | 138 | 142 | 159 | 165 | 200 | 210 | 250 |
| Cholesterol Level, $y$ | 170 | 160 | 173 | 181 | 201 | 192 | 240 | 290 |

For this data, $\Sigma x = 1394$, $\Sigma y = 1607$, $\Sigma xy = 291,990$, and $\Sigma x^2 = 255,214$.

(a) Find the equation of the least squares line.

(b) Predict the cholesterol level for a person whose blood sugar level is 190.

**Blood Vessel Volume**

98. A length of blood vessel is measured as 2.7 cm, with the radius measured as .7 cm. If each of these measurements could by off by .1 cm, estimate the maximum possible error in the volume of the vessel.

# 8 Differential Equations

Suppose that an economist wants to develop an equation that will forecast interest rates. By studying data on previous changes in interest rates, she hopes to find a relationship between the level of interest rates and their rate of change. A function giving the rate of change of interest rates would be the derivative of the function describing the level of interest rates. A **differential equation** is an equation that involves an unknown function, where $y = f(x)$, and a finite number of its derivatives. Solving the differential equation for $y$ would give the unknown function to be used for forecasting interest rates.

Differential equations have been important in the study of physical science and engineering since the eighteenth century. Among the pioneers in the field of differential equations was the French mathematician Alexis Claude Clairaut (1713–1765). A particular type of equation studied in elementary courses on differential equations bears his name.

More recently, differential equations have become useful in social sciences, life sciences, and economics for solving problems about population growth, ecological balance, and interest rates. In this chapter, we will introduce some methods for solving differential equations and give examples of their applications.

## ▤ 8.1  SOLUTIONS OF DIFFERENTIAL EQUATIONS

Usually a solution of an equation is a *number*. A solution of a differential equation, however, is a *function*. For example, the solutions of a differential equation such as

$$\frac{dy}{dx} = 3x^2 - 2x \tag{1}$$

consist of all expressions for $y$ that satisfy the equation. Since the left side of the equation is the derivative of $y$ with respect to $x$, we can solve the equation for $y$ by finding an antiderivative on each side. On the left, the antiderivative is $y + C_1$. On the right side,

$$\int (3x^2 - 2x)dx = x^3 - x^2 + C_2.$$

The solutions of equation (1) are given by

$$y + C_1 = x^3 - x^2 + C_2$$

or

$$y = x^3 - x^2 + C_2 - C_1.$$

Replacing the constant $C_2 - C_1$ with the single constant $C$ gives

$$y = x^3 - x^2 + C. \tag{2}$$

(From now on we will add just one constant, with the understanding that it represents the difference between the two constants obtained in the two integrations.)

Each different value of $C$ in equation (2) leads to a different solution of equation (1), showing that a differential equation can have an infinite number of solutions. Equation (2) is the **general solution** of the differential equation (1). Some of the solutions of equation (1) are graphed in Figure 1.

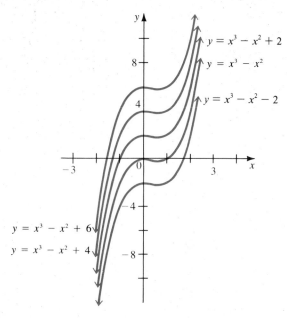

**FIGURE 1**

The simplest kind of differential equation has the form

$$\frac{dy}{dx} = f(x).$$

Since equation (1) has this form, the solution of equation (1) suggests the following generalization.

**GENERAL SOLUTION OF**
$$\frac{dy}{dx} = f(x)$$

The general solution of the differential equation $dy/dx = f(x)$ is

$$y = \int f(x)\ dx.$$

## EXAMPLE 1

The time-dating of dairy products depends on the solution of a differential equation. Suppose that $y$ is the number of bacteria (in thousands) present in a given product at time $t$ (in days) and

$$\frac{dy}{dt} = 10t. \tag{3}$$

Further, suppose at $t = 0$, $y = 50$ (in thousands). Find an equation for $y$ in terms of $t$.

Solve the differential equation $dy/dt = 10t$ by using the general solution given above.

$$y = \int 10t \, dt.$$

$$y = \frac{10t^2}{2} + C$$

$$y = 5t^2 + C$$

The result is the general solution of equation (3). Because of the given information that $y = 50$ when $t = 0$, a specific value of $C$ can be found here. Replacing $t$ with 0 and $y$ with 50 in the general solution gives

$$50 = 5 \cdot 0^2 + C$$

$$50 = C.$$

Let $C = 50$ in the general solution to get

$$y = 5t^2 + 50. \quad \blacksquare$$

In Example 1, the given information was used to produce a solution with a specific value of $C$. Such a solution is called a **particular solution** of the given differential equation. The given information, $y = 50$ when $t = 0$, is called an **initial condition** because $t = 0$.

Sometimes a differential equation must be rewritten in the form

$$\frac{dy}{dx} = f(x)$$

before it can be solved.

**■ EXAMPLE 2**

Find the particular solution of

$$\frac{dy}{dx} - 2x = 5,$$

given that $y = 2$ when $x = -1$.

Add $2x$ to both sides of the equation to get

$$\frac{dy}{dx} = 2x + 5.$$

The general solution is

$$y = \frac{2x^2}{2} + 5x + C = x^2 + 5x + C.$$

Substituting 2 for $y$ and $-1$ for $x$ gives

$$2 = (-1)^2 + 5(-1) + C$$
$$C = 6.$$

The particular solution is $y = x^2 + 5x + 6$. ▬

▬ EXAMPLE 3

The population $P$ of a flock of birds is growing exponentially so that

$$\frac{dP}{dx} = 20e^{.05x},$$

where $x$ is time in years. Find $P$ in terms of $x$ if there were 20 birds in the flock initially.

Solve the differential equation:

$$P = \int 20e^{.05x}\, dx = \frac{20}{.05}e^{.05x} + C = 400e^{.05x} + C.$$

Since $P$ is 20 when $x$ is 0,

$$20 = 400e^0 + C$$
$$-380 = C,$$

and

$$P = 400e^{.05x} - 380. \quad ▬$$

▬ 8.1 EXERCISES

*Find general solutions for the following differential equations.*

**1.** $\dfrac{dy}{dx} = x^2$

**2.** $\dfrac{dy}{dx} = -x + 2$

**3.** $\dfrac{dy}{dx} = -2x + 3x^2$

**4.** $\dfrac{dy}{dx} = 6x^2 - 4x$

**5.** $\dfrac{dy}{dx} = e^x$

**6.** $\dfrac{dy}{dx} = e^{3x}$

**7.** $\dfrac{dy}{dx} = 2e^{-x}$

**8.** $\dfrac{dy}{dx} = 3e^{-2x}$

**9.** $3\dfrac{dy}{dx} = -4x^2$

**10.** $4\dfrac{dy}{dx} - x = 0$

**11.** $3x^3 - 2\dfrac{dy}{dx} = 0$

**12.** $3x^2 - 3\dfrac{dy}{dx} = 2$

*Find particular solutions for the following differential equations.*

**13.** $\dfrac{dy}{dx} + 2x = 3x^2$;   $y = 2$ when $x = 0$

**14.** $\dfrac{dy}{dx} = 4x + 3$;   $y = -4$ when $x = 0$

**15.** $\dfrac{dy}{dx} = 3x^2 - 4x + 2$;   $y = 3$ when $x = -1$

**16.** $\dfrac{dy}{dx} = 4x^3 - 3x^2 + x$;   $y = 0$ when $x = 1$

**17.** $\dfrac{dy}{dx} = \dfrac{1}{x}$;   $y = 10$ when $x = 1$

**18.** $\dfrac{dy}{dx} = \dfrac{1}{x - 2}$;   $y = 50$ when $x = 3$

**19.** $2\dfrac{dy}{dx} = 4xe^{-x}$;   $y = 42$ when $x = 0$

**20.** $x\dfrac{dy}{dx} = x^2e^{3x}$;   $y = \dfrac{8}{9}$ when $x = 0$

# ▦ APPLICATIONS

## BUSINESS AND ECONOMICS

**Profit**   **21.** The marginal profit of a certain company is given by

$$\frac{dy}{dx} = \frac{100}{32 - 4x},$$

where $x$ represents the amount of money in thousands of dollars that the company spends on advertising.

Find the profit for each of the following advertising expenditures if the profit is $1000 when nothing is spent on advertising.

(a) $3000     (b) $5000

(c) Can advertising expenditures ever reach $8000 according to this model? Why?

**Sales**   **22.** A company has found the rate of change of sales during an advertising campaign is

$$\frac{dy}{dx} = 30e^{.03x},$$

where $y$ is the sales in thousands of dollars at time $x$ in months. If the sales are $200,000 when $x = 0$, find the sales at the end of 6 months.

**Elasticity of Demand**   *Elasticity of demand was discussed in the chapter on applications of the derivative, where it was defined as*

$$E = -\frac{p}{x} \cdot \frac{dx}{dp},$$

*for demand $x$ and price $p$. Find the general demand equation $x = f(p)$ for each of the following elasticity functions. (Hint: Set each elasticity function equal to $-\dfrac{p}{x} \cdot \dfrac{dx}{dp}$, then solve for $x$.)*

**23.** $E = \dfrac{4p^2}{x}$          **24.** $E = \dfrac{-1000}{px}$

## LIFE SCIENCES

**Population Growth**   **25.** Assume that the rate of change of the population of a certain city is given by

$$\frac{dy}{dt} = 6000e^{.06t},$$

where $y$ is the population at time $t$, measured in years. If the population was 100,000 in 1980 (assume $t = 0$ in 1980), predict the population in the year 2000.

**Bacterial Population Decline**   **26.** The average rate of decline of a bacteria population after the introduction of a bactericide is given by

$$\frac{dy}{dt} = \frac{-200}{1 + 2t},$$

where $t$ is time in hours after the bactericide is introduced. If there are 10,000 bacteria initially, what is the population at each of the following times?

(a) 4 hours     (b) 12 hours

## 8.2   SEPARATION OF VARIABLES

Interest on an investment is sometimes compounded continuously, which means that the investment grows at a rate proportional to the amount of money present.* If $A$ is the amount in an account at time $t$, then for some constant $k$, the differential equation

$$\frac{dA}{dt} = kA$$

gives the rate of growth of $A$ with respect to $t$. This differential equation is different from those discussed in the previous section, which had the form

$$\frac{dy}{dx} = f(x).$$

A more general differential equation is one that can be written in the form

$$\frac{dy}{dx} = \frac{f(x)}{g(y)}.$$

Using differentials and multiplying on both sides by $g(y)\ dx$ gives

$$g(y)\ dy = f(x)\ dx.$$

In this form all terms involving $y$ (including $dy$) are on one side of the equation, and all terms involving $x$ (and $dx$) are on the other side. A differential equation in this form is said to be **separable,** since the variables $x$ and $y$ can be separated. A separable differential equation may be solved by integrating each side.

### ■ EXAMPLE 1

Find the general solution of $y\ \dfrac{dy}{dx} = x^2$.

Begin by separating the variables to get

$$y\ dy = x^2\ dx.$$

The general solution is found by taking antiderivatives on each side.

$$\int y\ dy = \int x^2\ dx$$

$$\frac{y^2}{2} = \frac{x^3}{3} + C$$

$$y^2 = \frac{2}{3}x^3 + 2C = \frac{2}{3}x^3 + K$$

The constant $K$ was substituted for $2C$ in the last step. The solution is left in implicit form, not solved explicitly for $y$.   ■

---

*See Section 6 in the chapter on exponential and logarithmic functions.

### ▬▬ EXAMPLE 2

Find the general solution of $\dfrac{dy}{dx} = ky$, where $k$ is a constant.

Separating variables leads to

$$\frac{1}{y}\, dy = k\, dx.$$

To solve this equation, take antiderivatives on each side.

$$\int \frac{1}{y}\, dy = \int k\, dx$$

$$\ln |y| = kx + C$$

Use the definition of logarithm to write the equation in exponential form as

$$|y| = e^{kx+C}.$$

By properties of exponents,

$$|y| = e^{kx} e^{C}.$$

Finally, use the definition of absolute value to get

$$y = e^{kx}\, e^{C} \qquad \text{or} \qquad y = -e^{kx}\, e^{C}.$$

Since $e^{C}$ and $-e^{C}$ are constants, replace them with the constant $M$, which may have any nonzero real-number value, to get the single equation

$$y = Me^{kx}.$$

This equation, $y = Me^{kx}$, defines the exponential growth or decay function that was discussed previously (in the chapter on exponential and logarithmic functions).  ▬

Recall that equations of the form $y = Me^{kx}$ arise in situations where the rate of change of a quantity is proportional to the amount present at time $x$; that is, where

$$\frac{dy}{dx} = ky.$$

The constant $k$ is called the **growth rate constant,** while $M$ represents the amount present at time $x = 0$. (A positive value of $k$ indicates growth, while a negative value of $k$ indicates decay.)

Applying the results of Example 2 to the equation discussed at the beginning of this section,

$$\frac{dA}{dt} = kA,$$

shows that the amount in the account at time $t$ is

$$A = A_0 e^{kt},$$

where $A_0$ is the amount originally invested.

As a model of population growth, the equation $y = Me^{kx}$ is not realistic over the long run for most populations. As shown by graphs of functions of the form $y = Me^{kx}$, with both $M$ and $k$ positive, growth would be unbounded. Additional factors, such as space restrictions or a limited amount of food, tend to inhibit growth of populations as time goes on. In an alternative model that assumes a maximum population of size $N$, the rate of growth of a population is proportional to how close the population is to that maximum. These assumptions lead to the differential equation

$$\frac{dy}{dx} = k(N - y),$$

the limited growth function mentioned in the earlier chapter. Graphs of limited growth functions look like the graph in Figure 2, where $y_0$ is the initial population.

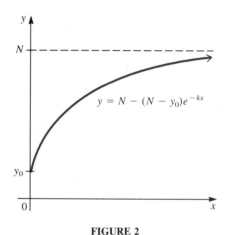

**FIGURE 2**

### ■ EXAMPLE 3

A certain area can support no more than 4000 mountain goats. There are 1000 goats in the area at present, with a growth constant of .20.

**(a)** Write a differential equation for the rate of growth of this population.

Let $N = 4000$ and $k = .20$. The rate of growth of the population is given by

$$\frac{dy}{dx} = .20(4000 - y).$$

To solve for $y$, first separate the variables.

$$\frac{dy}{4000 - y} = .2\, dx$$

$$\int \frac{dy}{4000 - y} = \int .2\, dx$$

$$-\ln(4000 - y) = .2x + C$$

$$\ln(4000 - y) = -.2x - C$$

$$4000 - y = e^{-.2x - C} = (e^{-.2x})(e^{-C})$$

The absolute value bars are not needed for $\ln(4000 - y)$ since $y$ must be less than 4000 for this population so that $4000 - y$ is always nonnegative. Let $e^{-C} = B$. Then

$$4000 - y = Be^{-.2x}$$

$$y = 4000 - Be^{-.2x}.$$

Find $B$ by using the fact that $y = 1000$ when $x = 0$.

$$1000 = 4000 - B$$

$$B = 3000$$

Notice that the value of $B$ is the difference between the maximum population and the initial population. Substituting 3000 for $B$ in the equation for $y$ gives

$$y = 4000 - 3000e^{-.2x}.$$

**(b)** What will the goat population be in 5 years?

In 5 years, the population will be

$$y = 4000 - 3000e^{(.2)(5)} = 4000 - 3000e^{-1}$$

$$= 4000 - 1103.6 = 2896.4,$$

or about 2900 goats. ∎

*Logistic Growth*   Let $y$ be the size of a certain population at time $x$. In the standard model for unlimited growth,

$$\frac{dy}{dx} = ky, \tag{1}$$

the rate of growth is proportional to the current population size. The constant $k$, the growth rate constant, is the difference between the birth and death rates of the population. The unlimited growth model predicts that the population's growth rate is a constant, $k$.

Growth usually is not unlimited, however, and the population's growth rate is usually not constant because the population is limited by environmental factors to a maximum size $N$, called the **carrying capacity** of the environment for the species. In the limited growth model given above,

$$\frac{dy}{dx} = k(N - y),$$

the rate of growth is proportional to the remaining room for growth, $N - y$. In the **logistic growth model**

$$\frac{dy}{dx} = \frac{k}{N}(N - y)y, \tag{2}$$

the rate of growth is proportional to both the current population size $y$ and the remaining room for growth $N - y$. Equation (2) is called the **logistic equation.** The logistic growth model predicts that the population's growth rate is a function of the population size $y$:

$$\frac{dy}{dx} = ky - \frac{k}{N}y^2.$$

As $y \to 0$, $y^2$ is insignificant when compared to $y$, and so

$$\frac{dy}{dx} = ky - \frac{k}{N}y^2 \approx ky.$$

That is, the growth of the population is essentially unlimited when $y$ is small. On the other hand, when $y \to N$,

$$\frac{dy}{dx} = ky - \frac{k}{N}y^2$$

$$= \left(k - \frac{k}{N}y\right)y$$

$$= \left(\frac{kN}{N} - \frac{k}{N}y\right)y$$

$$= \frac{k}{N}(N - y)y \to 0.$$

That is, population growth levels off as $y$ nears the maximum population size $N$. Thus, the logistic equation

$$\frac{dy}{dx} = ky - \frac{k}{N}y^2$$

is the unlimited growth equation (1) with a damping term $ky^2/N$ subtracted to account for limiting environmental factors when $y$ nears $N$. Let $y_0$ denote the initial population size. Under the assumption $0 < y < N$, the general solution of equation (2) is

$$y = \frac{N}{1 + be^{-kx}} \tag{3}$$

where $b = (N - y_0)/y_0$ (see Exercise 23). This solution, called a **logistic curve,** is shown in Figure 3.

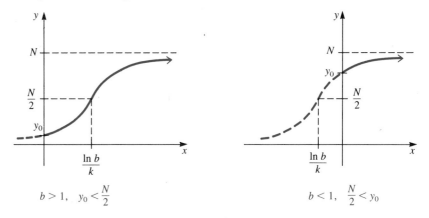

$$b > 1, \quad y_0 < \frac{N}{2}$$

$$b < 1, \quad \frac{N}{2} < y_0$$

**FIGURE 3**

As expected, the logistic curve begins exponentially and subsequently levels off. Another important feature is the point of inflection $((\ln b)/k, N/2)$, at which the growth rate $dy/dx$ is a maximum (see Exercise 25).

Logistic equations arise frequently in the study of populations. In about 1840 the Belgian sociologist P. F. Verhurst fitted a logistic curve to United States census figures and made predictions about the population that were subsequently proven to be quite accurate. American biologist Raymond Pearl (circa 1920) found that the growth of a population of fruit flies in a limited space could be modeled by the logistic equation

$$\frac{dy}{dx} = .2y - \frac{.2}{1035}y^2.$$

Logistic growth is an example of how a model is modified over time as new insights occur. The model for population growth changed from the early exponential curve $y = Me^{kx}$ to the logistic curve

$$y = \frac{N}{1 + be^{-kx}}.$$

Many other quantities besides population grow logistically. That is, their initial rate of growth is slow, but as time progresses, their rate of growth increases to a maximum value and subsequently begins to decline and to approach zero.

### ▰▰ EXAMPLE 4

Rapid technological advancements in the last 20 years have made many products obsolete practically overnight. J. C. Fisher and R. H. Pry* successfully

---

*Reprinted by permission of the publisher from "A Simple Substitution Model of Technological Change" by J. C. Fisher and R. H. Pry, *Technological Forecasting and Social Change,* Vol. 3, 1971–1972. Copyright © 1972 by Elsevier Science Publishing Co., Inc.

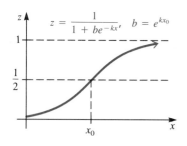

$$z = \frac{1}{1 + be^{-kx}}, \quad b = e^{kx_0}$$

**FIGURE 4**

described the phenomenon of a technically superior new product replacing another product by the logistic equation

$$\frac{dz}{dx} = k(1 - z)z, \tag{4}$$

where $z$ is the market share of the new product and $1 - z$ is the market share of the other product. The new product will initially have little or no market share; that is, $z_0 \approx 0$. Thus, the constant $b$ in equation (3) will have to be determined in a different way. Let $x_0$ be the time at which $z = 1/2$. Under the assumption $0 < z < 1$, the general solution of equation (4) is

$$z = \frac{1}{1 + be^{-kx}}$$

where $b = e^{kx_0}$ (see Exercise 24). This solution is shown in Figure 4.

The market share of the new product will be growing most rapidly when the new product has captured exactly half the market, and the market share of the older product will be shrinking most rapidly at the same time. Notice that the logistic equation (2) can be transformed into the simpler logistic equation (4) by the change of variable $z = y/N$. ∎

---

## 8.2 EXERCISES

*Find general solutions for the following differential equations.*

**1.** $y\dfrac{dy}{dx} = x$

**2.** $y\dfrac{dy}{dx} = x^2 - 1$

**3.** $\dfrac{dy}{dx} = 2xy$

**4.** $\dfrac{dy}{dx} = x^2y$

**5.** $\dfrac{dy}{dx} = 3x^2y - 2xy$

**6.** $(y^2 - y)\dfrac{dy}{dx} = x$

**7.** $\dfrac{dy}{dx} = \dfrac{y}{x}, \; x > 0$

**8.** $\dfrac{dy}{dx} = \dfrac{y}{x^2}$

**9.** $\dfrac{dy}{dx} = y - 5$

**10.** $\dfrac{dy}{dx} = 3 - y$

**11.** $\dfrac{dy}{dx} = y^2e^x$

**12.** $\dfrac{dy}{dx} = \dfrac{e^x}{e^y}$

*Find particular solutions for the following equations.*

**13.** $\dfrac{dy}{dx} = \dfrac{x^2}{y}; \quad y = 3$ when $x = 0$

**14.** $x^2\dfrac{dy}{dx} = y; \quad y = -1$ when $x = 1$

**15.** $(2x + 3)y = \dfrac{dy}{dx}; \quad y = 1$ when $x = 0$

**16.** $x\dfrac{dy}{dx} - y\sqrt{x} = 0; \quad y = 1$ when $x = 0$

**17.** $\dfrac{dy}{dx} = \dfrac{y^2}{x}; \quad y = 5$ when $x = e$

**18.** $\dfrac{dy}{dx} = x^{1/2}y^2; \quad y = 12$ when $x = 4$

**19.** $\dfrac{dy}{dx} = \dfrac{2x + 1}{y - 3}; \quad y = 4$ when $x = 0$

**20.** $\dfrac{dy}{dx} = \dfrac{x^2 + 5}{2y - 1}; \quad y = 11$ when $x = 0$

**21.** $\dfrac{dy}{dx} = (y - 1)^2 e^x$;   $y = 2$ when $x = 0$

**22.** $\dfrac{dy}{dx} = (x + 2)^2 e^y$;   $y = 0$ when $x = 1$

**23.** Solve the logistic equation (2) in this section by observing that

$$\frac{1}{y} + \frac{1}{N - y} = \frac{N}{(N - y)y}.$$

(a) Assume $0 < y < N$. Verify that $b = (N - y_0)/y_0$, where $y_0$ is the initial population size.

(b) Assume $0 < N < y$ for all y. Verify that $b = (y_0 - N)/y_0$.

**24.** Suppose that $0 < z < 1$ for all $z$. Solve the logistic equation (4) as in Exercise 23. Verify that $b = e^{kx_0}$, where $x_0$ is the time at which $z = 1/2$.

**25.** Suppose that $0 < y_0 < N$. Let $b = (N - y_0)/y_0$, and let

$$y(x) = \frac{N}{1 + be^{-kx}} \qquad \text{for all } x.$$

Show the following.

(a) $0 < y(x) < N$ for all $x$.

(b) The lines $y = 0$ and $y = N$ are horizontal asymptotes of the graph.

(c) $y(x)$ is an increasing function.

(d) $((\ln b)/k, N/2)$ is a point of inflection of the graph.

(e) $dy/dx$ is a maximum at $x_0 = (\ln b)/k$.

**26.** Suppose that $0 < N < y_0$. Let $b = (y_0 - N)/y_0$ and let

$$y(x) = \frac{N}{1 - be^{-kx}} \qquad \text{for all } x \neq \frac{\ln b}{k}.$$

See the figure.

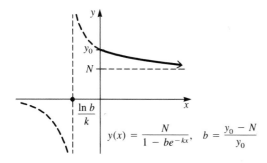

$$y(x) = \frac{N}{1 - be^{-kx}}, \quad b = \frac{y_0 - N}{y_0}$$

Show the following.

(a) $0 < b < 1$

(b) The lines $y = 0$ and $y = N$ are horizontal asymptotes of the graph.

(c) The line $x = (\ln b)/k$ is a vertical asymptote of the graph.

(d) $y(x)$ is decreasing on $((\ln b)/k, +\infty)$ and on $(-\infty, (\ln b)/k)$.

(e) $y(x)$ is concave upward on $((\ln b)/k, +\infty)$ and concave downward on $(-\infty, (\ln b)/k)$.

## ▦ APPLICATIONS

### BUSINESS AND ECONOMICS

**Marginal Cost**   **27.** Suppose the marginal cost of producing $x$ copies of a book is given by

$$\frac{dy}{dx} = \frac{50}{y\sqrt{x}}.$$

If $y = 150$ when $x = 4$, find the costs of producing the following numbers of books.

**(a)** 100 books       **(b)** 400 books       **(c)** 625 books

**Sales Decline**   **28.** Sales (in thousands) of a certain product are declining exponentially, with a decay constant of 25% a year.

**(a)** Write a differential equation to express the rate of sales decline.

**(b)** Find a general solution to the equation in part (a).

**Inflation**   **29.** If inflation grows continuously at a rate of 6% per year, how long will it take for $1 to lose half its value?

**Gross National Product**   **30.** Suppose that the gross national product (GNP) of a particular country increases exponentially, with a growth constant of 2% per year. Ten years ago, the GNP was $10^5$ dollars. What will be the GNP in 5 years?

**Bankruptcy**   **31.** In a certain area, 1500 small business firms are threatened by bankruptcy. Assume the rate of change in the number of bankruptcies is proportional to the number of small firms that are not yet bankrupt. If the growth constant is 6%, and if 100 firms are bankrupt initially, how many will be bankrupt in 2 years?

**New Car Sales**   **32.** The sales of a new model of car are expected to follow the logistic curve

$$y = \frac{100,000}{1 + 100e^{-x}},$$

where $x$ is the number of months since the introduction of the model. Find the following.

**(a)** The initial sales level

**(b)** The maximum sales level

**(c)** The month in which the maximum sales level will be achieved

**(d)** The rate at which sales are growing 6 months after the introduction of the model

### LIFE SCIENCES

**Tracer Dye**   **33.** The amount of a tracer dye injected into the bloodstream decreases exponentially, with a decay constant of 3% per minute. If 6 cc are present initially, how many cc are present after 10 min? (Here $k$ will be negative.)

**Bacterial Population**   **34.** The rate at which the number of bacteria (in thousands) in a culture is changing after the introduction of a bactericide is given by

$$\frac{dy}{dx} = 50 - y,$$

where $y$ is the number of bacteria (in thousands) present at time $x$. Find the number of bacteria present at each of the following times if there were 1000 thousand bacteria present at time $x = 0$.

**(a)** $x = 2$       **(b)** $x = 5$       **(c)** $x = 10$       **(d)** $x = 15$

**Fish Population** **35.** An isolated fish population is limited to 5000 by the amount of food available. If there are now 150 fish and the population is growing with a growth constant of 1% a year, find the expected population at the end of 5 years.

**Mite Population** **36.** A population of mites (in hundreds) increases exponentially, with a growth constant of 5% per week. At the beginning of an observation period there were 3000 mites. How many are present 4 weeks later?

## SOCIAL SCIENCES

**Spread of a Rumor** **37.** Suppose the rate at which a rumor spreads—that is, the number of people who have heard the rumor over a period of time—increases with the number of people who have heard it. If $y$ is the number of people who have heard the rumor, then

$$\frac{dy}{dt} = ky,$$

where $t$ is the time in days.

(a) If $y$ is 1 when $t = 0$, and $y$ is 5 when $t = 2$, find $k$.

Using the value of $k$ from part (a), find $y$ for each of the following times.

(b) $t = 3$    (c) $t = 5$    (d) $t = 10$

**Worker Productivity** **38.** A company has found that the rate at which a person new to the assembly line produces items is

$$\frac{dy}{dx} = 7.5e^{-.3y},$$

where $x$ is the number of days the person has worked on the line. How many items can a new worker be expected to produce on the eighth day if he produces none when $x = 0$?

## PHYSICAL SCIENCES

**Radioactive Decay** **39.** The amount of a radioactive substance decreases exponentially, with a decay constant of 5% a month.

(a) Write a differential equation to express the rate of change.

(b) Find a general solution to the differential equation from part (a).

(c) If there are 90 grams at the start of the decay process, find a particular solution for the differential equation from part (a).

(d) Find the amount left at 10 months.

# 8.3  LINEAR FIRST-ORDER DIFFERENTIAL EQUATIONS

Recall that $f^{(n)}(x)$ represents the $n$th derivative of $f(x)$, and that $f^{(n)}(x)$ is called an *nth-order* derivative. By this definition, the derivative $f'(x)$ is first-order, $f''(x)$ is second-order, and so on. The **order of a differential equation** is that of the highest-order derivative in the equation. In this section only first-order differential equations are discussed.

A **linear first-order differential equation** is an equation of the form

$$y' + P(x)y = Q(x).$$

Many useful models produce such equations. In this section we develop a general method for solving first-order linear differential equations. For simplicity, we shall use the notation $y'$ for the derivative in this section.

The first step in solving the equation

$$xy' + 6y + 2x^4 = 0, \tag{1}$$

for example, is to divide both sides of the equation by $x$ and rearrange the terms to get the linear differential equation

$$y' + \frac{6}{x}y = -2x^3.$$

This equation does not have separable variables and cannot be solved by the methods discussed so far. Instead, multiply both sides of the equation by $x^6$ to get

$$x^6y' + 6x^5y = -2x^9. \tag{2}$$

On the left, $6x^5$, the coefficient of $y$, is the derivative of $x^6$, the coefficient of $y'$. Recall the product rule for derivatives:

$$D_x(uv) = uv' + u'v.$$

If $u = x^6$ and $v = y$, the product rule gives

$$D_x(x^6y) = x^6y' + 6x^5y,$$

which is the left side of equation (2). Substituting $D_x(x^6y)$ for the left side of equation (2) gives

$$D_x(x^6y) = -2x^9.$$

Assuming $y = f(x)$, as usual, both sides of this equation can be integrated with respect to $x$ and the result solved for $y$ to get

$$x^6y = \int -2x^9 dx = -2\left(\frac{x^{10}}{10}\right) + C = -\frac{x^{10}}{5} + C$$

$$y = -\frac{x^4}{5} + \frac{C}{x^6}. \tag{3}$$

Equation (3) is the general solution of equation (2) and therefore of equation (1).

This procedure has given us a solution, but what motivated our choice of the multiplier $x^6$? To see where $x^6$ came from, let $I(x)$ represent the multiplier, and multiply both sides of the general equation

$$y' + P(x)y = Q(x)$$

by $I(x)$:

$$I(x)y' + I(x)P(x)y = I(x)Q(x). \tag{4}$$

The method illustrated above will work only if the left side of the equation is the derivative of the product function $I(x) \cdot y$, which is

$$I(x)y' + I'(x)y. \tag{5}$$

Comparing the coefficients of $y$ in equations (4) and (5) shows that $I(x)$ must satisfy

$$I'(x) = I(x)P(x),$$

or

$$\frac{I'(x)}{I(x)} = P(x).$$

Integrating both sides of this last equation gives

$$\ln |I(x)| = \int P(x)dx + C$$
$$|I(x)| = e^{\int P(x)dx + C}$$

or

$$I(x) = \pm e^C e^{\int P(x)dx}.$$

Only one value of $I(x)$ is needed, so let $C = 0$, so that $e^C = 1$, and use the positive result, giving

$$I(x) = e^{\int P(x)dx}.$$

In summary, choosing $I(x)$ as $e^{\int P(x)dx}$ and multiplying both sides of a linear first-order differential equation by $I(x)$ puts the equation in a form that can be solved by integration.

**INTEGRATING FACTOR**

> $I(x) = e^{\int P(x)dx}$ is called an **integrating factor** for the differential equation $y' + P(x)y = Q(x)$.

For equation (1) above, written as the linear differential equation

$$y' + \frac{6}{x}y = -2x^3,$$

$P(x) = 6/x$, and the integrating factor is

$$I(x) = e^{\int (6/x)dx} = e^{6 \ln |x|} = e^{\ln |x|^6} = e^{\ln x^6} = x^6.$$

This last step used the fact that, for all $a$, $e^{\ln a} = a$.

In summary, we solve a linear first-order differential equation with the following steps.

**SOLVING A LINEAR FIRST-ORDER DIFFERENTIAL EQUATION**

>
> 1. Put the equation in the linear form $y' + P(x)y = Q(x)$.
> 2. Find the integrating factor $I(x) = e^{\int P(x)dx}$.
> 3. Multiply each term of the equation from Step 1 by $I(x)$.
> 4. Replace the sum of terms on the left with $D_x[I(x)y]$.
> 5. Integrate both sides of the equation and solve for $y$ if possible.

■■■ **EXAMPLE 1**

Give the general solution of $y' + 2xy = x$.

*Step 1*  This equation is already in the required form.

*Step 2*  The integrating factor is

$$I(x) = e^{\int 2x \, dx} = e^{x^2}.$$

*Step 3*  Multiplying each term by $e^{x^2}$ gives

$$e^{x^2}y' + 2xe^{x^2}y = xe^{x^2}.$$

*Step 4*  The sum of terms on the left can now be replaced with $D_x(e^{x^2}y)$, to get

$$D_x(e^{x^2}y) = xe^{x^2}.$$

*Step 5*  Integrating on both sides leads to the general solution

$$e^{x^2}y = \int xe^{x^2}dx = \frac{1}{2}e^{x^2} + C$$

$$y = \frac{1}{2} + Ce^{-x^2}. \quad ■$$

■■■ **EXAMPLE 2**

Solve $2y' - 6y - e^x = 0$ if $y = 5$ when $x = 0$.

Write the equation in the required form:

$$y' - 3y = \frac{1}{2}e^x.$$

The integrating factor is $\quad I(x) = e^{\int(-3)dx} = e^{-3x}.$

Multiplying each term by $I(x)$ gives

$$e^{-3x}y' - 3e^{-3x}y = \frac{1}{2}e^x e^{-3x}$$

$$e^{-3x}y' - 3e^{-3x}y = \frac{1}{2}e^{-2x}.$$

The left side can now be replaced by $D_x(e^{-3x}y)$ to get

$$D_x(e^{-3x}y) = \frac{1}{2}e^{-2x}.$$

Integrating on both sides and solving for $y$ gives

$$e^{-3x}y = \int \frac{1}{2}e^{-2x}dx = \frac{1}{2}\left(-\frac{1}{2}\right)e^{-2x} + C.$$

Now, multiply both sides by $e^{3x}$ to get

$$y = -\frac{1}{4}e^x + Ce^{3x},$$

the general solution. Find the particular solution by substituting 0 for $x$ and 5 for $y$:

$$5 = -\frac{1}{4}e^0 + Ce^0 = -\frac{1}{4} + C$$

or
$$\frac{21}{4} = C,$$

which leads to the particular solution

$$y = -\frac{1}{4}e^x + \frac{21}{4}e^{3x}. \quad \blacksquare$$

## ▬ EXAMPLE 3*

Suppose glucose is infused into a patient's bloodstream at a constant rate of $a$ grams per minute. At the same time, glucose is removed from the bloodstream at a rate proportional to the amount of glucose present. Then the amount of glucose, $G(t)$, present at time $t$ satisfies

$$\frac{dG}{dt} = a - KG.$$

for some constant $K$. Solve this equation for $G$. Does the glucose concentration eventually reach a constant? That is, what happens to $G$ as $t \rightarrow +\infty$?

   The equation can be written in the form of the linear first-order differential equation

$$G' + KG = a. \tag{6}$$

The integrating factor is

$$I(t) = e^{\int K\,dt} = e^{Kt}.$$

Multiply both sides of equation (6) by $I(t) = e^{Kt}$

$$e^{Kt}G' + Ke^{Kt}G = ae^{Kt}$$

Write the left side as $D_t(e^{Kt}G)$ and solve for $G$ by integrating on each side.

$$D_t(e^{Kt}G) = ae^{Kt}$$

$$e^{Kt}G = \int ae^{Kt}dt = \frac{a}{K}e^{Kt} + C$$

Multiply both sides by $e^{-Kt}$ to get

$$G = \frac{a}{K} + Ce^{-Kt}.$$

---

*From *Ordinary Differential Equations with Applications*, Larry C. Andrews, Scott, Foresman and Company, 1982, p. 79.

As $t \rightarrow +\infty$,

$$\lim_{t \to +\infty} G = \lim_{t \to +\infty}\left(\frac{a}{K} + Ce^{-Kt}\right) = \lim_{t \to +\infty}\left(\frac{a}{K} + \frac{C}{e^{Kt}}\right) = \frac{a}{K}.$$

Thus, the glucose concentration stabilizes at $a/K$. ▬

## ▬ 8.3 EXERCISES

*Solve the following first-order differential equations.*

**1.** $y' + 2y = 5$

**2.** $y' + 4y = 10$

**3.** $y' + xy = 3x$

**4.** $y' + 2xy = x$

**5.** $xy' - y - x = 0, \quad x > 0$

**6.** $xy' + 2xy - x^2 = 0, \quad x > 0$

**7.** $2y' - 2xy - x = 0$

**8.** $3y' + 6xy + x = 0$

**9.** $xy' + 2y = x^2 + 3x, \quad x > 0$

**10.** $x^2y' + xy = x^3 - 2x^2, \quad x > 0$

**11.** $y - x\dfrac{dy}{dx} = x^3, \quad x > 0$

**12.** $2xy + x^3 = x\dfrac{dy}{dx}$

*Solve each differential equation, subject to the given initial condition.*

**13.** $y' + y = 2e^x; \quad y = 100$ when $x = 0$

**14.** $y' + 2y = e^{3x}; \quad y = 50$ when $x = 0$

**15.** $\dfrac{dy}{dx} - xy - x = 0; \quad y = 10$ when $x = 1$

**16.** $x\dfrac{dy}{dx} - 3y + 2 = 0; \quad y = 8$ when $x = 1$

**17.** $x\dfrac{dy}{dx} + 5y = x^2; \quad y = 12$ when $x = 2$

**18.** $2\dfrac{dy}{dx} - 4xy = 5x; \quad y = 10$ when $x = 1$

**19.** $xy' + (1 + x)y = 3; \quad y = 50$ when $x = 4$

**20.** $y' + 2xy - e^{-x^2} = 0; \quad y = 100$ when $x = 0$

## ▬ APPLICATIONS

### LIFE SCIENCES

**Population Growth**   **21.** The logistic equation introduced in Section 2,

$$\frac{dy}{dx} = \frac{k}{N}(N - y)y, \tag{*}$$

can be written as

$$\frac{dy}{dx} = cy - py^2,$$

where $c$ and $p$ are positive constants. Although this is a nonlinear differential equation, it can be reduced to a linear equation by a suitable substitution for the variable $y$.

**(a)** Letting $y = 1/z$ and $y' = -z'/z^2$, rewrite equation (*) in terms of $z$ and solve.

**(b)** Let $z(0) = 1/y_0$ in part (a) and find a particular solution for $y$.

**(c)** Find the limit of $y$ as $x \rightarrow +\infty$. This is the saturation level of the population.

### SOCIAL SCIENCES

**Immigration and Emigration**
*If population is changed either by immigration or emigration, the population model discussed in Section 2 is modified to*

$$\frac{dy}{dt} = ky + f(t),$$

*where y is the population at time t and f(t) is some (other) function of t that describes the net effect of the emigration/immigration. Assume k = .02 and y = 10,000 when t = 0. Solve this differential equation for y, given the following functions f(t).*

**22.** $f(t) = e^t$      **23.** $f(t) = e^{-t}$      **24.** $f(t) = -t$      **25.** $f(t) = t$

### PHYSICAL SCIENCES

**Newton's Law of Cooling**
*Newton's law of cooling states that the rate of change of temperature of an object is proportional to the difference in temperature between the object and the surrounding medium. Thus, if T is the temperature of the object after t hr, and $T_F$ is the (constant) temperature of the surrounding medium, then*

$$\frac{dT}{dt} = -k(T - T_F),$$

*where k is a constant. Exercises 26–28 use this equation.*

**26.** Show that the solution of this equation is

$$T = ce^{-kt} + T_F,$$

where $c$ is a constant.

**27.** Suppose a container of soup with a temperature of 180°F is put into a refrigerator where the temperature is 40°F. When checked 1/2 hr later, the temperature of the soup is 110°F.

    **(a)** Find an equation for the temperature of the soup in terms of the number of hours it has been in the refrigerator.

    **(b)** Find the temperature of the soup after 1 hr.

    **(c)** Find the temperature of the soup after 5 hr.

**28.** A ceramic jar is taken from a kiln at a temperature of 50°C and set to cool in a room where the temperature is 22°C. In 1 hr, the temperature of the jar is 35°C.

    **(a)** Find an equation for the temperature of the jar at time $t$.

    **(b)** Find the temperature of the jar after 2 hr.

## ▤ 8.4 EULER'S METHOD

Applications sometimes involve differential equations such as

$$\frac{dy}{dx} = \frac{x + y}{y}$$

that cannot be solved by the methods discussed so far, but approximate solutions to these equations often can be found by numerical methods. For many applications, these numerical methods are quite adequate. In this section we introduce Euler's method, which is only one of numerous mathematical contributions made by Leonhard Euler (1707–1783) of Switzerland. (His name is pronounced "oiler".) He also introduced the $f(x)$ notation used throughout this text. In spite of becoming blind during his later years, he was the most prolific mathematician of his era. He published in all mathematical fields. The equation

$$e^{i\theta} = \cos\theta + i\sin\theta,$$

which shows how concepts from three branches of mathematics are interrelated, was developed by him. The number $e$ (in honor of Euler) is the base for the natural logarithm, the number $i$ represents the imaginary unit $\sqrt{-1}$, and the symbols cos and sin are found in the study of trigonometry. (Trigonometric applications are given in Chapter 10 of this book).

**Euler's method** of solving differential equations gives approximate solutions to differential equations involving $y = f(x)$ where the initial values of $x$ and $y$ are known, that is, equations of the form

$$y' = g(x, y), \quad \text{with } f(x_0) = y_0.$$

Geometrically, Euler's method approximates the graph of the solution $y = f(x)$ with a polygonal line whose first segment is tangent to the curve at the point $(x_0, y_0)$, as shown in Figure 5.

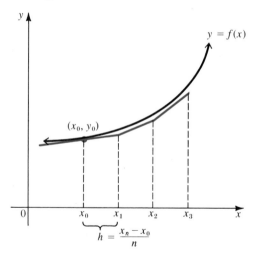

**FIGURE 5**

To use Euler's method, divide the interval from $x_0$ to another point $x_n$ into $n$ subintervals, each of equal width (see Figure 5). The width of each subinterval is $h = (x_n - x_0)/n$. Since the first segment of the polygonal line goes through $(x_0, y_0)$ and is tangent to the curve $y = f(x)$ at $(x_0, y_0)$, the slope of the line segment from $(x_0, y_0)$ to $(x_1, y_1)$ is $f'(x_0) = y' = g(x_0, y_0)$.

The equation of this line is

$$y - y_0 = m(x - x_0) \quad \text{or} \quad y - y_0 = g(x_0, y_0)(x - x_0),$$

from which

$$y = y_0 + g(x_0, y_0)(x - x_0).$$

The other endpoint, $(x_1, y_1)$, is also on the line and satisfies the equation, so

$$y_1 = y_0 + g(x_0, y_0)(x_1 - x_0).$$

Since each subinterval has width $h$, $x_1 - x_0 = h$, and

$$y_1 = y_0 + g(x_0, y_0)h$$

is an approximation of $f(x_1)$.

To find an approximation to $f(x_2)$, use the second line segment. Since $y_1 \approx f(x_1)$, the slope of the curve at $x_1$ is close to the slope of the second line segment at $x_1$. That is, $f'(x_1) \approx g(x_1, y_1)$, and

$$y - y_1 = g(x_1, y_1)(x - x_1)$$
$$y = y_1 + g(x_1, y_1)(x - x_1).$$

Then, at $(x_2, y_2)$,

$$y_2 = y_1 + g(x_1, y_1)(x_2 - x_1)$$

or

$$y_2 = y_1 + g(x_1, y_1)h,$$

since $x_2 - x_1 = h$.

Continuing in this way,

$$y_{i+1} = y_i + g(x_i, y_i)h$$

is an approximation of $f(x_{i+1})$. A summary of Euler's method follows.

**EULER'S METHOD**

Let $y = f(x)$ be the solution of the differential equation

$$y' = g(x, y), \quad \text{with } f(x_0) = y_0,$$

for $x_0 \le x \le x_n$. Let $x_{i+1} - x_i = h$ and

$$y_{i+1} = y_i + g(x_i, y_i)h,$$

for $0 \le i \le n$. Then

$$f(x_{i+1}) \approx y_{i+1}.$$

As the following examples will show, the accuracy of the approximation varies for different functions. As $h$ gets smaller, however, the approximation improves.

### ■ EXAMPLE 1

Use Euler's method to approximate the solution of $y' + 2xy = x$, with $f(0) = 1.5$, for [0, 1]. Use $h = .1$.

The general solution of this equation was found in Example 1 of the last section, so the results using Euler's method can be compared with the actual solution. Begin by writing the differential equation in the required form as

$$y' = x - 2xy, \quad \text{so that} \quad g(x, y) = x - 2xy.$$

Since $x_0 = 0$ and $y_0 = 1.5$, $g(x_0, y_0) = 0 - 2(0)(1.5) = 0$, and

$$y_1 = 1.5 + 0(.1) = 1.5.$$

Now $x_1 = .1$, $y_1 = 1.5$ and $g(x_1, y_1) = .1 - 2(.1)(1.5) = -.2$. Then

$$y_2 = 1.5 + (-.2)(.1) = 1.48.$$

The ten values for $x_i$ and $y_i$ for $0 \le i \le 10$ are shown in Table 1, together with the actual values using the result from Example 1 in the last section. (Since the result was only a general solution, replace $x$ with 0 and $y$ with 1.5 to get the particular solution $y = 1/2 + e^{-x^2}$.)

**Table 1**

| | Euler's Method | Actual Solution | Difference |
|---|---|---|---|
| $x_i$ | $y_i$ | $f(x_i)$ | $y_i - f(x_i)$ |
| 0 | 1.5 | 1.5 | 0 |
| .1 | 1.5 | 1.4900498 | .0099502 |
| .2 | 1.48 | 1.4607894 | .0192106 |
| .3 | 1.4408 | 1.4139312 | .0268688 |
| .4 | 1.384352 | 1.3521438 | .0322082 |
| .5 | 1.3136038 | 1.2788008 | .034803 |
| .6 | 1.232243 | 1.1976763 | .0345667 |
| .7 | 1.1443742 | 1.1126264 | .0317102 |
| .8 | 1.0541619 | 1.0272924 | .0268695 |
| .9 | .96549595 | .94485807 | .02063788 |
| 1 | .88170668 | .86787944 | .01382724 |

The results in Table 1 look quite good. Figure 6 shows that the polygonal line follows the actual graph of $f(x)$ closely. ▬

**FIGURE 6**

Euler's method produces a very good approximation for this differential equation because the slope of the solution $f(x)$ is not steep in the interval under investigation. The next example shows that such good results cannot always be expected.

## ▬ EXAMPLE 2

Use Euler's method to solve $y' = 3y + (1/2)e^x$, with $y(0) = 5$, for $[0, 1]$, using 10 subintervals.

This is the differential equation of Example 2 in the last section. The general solution found there, with the initial condition given above, leads to the particular solution

$$y = -\frac{1}{4}e^x + \frac{21}{4}e^{3x}.$$

To solve by Euler's method, start with $g(x, y) = 3y + (1/2)e^x$, $x_0 = 0$, and $y_0 = 5$. For $n = 10$, $h = (1 - 0)/10 = .1$ again, and

$$y_{i+1} = y_i + g(x_i, y_i)h = y_i + \left(3y_i + \frac{1}{2}e^{x_i}\right)h.$$

For $y_1$, this gives

$$y_1 = y_0 + \left(3y_0 + \frac{1}{2}e^{x_0}\right)h$$

$$= 5 + \left[3(5) + \frac{1}{2}(e^0)\right](.1)$$

$$= 5 + [15.5](.1) = 6.55.$$

Similarly,
$$y_2 = 6.55 + [3(6.55) + \frac{1}{2}e^{.1}](.1)$$

$$= 6.55 + [19.65 + .55258546](.1)$$

$$= 8.57025855.$$

These and the remaining values for the interval $[0, 1]$ are shown in Table 2.

Table 2  Approximate Solution Using $h = .1$

| | Euler's Method | Actual Solution | Difference |
|---|---|---|---|
| $x_i$ | $y_i$ | $f(x_i)$ | $y_i - f(x_i)$ |
| 0 | 5 | 5 | 0 |
| .1 | 6.55 | 6.8104660 | − .260466 |
| .2 | 8.570259 | 9.2607730 | − .690514 |
| .3 | 11.20241 | 12.575452 | − 1.373042 |
| .4 | 14.63062 | 17.057658 | − 2.427038 |
| .5 | 19.09440 | 23.116687 | − 4.022287 |
| .6 | 24.90516 | 31.305120 | − 6.399960 |
| .7 | 32.46781 | 42.368954 | − 9.901144 |
| .8 | 42.30884 | 57.315291 | − 15.006451 |
| .9 | 55.11277 | 77.503691 | − 22.390921 |
| 1 | 71.76958 | 104.76950 | − 32.999920 |

In this example the absolute value of the differences grows very rapidly as $x_i$ gets farther from $x_0$. See Figure 7. These large differences come from the term $e^{3x}$ in the solution; this term grows very quickly as $x$ increases. ▬

**FIGURE 7**

As these examples show, numerical methods may produce large errors. The error often can be reduced by using more subintervals of smaller width, letting $n = 100$ or $1000$, for example. Approximations for the function in Example 2 with $n = 100$ and $h = (1 - 0)/100 = .01$ are shown in Table 3. The approximations are considerably improved.

Table 3   Approximate Solution Using $h = .01$

| | Euler's Method | Actual Solution | Difference |
|---|---|---|---|
| $x_i$ | $y_i$ | $f(x_i)$ | $y_i - f(x_i)$ |
| 0 | 5 | 5 | 0 |
| .1 | 6.779419 | 6.810466 | −.031047 |
| .2 | 9.177102 | 9.260773 | −.083671 |
| .3 | 12.40634 | 12.575452 | −.169112 |
| .4 | 16.75386 | 17.057658 | −.303798 |
| .5 | 22.60505 | 23.116687 | −.511637 |
| .6 | 30.47795 | 31.305120 | −.827170 |
| .7 | 41.06884 | 42.368954 | −1.30011 |
| .8 | 55.31358 | 57.315291 | −2.001711 |
| .9 | 74.46998 | 77.503691 | −3.033711 |
| 1 | 100.22860 | 104.769498 | −4.540898 |

It is possible to improve Euler's method to increase its accuracy. Other numerical methods are also used, in particular the Runge-Kutta method, the most commonly used technique, which is more accurate than Euler's method. These approaches are more complex than Euler's method and we will not discuss them here.*

## 8.4 EXERCISES

*Use Euler's method to approximate the indicated function value to three decimal places using $h = .1$.*

**1.** $y' = x^2 + y^2$;  $f(0) = 1$;   find $f(.5)$

**2.** $y' = xy + 2$;  $f(0) = 0$;   find $f(.5)$

**3.** $y' = 1 + y$;  $f(0) = 2$;   find $f(.6)$

**4.** $y' = x + y^2$;  $f(0) = 0$;   find $f(.6)$

**5.** $y' = x + \sqrt{y}$;  $f(0) = 1$;   find $f(.4)$

**6.** $y' = 1 + \dfrac{y}{x}$;  $f(1) = 0$;   find $f(1.4)$

**7.** $y' = 1 - e^{-x}$;  $f(0) = 0$;   find $f(.5)$

**8.** $y' = e^{-y} + x$;  $f(0) = 0$;   find $f(.5)$

*Use Euler's method to approximate the indicated function value to three decimal places using $h = .1$. Next, solve the differential equation and find the indicated function value to three decimal places. Compare the result with the approximation.*

**9.** $y' = -4 + x$;  $f(0) = 1$;   find $f(.4)$

**10.** $y' = 4x + 3$;  $f(1) = 0$;   find $f(1.5)$

**11.** $y' = x^2$;  $f(0) = 2$;   find $f(.5)$

**12.** $y' = \dfrac{1}{x}$;  $f(1) = 1$;   find $f(1.4)$

**13.** $y' = 2xy$;  $f(1) = 1$;   find $f(1.6)$

**14.** $y' = x^2y$;  $f(0) = 1$;   find $f(.6)$

**15.** $y' = x^2 - x$;  $f(0) = 0$;   find $f(.3)$

**16.** $y' = \dfrac{x}{y}$;  $f(0) = 2$;   find $f(.3)$

*In each of the following, construct a table like the ones in the examples for $0 \le x \le 1$ with $h = .2$.*

**17.** $y' = \sqrt[3]{x}$;  $f(0) = 0$

**18.** $y' = y$;  $f(0) = 1$

**19.** $y' = 1 - y$;  $f(0) = 0$

**20.** $y' = x - xy$;  $f(0) = .5$

*Solve each differential equation and graph the function $y = f(x)$ and the polygonal approximation on the same axes. (The approximations were found in Exercises 17–20.)*

**21.** $y' = \sqrt[3]{x}$;  $f(0) = 0$

**22.** $y' = y$;  $f(0) = 1$

**23.** $y' = 1 - y$;  $f(0) = 0$

**24.** $y' = x - xy$;  $f(0) = .5$

---

*For a discussion of other methods, see Elementary Differential Equations,* William E. Boyce and Richard C. DiPrima, John Wiley and Sons, Inc., 1977.

▄▄▄ APPLICATIONS

*Solve Exercises 25–30 by using Euler's method.*

## BUSINESS AND ECONOMICS

**Bankruptcy**    **25.** Suppose 150 small business firms are threatened by bankruptcy.* If $y$ is the number bankrupt by time $t$, then $150 - y$ is the number not yet bankrupt by time $t$. The rate of change of $y$ depends on both $y$ and $150 - y$. Let 1982 correspond to $t = 0$. Assume 20 firms are bankrupt at $t = 0$.

   **(a)** Write a differential equation using the given information. Use $-.02$ for the constant of proportionality.

   **(b)** Approximate the number of firms that are bankrupt in 1986, using $h = 1$.

## LIFE SCIENCES

**Algae Growth**    **26.** The phosphate compounds found in many detergents are highly water soluble and are excellent fertilizers for algae. Assume that there are 3000 algae present at time $t = 0$ and conditions will support at most 100,000 algae. Assume that the rate of growth of algae, in the presence of sufficient phosphates, is proportional both to the number present, in thousands, and to the difference between 100,000 and the number present (in thousands).

   **(a)** Write a differential equation using the given information. Use $.01$ for the constant of proportionality.

   **(b)** Approximate the number present when $t = 2$ using $h = .5$.

**Immigration**    **27.** An island is colonized by immigration from the mainland, where there are 100 species. Let the number of species on the island at time $t$ (in years) equal $y$, where $y = f(t)$. Suppose the rate at which new species immigrate to the island is

$$\frac{dy}{dt} = .02(100 - y^{1/2}).$$

Use Euler's method with $h = .5$ to approximte $y$ when $t = 5$ if there were 10 species initially.

**Insect Population**    **28.** A population of insects, $y$, living in a circular colony grows at a rate

$$\frac{dy}{dt} = .05y - .1y^{1/2},$$

where $t$ is time in weeks. If there were 60 insects initially, use Euler's method with $h = 1$ to approximate the number of insects after 6 wk.

**Whale Population**    **29.** Under certain conditions a population may exhibit a polynomial growth rate function. A population of blue whales is growing according to the function

$$\frac{dy}{dt} = -y + .02y^2 + .003y^3.$$

Here $y$ is the population in thousands and $t$ is measured in years. Use Euler's method with $h = 1$ to approximate the population in 4 yr if the initial population is 15,000.

*From *An Introduction to Mathematical Modeling,* Edward A. Bender, John Wiley and Sons, Inc., 1978.

## SOCIAL SCIENCES

**Spread of Rumors**   **30.** A rumor spreads through a community of 500 people at the rate

$$\frac{dN}{dt} = .2(500 - N)N^{1/2},$$

where $N$ is the number of people who have heard the rumor at time $t$ in hours. Use Euler's method with $h = .5$ to find the number who have heard the rumor after 3 hr, if only 2 people heard it initially.

## FOR THE COMPUTER

*Use Euler's method to approximate $y$ to three decimal places for each function, when $x = 1.5$. Use $h = .1$.*

**31.** $y' = e^{.02x} + y;$  $f(0) = .2$

**32.** $y' = e^x - e^{-y};$  $f(1) = .8$

**33.** $y' = \dfrac{x - y}{x + y};$  $f(1) = 1.5$

**34.** $y' = \dfrac{x^2 + y}{y};$  $f(0) = 2.3$

---

## ▓▓ 8.5  APPLICATIONS OF DIFFERENTIAL EQUATIONS

The Austrian mathematician A. J. Lotka (1880–1949) and the Italian mathematician Vito Volterra (1860–1940) proposed the following simple model for the way in which the fluctuations of populations of a predator and its prey affect each other.* Let $x = f(t)$ denote the population of the predator and $y = g(t)$ denote the population of the prey at time $t$. The predator might be a type of insect and its prey a type of scale, or the predator might be a type of bird and the prey a type of insect.

Assume that if there were no predators present, the population of prey would increase at a rate $py$ proportional to their number, but that the predators consume the prey at a rate $qxy$ proportional to the product of the number of prey and the number of predators. The net rate of change $dy/dt$ of $y$ is the rate of increase of the prey minus the rate at which the prey are eaten, that is,

$$\frac{dy}{dt} = py - qxy, \tag{1}$$

with positive constants $p$ and $q$.

Assume that if there were no prey, the predators would starve and their population would *decrease* at a rate $rx$ proportional to their number, but that in the presence of prey the rate of growth of the population of predators is increased by an amount $sxy$. These assumptions give a second differential equation.

$$\frac{dx}{dt} = -rx + sxy, \tag{2}$$

with additional positive constants $r$ and $s$.

---

*From *Calculus and Analytic Geometry,* Third Edition, Al Shenk, Scott, Foresman and Company, 1984.

Equations (1) and (2) form a system of differential equations known as the **Lotka-Volterra equations.** They cannot be solved for $x$ and $y$ as functions of $t$, but an equation relating the variables $x$ and $y$ can be found. Dividing equation (1) by equation (2) gives the separable differential equation

$$\frac{dy}{dx} = \frac{dy/dt}{dx/dt} = \frac{py - qxy}{-rx + sxy}$$

or

$$\frac{dy}{dx} = \frac{y(p - qx)}{x(sy - r)}. \tag{3}$$

Equation (3) is solved for specific values of the constants $p$, $q$, $r$, and $s$ in the next example.

### ■ EXAMPLE 1

Suppose that $x = f(t)$ (hundreds of predators) and $y = g(t)$ (thousands of prey) satisfy the Lotka-Volterra equations (1) and (2) with $p = 3$, $q = 1$, $r = 4$, and $s = 2$. Suppose that at a time when there are 100 predators ($x = 1$), there are 1000 prey ($y = 1$). Find an equation relating $x$ and $y$.

With the given values of the constants, $p$, $q$, $r$, and $s$, equation (3) reads

$$\frac{dy}{dx} = \frac{y(3 - x)}{x(2y - 4)}.$$

Separating the variables yields

$$\frac{2y - 4}{y}dy = \frac{3 - x}{x}dx$$

or

$$\int\left(2 - \frac{4}{y}\right)dy = \int\left(\frac{3}{x} - 1\right)dx.$$

Evaluating the integrals gives

$$2y - 4 \ln y = 3 \ln x - x + C.$$

(It is not necessary to use absolute value for the logarithms since $x$ and $y$ are positive.) Use the initial conditions $x = 1$ and $y = 1$ to find $C$.

$$2 - 4(0) = 3(0) - 1 + C$$
$$C = 3$$

The desired equation is

$$x + 2y - 3 \ln x - 4 \ln y = 3. \tag{4}$$

A graph of equation (4), in Figure 8, shows that the variations in the populations of the predator and prey are cyclic. The population of prey tends to rise when the population of predators is low. Then, with the resulting abundance of food, the population of predators increases. As a result, the population of prey decreases again, and this forces the population of predators to decline because of the lack of food. The pattern repeats indefinitely. ■

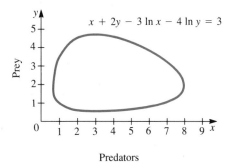

**FIGURE 8**

***Arms Race***    Lewis F. Richardson* proposed a pair of linear first-order differential equations as a model of an arms race between two rival countries $X$ and $Y$. In order to simplify the situation, the following assumptions are made.

**1.** $X$ and $Y$ are roughly equal in size.
**2.** $T_0$, the annual trade flow from $X$ to $Y$ in dollars, is a constant.
**3.** $U_0$, the annual trade flow from $Y$ to $X$ in dollars, is a constant.
**4.** The arms race between $X$ and $Y$ is not affected by the policies of other countries.

Let

$$t = \text{the time in years,}$$
$$x = \text{the armaments budget of } X \text{ in year } t \text{ in}$$
$$\text{dollars, minus } T_0,$$
$$\text{and} \quad y = \text{the armaments budget of } Y \text{ in year } t \text{ in}$$
$$\text{dollars, minus } U_0.$$

Richardson assumed that $x$ stimulates the growth of $y$, but the sheer size of $x$ would eventually inhibit its own growth. He made similar assumptions about $y$. These assumptions led him to the following model:

$$\frac{dx}{dt} = ky - mx + g \quad \text{and} \quad \frac{dy}{dt} = kx - my + h, \tag{5}$$

where $k$, $m$, $g$, and $h$ are constants. The constants $g$ and $h$ account for the flow of trade between the two nations and their underlying attitudes toward each other. For example, if the two countries have long-standing grievances and ideological differences that outweigh the value of the constant trade flow, then $g > 0$, $h > 0$, and the armament budgets will grow even when $x = 0 = y$.

---

*Arms and Insecurity* by Lewis F. Richardson (Pittsburgh: The Boxwood Press, 1960.)

On the other hand, if the constant trade flow is the dominant factor, then $g < 0$, $h < 0$, and the armament budgets will shrink when $x = 0 = y$. Richardson found that the constant $k$ was roughly proportional to the size of the country, but the constant $m$ was roughly the same for all countries.

Richardson set $z = x + y$ and added the two equations in (5), obtaining

$$
\begin{aligned}
\frac{dz}{dt} &= \frac{dx}{dt} + \frac{dy}{dt} \\
&= ky - mx + g + kx - my + h \\
&= k(x + y) - m(x + y) + g + h \\
&= kz - mz + g + h
\end{aligned}
$$

$$
\frac{dz}{dt} = g + h - (m - k)z. \tag{6}
$$

Solving this differential equation as in Example 3 in Section 8.3 gives the general solution of (6) as

$$
z = \frac{g + h}{m - k} + Ce^{-(m - k)t}. \tag{7}
$$

Richardson tested his model on the growth of armaments in the years preceding World War I for

$$X = \text{the bloc consisting of Russia and France,}$$

and $$Y = \text{the bloc consisting of Germany and Austro-Hungary.}$$

He found a remarkably good fit of (6) and (7) to the data.

R. Taagepera, et al.,* proposed the following model for the arms race (within a finite time span) between Israel and the Arab countries:

$$
\frac{dx}{dt} = kx \quad \text{and} \quad \frac{dy}{dt} = my, \tag{8}
$$

where $k$ and $m$ are constants. By Example 2 in Section 8.2, the general solution of (8) is

$$
x = x_0 e^{kt} \quad \text{and} \quad y = y_0 e^{mt}. \tag{9}
$$

This model asserts that $x$ stimulates the growth of $x$, but not of $y$, and vice versa. Here we are supposing that self-stimulation is the controlling factor, whereas the Richardson model presupposes mutual stimulation. Taagepera et al. found that the self-stimulation model (8)–(9) fitted the Israeli-Arab arms race better than the Richardson model (5)–(7). This result is very surprising, for it challenges the common assumption that the driving force in any arms race is mutual stimulation.

***Mixing Problems***   The mixing of two solutions can lead to a first-order differential equation, as the next example shows.

---

*R. Taagepera, G. M. Shiffler, R. T. Perkins, and D. L. Wagner, "Soviet-American and Israeli-Arab arms races and the Richardson model," *General Systems* 20 (1975), 151–58.

4 gal/min

3 gal/min

**FIGURE 9**

### ■ EXAMPLE 2*

Suppose a tank contains 100 gallons of a solution of dissolved salt and water, which is kept uniform by stirring. If pure water is allowed to flow into the tank at the rate of 4 gallons per minute, and the mixture flows out at the rate of 3 gallons per minute (Figure 9), how much salt will remain in the tank after $t$ minutes if 15 pounds of salt are in the mixture initially?

Let the amount of salt present in the tank at any specific time be $y = f(t)$. The net rate at which $y$ changes is given by

$$\frac{dy}{dt} = (\text{Rate of Salt In}) - (\text{Rate of Salt Out}).$$

Since pure water is coming in, the rate of salt entering the tank is zero. The rate at which salt is leaving the tank is the product of the amount of salt per gallon (in $V$ gallons) and the number of gallons per minute leaving the tank.

$$\text{Rate of Salt Out} = \left(\frac{y}{V} \text{ lb/gal}\right)(3 \text{ gal/min}).$$

The differential equation, therefore, can be written as

$$\frac{dy}{dt} = -\frac{3y}{V}, \quad y(0) = 15,$$

where $y(0)$ is the initial amount of salt in the solution. We must take into account the fact that the volume $V$ of the mixture is not constant but is determined by

$$\frac{dV}{dt} = (\text{Rate of Liquid In}) - (\text{Rate of Liquid Out}) = 4 - 3 = 1,$$

or $\quad \dfrac{dV}{dt} = 1,$

from which

$\quad V(t) = t + C_1.$

Because the volume is known to be 100 at time $t = 0$, we have $C_1 = 100$, and

$$\frac{dy}{dt} = \frac{-3y}{t + 100}, \quad y(0) = 15,$$

a variables separable equation with solution

$$\frac{dy}{y} = \frac{-3}{t + 100} dt$$

$$\ln y = -3 \ln(t + 100) + C.$$

---

*This example is taken from *Ordinary Differential Equations with Applications*, Larry C. Andrews, Scott, Foresman and Company, 1982, pp. 77–78.

Since $y = 15$ when $t = 0$,

$$\ln 15 = -3 \ln 100 + C$$
$$\ln 15 + 3 \ln 10^2 = C$$
$$\ln(15 \times 10^6) = C.$$

Finally,

$$\ln y = \ln(t + 100)^{-3} + \ln(15 \times 10^6)$$
$$= \ln[(t + 100)^{-3}(15 \times 10^6)]$$
$$y = \frac{15 \times 10^6}{(t + 100)^3}. \ \blacksquare$$

***Continuous Deposits***    An amount of money $A$ invested at an annual interest rate $r$, compounded continuously, grows according to the differential equation

$$\frac{dA}{dt} = rA,$$

where $t$ is time in years. Suppose regular deposits are made to the account at frequent intervals at a rate of $D$ dollars per year. For simplicity, assume these deposits to be continuous. The differential equation for the growth of the account then becomes

$$\frac{dA}{dt} = rA + D.$$

■■■■ **EXAMPLE 3**

When Michael was born, his grandfather arranged to deposit $5000 in an account for him at 8% annual interest compounded continuously. Grandfather plans to add to the account "continuously" at the rate of $1000 a year. How much will be in the account when Michael is 18?

Here $r = .08$ and $D = 1000$, so the differential equation is

$$\frac{dA}{dt} = .08A + 1000.$$

Separate the variables and integrate on each side.

$$\frac{1}{.08A + 1000} dA = dt$$

$$\frac{1}{.08} \ln(.08A + 1000) = t + C$$

$$\ln(.08A + 1000) = .08t + K \qquad\qquad K = .08C$$

$$.08A + 1000 = e^{.08t + K}$$

$$.08A = -1000 + e^{.08t} e^K$$

$$.08A = -1000 + Me^{.08t} \qquad M = e^K$$

$$A = -12,500 + \frac{M}{.08} e^{.08t}$$

Use the fact that the initial amount deposited was $5000 to find $M$.

$$5000 = -12,500 + \frac{M}{.08} e^{(.08)0}$$

$$5000 = -12,500 + \frac{M}{.08} (1)$$

$$1400 = M$$

$$A = -12,500 + \frac{1400}{.08} e^{.08t}$$

$$A = -12,500 + 17,500e^{.08t}$$

When Michael is 18, $t = 18$, and the amount in the account will be

$$A = -12,500 + 17,500e^{(.08)18}$$
$$= -12,500 + 17,500e^{1.44}$$
$$= 61,362.18,$$

or about $61,400. ▃▃

*Epidemics*   Under certain conditions, the spread of a contagious disease can be described with the logistic growth model as in the next example.

## ▃▃ EXAMPLE 4

Consider a population of size $N$ that satisfies the following conditions.

1. Initially there is only one infected individual.
2. All uninfected individuals are susceptible, and infection occurs when an uninfected individual contacts an infected individual.
3. Contact between any two individuals is just as likely as contact between any other two individuals.
4. Infected individuals remain infectious.

Let $t$ = the time in days and $y$ = the number of individuals infected at time $t$. At any moment there are $(N - y)y$ possible contacts between an uninfected individual and an infected individual. Thus, it is reasonable to assume that the rate of spread of the disease satisfies the logistic equation (discussed in Section 8.2)

$$\frac{dy}{dt} = a(N - y)y. \tag{10}$$

As shown in Section 8.2, the general solution of equation (10) is

$$y = \frac{N}{1 + be^{-kt}}, \tag{11}$$

where $b = (N - y_0)/y_0$ and $k = aN$. Since just one individual is infected initially, $y_0 = 1$ here. Substituting these values into equation (11) gives

$$y = \frac{N}{1 + (N - 1)e^{-aNt}} \tag{12}$$

as the general solution of equation (10).

The infection rate will be a maximum when its derivative is 0, that is, when $d^2y/dt^2 = 0$. Since

$$\frac{dy}{dt} = a(N - y)y,$$

we have

$$\frac{d^2y}{dt^2} = a[(N - y)(1) + y(-1)]$$

$$= aN - 2ay.$$

Set $d^2y/dt^2 = 0$ to get

$$0 = aN - 2ay$$

$$2ay = aN$$

$$y = \frac{N}{2}.$$

The maximum infection rate occurs when exactly half the total population is still uninfected and equals

$$\frac{dy}{dt} = a\left(N - \frac{N}{2}\right)\frac{N}{2} = \frac{aN^2}{4}.$$

Letting $y = N/2$ in equation (12) and solving for $t$ shows that the maximum infection rate occurs at time

$$t_m = \frac{\ln(N - 1)}{aN}.$$

The graph of $dy/dt$, shown in Figure 10, is called the *epidemic curve*. It is symmetric about the line $t = t_m$. ▬

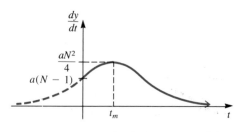

**FIGURE 10**

## 8.5 EXERCISES    APPLICATIONS

### BUSINESS AND ECONOMICS

**Continuous Deposits**

1. Jessie deposits $2000 in an IRA at 6% interest compounded continuously for her retirement in ten years. She intends to make continuous deposits at the rate of $2000 a year until she retires. How much will she have accumulated at that time?

**Continuous Deposits**

2. In Exercise 1, how long will it take Jessie to accumulate $20,000 in her retirement account?

**Continuous Deposits**

3. To provide for a future expansion, a company plans to make continuous deposits to a savings account at the rate of $10,000 per year, with no initial deposit. They want to accumulate $70,000. How long will it take if the account earns 7% interest compounded continuously?

**Continuous Deposits**

4. Suppose the company in Exercise 3 wants to accumulate $50,000 in 3 years. Find the approximate yearly deposit that will be required.

### LIFE SCIENCES

**Competing Species**

5. The system of equations

$$\frac{dy}{dt} = 3y - 2xy$$

$$\frac{dx}{dt} = -2x + 3xy$$

describes the influence of the populations of two competing species on their growth rates.

(a) Following Example 1 in the text, find an equation relating $x$ and $y$, assuming $y = 2$ when $x = 1$.

(b) Find values of $x$ and $y$ so that both populations are constant. (*Hint:* Set both differential equations equal to 0.)

**Symbiotic Species**

6. When two species coexist in a symbiotic (dependent) relationship, they either increase together or decrease together. Typical equations for the growth rates of two such species might be

$$\frac{dx}{dt} = -3x + 4xy$$

$$\frac{dy}{dt} = -3y + xy.$$

(a) Find an equation relating $x$ and $y$ if $x = 3$ when $y = 1$.

(b) Find values of $x$ and $y$ so that both populations are constant. (See Exercise 1.)

**Spread of an Epidemic**

7. The native Hawaiians lived for centuries in isolation from other peoples. When foreigners finally came to the islands they brought with them diseases such as measles, whooping cough, and smallpox, which decimated the population. Suppose such an island has a native population of 5000, and a sailor from a visiting ship introduces measles, which has an infection rate of .00005.

(a) Write an equation for the number of natives who remain uninfected. Let $t$ represent time in days.

(b) How many are uninfected after 30 days?

(c) How many are uninfected after 50 days?

(d) When will the maximum infection rate occur?

**Spread of an Epidemic**    **8.** In Example 4, the number of infected individuals is given by equation (12).

(a) Show that the number of uninfected individuals is given by

$$N - y = \frac{N(N - 1)}{N - 1 + e^{aNt}}.$$

(b) Graph the equation in (a) and equation (12) on the same axes when $N = 100$ and $a = .01$.

(c) Find the common inflection point of the two graphs.

(d) What is the significance of the common inflection point?

(e) What are the limiting values of $y$ and $N - y$?

**Spread of an Epidemic**    **9.** An influenza epidemic spreads at a rate proportional to the product of the number of people infected and the number not yet infected. Assume that 50 people are infected at the beginning of the epidemic in a community of 10,000 people, and 300 are infected 10 days later.

(a) Write an equation for the number of people infected, $y$, after $t$ days.

(b) When will half the community be infected?

**Spread of an Epidemic**    **10.** The Gompertz growth law,

$$\frac{dy}{dt} = kye^{-at},$$

for constants $k$ and $a$, is another model used to describe the growth of an epidemic. Repeat Exercise 9, using this differential equation with $a = .02$.

**Spread of Gonorrhea**    **11.** Gonorrhea is spread by sexual contact, takes 3 to 7 days to incubate, and can be treated with antibiotics. There is no evidence that a person ever develops immunity. One model proposed for the rate of change in the number of men infected by this disease is

$$\frac{dy}{dt} = -ay + b(f - y)Y,$$

where $y$ is the fraction of men infected, $f$ is the fraction of men who are promiscuous, $Y$ is the fraction of women infected, and $a$ and $b$ are appropriate constants.*

(a) Assume $a = 1$, $b = 1$, $f = .5$, $Y = .01$, and solve for $y$ using $y = .02$ when $t$ is 0 as an initial condition. Round your answer to three decimal places.

(b) A comparable model for women is

$$\frac{dY}{dt} = -AY + B(F - Y)y,$$

where $F$ is the fraction of women who are promiscuous and $A$ and $B$ are constants. Assume $A = 1$, $B = 1$, $y = .1$, and $F = .03$, and solve for $Y$, using $Y = .01$ as an initial condition.

---

*From *An Introduction to Mathematical Modeling* by Edward A. Bender. Copyright © 1978 by John Wiley and Sons., Inc. Reprinted by permission.

## SOCIAL SCIENCES

**Spread of a Rumor**   *The equation developed in the text for the spread of an epidemic also can be used to describe diffusion of information. In a population of size N, let y be the number who have heard a particular piece of information. Then*

$$\frac{dy}{dt} = a(N - y)y$$

*for a positive constant a. Use this model in Exercises 12–14.*

12. Suppose a rumor starts among 3 people in a certain office building. That is, $y_0 = 3$. Suppose 500 people work in the building and 50 people have heard the rumor in 2 days. Using equation (11), write an equation for the number who have heard the rumor in $t$ days. How many people will have heard the rumor in 5 days?

13. A rumor spreads at a rate proportional to the product of the number of people who have heard it and the number who have not heard it. Assume that 5 people in an office with 50 employees heard the rumor initially, and 15 people have heard it 3 days later.

    (a) Write an equation for the number, $y$, of people who have heard the rumor in $t$ days.

    (b) When will 30 employees have heard the rumor?

14. A news item is heard on the late news by 5 of the 100 people in a small community. By the end of the next day 20 people have heard the news. Using equation (11), write an equation for the number of people who have heard the news in $t$ days. How many have heard the news after 3 days?

15. Repeat Exercise 13 using the Gompertz growth law,

$$\frac{dy}{dt} = kye^{-at},$$

    for constants $k$ and $a$, with $a = .1$.

## PHYSICAL SCIENCES

**Salt Concentration**   16. A tank holds 100 gal of water that contains 20 lb of dissolved salt. A brine (salt) solution is flowing into the tank at the rate of 2 gal/min while the solution flows out of the tank at the same rate. The brine solution entering the tank has a salt concentration of 2 lb/gal.

    (a) Find an expression for the amount of salt in the tank at any time.

    (b) How much salt is present after 1 hr?

    (c) As time increases, what happens to the salt concentration?

17. Solve Exercise 16 if the brine solution is introduced at the rate of 3 gal/min while the rate of outflow remains the same.

18. Solve Exercise 16 if the brine solution is introduced at the rate of 1 gal/min while the rate of outflow stays the same.

19. Solve Exercise 16 if pure water is added instead of brine.

**Chemical in a Solution**   20. Five grams of a chemical is dissolved in 100 liters of alcohol. Pure alcohol is added at the rate of 2 liters/min and at the same time the solution is being drained at the rate of 1 liter/min.

    (a) Find an expression for the amount of the chemical in the mixture at any time.

    (b) How much of the chemical is present after 30 min?

**21.** Solve Exercise 20 if a 10% solution of the same mixture is added instead of pure alcohol.

**Soap Concentration**   **22.** A prankster puts 4 lb of soap in a fountain that contains 200 gal of water. To clean up the mess a city crew runs clear water into the fountain at the rate of 8 gal/min allowing the excess solution to drain off at the same rate. How long will it be before the amount of soap in the mixture is reduced to 1 lb?

---

**KEY WORDS**

| | | | |
|---|---|---|---|
| | differential equation | | logistic curve |
| **8.1** | general solution | **8.3** | linear first-order differential |
| | particular solution | | equation |
| | initial condition | | integrating factor |
| **8.2** | separable differential equation | **8.4** | Euler's method |
| | growth rate constant | **8.5** | Lotka-Volterra equations |
| | carrying capacity | | mixing problems |
| | limited growth function | | continuous deposits |
| | logistic growth | | epidemic curve |

---

## CHAPTER 8        REVIEW EXERCISES

*Find general solutions for the following differential equations.*

**1.** $\dfrac{dy}{dx} = 2x^3 + 6x$    **2.** $\dfrac{dy}{dx} = x^2 + 5x^4$    **3.** $\dfrac{dy}{dx} = 4e^x$    **4.** $\dfrac{dy}{dx} = \dfrac{1}{2x + 3}$

*Find particular solutions for the following differential equations.*

**5.** $\dfrac{dy}{dx} = x^2 - 5x; \quad y = 1$ when $x = 0$

**6.** $\dfrac{dy}{dx} = 4x^3 + 2; \quad y = 3$ when $x = 1$

**7.** $\dfrac{dy}{dx} = 5(e^{-x} - 1); \quad y = 17$ when $x = 0$

**8.** $\dfrac{dy}{dx} = \dfrac{x}{x^2 - 3}; \quad y = 52$ when $x = 2$

*Find general solutions for the following differential equations.*

**9.** $\dfrac{dy}{dx} = \dfrac{3x + 1}{y}$    **10.** $\dfrac{dy}{dx} = \dfrac{e^x + x}{y - 1}$    **11.** $\dfrac{dy}{dx} = \dfrac{2y + 1}{x}$    **12.** $\dfrac{dy}{dx} = \dfrac{3 - y}{e^x}$

*Find particular solutions for the following differential equations. (Some solutions may give y implicitly.)*

**13.** $(5 - 2x)y = \dfrac{dy}{dx}; \quad y = 2$ when $x = 0$

**14.** $\sqrt{x}\,\dfrac{dy}{dx} = xy; \quad y = 4$ when $x = 1$

**15.** $\dfrac{dy}{dx} = \dfrac{1 - 2x}{y + 3}; \quad y = 16$ when $x = 0$

**16.** $\dfrac{dy}{dx} = (3x + 2)^2\, e^y; \quad y = 0$ when $x = 0$

*Find the general solutions of the following differential equations. (Assume y' = dy/dx.)*

**17.** $y' + 5y = 12$

**18.** $y' + xy = 4x$

**19.** $3y' + xy - x = 0$

**20.** $x\,\dfrac{dy}{dx} + y - e^x = 0$

*Solve each differential equation subject to the given condition.*

**21.** $y' + x^2y = x^2$;   $y = 8$ when $x = 0$

**22.** $e^x y' - e^x y = x^2 - 1$;   $y = 42$ when $x = 0$

**23.** $x \dfrac{dy}{dx} - 2x^2 y + 3x^2 = 0$;   $y = 15$ when $x = 0$

**24.** $x^2 \dfrac{dy}{dx} + 4xy - e^{2x^3} = 0$;   $y = e^2$ when $x = 1$

*Use Euler's method to approximate the indicated function value for $y = f(x)$ to three decimal places using $h = .2$.*

**25.** $y' = x + y^{-1}$;   $f(0) = 1$; find $f(1)$

**26.** $y' = e^x + 1$;   $f(0) = 1$; find $f(.6)$

**27.** Let $y = f(x)$ and $y' = (x/2) + 4$ with $f(0) = 0$. Use Euler's method with $h = .1$ to approximate $f(.3)$ to three decimal places. Then solve the differential equation and find $f(.3)$ to three decimal places. Also, find $y_3 - f(x_3)$.

**28.** Let $y = f(x)$ and $y' = 3 + \sqrt{y}$, with $f(0) = 0$. Construct a table like the ones in Section 8.4, Example 2, for $[0, 1]$, with $h = .2$. Then graph the polygonal approximation of the graph of $y = f(x)$.

 APPLICATIONS

## BUSINESS AND ECONOMICS

**Marginal Sales**   **29.** The marginal sales (in hundreds of dollars) of a computer software company are given by

$$\frac{dy}{dx} = 5e^{.2x},$$

where $x$ is the number of months the company has been in business. Assume that sales were 0 initially.

**(a)** Find the sales after 6 months.

**(b)** Find the sales after 12 months.

**Continuous Withdrawals**   **30.** A deposit of \$10,000 is made to a savings account at 5% interest compounded continuously. Assume that continuous *withdrawals* of \$1000 per year are made.

**(a)** Write a differential equation to describe the situation.

**(b)** How much will be left in the account after 1 year?

**31.** In Exercise 30, approximately how long will it take to use up the account?

## LIFE SCIENCES

**Effect of Insecticide**   **32.** After use of an experimental insecticide, the rate of decline of an insect population is

$$\frac{dy}{dt} = \frac{-10}{1 + 5t},$$

where $t$ is the number of hours after the insecticide is applied. Assume that there were 50 insects initially.

**(a)** How many are left after 24 hours?

**(b)** How long will it take for the entire population to die?

**Growth of a Mite Population**  33. A population of mites grows at a rate proportional to the number present, $y$. If the growth constant is 10% and 120 mites are present at time $t = 0$, in weeks, find the number present after 6 wk.

**Competing Species**  34. Find an equation relating $x$ to $y$ given the following equations, which describe the interaction of two competing species and their growth rates.

$$\frac{dx}{dt} = .2x - .5xy$$

$$\frac{dy}{dt} = -.3y + .4xy$$

Find the values of $x$ and $y$ for which both growth rates are 0.

**Smoke Content in a Room**  35. The air in a meeting room with 15,000 cubic feet has a smoke content of 20 ppm (parts per million). An air conditioner is turned on, which brings fresh air (with no smoke) into the room at a rate of 1200 cubic feet per minute and forces the smoky air out at the same rate. How long will it take to reduce the smoke content to 5 ppm?

**Smoke Content in a Room**  36. In Exercise 35, how long will it take to reduce the smoke content to 10 ppm if smokers in the room are adding smoke at the rate of 5 ppm per minute?

**Spread of Influenza**  37. A small, isolated mountain community with a poulation of 700 is visited by an outsider who carries influenza. After 6 weeks, 300 people are uninfected.

(a) Write an equation for the number of people who remain uninfected at time $t$ in weeks.

(b) Find the number still uninfected after 7 weeks.

(c) When will the maximum infection rate occur?

## SOCIAL SCIENCES

**Production Rate**  38. The rate at which a new worker in a certain factory produces items is given by

$$\frac{dy}{dx} = .2(125 - y),$$

where $y$ is the number of items produced by the worker per day, $x$ is the number of days worked, and the maximum production per day is 125 items. Assume the worker produced 20 items the first day on the job ($x = 0$).

(a) Find the number of items the new worker will produce in 10 days.

(b) According to the function that is the solution of the differential equation, can the worker ever produce 125 items in a day?

**Spread of a Rumor**  39. A rumor spreads through the offices of a company with 100 employees, starting in a meeting with 4 people. After 3 days 15 people have heard the rumor.

(a) Write an equation for the number of people who have heard the rumor in $x$ days.

(b) How many people have heard the rumor in 5 days?

**Population Growth**  40. Let

$$y = \frac{N}{1 + be^{-kx}}.$$

If $y$ is $y_1$, $y_2$, and $y_3$ at times $x_1$, $x_2$, and $x_3 = 2x_2 - x_1$ (that is, at three equally spaced times), then prove that

$$N = \frac{1/y_1 + 1/y_3 - 2/y_2}{1/(y_1 y_3) - 1/y_2^2}.$$

**Population Growth**  *In the following table of United States census figures, y is population in millions.*

| Year | y | | Year | y |
|------|-----|---|------|-----|
| 1790 | 4 | | 1890 | 63 |
| 1800 | 5 | | 1900 | 76 |
| 1810 | 7 | | 1910 | 92 |
| 1820 | 10 | | 1920 | 106 |
| 1830 | 13 | | 1930 | 123 |
| 1840 | 17 | | 1940 | 132 |
| 1850 | 23 | | 1950 | 151 |
| 1860 | 31 | | 1960 | 179 |
| 1870 | 40 | | 1970 | 204 |
| 1880 | 50 | | 1980 | 227 |

**41.** Use Exercise 40 and the above data to find the following.

(a) Find $N$ using the years 1800, 1850, 1900.

(b) Find $N$ using the years 1850, 1900, and 1950.

(c) Find $N$ using the years 1870, 1920, and 1970.

(d) Explain why different values of $N$ were obtained in (a)–(c). What does this suggest about the validity of this model and others?

**42.** Given the value of $N$ found in Exercise 41(c), assume $y_0 = 40$.

(a) Find $b$ and $k$ to two decimal places using the population of the United States in 1920 or 1970 and

$$y = \frac{N}{1 + be^{-kx}}.$$

(b) Estimate the population of the United States in 1980 and compare your estimate to the actual population in 1980.

(c) Predict the populations of the United States in 1990 and 2000.

## PHYSICAL SCIENCES

**Newton's Law of Cooling**  **43.** A roast at a temperature of 40° is put in a 300° oven. After 1 hr the roast has reached a temperature of 150°. Newton's law of cooling states that

$$\frac{dT}{dt} = k(T - T_F),$$

where $T$ is the temperature of an object, the surrounding medium has temperature $T_F$ at time $t$, and $k$ is a constant. Use Newton's law to find the temperature of the roast after 2 hr.

**44.** In Exercise 43, how long does it take for the roast to reach a temperature of 250°?

## EXTENDED APPLICATION

## POLLUTION OF THE GREAT LAKES*

Industrial nations are beginning to face the problems of water pollution. Lakes present a problem, because a polluted lake contains a considerable amount of water that must somehow be cleaned. The main cleanup mechanism is the natural process of gradually replacing the water in the lake. This application deals with pollution in the Great Lakes. The basic idea is to regard the flow in the Great Lakes as a mixing problem.

We make the following assumptions:

1. Rainfall and evaporation balance each other, so the average rates of inflow and outflow are equal.
2. The average rates of inflow and outflow do not vary much seasonally.
3. When water enters the lake, perfect mixing occurs, so that the pollutants are uniformly distributed.
4. Pollutants are not removed from the lake by decay, sedimentation, or in any other way except outflow.
5. Pollutants flow freely out of the lake; they are not retained as DDT is.

The first two are valid assumptions; however, the last three are questionable.

We will use the following variables in the discussion to follow.

$$V = \text{volume of the lake}$$
$$P_L = \text{pollution concentration in the lake at time } t$$
$$P_i = \text{pollution concentration in the inflow to the lake at time } t$$
$$r = \text{rate of flow}$$
$$t = \text{time in years}$$

By the assumptions stated above, the net change in total pollutants during the time interval $\Delta t$ is (approximately)

$$VP_L = (P_i - P_L)(r\,\Delta t).$$

Dividing this equation by $\Delta t$ and taking the limit as $\Delta t \to 0$, we get the differential equation

$$P'_L = \frac{(P_i - P_L)r}{V}.$$

Since we are treating $V$ and $r$ as constants, we replace $r/V$ with $k$, so the equation can be written as the first-order linear equation

$$P'_L + kP_L = kP_i.$$

The solution is

$$P_L(t) = e^{-kt}\left[P_L(0) + k\int_0^t P_i(x)e^{kx}dx\right]. \tag{1}$$

Figure 11 shows values of $1/k$ for each lake (except Huron) measured in years. *If the model is reasonable,* the numbers in the figure can be used in equation (1) to determine the effect of various pollution abatement schemes. Lake Ontario is excluded from the discussion because about 84% of its inflow comes from Erie and can be controlled only indirectly.

The fastest possible cleanup will occur if all pollution inflow ceases. This means that $P_i = 0$. In this case, equation (1) leads to

$$t = \frac{1}{k}\ln\left(\frac{P_L(0)}{P_L(t)}\right).$$

---

*From *An Introduction to Mathematical Modeling* by Edward A. Bender. Copyright © 1978 by John Wiley & Sons, Inc. Reprinted by permission.

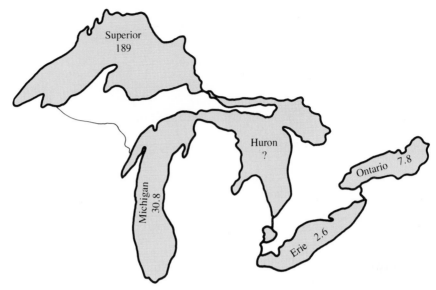

**FIGURE 11**

From this we can tell the length of time necessary to reduce pollution to a given percentage of its present level. For example, from the figure, for Lake Superior $1/k = 189$. Thus, to reduce pollution to 50% of its present level, $P_L(0)$, we want

$$\frac{P_L(t)}{P_L(0)} = .5 \quad \text{or} \quad \frac{P_L(0)}{P_L(t)} = 2,$$

from which

$$t = 189 \ln 2 \approx 131.$$

The following figures, representing years, were found in this way.

| Lake | 50% | 20% | 10% | 5% |
|------|-----|-----|-----|-----|
| Erie | 2 | 4 | 6 | 8 |
| Michigan | 21 | 50 | 71 | 92 |
| Superior | 131 | 304 | 435 | 566 |

Fortunately, the pollution in Lake Superior is quite low at present.

As mentioned before, assumptions 3, 4, and 5 are questionable. For persistent pollutants like DDT, the estimated cleanup times may be too low. For other pollutants, how assumptions 4 and 5 affect cleanup times is unclear. However, the values of $1/k$ given in the figure probably provide rough lower bounds for the cleanup times of persistent pollutants.

**EXERCISES**

1. Calculate the number of years to reduce pollution in Lake Erie to each of the following levels.
   **(a)** 40%  **(b)** 30%
2. Repeat Exercise 1 for Lake Michigan.
3. Repeat Exercise 1 for Lake Superior.

# 9

# Probability and Calculus

In recent years, probability has become increasingly useful in fields ranging from manufacturing to medicine, as well as in all types of research. The foundations of probability were laid in the seventeenth century by Blaise Pascal (1623–1662) and Pierre de Fermat (1601–1665), who investigated *the problem of the points*. This problem dealt with the fair distribution of winnings in an interrupted game of chance between two equally matched players whose scores were known at the time of the interruption.

Probability has advanced from a study of gambling to a well-developed, deductive mathematical system. In this chapter we give a brief introduction to the use of calculus in probability.

## ▦ 9.1 CONTINUOUS PROBABILITY MODELS

Suppose that a bank is studying the transaction times of its tellers. The lengths of time spent on observed transactions, rounded to the nearest minute, are shown in Table 1.

**Table 1**

| Time | 1 | 2 | 3 | 4 | 5 | 6 | 7 | 8 | 9 | 10 | |
|------|---|---|---|---|---|---|---|---|---|----|---|
| Frequency | 3 | 5 | 9 | 12 | 15 | 11 | 10 | 6 | 3 | 1 | (Total: 75) |

The table shows, for example, that 9 of the 75 transactions in the study took 3 minutes, 15 transactions took 5 minutes, and 1 transaction took 10 minutes. Because the time for any particular transaction is a random event, the number of minutes for a transaction is called a **random variable.** The frequencies can be converted to probabilities by dividing each frequency by the total number of transactions (75) to get the results shown in Table 2.*

**Table 2**

| Time | 1 | 2 | 3 | 4 | 5 | 6 | 7 | 8 | 9 | 10 |
|------|---|---|---|---|---|---|---|---|---|----|
| Probability | .04 | .07 | .12 | .16 | .20 | .15 | .13 | .08 | .04 | .01 |

---

*One definition of the *probability of an event* is the number of outcomes that favor the event divided by the total number of outcomes in an experiment.

Because each value of the random variable is associated with just one probability, Table 2 defines a function. Such a function is called a **probability distribution function** or simply a **probability function.** The special properties of a probability distribution function are given below.

**PROBABILITY FUNCTION OF A RANDOM VARIABLE**

If the function $f(x)$ is a probability distribution function with domain $\{x_1, x_2, \ldots, x_n\}$, then for $1 \le i \le n$,

$$0 \le f(x_i) \le 1,$$

and

$$f(x_1) + f(x_2) + \cdots + f(x_n) = 1.$$

The information in Table 2 can be displayed graphically with a special kind of bar graph called a **histogram.** The bars of a histogram have the same width, and their heights are determined by the probabilities of the various values of the random variable.

The probability distribution function in Table 2 is a **discrete probability function** because it has a finite domain—the integers from 1 to 10, inclusive. Many distributions, however, have infinite domains. For example, the distribution of heights (in inches) of college women includes infinitely many possible measurements, such as 53, 54.2, 66.5, 72.33. . . , and so on. Probability distribution functions with infinite domains are called *continuous probability distributions*.

**CONTINUOUS PROBABILITY DISTRIBUTION**

A **continuous random variable** can take on any value in some interval of real numbers. The distribution of this random variable is called a **continuous probability distribution.**

In the bank example discussed earlier in this section, it would have been possible to time the teller transactions with greater precision—to the nearest tenth of a minute or even to the nearest second (one-sixtieth of a minute) if desired. Theoretically, at least, $t$ could take on any positive real-number value between, say, 0 and 11 minutes. The graph of the probabilities $f(t)$ of these transaction times can be thought of as the continuous curve shown in Figure 1. As indicated in Figure 1, the curve was derived from Table 2 by connecting the points at the tops of the bars in the corresponding histogram and smoothing the resulting polygon into a curve.

For a discrete probability function, the area of each bar (or rectangle) gives the probability of a particular transaction time. Thus, by considering the possible transaction times $t$ as all the real numbers between 0 and 11, the area under the curve of Figure 2 between any two values of $t$ can be interpreted as the probability that a transaction time will be between those two numbers. For example, the shaded region in Figure 2 corresponds to the probability that $t$ is between $a$ and $b$, written $P(a \le t \le b)$.

**FIGURE 1**

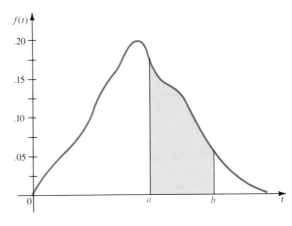

**FIGURE 2**

It was shown earlier that the definite integral of a continuous function $f(x)$, where $f(x) \geq 0$, gives the area under the graph of $f(x)$ from $x = a$ to $x = b$. If a function $f(x)$ can be found to describe a continuous probability function, then the definite integral can be used to find the area under the curve from $a$ to $b$. Since that area represents the probability that $x$ will be between $a$ and $b$,

> if $x$ is a continuous probability function in $[a, b]$, then
>
> $$P(a \leq x \leq b) = \int_a^b f(x) \, dx.$$

A function $f(x)$ that describes a continuous probability distribution is called a *probability density function*. Such a function must satisfy the following conditions.

**PROBABILITY DENSITY FUNCTION**

> The function $f(x)$ is a **probability density function** of a random variable $x$ in the interval $[a, b]$ if
>
> **1.** $\int_a^b f(x) \, dx = 1$, and
>
> **2.** $f(x) \geq 0$ for all $x$ in the interval $[a, b]$.

Intuitively, condition 1 says that the total probability for the interval must be 1; *something* must happen. Condition 2 says that the probability of a particular event can never be negative.

### ■ EXAMPLE 1

**(a)** Show that the function defined by $f(x) = (3/26)x^2$ is a probability density function for the interval $[1, 3]$.

First, show that condition 1 holds.

$$\int_1^3 \frac{3}{26}x^2 \, dx = \frac{3}{26}\left(\frac{x^3}{3}\right)\Big|_1^3 = \frac{3}{26}\left(9 - \frac{1}{3}\right) = 1$$

Next, show that condition 2 holds; that is, $f(x) \geq 0$ for the interval $[1, 3]$. Since $x^2$ is always positive, $f(x)$ is a probability density function.

**(b)** Find the probability that $x$ will be between 1 and 2.

The desired probability is given by the area under the graph of $f(x)$ between $x = 1$ and $x = 2$, as shown in Figure 3. The area is found by using a definite integral.

$$P(1 \leq x \leq 2) = \int_1^2 \frac{3}{26}x^2 \, dx = \frac{3}{26}\left(\frac{x^3}{3}\right)\Big|_1^2 = \frac{7}{26} \quad ■$$

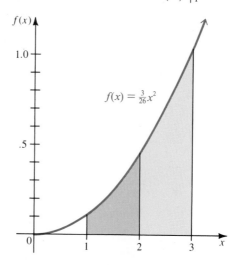

**FIGURE 3**

### ■ EXAMPLE 2

Is the function defined by $f(x) = 3x^2$ a probability density function for the interval $[0, 4]$? If not, convert it to one.

First,

$$\int_0^4 3x^2 \, dx = x^3 \Big|_0^4 = 64.$$

Since the integral is not equal to 1, the function is not a probability density function. To convert it to one, multiply $f(x)$ by $\frac{1}{64}$. The function defined by $\frac{3}{64}x^2$ for [0, 4] will be a probability density function, since

$$\int_0^4 \frac{3}{64}x^2 = 1,$$

and $(3/64)\,x^2 \geq 0$ for all $x$ in [0, 4]. ■

An important distinction is made between a discrete probability distribution and a probability density function (which is continuous). In a discrete distribution, the probability that the random variable $x$ will assume a specific value is given in the distribution for every possible value of $x$. In a probability density function, however, the probability that $x$ equals a specific value, say $c$, is

$$P(x = c) = \int_c^c f(x)\,dx = 0.$$

For a probability density function, only probabilities of *intervals* can be found. For example, suppose the random variable is the annual rainfall for a given region. The amount of rainfall in one year can take on any value within some continuous interval that depends on the region; however, the probability that the rainfall in a given year will be some specific amount, say 33.25 inches, is actually zero.

The above definition of a probability density function is extended to intervals such as $(-\infty, b]$, $[a, +\infty)$, or $(-\infty, +\infty)$ by using improper integrals as follows.

**PROBABILITY DENSITY FUNCTIONS ON $(-\infty, +\infty)$**

If $f(x)$ is a probability density function for a continuous random variable $x$ on $(-\infty, +\infty)$, then

$$P(x \leq b) = \int_{-\infty}^b f(x)\,dx,$$

$$P(x \geq a) = \int_a^{+\infty} f(x)\,dx,$$

$$P(-\infty \leq x \leq +\infty) = \int_{-\infty}^{+\infty} f(x)\,dx.$$

The total area under the graph of a probability density function of this type must still equal 1.

### ■ EXAMPLE 3

Suppose the random variable $x$ is the distance (in kilometers) from a given point to the nearest bird's nest, with the probability density function of the distribution given by $f(x) = 2xe^{-x^2}$ for $x \geq 0$.

**(a)** Show that $f(x)$ is a probability density function.

Since $e^{-x^2} = 1/e^{x^2}$ is always positive, and $x \geq 0$,

$$f(x) = 2xe^{-x^2} \geq 0.$$

Use substitution to evaluate the definite integral $\int_0^{+\infty} 2xe^{-x^2} \, dx$. Let $u = -x^2$, so that $du = -2x \, dx$, and

$$\int 2xe^{-x^2} \, dx = -\int e^{-x^2}(-2x \, dx) = -\int e^u \, du = -e^u = -e^{-x^2}.$$

Then

$$\int_0^{+\infty} 2xe^{-x^2} \, dx = \lim_{a \to +\infty} \int_0^a 2xe^{-x^2} \, dx = \lim_{a \to +\infty} (-e^{-x^2}) \Big|_0^a$$

$$= \lim_{a \to +\infty} \left( -\frac{1}{e^{a^2}} + e^0 \right) = 0 + 1 = 1.$$

The function defined by $f(x) = 2xe^{-x^2}$ satisfies the two conditions required of a probability density function.

**(b)** Find the probability that there is a bird's nest within .5 kilometers of the given point.

Find $P(x \leq .5)$ where $x \geq 0$. This probability is given by

$$P(0 \leq x \leq .5) = \int_0^{.5} 2xe^{-x^2} \, dx.$$

Now evaluate the integral.

$$P(0 \leq x \leq .5) = \int_0^{.5} 2xe^{-x^2} \, dx = (-e^{-x^2}) \Big|_0^{.5}$$

$$= -e^{-(.5)^2} - (-e^0) = -e^{-.25} + 1$$

$$\approx -.78 + 1 = .22$$

The probability that a bird's nest will be found within .5 kilometers of the point is about .22. ■

---

## 9.1 EXERCISES

*Decide whether the functions defined as follows are probability density functions on the indicated intervals. If not, tell why.*

**1.** $f(x) = \dfrac{1}{9}x - \dfrac{1}{18}$;   [2, 5]

**2.** $f(x) = \dfrac{1}{3}x - \dfrac{1}{6}$;   [3, 4]

**3.** $f(x) = \dfrac{3}{63}x^2$;   [1, 4]

**4.** $f(x) = \dfrac{3}{98}x^2$;   [3, 5]

**5.** $f(x) = 4x^3$;   [0, 3]

**6.** $f(x) = \dfrac{x^3}{81}$;   [0, 3]

**7.** $f(x) = \dfrac{x^2}{16}$; $[-2, 2]$

**8.** $f(x) = 2x^2$; $[-1, 1]$

**9.** $f(x) = 2x$; $[-2, \sqrt{5}]$

**10.** $f(x) = 4x^3$; $[-1, \sqrt[4]{2}]$

*Find a value of k that will make f a probability density function on the indicated interval.*

**11.** $f(x) = kx^{1/2}$; $[1, 4]$

**12.** $f(x) = kx^{3/2}$; $[4, 9]$

**13.** $f(x) = kx^2$; $[0, 5]$

**14.** $f(x) = kx^2$; $[-1, 2]$

**15.** $f(x) = kx$; $[0, 3]$

**16.** $f(x) = kx$; $[2, 3]$

**17.** $f(x) = kx$; $[1, 5]$

**18.** $f(x) = kx$; $[0, 4]$

**19.** $f(x) = kx^2$; $[1, 3]$

**20.** $f(x) = kx^3$; $[2, 4]$

*Show that each function defined as follows is a probability density function on the given interval, then find the indicated probabilities.*

**21.** $f(x) = \dfrac{1}{2}(1 + x)^{-3/2}$; $[0, +\infty)$

   **(a)** $P(0 \le x \le 2)$   **(b)** $P(1 \le x \le 3)$   **(c)** $P(x \ge 5)$

**22.** $f(x) = e^{-x}$; $[0, +\infty)$

   **(a)** $P(0 \le x \le 1)$   **(b)** $P(1 \le x \le 2)$   **(c)** $P(x \le 2)$

**23.** $f(x) = (1/2)e^{-x/2}$; $[0, +\infty)$

   **(a)** $P(0 \le x \le 1)$   **(b)** $P(1 \le x \le 3)$   **(c)** $P(x \ge 2)$

**24.** $f(x) = \dfrac{20}{(x + 20)^2}$; $[0, +\infty)$

   **(a)** $P(0 \le x \le 1)$   **(b)** $P(1 \le x \le 5)$   **(c)** $P(x \ge 5)$

---

## ▆ APPLICATIONS

### BUSINESS AND ECONOMICS

**Life Span of a Computer Part**   **25.** The life in months of a certain electronic computer part has a probability density function defined by

$$f(x) = \frac{1}{2}e^{-x/2} \quad \text{for } [0, +\infty).$$

Find the probability that a randomly selected component will last each of the following lengths of time.

**(a)** At most 12 months

**(b)** Between 12 and 20 months

**Machine Life**   **26.** A machine has a useful life of 4 to 9 years, and its life in years is a probability density function defined by

$$f(x) = \frac{1}{11}\left(1 + \frac{3}{\sqrt{x}}\right).$$

Find the following probabilities for the life of such a machine selected at random.

**(a)** $P(x \ge 6)$   **(b)** $P(x \le 5)$   **(c)** $P(4 \le x \le 7)$

## LIFE SCIENCES

**Petal Length**    27. The length of a petal on a certain flower varies from 1 cm to 4 cm and is a probability density function defined by

$$f(x) = \frac{1}{2\sqrt{x}}.$$

Find the probability that the length of a randomly selected petal will be as follows.

(a) Greater than or equal to 3 cm

(b) Less than or equal to 2 cm

(c) Between 2 cm and 3 cm

**Clotting Time of Blood**    28. The clotting time of blood is a random variable $x$ with values from 1 to 20 sec and probability density function defined by

$$f(x) = \frac{1}{(\ln 20)x}.$$

Find the following probabilities for a person selected at random.

(a) $P(1 \leq x \leq 5)$    (b) $P(x \geq 10)$

## SOCIAL SCIENCES

**Time to Learn a Task**    29. The time $x$ required for a person to learn a certain task is a random variable with probability density function defined by

$$f(x) = \frac{8}{7(x - 2)^2}.$$

The time required to learn the task is between 3 and 10 min. Find the probability that a randomly selected person will learn the task in the following lengths of time.

(a) Less than 4 min    (b) More than 5 min

## PHYSICAL SCIENCES

**Annual Rainfall**    30. The annual rainfall in a remote Middle Eastern country varies from 0 to 5 inches and is a probability density function defined by

$$f(x) = \frac{1}{15}\left(x + \frac{1}{2}\right).$$

Find the following probabilities for the annual rainfall in a randomly selected year.

(a) $P(x \geq 3)$    (b) $P(x \leq 2)$    (c) $P(1 \leq x \leq 4)$

**Temperature**    31. The minimum temperature in March (in degrees Celsius) on a North Atlantic island is a random variable with probability density function defined by

$$f(x) = \frac{1}{2} \quad \text{for } [1, 3].$$

Find the probability that the temperature on a given day in March will be as follows.

(a) Greater than or equal to 2°C

(b) Less than or equal to 1.5°C

(c) Between 1°C and 2.5°C, inclusive

GENERAL

**Length of a Telephone Call**   **32.** The length of a telephone call (in minutes) $x$ for a certain town is a continuous random variable with probability density function defined by

$$f(x) = 3x^{-4} \quad \text{for } [1, +\infty)$$

Find the following probabilities.

(a) $P(1 \le x \le 2)$     (b) $P(3 \le x \le 5)$     (c) $P(x \ge 3)$

---

## ▋ 9.2   EXPECTED VALUE AND VARIANCE OF CONTINUOUS RANDOM VARIABLES

It often is useful to have a single number, a typical or "average" number, that represents a random variable. The *mean* or *expected value* for a discrete random variable is found by multiplying each value of the random variable by its corresponding probability, as follows.

**EXPECTED VALUE**

> Suppose the random variable $x$ can take on the $n$ values $x_1$, $x_2$, $x_3$, $\cdots$, $x_n$. Also, suppose the probabilities that each of these values occurs are respectively $p_1$, $p_2$, $p_3$, $\cdots$, $p_n$. Then the **mean**, or **expected value**, of the random variable is
>
> $$\mu = x_1 p_1 + x_2 p_2 + x_3 p_3 + \cdots + x_n p_n = \sum_{i=1}^{n} x_i p_i.$$

This definition can be extended to continuous random variables by using definite integrals. Suppose a continuous random variable has probability function $f(x)$ on $[a, b]$. We can divide the interval from $a$ to $b$ into $n$ subintervals of length $\Delta x$, where $\Delta x = (b - a)/n$. In the $i$th subinterval, the probability that the random variable takes the value $x_i$ is $f(x_i)\Delta x_i$, and so

$$\mu = \sum_{i=1}^{n} x_i (fx_i)\Delta x_i.$$

As $n \to +\infty$, the limit of this sum gives the expected value

$$\mu = \int_{a}^{b} x\, f(x)dx.$$

The **variance** of a probability distribution is a measure of the *spread* of the values of the distribution. For a discrete distribution, the variance is found by taking the expected value of the squares of the differences of the values of the random variable and the mean. If the random variable $x$ takes the values $x_1$, $x_2$, $x_3$, . . . , $x_n$ with respective probabilities $p_1$, $p_2$, $p_3$, . . . , $p_n$ and mean $\mu$, then the variance of $x$ is

$$\text{Var}(x) = \sum_{i=1}^{n} (x_i - \mu)^2 p_i.$$

The **standard deviation** of $x$ is defined as

$$\sigma = \sqrt{\text{Var}(x)}.$$

Like the mean or expected value, the variance of a continuous random variable is an integral.

$$\text{Var}(x) = \int_a^b (x - \mu)^2 f(x) dx$$

To find the standard deviation of a continuous probability distribution, like that of a discrete distribution, we find the square root of the variance. The formulas for the expected value, variance, and standard deviation of a continuous probability distribution are summarized below.

**EXPECTED VALUE, VARIANCE, AND STANDARD DEVIATION**

If $x$ is a continuous random variable with probability density function $f(x)$ on $[a, b]$, then the expected value of $x$ is

$$E(x) = \mu = \int_a^b xf(x) \, dx.$$

The variance of $x$ is

$$\text{Var}(x) = \int_a^b (x - \mu)^2 f(x) \, dx,$$

and the standard deviation of $x$ is

$$\sigma = \sqrt{\text{Var}(x)}.$$

Geometrically, the expected value (or mean) of a probability distribution represents the "center" of the distribution. If a fulcrum were placed at $\mu$ on the $x$-axis the figure would be in balance. See Figure 4.

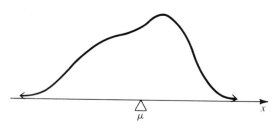

**FIGURE 4**

The variance or standard deviation of a probability distribution indicates how closely the values of the distribution cluster about the mean. These measures are most useful for comparing different distributions, as in Figure 5.

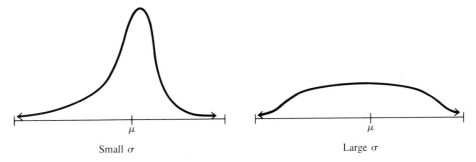

Small σ                    Large σ

**FIGURE 5**

### ◼ EXAMPLE 1

Find the expected value and variance of the random variable with probability density function defined by $f(x) = (3/26)x^2$ on $[1, 3]$.

By the definition of expected value given above,

$$\mu = \int_1^3 xf(x)\, dx$$

$$= \int_1^3 x\left(\frac{3}{26}x^2\right) dx$$

$$= \frac{3}{26}\int_1^3 x^3\, dx$$

$$= \frac{3}{26}\left(\frac{x^4}{4}\right)\bigg|_1^3 = \frac{3}{104}(81 - 1) = \frac{30}{13},$$

or about 2.31.

The variance is

$$\text{Var}(x) = \int_1^3 \left(x - \frac{30}{13}\right)^2 \left(\frac{3}{26}x^2\right) dx$$

$$= \int_1^3 \left(x^2 - \frac{60}{13}x + \frac{900}{169}\right)\left(\frac{3}{26}x^2\right) dx$$

$$= \frac{3}{26}\int_1^3 \left(x^4 - \frac{60}{13}x^3 + \frac{900}{169}x^2\right) dx$$

$$= \frac{3}{26}\left(\frac{x^5}{5} - \frac{60}{13}\cdot\frac{x^4}{4} + \frac{900}{169}\cdot\frac{x^3}{3}\right)\bigg|_1^3$$

$$= \frac{3}{26}\left[\left(\frac{243}{5} - \frac{60(81)}{52} + \frac{900(27)}{169(3)}\right) - \left(\frac{1}{5} - \frac{60}{52} + \frac{300}{169}\right)\right]$$

$$\approx .259.$$

From the variance, the standard deviation is $\sigma \approx \sqrt{.259} \approx .51$. The expected value and standard deviation are shown on the graph of the probability density function in Figure 6.

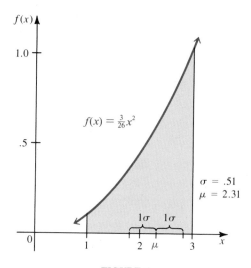

**FIGURE 6**

Calculating the variance in the last example was a messy job. An alternative version of the formula for the variance is easier to compute. This alternative formula is derived as follows.

$$
\begin{aligned}
\text{Var}(x) &= \int_a^b (x - \mu)^2 f(x) \, dx \\
&= \int_a^b (x^2 - 2\mu x + \mu^2) f(x) \, dx \\
&= \int_a^b x^2 f(x) \, dx - 2\mu \int_a^b x f(x) \, dx + \mu^2 \int_a^b f(x) \, dx
\end{aligned}
\tag{1}
$$

By definition,

$$
\int_a^b x f(x) \, dx = \mu,
$$

and, since $f(x)$ is a probability density function,

$$
\int_a^b f(x) \, dx = 1.
$$

Substitute back into equation (1) to get the alternative formula,

$$
\text{Var}(x) = \int_a^b x^2 f(x) \, dx - 2\mu^2 + \mu^2 = \int_a^b x^2 f(x) \, dx - \mu^2.
$$

**ALTERNATIVE FORMULA FOR VARIANCE**

If $x$ is a random variable with probability density function $f(x)$ on $[a,b]$, and if $E(x) = \mu$, then

$$\textbf{Var}(x) = \int_a^b x^2 f(x) \; dx - \mu^2.$$

It is important to note that the term $\mu^2$ comes *after* the differential $dx$ and so is *not* integrated.

### ▬▬ EXAMPLE 2

Use the alternative formula for variance to compute the variance of the random variable $x$ with probability density function defined by $f(x) = 3/x^4$ for $x \geq 1$.

To find the variance, first find the expected value.

$$\mu = \int_1^{+\infty} xf(x) \; dx = \int_1^{+\infty} x \cdot \frac{3}{x^4} \; dx = \int_1^{+\infty} \frac{3}{x^3} \; dx$$

$$= \lim_{a \to +\infty} \int_1^a \frac{3}{x^3} \; dx = \lim_{a \to +\infty} \left( \frac{3}{-2x^2} \right) \bigg|_1^a = \frac{3}{2}, \quad \text{or} \quad 1.5$$

Now find the variance by the alternative formula for variance.

$$\text{Var}(x) = \int_1^{+\infty} x^2 \left( \frac{3}{x^4} \right) dx - \left( \frac{3}{2} \right)^2$$

$$= \int_1^{+\infty} \frac{3}{x^2} \; dx - \frac{9}{4}$$

$$= \lim_{a \to +\infty} \int_1^a \frac{3}{x^2} \; dx - \frac{9}{4}$$

$$= \lim_{a \to +\infty} \left( \frac{-3}{x} \right) \bigg|_1^a - \frac{9}{4}$$

$$= 3 - \frac{9}{4} = \frac{3}{4}, \quad \text{or} \quad .75 \quad \text{▬▬}$$

### ▬▬ EXAMPLE 3

A recent study has shown that on any given day the proportion of airline passengers who rent a car at their destination is a random variable with probability density function $f(x) = 6x - 6x^2$, for $0 \leq x \leq 1$.

**(a)** Find and interpret the expected value for this distribution.

The expected value is

$$\mu = \int_0^1 x(6x - 6x^2) \; dx = \int_0^1 (6x^2 - 6x^3) \; dx$$

$$= \left( 2x^3 - \frac{3}{2}x^4 \right) \bigg|_0^1 = \frac{1}{2}, \quad \text{or} \quad .5.$$

This result indicates that over a long period of time, about 1/2 of all airline passengers rent a car at their destination.

**(b)** Compute the standard deviation.

First compute the variance. We use the alternative formula.

$$\text{Var}(x) = \int_0^1 x^2(6x - 6x^2)dx - \left(\frac{1}{2}\right)^2$$

$$= \int_0^1 (6x^3 - 6x^4)\, dx - \left(\frac{1}{2}\right)^2$$

$$= \left(\frac{3}{2}x^4 - \frac{6}{5}x^5\right)\Big|_0^1 - \frac{1}{4}$$

$$= \frac{3}{10} - \frac{1}{4} = \frac{1}{20} = .05$$

The standard deviation is $\sigma = \sqrt{.05} \approx .22.$ ▬

## 9.2 EXERCISES

*In Exercises 1–8 a probability density function of a random variable is defined. Find the expected value, the variance, and the standard deviation. Round answers to the nearest hundredth.*

**1.** $f(x) = \frac{1}{4};$ $[3, 7]$

**2.** $f(x) = \frac{1}{10};$ $[0, 10]$

**3.** $f(x) = \frac{x}{8} - \frac{1}{4};$ $[2, 6]$

**4.** $f(x) = 2(1 - x);$ $[0, 1]$

**5.** $f(x) = 1 - \frac{1}{\sqrt{x}};$ $[1, 4]$

**6.** $f(x) = \frac{1}{11}\left(1 + \frac{3}{\sqrt{x}}\right);$ $[4, 9]$

**7.** $f(x) = 4x^{-5};$ $[1, +\infty)$

**8.** $f(x) = 3x^{-4};$ $[1, +\infty)$

*In Exercises 9–12, the probability density function of a random variable is defined.*

**(a)** *Find the expected value to the nearest hundredth.*

**(b)** *Find the variance to the nearest hundredth.*

**(c)** *Find the standard deviation. Round to the nearest hundredth.*

**(d)** *Find the probability that the random variable has a value greater than the mean.*

**(e)** *Find the probability that the value of the random variable is within one standard deviation of the mean.*

**9.** $f(x) = \frac{\sqrt{x}}{18};$ $[0, 9]$

**10.** $f(x) = \frac{x^{-1/3}}{6};$ $[0, 8]$

**11.** $f(x) = \frac{1}{2}x;$ $[0, 2]$

**12.** $f(x) = \frac{3}{2}(1 - x^2);$ $[0, 1]$

If x is a random variable with probability density function f(x) on [a, b], then the
**median** of x is the number m such that

$$\int_a^m f(x) \, dx = \frac{1}{2}.$$

(a) Find the median of each random variable for the probability density functions in
Exercises 13–18.

(b) In each case, find the probability that the random variable is between the expected
value (mean) and the median. The expected value for each of these functions was
found in Exercises 1–8.

**13.** $f(x) = \frac{1}{4}$;  [3, 7]

**14.** $f(x) = \frac{1}{10}$;  [0, 10]

**15.** $f(x) = \frac{x}{8} - \frac{1}{4}$;  [2, 6]

**16.** $f(x) = 2(1 - x)$;  [0, 1]

**17.** $f(x) = 4x^{-5}$;  [1, +∞)

**18.** $f(x) = 3x^{-4}$;  [1, +∞)

---

## ▦ APPLICATIONS

### BUSINESS AND ECONOMICS

**Life of a Light Bulb**  **19.** The life (in hours) of a certain kind of light bulb is a random variable with probability density function defined by

$$f(x) = \frac{1}{58\sqrt{x}} \quad \text{for } [1, 900].$$

(a) What is the expected life of such a bulb?

(b) Find $\sigma$.

(c) Find the probability that one of these bulbs lasts longer than one standard
deviation above the mean.

**Machine Life**  **20.** The life in years of a certain machine is a random variable with probability density
function defined by

$$f(x) = \frac{1}{11}\left(1 + \frac{3}{\sqrt{x}}\right) \quad \text{for } [4, 9].$$

(a) Find the mean life of this machine.

(b) Find the standard deviation of the distribution.

(c) Find the probability that a particular machine of this kind will last longer than
the mean.

**Life of an Automobile Part**  **21.** A certain automobile part has a life span in months with probability density function defined by

$$f(x) = \frac{1}{2}e^{-x/2} \quad \text{for } [0, +∞).$$

(a) Find the expected life of this part.

(b) Find the standard deviation of the distribution.

(c) Find the probability that one of these parts lasts less than the mean number of
months.

## LIFE SCIENCES

**Blood Clotting Time**    **22.** The clotting time of blood, in seconds, is a random variable with probability density function defined by

$$f(x) = \frac{1}{(\ln 20)x} \quad \text{for } [1, 20].$$

(a) Find the mean clotting time.

(b) Find the standard deviation.

(c) Find the probability that a person's blood clotting time is within one standard deviation of the mean.

**Length of a Leaf**    **23.** The length of a leaf on a tree is a random variable with probability density function defined by

$$f(x) = \frac{3}{32}(4x - x^2) \quad \text{for } [0, 4].$$

(a) What is the expected leaf length?

(b) Find $\sigma$ for this distribution.

(c) Find the probability that the length of a given leaf is within one standard deviation of the expected value.

**Petal Length**    **24.** The length in centimeters of a petal on a certain flower is a random variable with probability density function defined by

$$f(x) = \frac{1}{2\sqrt{x}} \quad \text{for } [1, 4].$$

(a) Find the expected petal length.

(b) Find the standard deviation.

(c) Find the probability that a petal selected at random has a length more than two standard deviations above the mean.

## PHYSICAL SCIENCES

**Annual Rainfall**    **25.** The annual rainfall in a remote Middle Eastern country is a random variable with probability density function defined by

$$f(x) = \frac{1}{15}\left(x + \frac{1}{2}\right) \quad \text{for } [0, 5].$$

(a) Find the mean annual rainfall.

(b) Find the standard deviation.

(c) Find the probability of a year with rainfall less than two standard deviations below the mean.

**Temperature**    **26.** The probability density function for the minimum March temperature (in degrees Celsius) at a North Atlantic island is defined by

$$f(x) = \frac{1}{2} \quad \text{for } [1, 3].$$

(a) Find the expected minimum March temperature.

(b) Find the standard deviation.

(c) Find the probability that the minimum March temperature will vary from the mean by less than two standard deviations.

## 9.3   SPECIAL PROBABILITY DENSITY FUNCTIONS

In practice, it is not feasible to construct a probability density function for each experiment. Instead, a researcher uses one of several probability density functions that are well known, matching the shape of the experimental distribution to one of the known distributions. In this section we discuss some of the most commonly used probability distributions.

***Uniform Distribution***    The probability density function for the **uniform distribution** is defined by

$$f(x) = \frac{1}{b - a} \quad \text{for } [a, b],$$

where $a$ and $b$ are nonnegative real numbers. The graph of $f(x)$ is shown in Figure 7.

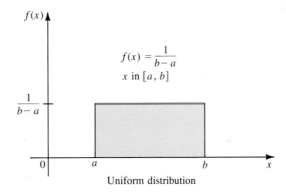

$$f(x) = \frac{1}{b - a}$$
$$x \text{ in } [a, b]$$

Uniform distribution

**FIGURE 7**

Since $b - a$ is positive, $f(x) \geq 0$, and

$$\int_a^b \frac{1}{b - a}\, dx = \frac{1}{b - a} x \Big|_a^b = \frac{1}{b - a}(b - a) = 1.$$

Therefore, the function is a probability density function.

The expected value for the uniform distribution is

$$\mu = \int_a^b \left(\frac{1}{b - a}\right) x\, dx = \left(\frac{1}{b - a}\right)\frac{x^2}{2}\Big|_a^b$$

$$= \frac{1}{2(b - a)}(b^2 - a^2) = \frac{1}{2}(b + a).$$

The variance is given by

$$\text{Var}(x) = \int_a^b \left(\frac{1}{b-a}\right) x^2 \, dx - \left(\frac{b+a}{2}\right)^2$$

$$= \left(\frac{1}{b-a}\right) \frac{x^3}{3} \Big|_a^b - \frac{(b+a)^2}{4}$$

$$= \frac{1}{3(b-a)}(b^3 - a^3) - \frac{1}{4}(b+a)^2$$

$$= \frac{b^2 + ab + a^2}{3} - \frac{b^2 + 2ab + a^2}{4}$$

$$= \frac{b^2 - 2ab + a^2}{12}$$

Thus,

$$\text{Var}(x) = \frac{1}{12}(b-a)^2,$$

and

$$\sigma = \frac{1}{\sqrt{12}}(b-a).$$

These properties of the uniform distribution are summarized below.

**UNIFORM DISTRIBUTION**

If $x$ is a random variable with probability density function

$$f(x) = \frac{1}{b-a} \quad \text{for } [a, b],$$

then

$$\mu = \frac{1}{2}(b+a) \quad \text{and} \quad \sigma = \frac{1}{\sqrt{12}}(b-a).$$

■ **EXAMPLE 1**

A couple is planning to vacation in San Francisco. They have been told that the maximum daily temperature during the time they plan to be there ranges from 15°C to 27°C. Assume that the probability of any temperature between 15°C and 27°C is equally likely for any given day during the specified time period.

**(a)** What is the probability that the maximum temperature on the day they arrive will be greater than 24°C?

If the random variable $t$ represents the maximum temperature on a given day, then the uniform probability density function for $t$ is defined by $f(t) = 1/12$ for the interval $[15, 27]$. By definition,

$$P(t > 24) = \int_{24}^{27} \frac{1}{12} \, dt = \frac{1}{12}t \Big|_{24}^{27} = \frac{1}{4}.$$

**(b)** What average maximum temperature can they expect?

The expected maximum temperature is

$$\mu = \frac{1}{2}(27 + 15) = 21,$$

or 21°C.

**(c)** What is the probability that the maximum temperature on a given day will be at least one standard deviation below the mean?

First find $\sigma$:

$$\sigma = \frac{1}{\sqrt{12}}(27 - 15) = \frac{12}{\sqrt{12}} = \sqrt{12} = 2\sqrt{3} \approx 3.5.$$

One standard deviation below the mean indicates a temperature of $21 - 3.5 = 17.5$°C.

$$P(T \le 17.5°) = \int_{15}^{17.5} \frac{1}{12} \, dt = \frac{1}{12} t \Big|_{15}^{17.5} \approx .21$$

The probability is .21 that the temperature will not exceed 17.5°C. ▄

***Exponential Distribution*** The probability density function for the **exponential distribution** is defined by

$$f(x) = ae^{-ax} \quad \text{for } [0, +\infty),$$

where $a$ is a positive real number. The graph of $f(x)$ is shown in Figure 8.

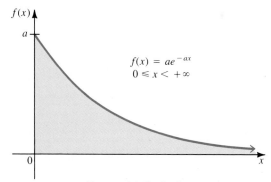

Exponential distribution

**FIGURE 8**

Here $f(x) \ge 0$, since $e^{-ax}$ and $a$ are both positive for all values of $x$. Also,

$$\int_0^{+\infty} ae^{-ax} \, dx = \lim_{b \to +\infty} \int_0^b ae^{-ax} \, dx$$

$$= \lim_{b \to +\infty} (-e^{-ax}) \Big|_0^b = \lim_{b \to +\infty} \left( \frac{-1}{e^{ab}} + \frac{1}{e^0} \right) = 1,$$

so the function is a probability density function.

The expected value and standard deviation of the exponential distribution can be found using integration by parts. The results are given below.

EXPONENTIAL
DISTRIBUTION

If $x$ is a random variable with probability density function

$$f(x) = ae^{-ax} \quad \text{for } [0, +\infty),$$

then

$$\mu = \frac{1}{a} \quad \text{and} \quad \sigma = \frac{1}{a}.$$

## EXAMPLE 2

Suppose the useful life (in hours) of a flashlight battery is the random variable $t$, with probability density function given by the exponential distribution

$$f(t) = \frac{1}{20}e^{-t/20} \quad \text{for } t \geq 0.$$

**(a)** Find the probability that a particular battery, selected at random, has a useful life of less than 100 hours.

The probability is given by

$$P(t \leq 100) = \int_0^{100} \frac{1}{20}e^{-t/20} \, dt = \frac{1}{20}\left(-20e^{-t/20}\right)\Big|_0^{100}$$

$$= -(e^{-100/20} - e^0) = -(e^{-5} - 1)$$

$$\approx 1 - .0067 = .9933.$$

**(b)** Find the expected value and standard deviation of the distribution.

Use the formulas given above. Both $\mu$ and $\sigma$ equal $1/a$, and since $a = 1/20$ here,

$$\mu = 20 \quad \text{and} \quad \sigma = 20.$$

This means that the average life of a battery is 20 hours and no battery lasts less than one standard deviation below the mean.

**(c)** What is the probability that a battery will last more than 40 hours?

The probability is given by

$$P(t > 40) = \int_{40}^{+\infty} \frac{1}{20}e^{-t/20} \, dt = \lim_{b \to +\infty} \left(-e^{-t/20}\right)\Big|_{40}^{b} = \frac{1}{e^2} \approx .1353,$$

or about 14%. ▬

*Normal Distribution*  The **normal distribution,** with its well-known bell-shaped graph, is undoubtedly the most important probability density function. It is widely used in various applications of statistics. The probability density function for the normal distribution has the following characteristics.

**NORMAL DISTRIBUTION**

If $\mu$ and $\sigma$ are real numbers, $\sigma \geq 0$, and if $x$ is a random variable with probability density function defined by

$$f(x) = \frac{1}{\sigma\sqrt{2\pi}} e^{-(x-\mu)^2/(2\sigma^2)} \quad \text{for } (-\infty, +\infty),$$

then

$$E(x) = \mu \quad \text{and} \quad \text{Var}(x) = \sigma^2, \quad \text{with } \textbf{standard deviation } \sigma.$$

Advanced techniques can be used to show that

$$\int_{-\infty}^{+\infty} \frac{1}{\sigma\sqrt{2\pi}} e^{-(x-\mu)^2/(2\sigma^2)} \, dx = 1.$$

Deriving the expected value and standard deviation for the normal distribution also requires techniques beyond the scope of this text.

Each normal probability distribution has associated with it a bell-shaped curve, called a **normal curve,** such as the one in Figure 9. Each normal curve is symmetric about a vertical line through the mean, $\mu$. Vertical lines at points $+1\sigma$ and $-1\sigma$ from the mean show the inflection points of the graph. A normal curve never touches the $x$-axis; it extends indefinitely in both directions.

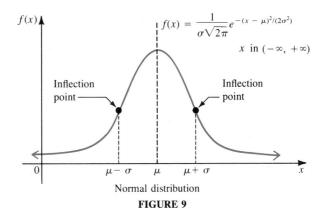

Normal distribution
**FIGURE 9**

The development of the normal curve is credited to the Frenchman Abraham de Moivre (1667–1754). Three of his publications dealt with probability and associated topics: *Annuities upon Lives* (which contributed to the development of actuarial studies), *Doctrine of Chances,* and *Miscellanea Analytica.*

Many different normal curves have the same mean. In such cases, a larger value of $\sigma$ produces a "flatter" normal curve, while smaller values of $\sigma$ produce more values near the mean, resulting in a "taller" normal curve. See Figure 10.

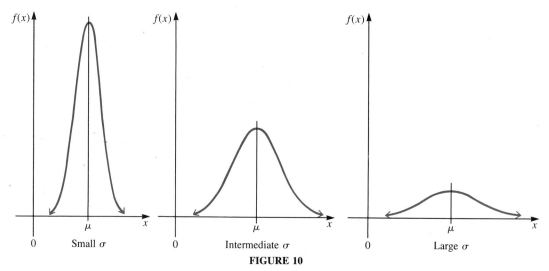

FIGURE 10

It would be far too much work to calculate values for the normal probability distribution for various values of $\mu$ and $\sigma$. Instead, values are calculated for the **standard normal distribution,** which has $\mu = 0$ and $\sigma = 1$. The graph of the standard normal distribution is shown in Figure 11.

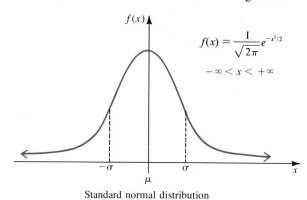

$$f(x) = \frac{1}{\sqrt{2\pi}} e^{-x^2/2}$$

$$-\infty < x < +\infty$$

Standard normal distribution

**FIGURE 11**

Probabilities for the standard normal distribution come from the definite integral

$$\int_a^b \frac{1}{\sqrt{2\pi}} e^{-x^2/2} \, dx.$$

Since $f(x) = e^{-x^2/2}$ does not have an antiderivative that can be expressed in terms of functions used in this course, numerical methods are used to find values of this definite integral. A table in the Appendix of this book gives areas under the standard normal curve, along with a sketch of the curve. Each value in this table is the total area under the standard normal curve to the left of the number $z$.

### ■ EXAMPLE 3

Find the following areas from the table for the standard normal curve.

**(a)** The area of the region to the left of $z = .89$

Look up .89 in the normal curve table. The corresponding area is .8133. Thus, the shaded area shown in Figure 12 is .8133. This area represents 81.33% of the total area under the normal curve.

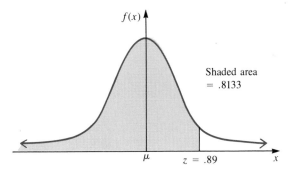

Shaded area = .8133

FIGURE 12

**(b)** The area of the region to the right of $z = .89$

From part (a) of this example, the area to the left of $z = .89$ is .8133. The total area under the normal curve is 1, so the area to the *right* of $z = .89$ is

$$1 - .8133 = .1867.$$

See Figure 13, where the shaded area represents 18.67% of the total area under the normal curve.

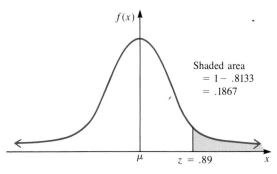

Shaded area
= 1 - .8133
= .1867

FIGURE 13

**(c)** The area of the region between $z = -.54$ and $z = 1.07$

To find this area, which is shaded in Figure 14, start with the area to the left of $z = 1.07$ and subtract the area to the left of $z = -.54$. See Figure 15. This gives $.8577 - .2946 = .5631$. ▬

**FIGURE 14**

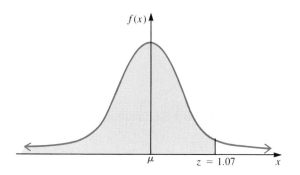

**FIGURE 15**

If a normal distribution does not have $\mu = 0$ and $\sigma = 1$, we use the following theorem, which is stated without proof.

*z*-SCORES THEOREM

> Suppose a normal distribution has mean $\mu$ and standard deviation $\sigma$. The area under the associated normal curve that is to the left of the value $x$ is exactly the same as the area to the left of
> $$z = \frac{x - \mu}{\sigma}$$
> for the standard normal curve.

Using this result, the table can be used for *any* normal distribution, regardless of the values of $\mu$ and $\sigma$. The number $z$ in the theorem is called a **z-score.**

### ■ EXAMPLE 4

A normal distribution has mean 35 and standard deviation 5.9. Find the following areas under the associated normal curve.

**(a)** The area of the region to the left of 40

Find the appropriate $z$-score using $x = 40$, $\mu = 35$, and $\sigma = 5.9$. Round to the nearest hundredth.

$$z = \frac{40 - 35}{5.9} = \frac{5}{5.9} \approx .85.$$

From the table, the desired area is .8023.

**(b)** The area of the region to the right of 32

$$z = \frac{32 - 35}{5.9} = \frac{-3}{5.9} \approx -.51$$

The area to the *left* of $z = -.51$ is .3050, so the area to the *right is* $1 - .3050 = .6950$.

**(c)** The area of the region between 30 and 33

Find $z$-scores for both values.

$$z = \frac{30 - 35}{5.9} = \frac{-5}{5.9} \approx -.85 \quad \text{and} \quad z = \frac{33 - 35}{5.9} = \frac{-2}{5.9} \approx -.34$$

Start with the area to the left of $z = -.34$ and subtract the area to the left of $z = -.85$:

$$.3669 - .1977 = .1692. \quad ■$$

The $z$-scores are actually standard deviation multiples—that is, a $z$-score of 2.5 corresponds to a value 2.5 standard deviations above the mean. For example, looking up $z = 1.00$ and $z = -1.00$ in the table shows that

$$.8413 - .1587 = .6826,$$

with 68.26% of the area under a normal curve within one standard deviation of the mean. Also, using $z = 2.00$ and $z = -2.00$,

$$.9772 - .0228 = .9544,$$

with 95.44% of the area within two standard deviations of the mean. These results, summarized in Figure 16, can be used to get a quick estimate of results when working with normal curves.

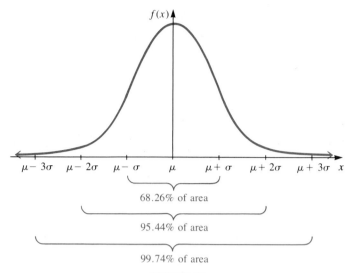

68.26% of area

95.44% of area

99.74% of area

**FIGURE 16**

---

### 9.3 EXERCISES

*For Exercises 1–6, find each of the following:*

**(a)** *the mean of the distribution;*

**(b)** *the standard deviation of the distribution;*

**(c)** *the probability that the random variable is between the mean and one standard deviation above the mean.*

**1.** The length (in centimeters) of the leaf of a certain plant is a continuous random variable with probability density function defined by

$$f(x) = \frac{5}{4} \quad \text{for [4, 4.8]}.$$

**2.** The price of an item in dollars is a continuous random variable with probability density function defined by

$$f(x) = 2 \quad \text{for [1.25, 1.75]}.$$

**3.** The length of time in years until a particular radioactive particle decays is a random varible $t$ with probability density function defined by

$$f(t) = .03e^{-.03t} \quad \text{for [0, } +\infty).$$

**4.** The length of time (in years) that a seedling tree survives is a random variable $t$ with probability density function defined by

$$f(t) = .05e^{-.05t} \quad \text{for [0, } +\infty).$$

**5.** The length of time (in days) required to learn a certain task is a random variable $t$ with probability density function defined by

$$f(t) = e^{-t} \quad \text{for [0, } +\infty).$$

**6.** The distance (in meters) that seeds are dispersed from a certain kind of plant is a random variable $x$ with probability density function defined by

$$f(x) = .1e^{-.1x} \quad \text{for } [0, +\infty).$$

*Find the percent of the area under a normal curve between the mean and the number of standard deviations above the mean given in Exercises 7–10.*

**7.** 3.50 **8.** 1.68 **9.** .45 **10.** 1.71

*Find the percent of the total area under the normal curve between the z-scores given in Exercises 11–14.*

**11.** 1.28 and 2.05 **12.** $-2.13$ and $-.04$ **13.** $-3.05$ and 1.27 **14.** $-.53$ and .53

*Find a z-score satisfying the conditions given in Exercises 15–18. (Hint: Use the table backwards.)*

**15.** 10% of the total area is to the left of $z$   **16.** 2% of the total area is to the left of $z$

**17.** 18% of the total area is to the right of $z$   **18.** 22% of the total area is to the right of $z$

**19.** Find an expression for the median $m$ of the uniform distribution.

**20.** Find an expression for the median $m$ of the exponential distribution.

---

### ▤ APPLICATIONS

#### BUSINESS AND ECONOMICS

**Insurance Sales** **21.** The amount of insurance (in thousands of dollars) sold in a day by a particular agent is uniformly distributed over the interval [10, 85].

   **(a)** What amount of insurance does the agent sell on an average day?

   **(b)** Find the probability that the agent sells more than \$50,000 of insurance on a particular day.

**Fast-Food Outlets** **22.** The number of new fast-food outlets opening during June in a certain city is exponentially distributed with a mean of 5.

   **(a)** Give the probability density function for this distribution.

   **(b)** What is the probability that the number of outlets opening is between 2 and 6?

**Sales Expense** **23.** A salesperson's monthly expenses (in thousands of dollars) are exponentially distributed with an average of 4.25 (thousand).

   **(a)** Give the probability density function for the expenses.

   **(b)** Find the probability that the expenses are more than \$10,000.

*In Exercises 24–26, assume a normal distribution.*

**Machine Accuracy** **24.** A machine that fills quart bottles with apple juice averages 32.8 oz per bottle, with a standard deviation of 1.1 oz. What is the probability that the amount of juice in a bottle is as follows?

   **(a)** Less than a quart

   **(b)** At least 1 oz more than a quart

**Machine Accuracy**   **25.** A machine produces screws with a mean length of 2.5 cm and a standard deviation of .2 cm. Find the probability that a screw produced by this machine has length as follows.

(a) Greater than 2.7 cm

(b) Within 1.2 standard deviations of the mean

**Customer Expenditures**   **26.** Customers at a certain pharmacy spend an average of $54.40, with a standard deviation of $13.50. What are the largest and smallest amounts spent by the middle 50% of these customers?

## LIFE SCIENCES

**Insect Life Span**   **27.** The life span of a certain insect in days is uniformly distributed over the interval [20, 36].

(a) What is the expected life of this insect?

(b) Find the probability that one of these insects, randomly selected, lives longer than 30 days.

**Emergency Room Patients**   **28.** The number of emergency room patients at a certain hospital is uniformly distributed over the interval [8, 25].

(a) What is the expected number of patients on any given day?

(b) What is the probability that the number of patients will be within one standard deviation of the mean?

**Location of a Bee Swarm**   **29.** A swarm of bees is released from a certain point. The proportion of the swarm located at least 2 m from the point of release after 1 hr is a random variable that is exponentially distributed with $a = 2$ over the interval $[0, +\infty)$.

(a) Find the expected proportion under the given conditions.

(b) Find the probability that fewer than 1/3 of the bees are located at least 2 m from the release point after 1 hr.

**Digestion Time**   **30.** The digestion time in hours of a fixed amount of food is exponentially distributed, with $a = 1$.

(a) Find the mean digestion time.

(b) Find the probability that the digestion time is less than 30 min.

**Pygmy Heights**   **31.** The average height of a member of a certain tribe of pygmies is 3.2 ft, with a standard deviation .2 ft. What are the largest and smallest heights of the middle 50% of this population?

## PHYSICAL SCIENCES

**Rainfall**   **32.** The rainfall in inches in a certain region is uniformly distributed over the interval [32, 44].

(a) What is the expected number of inches of rainfall?

(b) What is the probability that the rainfall will be between 38 and 40 inches?

## FOR THE COMPUTER

*Use Simpson's rule with $n = 100$ to approximate the following integrals.*

**33.** $\int_0^{50} .5e^{-.5x}\, dx$     **34.** $\int_0^{50} .5xe^{-.5x}\, dx$     **35.** $\int_0^{50} .5x^2e^{-.5x}\, dx - 4$

**36.** Use your results from Exercises 34 and 35 to verify that, for the exponential distribution, $\mu = 1/a$ and $\sigma = 1/a$.

**37.** The standard normal probability density function is defined by

$$f(x) = \int_{-\infty}^{+\infty} \frac{1}{\sqrt{2\pi}} e^{-x^2/2} \, dx.$$

Use Simpson's rule with $n = 100$ and the formulas for the mean and standard deviation of a probability density function to approximate the following for the standard normal probability distribution. Use limits of $-4$ and $4$ (instead of $-\infty$ and $+\infty$).

(a) The mean    (b) The standard deviation

---

### KEY WORDS

| | |
|---|---|
| **9.1** random variable | **expected value** |
| probability distribution function | **variance** |
| histogram | **standard deviation** |
| discrete probability function | **9.3** uniform distribution |
| continuous random variable | exponential distribution |
| continuous probability | normal distribution |
| distribution | normal curve |
| probability density function | standard normal distribution |
| **9.2** mean | z-score |

---

## CHAPTER 9        REVIEW EXERCISES

*Decide whether each function defined as follows is a probability density function for the given interval.*

**1.** $f(x) = \dfrac{1}{27}(2x + 4);$   $[1, 4]$

**2.** $f(x) = \sqrt{x};$   $[4, 9]$

**3.** $f(x) = .1;$   $[0, 10]$

**4.** $f(x) = e^{-x};$   $[0, +\infty)$

*In Exercises 5 and 6, find a value of k that will make f(x) define a probability density function for the indicated interval.*

**5.** $f(x) = k\sqrt{x};$   $[1, 4]$

**6.** $f(x) = kx^2;$   $[0, 3]$

**7.** The probability density function of a random variable $x$ is defined by

$$f(x) = 1 - \frac{1}{\sqrt{x - 1}} \quad \text{for } [2, 5].$$

Find the following probabilities.

(a) $P(x \geq 3)$    (b) $P(x \leq 4)$    (c) $P(3 \leq x \leq 4)$

**8.** The probability density function of a random variable $x$ is defined by

$$f(x) = \frac{1}{10} \quad \text{for } [10, 20].$$

Find the following probabilities.

(a) $P(x \leq 12)$    (b) $P(x \geq 31/2)$    (c) $P(10.8 \leq x \leq 16.2)$

*For the probability density functions defined in Exercises 9–12, find the expected value, the variance, and the standard deviation.*

**9.** $f(x) = \dfrac{1}{5}$;  [4, 9]

**10.** $f(x) = \dfrac{2}{9}(x - 2)$;   [2, 5]

**11.** $f(x) = \dfrac{1}{7}\left(1 + \dfrac{2}{\sqrt{x}}\right)$;   [1, 4]

**12.** $f(x) = 5x^{-6}$;   [1, +∞)

**13.** The probability density function of a random variable is defined by $f(x) = 4x - 3x^2$ for [0, 1].
   Find each of the following for the distribution:
   **(a)** the mean;     **(b)** the standard deviation;
   **(c)** the probability that the value of the random variable will be less than the mean;
   **(d)** the probability that the value of the random variable will be within one standard deviation of the mean.

**14.** Find the median of the random variable of Exercise 13. (See Exercises 13–18 in Section 2.) Then find the probability that the value of the random variable will be between the median and the mean of the distribution.

*For Exercises 15–16, find **(a)** the mean of the distribution, **(b)** the standard deviation of the distribution, and **(c)** the probability that the value of the random variable is within one standard deviation of the mean.*

**15.** $f(x) = \dfrac{5}{112}(1 - x^{-3/2})$   for [1, 25]

**16.** $f(x) = .01e^{-.01x}$   for [0, +∞)

*In Exercises 17–22, find the percent of the area under a normal curve for each of the following.*

**17.** The region to the right of $z = 1.53$

**18.** The region to the left of $z = -.49$

**19.** The region between $z = -1.47$ and $z = 1.03$

**20.** The region between $z = -.98$ and $z = -.15$

**21.** The region that is up to 2.5 standard deviations above the mean

**22.** The region that is up to 1.2 standard deviations below the mean

**23.** Find a $z$-score so that 21% of the area under the normal curve is to the left of $z$.

**24.** Find a $z$-score so that 52% of the area under the normal curve is to the right of $z$.

## ▦ APPLICATIONS

### BUSINESS AND ECONOMICS

**Mutual Funds**   **25.** The price per share in dollars of a particular mutual fund is a random variable $x$ with probability density function defined by

$$f(x) = \frac{3}{4}(x^2 - 16x + 65)   \text{for [8, 9].}$$

Find the probability that the price will be less than $8.50.

**Machine Repairs**  26. The time (in years) until a certain machine requires repairs is a random variable $t$ with probability density function defined by

$$f(t) = \frac{5}{112}(1 - t^{-3/2}) \quad \text{for } [1, 25].$$

Find the probability that no repairs are required in the first three years by finding the probability that a repair will be needed in years 4 through 25.

**Product Repairs**  27. The number of repairs required by a new product each month is exponentially distributed, with an average of 8.

(a) What is the probability density function for this distribution?

(b) Find the expected number of repairs per month.

(c) Find the standard deviation.

(d) What is the probability that the number of repairs per month will be between 5 and 10?

**Retail Outlets**  28. The number of new outlets for a clothing manufacturer is an exponential distribution with probability density function defined by

$$f(x) = \frac{1}{6}e^{-x/6} \quad \text{for } [0, +\infty).$$

Find each of the following for this distribution.

(a) The mean

(b) The standard deviation

(c) The probability that the number of new outlets will be greater than the mean

**Useful Life of an Appliance Part**  29. The useful life of a certain appliance part (in hundreds of hours) is 46.2, with a standard deviation of 15.8. Find the probability that one such part would last for at least 6000 (60 hundred) hr. Assume a normal distribution.

## LIFE SCIENCES

**Movement of a Released Animal**  30. The distance (in meters) that a certain animal moves away from a release point is a random variable with probability density function defined by

$$f(x) = .01e^{-.01x} \quad \text{for } [0, +\infty).$$

Find the probability that the animal will move no farther than 100 m away.

**Weight Gain of Rats**  31. The weight gain (in grams) of rats fed a certain vitamin supplement is a continuous random variable with probability density function defined by

$$f(x) = \frac{8}{7}x^{-2} \quad \text{for } [1, 8].$$

(a) Find the mean of the distribution.

(b) Find the standard deviation of the distribution.

(c) Find the probability that the value of the random variable is within one standard deviation of the mean.

**Body Temperature of a Bird**  32. The body temperature (in degees Celsius) of a particular species of bird is a continuous random variable with probability density function defined by

$$f(x) = \frac{6}{15,925}(x^2 + x) \quad \text{for } [20, 25].$$

(a) What is the expected body temperature of this species?

(b) Find the probability of a body temperature below the mean.

**Snowfall**      **33.** The snowfall in inches in a certain area is uniformly distributed over the interval [2, 40].

(a) What is the expected snowfall?

(b) What is the probability of getting more than 20 inches of snow?

**Heart Muscle Tension**      **34.** In a pilot study on tension of the heart muscle in dogs, the mean developed tension was 2.4 g, with a standard deviation of .4 g. Find the probability of a tension of less than 1.9 g. Assume a normal distribution.

**Average Birth Weight**      **35.** The average birth weight of infants in the United States is 7.8 lb, with a standard deviation of 1.1 lb. Assuming a normal distribution, what is the probability that a newborn will weigh more than 9 lb?

## GENERAL

**State-Run Lotteries**      **36.** The average state "take" on lotteries is 40%, with a standard deviation of 13%. Assuming a normal distribution, what is the probability that a state-run lottery will have a "take" of more than 50%?

## EXTENDED APPLICATION

## ■ A CROP-PLANTING MODEL*

Many firms in food processing, seed production, and similar industries face a problem every year deciding how many acres of land to plant in each of various crops. Demand for the crop is unknown, as is the actual yield per acre. In this application, we set up a mathematical model for determining the optimum number of acres to plant in a crop.

This model is designed to tell the company the number of acres of seed that it should plant. The model uses the following variables:

$D$ = number of tons of seed demanded;

$f(D)$ = continuous probability density function for the quantity of seed demanded, $D$;

$X$ = quantity of seed produced per acre of land;

$Q$ = quantity of seed carried over in inventory from previous years;

$S$ = selling price per ton of seed;

$C_p$ = variable costs of production, marketing, etc., per ton of seed;

$C_c$ = cost to carry over a ton of seed from previous years;

$A$ = number of acres of land to be planted;

$C_A$ = variable cost per acre of land contracted;

$T$ = total number of tons of seed available for sale;

$a$ = lower limit of the domain of $f(D)$;

and   $b$ = upper limit of the domain of $f(D)$.

To find the optimum number of acres to plant, it is necessary to calculate the expected value of the profit from the planting of $A$ acres.

*Based on work by David P. Rutten, Senior Mathematician, The Upjohn Company, Kalamazoo, Michigan. Reprinted with permission.

Based on the definition of the variables above, the total number of tons of seed that will be available for sale is given by the product of the number of acres planted, $A$, and the yield per acre, $X$, added to the carryover, $Q$. If $T$ represents this total, then

$$T = AX + Q.$$

The variable here is $A$; we assume $X$ and $Q$ are known and fixed.

The expected profit can be broken down into several parts. The first portion comes from multiplying the profit per ton and the average number of tons demanded. The profit per ton is found by subtracting the variable cost per ton, $C_p$, from the selling price per ton, $S$:

$$\text{Profit per Ton} = S - C_p.$$

The average number of tons demanded for our interval of concern is given by

$$\int_a^T D \cdot f(D)dD.$$

Thus, this portion of the expected profit is

$$(S - C_p) \cdot \int_a^T D \cdot f(D)dD. \tag{1}$$

A second portion of expected profit is found by multiplying the profit per ton, $S - C_p$, the total number of tons available, $T$ (recall that this is a variable), and the probability that $T$ or more tons will be demanded by the marketplace.

$$(S - C_p)(T)\int_T^b f(D)dD \tag{2}$$

If $T$ is greater than $D$, there will be costs associated with carrying over the excess seed. The expected value of these costs is given by the product of the carrying cost per ton, $C_c$, and the number of tons to be carried over, or

$$-C_c\int_a^T (T - D)f(D)dD. \tag{3}$$

The minus sign shows that these costs reduce profit. If $T < D$, this term would be omitted. Finally, the total cost of producing the seed is given by the product of the variable cost per acre and the number of acres:

$$-C_A \cdot A. \tag{4}$$

The expected profit is the sum of the expressions in (1)–(4), or

$$\text{Expected Profit} = (S - C_p) \cdot \int_a^T D \cdot f(D)dD + (S - C_p)(T)\int_T^b f(D)dD$$

$$-C_c\int_a^T (T - D)f(D)dD - C_A \cdot A. \tag{5}$$

As an example, suppose that we have the following information for a particular crop.

$$\text{Probability Density Function} = f(D) = \frac{1}{1000} \quad \text{for } 500 \le D \le 1500 \text{ tons}$$

$$a = 500$$
$$b = 1500$$
$$\text{Selling Price} = S = \$10,000 \text{ per ton}$$
$$\text{Variable Cost} = C_p = \$5000 \text{ per ton}$$
$$\text{Carrying Cost} = C_c = \$3000 \text{ per ton}$$
$$\text{Variable Cost per Acre} = C_A = \$100$$
$$\text{Inventory Carryover} = Q = 200 \text{ tons}$$
$$\text{Yield per Acre} = X = .1 \text{ ton}$$
$$T = AX + Q = .1A + 200$$

Substitute all this into equation (5).

$$\text{Expected Profit} = (10{,}000 - 5000)\int_{500}^{.1A+200} D \cdot \frac{1}{1000}\, dD$$

$$+ (10{,}000 - 5000)(.1A + 200)\int_{.1A+200}^{1500} \frac{1}{1000}\, dD$$

$$- 3000\int_{500}^{.1A+200} (.1A + 200 - D) \cdot \frac{1}{1000}\, dD - 100A$$

Simplify all this.

$$\text{Expected Profit} = \frac{5000}{1000} \cdot \frac{D^2}{2}\Big|_{500}^{.1A+200} + (5000)(.1A + 200)\frac{D}{1000}\Big|_{.1A+200}^{1500}$$

$$- 3000\left(\frac{.1AD}{1000} + \frac{200D}{1000} - \frac{D^2}{2000}\right)\Big|_{500}^{.1A+200} - 100A$$

$$= \frac{5}{2}(.1A + 200)^2 - \frac{5}{2} \cdot 500^2 + 5(.1A + 200)1500$$

$$- 5(.1A + 200)(.1A + 200) - .3A(.1A + 200)$$

$$- 600(.1A + 200) + \frac{3}{2}(.1A + 200)^2 + 150A$$

$$+ 600(500) - \frac{3}{2}(500)^2 - 100A$$

$$= -(.1A + 200)^2 + 6900(.1A + 200) - .3A(.1A + 200)$$

$$+ 50A - 700{,}000$$

To find the maximum expected profit, take the derivative of this function with respect to $A$ and then place it equal to 0.

$$D_A[\text{Expected Profit}] = -2(.1A + 200)(.1) + 690 - .06A - 60 + 50$$
$$= -.02A - 40 + 690 - .06A - 10$$
$$= -.08A + 640$$

Place this derivative equal to 0.

$$-.08A + 640 = 0$$
$$8000 = A$$

Planting 8000 acres will give maximum profit.

# 10

# The Trigonometric Functions

Throughout this book we have discussed many different types of functions, including linear, quadratic, exponential, and logarithmic functions. In this chapter we introduce the *trigonometric functions,* which differ in a fundamental way from those previously studied: the trigonometric functions describe periodic or repetitive relationships.

An example of a periodic relationship is given by an electrocardiogram (EKG), a graph of a human heartbeat. The EKG in Figure 1 shows electrical impulses from a heart.* Each small square represents .04 second, and each large square represents .2 second. How often does this heart beat?

**FIGURE 1**

Trigonometric functions describe many natural phenomena and are important in the study of optics, heat, electronics, acoustics, and seismology. Also, many algebraic functions have integrals involving trigonometric functions.

## 10.1   DEFINITIONS OF THE TRIGONOMETRIC FUNCTIONS

The angle is one of the basic concepts of trigonometry. The definition of an angle depends on that of a ray: a **ray** is the portion of a line that starts at a given point and continues indefinitely in one direction. For example, Figure 2 shows a line through the two points $A$ and $B$. The portion of line $AB$ that starts at $A$ and continues through and past $B$ is called ray $AB$. Point $A$ is the **endpoint** of the ray.

**FIGURE 2**

*EKG courtesy of Nancy Schiller, Glendale, Arizona.

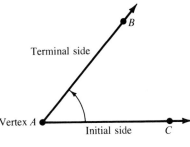

**FIGURE 3**

An **angle** is formed by rotating a ray about its endpoint. The initial position of the ray is called the **initial side** of the angle, and the endpoint of the ray is called the **vertex** of the angle. The location of the ray at the end of its rotation is called the **terminal side** of the angle. Figure 3 shows the initial and terminal sides of an angle with vertex *A*.

An angle can be named by its vertex. For example, the angle in Figure 3 can be called angle *A*. An angle also can be named by using three letters. For example, the angle in Figure 3 could be named angle *BAC* or angle *CAB*. (The vertex is always the middle letter.)

***Degree Measure*** The sizes of angles are often indicated in *degrees*. Degree measure has remained unchanged since the Babylonians developed it 4000 years ago. In degree measure, 360 degrees represents a complete rotation of a ray. **One degree,** written 1°, is 1/360 of a rotation. Also, 90° is 90/360 or 1/4 of a rotation, and 180° is 180/360 or 1/2 of a rotation.

There are various theories about why the Babylonians chose to divide a complete revolution into 360 parts. One theory suggests that 360 was chosen because there are approximately 360 days in one calendar year. Another theory says that because the Babylonians used a sexagesimal (base sixty) system of numeration (as opposed to our familiar decimal, or base ten, system), they used 360 since it is a multiple of 60. In any event, the practice has continued for thousands of years.

An angle having degree measure between 0° and 90° is called an **acute angle.** An angle of 90° is a **right angle.** An angle having measure more than 90° and less than 180° is an **obtuse angle,** while an angle of 180° is a **straight angle.**

An angle is in **standard position** if its vertex is at the origin of a coordinate system and its initial side is along the positive *x*-axis. The two angles in Figure 4 are in standard position. An angle in standard position is said to lie in the quadrant where its terminal side lies. For example, an acute angle in standard position is in quadrant I, while an obtuse angle is in quadrant II.

**FIGURE 4**

**FIGURE 5**

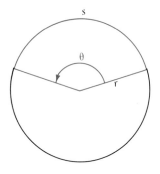

**FIGURE 6**

***Radian Measure***    While degree measure is best for some applications, this system of angle measurement is not the best for calculus. To keep the formulas for derivatives as simple as possible, it is better to measure angles with *radian measure*. To see how this alternative system for measuring angles is obtained, look at angle $\theta$ (the Greek letter *theta*) in Figure 5. The angle $\theta$ is in standard position; Figure 5 also shows a circle of radius $r$, centered at the origin.

The vertex of $\theta$ is at the center of the circle in Figure 5. Angle $\theta$ cuts a piece of the circle called an **arc.** Because the length of this arc is equal to the radius of the circle, angle $\theta$ has a measure of one radian. **One radian** is the measure of an angle that has its vertex at the center of a circle and that cuts an arc on the circle equal in length to the radius of the circle. The term "radian" comes from the phrase "radial angle." Two nineteenth-century scientists, mathematician Thomas Muir and physicist James Thomson, are credited with the development of the radian as a unit of angular measure.

Generalizing, the radian measure of a central angle $\theta$ (see Figure 6) cutting off an arc of length $s$ in a circle of radius $r$ is defined as follows:

$$\text{Radian Measure of } \theta = \frac{\text{Length of Arc}}{\text{Radius}} = \frac{s}{r}.$$

In this definition, the units of measure of the length of the arc and the radius "cancel," leaving a quotient without units. For this reason, a real number can be thought of as the radian measure of some angle.

Since the circumference of a circle is $2\pi$ times the radius of the circle, the radius could be marked off $2\pi$ times around the circle. Therefore, an angle of $360°$, that is, a complete circle, cuts off an arc equal in length to $2\pi$ times the radius of the circle, or

$$360° = 2\pi \text{ radians.}$$

This result gives a basis for comparing degree and radian measure.

Since an angle of $180°$ is half the size of an angle of $360°$, an angle of $180°$ would have half the radian measure of an angle of $360°$, or

$$180° = \frac{1}{2}(2\pi) \text{ radians} = \pi \text{ radians.}$$

**180° = π radians**

Since $\pi$ radians $= 180°$, divide both sides by $\pi$ to find the degree measure of one radian.

**1 RADIAN**

$$1 \text{ radian} = \left(\frac{180°}{\pi}\right)$$

This quotient is approximately $57.29578°$. Since $180° = \pi$ radians, we can find the radian measure of one degree by dividing by $180°$ on both sides.

**1 DEGREE**

$$1° = \frac{\pi}{180} \text{ radians}$$

One degree is approximately equal to .0174533 radians.

### ■ EXAMPLE 1

Convert degree measures to radians and radian measures to degrees.

**(a)** 45°

Since $1° = \pi/180$ radians,

$$45° = 45\left(\frac{\pi}{180}\right) \text{ radians} = \frac{45\pi}{180} \text{ radians} = \frac{\pi}{4} \text{ radians}.$$

The word *radian* is often omitted, so the answer could be written as just $45° = \pi/4$.

**(b)** $240° = 240\left(\frac{\pi}{180}\right) = \frac{4\pi}{3}$

**(c)** $\frac{9\pi}{4}$

Since 1 radian $= 180°/\pi$,

$$\frac{9\pi}{4} \text{ radians} = \frac{9\pi}{4}\left(\frac{180°}{\pi}\right) = 405°.$$

**(d)** $\frac{11\pi}{3} \text{ radians} = \frac{11\pi}{3}\left(\frac{180°}{\pi}\right) = 660°$ ■

***The Trigonometric Functions***   The origins of trigonometry, while somewhat obscure, are tied closely to the development of another science: astronomy. The early foundations of trigonometry consisted of tables of shadow lengths corresponding to various times of day, and lengths of chords of arbitrary circular arcs.* Based upon the work of the Greek astronomer Hipparchus, the *Almagest* of Claudius Ptolemy was a source of information for astronomers such as Copernicus and Kepler. This work consisted of tables of chords and identities that could be used for computing chord lengths. With the invention of the calculus, trigonometry became more than simply a tool for calculations. The work of Newton, Johann Bernoulli (1667–1748), and Euler paved the way for new applications by expanding the study of trigonometry into a theory of functions of real numbers with no dependence on angles or arc lengths.

---

*A *chord* is a line segment in a circle with endpoints on the circle itself. (The diameter of a circle is the length of a chord that passes through the center of the circle.)

To define the six basic trigonometric functions, we start with an angle $\theta$ in standard position, as shown in Figure 7. Next, we choose an arbitrary point $P$ having coordinates $(x, y)$, located on the terminal side of angle $\theta$. (The point $P$ must not be the vertex of $\theta$).

Drawing a line segment perpendicular to the $x$-axis from $P$ to point $Q$ sets up a right triangle having vertices at $O$ (the origin), $P$, and $Q$. The distance from $P$ to $O$ is $r$. Since the distance from $P$ to $O$ can never be negative, $r > 0$. The six **trigonometric functions** of angle $\theta$ are defined as follows.

**TRIGONOMETRIC FUNCTIONS**

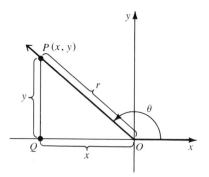

Let $(x, y)$ be a point other than the origin on the terminal side of an angle $\theta$ in standard position. Let $r$ be the distance from the origin to $(x, y)$. Then

$$\text{sine } \theta = \sin \theta = \frac{y}{r} \qquad \text{cosecant } \theta = \csc \theta = \frac{r}{y}.$$

$$\text{cosine } \theta = \cos \theta = \frac{x}{r} \qquad \text{secant } \theta = \sec \theta = \frac{r}{x}$$

$$\text{tangent } \theta = \tan \theta = \frac{y}{x} \qquad \text{cotangent } \theta = \cot \theta = \frac{x}{y}$$

**FIGURE 7**

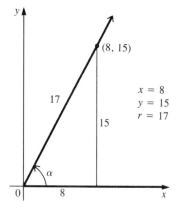

**FIGURE 8**

## ▰ EXAMPLE 2

The terminal side of an angle $\alpha$ (the Greek letter alpha) goes through the point $(8, 15)$. Find the values of the six trigonometric functions of angle $\alpha$.

Figure 8 shows angle $\alpha$ and the triangle formed by dropping a perpendicular from the point $(8, 15)$. To reach the point $(8, 15)$, begin at the origin and go 8 units to the right and 15 units up, so that $x = 8$ and $y = 15$. To find the radius $r$, use the Pythagorean theorem*: in a triangle with a right angle, if the longest side of the triangle is $r$ and the shorter sides are $x$ and $y$, then

$$r^2 = x^2 + y^2$$

or

$$r = \sqrt{x^2 + y^2}.$$

(Recall that $\sqrt{a}$ represents the *positive* square root of $a$.)

Substituting the known values $x = 8$ and $y = 15$ in the equation gives

$$r = \sqrt{8^2 + 15^2} = \sqrt{64 + 225} = \sqrt{289} = 17.$$

---

*While one of the most famous theorems in mathematics is named after the Greek mathematician Pythagoras, there is much evidence that the relationship between the sides of a right triangle was known long before his time. The Babylonian mathematical tablet identified as *Plimpton 322* has been determined to be essentially a list of *Pythagorean triples*—sets of three numbers $a$, $b$, and $c$ that satisfy the equation $a^2 + b^2 = c^2$.

We have $x = 8$, $y = 15$, and $r = 17$. The values of the six trigonometric functions of angle $\alpha$ are found by using the definitions given above.

$$\sin \alpha = \frac{y}{r} = \frac{15}{17} \qquad \tan \alpha = \frac{y}{x} = \frac{15}{8} \qquad \sec \alpha = \frac{r}{x} = \frac{17}{8}$$

$$\cos \alpha = \frac{x}{r} = \frac{8}{17} \qquad \cot \alpha = \frac{x}{y} = \frac{8}{15} \qquad \csc \alpha = \frac{r}{y} = \frac{17}{15} \quad ▬$$

### ▬ EXAMPLE 3

Find the values of the six trigonometric functions for an angle of $\pi/2$.

Select any point on the terminal side of an angle of measure $\pi/2$ radians (or 90°). See Figure 9. Selecting the point $(0, 1)$ gives $x = 0$ and $y = 1$. Check that $r = 1$ also. Then

$$\sin \frac{\pi}{2} = \frac{1}{1} = 1 \qquad \tan \frac{\pi}{2} = \frac{1}{0} \text{ (undefined)} \qquad \sec \frac{\pi}{2} = \frac{1}{0} \text{ (undefined)}$$

$$\cos \frac{\pi}{2} = \frac{0}{1} = 0 \qquad \cot \frac{\pi}{2} = \frac{0}{1} = 0 \qquad \csc \frac{\pi}{2} = \frac{1}{1} = 1. \quad ▬$$

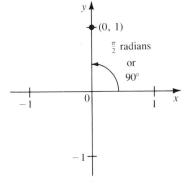

**FIGURE 9**

Methods similar to the procedure in Example 3 can be used to find the values of the six trigonometric functions for the angles with measures 0, $\pi$, and $3\pi/2$. These results are summarized in the following table. The table shows that the results for $2\pi$ are the same as those for 0.

| $\theta$ in radians | $\theta$ in degrees | $\sin \theta$ | $\cos \theta$ | $\tan \theta$ | $\cot \theta$ | $\sec \theta$ | $\csc \theta$ |
|---|---|---|---|---|---|---|---|
| 0 | 0° | 0 | 1 | 0 | Undefined | 1 | Undefined |
| $\pi/2$ | 90° | 1 | 0 | Undefined | 0 | Undefined | 1 |
| $\pi$ | 180° | 0 | $-1$ | 0 | Undefined | $-1$ | Undefined |
| $3\pi/2$ | 270° | $-1$ | 0 | Undefined | 0 | Undefined | $-1$ |
| $2\pi$ | 360° | 0 | 1 | 0 | Undefined | 1 | Undefined |

***Special Angles*** The values of the trigonometric functions for most angles must be found in a table or by using a calculator with trigonometric keys. For a few angles called *special angles*, however, the function values can be found exactly. These values are found with the aid of two kinds of right triangles that will be described in this section.

**30°–60°–90° TRIANGLE**

In a right triangle having angles of 30°, 60°, and 90°, the hypotenuse is always twice as long as the shortest side, and the middle side has a length that is $\sqrt{3}$ times as long as that of the shortest side. Also, the shortest side is opposite the 30° angle.

## ■■ EXAMPLE 4

Find the values of the trigonometric functions for an angle of $\pi/6$ radians.

Since $\pi/6$ radians $= 30°$, find the necessary values by placing a 30° angle in standard position, as in Figure 10. Choose a point $P$ on the terminal side of the angle so that $r = 2$. From the description of 30°–60°–90° triangles, $P$ will have coordinates $(\sqrt{3}, 1)$, with $x = \sqrt{3}$, $y = 1$, and $r = 2$. Using the definitions of the trigonometric functions gives the following results.

$$\sin \frac{\pi}{6} = \frac{1}{2} \qquad \tan \frac{\pi}{6} = \frac{1}{\sqrt{3}} = \frac{\sqrt{3}}{3} \qquad \sec \frac{\pi}{6} = \frac{2}{\sqrt{3}} = \frac{2\sqrt{3}}{3}$$

$$\cos \frac{\pi}{6} = \frac{\sqrt{3}}{2} \qquad \cot \frac{\pi}{6} = \sqrt{3} \qquad \csc \frac{\pi}{6} = 2 \quad ■■$$

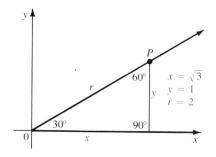

**FIGURE 10**

We can find the trigonometric function values for 45° angles by using the properties of a right triangle having two sides of equal length.

**45°–45°–90° TRIANGLE**

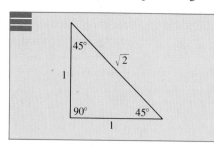

In a 45°–45° right triangle, the hypotenuse has a length that is $\sqrt{2}$ times as long as the length of either of the shorter (equal) sides.

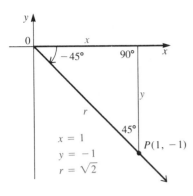

**FIGURE 11**

### ■■ EXAMPLE 5

Find the trigonometric function values for an angle of $-\pi/4$.

Place an angle of $-\pi/4$ radians, or $-45°$, in standard position, as in Figure 11. Choose point $P$ on the terminal side so that $r = \sqrt{2}$. By the description of $45°$–$45°$–$90°$ triangles, $P$ has coordinates $(1, -1)$, with $x = 1$, $y = -1$, and $r = \sqrt{2}$.

$$\sin -\frac{\pi}{4} = -\frac{1}{\sqrt{2}} = -\frac{\sqrt{2}}{2} \qquad \tan -\frac{\pi}{4} = -1 \qquad \sec -\frac{\pi}{4} = \sqrt{2}$$

$$\cos -\frac{\pi}{4} = \frac{1}{\sqrt{2}} = \frac{\sqrt{2}}{2} \qquad \cot -\frac{\pi}{4} = -1 \qquad \csc -\frac{\pi}{4} = -\sqrt{2} \quad ■■$$

For angles other than the special angles of $30°$, $45°$, $60°$, and their multiples, use a calculator or a table of trigonometric functions like the one in the Appendix of this book. Use the degree column or the radian column, as appropriate.

### ■■ EXAMPLE 6

Use the table of trigonometric functions or a calculator to verify the following results.

**(a)** $\sin 10° = .1736$

**(b)** $\cos 48° = .6691$

**(c)** $\tan 82° = 7.1154$

**(d)** $\sin .2618 = .2588$

**(e)** $\tan 1.2043 = 2.6053$

**(f)** $\sec .7679 = 1.3901$  ■■

***Graphs of the Trigonometric Functions***    Because of the way the trigonometric functions are defined (using a circle), the same function values will be obtained for any two angles that differ by $2\pi$ radians (or $360°$). That is,

$$\sin(x + 2\pi) = \sin x \qquad \text{and} \qquad \cos(x + 2\pi) = \cos x$$

for any value of $x$. Because of this property, the sine and cosine functions are *periodic functions*.

**PERIODIC FUNCTION**

■■ A function $y = f(x)$ is **periodic** if there exists a positive real number $a$ such that

$$f(x) = f(x + a)$$

for all values of $x$ in the domain of the function. The smallest possible value of $a$ is called the **period** of the function.

When graphing, it is customary to use $x$ (rather than $\theta$) for the domain elements, as we did with earlier functions, and to write $y = \sin x$ instead of $y = \sin \theta$. Because sine is periodic, with period $2\pi$, we can graph $y = \sin x$ by finding $y$ for values of $x$ between 0 and $2\pi$. This portion of the graph can then be repeated as many times as necessary.

Think of a point moving counterclockwise around a circle, tracing out an arc for angle $x$. The value of $\sin x$ gradually increases from 0 to 1 as $x$ increases from 0 to $\pi/2$. The values of $\sin x$ then decrease back to 0 as $x$ goes from $\pi/2$ to $\pi$. For $\pi < x < 2\pi$, $\sin x$ is negative. A few typical values from these intervals are given in the following table, where decimals have been rounded to the nearest tenth.

| $x$ | 0 | $\pi/4$ | $\pi/2$ | $3\pi/4$ | $\pi$ | $5\pi/4$ | $3\pi/2$ | $7\pi/4$ | $2\pi$ |
|-----|---|---------|---------|----------|-------|----------|----------|----------|--------|
| $\sin x$ | 0 | .7 | 1 | .7 | 0 | $-.7$ | $-1$ | $-.7$ | 0 |

Plotting the points from the table of values and connecting them with a smooth line gives the solid portion of the graph in Figure 12. Since $y = \sin x$ is periodic, the graph continues in both directions indefinitely, as suggested with dashed lines.

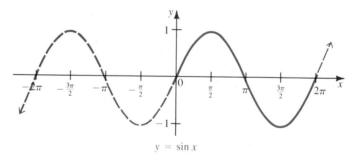

$y = \sin x$

**FIGURE 12**

The graph of $y = \cos x$ in Figure 13 can be found in much the same way. Again, the period is $2\pi$.

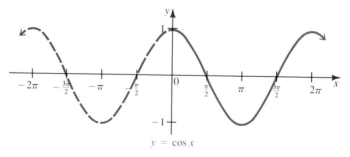

$y = \cos x$

**FIGURE 13**

Finally, Figure 14 shows the graph of $y = \tan x$. Since $\tan x$ is undefined (because of zero denominators) for $x = \pi/2, 3\pi/2, -\pi/2$, and so on, the graph has vertical asymptotes at these values. As the graph suggests, the tangent function is periodic, with a period of $\pi$.

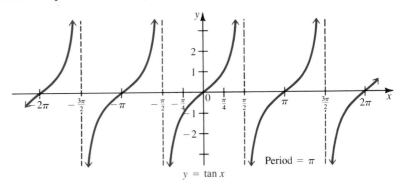

**FIGURE 14**

The graphs of the other trigonometric functions are not as useful as these three, so they are not given here.

## 10.1 EXERCISES

*Convert the following degree measures to radians. Leave answers as multiples of $\pi$.*

**1.** $60°$      **2.** $90°$      **3.** $150°$      **4.** $135°$

**5.** $210°$      **6.** $300°$      **7.** $390°$      **8.** $480°$

*Convert the following radian measures to degrees.*

**9.** $\dfrac{7\pi}{4}$      **10.** $\dfrac{2\pi}{3}$      **11.** $\dfrac{11\pi}{6}$      **12.** $\dfrac{-\pi}{4}$

**13.** $\dfrac{8\pi}{5}$      **14.** $\dfrac{7\pi}{10}$      **15.** $\dfrac{4\pi}{15}$      **16.** $5\pi$

*Find the values of the six trigonometric functions for the angles in standard position having the points in Exercises 17–20 on their terminal sides.*

**17.** $(-3, 4)$      **18.** $(-12, -5)$      **19.** $(6, 8)$      **20.** $(-7, 24)$

*In quadrant I, x, y, and r are all positive, so that all six trigonometric functions have positive values. In quadrant II, x is negative and y is positive (r is always positive). Thus, in quadrant II, sine is positive, cosine is negative, and so on. For Exercises 21–24, complete the following table of values for the signs of the trigonometric functions.*

| | Quadrant of $\theta$ | $\sin \theta$ | $\cos \theta$ | $\tan \theta$ | $\cot \theta$ | $\sec \theta$ | $\csc \theta$ |
|---|---|---|---|---|---|---|---|
| **21.** | I | | | | | | |
| **22.** | II | | | | | | |
| **23.** | III | | | | | | |
| **24.** | IV | | | | | | |

*For Exercises 25–32, complete the following chart. Do not use tables or a calculator.*

|   | $\theta$ | $\sin\theta$ | $\cos\theta$ | $\tan\theta$ | $\cot\theta$ | $\sec\theta$ | $\csc\theta$ |
|---|---|---|---|---|---|---|---|
| **25.** | 30° | 1/2 | $\sqrt{3}/2$ | | | $2\sqrt{3}/3$ | |
| **26.** | 45° | | | 1 | 1 | | |
| **27.** | 60° | | 1/2 | $\sqrt{3}$ | | 2 | |
| **28.** | 120° | $\sqrt{3}/2$ | | $-\sqrt{3}$ | | | $2\sqrt{3}/3$ |
| **29.** | 135° | $\sqrt{2}/2$ | $-\sqrt{2}/2$ | | | $-\sqrt{2}$ | $\sqrt{2}$ |
| **30.** | 150° | | $-\sqrt{3}/2$ | $-\sqrt{3}/3$ | | | 2 |
| **31.** | 210° | $-1/2$ | | $\sqrt{3}/3$ | $\sqrt{3}$ | | $-2$ |
| **32.** | 240° | $-\sqrt{3}/2$ | $-1/2$ | | | $-2$ | $-2\sqrt{3}/3$ |

*Find the following function values without using a calculator or a table.*

**33.** $\sin\dfrac{\pi}{3}$   **34.** $\cos\dfrac{\pi}{6}$   **35.** $\tan\dfrac{\pi}{4}$   **36.** $\cot\dfrac{\pi}{3}$

**37.** $\sec\dfrac{\pi}{6}$   **38.** $\sin\dfrac{\pi}{2}$   **39.** $\cos 3\pi$   **40.** $\sec\pi$

**41.** $\sin\dfrac{4\pi}{3}$   **42.** $\tan\dfrac{3\pi}{4}$   **43.** $\csc\dfrac{5\pi}{4}$   **44.** $\cos 5\pi$

**45.** $\tan\left(-\dfrac{\pi}{3}\right)$   **46.** $\cot\left(-\dfrac{2\pi}{3}\right)$   **47.** $\sin\left(-\dfrac{7\pi}{6}\right)$   **48.** $\cos\left(-\dfrac{\pi}{6}\right)$

*Use the table of trigonometric functions or a calculator to find the following function values.*

**49.** $\sin 39°$   **50.** $\cos 58°$   **51.** $\tan 82°$   **52.** $\tan 54°$

**53.** $\sin .4014$   **54.** $\tan 1.0123$   **55.** $\cos 1.4137$   **56.** $\sin 1.5359$

*Graph each function defined as follows over a two-period interval.*

**57.** $y = 2\sin x$   **58.** $y = 2\cos x$   **59.** $y = \cos 2x$   **60.** $y = \sin 2x$

**61.** $y = -\sin x$   **62.** $y = -\dfrac{1}{2}\cos x$   **63.** $y = \dfrac{1}{2}\tan x$   **64.** $y = -3\tan x$

████ APPLICATIONS

## BUSINESS AND ECONOMICS

**Sales**   **65.** Sales of snow blowers are seasonal. Suppose the sales of snow blowers in one region of the country are approximated by

$$S(t) = 500 + 500\cos\frac{\pi}{6}t,$$

where $t$ is time in months, with $t = 0$ corresponding to November.
Find sales in each of the following months.

**(a)** November    **(b)** January    **(c)** February    **(d)** May    **(e)** August

**(f)** Graph $y = S(t)$.

## PHYSICAL SCIENCES

**Light Rays**    *When a light ray travels from one medium, such as air, to another medium, such as water or glass, the speed of the light changes, and the direction that the ray is traveling changes. (This is why a fish under water is in a different position from the place at which it appears to be.) These changes are given by Snell's law,*

$$\frac{c_1}{c_2} = \frac{\sin \theta_1}{\sin \theta_2},$$

*where $c_1$ is the speed in the first medium, $c_2$ is the speed in the second medium, and $\theta_1$ and $\theta_2$ are the angles shown in the figure.*

Medium 1    $\theta_1$

Medium 2    $\theta_2$

If this medium is less dense, light travels at a faster speed, $c_1$.

If this medium is more dense, light travels at a slower speed, $c_2$.

*In Exercises 66 and 67, assume that $c_1 = 3 \times 10^8$ meters per second, and find the speed of light in the second medium.*

**66.** $\theta_1 = 39°, \theta_2 = 28°$        **67.** $\theta_1 = 46°, \theta_2 = 31°$

**Sound**    *Pure sounds produce single sine waves on an oscilloscope. Find the period of each sine wave in the photographs in Exercises 68 and 69. On the vertical scale each square represents .5 and on the horizontal scale each square represents 30°.*

**68.**

**69.**

**Air Pollution** **70.** The amount of pollution in the air fluctuates with the seasons. It is lower after heavy spring rains and higher after periods of little rain. In addition to this seasonal fluctuation, the long-term trend is upward. An idealized graph of this situation is shown in the figure. Trigonometric functions can be used to describe the fluctuating part of the pollution levels. Powers of the number $e$ can be used to show the long-term growth. In fact, the pollution level in a certain area might be given by

$$P(t) = 7(1 - \cos 2\pi t)(t + 10) + 100e^{.2t},$$

where $t$ is time in years, with $t = 0$ representing January 1 of the base year. Thus, July 1 of the same year would be represented by $t = .5$, while October 1 of the following year would be represented by $t = 1.75$. Find the pollution levels on the following dates.

(a) January 1, base year

(b) July 1, base year

(c) January 1, following year

(d) July 1, following year

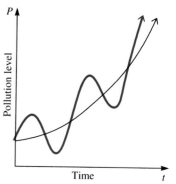

**Temperature** **71.** The maximum afternoon temperature in degrees Fahrenheit in a given city is approximated by

$$T(x) = 60 - 30 \cos(x/2),$$

where $x$ represents the month, with $x = 0$ representing January, $x = 1$ representing February, and so on. Use a calculator with trigonometric function keys to find the maximum afternoon temperature for each of the following months.

(a) January    (b) March    (c) October    (d) June    (e) August

**Temperature** **72.** A mathematical model for the temperature in Fairbanks is

$$T(x) = 37 \sin\left[\frac{2\pi}{365}(x - 101)\right] + 25,*$$

where $T(x)$ is the temperature in degrees Celsius on day $x$, with $x = 0$ corresponding to January 1 and $x = 365$ corresponding to December 31.

Use a calculator with trigonometric function keys or the table of trigonometric functions in the Appendix to estimate the temperature on each of the following days.

(a) March 1 (Day 60)    (b) April 1 (Day 91)    (c) Day 101    (d) Day 150

(e) Find maximum and minimum values of $T$.

*Barbara Lando and Clifton Lando, "Is the Graph of Temperature Variation a Sine Curve?" *The Mathematics Teacher* 70 (September 1977): 534–37.

## 10.2   DERIVATIVES OF TRIGONOMETRIC FUNCTIONS

In this section we derive formulas for the derivatives of some of the trigonometric functions. All these derivatives can be found from the formula for the derivative of $y = \sin x$.

We will need to use the following identities, which are listed without proof, to find the derivatives of the trigonometric functions.

**BASIC IDENTITIES**

$$\sin^2 x + \cos^2 x = 1$$

$$\tan x = \frac{\sin x}{\cos x}$$

$$\sin(x + y) = \sin x \cos y + \cos x \sin y$$

$$\sin(x - y) = \sin x \cos y - \cos x \sin y$$

$$\cos(x + y) = \cos x \cos y - \sin x \sin y$$

$$\cos(x - y) = \cos x \cos y + \sin x \sin y$$

The derivative of $y = \sin x$ also depends on the value of

$$\lim_{x \to 0} \frac{\sin x}{x}.$$

To estimate this limit, find the quotient $(\sin x)/x$ for various values of $x$ close to 0. (Be sure that your calculator is set for radian measure.)

| $x$ | .1 | .01 | .001 | − .001 | − .01 | − .1 |
|---|---|---|---|---|---|---|
| $\dfrac{\sin x}{x}$ | .998 | .99998 | .99999983 | .99999983 | .99998 | .998 |

*x approaches 0.*

*Values approach 1.*

These results suggest that

$$\lim_{x \to 0} \frac{\sin x}{x} = 1.$$

In Example 1, this limit is used to obtain another limit. Then the derivative of $y = \sin x$ can be found.

### ■ EXAMPLE 1

Find $\lim\limits_{h \to 0} \dfrac{\cos h - 1}{h}$.

Use the limit above and some trigonometric identities.

$$\lim_{h \to 0} \frac{\cos h - 1}{h} = \lim_{h \to 0} \frac{(\cos h - 1)}{h} \cdot \frac{(\cos h + 1)}{(\cos h + 1)}$$

$$= \lim_{h \to 0} \frac{\cos^2 h - 1}{h(\cos h + 1)}$$

$$= \lim_{h \to 0} \frac{-\sin^2 h}{h(\cos h + 1)} \qquad \cos^2 h = 1 - \sin^2 h$$

$$= \lim_{h \to 0} (-\sin h)\left(\frac{\sin h}{h}\right)\left(\frac{1}{\cos h + 1}\right)$$

$$= 0(1)\left(\frac{1}{1 + 1}\right)$$

$$\lim_{h \to 0} \frac{\cos h - 1}{h} = 0 \quad ■$$

We can now find the derivative of $y = \sin x$ by using the general definition for the derivative of a function $f$ given in Chapter 2:

$$f'(x) = \lim_{h \to 0} \frac{f(x + h) - f(x)}{h},$$

provided this limits exists. By this definition, the derivative of $f(x) = \sin x$ is

$$f'(x) = \lim_{h \to 0} \frac{\sin(x + h) - \sin x}{h}$$

$$= \lim_{h \to 0} \frac{\sin x \cdot \cos h + \cos x \cdot \sin h - \sin x}{h} \qquad \text{Identity for } \sin(x + h)$$

$$= \lim_{h \to 0} \frac{(\sin x \cdot \cos h - \sin x) + \cos x \cdot \sin h}{h} \qquad \text{Rearranging terms}$$

$$= \lim_{h \to 0} \frac{\sin x(\cos h - 1) + \cos x \cdot \sin h}{h} \qquad \text{Factoring}$$

$$= \lim_{h \to 0} \left(\sin x \frac{\cos h - 1}{h}\right) + \lim_{h \to 0} \left(\cos x \frac{\sin h}{h}\right)$$

$$= (\sin x)(0) + (\cos x)(1)$$

$$f'(x) = \cos x.$$

This result is summarized below.

**DERIVATIVE OF $\sin x$**

$$D_x(\sin x) = \cos x$$

Recall that the symbol $D_x[f(x)]$ means the derivative of $f(x)$ with respect to $x$.

We can use the chain rule to find derivatives of other sine functions, as shown in the following examples.

### ■ EXAMPLE 2

Find the derivatives of the functions defined as follows.

**(a)** $y = \sin 6x$

By the chain rule,

$$y' = (\cos 6x) \cdot D_x(6x)$$
$$= (\cos 6x) \cdot 6$$
$$y' = 6 \cos 6x.$$

**(b)** $y = 5 \sin(9x^2 + 2)$

$$y' = [5 \cos(9x^2 + 2)] \cdot D_x(9x^2 + 2)$$
$$= 18x[5 \cos(9x^2 + 2)]$$
$$y' = 90x \cos(9x^2 + 2) \quad ■$$

### ■ EXAMPLE 3

Find $D_x(\sin^4 x)$.

The expression $\sin^4 x$ means $(\sin x)^4$. By the generalized power rule,

$$D_x(\sin^4 x) = 4 \cdot \sin^3 x \cdot D_x(\sin x)$$
$$= 4 \sin^3 x \cos x. \quad ■$$

The derivative of $y = \cos x$ is found from trigonometric identities and the fact that $D_x(\sin x) = \cos x$. First, use the identity for $\sin(x - y)$ to get

$$\sin\left(\frac{\pi}{2} - x\right) = \sin\frac{\pi}{2} \cdot \cos x - \cos\frac{\pi}{2} \cdot \sin x$$
$$= 1 \cdot \cos x - 0 \cdot \sin x$$
$$= \cos x.$$

In the same way, $\cos\left(\dfrac{\pi}{2} - x\right) = \sin x$. Therefore,

$$D_x(\cos x) = D_x\left[\sin\left(\frac{\pi}{2} - x\right)\right].$$

By the chain rule,

$$D_x\left[\sin\left(\frac{\pi}{2} - x\right)\right] = \cos\left(\frac{\pi}{2} - x\right) \cdot D_x\left(\frac{\pi}{2} - x\right)$$

$$= \cos\left(\frac{\pi}{2} - x\right) \cdot (-1) \qquad \pi/2 \text{ is constant}$$

$$= -\cos\left(\frac{\pi}{2} - x\right)$$

$$= -\sin x.$$

**DERIVATIVE OF cos** $x$

$$D_x(\cos x) = -\sin x$$

### EXAMPLE 4

Find the following derivatives.

**(a)** $D_x[\cos(3x)] = -\sin(3x) \cdot D_x(3x) = -3 \sin 3x$

**(b)** $D_x(\cos^4 x) = 4 \cos^3 x \cdot D_x(\cos x) = 4 \cos^3 x(-\sin x)$
$$= -4 \sin x \cos^3 x$$

**(c)** $D_x(3x \cdot \cos x)$

Use the product rule.

$$D_x(3x \cdot \cos x) = 3x(-\sin x) + (\cos x)(3)$$

$$= -3x \sin x + 3 \cos x$$

As mentioned in the list of basic identities at the beginning of this section, $\tan x = (\sin x)/\cos x$. The derivative of $y = \tan x$ can be found by using the quotient rule to find the derivative of $y = (\sin x)/\cos x$.

$$D_x(\tan x) = D_x\left(\frac{\sin x}{\cos x}\right) = \frac{\cos x \cdot D_x(\sin x) - \sin x \cdot D_x(\cos x)}{\cos^2 x}$$

$$= \frac{\cos x(\cos x) - \sin x(-\sin x)}{\cos^2 x}$$

$$= \frac{\cos^2 x + \sin^2 x}{\cos^2 x}$$

$$= \frac{1}{\cos^2 x}$$

$$= \sec^2 x$$

The last step follows from the definitions of the trigonometric functions, which could be used to show that $1/\cos x = \sec x$. A similar calculation leads to the derivative of $\cot x$.

DERIVATIVES OF tan $x$
AND cot $x$

$$D_x(\tan x) = \sec^2 x$$
$$D_x(\cot x) = -\csc^2 x$$

## EXAMPLE 5

Find the derivatives of the following functions.

(a)  $D_x(\tan 9x) = \sec^2 9x \cdot D_x(9x) = 9 \sec^2 9x$

(b)  $D_x(\cot^6 x) = 6 \cot^5 x \cdot D_x(\cot x) = -6 \cot^5 x \csc^2 x$

(c)  $D_x(\ln |6 \tan x|) = \dfrac{D_x(6 \tan x)}{6 \tan x} = \dfrac{6 \sec^2 x}{6 \tan x} = \dfrac{\sec^2 x}{\tan x}$

## 10.2 EXERCISES

*Find the derivatives of the functions defined as follows.*

**1.** $y = 2 \sin 6x$

**2.** $y = -\cos 4x$

**3.** $y = 12 \tan (9x + 1)$

**4.** $y = -3 \cos (8x^2 + 2)$

**5.** $y = \cos^4 x$

**6.** $y = -9 \sin^5 x$

**7.** $y = \tan^5 x$

**8.** $y = 2 \cot^4 x$

**9.** $y = -5x \cdot \sin 4x$

**10.** $y = 6x \cdot \cos 3x$

**11.** $y = \dfrac{\sin x}{x}$

**12.** $y = \dfrac{\tan x}{2x + 4}$

**13.** $y = \sin e^{5x}$

**14.** $y = \cos 4e^{2x}$

**15.** $y = e^{\sin x}$

**16.** $y = -8e^{\tan x}$

**17.** $y = \sin (\ln 4x^2)$

**18.** $y = \cos(\ln |2x^3|)$

**19.** $y = \ln |\sin x^2|$

**20.** $y = \ln |\tan^2 x|$

**21.** $y = \dfrac{2 \sin x}{3 - 2 \sin x}$

**22.** $y = \dfrac{4 \cos x}{2 - \cos x}$

**23.** $y = \sqrt{\dfrac{\sin x}{\sin 3x}}$

**24.** $y = \sqrt{\dfrac{\cos 4x}{\cos x}}$

## APPLICATIONS

### BUSINESS AND ECONOMICS

Revenue **25.** The revenue received from the sale of electric fans is seasonal, with maximum revenue in the summer. Let the revenue received from the sale of fans be approximated by

$$R(x) = 100 \cos 2\pi x,$$

where $x$ is time in years, measured from July 1.

(a)  Find $R'(x)$.

(b)  Find $R'(x)$ for August 1. (*Hint:* August 1 is 1/12 of a year from July 1.)

(c)  Find $R'(x)$ for January 1.

(d)  Find $R'(x)$ for June 1.

## PHYSICAL SCIENCES

**Motion of a Particle**    **26.** A particle moves along a straight line. The distance of the particle from the origin at time $t$ is given by

$$s(t) = \sin t + 2 \cos t.$$

Find the velocity at each of the following times.

**(a)** $t = 0$    **(b)** $t = \pi/2$    **(c)** $t = \pi$

Find the acceleration at each of the following times.

**(d)** $t = 0$    **(e)** $t = \pi/2$    **(f)** $t = \pi$

## LIFE SCIENCES

**Swing of a Runner's Arm**    **27.** A runner's arm swings rhythmically according to the equation

$$y = \frac{\pi}{8} \cos 3\pi\left(t - \frac{1}{3}\right),$$

where $y$ denotes the angle between the actual position of the upper arm and the downward vertical position as shown in the figure,* and where $t$ denotes time in seconds.

Schematic diagrams of a rhythmically moving arm. The upper arm $RO$ rotates back and forth about the point $R$; the position of the arm is measured by the angle $y$ between the actual position and the downward vertical position. The arrow labeled $F$ denotes the effective component of the force exerted by the muscle as the arm moves.

**(a)** Make a rough sketch of the graph of $y$ as a function of $t$.

**(b)** Calculate the velocity and the acceleration of the arm.

**(c)** Verify that the angle $y$ and the acceleration $d^2y/dt^2$ are related by the differential equation

$$\frac{d^2y}{dt^2} + 9\pi^2 y = 0.$$

*Art for Exercises 27 and 28 from *Calculus for the Life Sciences* by Rodolfo De Sapio. Copyright © 1976, 1978 by W. H. Freeman and Company. Reprinted by permission.

(d) Apply the fact that the force exerted by the muscle as the arm swings is proportional to the acceleration of $y$, with a positive constant of proportionality, to find the direction of the force (counterclockwise or clockwise) at $t = 1$ sec, $t = 4/3$ sec, and $t = 5/3$ sec. What is the position of the arm at each of these times?

**Swing of a Jogger's Arm**    **28.** A jogger's arm swings according to the equation

$$y = \frac{1}{5} \sin \pi(t - 1).$$

Proceed as directed in (a), (b), (c), and (d) of the preceding exercise with the following exceptions. In (c) replace the differential equation with

$$\frac{d^2y}{dt^2} + \pi^2 y = 0,$$

and in (d) consider the times $t = 1.5$ sec, $t = 2.5$ sec, and $t = 3.5$ sec.

## 10.3 INTEGRATION

Any differentiation formula leads to a corresponding formula for integration. In particular, the formulas of the last section lead to the following indefinite integrals.

**BASIC TRIGONOMETRIC INTEGRALS**

$$\int \sin x \, dx = -\cos x + C \qquad \int \cos x \, dx = \sin x + C$$

$$\int \sec^2 x \, dx = \tan x + C \qquad \int \csc^2 x \, dx = -\cot x + C$$

### EXAMPLE 1

Find each integral.

**(a)** $\int \sin 7x \, dx$

Use substitution. Let $u = 7x$, so that $du = 7dx$. Then

$$\int \sin 7x \, dx = \frac{1}{7} \int \sin 7x(7dx)$$

$$= \frac{1}{7} \int \sin u \, du$$

$$= -\frac{1}{7} \cos u + C$$

$$= -\frac{1}{7} \cos 7x + C.$$

**(b)** $\int \cos \frac{2}{3}x \, dx = \frac{3}{2} \int \cos \frac{2}{3}x \left(\frac{2}{3}dx\right) = \frac{3}{2} \sin \frac{2}{3}x + C$

**(c)** $\int \sin^2 x \cos x \, dx$

Let $u = \sin x$, with $du = \cos x \, dx$. This gives

$$\int \sin^2 x \cos x \, dx = \int u^2 \, du = \frac{1}{3}u^3 + C.$$

Replacing $u$ with $\sin x$ gives

$$\int \sin^2 x \cos x \, dx = \frac{1}{3}\sin^3 x + C.$$

**(d)** $\int \dfrac{\sin x}{\sqrt{\cos x}} dx$

Rewrite the integrand as

$$\int (\cos x)^{-1/2} \sin x \, dx.$$

If $u = \cos x$, then $du = -\sin x \, dx$, with

$$\int (\cos x)^{-1/2} \sin x \, dx = -\int (\cos x)^{-1/2}(-\sin x \, dx)$$
$$= -\int u^{-1/2} \, du$$
$$= -2u^{1/2} + C$$
$$= -2\cos^{1/2} x + C.$$

**(e)** $\int \sec^2 12x \, dx = \dfrac{1}{12}\int \sec^2 12x(12dx) = \dfrac{1}{12}\tan 12x + C$ ▬

As in Chapter 5, we can find the area under a curve by setting up an appropriate definite integral.

## ▬ EXAMPLE 2

Find the shaded area in Figure 15.

The shaded area in Figure 15 is bounded by $y = \cos x$, $y = 0$, $x = -\pi/2$, and $x = \pi/2$. By the fundamental theorem of calculus, this area is given by

$$\int_{-\pi/2}^{\pi/2} \cos x \, dx = \sin x \Big|_{-\pi/2}^{\pi/2}$$

$$= \sin \frac{\pi}{2} - \sin\left(-\frac{\pi}{2}\right)$$

$$= 1 - (-1)$$

$$= 2.$$

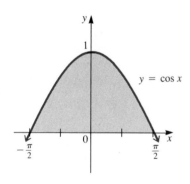

**FIGURE 15**

By symmetry, the same area could be found by evaluating

$$2\int_0^{\pi/2} \cos x \, dx. \quad \blacksquare$$

The method of integration by parts discussed in Chapter 6 is often useful for finding certain integrals involving trigonometric functions.

### ■ EXAMPLE 3

Find $\int 2x \sin x \, dx$.

Let $u = 2x$ and $dv = \sin x \, dx$. Then $du = 2 \, dx$ and $v = -\cos x$. Use the formula for integration by parts,

$$\int u \, dv = uv - \int v \, du,$$

to get

$$\int 2x \sin x \, dx = -2x \cos x - \int (-\cos x)2 \, dx$$

$$= -2x \cos x + 2\int \cos x \, dx$$

$$= -2x \cos x + 2 \sin x + C.$$

Check the result by differentiating. ■

As mentioned earlier, $\tan x = (\sin x)/\cos x$, so that

$$\int \tan x \, dx = \int \frac{\sin x}{\cos x} dx.$$

To find $\int \tan x \, dx$, let $u = \cos x$, with $du = -\sin x \, dx$. Then

$$\int \tan x \, dx = \int \frac{\sin x}{\cos x} dx = -\int \frac{du}{u} = -\ln |u| + C.$$

Replacing $u$ with $\cos x$ gives the formula for integrating $\tan x$. The integral for $\cot x$ is found in a similar way.

**INTEGRALS OF tan $x$ AND cot $x$**

$$\int \tan x \, dx = -\ln |\cos x| + C$$

$$\int \cot x \, dx = \ln |\sin x| + C$$

### ■ EXAMPLE 4

**(a)** $\int \tan 6x \, dx = \frac{1}{6}\int \tan 6x(6 \, dx) = -\frac{1}{6} \ln |\cos 6x| + C$

**(b)** $\int x \cdot \cot x^2 \, dx = \frac{1}{2}\int (\cot x^2)(2x \, dx) = \frac{1}{2} \ln |\sin x^2| + C$ ■

## 10.3 EXERCISES

*Find the following integrals.*

**1.** $\int \cos 5x \, dx$

**2.** $\int \sin 8x \, dx$

**3.** $\int (5 \cos x + 2 \sin x) dx$

**4.** $\int (7 \sin x - 8 \cos x) dx$

**5.** $\int x \sin x^2 \, dx$

**6.** $\int 2x \cdot \cos x^2 \, dx$

**7.** $-\int 6 \sec^2 2x \, dx$

**8.** $-\int 2 \csc^2 8x \, dx$

**9.** $\int \sin^7 x (\cos x) \, dx$

**10.** $\int \sin^6 x (\cos x) \, dx$

**11.** $\int \sqrt{\sin x} (\cos x) \, dx$

**12.** $\int \dfrac{\cos x}{\sqrt{\sin x}} \, dx$

**13.** $\int \dfrac{\sin x}{1 + \cos x} \, dx$

**14.** $\int \dfrac{\cos x}{1 - \sin x} \, dx$

**15.** $\int x^5 \cos x^6 \, dx$

**16.** $\int (x + 2)^4 \sin(x + 2)^5 \, dx$

**17.** $\int \tan \dfrac{1}{4} x \, dx$

**18.** $\int \cot \left( -\dfrac{3}{8} x \right) dx$

**19.** $\int x^2 \cot x^3 \, dx$

**20.** $\int \dfrac{x}{4} \cdot \tan \left( \dfrac{x}{4} \right)^2 dx$

**21.** $\int e^x \cdot \sin e^x \, dx$

**22.** $\int e^{-x} \cdot \tan e^{-x} \, dx$

*Evaluate the following definite integrals.*

**23.** $\displaystyle\int_0^{\pi/4} \sin x \, dx$

**24.** $\displaystyle\int_{-\pi/2}^0 \cos x \, dx$

**25.** $\displaystyle\int_0^{\pi/3} \tan x \, dx$

**26.** $\displaystyle\int_{\pi/4}^{\pi/2} \cot x \, dx$

**27.** $\displaystyle\int_{\pi/2}^{2\pi/3} \cos x \, dx$

**28.** $\displaystyle\int_{\pi/4}^{3\pi/4} \sin x \, dx$

*Use integration by parts to find each integral.*

**29.** $\int -6x \cos 5x \, dx$

**30.** $\int 9x \sin 2x \, dx$

**31.** $\int 8x \sin x \, dx$

**32.** $\int -11x \cos x \, dx$

**33.** $\int -6x \cos 8x \, dx$

**34.** $\int 10x \sin \dfrac{1}{2} x \, dx$

*Find each improper integral (if you have studied improper integrals).*

**35.** $\displaystyle\int_0^{+\infty} e^{-x} \sin x \, dx$

**36.** $\displaystyle\int_0^{+\infty} e^{-x} \cos x \, dx$

## APPLICATIONS

### BUSINESS AND ECONOMICS

**Sales**  **37.** Sales of snow blowers are seasonal. Suppose the sales of snow blowers in one region of the country are approximated by

$$S(t) = 500 + 500 \cos \left( \dfrac{\pi}{6} t \right),$$

where $t$ is time in months, with $t = 0$ corresponding to November. The figure at the top of the next page shows a graph of $S$. Use a definite integral to find total sales over a year.

$S(t) = 500 + 500 \cos\left(\frac{\pi}{6} t\right)$

**EXERCISE 37**

## LIFE SCIENCES

**Migratory Animals** 38. The number of migratory animals (in hundreds) counted at a certain checkpoint is given by

$$T(t) = 50 + 50 \cos\left(\frac{\pi}{6}t\right),$$

where $t$ is time in months, with $t = 0$ corresponding to July. The figure shows a graph of $T$. Use a definite integral to find the number of animals passing the checkpoint in a year.

$T(t) = 50 + 50 \cos\left(\frac{\pi}{6} t\right)$

**EXERCISE 38**

## 10.4 INVERSE TRIGONOMETRIC FUNCTIONS

To solve the equation $y = e^x$ for $x$, we introduced a new function, the logarithmic function. Similarly, we introduce the *inverse sine* function in order to solve the equation $y = \sin x$ for $x$. The inverse sine function determines an angle from the sine of the angle. For example, since $\sin(\pi/2) = 1$, the inverse sine function must start with 1 and produce the result $\pi/2$. One difficulty is that there are many values of $x$ such that $\sin x = 1 : \pi/2$ is one, as are $5\pi/2$, $9\pi/2$, and $13\pi/2$. To be sure that the inverse sine produces only one answer, it is necessary to restrict the range of the inverse sine function. The range is commonly restricted to the interval $[-\pi/2, \pi/2]$. With this restriction, the inverse sine function is defined as follows.

**INVERSE SINE**

> Let $x = \sin y$, for $y$ in $[-\pi/2, \pi/2]$. Then
> $$y = \sin^{-1} x.$$

### ◼ EXAMPLE 1

Find each of the following.

**(a)** $\sin^{-1}(1/2)$

Let $y = \sin^{-1}(1/2)$. Then by the definition of the inverse sine function, $\sin y = 1/2$. Since $\sin(\pi/6) = 1/2$, and $\pi/6$ is in $[-\pi/2, \pi/2]$,

$$\sin^{-1}\frac{1}{2} = \frac{\pi}{6}.$$

**(b)** $\sin^{-1}\left(-\dfrac{\sqrt{3}}{2}\right) = -\dfrac{\pi}{3}$

**(c)** $\sin^{-1} 1 = \dfrac{\pi}{2}$ ◼

A graph of $y = \sin^{-1} x$ is shown in Figure 16. (Sometimes the notation $y = \arcsin x$ is used for $y = \sin^{-1} x$.)

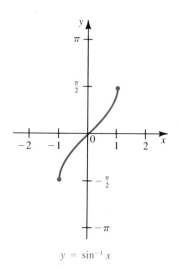

$$y = \sin^{-1} x$$

**FIGURE 16**

For each of the other trigonometric functions, an inverse function can be defined by putting a suitable restriction on the range of the inverse function. The three most commonly used **inverse trigonometric functions** and their ranges are listed below.

**INVERSE TRIGONOMETRIC FUNCTIONS**

$$y = \sin^{-1} x, \qquad -\frac{\pi}{2} \le y \le \frac{\pi}{2}$$

$$y = \cos^{-1} x, \qquad 0 \le y \le \pi$$

$$y = \tan^{-1} x, \qquad -\frac{\pi}{2} < y < \frac{\pi}{2}$$

Graphs of $y = \cos^{-1} x$ and $y = \tan^{-1} x$ are shown in Figures 17 and 18, respectively.

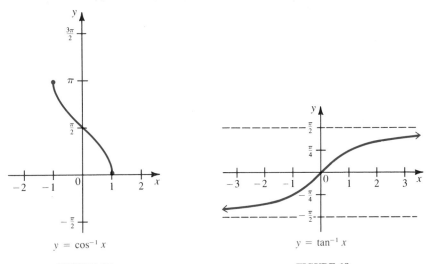

$y = \cos^{-1} x$

**FIGURE 17**

$y = \tan^{-1} x$

**FIGURE 18**

### EXAMPLE 2

Find $\cos^{-1} (-\sqrt{2}/2)$.

The values of $\cos^{-1} x$ are in quadrants I and II, according to the definition given above. Since $-\sqrt{2}/2$ is negative, $\cos^{-1} (-\sqrt{2}/2)$ is restricted to quadrant II. Let $y = \cos^{-1} (-\sqrt{2}/2)$. Then $\cos y = -\sqrt{2}/2$. In quadrant II, $\cos (3\pi/4) = -\sqrt{2}/2$, so

$$\cos^{-1} \left( -\frac{\sqrt{2}}{2} \right) = \frac{3\pi}{4}. \quad \blacksquare$$

***Derivatives of Inverse Trigonometric Functions*** We found that the derivatives of the trigonometric functions were themselves trigonometric functions. The derivatives of the inverse trigonometric functions, however, are *algebraic functions*. To find the derivative of $y = \sin^{-1} x$, use the definition of $y = \sin^{-1} x$ to write

$$x = \sin y.$$

Use implicit differentiation to find $D_x y$. Taking the derivative with respect to $x$ on both sides of $x = \sin y$ gives

$$D_x(x) = (\cos y)D_x y.$$

Since $D_x(x) = 1$,

$$1 = (\cos y)D_x y$$

or $$D_x y = \frac{1}{\cos y}.$$

Since $\sin^2 y + \cos^2 y = 1$, $\cos y = \sqrt{1 - \sin^2 y}$, and

$$D_x y = \frac{1}{\sqrt{1 - \sin^2 y}}.$$

Replace $\sin y$ with $x$ to get

$$D_x y = \frac{1}{\sqrt{1 - x^2}}.$$

Choose the positive square root since $\cos y > 0$ for $y$ in $[-\pi/2, \pi/2]$. This result is summarized below.

**DERIVATIVE OF $\sin^{-1} x$**

$$D_x(\sin^{-1} x) = \frac{1}{\sqrt{1 - x^2}}$$

### EXAMPLE 3

Find each derivative.

**(a)** $y = \sin^{-1} 9x$

Use the above formula with the chain rule to get

$$D_x(\sin^{-1} 9x) = \frac{1}{\sqrt{1 - (9x)^2}} D_x(9x) = \frac{9}{\sqrt{1 - 81x^2}}.$$

**(b)** $D_x[\sin^{-1}(5x^2 + 1)] = \dfrac{1}{\sqrt{1 - (5x^2 + 1)^2}} D_x(5x^2 + 1)$

$$= \dfrac{10x}{\sqrt{1 - (5x^2 + 1)^2}} \quad \blacksquare$$

Starting with $y = \cos^{-1} x$ and $y = \tan^{-1} x$ and again using implicit differentiation gives the following derivatives.

**DERIVATIVES OF**
$\cos^{-1} x$ **AND** $\tan^{-1}x$

$$D_x(\cos^{-1} x) = \dfrac{-1}{\sqrt{1 - x^2}}$$

$$D_x(\tan^{-1} x) = \dfrac{1}{1 + x^2}$$

### EXAMPLE 4

Find each derivative.

**(a)** $y = \cos^{-1}(3x^2)$

Using the formula for the derivative of inverse cosine, together with the chain rule, gives

$$D_x[\cos^{-1}(3x^2)] = \dfrac{-1}{\sqrt{1 - (3x^2)^2}}(6x) = \dfrac{-6x}{\sqrt{1 - 9x^4}}.$$

**(b)** $D_x[\tan^{-1}(8x^2)] = \dfrac{1}{1 + (8x^2)^2} D_x(8x^2) = \dfrac{16x}{1 + 64x^4}$

**(c)** $D_x[\tan^{-1}(e^{2x})] = \dfrac{1}{1 + (e^{2x})^2} D_x(e^{2x}) = \dfrac{2e^{2x}}{1 + e^{4x}} \quad \blacksquare$

***Integrals***   The three derivatives above lead to the following integrals.

$$\int \dfrac{1}{\sqrt{1 - x^2}}\, dx = \sin^{-1} x + C$$

$$\int \dfrac{-1}{\sqrt{1 - x^2}}\, dx = \cos^{-1} x + C$$

$$\int \dfrac{1}{1 + x^2}\, dx = \tan^{-1} x + C$$

The second integral is seldom used.

■■■ **EXAMPLE 5**

Find each integral.

(a) $\int \dfrac{2x}{\sqrt{1 - x^4}}\, dx$

Let $u = x^2$ to get $du = 2x\, dx$, so that

$$\int \frac{2x}{\sqrt{1 - x^4}}\, dx = \int \frac{2x\, dx}{\sqrt{1 - (x^2)^2}}$$

$$= \int \frac{du}{\sqrt{1 - u^2}}$$

$$= \sin^{-1} u + C$$

$$= \sin^{-1} (x^2) + C.$$

(b) $\int \dfrac{e^x}{1 + e^{2x}}\, dx$

Let $u = e^x$, with $du = e^x\, dx$. Then

$$\int \frac{e^x}{1 + e^{2x}}\, dx = \int \frac{e^x\, dx}{1 + (e^x)^2} = \int \frac{du}{1 + u^2}$$

$$= \tan^{-1} u + C$$

$$= \tan^{-1} e^x + C. \quad ■■$$

═══ **10.4 EXERCISES**

*Give the value of y in radians.*

**1.** $y = \sin^{-1} (-\sqrt{3}/2)$     **2.** $y = \cos^{-1} (\sqrt{3}/2)$     **3.** $y = \tan^{-1} 1$     **4.** $y = \tan^{-1} (-1)$

**5.** $y = \sin^{-1} (-1)$     **6.** $y = \cos^{-1} (-1)$     **7.** $y = \cos^{-1} (1/2)$     **8.** $y = \sin^{-1} (-\sqrt{2}/2)$

**9.** $y = \cos^{-1} (-\sqrt{2}/2)$     **10.** $y = \tan^{-1} (\sqrt{3}/3)$     **11.** $y = \tan^{-1} (-\sqrt{3})$     **12.** $y = \cos^{-1} (-1/2)$

*Use the table of trigonometric functions or a calculator to give each value in degrees.*

**13.** $\sin^{-1} (-.1392)$     **14.** $\cos^{-1} (-.1392)$     **15.** $\cos^{-1} (-.8988)$     **16.** $\sin^{-1} .7880$

**17.** $\cos^{-1} .9272$     **18.** $\tan^{-1} 1.7321$     **19.** $\tan^{-1} 1.111$     **20.** $\sin^{-1} .8192$

**21.** $\tan^{-1} (-.9004)$     **22.** $\tan^{-1} (-.2867)$     **23.** $\sin^{-1} .9272$     **24.** $\cos^{-1} .4384$

*Find the derivative of each inverse trigonometric function.*

**25.** $y = \sin^{-1} 12x$     **26.** $y = \cos^{-1} 10x$     **27.** $y = \tan^{-1} 3x$     **28.** $y = \sin^{-1} (1/x)$

**29.** $y = \cos^{-1}(-2/x)$     **30.** $y = \cos^{-1} \sqrt{x}$     **31.** $y = \tan^{-1} (\ln |7x|)$     **32.** $y = \tan^{-1} (\ln |x + 2|)$

**33.** $y = \ln |\tan^{-1} (x + 1)|$     **34.** $y = \ln |\tan^{-1}(3x - 5)|$

*A painting 1 m high and 3 m from the floor will cut off an angle θ to an observer, where*

$$\theta = \tan^{-1} \frac{x}{x^2 + 2}.$$

*We assume the observer is x m from the wall displaying the painting and that the eyes of the observer are 2 m above the ground. (See the figure.) Find the value of θ for each of the following values of x. Round to the nearest degree.*

**35.** 1

**36.** 2

**37.** 3

**38.** 4

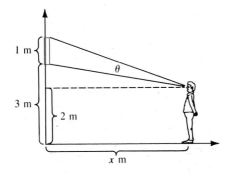

*Find each integral.*

**39.** $\int \dfrac{x^3}{1 + x^8}\, dx$

**40.** $\int \dfrac{x^2}{\sqrt{1 - x^6}}\, dx$

**41.** $\int \dfrac{e^x}{\sqrt{1 - e^{2x}}}\, dx$

**42.** $\int \dfrac{-e^{2x}}{1 + e^{4x}}\, dx$

**43.** $\int \dfrac{4}{\sqrt{1 - 9x^2}}\, dx$

**44.** $\int \dfrac{\cos x}{1 + \sin^2 x}\, dx$

**45.** $\int \dfrac{1/x}{1 + (\ln x)^2}\, dx$

**46.** $\int \dfrac{1}{\sqrt{25 - x^2}}\, dx$

## 10.5   APPLICATIONS

As mentioned earlier in this chapter, trigonometric functions are used in a wide variety of settings. Some of these applications are presented in this section.

### EXAMPLE 1

Suppose a light source is hanging at the end of a pendulum, as shown in Figure 19. As the pendulum swings back and forth, the light will oscillate on the floor. If a roll of photographic paper is moved at a constant speed under the light, a sine curve will be traced on the paper. This curve will have an equation of the form

$$y = A \sin(Bt + C)$$

for some constants *A*, *B*, and *C* and time *t*. Different values of *A* cause the range of the graph to stretch or contract vertically, while $B$ stretches or contracts the function horizontally, and $C$ shifts the whole graph horizontally. In $y = A \sin (Bt + C)$, the *amplitude* of the graph is $|A|$, the *period* is $2\pi/|B|$, and the *phase shift* is the value of $t$ that makes $Bt + C = 0$.   ▬

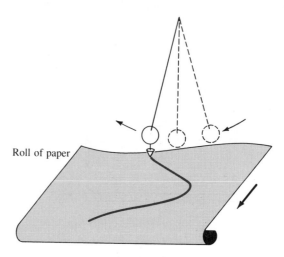

Roll of paper

**FIGURE 19**

Motion of this type, called **simple harmonic motion,** occurs in such diverse areas as sound and wave mechanics. In specific examples, calculus can be used to find the maximum and minimum values of $y$ and the $x$-values where these occur. (See Exercises 7–12 and Exercise 24 in this section.)

### ▬ EXAMPLE 2

Many biological populations, both plant and animal, experience seasonal growth. For example, an animal population might flourish during the spring and summer and die back in the fall. If $f(t)$ represents the population at time $t$, then the rate of change of the population with respect to time might be governed by the differential equation

$$\frac{df}{dt} = c \cdot f(t) \cdot \cos t.$$   $c$ is a positive constant

Since $\cos t$ ranges from $-1$ to $1$, $df/dt$ will change from negative to positive, so that $f(t)$ is both increasing and decreasing. The solution of the differential equation above is

$$f(t) = f(0)e^{c \, \sin t},$$

where $f(0)$ is the size of the population when $t = 0$.   ▬

### ▰ EXAMPLE 3

Mathematical models of ground temperature variation usually involve Fourier series or other sophisticated methods. However, Mary Kay Corbitt and C. H. Edwards, Jr., have given the elementary model

$$u(x, \ t) = T_0 + A_0 \ e^{-ax} \cos (wt - ax) \qquad \textbf{(1)}$$

for the temperature $u(x, \ t)$ at a given location at a given time $t$ and a given depth $x$ beneath the earth's surface.* In equation (1),

$\qquad T_0 = $ the annual average surface temperature;

$\qquad A_0 = $ the amplitude of the seasonal surface temperature variation;

$\qquad w = $ a constant chosen to make the period one year in length;

$\qquad k = $ the thermal conductivity of the ground;

and $\qquad a = \sqrt{w/2k}$.

Figure 20 shows $u(0, \ t)$, which models the temperature at the earth's surface.

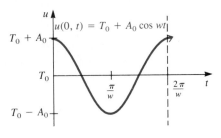

**FIGURE 20**

In deriving equation (1), geothermal heat was ignored and the density of the earth was assumed to be uniform so that $k$ is constant and the annual average temperature *at any depth* is $T_0$. The factor $e^{-ax}$ in equation (1) accounts for the relationship between temperature at depth $x$ and surface temperature. The phase shift $-ax$ in equation (1) accounts for the seasons at depth $x$ lagging behind those at the surface. ▰

---

### ▰ 10.5 EXERCISES

*In Exercises 1–6, recall that the slope of the tangent line to a graph is given by the derivative of the function. Find the slope of the tangent line to the graph of each equation at the given point.*

**1.** $y = \sin x$;  $x = 0$

**2.** $y = \sin x$;  $x = \pi/4$

**3.** $y = \cos x$;  $x = \pi/2$

**4.** $y = \cos x$;  $x = -\pi/4$

**5.** $y = \tan x$;  $x = 0$

**6.** $y = \cot x$;  $x = \pi/2$

---

*From "Mathematical Modeling and Cool Buttermilk in the Summer" by Mary Kay Corbitt and C. H. Edwards, Jr. in *Applications In School Mathematics—1979 Yearbook,* edited by Sidney Sharron and Robert E. Hays. Copyright © 1979 by The National Council of Teachers of Mathematics, Inc. Reprinted by permission.

*Find the maximum and minimum values of y for the following equations of simple harmonic motion.*

**7.** $y = 2 \sin (3x - 1)$

**8.** $y = 4 \sin (8x + 2)$

**9.** $y = -9 \sin (5x + 4)$

**10.** $y = -3 \sin (2x - 7)$

**11.** $y = \frac{1}{2} \sin \left( \frac{1}{4} x - 2 \right)$

**12.** $y = -\frac{2}{3} \sin \left( \frac{3}{5} x - \frac{1}{2} \right)$

---

### ≣ APPLICATIONS

### LIFE SCIENCES

**Seasonal Growth**     *Example 2 gave the following mathematical model for seasonal growth:*

$$f(t) = f(0)e^{c \sin t}.$$

*Suppose that $f(0) = 1000$ and $c = 2$. Find the function values in Exercises 13–22 using a calculator with trigonometric functions and powers of e.*

**13.** $f(.2)$     **14.** $f(.4)$     **15.** $f(.5)$     **16.** $f(.8)$     **17.** $f(1)$

**18.** $f(1.4)$     **19.** $f(1.8)$     **20.** $f(2.3)$     **21.** $f(3)$     **22.** $f(3.1)$

**23.** Graph $f(t)$ using the information from Exercises 13–22.

**24.** Find the maximum and minimum values of $f(t)$ and the values of $t$ where they occur.

**Blood Vessel System**     *The body's system of blood vessels is made up of arteries, arterioles, capillaries, and veins. The transport of blood from the heart through all organs of the body and back to the heart should be as efficient as possible. One way this can be done is by having large enough blood vessels to avoid turbulence, with blood cells small enough to minimize viscosity.\**

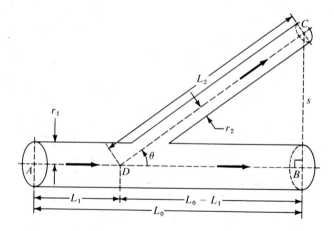

---

\*Art from *Introduction to Mathematics for Life Scientists* by Edward Batschelet. Copyright © 1971 by Springer-Verlag New York, Inc. Reprinted by permission.

*In Exercises 25–38 we will find the value of angle θ (see the figure) such that total resistance to the flow of blood is minimized. Assume that a main vessel of radius $r_1$ runs along the horizontal line from A to B. A side artery, of radius $r_2$, heads for a point C. Choose point B so that CB is perpendicular to AB. Let CB = s and let D be the point where the axis of the branching vessel cust the axis of the main vessel.*

*According to Poiseuille's law, the resistance R in the system is proportional to the length L of the vessel and inversely proportional to the fourth power of the radius r. That is,*

$$R = k \cdot \frac{L}{r^4}, \tag{1}$$

*where k is a constant determined by the viscosity of the blood. Let AB = $L_0$, AD = $L_1$, and DC = $L_2$.*

**25.** Use right triangle *BDC* and find $\sin \theta$.

**26.** Solve the result of Exercise 25 for $L_2$.

**27.** Find $\cot \theta$ in terms of $s$ and $L_0 - L_1$.

**28.** Solve the result of Exercise 27 for $L_1$.

**29.** Write an expression similar to equation (1) for the resistance $R_1$ along *AD*.

**30.** Write a formula for the resistance along *DC*.

**31.** The total resistance *R* is given by the sum of the resistances along *AD* and *DC*. Use your answers to Exercises 29 and 30 to write an expression for *R*.

**32.** In your formula for *R*, replace $L_1$ with the result of Exercise 28 and $L_2$ with the result of Exercise 26.

**33.** Find $R'$. (Remember that $k$, $L_1$, $L_0$, $s$, $r_1$, and $r_2$ are constants.)

**34.** Place $R'$ equal to 0.

**35.** Multiply through by $(\sin^2 \theta)/s$.

**36.** Solve for $\cos \theta$.

**37.** Suppose $r_1 = 1$ cm and $r_2 = 1/4$ cm. Find $\cos \theta$ and then find $\theta$. (Use a calculator or the table of trigonometric functions.)

**38.** Find $\theta$ if $r_1 = 1.4$ cm and $r_2 = .8$ cm.

**Ground Temperature**  *In Exercises 39–41, use the ground temperature model in Example 3 and assume that $T_0 = 16°C$ and $A_0 = 11°C$ at a certain location. Also assume that a = .00706 in cgs (centimeter-gram-second) units.*

**39.** At what minimum depth $x$ in centimeters is the amplitude of $u(x, t)$ at most 1°C?

**40.** Suppose that we wish to construct a cellar to keep wine at a temperature between 14°C and 18°C. What minimum depth in centimeters will accomplish this?

**41.** At what minimum depth $x$ does the ground temperature model predict that it will be winter when it is summer at the surface, and vice versa? That is, when will the phase shift correspond to 1/2 year?

**42.** Show that the ground temperature model

$$u(x, t) = T_0 + A_0 e^{-ax} \cos(wt - ax)$$

satisfies the *heat equation*

$$\frac{\partial u}{\partial t} = k \frac{\partial^2 u}{\partial x^2}.$$

═══ **KEY WORDS** ═══    10.1  **angle**                                  **sine**
                               **ray**                                    **cosine**
                               **vertex**                                 **tangent**
                               **degree measure**                         **cotangent**
                               **acute angle**                            **secant**
                               **right angle**                            **cosecant**
                               **obtuse angle**                           **special angles**
                               **straight angle**                         **periodic functions**
                               **standard position**                      **period**
                               **arc**                            10.4  **inverse trigonometric functions**
                               **radian measure**                 10.5  **simple harmonic motion**
                               **trigonometric functions**

═══ **CHAPTER 10**          **REVIEW EXERCISES** ══════════════════════════

*Convert the following degree measures to radians. Leave answers as multiples of $\pi$.*

**1.** $90°$        **2.** $120°$        **3.** $210°$        **4.** $270°$        **5.** $360°$        **6.** $420°$

*Convert the following radian measures to degrees.*

**7.** $7\pi$    **8.** $\dfrac{3\pi}{4}$    **9.** $\dfrac{9\pi}{20}$    **10.** $\dfrac{7\pi}{15}$    **11.** $\dfrac{13\pi}{20}$    **12.** $\dfrac{11\pi}{15}$

*Find each function value without using tables or a calculator.*

**13.** $\sin 60°$        **14.** $\tan 120°$        **15.** $\cos(-30°)$        **16.** $\sec 45°$        **17.** $\csc 120°$

**18.** $\cot 300°$    **19.** $\sin \dfrac{\pi}{6}$    **20.** $\cos \dfrac{2\pi}{3}$    **21.** $\sec \dfrac{5\pi}{4}$    **22.** $\csc \dfrac{7\pi}{3}$

*Find each function value.*

**23.** $\sin 47°$        **24.** $\cos 59°$        **25.** $\tan 81°$        **26.** $\sin(-32°)$

**27.** $\sin 1.4661$        **28.** $\cos .3142$        **29.** $\cos .5934$        **30.** $\tan 1.2915$

*Graph one period of each function.*

**31.** $y = 3\cos x$    **32.** $y = \dfrac{1}{2}\tan x$    **33.** $y = -\tan x$    **34.** $y = -2\sin x$

*Find the derivative of each function.*

**35.** $y = -4\sin 7x$        **36.** $y = 6\tan 3x$        **37.** $y = \tan(4x^2 + 3)$        **38.** $y = \cot(9 - x^2)$

**39.** $y = 3\cos^6 x$        **40.** $y = 2\sin^4(4x^2)$        **41.** $y = \cot(4x^5)$        **42.** $y = \cos(1 + x^2)$

**43.** $y = x^2 \cdot \cos x$    **44.** $y = e^{-x}\sin x$    **45.** $y = \dfrac{\sin x - 1}{\sin x + 1}$    **46.** $y = \dfrac{\cos^2 x}{1 - \cos x}$

**47.** $y = \dfrac{x - 2}{\sin x}$    **48.** $y = \dfrac{\tan x}{1 + x}$    **49.** $y = \ln|\cos x|$    **50.** $y = \ln|5\sin x|$

*Find each integral.*

**51.** $\int \sin 2x \, dx$        **52.** $\int \cos 3x \, dx$        **53.** $\int \tan 9x \, dx$

**54.** $\int \sec^2 5x \, dx$        **55.** $\int 5\sec^2 x \, dx$        **56.** $\int 4\csc^2 x \, dx$

**57.** $\int x \sin 3x^2 \, dx$

**58.** $\int 5x \cos 2x^2 \, dx$

**59.** $\int \sqrt{\cos x} \cdot \sin x \, dx$

**60.** $\int \sin^4 x \cos x \, dx$

**61.** $\int x \tan 11x^2 \, dx$

**62.** $\int x^2 \cot 8x^3 \, dx$

**63.** $\int (\sin x)^{5/2} \cos x \, dx$

**64.** $\int (\cos x)^{-4/3} \sin x \, dx$

*Find each definite integral.*

**65.** $\int_0^{\pi/2} \cos x \, dx$

**66.** $\int_{\pi/2}^{\pi} \sin x \, dx$

**67.** $\int_0^{2\pi} (10 + 10 \cos x) dx$

**68.** $\int_0^{2\pi} (5 + 5 \sin x) dx$

*Give each value in radians.*

**69.** $\sin^{-1}(-1/2)$

**70.** $\tan^{-1}(-1)$

**71.** $\cos^{-1} 1$

**72.** $\sin^{-1} \sqrt{3}/2$

**73.** $\sin^{-1}(-1)$

**74.** $\tan^{-1} \sqrt{3}$

*Find the derivative of each function.*

**75.** $y = \sin^{-1}(-3x)$

**76.** $y = \tan^{-1}(1 + x)$

**77.** $y = \tan^{-1}(5x^2)$

**78.** $y = \sin^{-1}(2 - x)$

*Find each integral.*

**79.** $\int \dfrac{3}{\sqrt{1 - x^2}} \, dx$

**80.** $\int \dfrac{4}{1 + x^2} \, dx$

**81.** $\int \dfrac{-6x^4}{1 + x^{10}} \, dx$

**82.** $\int \dfrac{-12x^2}{\sqrt{1 - x^6}} \, dx$

---

## ▤ APPLICATIONS

### LIFE SCIENCES

**Blood Pressure**  **83.** A person's blood pressure at time $t$ in seconds is given by

$$P(t) = 90 + 15 \sin 144\pi t.$$

Find the maximum and minimum values of $P$ on the interval [0, 1/72]. Graph one period of $y = P(t)$.

### PHYSICAL SCIENCES

**Voltage**  **84.** The voltage $E$ in an electrical circuit is given by $E = 5 \cos 120\pi t$, where $t$ is time in seconds. Find the maximum and minimum voltage in the interval (0, 1/60].

---

## ▤ EXTENDED APPLICATION

## ▤ LIGHT INTENSITY IN WATER*

In this application, we derive an elementary model for the light intensity in a body of water $B$ on the surface of the earth. To simplify the situation, assume that the light is all the same color, that the stream of light from the sun is constant, and that the sun is directly overhead 6 hours after dawn. See Figure 21. Let

$$x = \text{the depth below the surface of } B \text{ in feet;}$$
$$t = \text{the time after dawn in hours;}$$

and                $i(x, t) = \text{the light intensity at depth } x \text{ in } B \text{ at time } t.$

---

*Supplied by T. B. Muenzenberger, Kansas State University, Manhattan, Kansas.

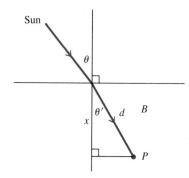

**FIGURE 21**

Suppose that a particle of light first travels from the sun to $B$ along a path that makes an angle $\theta$ with respect to the vertical, as shown in Figure 21. Then the particle travels in $B$ a distance $d$ along a path that makes an angle $\theta'$ with respect to the vertical and arrives at time $t$ at a point $P$ that is at depth $x$. If the scattering of light in the sky and the water is ignored, then there is only one direction from which light can arrive at $P$ at time $t$.

When $t = 0$ (at dawn), $\theta = \pi/2$, and since the sun is directly overhead at time $t = 6$, $\theta = 0$ when $t = 6$. Thus,

$$\frac{\pi}{2} - \theta = \frac{t\pi}{12} \quad \text{for} \quad 0 \le t \le 6 \quad \text{and} \quad \frac{\pi}{2} \ge \theta \ge 0.$$

By Snell's law (see 10.1 Exercises),

$$\sin \theta = \frac{4}{3} \sin \theta',$$

where 4/3 is the index of refraction for the transmission of light from air into water. Since $\cos \theta' = x/d$,

$$d = \frac{x}{\cos \theta'}$$

$$= \frac{x}{\sqrt{1 - \sin^2 \theta'}}$$

$$= \frac{x}{\sqrt{1 - \dfrac{9}{16} \sin^2 \theta}}$$

$$= \frac{x}{\sqrt{1 - \dfrac{9}{16} \cos^2 \left( \dfrac{\pi}{2} - \theta \right)}}$$

$$= \frac{x}{\sqrt{1 - \dfrac{9}{16} \cos^2 \left( \dfrac{t\pi}{12} \right)}}.$$

Now suppose that the ight intensity varies trigonometrically with the time and decays exponentially with the distance $d$ traveled in water. This can be expressed as

$$i(x, t) = i_0 \sin \frac{t\pi}{12} \cdot e^{-\alpha d}$$

$$= i_0 \sin \frac{t\pi}{12} \cdot e^{-\alpha x/\sqrt{1 - (9/16) \cos^2 (t\pi/12)}}.$$

The model makes the following predictions:

the light intensity at dawn is $i(x,0) = 0$;
the light intensity at noon at the surface of $B$ is $i(0, 6) = i_0$;
the light intensity at noon is $i(x, 6) = i_0 \, e^{-\alpha x}$;

the light intensity at the surface is $i(0, t) = i_0 \sin \dfrac{t\pi}{12}$.

All of these predictions are reasonable.

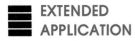

**EXERCISES**

1. Use the light intensity model to predict the light intensity at noon 5 ft below the surface of *B*.

2. Predict the light intensity at the surface 3 hr after dawn.

3. Use the light intensity model $i(x, t)$ to calculate and interpret the following.

    **(a)** $\dfrac{\partial i}{\partial x}$    **(b)** $\dfrac{\partial i}{\partial t}$

**EXTENDED APPLICATION**

## THE MATHEMATICS OF A HONEYCOMB*

The bee's cell is a regular hexagonal prism with one open end and one trihedral apex (see Figure 22[†]). We may construct the surface by starting with a regular hexagonal base *abcdef* with side *s*. Over the base we raise a right prism of height *h* and top *ABCDEF*. The corners *B*, *D*, and *F* are cut off by planes through the lines *AC*, *CE*, and *EA*, meeting in a point *V* on the axis *VN* of the prism, and intersecting *Bb*, *Dd*, and *Ff* in *X*, *Y*, and *Z*. The three cutoff pieces are the tetrahedra *ABCX*, *CDEY*, and *EFAZ*. We put these pieces on top of the remaining solid such that *X*, *Y*, and *Z* coincide with *V*. Hereby, the lines *AC*, *CE*, and *EA* act as "hinges". The faces *AXCV*, *CYEV*, *EZAV* are rhombuses—that is, quadrilaterals with equal sides. The new body is the bee's cell and has the same volume as the original prism. The hexagonal base *abcdef* is the open end.

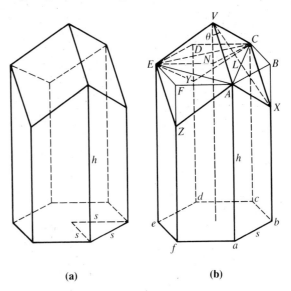

(a)                          (b)

**FIGURE 22**

The bees form the faces by using wax. When the volume is given, it is economic to spare wax and therefore to choose the angle of inclination, $\theta = NVX$, in such a way that the *surface of the bee's cell is minimized.*

---

The problem can be solved mathematically as follows. Let $L$ be the intersection of $CA$ and $VX$. Then $L$ bisects the segment $NB$ and, hence, $NL = s/2$. The segment $CL$ is the height of the equilateral triangle $BCN$. Therefore,

$$CL = \frac{s}{2}\sqrt{3}. \tag{1}$$

In the triangle $NLV$ we have the relationship

$$VL = \frac{s}{2\sin\theta}. \tag{2}$$

The rhombus $AXCV$ has its center in $L$ and consists of four congruent right triangles with legs equal to $CL$ and $VL$. Therefore, from (1) and (2) we get

$$\text{area } AXCV = 4 \cdot \frac{1}{2} \cdot \frac{s}{2}\sqrt{3} \cdot \frac{s}{2\sin\theta} = \frac{s^2\sqrt{3}}{2\sin\theta}.$$

The surface of the bee's cell contains three such areas.

The six lateral faces of the bee's cell, such as $abXA$, are congruent trapezoids. Since $BX = VN$, we obtain from triangle $VNL$

$$BX = \frac{s}{2}\cot\theta.$$

Hence,

$$\text{Area } abXA = \frac{s}{2}(aA + bX) = \frac{s}{2}(h + h - BX) = hs - \frac{s^2}{4}\cot\theta.$$

The total area made of wax amounts to

$$6hs - \frac{3}{2}s^2\cot\theta + \frac{3}{2}\frac{s^2\sqrt{3}}{\sin\theta}.$$

This area is a function of the variable angle $\theta$ and, thus, we denote it by $f(\theta)$. We may rewrite $f(\theta)$ in the form

$$f(\theta) = 6hs + \frac{3}{2}s^2\left(-\cot\theta + \frac{\sqrt{3}}{\sin\theta}\right).$$

Only the expression in the parentheses contains the variable $\theta$. Some numerical values rounded off to two decimals are given in the following table.

| $\theta$ | $-\cot\theta + \dfrac{\sqrt{3}}{\sin\theta}$ |
|---|---|
| 10° | 4.30 |
| 20° | 2.32 |
| 30° | 1.73 |
| 40° | 1.50 |
| 50° | 1.42 |
| 60° | 1.42 |
| 70° | 1.48 |
| 80° | 1.58 |
| 90° | 1.73 |

The minimum of $f(\theta)$ is reached somewhere between $\theta = 50°$ and $\theta = 60°$. To get the optimal angle, say $\theta_0$, we differentiate $f(\theta)$.

$$f'(\theta) = \frac{3}{2}s^2\left(\frac{1}{\sin^2\theta} - \frac{\sqrt{3}\cos\theta}{\sin^2\theta}\right) \qquad (3)$$

The derivative vanishes if, and only if,

$$1 = \sqrt{3}\cos\theta. \qquad (4)$$

Hence, $\cos\theta_0 = 1/\sqrt{3} = .57735$ and $\theta_0 = 54.7°$. Notice that the optimal angle $\theta_0$ is independent of the choice of $s$ and $h$.

It is worth comparing the result with the actual angle chosen by the bees. It is difficult to measure this angle; however, the average of all measurements does not differ significantly from the theoretical value $\theta_0 = 54.7°$. Therefore, the bees strongly prefer the optimal angle. It is unlikely that the result is due to chance. We may rather suppose that selection pressure had an effect on the angle $\theta$.

## EXERCISES

**1.** Show how equation (3) above was found.

**2.** Show how the result given in equation (4) was obtained.

# 11 Sequences and Series

A function whose domain is the set of natural numbers, such as

$$a(n) = 2n \quad \text{for } n = 1, 2, 3, 4, \ldots$$

is a **sequence.** The sequence $a(n) = 2n$ can be written by listing its *terms,* 2, 4, 6, 8, $\cdots$, $2n \cdots$. The letter $n$ is used instead of $x$ as a variable to emphasize the fact that the domain includes only natural numbers. For the same reason, $a$ is used instead of $f$ to name the function.

Sequences have many different applications; one example is the sequence of payments used to pay off a car loan or a home mortgage. (For most practical problems, the domain is a *subset* of the set of natural numbers.) This use of sequences is discussed in Section 11.2.

The indicated sum of the terms of a sequence is called a **series.** One main application of series is in finding approximate values of a function. For example, computers and calculators find values of exponential, logarithmic, and trigonometric functions by approximating these functions with polynomials. When you enter a number in your calculator and push the $e^x$ key or the ln $x$ key, the calculator actually evaluates a polynomial for your value of $x$, producing an excellent approximation for the true value of $e^x$ or ln $x$. Methods for finding these approximating polynomials, called *Taylor polynomials,* are given in this chapter.

## ▬ 11.1  GEOMETRIC SEQUENCES

In our definition of sequence we used the example $a(n) = 2n$. The range values of this sequence function,

$$a(1) = 2, \quad a(2) = 4, \quad a(3) = 6, \ldots,$$

are called the **elements** or **terms** of the sequence. Instead of writing $a(5)$ for the fifth term of a sequence, it is customary to write $a_5$; for the sequence above

$$a_5 = 10.$$

In the same way, for the sequence above, $a_1 = 2$, $a_2 = 4$, $a_8 = 16$, $a_{20} = 40$, and $a_{51} = 102$.

The symbol $a_n$ is used for the **general** or **nth term** of a sequence. For example, for the sequence 4, 7, 10, 13, 16, . . . the general term might be given by $a_n = 1 + 3n$. This formula for $a_n$ can be used to find any term of the sequence that might be needed. For example, the first three terms of the sequence are

$$a_1 = 1 + 3(1) = 4, \qquad a_2 = 1 + 3(2) = 7, \quad \text{and} \quad a_3 = 1 + 3(3) = 10.$$

Also, $a_8 = 25$ and $a_{12} = 37$.

### EXAMPLE 1

Find the first four terms of the sequence having general term $a_n = -4n + 2$.
Replace $n$, in turn, with 1, 2, 3, and 4.

$$\text{If } n = 1, \quad a_1 = -4(1) + 2 = -4 + 2 = -2.$$

$$\text{If } n = 2, \quad a_2 = -4(2) + 2 = -8 + 2 = -6.$$

$$\text{If } n = 3, \quad a_3 = -4(3) + 2 = -12 + 2 = -10.$$

$$\text{If } n = 4, \quad a_4 = -4(4) + 2 = -16 + 2 = -14.$$

The first four terms of this sequence are $-2$, $-6$, $-10$, and $-14$. ■

A sequence in which each term after the first is found by *multiplying* the preceding term by the same number is called a **geometric sequence.** The ratio of any two consecutive terms,

$$\frac{a_{n+1}}{a_n}, \quad \text{where } n \geq 1,$$

is a constant, $r$, called the **common ratio.** For example,

$$3, -6, 12, -24, 48, -96, \ldots$$

is a geometric sequence in which each term after the first is found by multiplying the preceding term by the number $-2$, the common ratio.

If $a$ is the first term of a geometric sequence and $r$ is the common ratio, then the second term is given by $a_2 = ar$ and the third term by $a_3 = a_2 r = ar^2$. Also, $a_4 = ar^3$ and $a_5 = ar^4$. These results are generalized below.

**GENERAL TERM OF A GEOMETRIC SEQUENCE**

If a geometric sequence has first term $a$ and common ratio $r$, then
$$a_n = ar^{n-1}.$$

### EXAMPLE 2

Find the indicated term for each of the following geometric sequences.

**(a)** Find $a_7$ for 6, 24, 96, 384, . . . .

Here $a = a_1 = 6$. To find $r$, choose any term except the first and divide it by the preceding term. Choosing 96 and the preceding term 24 gives

$$r = \frac{96}{24} = 4.$$

Now use the formula for $a_n$ with $n = 7$, $a = 6$, and $r = 4$.

$$a_7 = 6(4)^{7-1} = 6(4)^6 = 6(4096) = 24{,}576$$

**(b)** Find $a_6$ for 8, $-16$, 32, $-64$, 128, . . . .

In this sequence $n = 6$, $a = a_1 = 8$, and $r = -2$, so

$$a_6 = 8(-2)^{6-1} = 8(-2)^5 = 8(-32) = -256. \quad ■$$

In the next section we will need to know how to find the sum of the first *n* terms of a geometric sequence. To get a general rule for finding such a sum, begin by writing the sum $S_n$ of the first *n* terms of a geometric sequence with first term $a = a_1$ and common ratio *r* as

$$S_n = a_1 + a_2 + a_3 + \cdots + a_n.$$

Since $a_n = ar^{n-1}$, this sum also can be written as

$$S_n = a + ar + ar^2 + \cdots ar^{n-1}. \tag{1}$$

If $r = 1$, all the terms are equal to *a*, and $S_n = n \cdot a$, the correct result for this case. If $r \neq 1$, multiply both sides of equation (1) by *r*, obtaining

$$rS_n = ar + ar^2 + ar^3 + \cdots + ar^n. \tag{2}$$

Now subtract corresponding sides of equation (1) from equation (2):

$$rS_n - S_n = ar^n - a.$$

Factoring yields

$$S_n(r - 1) = a(r^n - 1),$$

and dividing by $r - 1$ on both sides gives

$$S_n = \frac{a(r^n - 1)}{r - 1}.$$

This result is summarized below.

**SUM OF THE FIRST *n* TERMS OF A GEOMETRIC SEQUENCE**

> If a geometric sequence has first term *a* and common ratio *r*, then the sum of the first *n* terms, $S_n$, is given by
>
> $$S_n = \frac{a(r^n - 1)}{r - 1}, \quad \text{where } r \neq 1.$$

### ▰ EXAMPLE 3

Find the sum of the first six terms of the geometric sequence 3, 12, 48, . . . .

Here $a = a_1 = 3$ and $r = 4$. Find $S_6$, the sum of the first six terms, by the formula above.

$$S_6 = \frac{3(4^6 - 1)}{4 - 1} \qquad \text{Let } n = 6, a = 3, r = 4$$

$$= \frac{3(4096 - 1)}{3}$$

$$= 4095 \quad ▰$$

▬ **EXAMPLE 4**

A child saves 1 cent on Sunday, 2 cents on Monday, 4 cents on Tuesday, and so on, doubling the amount each day. How much will the child have saved in one week?

This is a geometric sequence with $a = 1$ and $r = 2$. Since a week has 7 days, find $S_7$.

$$S_7 = \frac{1(2^7 - 1)}{2 - 1} = 127$$

In one week, the child will have saved 127 cents, or \$1.27. ▬

## ▬ 11.1 EXERCISES

*List the first n terms of each of the following geometric sequences.*

**1.** $a_1 = 2, r = 3, n = 4$

**2.** $a_1 = 4, r = 2, n = 5$

**3.** $a_1 = 1/2, r = 4, n = 4$

**4.** $a_1 = 2/3, r = 6, n = 3$

**5.** $a_3 = 6, a_4 = 12, n = 5$

**6.** $a_2 = 9, a_3 = 3, n = 4$

*Find $a_5$ and $a_n$ for the following geometric sequences.*

**7.** $a_1 = 4, r = 3$

**8.** $a_1 = 8, r = 4$

**9.** $a_1 = -3, r = -5$

**10.** $a_1 = -4, r = -2$

**11.** $a_2 = 3, r = 2$

**12.** $a_3 = 6, r = 3$

**13.** $a_4 = 64, r = -4$

**14.** $a_4 = 81, r = -3$

*For each sequence in Exercises 15–22 that is geometric, find r and $a_n$.*

**15.** 6, 12, 24, 48, . . .

**16.** 4, 16, 64, 256, . . .

**17.** 3/4, 3/2, 3, 6, 12, . . .

**18.** $-7, -5, -3, -1, 1, 3, \ldots$

**19.** 4, 8, $-16$, 32, 64, $-128$, . . .

**20.** 6, 8, 10, 12, 14, . . .

**21.** $-5/8, 5/12, -5/18, 5/27, \ldots$

**22.** 7/4, $-7/12$, 7/36, $-7/108$, . . .

*Find the sum of the first five terms of each geometric sequence.*

**23.** 3, 6, 12, 24, . . .

**24.** 5, 20, 80, 320, . . .

**25.** 12, $-6$, 3, $-3/2$, . . .

**26.** 18, $-3$, 1/2, $-1/12$, . . .

**27.** $a_1 = 3, r = -2$

**28.** $a_1 = -4, r = 3$

**29.** $a_1 = 8.423, r = 2.859$

**30.** $a_1 = -3.772, r = -1.553$

*Sums of the terms of a sequence are often written in* sigma *notation. A sum in the form*

$$\sum_{i=1}^{n} c(b^i)$$

*(where $\Sigma$ is the capital Greek letter* sigma*) represents the sum of the first n terms of a geometric sequence with first term $a = cb$ and common ratio b. Use the formula for the sum of the first n terms of a geometric sequence to evaluate the following sums.*

**31.** $\sum_{i=1}^{4} 3(2^i)$

**32.** $\sum_{i=1}^{3} 2(3^i)$

**33.** $\sum_{i=1}^{4} \frac{1}{2}(4^i)$

**34.** $\sum_{i=1}^{4} \frac{3}{2}(2^i)$

**35.** $\sum_{i=1}^{4} \frac{4}{3}(3^i)$

**36.** $\sum_{i=1}^{4} \frac{5}{3}(3^i)$

**37.** $\sum_{i=1}^{8} 64\left(\frac{1}{2}\right)^i$

**38.** $\sum_{i=1}^{6} 81\left(\frac{2}{3}\right)^i$

APPLICATIONS

### BUSINESS AND ECONOMICS

**Depreciation** **39.** A certain machine annually loses 30% of the value it had at the beginning of that year. If its initial value is $10,000, use a calculator to find its value at the following times.

**(a)** The end of the fifth year     **(b)** The end of the eighth year

**Income** **40.** An oil well produced $4,000,000 of income its first year. Each year thereafter, the well produced half as much income as the previous year. What is the total amount of income produced by the well in 6 yr?

**Savings** **41.** Suppose you could save $1 on January 1, $2 on January 2, $4 on January 3, and so on. What amount would you save on January 31? What would be the total amount of your savings during January?

**Depreciation** **42.** Each year a machine loses 20% of the value it had at the beginning of the year. Find the value of the machine at the end of 6 yr if it cost $100,000 new.

### LIFE SCIENCES

**Population** **43.** The population of a certain colony of bacteria increases by 10% each hour. After 5 hr, what is the percent increase in the population over the initial population?

### PHYSICAL SCIENCES

**Radioactive Decay** **44.** The half-life of a radioactive substance is the time it takes for half the substance to decay. Suppose the half-life of a substance is 3 yr and that $10^{15}$ molecules of the substance are present initially. How many molecules will be unchanged after 15 yr?

**Rotation of a Wheel** **45.** A bicycle wheel rotates 400 times per minute. If the rider removes his or her feet from the pedals, the wheel will start to slow down. Each minute, it will rotate only 3/4 as many times as in the preceding minute. How many times will the wheel rotate in the fifth minute after the rider's feet are removed from the pedals?

### GENERAL

**Thickness of a Paper Stack** **46.** A piece of paper is .008 inch thick. Suppose the paper is folded in half, so that its thickness doubles, for 12 times in a row. How thick is the final stack of paper?

## 11.2 ANNUITIES: AN APPLICATION OF SEQUENCES

Suppose $1500 is deposited at the end of each year for the next six years in an account paying 8% per year, compounded annually. How much will be in the account after six years?

Such a sequence of equal payments made at equal periods of time is called an **annuity.** If each payment is made at the end of a period, and if the frequency of payments is the same as the frequency of compounding, the annuity is called an **ordinary annuity.** The time between payments is the **payment**

**period,** and the time from the beginning of the first period to the end of the last period is called the **term** of the annuity. The **amount** of the annuity, the final sum on deposit in the account, is defined as the sum of the compound amounts of all the payments, compounded to the end of the term.

Figure 1 shows the annuity described above. To find the amount of this annuity, look at each of the $1500 payments separately. The first of these payments will produce a compound amount of

$$1500(1 + .08)^5 = 1500(1.08)^5$$

at the end of 6 years. Use 5 as the exponent instead of 6 because the money is deposited at the *end* of the first year and thus earns interest for only five years.

Term of the annuity

End of year

| 1 | 2 | 3 | 4 | 5 | 6 |

Period 1   Period 2   Period 3   Period 4   Period 5   Period 6

$1500     $1500     $1500     $1500     $1500     $1500

— The $1500 is paid at the end of the year.

**FIGURE 1**

The second payment of $1500 will produce a compound amount (at the end of 5 years) of $1500(1.08)^4$. As shown in Figure 2, the total amount of the annuity is

$$1500(1.08)^5 + 1500(1.08)^4 + 1500(1.08)^3 + 1500(1.08)^2$$
$$+ 1500(1.08)^1 + 1500. \tag{1}$$

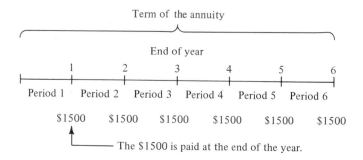

| Year | 1 | 2 | 3 | 4 | 5 | 6 |

Deposit   $1500   $1500   $1500   $1500   $1500   $1500

$1500.00
1620.00
1749.60
1889.57
2040.73
2203.99

Total   $11,003.89

**FIGURE 2**

(The last payment earns no interest.) Reversing the order of the terms, so the last term is first, shows that (1) is the sum of the terms of a geometric sequence with $a = 1500$, $r = 1.08$, and $n = 6$. Using the formula for the sum of the first $n$ terms of a geometric sequence gives

$$1500 + 1500(1.08)^1 + 1500(1.08)^2 + 1500(1.08)^3$$
$$+ 1500(1.08)^4 + 1500(1.08)^5 = \frac{1500(1.08^6 - 1)}{1.08 - 1}$$
$$\approx \frac{1500(1.58687 - 1)}{.08}$$
$$= \$11,003.89.$$

To generalize this result, suppose that $R$ dollars are paid into an account at the end of each period for $n$ periods, at a rate of interest $i$ per period. The first payment of $R$ dollars will produce a compound amount of $R(1 + i)^{n-1}$ dollars, the second payment will produce $R(1 + i)^{n-2}$ dollars, and so on; the final payment earns no interest and contributes just $R$ dollars to the total. If $S$ represents the future value (or sum) of the annuity, then

$$S = R(1 + i)^{n-1} + R(1 + i)^{n-2} + R(1 + i)^{n-3} + \cdots + R(1 + i) + R,$$

or, written in reverse order.

$$S = R + R(1 + i)^1 + R(1 + i)^2 + \cdots + R(1 + i)^{n-1}.$$

This is the sum of the first $n$ terms of the geometric sequence having first term $R$ and common ratio $1 + i$. Using the formula for the sum of the first $n$ terms of a geometric sequence gives

$$S = \frac{R[(1 + i)^n - 1]}{(1 + i) - 1} = \frac{R[(1 + i)^n - 1]}{i} = R\left[\frac{(1 + i)^n - 1}{i}\right].$$

The quantity in brackets is commonly written $s_{\overline{n}|i}$ (read "$s$-angle-$n$-at-$i$"), so that

$$S = R \cdot s_{\overline{n}|i}.$$

Values of $s_{\overline{n}|i}$ can be found by a calculator or from the "Amount of an Annuity" table in the Appendix of this book.

To check this result, go back to the example that began the section: an annuity of $1500 at the end of each year for 6 years, with interest of 8% per year compounded annually. Using the table gives the number 7.33593. Multiply this number by 1500:

$$S = 1500(7.33593) \approx 11,003.90.$$

or $11,003.90. This result differs by 1 cent from the result given in Figure 2, due to rounding errors.

Our work with annuities can be summarized as follows.

AMOUNT OF AN
ANNUITY

> The amount $S$ of an annuity of payments of $R$ dollars each, made at the end of each period for $n$ consecutive interest periods at a rate of interest $i$ per period, is given by
>
> $$S = R\left[\frac{(1 + i)^n - 1}{i}\right] \qquad \text{or} \qquad S = R \cdot s_{\overline{n}|i}.$$

### ▬ EXAMPLE 1

Reggie Theus is an athlete who feels that his playing career will last 7 more years. To prepare for his future, he deposits \$22,000 at the end of each year for 7 years in an account paying 8% compounded annually. How much will he have on deposit after 7 years?

His payments form an ordinary annuity with $R = 22,000$, $n = 7$, and $i = .08$. The amount of this annuity is (by the formula above)

$$S = 22,000\left[\frac{(1.08)^7 - 1}{.08}\right].$$

From the table, the number in brackets, $s_{\overline{7}|.08}$, is 8.92280, so that

$$S = 22,000(8.92280) = 196,301.60,$$

or \$196,301.60.  ▬

### ▬ EXAMPLE 2

Suppose \$1000 is deposited at the end of each six-month period for 5 years in an account paying 6% per year compounded semiannually. Find the amount of the annuity.

Interest of 6%/2 = 3% is earned semiannually. In 5 years there are 5 × 2 = 10 semiannual periods. We need to find $s_{\overline{10}|.03}$ from the table. Looking in the 3% column and down 10 periods gives $s_{\overline{10}|.03} = 11.46388$. Thus, the \$1000 deposits will produce a total of

$$S = 1000(11.46388) = 11,463.88,$$

or \$11,463.88.  ▬

The formula for $S$ involves the variables $R$, $i$, and $n$. The next example shows how to solve for one of these other variables.

### ▬ EXAMPLE 3

Hong Le wants to buy an expensive video camera three years from now. She plans to deposit an equal amount at the end of each quarter for three years in order to accumulate enough money to pay for the camera. Le expects the camera to cost \$2400 at that time. The bank pays 6% interest per year compounded quarterly. Find the amount of each of the twelve equal deposits she must make.

This example describes an ordinary annuity with $S = 2400$, $i = .015$ ($6\%/4 = 1.5\%$), and $n = 3 \cdot 4 = 12$ periods. The unknown here is the amount of each payment, $R$. By the formula for the amount of an annuity given above,

$$2400 = R \cdot s_{\overline{12}|.015}.$$

From the amount of annuity table in the Appendix,

$$2400 = R(13.04121)$$
$$R = 184.03, \quad \text{Dividing both sides by 13.04121}$$

or $184.03. ▬

***Sinking Fund*** A **sinking fund** is a fund set up to receive periodic payments; these periodic payments plus the interest on them are designed to produce a given total at some time in the future. As an example, a corporation might set up a sinking fund to receive money that will be needed to pay off a loan in the future. The deposits in Examples 1 and 3 form sinking funds.

▬ **EXAMPLE 4**

The Aquilars are close to retirement. They agree to sell an antique urn to a local museum for $17,000. Their tax adviser suggests that they defer receipt of this money until they retire, 5 years in the future. (At that time, they might well be in a lower tax bracket.) The museum agrees to pay them 10% per year for 5 years, and then pay them the $17,000 in a lump sum. Find the amount of each payment the museum must make into a sinking fund so that it will have the necessary $17,000 in 5 years. Assume that the museum can earn 8% compounded annually on its money and that the payments are made annually.

These payments make up an ordinary annuity. The annuity will amount to $17,000 in 5 years at 8% compounded annually, so

$$17,000 = R \cdot s_{\overline{5}|.08}$$

$$R = \frac{17,000}{s_{\overline{5}|.08}} = \frac{17,000}{5.86660} \approx 2897.76,$$

or $2897.76. If the museum deposits $2897.76 at the end of each year for 5 years in an account paying 8% compounded annually, it will have the needed $17,000. This result is shown in the following table. In other cases, the last payment might differ slightly from the others.

| Payment Number | Amount of Deposit | Interest Earned | Total in Account |
|---|---|---|---|
| 1 | $2897.76 | $0 | $2897.76 |
| 2 | $2897.76 | $231.82 | $6027.34 |
| 3 | $2897.76 | $482.19 | $9407.29 |
| 4 | $2897.76 | $752.58 | $13.057.63 |
| 5 | $2897.76 | $1044.61 | $17,000.00 ▬ |

***Present Value of an Annuity***   As shown above, if a deposit of $R$ dollars is made at the end of each period for $n$ periods, at a rate of interest $i$ per period, then the account will contain

$$S = R \cdot s_{\overline{n}|i} = R\left[\frac{(1 + i)^n - 1}{i}\right]$$

dollars after $n$ periods. Now suppose we want to find the *lump sum P* that must be deposited today at a rate of interest $i$ per period in order to produce the same amount $S$ after $n$ periods.

First recall that $P$ dollars deposited today will amount to $P(1 + i)^n$ dollars after $n$ periods at a rate of interest $i$ per period. This amount, $P(1 + i)^n$, should be the same as $S$, the amount of the annuity. Substituting $P(1 + i)^n$ for $S$ in the formula above gives

$$P(1 + i)^n = R\left[\frac{(1 + i)^n - 1}{i}\right].$$

To solve this equation for $P$, multiply both sides of the equation by $(1 + i)^{-n}$.

$$P = R(1 + i)^{-n}\left[\frac{(1 + i)^n - 1}{i}\right]$$

Use the distributive property and the fact that $(1 + i)^{-n}(1 + i)^n = 1$.

$$P = R\left[\frac{(1 + i)^{-n}(1 + i)^n - (1 + i)^{-n}}{i}\right] = R\left[\frac{1 - (1 + i)^{-n}}{i}\right]$$

The amount $P$ is called the **present value of the annuity.** The quantity in brackets is abbreviated as $a_{\overline{n}|i}$.

**PRESENT VALUE OF AN ANNUITY**

> The present value $P$ of an annuity of payments of $R$ dollars each, made at the end of each period for $n$ consecutive interest periods at a rate of interest $i$ per period is given by
>
> $$P = R\left[\frac{1 - (1 + i)^{-n}}{i}\right] \qquad \text{or} \qquad P = R \cdot a_{\overline{n}|i}.$$

Values of $a_{\overline{n}|i}$ are given in the "Present Value of an Annuity" table in the Appendix.

■■■ EXAMPLE 5

What lump sum deposited today at 6% interest compounded annually will yield the same total amount as payments of $1500 at the end of each year for 12 years, also at 6% compounded annually?

Find the present value of an annuity of $1500 per year for 12 years at 6% compounded annually. From the present value table, $a_{\overline{12}|.06} = 8.38384$, so

$$P = 1500(8.38384) = 12,575.76,$$

or $12,575.76. A lump sum deposit of $12,575.76 today at 6% compounded annually will yield the same total after 12 years as deposits of $1500 at the end of each year for 12 years at 6% compounded annually.

Check this result as follows. The compound amount in 12 years of a deposit of $12,575.76 today at 6% compounded annually can be found by the formula $A = P(1 + i)^n$. From the compound interest table, $12,575.76 will produce a total of

$$(12,575.76)(2.01220) \approx 25,304.94,$$

or $25,304.94. On the other hand, from the table for the amount of an annuity, deposits of $1500 at the end of each year for 12 years, at 6% compounded annually, give an amount of

$$1500(16.86994) = 25,304.91,$$

or $25,304.91. (The difference of 3 cents is due to rounding errors.)

In summary, there are two ways to have $25,304.91 in 12 years at 6% compounded annually—a single deposit of $12,575.76 today, or payments of $1500 at the end of each year for 12 years.    ▬

The formula above can be used if the lump sum is known and the periodic payment of the annuity must be found. The next example shows how to do this.

### ▬ EXAMPLE 6

A used car costs $6000. After a down payment of $1000, the balance will be paid off in 36 monthly payments with interest of 12% per year, compounded monthly. Find the amount of each payment.

A single lump sum payment of $5000 today would pay off the loan, so $5000 is the present value of an annuity of 36 monthly payments with interest of 12%/12 = 1% per month. We can find $R$, the amount of each payment, by using the formula

$$P = R \cdot a_{\overline{n}|i}$$

and replacing $P$ with 5000, $n$ with 36, and $i$ with .01. From the present value table, $a_{\overline{36}|.01} = 30.10751$, so

$$5000 = R(30.10751)$$
$$R \approx 166.07$$

or $166.07. Monthly payments of $166.07 each will be needed.    ▬

***Amortization***    A loan is **amortized** if both the principal and interest are paid by a sequence of equal periodic payments. In Example 6 above, a loan of $5000 at 12% interest compounded monthly could be amortized by paying $166.07 per month for 36 months, or (it turns out) $131.67 per month for 48 months.

## ▰ EXAMPLE 7

A speculator agrees to pay \$15,000 for a parcel of land. Payments will be made twice each year for 4 years at an interest rate of 12% compounded semiannually.

**(a)** Find the amount of each payment.

If the speculator were to pay \$15,000 immediately, there would be no need for any payments at all. Thus, \$15,000 is the present value of an annuity of $R$ dollars, with $2 \cdot 4 = 8$ periods, and $i = 12\%/2 = 6\% = .06$ per period. If $P$ is the present value of an annuity,

$$P = R \cdot a_{\overline{n}|i}.$$

In this example, $P = 15,000$, with

$$15,000 = R \cdot a_{\overline{8}|.06}$$

or

$$R = \frac{15,000}{a_{\overline{8}|.06}}.$$

From the present value table, $a_{\overline{8}|.06} = 6.20979$, so

$$R = \frac{15,000}{6.20979} \approx 2415.54,$$

or \$2415.54. Each payment is \$2415.54.

**(b)** Find the portion of the first payment that is applied to the reduction of the debt.

Interest is 12% per year, compounded semiannually. During the first period, the entire \$15,000 is owed. Interest on this amount for 6 months (1/2 year) is found by the formula for simple interest, $I = Prt$, so that

$$I = 15,000(.12)\left(\frac{1}{2}\right) = 900,$$

or \$900. At the end of 6 months, the speculator makes a payment of \$2415.54; since \$900 of this represents interest, a total of

$$\$2415.54 - \$900 = \$1515.54$$

is applied to the reduction of the original debt.

**(c)** Find the balance due after 6 months.

The original balance due is \$15,000. After 6 months, \$1515.54 is applied to reduction of the debt. The debt owed after 6 months is

$$\$15,000 - \$1515.54 = \$13,484.46.$$

**(d)** How much interest is owed for the second 6-month period? How much will be applied to the debt?

A total of $13,484.46 is owed for the second 6 months. Interest on this amount is

$$I = 13,484.46(.12)\left(\frac{1}{2}\right) \approx 809.07,$$

or $809.07. A payment of $2415.54 is made at the end of this period; a total of

$$\$2415.54 - \$809.07 = \$1606.47$$

is applied to the reduction of the debt.

Continuing this process gives the *amortization schedule* shown below. As the schedule shows, each payment is the same, except perhaps for a small adjustment in the final payment. Payment 0 represents the original amount of the loan.

| Payment Number | Amount of Payment | Interest for Period | Portion to Principal | Principal at End of Period |
|---|---|---|---|---|
| 0 | — | — | — | $15,000.00 |
| 1 | $2415.54 | $900 | $1515.54 | $13,484.46 |
| 2 | $2415.54 | $809.07 | $1606.47 | $11,877.99 |
| 3 | $2415.54 | $712.68 | $1702.86 | $10,175.13 |
| 4 | $2415.54 | $610.51 | $1805.03 | $8370.10 |
| 5 | $2415.54 | $502.21 | $1913.33 | $6456.77 |
| 6 | $2415.54 | $387.41 | $2028.13 | $4428.64 |
| 7 | $2415.54 | $265.72 | $2149.82 | $2278.82 |
| 8 | $2415.55 | $136.73 | $2278.82 | $0 |

### ■ EXAMPLE 8

The Millers buy a house for $74,000, with a down payment of $16,000. Interest is charged at 10.25% per year for 30 years. Find the amount of each monthly payment to amortize the loan.

Here, the present value, $P$, is 58,000 (or 74,000 − 16,000). Also, $i = .1025/12 \approx .0085416667$, and $n = 12 \cdot 30 = 360$. $R$ must be found. From the formula for the present value of an annuity,

$$58,000 = R \cdot a \,\overline{_{360}}|_{.0085416667}$$

$$= R\left[\frac{1 - (1 + .0085416667)^{-360}}{.0085416667}\right].$$

Use a financial calculator or a calculator with a $y^x$ key to get

$$58,000 = R\left[\frac{1 - .0467967507}{.0085416667}\right]$$

$$= R\left[\frac{.9532032493}{.0085416667}\right],$$

or

$$R \approx 519.74.$$

Monthly payments of $519.74 will be required to amortize the loan.

## 11.2 EXERCISES

*Find the value of each of the following ordinary annuities. (Interest is compounded annually.)*

**1.** $R = 100$, $i = .06$, $n = 10$    **2.** $R = 1000$, $i = .06$, $n = 12$    **3.** $R = 10,000$, $i = .05$, $n = 19$

**4.** $R = 100,000$, $i = .08$, $n = 23$    **5.** $R = 8500$, $i = .06$, $n = 30$    **6.** $R = 11,200$, $i = .08$, $n = 25$

*Find the value of each ordinary annuity based on the information given.*

**7.** $R = 9200$, 16% interest compounded semiannually for 7 years

**8.** $R = 3700$, 12% interest compounded semiannually for 11 years

**9.** $R = 800$, 12% interest compounded semiannually for 12 years

**10.** $R = 4600$, 16% interest compounded quarterly for 9 years

*Find the periodic payments that will amount to the given sums under the given conditions.*

**11.** $A = \$10,000$; interest is 8% compounded annually; payments are made at the end of each year for 12 years

**12.** $A = \$100,000$; interest is 16% compounded semiannually; payments are made at the end of each semiannual period for 9 years

**13.** $A = \$50,000$; interest is 12% compounded quarterly; payments are made at the end of each quarter for 8 years

*Find the present value of each of the following ordinary annuities.*

**14.** Payments of $1000 are made annually for 9 years at 8% compounded annually.

**15.** Payments of $5000 are made annually for 11 years at 6% compounded annually.

**16.** Payments of $890 are made annually for 16 years at 8% compounded annually.

**17.** Payments of $1400 are made annually for 8 years at 8% compounded annually.

**18.** Payments of $10,000 are made semiannually for 15 years at 10% compounded semiannually.

**19.** Payments of $50,000 are made quarterly for 10 years at 8% compounded quarterly.

*For Exercises 20–23, find the lump sum deposited today that will yield the same total amount as payments of $10,000 at the end of each year for 15 years, at each of the following interest rates. Interest is compounded annually.*

**20.** 4%                **21.** 5%                **22.** 6%                **23.** 8%

*Find the payments necessary to amortize the loans in Exercises 24–29.*

**24.** $1000, 8% compounded annually, 9 annual payments

**25.** $2500, 16% compounded quarterly, 6 quarterly payments

**26.** $41,000, 12% compounded semiannually, 10 semiannual payments

**27.** $90,000, 8% compounded annually, 12 annual payments

**28.** $5500, 18% compounded monthly, 24 monthly payments

**29.** $45,000, 18% compounded monthly, 36 monthly payments

## ▤ APPLICATIONS

### BUSINESS AND ECONOMICS

**Amount of an Annuity**   **30.** Pat Dillon deposits $12,000 at the end of each year for 9 years in an account paying 8% interest compounded annually. Find the final amount she will have on deposit.

**Amount of an Annuity**   **31.** Pat's brother-in-law works in a bank that pays 6% compounded annually. If she deposits her money in this bank instead of the one in Exercise 30, how much will she have in her account?

**Amount of an Annuity**   **32.** Referring to Exercises 30 and 31, how much would Pat lose over 9 years by using her brother-in-law's bank instead of the bank in Exercise 30?

**Sinking Fund**   **33.** Ray Berkowitz needs $10,000 in 8 years. What amount should he deposit at the end of each quarter at 16% compounded quarterly to accumulate the $10,000?

**Sinking Fund**   **34.** Find Berkowitz's quarterly deposit (see Exercise 33) if the money is deposited at 12% compounded quarterly.

**Sinking Fund**   **35.** Barb Silverman wants to buy a car that she estimates will cost $18,000 in 6 years. How much money must she deposit at the end of each quarter at 12% interest compounded quarterly in order to have enough in 6 years to pay for her car?

**Sinking Fund**   **36.** Harv's Meats will need to buy a new deboner machine in 4 years. At that time Harv expects the machine to cost $12,000. In order to accumulate enough money to pay for the machine, Harv decides to deposit a sum of money at the end of each six-month period in an account paying 16% compounded semiannually. How much should each payment be?

**Individual Retirement Accounts**   *Despite recent tax law changes, Individual Retirement Accounts (IRAs) are still available to some workers. Under these plans, a worker whose income does not exceed certain limits can deposit $2000 annually, with taxes deferred on the principal and interest. To attract depositors, banks have been advertising the amount that would accumulate by retirement. Suppose a person 40 years old deposits $2000 per year until age 65. Find the total in the account with the interest rates stated in Exercises 37–40. Assume semiannual compounding with payments of $1000 made at the end of each semiannual period.*

**37.** 6%                **38.** 8%                **39.** 10%                **40.** 12%

**Sinking Fund**   **41.** A firm must pay off $40,000 worth of bonds in 7 years. Find the amount of each annual payment to be made into a sinking fund, if the money earns 6% compounded annually.

**Sinking Fund**   **42.** What payment should be made in Exercise 41 if the firm can get interest of 8% compounded annually?

**Sinking Fund**   **43.** Dawn Sullivan sells some land in Nevada. She will be paid a lump sum of $60,000 in 7 years. Until then, the buyer pays 8% interest, compounded quarterly.

    **(a)** Find the amount of each quarterly interest payment.

    **(b)** The buyer sets up a sinking fund so that enough money will be present to pay off the $60,000. The buyer wants to make semiannual payments into the sinking fund; the account pays 6% compounded semiannually. Find the amount of each payment into the fund.

    **(c)** Prepare a table showing the amount in the sinking fund after each deposit.

**Sinking Fund**   **44.** Jeff Reschke bought a rare stamp for his collection. He agreed to pay a lump sum of $4000 after 5 years. Until then, he pays 6% interest, compounded semiannually.

    **(a)** Find the amount of each semiannual interest payment.

    **(b)** Reschke sets up a sinking fund so that enough money will be present to pay off the $4000. He wants to make annual payments into the fund. The account pays 8% compounded annually. Find the amount of each payment into the fund.

    **(c)** Prepare a table showing the amount in the sinking fund after each deposit.

**Present Value of an Annuity**   **45.** In his will the late Mr. Hudspeth said that each child in his family could have an annuity of $2000 at the end of each year for 9 years, or the equivalent present value. If money can be deposited at 8% compounded annually, what is the present value?

**Lottery winnings**   **46.** In the "Million Dollar Lottery," a winner is paid a million dollars at the rate of $50,000 per year for 20 years. Assume that these payments form an ordinary annuity and that the lottery managers can invest money at 6% compounded annually. Find the lump sum that the management must put away to pay off the "million dollar" winner.

**Car Payments**   **47.** Lynn Meyers buys a new car costing $6000. She agrees to make payments at the end of each month for 4 years. If she pays 12% interest, compounded monthly, what is the amount of each payment? Find the total amount of interest Meyers will pay.

**Present Value of an Annuity**   **48.** What lump sum deposited today at 5% compounded annually for 8 years will provide the same amount as $1000 deposited at the end of each year for 8 years at 6% compounded annually?

**House Payments**   *Find the monthly house payment necessary to amortize each of the loans in Exercises 49–52. You will need a financial calculator or one with a $y^x$ key.*

    **49.** $49,560 at 15.75% for 25 years       **50.** $70,892 at 14.11% for 30 years

    **51.** $53,762 at 16.45% for 30 years       **52.** $96,511 at 15.57% for 25 years

**Annuity**   **53.** When Ms. Thompson died, she left $25,000 to her husband, which he deposited at 6% compounded annually. He wants to make annual withdrawals from the account so that the money (principal and interest) is gone in exactly 8 years.

    **(a)** Find the amount of each withdrawal.

    **(b)** Find the amount of each withdrawal if the money must last 12 years.

**Annuity** **54.** The trustees of a college have accepted a gift of $150,000. The donor has directed the trustees to deposit the money in an account paying 6% per year, compounded semiannually. The trustees may withdraw an equal amount of money at the end of each 6-month period; the money must last 5 years.

**(a)** Find the amount of each withdrawal.

**(b)** Find the amount of each withdrawal if the money must last 7 years.

## FOR THE COMPUTER

**Amortization** **55.** An insurance firm pays $4000 for a new printer for its computer. It amortizes the loan for the printer in 4 annual payments at 8% compounded annually. Prepare an amortization schedule for this machine.

**Amoritzation** **56.** Certain large semitrailer trucks cost $72,000 each. Ace Trucking buys such a truck and agrees to pay for it with a loan that will be amortized with 9 semiannual payments at 16% compounded semiannually. Prepare an amortization schedule for this truck.

**Amortization** **57.** A typewriter manufacturer charges $1048 for a self-correcting typewriter. A firm of tax accountants buys 8 of these machines. They make a down payment of $1200 and agree to amortize the balance with monthly payments at 18% compounded monthly for 4 years. Prepare an amortization schedule showing the first six payments.

**Amortization** **58.** When Denise Sullivan opened her law office, she bought $14,000 worth of law books and $7200 worth of office furniture. She paid $1200 down and agreed to amortize the balance with semiannual payments for 5 years at 8% compounded semiannually. Prepare an amortization schedule for this purchase.

## ▦ 11.3 TAYLOR POLYNOMIALS AT 0

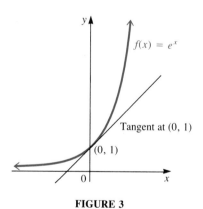

**FIGURE 3**

As mentioned in the introduction to this chapter, exponential, logarithmic, and trigonometric functions can be approximated closely by polynomials. These approximating polynomials are called **Taylor polynomials** after British mathematician Brook Taylor (1685–1731), who published his work on them in 1715.

One of the most useful nonpolynomial functions is the exponential function $f(x) = e^x$ Let us begin our discussion of Taylor polynomials by finding polynomials that approximate $e^x$ for values of $x$ close to 0. As a first approximation to $e^x$, choose the straight line that is tangent to the graph of $f(x) = e^x$ at the point (0, 1). (See Figure 3.) Since the slope of a tangent line is given by the derivative of the function, and the derivative of $f(x) = e^x$ is $f'(x) = e^x$, the slope of the tangent line at $x = 0$ is $f'(0) = e^0 = 1$.

The tangent line goes through $(0, f(0)) = (0, 1)$ and has slope $f'(0) = 1$. By the point-slope form of the equation of a line, the equation of the tangent line is

$$y - y_1 = m(x - x_1)$$
$$y - f(0) = f'(0)(x - 0)$$
$$y = f(0) + f'(0) \cdot x,$$

or, after substituting 1 for $f(0)$ and 1 for $f'(0)$,

$$y = 1 + x.$$

If $P_1(x)$ is used to represent $1 + x$, then

$$P_1(x) = 1 + x$$

is called the **Taylor polynomial of degree 1** for $f(x) = e^x$ at $x = 0$.

To be useful, $P_1(x)$ should approximate $e^x$ for values of $x$ close to 0. To check the accuracy of this approximation, compare values of $P_1(x)$ and values of $e^x$, for $x$ close to 0, in the following table.

| $x$ | $P_1(x) = 1 + x$ | $f(x) = e^x$ |
|---|---|---|
| $-1$ | 0 | .3678794412 |
| $-.1$ | .9 | .904837418 |
| $-.01$ | .99 | .9900498337 |
| $-.001$ | .999 | .9990004998 |
| 0 | 1 | 1 |
| .001 | 1.001 | 1.0010005 |
| .01 | 1.01 | 1.010050167 |
| .1 | 1.1 | 1.105170918 |
| 1 | 2 | 2.718281828 |

This table agrees with the graph in Figure 3: the polynomial $P_1(x)$ is a good approximation for $f(x) = e^x$ only when $x$ is close to 0.

For the polynomial $P_1(x)$,

$$P_1(0) = f(0) \quad \text{and} \quad P_1'(0) = f'(0);$$

that is, $P_1(x)$ and $f(x)$ are equal at 0 and their derivatives are equal at 0. A better approximation could be found with a curve. Since $P_1(x)$ is first-degree, we use a second-degree polynomial and require the second derivatives to be equal at 0. If $P_1(x)$ is written as

$$P_1(x) = a_0 + a_1 x,$$

with $a_0 = f(0)$ and $a_1 = f'(0)$, then a second-degree polynomial can be written as

$$P_2(x) = a_0 + a_1 x + a_2 x^2,$$

where $a_0 = f(0)$, $a_1 = f'(0)$, and $a_2 = f''(0)$.

Just as above, we want

$$P_2(0) = f(0) \quad \text{and} \quad P_2'(0) = f'(0),$$

but we also want

$$P_2''(0) = f''(0).$$

Since $P_2(0) = a_0$, and $f(0) = 1$, $a_0 = 1$. Also, $P_2'(x) = a_1 + 2a_2 x$, so $P_2'(0) = a_1$. Since $f'(0) = e^0 = 1$, we also must have $a_1 = 1$.

Finally, $P_2''(x) = 2a_2$. Since $f''(x) = e^x$, $P_2''(0) = 2a_2$ and $f''(0) = 1$, so that

$$2a_2 = 1$$

$$a_2 = \frac{1}{2}.$$

Our second approximation, the **Taylor polynomial of degree 2** for $f(x) = e^x$ at $x = 0$, is thus

$$P_2(x) = 1 + x + \frac{1}{2}x^2.$$

A graph of this polynomial, along with the graph of $f(x) = e^x$, is shown in Figure 4.

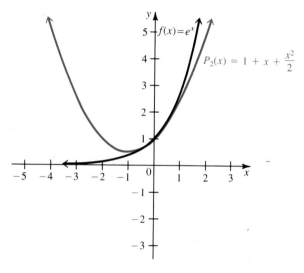

**FIGURE 4**

As above, the accuracy of this approximation can be checked with a table comparing values of $P_2(x)$ with those of $P_1(x)$ and $f(x)$.

| $x$ | $P_1(x) = 1 + x$ | $P_2(x) = 1 + x + \frac{1}{2}x^2$ | $f(x) = e^x$ |
|---|---|---|---|
| $-1$ | 0 | .5 | .3678794412 |
| $-.1$ | .9 | .905 | .904837418 |
| $-.01$ | .99 | .99005 | .9900498337 |
| $-.001$ | .999 | .9990005 | .9990004998 |
| 0 | 1 | 1 | 1 |
| .001 | 1.001 | 1.0010005 | 1.0010005 |
| .01 | 1.01 | 1.01005 | 1.010050167 |
| .1 | 1.1 | 1.105 | 1.105170918 |
| 1 | 2 | 2.5 | 2.718281828 |

While the approximations provided by $P_2(x)$ are better than those provided by $P_1(x)$, they are still accurate only for values of $x$ close to 0. A better approximation would be given by a polynomial $P_3(x)$ that equals $f(x)$ when $x = 0$, and also has the first, second, and *third* derivatives of $P_3(x)$ and $f(x) = e^x$ equal when $x = 0$. If we let

$$P_3(x) = a_0 + a_1x + a_2x^2 + a_3x^3,$$

we can find the first three derivatives:*

$$P_3^{(1)}(x) = a_1 + 2a_2x + 3a_3x^2$$
$$P_3^{(2)}(x) = 2a_2 + 6a_3x$$
$$P_3^{(3)}(x) = 6a_3.$$

Letting $x = 0$ in $P_3(x)$ and in each derivative, in turn, yields

$$P_3(0) = a_0, \qquad P_3^{(1)}(0) = a_1, \qquad P_3^{(2)}(0) = 2a_2, \qquad P_3^{(3)}(0) = 6a_3.$$

Since $f(0) = 1, f^{(1)}(0) = 1, f^{(2)}(0) = 1,$ and $f^{(3)}(0) = 1$ for $f(x) = e^x$,

$$a_0 = 1, \qquad a_1 = 1, \qquad a_2 = \frac{1}{2}, \quad \text{and} \quad a_3 = \frac{1}{6},$$

with
$$P_3(x) = 1 + x + \frac{1}{2}x^2 + \frac{1}{6}x^3.$$

A graph of $f(x) = e^x$ and $P_3(x)$ is shown in Figure 5.

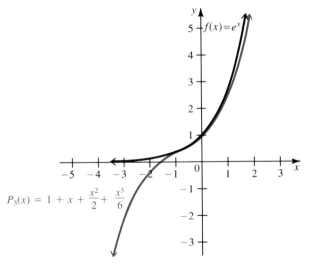

$$P_3(x) = 1 + x + \frac{x^2}{2} + \frac{x^3}{6}$$

**FIGURE 5**

---

*Recall that $P_3^{(n)}(x)$ represents the *n*th derivative of $P_3(x)$.

To generalize the work above, let $f(x) = e^x$ be approximated by

$$P_n(x) = a_0 + a_1 x + a_2 x^2 + \cdots + a_n x^n,$$

where

$$P_n(0) = f(0),$$

$$P_n^{(n)}(0) = f^{(1)}(0)$$

$$\cdot$$
$$\cdot$$
$$\cdot$$

$$P_n^{(1)}(0) = f^{(n)}(0).$$

Taking derivatives of $P_n(x)$ gives

$$P_n^{(1)}(x) = a_1 + 2a_2 x + 3a_3 x^2 + \cdots + n \cdot a_n x^{n-1}$$

$$P_n^{(2)}(x) = 2a_2 + 2 \cdot 3a_3 x + \cdots + n(n-1)a_n x^{n-2}$$

$$P_n^{(3)}(x) = 2 \cdot 3a_3 + \cdots + n(n-1)(n-2)a_n x^{n-3}$$

$$\cdot$$
$$\cdot$$
$$\cdot$$

$$P_n^{(n)}(x) = n(n-1)(n-2)(n-3) \cdots 3 \cdot 2 \cdot 1 \cdot a_n = n!a_n.$$

The symbol $n!$ (read "*n*-factorial") is used for the product

$$n(n-1)(n-2)(n-3) \cdots 3 \cdot 2 \cdot 1.$$

For example, $3! = 3 \cdot 2 \cdot 1 = 6$, while $5! = 120$. By convention, $0! = 1$.*
If we use factorials and replace $x$ with 0, the various derivatives of $P_n(x)$ become

$$P_n^{(1)}(0) = 1!a_1$$

$$P_n^{(2)}(0) = 2!a_2$$

$$P_n^{(3)}(0) = 3!a_3$$

$$\cdot$$
$$\cdot$$

$$P_n^{(n)}(0) = n!a_n.$$

---

*The symbol $n!$ for the product

$$n(n-1)(n-2)(n-3) \cdots 3 \cdot 2 \cdot 1$$

came into use during the late nineteenth century, although it was by no means the only symbol for *n*-factorial. Another popular symbol was $\underline{|n}$. The exclamation point notation has won out, probably because it is more convenient for printers of textbooks.

For every value of $n$, $f^{(n)}(0) = 1$. Setting corresponding derivatives equal gives

$$1!a_1 = 1$$
$$2!a_2 = 1$$
$$3!a_3 = 1$$
$$\vdots$$
$$n!a_n = 1,$$

from which

$$a_1 = \frac{1}{1!}, \quad a_2 = \frac{1}{2!}, \quad a_3 = \frac{1}{3!}, \quad \cdots, \quad a_n = \frac{1}{n!}.$$

Finally, the **Taylor polynomial of degree $n$** for $f(x) = e^x$ at $x = 0$ is

$$P_n(x) = 1 + \frac{1}{1!} x + \frac{1}{2!} x^2 + \frac{1}{3!} x^3 + \cdots + \frac{1}{n!} x^n.$$

Using the convention that the *zero-th derivative* of $y = f(x)$ is just $f$ itself, and using $\Sigma$ to represent a sum, we can write this result in the following way.

**TAYLOR POLYNOMIAL
FOR $f(x) = e^x$**

> The Taylor polynomial of degree $n$ for $f(x) = e^x$ at $x = 0$ is
> $$P_n(x) = \sum_{i=0}^{n} \frac{f^{(i)}(0)}{i!} x^i.$$

Taylor polynomials of degree up to 10 for $f(x) = e^x$ are shown in Figure 6.

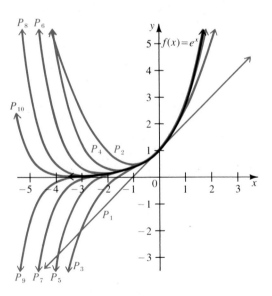

**FIGURE 6**

### ▬▬ EXAMPLE 1

Use a Taylor polynomial of degree five to approximate $e^{-.2}$.

In the work above, we found Taylor polynomials for $e^x$ at $x = 0$. As the graphs in Figure 6 suggest, these polynomials can be used to find approximate values of $e^x$ for values of $x$ "near" 0. The Taylor polynomial of degree five for $f(x) = e^x$ is

$$P_5(x) = 1 + \frac{1}{1!}x + \frac{1}{2!}x^2 + \frac{1}{3!}x^3 + \frac{1}{4!}x^4 + \frac{1}{5!}x^5.$$

Replacing $x$ with $-.2$ gives

$$1 + \frac{1}{1!}(-.2) + \frac{1}{2!}(-.2)^2 + \frac{1}{3!}(-.2)^3 + \frac{1}{4!}(-.2)^4 + \frac{1}{5!}(-.2)^5 \approx .8187307.$$

Using a calculator to evaluate $e^{-.2}$ directly gives .8187308, showing that this approximation is correct to six decimal places.  ▬▬

Generalizing our work in finding the Taylor polynomials for $f(x) = e^x$ leads to the following definition of Taylor polynomials for any appropriate function $f$.

**TAYLOR POLYNOMIAL**
**OF DEGREE $n$**

> ▬▬ Let $f$ be a function that can be differentiated $n$ times at 0. The **Taylor polynomial of degree $n$ for $f$ at 0** is
>
> $$P_n(x) = f(0) + \frac{f^{(1)}(0)}{1!}x + \frac{f^{(2)}(0)}{2!}x^2 + \frac{f^{(3)}(0)}{3!}x^3 + \cdots + \frac{f^{(n)}(0)}{n!}x^n.$$

(Because of the $f(0)$ term, a Taylor polynomial of degree $n$ has $n + 1$ terms.)

### ▬▬ EXAMPLE 2

Let $f(x) = \sqrt{x + 1}$. Find the Taylor polynomial of degree 4 at $x = 0$.

To find the Taylor polynomial of degree 4, use the first four derivatives of $f$, evaluated at 0. Arrange the work as follows.

| Derivative | Value at 0 |
|---|---|
| $f(x) = \sqrt{x + 1} = (x + 1)^{1/2}$ | $f(0) = 1$ |
| $f^{(1)}(x) = \frac{1}{2}(x + 1)^{-1/2} = \frac{1}{2(x + 1)^{1/2}}$ | $f^{(1)}(0) = \frac{1}{2}$ |
| $f^{(2)}(x) = -\frac{1}{4}(x + 1)^{-3/2} = \frac{-1}{4(x + 1)^{3/2}}$ | $f^{(2)}(0) = -\frac{1}{4}$ |
| $f^{(3)}(x) = \frac{3}{8}(x + 1)^{-5/2} = \frac{3}{8(x + 1)^{5/2}}$ | $f^{(3)}(0) = \frac{3}{8}$ |
| $f^{(4)}(x) = -\frac{15}{16}(x + 1)^{-7/2} = \frac{-15}{16(x + 1)^{7/2}}$ | $f^{(4)}(0) = -\frac{15}{16}$ |

Now use the definition of a Taylor polynomial.

$$P_4(x) = f(0) + \frac{f^{(1)}(0)}{1!}x + \frac{f^{(2)}(0)}{2!}x^2 + \frac{f^{(3)}(0)}{3!}x^3 + \frac{f^{(4)}(0)}{4!}x^4$$

$$= 1 + \frac{1/2}{1!}x + \frac{-1/4}{2!}x^2 + \frac{3/8}{3!}x^3 + \frac{-15/16}{4!}x^4$$

$$= 1 + \frac{1}{2}x - \frac{1}{8}x^2 + \frac{1}{16}x^3 - \frac{5}{128}x^4 \quad \blacksquare$$

### ■■■ EXAMPLE 3

Use the result of Example 2 to approximate $\sqrt{.9}$.

To approximate $\sqrt{.9}$, we must evaluate $f(-.1) = \sqrt{-.1 + 1} = \sqrt{.9}$. Using $P_4(x)$ from Example 2, with $x = -.1$, gives

$$P_4(-.1) = 1 + \frac{1}{2}(-.1) - \frac{1}{8}(-.1)^2 + \frac{1}{16}(-.1)^3 - \frac{5}{128}(-.1)^4$$

$$= 1 - .05 - .00125 - .0000625 - .000003906 = .948683594.$$

Thus, $\sqrt{.9} \approx .948683594$. A calculator gives a value of .9486832981 for the square root of .9. ■■

### ■■■ EXAMPLE 4

Find the Taylor polynomial of degree $n$ at $x = 0$ for

$$f(x) = \frac{1}{1 - x}.$$

As above, find the first $n$ derivatives, and evaluate each at 0.

| Derivative | Value at 0 |
|---|---|
| $f(x) = \dfrac{1}{1 - x} = (1 - x)^{-1}$ | $f(0) = 1$ |
| $f^{(1)}(x) = -1(1 - x)^{-2}(-1) = (1 - x)^{-2}$ | $f^{(1)}(0) = 1 = 1!$ |
| $f^{(2)}(x) = 2(1 - x)^{-3}$ | $f^{(2)}(0) = 2 = 2!$ |
| $f^{(3)}(x) = 3!(1 - x)^{-4}$ | $f^{(3)}(0) = 3!$ |
| $f^{(4)}(x) = 4!(1 - x)^{-5}$ | $f^{(4)}(0) = 4!$ |

Continuing this process,

$$f^{(n)}(x) = n!(1 - x)^{-1-n} \quad \text{and} \quad f^{(n)}(0) = n!.$$

By the definition of Taylor polynomials,

$$P_n(x) = 1 + \frac{1!}{1!}x + \frac{2!}{2!}x^2 + \frac{3!}{3!}x^3 + \frac{4!}{4!}x^4 + \cdots + \frac{n!}{n!}x^n$$

$$= 1 + x + x^2 + x^3 + x^4 + \cdots + x^n. \quad \blacksquare$$

## ■■ EXAMPLE 5

Use a Taylor polynomial of degree 4 to approximate $1/.98$.

Use the function $f$ from Example 4, with $x = .02$, to get

$$f(.02) = \frac{1}{1 - .02} = \frac{1}{.98}.$$

Based on the result obtained in Example 4,

$$P_4(x) = 1 + x + x^2 + x^3 + x^4,$$

with
$$\begin{aligned} P_4(.02) &= 1 + (.02) + (.02)^2 + (.02)^3 + (.02)^4 \\ &= 1 + .02 + .0004 + .000008 + .00000016 \\ &= 1.02040816. \end{aligned}$$

A calculator gives $1/.98 = 1.020408163$. ■

## ■■ EXAMPLE 6

Find a Taylor polynomial of degree 9 at $x = 0$ for $y = \sin x$.

Work as follows.

| Derivative | Value at 0 |
|---|---|
| $f(x) = \sin x$ | $f(0) = 0$ |
| $f^{(1)}(x) = \cos x$ | $f^{(1)}(0) = 1$ |
| $f^{(2)}(x) = -\sin x$ | $f^{(2)}(0) = 0$ |
| $f^{(3)}(x) = -\cos x$ | $f^{(3)}(0) = -1$ |
| $f^{(4)}(x) = \sin x$ | $f^{(4)}(0) = 0$ |

We're back where we started—these derivatives will repeat this pattern indefinitely. Using these derivatives and the definition, we can write the Taylor polynomial of degree 9 at $x = 0$ for $f(x) = \sin x$ as

$$P_9(x) = 0 + \frac{1}{1!}x + \frac{0}{2!}x^2 + \frac{(-1)}{3!}x^3 + \frac{0}{4!}x^4 + \cdots + \frac{1}{9!}x^9$$

or
$$P_9(x) = x - \frac{1}{3!}x^3 + \frac{1}{5!}x^5 - \frac{1}{7!}x^7 + \frac{1}{9!}x^9.$$

Figure 7 shows a graph of $y = \sin x$, along with the graphs of 11 Taylor polynomials at 0. ■

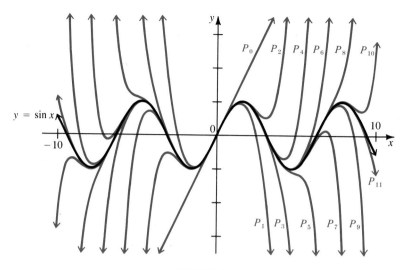

**FIGURE 7**

## 11.3 EXERCISES

*For the functions defined as follows, find the Taylor polynomials of degree 4 at 0.*

**1.** $f(x) = e^{-2x}$

**2.** $f(x) = e^{3x}$

**3.** $f(x) = e^{x+1}$

**4.** $f(x) = e^{-x}$

**5.** $f(x) = \sqrt{x + 9}$

**6.** $f(x) = \sqrt{x + 16}$

**7.** $f(x) = \sqrt[3]{x - 1}$

**8.** $f(x) = \sqrt[3]{x + 8}$

**9.** $f(x) = \sqrt[4]{x + 1}$

**10.** $f(x) = \sqrt[4]{x + 16}$

**11.** $f(x) = \ln(1 - x)$

**12.** $f(x) = \ln(1 + 2x)$

**13.** $f(x) = \ln(1 + 2x^2)$

**14.** $f(x) = \ln(1 - x^3)$

**15.** $f(x) = \cos x$

**16.** $f(x) = \sin 3x$

**17.** $f(x) = xe^{-x}$

**18.** $f(x) = x^2 e^x$

**19.** $f(x) = (9 - x)^{3/2}$

**20.** $f(x) = (1 + x)^{3/2}$

**21.** $f(x) = \dfrac{1}{1 + x}$

**22.** $f(x) = \dfrac{1}{x - 1}$

**23.** $f(x) = x \sin x$

**24.** $f(x) = x \cos x$

*Use Taylor polynomials of degree 4 at $x = 0$, found in Exercises 1–24 above, to approximate the quantities in Exercises 25–40. Round answers to four decimal places.*

**25.** $e^{-.02}$

**26.** $e^{.03}$

**27.** $e^{1.01}$

**28.** $e^{-.04}$

**29.** $\sqrt{8.9}$

**30.** $\sqrt{16.01}$

**31.** $\sqrt[3]{-1.02}$

**32.** $\sqrt[3]{7.98}$

**33.** $\sqrt[4]{1.03}$

**34.** $\sqrt[4]{15.97}$

**35.** $\ln .99$

**36.** $\ln 1.04$

**37.** $\ln 1.0002$

**38.** $\ln .999$

**39.** $\cos .01$

**40.** $\sin .03$

### ■ APPLICATIONS

*In Exercises 41–44 and 46–47, use a Taylor polynomial of degree 2 at $x = 0$ to approximate the desired value. Compare your answers with the results obtained by direct substitution.*

### BUSINESS AND ECONOMICS

**Profit**   **41.** The profit (in thousands of dollars) when $x$ thousand tons of apples are sold is

$$P(x) = \frac{20 + x^2}{50 + x}.$$

Find $P(.3)$.

**Profit**   **42.** The profit (in thousands of dollars) from the sale of $x$ thousand packages of note paper is

$$P(x) = \ln(100 + 3x).$$

Find $P(.6)$ if $\ln 100$ is given as 4.605.

**Cost**   **43.** For a certain electronic part, the cost to make the part declines as more parts are made. Suppose that the cost in dollars to manufacture the $x$th part is

$$C(x) = e^{-x/50}.$$

Find $C(5)$.

**Revenue**   **44.** Revenue from selling agricultural products often increases at a slower and slower rate as more of the products are sold. Suppose the revenue from the $x$th unit of a product is

$$R(x) = 500 \ln\left(4 + \frac{x}{50}\right).$$

Find $R(10)$ if $\ln 4$ is given as 1.386.

**Replacement Time**
**for a Part**   **45.** A book on management science gives the equation

$$\frac{e^{\lambda N}}{\lambda} - N = \frac{1}{\lambda} + k$$

to determine $N$, the time until a particular part can be expected to be replaced. ($\lambda$ and $k$ are constants for a particular machine.) To find a useful approximate value for $N$, go through the following steps.
**(a)** Find a Taylor polynomial of degree 2 at $N = 0$ for $e^{\lambda N}$.
**(b)** Substitute this polynomial into the given equation.
**(c)** Solve for $N$.

### LIFE SCIENCES

**Population Size**   **46.** The number of animals of a certain species newly introduced into an area is given by

$$500 - 250 \cos x,$$

where $x$ represents time in months since the species was introduced. Find $N(.5)$.

**Amount of a Drug in the Bloodstream**

**47.** The amount in milliliters of a certain drug in the bloodstream $x$ minutes after being administered is

$$A(x) = \frac{6x}{1 + 10x}.$$

Find $A(.05)$

**Species Survival**

**48.** According to a text on species survival, the probability $P$ that a certain species survives is given by
$$P = 1 - e^{-(1+k)P},$$

where $k$ is a constant. Use a Taylor polynomial to show that if $k$ is small, $P$ is approximately $2k$.

## 11.4  INFINITE SERIES

A repeating decimal such as .66666 . . . is really the "sum" of an infinite number of terms:

$$.66666 \ldots = .6 + .06 + .006 + .0006 + \cdots$$
$$= \frac{6}{10} + \frac{6}{10^2} + \frac{6}{10^3} + \frac{6}{10^4} + \cdots.$$

In this section we will show how an infinite number of terms can sometimes be "added" to get an infinite "sum" by a limit process. To do this, we need the following definition.

**INFINITE SERIES**

An **infinite series** is an expression of the form

$$a_1 + a_2 + a_3 + a_4 + \cdots + a_n + \cdots = \sum_{i=1}^{+\infty} a_i.$$

To find the sum $a_1 + a_2 + a_3 + a_4 + \cdots + a_n + \cdots$, first find the sum $S_n$ of the first $n$ terms, called the **$n$th partial sum.** For example,

$$S_1 = a_1$$
$$S_2 = a_1 + a_2$$
$$S_3 = a_1 + a_2 + a_3$$
$$\vdots$$
$$S_n = a_1 + a_2 + a_3 + \cdots + a_n = \sum_{i=1}^{n} a_i.$$

▬▬ EXAMPLE 1

Find the first five partial sums for the sequence

$$1, \frac{1}{2}, \frac{1}{4}, \frac{1}{8}, \frac{1}{16}, \cdots$$

By the definition of partial sum,

$$S_1 = 1$$

$$S_2 = 1 + \frac{1}{2} = \frac{3}{2}$$

$$S_3 = 1 + \frac{1}{2} + \frac{1}{4} = \frac{7}{4}$$

$$S_4 = 1 + \frac{1}{2} + \frac{1}{4} + \frac{1}{8} = \frac{15}{8}$$

$$S_5 = 1 + \frac{1}{2} + \frac{1}{4} + \frac{1}{8} + \frac{1}{16} = \frac{31}{16}. \quad ▬▬$$

As $n$ gets larger, the partial sum $S_n = a_1 + a_2 + \cdots + a_n$ includes more and more terms from the infinite series. It is thus reasonable to define the *sum* of the infinite series as $\lim\limits_{n \to +\infty} S_n$, if it exists.

**SUM OF INFINITE SERIES**

▬▬ Let $S_n = a_1 + a_2 + a_3 + \cdots + a_n$ be the $n$th partial sum for the series $a_1 + a_2 + a_3 + \cdots + a_n + \cdots$. Suppose

$$\lim_{n \to +\infty} S_n = L$$

for some real number $L$. Then $L$ is called the **sum** of the infinite series $a_1 + a_2 + a_3 + \cdots + a_n + \cdots$, and the infinite series **converges.** If no such limit exists, then the infinite series has no sum and **diverges.**

***Infinite Geometric Series*** Some good examples of convergent and divergent series come from the study of infinite geometric series, which are the sums of the terms of geometric sequences, discussed in Section 11.1. For example,

$$1, \frac{1}{2}, \frac{1}{4}, \frac{1}{8}, \frac{1}{16}, \cdots, \frac{1}{2^n}, \cdots$$

is a geometric sequence with first term $a_1 = 1$ and common ratio $r = 1/2$. The first five partial sums for this sequence were found in Example 1 above. To find $S_n$, the $n$th partial sum, use the formula given in Section 11.1: the sum of the first $n$ terms of a geometric sequence having first term $a = a_1$ and common ratio $r$ is

$$S_n = \frac{a(r^n - 1)}{r - 1}.$$

For any value of $n$, $S_n$ can be found for the geometric sequence above by using the formula with $a = 1$ and $r = 1/2$.

$$S_n = \frac{a(r^n - 1)}{r - 1} = \frac{1\left[\left(\frac{1}{2}\right)^n - 1\right]}{\frac{1}{2} - 1}$$

$$= \frac{\left(\frac{1}{2}\right)^n - 1}{-\frac{1}{2}} = -2\left[\left(\frac{1}{2}\right)^n - 1\right] = 2\left[1 - \left(\frac{1}{2}\right)^n\right]$$

As $n$ gets larger and larger, that is, as $n \to +\infty$, the value of $(1/2)^n$ gets closer and closer to 0, so that

$$\lim_{n \to +\infty} \left(\frac{1}{2}\right)^n = 0.$$

Using properties of limits from Chapter 2,

$$\lim_{n \to +\infty} S_n = \lim_{n \to +\infty} 2\left[1 - \left(\frac{1}{2}\right)^n\right] = 2[1 - 0] = 2.$$

By the definition of the sum of an infinite series.

$$1 + \frac{1}{2} + \frac{1}{4} + \frac{1}{8} + \cdots = 2,$$

and

$$1 + \frac{1}{2} + \frac{1}{4} + \frac{1}{8} + \cdots \text{ converges.}$$

To generalize from this example, start with the formula for the sum of the first $n$ terms of a geometric sequence,

$$S_n = \frac{a(r^n - 1)}{r - 1}.$$

For values of $r$ in the interval $(-1, 1)$, and only values in this interval,

$$\lim_{n \to +\infty} r^n = 0,$$

with

$$\lim_{n \to +\infty} S_n = \lim_{n \to +\infty} \frac{a(r^n - 1)}{r - 1} = \frac{a(0 - 1)}{r - 1}$$

$$= \frac{a}{1 - r}.$$

**SUM OF A GEOMETRIC SERIES**

The infinite geometric series

$$a + ar + ar^2 + ar^3 + \cdots$$

converges, if $r$ is in $(-1, 1)$, to the sum

$$\frac{a}{1 - r}.$$

The series diverges if $r$ is not in $(-1, 1)$.

### EXAMPLE 2
Show that each of the following series converges. Find the sum.

**(a)** $3 + \dfrac{3}{8} + \dfrac{3}{64} + \dfrac{3}{512} + \cdots$

   This is a geometric series, with $a = a_1 = 3$ and $r = 1/8$. Since $r$ is in $(-1, 1)$, the series converges and has sum

$$\frac{a}{1 - r} = \frac{3}{1 - 1/8} = \frac{3}{7/8} = 3 \cdot \frac{8}{7} = \frac{24}{7}.$$

**(b)** $\dfrac{3}{4} - \dfrac{9}{16} + \dfrac{27}{64} - \dfrac{81}{256} + \cdots$

   This geometric series has $a = a_1 = 3/4$ and $r = -3/4$. Since $r$ is in $(-1, 1)$, the series converges. The sum of the series is

$$\frac{3/4}{1 - (-3/4)} = \frac{3/4}{1 + 3/4} = \frac{3/4}{7/4} = \frac{3}{7}.$$

### EXAMPLE 3
The series

$$1 + 1.1 + (1.1)^2 + (1.1)^3 + (1.1)^4 + \cdots + (1.1)^{n-1} + \cdots$$

is a geometric series with common ratio $r = 1.1$. Since $r > 1$, the series diverges. (The partial sums $S_n$ will eventually exceed any preassigned number, no matter how large.)

### EXAMPLE 4
Suppose a company spends $100,000 on payroll in a certain city. Suppose also that the employees of the company reside in the city. Assume that on the average the inhabitants of this city spend 80% of their income in the same city and save the remaining 20% or spend it elsewhere. Then 80% of the original $100,000, or $(.80)(\$100,000) = \$80,000$ also will be spent in that city. An

additional 80% of this $80,000, or $64,000, will in turn be spent in the city, as will 80% of the $64,000, and so on. These amounts, $100,000, $80,000, $64,000, $51,200, and so on, form an infinite series with $a = a_1 = $100,000$ and $r = .80$. The sum of these amounts is

$$\frac{a}{1 - r} = \frac{$100,000}{1 - .80} = $500,000.$$

The original $100,000 payroll leads to a total expenditure of $500,000 in the city. In economics, the quotient of these numbers, $500,000/$100,000 = 5$, is called the *multiplier*.  ▬

▬ **EXAMPLE 5**

Retinoblastoma is a kind of cancer of the eye in children. Medical researchers believe that the disease depends on a single dominant gene, say $A$. Let $a$ be the normal gene. It is believed that a fraction $m$ of the population, $m = 2 \times 10^{-5}$, per generation will experience *mutation,* a sudden unaccountable change, of $a$ into $A$. (We exclude the possibility of back mutations of $A$ into $a$.) With medical care, approximately 70% of those affected with the disease survive. According to past data, the survivors reproduce at half the normal rate. The net fraction of affected persons who produce offspring is thus $r = 35\% = .35$. Since gene $A$ is extremely rare, practically all the affected persons are of genotype $Aa$, so that we may neglect the few individuals of genotype $AA$. Starting with 0 inherited cases in an early generation,

$m = $ fraction of population with disease due to mutation in the $n$th generation;

$mr = $ fraction of population with disease due to mutation in the $(n - 1)$st generation;

$mr^2 = $ fraction of population with disease due to mutation in the $(n - 2)$nd generation; and

$mr^n = $ fraction of population with disease due to mutation in the original generation (number 0).

The total fraction $p_n$ of the population having the disease in the $n$th generation is thus

$$p_n = m + mr + mr^2 + \cdots + mr^n.$$

As $n \rightarrow +\infty$, use the formula for the sum of an infinite geometric series to find

$$\lim_{n \to +\infty} p_n = \frac{m}{1 - r} = \frac{2 \times 10^{-5}}{1 - (.35)} \approx 3.1 \times 10^{-5}.$$

The fraction of the population having retinoblastoma will finally stabilize at about $3 \times 10^{-5}$, about 50% more than the rate of mutation.  ▬

***The Harmonic Series*** Not all infinite series are geometric. One important non-geometric series is the **harmonic series,**

$$1 + \frac{1}{2} + \frac{1}{3} + \frac{1}{4} + \frac{1}{5} + \frac{1}{6} + \cdots + \frac{1}{n} + \cdots.$$

Since this series is not geometric, none of the methods for such series can be used with it. Instead, we use the following approach: we show that the partial sums eventually become larger than any given number, no matter how large. This shows that the series diverges. Start with some partial sums:

$$S_1 = 1,$$

$$S_2 = 1 + \frac{1}{2} = \frac{3}{2},$$

$$S_4 = 1 + \frac{1}{2} + \frac{1}{3} + \frac{1}{4} > S_2 + \left(\frac{1}{4} + \frac{1}{4}\right) = S_2 + \frac{1}{2},$$

$$S_8 = 1 + \frac{1}{2} + \frac{1}{3} + \frac{1}{4} + \frac{1}{5} + \frac{1}{6} + \frac{1}{7} + \frac{1}{8} > S_4 + \left(\frac{1}{8} + \frac{1}{8} + \frac{1}{8} + \frac{1}{8}\right) = S_4 + \frac{1}{2},$$

$$S_{16} = S_8 + \left(\frac{1}{9} + \frac{1}{10} + \frac{1}{11} + \frac{1}{12} + \frac{1}{13} + \frac{1}{14} + \frac{1}{15} + \frac{1}{16}\right)$$

$$> S_8 + \left(\frac{1}{16} + \frac{1}{16} + \frac{1}{16} + \frac{1}{16} + \frac{1}{16} + \frac{1}{16} + \frac{1}{16} + \frac{1}{16}\right) = S_8 + \frac{1}{2},$$

$$S_{32} > S_{16} + \frac{1}{2},$$

$$S_{64} > S_{32} + \frac{1}{2}.$$

As these results suggest, doubling the number of terms being added increases the value of the partial sum by at least 1/2. Doing this enough times would eventually give a result larger than any preassigned number. This means that

$$\lim_{n \to +\infty} S_n \text{ does not exist}$$

for the harmonic series; thus, the harmonic series diverges.

## 11.4 EXERCISES

*In Exercises 1–14, identify the geometric series that converge. Give the sum of all convergent series.*

**1.** $12 + 6 + 3 + \frac{3}{2} + \cdots$

**2.** $1 + .9 + .81 + .729 + \cdots$

**3.** $9 + 18 + 36 + 72 + \cdots$

**4.** $3 + 9 + 27 + 81 + \cdots$

**5.** $16 + 4 + 1 + \cdots$

**6.** $81 + 27 + 9 + 3 + 1 + \cdots$

**7.** $100 + 10 + 1 + \cdots$

**8.** $128 + 64 + 32 + \cdots$

**9.** $\frac{3}{4} + \frac{3}{8} + \frac{3}{16} + \cdots$

**10.** $\dfrac{4}{5} + \dfrac{2}{5} + \dfrac{1}{5} + \cdots$

**11.** $\dfrac{1}{3} - \dfrac{2}{9} + \dfrac{4}{27} - \dfrac{8}{81} + \cdots$

**12.** $1 + \dfrac{1}{1.01} + \dfrac{1}{(1.01)^2} + \cdots$

**13.** $e - 1 + \dfrac{1}{e} - \dfrac{1}{e^2} + \cdots$

**14.** $e + e^2 + e^3 + e^4 + \cdots$

*In Exercises 15–20, the nth term of a sequence is given. Calculate the first five partial sums.*

**15.** $a_n = \dfrac{1}{n}$

**16.** $a_n = \dfrac{1}{n + 1}$

**17.** $a_n = \dfrac{1}{2n + 5}$

**18.** $a_n = \dfrac{1}{3n - 1}$

**19.** $a_n = \dfrac{1}{(n + 1)(n + 2)}$

**20.** $a_n = \dfrac{1}{(n + 3)(2n + 1)}$

## ▬ APPLICATIONS

### BUSINESS AND ECONOMICS

**Production Orders**

**21.** A sugar factory receives an order for 1000 units of sugar. The production manager thus orders production of 1000 units of sugar. He forgets, however, that the production of sugar requires some sugar (to prime the machines, for example), and so he ends up with only 900 units. He then orders an additional 100 units, and receives only 90 units. A further order for 10 units produces 9 units. Finally seeing he is wrong, the manager decides to try mathematics. He views the production process as an infinite geometric series with $a_1 = 1000$ and $r = .1$. Using this, find the number of units of sugar that he should have ordered originally.

### PHYSICAL SCIENCES

**Distance**

**22.** Mitzi drops a ball from a height of 10 m and notices that on each bounce the ball returns to about 3/4 of its previous height. About how far will the ball travel before it comes to rest?

**Rotation of a Wheel**

**23.** After a person pedaling a bicycle removes his or her feet from the pedals, the wheel rotates 400 times the first minute. As it continues to slow down, each minute it rotates only 3/4 as many times as in the previous minute. How many times will the wheel rotate before coming to a complete stop?

**Pendulum Arc Length**

**24.** A pendulum bob swings through an arc 40 cm long on its first swing. Each swing thereafter, it swings only 80% as far as on the previous swing. How far will it swing altogether before coming to a complete stop?

### GENERAL

**Perimeter**

**25.** A sequence of equilateral triangles is constructed as follows: The first triangle has sides 2 m in length. To get the next triangle, midpoints of the sides of the previous triangle are connected. If this process could be continued indefinitely, what would be the total perimeter of all the triangles?

**Area**

**26.** What would be the total area of all the triangles of Exercise 25, disregarding the overlapping?

# 11.5 TAYLOR SERIES

The sum of the infinite geometric series having first term $a$ and common ratio $r$ is

$$\frac{a}{1 - r} \quad \text{for } r \text{ in } (-1, 1).$$

If the first term of an infinite geometric series is $a = 1$ and the common ratio is $x$, then the series is written

$$1 + x + x^2 + x^3 + x^4 + \cdots + x^{n-1} + \cdots.$$

If $x$ is in $(-1, 1)$, then by the formula for the sum of an infinite geometric series, the sum of this series is

$$\frac{1}{1 - x}.$$

That is,

$$\frac{1}{1 - x} = 1 + x + x^2 + x^3 + x^4 + \cdots \quad \text{for } x \text{ in } (-1, 1).$$

The interval $(-1, 1)$ is called the **interval of convergence** for the series. This series is not an approximation for $1/(1 - x)$; the series is actually *equal to* $1/(1 - x)$ for any $x$ in $(-1, 1)$.

Earlier in this chapter, we found that the Taylor polynomial of degree $n$ at $x = 0$ for $1/(1 - x)$ is

$$P_n(x) = 1 + x + x^2 + x^3 + x^4 + \cdots + x^n.$$

Since the series given above for $1/(1 - x)$ is just an extension of this Taylor polynomial, it seems natural to call the series a *Taylor series*.

**TAYLOR SERIES**

> If all derivatives of a function $f$ exist at 0, then the **Taylor series for $f$ at 0** is defined to be
>
> $$f(0) + f^{(1)}(0)x + \frac{f^{(2)}(0)}{2!} x^2 + \frac{f^{(3)}(0)}{3!} x^3 + \cdots$$
>
> for all values of $x$ for which the series converges.

The particular Taylor series at 0 is also called a **Maclaurin series.** Scotsman Colin Maclaurin (1698–1746) used this series in his work *Treatise of Fluxions,* published in 1742.

■ EXAMPLE 1

Find the Taylor series for $f(x) = e^x$ at 0.
   Work as in Section 3 of this chapter.

| Derivative | Value at 0 |
|---|---|
| $f(x) = e^x$ | $f(0) = 1$ |
| $f^{(1)}(x) = e^x$ | $f^{(1)}(0) = 1$ |
| $f^{(2)}(x) = e^x$ | $f^{(2)}(0) = 1$ |
| $f^{(3)}(x) = e^x$ | $f^{(3)}(0) = 1$ |
| $\vdots$ | $\vdots$ |
| $f^{(n)}(x) = e^x$ | $f^{(n)}(0) = 1$ |

Using the definition given in this section, the Taylor series for $f(x) = e^x$ is

$$1 + x + \frac{1}{2!}x^2 + \frac{1}{3!}x^3 + \frac{1}{4!}x^4 + \cdots + \frac{1}{n!}x^n + \cdots .$$

While we cannot prove it here, the interval of convergence is $(-\infty, +\infty)$. ■

The process for finding the interval of convergence for a given Taylor series is discussed in advanced calculus courses. The most common Taylor series are listed below, along with their intervals of convergence. (As is customary, these series are written so that the initial term is a "*zero-th*" term.)

**COMMON TAYLOR SERIES**

| $f(x)$ | Taylor Series | Interval of Convergence |
|---|---|---|
| ✳ $e^x$ | $1 + x + \dfrac{1}{2!}x^2 + \dfrac{1}{3!}x^3 + \cdots + \dfrac{1}{n!}x_n + \cdots$ | $(-\infty, +\infty)$ |
| $\ln(1 + x)$ | $x - \dfrac{x^2}{2} + \dfrac{x^3}{3} - \dfrac{x^4}{4} + \cdots + \dfrac{(-1)^n x^{n+1}}{n + 1} + \cdots$ | $(-1, 1]$ |
| $\dfrac{1}{1 - x}$ | $1 + x + x^2 + x^3 + \cdots + x^n + \cdots$ | $(-1, 1)$ |
| $\sin x$ | $x - \dfrac{x^3}{3!} + \dfrac{x^5}{5!} - \dfrac{x^7}{7!} + \cdots + \dfrac{(-1)^n x^{2n+1}}{(2n + 1)!} + \cdots$ | $(-\infty, +\infty)$ |
| $\cos x$ | $1 - \dfrac{x^2}{2!} + \dfrac{x^4}{4!} - \dfrac{x^6}{6!} + \cdots + \dfrac{(-1)^n x^{2n}}{(2n)!} + \cdots$ | $(-\infty, +\infty)$ |
| $\tan^{-1} x$ | $x - \dfrac{x^3}{3} + \dfrac{x^5}{5} - \dfrac{x^7}{7} + \cdots + \dfrac{(-1)^n x^{2n+1}}{2n + 1} + \cdots$ | $[-1, 1]$ |

***Operations on Taylor Series***    The first $n$ terms of a Taylor series form a polynomial. Because of this, we would expect many of the operations on polynomials to generalize to Taylor series; some properties of series concerning these operations are given in the following theorems.

**OPERATIONS ON
TAYLOR SERIES**

Let $f$ and $g$ be functions having Taylor series with

$$f(x) = a_0 + a_1x + a_2x^2 + a_3x^3 + \cdots + a_nx^n + \cdots$$

and

$$g(x) = b_0 + b_1x + b_2x^2 + b_3x^3 + \cdots + b_nx^n + \cdots.$$

1. The Taylor series for $f + g$ is

$$(a_0 + b_0) + (a_1 + b_1)x + (a_2 + b_2)x^2 + \cdots$$
$$+ (a_n + b_n)x^n + \cdots,$$

for all $x$ in the interval of convergence of both $f$ and $g$. (Convergent series may be added term by term.)

2. For a real number $c$, the Taylor series for $c \cdot f(x)$ is

$$c \cdot a_0 + c \cdot a_1x + c \cdot a_2x^2 + \cdots + c \cdot a_nx^n + \cdots,$$

for all $x$ in the interval of convergence of $f$.

3. For any positive integer $k$, the Taylor series for $x^k \cdot f(x)$ is

$$a_0x^k + a_1x^k \cdot x + a_2x^k \cdot x^2 + \cdots + a_nx^k \cdot x^n + \cdots$$
$$= a_0x^k + a_1x^{k+1} + a_2x^{k+2} + \cdots + a_nx^{k+n} + \cdots,$$

for all $x$ in the interval of convergence of $f$.

These properties follow from the properties of derivatives and from the definition of a Taylor series.

### ■ EXAMPLE 2

Find Taylor series for the following functions.

**(a)** $f(x) = 5e^x$

Use property (2) above, with $c = 5$, along with the Taylor series for $f(x) = e^x$ given earlier. The Taylor series for $5e^x$ is

$$5 \cdot 1 + 5 \cdot x + 5 \cdot \frac{1}{2!}x^2 + 5 \cdot \frac{1}{3!}x^3 + \cdots + 5 \cdot \frac{1}{n!}x^n + \cdots$$

$$= 5 + 5x + \frac{5}{2!}x^2 + \frac{5}{3!}x^3 + \cdots + \frac{5}{n!}x^n + \cdots$$

for all $x$ in $(-\infty, +\infty)$.

**(b)** $f(x) = x^3 \ln(1 + x)$

Use the Taylor series for $\ln(1 + x)$ with property (3). With $k = 3$, this gives the Taylor series for $x^3 \ln(1 + x)$.

$$x^3 \cdot x - x^3 \cdot \frac{1}{2}x^2 + x^3 \cdot \frac{1}{3}x^3 - x^3 \cdot \frac{1}{4}x^4 + \cdots + \frac{x^3(-1)^n \cdot x^{n+1}}{n+1} + \cdots$$

$$= x^4 - \frac{1}{2}x^5 + \frac{1}{3}x^6 - \frac{1}{4}x^7 + \cdots + \frac{(-1)^nx^{4+n}}{n+1} + \cdots \quad ■$$

To see why the properties are so useful, try writing the Taylor series for $x^3 \ln(1 + x)$ directly from the definition of a Taylor series.

The final property of Taylor series is perhaps the most useful of all.

**COMPOSITION WITH TAYLOR SERIES**

Let a function $f$ have a Taylor series
$$f(x) = a_0 + a_1 x + a_2 x^2 + a_3 x^3 + \cdots + a_n x^n + \cdots.$$
Then replacing each $x$ with $g(x)$ gives the Taylor series for $f[g(x)]$:
$$a_0 + a_1 g(x) + a_2[g(x)]^2 + a_3[g(x)]^3 + \cdots + a_n[g(x)]^n + \cdots.$$
The interval of convergence of this new series may be different from that of the first series.

### EXAMPLE 3

Find Taylor series for the following functions.

**(a)** $f(x) = e^{-x^2/2}$

We know that the Taylor series for $e^x$ is

$$1 + x + \frac{1}{2!} x^2 + \frac{1}{3!} x^3 + \cdots + \frac{1}{n!} x^n + \cdots$$

for all $x$ in $(-\infty, +\infty)$. Use the composition property above and replace each $x$ with $-x^2/2$ to get the Taylor series for $e^{-x^2/2}$.

$$1 + \left(-\frac{x^2}{2}\right) + \frac{1}{2!}\left(-\frac{x^2}{2}\right)^2 + \frac{1}{3!}\left(-\frac{x^2}{2}\right)^3 + \cdots + \frac{1}{n!}\left(-\frac{x^2}{2}\right)^n + \cdots$$

$$= 1 - \frac{1}{2} x^2 + \frac{1}{2!2^2} x^4 - \frac{1}{3!2^3} x^6 + \cdots + \frac{(-1)^n}{n!2^n} x^{2n} + \cdots$$

The Taylor series for $e^{-x^2/2}$ has the same interval of convergence, $(-\infty, +\infty)$, as the Taylor series for $e^x$.

**(b)** $f(x) = \dfrac{1}{1 + 4x}$

Write $1/(1 + 4x)$ as

$$\frac{1}{1 + 4x} = \frac{1}{1 - (-4x)},$$

which is $1/(1 - x)$ with $x$ replaced with $-4x$. Start with

$$\frac{1}{1 - x} = 1 + x + x^2 + x^3 + \cdots + x^n + \cdots,$$

which converges for $x$ in $(-1, 1)$, and replace each $x$ with $-4x$ to get

$$\frac{1}{1 + 4x} = \frac{1}{1 - (-4x)}$$
$$= 1 + (-4x) + (-4x)^2 + (-4x)^3 + \cdots + (-4x)^n + \cdots$$
$$= 1 - 4x + 16x^2 - 64x^3 + \cdots + (-1)^n 4^n x^n + \cdots.$$

The interval of convergence of the original series is $(-1, 1)$, or $-1 < x < 1$. Replacing $x$ with $-4x$ gives

$$-1 < -4x < 1 \qquad \text{or} \qquad \frac{1}{4} > x > -\frac{1}{4},$$

so that the interval of convergence of the new series is $(-1/4, 1/4)$.

**(c)** $f(x) = \dfrac{1}{2 - x^2}$

This function most nearly matches $1/(1 - x)$. To get 1 in the denominator, instead of 2, divide numerator and denominator by 2.

$$\frac{1}{2 - x^2} = \frac{1/2}{1 - x^2/2}$$

Thus, we can find the Taylor series for $1/(2 - x^2)$ by starting with the Taylor series for $1/(1 - x)$, multiplying each term by $1/2$, and replacing each $x$ with $x^2/2$.

$$\frac{1}{2 - x^2} = \frac{1/2}{1 - x^2/2}$$
$$= \frac{1}{2} \cdot 1 + \frac{1}{2} \cdot \left(\frac{x^2}{2}\right) + \frac{1}{2}\left(\frac{x^2}{2}\right)^2 + \frac{1}{2}\left(\frac{x^2}{2}\right)^3 + \cdots + \frac{1}{2}\left(\frac{x^2}{2}\right)^n + \cdots$$
$$= \frac{1}{2} + \frac{x^2}{4} + \frac{x^4}{8} + \frac{x^6}{16} + \cdots + \frac{x^{2n}}{2^{n+1}} + \cdots$$

The Taylor series for $1/(1 - x)$ is valid when $-1 < x < 1$. Replacing $x$ with $x^2/2$ gives

$$-1 < \frac{x^2}{2} < 1 \qquad \text{or} \qquad -2 < x^2 < 2.$$

This inequality is satisfied by any $x$ in the interval $(-\sqrt{2}, \sqrt{2})$. ■

While we do not go into detail in this book, the Taylor series we discuss may be differentiated and integrated term by term. This result is used in the next example.

### ■ EXAMPLE 4

The standard normal curve of statistics is given by

$$f(x) = \frac{1}{\sqrt{2\pi}} e^{-x^2/2}.$$

Find the area bounded by this curve and the lines $x = 0$, $x = 1$, and the $x$-axis.

The desired area is shown in Figure 8. By earlier methods, this area is given by the definite integral

$$\int_0^1 \frac{1}{\sqrt{2\pi}} e^{-x^2/2}\, dx = \frac{1}{\sqrt{2\pi}} \int_0^1 e^{-x^2/2}\, dx.$$

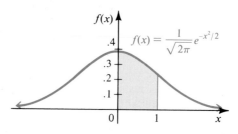

**FIGURE 8**

This integral cannot be evaluated by any method we have used, but recall that Example 3 gave the Taylor series for $f(x) = e^{-x^2/2}$:

$$e^{-x^2/2} = 1 - \frac{1}{2}x^2 + \frac{1}{2!2^2}x^4 - \frac{1}{3!2^3}x^6 + \cdots + \frac{(-1)^n}{n!2^n}x^{2n} + \cdots.$$

An approximation to $\int e^{-x^2/2}\, dx$ can be found by integrating this series term by term. Using, say, the first six terms of this series gives

$$\frac{1}{\sqrt{2\pi}} \int_0^1 e^{-x^2/2}\, dx \approx \frac{1}{\sqrt{2\pi}} \int_0^1 \left( 1 - \frac{1}{2}x^2 + \frac{1}{2!2^2}x^4 - \frac{1}{3!2^3}x^6 + \frac{1}{4!2^4}x^8 - \frac{1}{5!2^5}x^{10} \right) dx$$

$$= \frac{1}{\sqrt{2\pi}} \int_0^1 \left( 1 - \frac{1}{2}x^2 + \frac{1}{8}x^4 - \frac{1}{48}x^6 + \frac{1}{384}x^8 - \frac{1}{3840}x^{10} \right) dx$$

$$= \frac{1}{\sqrt{2\pi}} \left[ x - \frac{1}{6}x^3 + \frac{1}{40}x^5 - \frac{1}{336}x^7 + \frac{1}{3456}x^9 - \frac{1}{42,240}x^{11} \right]\Bigg|_0^1$$

$$= \frac{1}{\sqrt{2\pi}} \left[ 1 - \frac{1}{6} + \frac{1}{40} - \frac{1}{336} + \frac{1}{3456} - \frac{1}{42,240} - 0 \right]$$

$$\approx \frac{1}{\sqrt{2(3.1416)}} (.855623)$$

$$\approx .3413.$$

This result agrees with the value .3413 obtained from the normal curve table in the Appendix. We could have obtained a more accurate result by using more terms of the Taylor series. ▬

If $P$ dollars are deposited in an account that pays an interest rate $i$ compounded annually, then the compound interest formula

$$A = P(1 + i)^n \qquad (*)$$

gives the compound amount present in the account after $n$ years. The same formula can serve as a model for population growth (at least initially) or any other geometric growth. Based on this formula, how long will it take for $P$ to double?

Setting $A = 2P$ in equation (*) gives

$$2P = P(1 + i)^n.$$

Dividing both sides by $P$ and taking the natural logarithm of each side leads to

$$\ln 2 = n \ln(1 + i)$$

or

$$n = \frac{\ln 2}{\ln (1 + i)},$$

which gives the **doubling time** in years.

▬▬ EXAMPLE 5

Find the doubling time for an investment at each of the following interest rates.

**(a)** 9%

By the formula for doubling time, at an interest rate of 9%, money will double in

$$\frac{\ln 2}{\ln (1 + .09)} \approx 8 \text{ years.}$$

**(b)** 1%

At an interest rate of 1%, money will double in

$$\frac{\ln 2}{\ln (1 + .01)} \approx 70 \text{ years.} \quad ▬$$

Two other convenient formulas for the doubling time can be derived using a Taylor series. As shown in the list of common Taylor series,

$$\ln (1 + x) = x - \frac{x^2}{2} + \frac{x^3}{3} - \frac{x^4}{4} + \ldots \text{ for } x \text{ in } (-1, 1].$$

Further,

$$\ln(1 + x) = x - \frac{x^2}{2} + \frac{x^3}{3} - \frac{x^4}{4} + \cdots < x \qquad \text{for } x \text{ in } (-1, 1].$$

Therefore, for $0 < i < 1$, the doubling time,

$$n = \frac{\ln 2}{\ln (1 + i)} = \frac{\ln 2}{i - \dfrac{i^2}{2} + \dfrac{i^3}{3} - \dfrac{i^4}{4} + \cdots},$$

is just slightly larger than the quotient

$$\frac{\ln 2}{i} \approx \frac{.693}{i} = \frac{69.3}{100i}.$$

Since the actual value of

$$\ln (1 + i) = i - \frac{i^2}{2} + \frac{i^3}{3} - \frac{i^4}{4} + \cdots$$

is slightly smaller than $i$ for $0 < i < 1$, the quotients

$$\frac{70}{100i} \quad \text{and} \quad \frac{72}{100i}$$

give good approximations to the doubling time.

These two approximations are summarized below.

RULE OF 70
AND
RULE OF 72

**Rule of 70**   If a quantity is increasing at a constant rate $i$ compounded annually, where $.001 \le i \le .05$,

$$\text{Doubling Time} \approx \frac{70}{100i} \text{ years.}$$

**Rule of 72**   If a quantity is increasing at a constant rate $i$ compounded annually, where $.05 \le i \le .12$, then

$$\text{Doubling Time} \approx \frac{72}{100i} \text{ years.}$$

The rule of 70 is used by demographers because populations usually grow at rates less than 5%. The rule of 72 is preferred by economists and investors since money frequently grows at a rate between 5 and 12 percent.

### EXAMPLE 6

The rule of 70 predicts that at a growth rate of 1%, a population will double in 70 years, in agreement with Example 5(b). The rule of 72 predicts that at an interest rate of 9%, money will double in 8 years, in agreement with Example 5(a).

The following table gives the actual doubling time $n$ in years for various growth rates $i$, together with the approximate doubling times given by the rules of 70 and 72.

| $i$ | .001 | .005 | .01 | .02 | .03 | .04 | .05 | .06 | .07 | .08 | .09 | .10 | .11 | .12 |
|---|---|---|---|---|---|---|---|---|---|---|---|---|---|---|
| $n$ | 693 | 139 | 69.7 | 35.0 | 23.4 | 17.7 | 14.2 | 11.9 | 10.2 | 9.0 | 8.0 | 7.3 | 6.6 | 6.1 |
| $\dfrac{70}{100i}$ | 700 | 140 | 70 | 35 | 23.3 | 17.5 | 14 | 11.7 | 10 | 8.8 | 7.8 | 7.0 | 6.4 | 5.8 |
| $\dfrac{72}{100i}$ | 720 | 144 | 72 | 36 | 24 | 18 | 14.4 | 12 | 10.3 | 9 | 8 | 7.2 | 6.5 | 6 |

The last row in the table is particularly easy to compute because 72 has so many integral divisors. Therefore, the rule of 72 is frequently used by economists and investors for any interest rate $i$.

It can be shown that the rule of 70 will give the doubling time with an error of 2% or less if $.001 \le i \le .05$, and the rule of 72 will give the doubling time with a 2% error or less if $.05 \le i \le .12$. The table above shows the accuracy of the approximations, and the graph in Figure 9 shows that the graphs of

$$\frac{\ln 2}{\ln (1 + i)}, \quad \frac{70}{100i}, \quad \text{and} \quad \frac{72}{100i}$$

are virtually indistinguishable over the domains indicated above.  ■

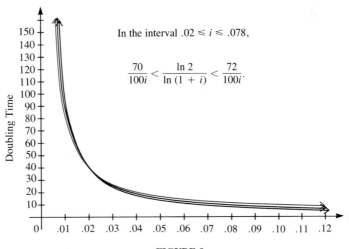

In the interval $.02 \le i \le .078$,

$$\frac{70}{100i} < \frac{\ln 2}{\ln (1 + i)} < \frac{72}{100i}.$$

**FIGURE 9**

## 11.5 EXERCISES

*Use the Taylor series given in this section to find the Taylor series for the functions defined as follows. (It is not necessary to give the general term for Exercises 19–20.) Give the interval of convergence for each series.*

**1.** $f(x) = \dfrac{5}{2 - x}$

**2.** $f(x) = \dfrac{-3}{4 - x}$

**3.** $f(x) = \dfrac{8x}{1 + 3x}$

**4.** $f(x) = \dfrac{7x}{1 + 2x}$

**5.** $f(x) = \dfrac{x^2}{4 - x}$

**6.** $f(x) = \dfrac{9x^4}{1 - x}$

**7.** $f(x) = \ln(1 + 4x)$

**8.** $f(x) = \ln\left(1 - \dfrac{x}{2}\right)$

**9.** $f(x) = e^{4x^2}$

**10.** $f(x) = e^{-3x^2}$

**11.** $f(x) = x^3 e^{-x}$

**12.** $f(x) = x^5 e^x$

**13.** $f(x) = \dfrac{2}{1 + x^2}$

**14.** $f(x) = \dfrac{6}{3 + x^2}$

**15.** $f(x) = \dfrac{e^x + e^{-x}}{2}$

**16.** $f(x) = \dfrac{e^x - e^{-x}}{2}$

**17.** $f(x) = \ln(1 + 2x^4)$

**18.** $f(x) = \ln(1 - 5x^2)$

**19.** $f(x) = \sin x + \cos x$

**20.** $f(x) = \cos x - \sin x$

**21.** $f(x) = \cos x^3$

**22.** $f(x) = \sin 2x^5$

**23.** $f(x) = \tan^{-1} x^2$

**24.** $f(x) = \tan^{-1} 8x^3$

**25.** Use the fact that

$$\frac{1 + x}{1 - x} = \frac{1}{1 - x} + \frac{x}{1 - x}$$

to find a Taylor series for $(1 + x)/(1 - x)$.

**26.** By properties of logarithms,

$$\ln\left(\frac{1 + x}{1 - x}\right) = \ln(1 + x) - \ln(1 - x).$$

Use this to find a Taylor series for $\ln[(1 + x)/(1 - x)]$.

**27.** Use the Taylor series for $e^x$ to suggest that

$$e^x \approx 1 + x + \frac{x^2}{2}$$

for all positive $x$.

**28.** Use the Taylor series for $e^{-x}$ to suggest that

$$e^{-x} \approx 1 - x + \frac{x^2}{2}$$

for all positive $x$.

*Use the method in Example 4 (with five terms of the appropriate Taylor series) to approximate the areas of the following regions.*

**29.** The region bounded by $f(x) = e^{x^2}$, $x = 0$, $x = 1/2$, and the $x$-axis

**30.** The region bounded by $f(x) = 1/(1 - x^2)$, $x = 0$, $x = 1/2$, and the $x$-axis

**31.** The region bounded by $f(x) = 1/(1 - \sqrt{x})$, $x = 1/4$, $x = 1/2$, and the $x$-axis

**32.** The region bounded by $f(x) = e^{\sqrt{x}}$, $x = 0$, $x = 1$, and the $x$-axis

As mentioned in Example 4, the equation of the standard normal curve is

$$f(x) = \frac{1}{\sqrt{2\pi}} e^{-x^2/2}.$$

Use the method in Example 4 with five terms of the Taylor series to approximate the area of the region bounded by the normal curve, the x-axis, x = 0, and the values of x in Exercises 33 and 34.

**33.** x = .5                            **34.** x = .75

Scientific calculators and computers usually determine the values of trigonometric functions by calculating polynomials. The polynomials

$$f(x) = x - \frac{1}{6}x^3 + \frac{1}{120}x^5 \quad and \quad g(x) = 1 - \frac{1}{2}x^2 + \frac{1}{24}x^4$$

give fairly good approximations for sin x and cos x, respectively, when x is close to zero. Use a calculator in radian mode to find each of the following. Compare this result to the value obtained by using the trigonometric function key.

**35.** $\sin .01 \approx f(.01)$          **36.** $\cos .01 \approx g(.01)$

**37.** $\cos .05 \approx g(.05)$          **38.** $\sin .05 \approx f(.05)$

## ▤ 11.6 L'HÔPITAL'S RULE

The study of calculus in this book began with a discussion of *limits*. For example,

$$\lim_{x \to 1} \frac{x^2 + 1}{x + 4} = \frac{2}{5},$$

or, in words, as x gets closer and closer to 1, the quotient $(x^2 + 1)/(x + 4)$ gets closer and closer to 2/5. In this section we will use derivatives to find limits of quotients of functions that could not easily be found before.

If we try to find

$$\lim_{x \to 1} \frac{x^2 - 1}{x - 1}$$

by evaluating the numerator and denominator at x = 1, we get

$$\frac{1^2 - 1}{1 - 1} = \frac{0}{0}.$$

an **indeterminate form.** Any attempt to assign a value to 0/0 leads to a meaningless result. The limit exists, however; as shown earlier, it is found by factoring.

$$\lim_{x \to 1} \frac{x^2 - 1}{x - 1} = \lim_{x \to 1} \frac{(x + 1)(x - 1)}{x - 1} = \lim_{x \to 1} (x + 1) = 2$$

As a second example,

$$\lim_{x \to 1} \frac{\ln x}{(x - 1)^2}$$

also leads to the indeterminate form 0/0. Selecting values of $x$ close to 1 and using a calculator gives the following table.

| $x$ | .99 | .999 | .9999 | 1.0001 | 1.001 | 1.01 |
|---|---|---|---|---|---|---|
| $\dfrac{\ln x}{(x - 1)^2}$ | $-100.5$ | $-1000.5$ | $-10{,}000.5$ | $9999.5$ | $999.5$ | $99.5$ |

As this table suggests, $\lim\limits_{x \to 1} (\ln x)/(x - 1)^2$ does not exist.

In the first example, trying to find $\lim\limits_{x \to 1} (x^2 - 1)/(x - 1)$ by evaluating the expression at $x = 1$ led to the indeterminate form 0/0, but factoring the expression led to the actual limit, 2. Evaluating $\lim\limits_{x \to 1} (\ln x)/(x - 1)^2$ in the second example led to the indeterminate form 0/0, but using a table of values showed that this limit did not exist. **L'Hôpital's rule** gives a quicker way to decide whether a quotient with the indeterminate form 0/0 has a limit.

**L'HÔPITAL'S RULE**

Let $f$ and $g$ be functions and let $a$ be a real number such that

$$\mathbf{\lim_{x \to a} f(x) = 0} \qquad \text{and} \qquad \mathbf{\lim_{x \to a} g(x) = 0.}$$

Let $f$ and $g$ have derivatives that exist at each point in some open interval containing $a$.

If $\lim\limits_{x \to a} \dfrac{f'(x)}{g'(x)} = L$, then $\lim\limits_{x \to a} \dfrac{f(x)}{g(x)} = L.$

If $\lim\limits_{x \to a} \dfrac{f'(x)}{g'(x)}$ does not exist because $\left| \dfrac{f'(x)}{g'(x)} \right|$ becomes large without bound for values of $x$ near $a$, then $\lim\limits_{x \to a} \dfrac{f(x)}{g(x)}$ also does not exist.

A partial proof of this rule is given at the end of this section. L'Hôpital's rule is another example of a mathematical misnomer. Although named after the Marquis de L'Hôpital (1661–1704), it was actually developed by Johann Bernoulli in a textbook published in 1696. L'Hôpital was a student of Bernoulli and published, with a financial arrangement, the works of his teacher under his own name.

### EXAMPLE 1

Find $\lim\limits_{x \to 1} \dfrac{x^2 - 1}{x - 1}$.

It is very important to first make sure that the conditions of L'Hôpital's rule are satisfied. Here

$$\lim_{x \to 1} (x^2 - 1) = 0 \quad \text{and} \quad \lim_{x \to 1} (x - 1) = 0.$$

Since the limits of both numerator and denominator are 0, the rule does apply. Now take the derivatives of both numerator and denominator separately. (Do *not* use the quotient rule for derivatives.)

$$\text{If } f(x) = x^2 - 1, \text{ then } f'(x) = 2x;$$
$$\text{if } g(x) = x - 1, \text{ then } g'(x) = 1.$$

Find the limit of the quotient of the derivatives.

$$\lim_{x \to 1} \frac{f'(x)}{g'(x)} = \lim_{x \to 1} \frac{2x}{1} = \lim_{x \to 1} 2x = 2$$

By L'Hôpital's rule, this result is the desired limit:

$$\lim_{x \to 1} \frac{x^2 - 1}{x - 1} = 2. \quad \blacksquare$$

### EXAMPLE 2

Find $\lim\limits_{x \to 1} \dfrac{\ln x}{(x - 1)^2}$.

Make sure that L'Hôpital's rule applies.

$$\lim_{x \to 1} \ln x = \ln 1 = 0 \quad \text{and} \quad \lim_{x \to 1} (x - 1)^2 = 0$$

Since the conditions of L'Hôpital's rule are satisfied, we can now take the derivatives of the numerator and denominator separately.

$$D_x(\ln x) = \frac{1}{x} \quad \text{and} \quad D_x[(x - 1)^2] = 2(x - 1)$$

Next, we find the limit of the quotient of these derivatives:

$$\lim_{x \to 1} \frac{1/x}{2(x - 1)} = \lim_{x \to 1} \frac{1}{2x(x - 1)} \quad \text{does not exist.}$$

By L'Hôpital's rule, this means that

$$\lim_{x \to 1} \frac{\ln x}{(x - 1)^2} \quad \text{does not exist.} \quad \blacksquare$$

Before looking at more examples of L'Hôpital's rule, consider the following summary.

**USING L'HÔPITAL'S RULE**

1. Be sure that $\lim\limits_{x \to a} f(x)/g(x)$ leads to the indeterminate form 0/0.
2. Take the derivatives of $f$ and $g$ separately.
3. Find the limit of $f'(x)/g'(x)$; this limit, if it exists, equals the limit of $f(x)/g(x)$.
4. If necessary, apply L'Hôpital's rule more than once.

### EXAMPLE 3

Find $\lim\limits_{x \to 0} \dfrac{x^3}{e^x - 1}$.

The limit in the numerator is 0, as is the limit in the denominator, so that L'Hôpital's rule applies. Taking derivatives separately in numerator and denominator gives

$$\lim_{x \to 0} \frac{3x^2}{e^x} = \frac{0}{e^0} = \frac{0}{1} = 0.$$

By L'Hôpital's rule,

$$\lim_{x \to 0} \frac{x^3}{e^x - 1} = 0.$$

### EXAMPLE 4

Find $\lim\limits_{x \to 0} \dfrac{e^x - x - 1}{x^2}$.

Find the limit in both numerator and denominator to verify that L'Hôpital's rule applies. Then take derivatives of both numerator and denominator separately.

$$\lim_{x \to 0} \frac{e^x - 1}{2x} = \frac{e^0 - 1}{2 \cdot 0} = \frac{1 - 1}{0} = \frac{0}{0}$$

The result is still the indeterminate form 0/0; use L'Hôpital's rule a second time. Taking derivatives of $e^x - 1$ and $2x$ gives

$$\lim_{x \to 0} \frac{e^x}{2} = \frac{e^0}{2} = \frac{1}{2}.$$

Finally, by L'Hôpital's rule,

$$\lim_{x \to 0} \frac{e^x - x - 1}{x^2} = \frac{1}{2}.$$

■ **EXAMPLE 5**

Find $\lim\limits_{x \to 1} \dfrac{x^2 - 1}{\sqrt{x}}$.

Taking derivatives of numerator and denominator separately gives

$$\lim_{x \to 1} \frac{2x}{(1/2)x^{-1/2}} = \lim_{x \to 1} 4x^{3/2} = 4 \cdot 1^{3/2} = 4 \cdot 1 = 4. \qquad \text{Incorrect}$$

Unfortunately, 4 is the wrong answer. What happened? We did not verify that the conditions of L'Hôpital's rule were satisfied. In fact,

$$\lim_{x \to 1} (x^2 - 1) = 0, \qquad \text{but} \qquad \lim_{x \to 1} \sqrt{x} = 1 \neq 0.$$

Since L'Hôpital's rule does not apply, we must use another method to find the limit. By substitution,

$$\lim_{x \to 1} \frac{x^2 - 1}{\sqrt{x}} = \frac{1^2 - 1}{\sqrt{1}} = \frac{0}{1} = 0. \quad ■$$

***Proof of L'Hôpital's Rule***   We cannot prove L'Hôpital's rule in general, but we can prove the theorem for the special case where $f$, $g$, $f'$, and $g'$ are continuous on some open interval containing $a$. With this assumption, the fact that

$$\lim_{x \to a} f(x) = 0 \qquad \text{and} \qquad \lim_{x \to a} g(x) = 0$$

means that both $f(a) = 0$ and $g(a) = 0$. Thus,

$$\lim_{x \to a} \frac{f(x)}{g(x)} = \lim_{x \to a} \frac{f(x) - f(a)}{g(x) - g(a)}, \qquad f(a) = 0 \text{ and } g(a) = 0$$

where we subtracted 0 in both numerator and denominator. Multiplying numerator and denominator by $1/(x - a)$ gives

$$\lim_{x \to a} \frac{f(x)}{g(x)} = \lim_{x \to a} \frac{\dfrac{f(x) - f(a)}{x - a}}{\dfrac{g(x) - g(a)}{x - a}}.$$

By a property of limits, this becomes

$$\lim_{x \to a} \frac{f(x)}{g(x)} = \frac{\lim\limits_{x \to a} \dfrac{f(x) - f(a)}{x - a}}{\lim\limits_{x \to a} \dfrac{g(x) - g(a)}{x - a}}.$$

By the definition of derivative, the limit in the numerator is $f'(a)$, and the limit in the denominator is $g'(a)$. From our assumption that both $f'$ and $g'$ are continuous, the quotient on the right above becomes

$$\frac{f'(a)}{g'(a)} = \frac{\lim_{x \to a} f'(x)}{\lim_{x \to a} g'(x)} = \lim_{x \to a} \frac{f'(x)}{g'(x)}.$$

Thus,

$$\lim_{x \to a} \frac{f(x)}{g(x)} = \lim_{x \to a} \frac{f'(x)}{g'(x)},$$

which is what we wanted to show.

## 11.6 EXERCISES

*Use l'Hôpital's rule where applicable to find each limit.*

**1.** $\lim\limits_{x \to 1} \dfrac{x^3 + x^2 - x - 1}{x^2 - x}$

**2.** $\lim\limits_{x \to 3} \dfrac{x^3 + x^2 - 11x - 3}{x^2 - 3x}$

**3.** $\lim\limits_{x \to 0} \dfrac{x^5 - 2x^3 + 4x^2}{8x^5 - 2x^2 + 5x}$

**4.** $\lim\limits_{x \to 0} \dfrac{8x^6 + 3x^4 - 9x}{9x^7 - 2x^4 + x^3}$

**5.** $\lim\limits_{x \to 2} \dfrac{\ln (x - 1)}{x - 2}$

**6.** $\lim\limits_{x \to 0} \dfrac{\ln (x + 1)}{x}$

**7.** $\lim\limits_{x \to 0} \dfrac{e^x - 1}{x^4}$

**8.** $\lim\limits_{x \to 0} \dfrac{e^{2x} - 1}{5x^2 - x}$

**9.** $\lim\limits_{x \to 0} \dfrac{xe^x}{e^x - 1}$

**10.** $\lim\limits_{x \to 0} \dfrac{xe^{-x}}{2e^{2x} - 2}$

**11.** $\lim\limits_{x \to 0} \dfrac{e^x}{2x^3 + 9x^2 - 11x}$

**12.** $\lim\limits_{x \to 0} \dfrac{e^x}{8x^5 - 3x^4}$

**13.** $\lim\limits_{x \to 0} \dfrac{\sqrt{2 + x} - \sqrt{2}}{x}$

**14.** $\lim\limits_{x \to 0} \dfrac{\sqrt{9 + x} - 3}{x}$

**15.** $\lim\limits_{x \to 4} \dfrac{\sqrt{x} - 2}{x - 4}$

**16.** $\lim\limits_{x \to 9} \dfrac{\sqrt{x} - 3}{x - 9}$

**17.** $\lim\limits_{x \to 8} \dfrac{\sqrt[3]{x} - 2}{x - 8}$

**18.** $\lim\limits_{x \to 27} \dfrac{\sqrt[3]{x} - 3}{x - 27}$

**19.** $\lim\limits_{x \to 1} \dfrac{x^9 - 1}{x - 1}$

**20.** $\lim\limits_{x \to 2} \dfrac{x^7 - 128}{x - 2}$

**21.** $\lim\limits_{x \to 0} \dfrac{e^x + e^{-x} - 2}{x}$

**22.** $\lim\limits_{x \to 0} \dfrac{e^x - 1 + x}{e^{-x} - 1 - x}$

**23.** $\lim\limits_{x \to 3} \dfrac{\sqrt{x^2 + 7} - 4}{x^2 - 9}$

**24.** $\lim\limits_{x \to 5} \dfrac{\sqrt{x^2 + 11} - 6}{x^2 - 25}$

**25.** $\lim\limits_{x \to 0} \dfrac{1 + \dfrac{1}{3}x - (1 + x)^{1/3}}{x^2}$

**26.** $\lim\limits_{x \to 2} \dfrac{2 - \sqrt{x + 2}}{2x^2 - 10x + 8}$

**27.** $\lim\limits_{x \to 0} \dfrac{\sqrt{1 + x} - \sqrt{1 - x}}{x}$

**28.** $\lim\limits_{x \to 0} \dfrac{\sqrt{3 - x} - \sqrt{3 + x}}{x}$

**29.** $\lim\limits_{x \to 0} \dfrac{\sqrt{x^2 - 5x + 4}}{x}$

**30.** $\lim\limits_{x \to 1} \dfrac{\sqrt{x^2 + 5x + 9}}{x - 1}$

**31.** $\lim\limits_{x \to 0} \dfrac{(5 + x) \ln (x + 1)}{e^x - 1}$

**32.** $\lim\limits_{x \to 0} \dfrac{(7 - x) \ln (1 - x)}{e^{-x} - 1}$

**33.** $\lim\limits_{x \to 0} \dfrac{\sin x}{x}$

**34.** $\lim\limits_{x \to 0} \dfrac{1 - \cos x}{x}$

**35.** $\lim\limits_{x \to 0} \dfrac{\sin x}{e^x - 1}$

**36.** $\lim\limits_{x \to 0} \dfrac{\sin x}{x^2}$

**37.** $\lim\limits_{x \to 0} \dfrac{x(\cos x - 1)}{\sin x - x}$

**38.** $\lim\limits_{x \to 0} \dfrac{x^3}{\sin x - x}$

**39.** $\lim\limits_{x \to 0} \dfrac{x - \tan x}{\sin x - x}$

**40.** $\lim\limits_{x \to 0} \dfrac{\sin x^2}{x^2 + 2x^3}$

*First get a common denominator; then find the limits that exist.*

**41.** $\lim\limits_{x \to 0} \left( \dfrac{1}{x} - \dfrac{1}{x^2} \right)$

**42.** $\lim\limits_{x \to 0} \left( \dfrac{1}{x^2} + \dfrac{1}{x^3} \right)$

**43.** $\lim\limits_{x \to 0} \left( \dfrac{1}{x} + \dfrac{1}{\sqrt[3]{x}} \right)$

**44.** $\lim\limits_{x \to 0} \left( \dfrac{1}{\sqrt[5]{x}} - \dfrac{1}{x} \right)$

**45.** $\lim\limits_{x \to 0} \left( \dfrac{1}{x} - \dfrac{1}{\sin x} \right)$

**46.** $\lim\limits_{x \to 0} \left( \dfrac{1}{\sin x} - \dfrac{1}{\sin 2x} \right)$

**47.** $\lim\limits_{x \to 1} \left( \dfrac{1}{\ln x} - \dfrac{1}{x - 1} \right)$

**48.** $\lim\limits_{x \to 1} \left( \dfrac{1}{\ln x} - \dfrac{1}{(\ln x)^2} \right)$

## 11.7 NEWTON'S METHOD

Given a function $f$, a number $r$ such that $f(r) = 0$ is called a **zero** of $f$. For example, if $f(x) = x^2 - 4x + 3$, then $f(3) = 0$ and $f(1) = 0$, so that both 3 and 1 are zeros of $f$. The zeros of linear and quadratic functions can be found with the methods of algebra. More complicated methods exist for finding zeros of third-degree or fourth-degree polynomial functions, but there is no general method for finding zeros of higher-degree polynomials.

In practical applications of mathematics, it is seldom necessary to find *exact* zeros of a function; usually a decimal approximation is all that is needed. In this section we introduce a method that can be used to find approximate zeros of many useful functions.

The zeros of a function $f$, where $f$ is continuous on some closed interval $[a, b]$, can be approximated as follows. Find $a$ and $b$ so that $f(a)$ and $f(b)$ are of opposite sign, one positive and one negative. As suggested by Figure 10, this means there must exist at least one value $c$ in the interval $(a, b)$ such that $f(c) = 0$. This number $c$ is a zero of the function $f$.

 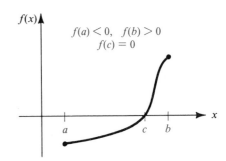

**FIGURE 10**

To find an approximate value for $c$, first make a guess for $c$. Let $c_1$ be the initial guess. (See Figure 11.) Then locate the point $(c_1, f(c_1))$ on the graph of $y = f(x)$ and identify the tangent line at this point. This tangent line will cut the $x$-axis at a point $c_2$. The number $c_2$ is often a better approximation to $c$ than was $c_1$.

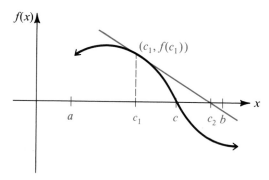

**FIGURE 11**

To locate $c_2$, first find the equation of the tangent line through $(c_1, f(c_1))$. The slope of this tangent line is $f'(c_1)$. The point-slope form of the equation of the tangent line is

$$y - f(c_1) = f'(c_1)(x - c_1).$$

At the point $x = c_2$, $y = 0$. Substituting into the equation of the tangent line gives

$$0 - f(c_1) = f'(c_1)(c_2 - c_1)$$

or

$$-\frac{f(c_1)}{f'(c_1)} = c_2 - c_1,$$

from which

$$c_2 = c_1 - \frac{f(c_1)}{f'(c_1)}.$$

If $f'(c_1)$ should be 0, the tangent line would be horizontal and not cut the $x$-axis. For this reason, assume $f'(c_1) \neq 0$. This new value, $c_2$, is usually a better approximation to $c$ than was $c_1$. To improve the approximation further, locate the tangent line to the curve at $(c_2, f(c_2))$. Let this tangent cut the $x$-axis at $c_3$. (See Figure 12). Find $c_3$ by a process similar to that used above: if $f'(c_2) \neq 0$,

$$c_3 = c_2 - \frac{f(c_2)}{f'(c_2)}.$$

The approximation to $c$ often can be improved by repeating this process as many times as desired. In general, if $c_n$ is an approximation to $c$, a better approximation $c_{n+1}$ frequently can be found by the following formula.

**NEWTON'S METHOD**

If $f'(c_n) \neq 0$, then

$$c_{n+1} = c_n - \frac{f(c_n)}{f'(c_n)}.$$

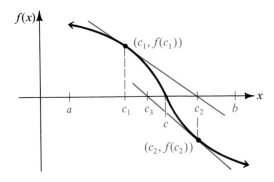

**FIGURE 12**

This process of first obtaining a rough approximation for $c$, then replacing it successively by approximations that are often better, is called **Newton's method,** named after Sir Isaac Newton, the codiscoverer of calculus. An early version of this method appeared in his work *Method of Fluxions,* published in 1736.

### ■ EXAMPLE 1

Approximate (to the nearest hundredth) a solution for the equation

$$3x^3 - x^2 + 5x - 12 = 0$$

in the interval $[1, 2]$.

Let $f(x) = 3x^3 - x^2 + 5x - 12$, so that $f'(x) = 9x^2 - 2x + 5$. Check that $f(1) < 0$ with $f(2) > 0$. Since $f(1)$ and $f(2)$ have opposite signs, there is a solution for the equation in the interval $(1, 2)$. As an initial guess, let $c_1 = 1$. A better guess, $c_2$, can be found as follows.

$$c_2 = c_1 - \frac{f(c_1)}{f'(c_1)} = 1 - \frac{-5}{12} = \frac{17}{12} = 1.42$$

A third approximation, $c_3$, can now be found.

$$c_3 = c_2 - \frac{f(c_2)}{f'(c_2)} = 1.42 - \frac{1.67}{20.3} = 1.34$$

In the same way,

$$c_4 = 1.34 - \frac{.122}{18.5} = 1.33 \quad \text{and} \quad c_5 = 1.33 - \frac{-.06}{18.3} = 1.33.$$

Since $c_4 = c_5 = 1.33$, to two decimal places of accuracy $x = 1.33$ is a solution of $3x^3 - x^2 + 5x - 12 = 0$. (The exact solution is 4/3.) ■

In Example 1 we had to go through five steps in order to get the degree of accuracy that we needed. The solutions of similar polynomial equations usually can be found in about as many steps, although other types of equations might require more steps.

In any case, if a solution can be found by Newton's method, it usually can be found by a computer in a few seconds. Some functions, such as the one in Figure 13, have solutions which cannot be found by Newton's method. Because of the symmetry of the graph in Figure 13, $c_1$ will always equal $c_3$ and $c_5$, while $c_2$ will equal $c_4$ and $c_6$. Newton's method would not give a solution here. Such cases are rare in practice.

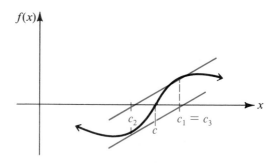

**FIGURE 13**

Newton's method also can be used to approximate the values of radicals, as shown by the next example.

### ▰▰ EXAMPLE 2
Approximate $\sqrt{12}$ to the nearest thousandth.

First, note that $\sqrt{12}$ is a solution of the equation $x^2 - 12 = 0$. Therefore, let $f(x) = x^2 - 12$, so that $f'(x) = 2x$. Since $3 < \sqrt{12} < 4$, use $c_1 = 3$ as the first approximation to $\sqrt{12}$. A better approximation is given by $c_2$:

$$c_2 = 3 - \frac{-3}{6} = 3.5.$$

Now find $c_3$ and $c_4$:    $c_3 = 3.5 - \frac{.25}{7} = 3.464,$

$$c_4 = 3.464 - \frac{-.0007}{6.928} = 3.464.$$

Since $c_3 = c_4 = 3.464$, to the nearest thousandth, $\sqrt{12} = 3.464$.

## ▰▰ 11.7 EXERCISES

*Use Newton's method to find a solution for each equation in the given interval. Find all solutions to the nearest hundredth. Use a calculator if available.*

1. $x^2 - 2x - 2 = 0$;   $[2, 3]$
2. $x^2 - 6x + 4 = 0$;   $[5, 6]$
3. $3x^2 + 2x - 1 = 0$;   $[0, 1]$
4. $6x^2 - 13x + 2 = 0$;   $[0, 1]$
5. $3x^3 - 5x^2 + 8x - 4 = 0$;   $[0, 1]$
6. $5x^3 - 2x^2 - 2x - 7 = 0$;   $[1, 2]$
7. $3x^3 - 4x^2 - 4x - 7 = 0$;   $[2, 3]$
8. $3x^3 - 14x^2 + 17x - 22 = 0$;   $[3, 4]$
9. $5x^3 + 6x^2 + 4x + 33 = 0$;   $[-3, -2]$
10. $7x^3 + 2x^2 - 30x - 27 = 0$;   $[2, 3]$
11. $x^3 - 3x^2 - 1 = 0$;   $[3, 4]$
12. $x^3 + 4x^2 + 6 = 0$;   $[-5, -4]$

**13.** $6x^3 - 13x^2 - 36x + 28 = 0$; $[-2, -1]$, $[0, 1]$, $[3, 4]$

**14.** $6x^3 + x^2 - 46x + 15 = 0$; $[0, 1]$, $[2, 3]$

**15.** $3x^4 - 5x^3 - 16x^2 - 19x - 11 = 0$; $[-2, -1]$, $[3, 4]$

**16.** $2x^4 + 7x^3 + 6x^2 + 7x - 6 = 0$; $[0, 1]$, $[-3, -2]$

*Use Newton's method to find each root to the nearest thousandth.*

**17.** $\sqrt{2}$  **18.** $\sqrt{3}$  **19.** $\sqrt{11}$  **20.** $\sqrt{15}$  **21.** $\sqrt{250}$

**22.** $\sqrt{300}$  **23.** $\sqrt[3]{9}$  **24.** $\sqrt[3]{15}$  **25.** $\sqrt[3]{100}$  **26.** $\sqrt[3]{121}$

*Use Newton's method to find the critical points for the functions defined as follows. Approximate them to the nearest hundredth. Decide whether each critical point leads to a relative maximum or a relative minimum.*

**27.** $f(x) = x^3 - 3x^2 - 18x + 4$

**28.** $f(x) = x^3 + 9x^2 - 6x + 4$

**29.** $f(x) = x^4 - 3x^3 + 6x - 1$

**30.** $f(x) = x^4 + 2x^3 - 5x + 2$

## ▓ APPLICATIONS

### BUSINESS AND ECONOMICS

**Break-Even Point** **31.** For a particular product, the revenue and cost functions are
$$R(x) = \sqrt{64 - x^2} \quad \text{and} \quad C(x) = x - 1.$$
Approximate the break-even point to the nearest hundredth.

**Manufacturing** **32.** A new manufacturing process produces savings of
$$S(x) = x^3 + 5x^2 + 9$$
dollars after $x$ years, with increased costs of
$$C(x) = x^2 + 40x + 20$$
dollars. For how many years, to the nearest hundredth, should the process be used?

### FOR THE COMPUTER

**True Annual Interest Rate** **33.** Federal government regulations require that people loaning money to consumers disclose the true annual interest rate of the loan. The formulas for calculating this interest rate are very complex. For example, suppose $P$ dollars is loaned, with the money to be repaid in $n$ monthly payments of $M$ dollars each. Then the true annual interest rate is found by solving the equation
$$\frac{1 - (1 + i)^{-n}}{i} - \frac{P}{M} = 0$$
for $i$, the monthly interest rate, and then multiplying $i$ by 12 to get the true annual rate. This equation can best be solved by Newton's method, as explained in the next few exercises.

(a) Let $f(i) = \dfrac{1 - (1 + i)^{-n}}{i} - \dfrac{P}{M}$. Find $f'(i)$.

(b) Form the quotient $f(i)/f'(i)$.

(c) Suppose that $P = \$4000$, $n = 24$, and $M = \$197$. Let the initial guess for $i$ be $i_1 = .01$. Use Newton's method and find $i_2$.

(d) Find $i_3$. (*Note:* For the accuracy required by federal law, it is usually sufficient to stop after two successive values of $i$ differ by no more than $10^{-7}$.)

*In Exercises 34 and 35, find $i_2$ and $i_3$.*

**34.** $P = \$600$, $M = \$57$, $n = 12$, $i_1 = .02$

**35.** $P = \$15,000$, $M = \$337$, $n = 60$, $i_1 = .01$

*Use Newton's method to find approximate solutions to four decimal places for the following equations.*

**36.** $x^5 - 3x^2 + 1 = 0$;  [0, 1]

**37.** $x^4 - 2x^2 - 1 = 0$;  [1, 2]

**38.** $x^5 + 5.7x^2 = 0$;  $[-2, -1]$

---

**KEY WORDS**

| | |
|---|---|
| **11.1** **sequence** | **11.3** **Taylor polynomial** |
| **series** | **11.4** **infinite series** |
| **element** | ***n*th partial sum** |
| **term** | **sum of an infinite series** |
| **general term** | **convergence** |
| ***n*th term** | **divergence** |
| **geometric sequence** | **harmonic series** |
| **common ratio** | **11.5** **interval of convergence** |
| **11.2** **annuity** | **Taylor series** |
| **ordinary annuity** | **Maclaurin series** |
| **payment period** | **doubling time** |
| **term of an annuity** | **rule of 70** |
| **amount of an annuity** | **rule of 72** |
| **sinking fund** | **11.6** **indeterminate form** |
| **present value of an annuity** | **L'Hôpital's rule** |
| **amortization** | **11.7** **Newton's method** |

---

## CHAPTER 11 REVIEW EXERCISES

*Find $a_5$ and $a_n$ for the following geometric sequences.*

**1.** $a_1 = -4$, $r = -2$     **2.** $a_1 = -1/2$, $r = 4$     **3.** $a_3 = 9$, $a_4 = 3$     **4.** $a_2 = -2/3$, $a_3 = 1/3$

*Find $S_5$ for the following geometric sequences.*

**5.** $a_1 = 6$, $r = 2$     **6.** $a_1 = -4$, $r = 3$     **7.** $a_2 = 9$, $r = 1/3$     **8.** $a_3 = 128$, $r = -1/2$

*Evaluate each sum.*

**9.** $\displaystyle\sum_{i=1}^{6} 2^i$          **10.** $\displaystyle\sum_{i=1}^{4} 3^i$          **11.** $\displaystyle\sum_{i=1}^{5} \frac{1}{2}(3^i)$          **12.** $\displaystyle\sum_{i=1}^{4} \frac{3}{4}(2^i)$

*Find the value of each annuity.*

**13.** $500 is deposited at the end of each 6-month period for 8 years; money earns 16% compounded semiannually

**14.** $1288 is deposited at the end of each year for 14 years; money earns 8% compounded annually

**15.** $4000 is deposited at the end of each quarter for 7 years; money earns 16% compounded quarterly

**16.** $233 is deposited at the end of each month for 4 years; money earns 18% compounded monthly

*Find the amount of each payment to be made into a sinking fund so that enough money will be available to pay off the indicated loan.*

**17.** $6500 loan, money earns 8% compounded annually, 6 annual payments

**18.** $57,000 loan, money earns 16% compounded semiannually, for 8½ years, semiannual payments

**19.** $233,188 loan, money earns 19.7% compounded quarterly, for 7¾ years, quarterly payments

**20.** $1,056,788 loan, money earns 18.12% compounded monthly, for 4½ years, monthly payments

*Find the present value of each ordinary annuity.*

**21.** Payments of $850 are made annually for 4 years at 8% compounded annually.

**22.** Payments of $1500 are made quarterly for 7 years, at 16% compounded quarterly.

**23.** Payments of $4210 are made semiannually for 8 years, at 18.6% compounded semiannually.

**24.** Payments of $877.34 are made monthly for 17 months, at 22.4% compounded monthly.

*Find the amount of the payment necessary to amortize each loan.*

**25.** $80,000 loan, 8% compounded annually, 9 annual payments

**26.** $3200 loan, 16% compounded quarterly, 10 quarterly payments

**27.** $32,000 loan, 19.4% compounded quarterly, 17 quarterly payments

**28.** $51,607 loan, 23.6% compounded monthly, 32 monthly payments

*Prepare amortization schedules for the following loans.*

**29.** $5000 at 10% compounded semiannually for 3 years

**30.** $12,500 at 12% compounded quarterly for 2 years

*Find Taylor polynomials of degree 4 at 0 for the functions defined as follows.*

**31.** $f(x) = e^{2-x}$  **32.** $f(x) = 5e^{2x}$  **33.** $f(x) = \sqrt{x+1}$  **34.** $f(x) = \sqrt[3]{x+27}$

**35.** $f(x) = \ln(2-x)$  **36.** $f(x) = \ln(3+2x)$  **37.** $f(x) = \cos 2x$  **38.** $f(x) = \sin 5x$

**39.** $f(x) = (1+x)^{2/3}$  **40.** $f(x) = x \cos 2x$

*Use Taylor polynomials of degree 4 at $x = 0$, found in Exercises 31–40 above, to approximate the quantities in Exercises 41–50. Round to four decimal places.*

**41.** $e^{1.99}$  **42.** $5e^{.02}$

**43.** $\sqrt{.98}$  **44.** $\sqrt[3]{26.98}$

**45.** $\ln 2.03$ ($\ln 2 = .69315$)  **46.** $\ln 2.96$ ($\ln 3 = 1.09861$)

**47.** cos .06        **48.** sin .05        **49.** $(1.01)^{2/3}$        **50.** .01 cos .02

*Identify the geometric series that converge. Give the sum of each convergent series.*

**51.** $9 - 6 + 4 - 8/3 + \cdots$

**52.** $2 + 1.4 + .98 + .686 + \cdots$

**53.** $3 + 9 + 27 + 81 + \cdots$

**54.** $4 + 4.8 + 5.76 + 6.912 + \cdots$

**55.** $\dfrac{2}{5} - \dfrac{2}{25} + \dfrac{2}{125} - \dfrac{2}{625} + \cdots$

**56.** $1 + \dfrac{1}{.99} + \dfrac{1}{(.99)^2} + \dfrac{1}{(.99)^3} + \cdots$

*In Exercises 57–58, the nth term of a sequence is given. Calculate the first five partial sums.*

**57.** $a_n = \dfrac{1}{2n - 1}$

**58.** $a_n = \dfrac{1}{(n + 2)(n + 3)}$

*Use the Taylor series given in the text to find the Taylor series for the functions defined as follows. Give the interval of convergence of each series.*

**59.** $f(x) = \dfrac{4}{3 - x}$

**60.** $f(x) = \dfrac{2x}{1 + 3x}$

**61.** $f(x) = \dfrac{x^2}{x + 1}$

**62.** $f(x) = \dfrac{3x^3}{2 - x}$

**63.** $f(x) = \ln(1 - 2x)$

**64.** $f(x) = \ln\left(1 + \dfrac{1}{3}x\right)$

**65.** $f(x) = e^{-2x^2}$

**66.** $f(x) = e^{-5x}$

**67.** $f(x) = 2x^3 e^{-3x}$

**68.** $f(x) = x^6 e^{-x}$

**69.** $f(x) = \cos x^2$

**70.** $f(x) = x \sin x^3$

**71.** $f(x) = \sin 3x^2$

**72.** $f(x) = \tan^{-1} 4x^2$

*Use L'Hôpital's rule, where applicable, to find each limit.*

**73.** $\displaystyle\lim_{x \to 2} \dfrac{x^3 - x^2 - x - 2}{x^2 - 4}$

**74.** $\displaystyle\lim_{x \to 0} \dfrac{x^3 - 4x^2 + 6x}{3x}$

**75.** $\displaystyle\lim_{x \to -5} \dfrac{x^3 - 3x^2 + 4x - 1}{x^2 - 25}$

**76.** $\displaystyle\lim_{x \to 0} \dfrac{\ln(3x + 1)}{x}$

**77.** $\displaystyle\lim_{x \to 0} \dfrac{5e^x - 5}{x^3 - 8x^2 + 7x}$

**78.** $\displaystyle\lim_{x \to 0} \dfrac{\sqrt{5 + x} - \sqrt{5}}{x}$

**79.** $\displaystyle\lim_{x \to 0} \dfrac{-xe^{2x}}{e^{2x} - 1}$

**80.** $\displaystyle\lim_{x \to 16} \dfrac{\sqrt{x} - 4}{x - 16}$

**81.** $\displaystyle\lim_{x \to 0} \dfrac{1 + 2x - (1 + x)^{1/2}}{x^3}$

**82.** $\displaystyle\lim_{x \to 0} \dfrac{\sqrt{5 + x} - \sqrt{5 - x}}{2x}$

**83.** $\displaystyle\lim_{x \to 0} \dfrac{\sin x}{6x}$

**84.** $\displaystyle\lim_{x \to 0} \dfrac{x^2}{\cos x - 1}$

*First get a common denominator; then find the limits that exist.*

**85.** $\displaystyle\lim_{x \to 0} \left(\dfrac{1}{2x} - \dfrac{1}{3x}\right)$

**86.** $\displaystyle\lim_{x \to 0} \left(\dfrac{1}{x} + \dfrac{1}{x^2}\right)$

**87.** $\displaystyle\lim_{x \to 0} \left(\dfrac{1}{x} - \dfrac{1}{\sqrt{x}}\right)$

**88.** $\displaystyle\lim_{x \to 0} \left(\dfrac{1}{\sqrt[3]{x}} + \dfrac{1}{\sqrt{x}}\right)$

*Use Newton's method to find a solution to the nearest hundredth for each equation in the given interval.*

**89.** $x^3 - 8x^2 + 18x - 12 = 0; \quad [4, 5]$

**90.** $3x^3 - 4x^2 - 4x - 7 = 0; \quad [2, 3]$

**91.** $x^4 + 3x^3 - 4x^2 - 21x - 21 = 0; \quad [2, 3]$

**92.** $x^4 + x^3 - 14x^2 - 15x - 15 = 0; \quad [3, 4]$

*Use Newton's method to approximate each radical to the nearest thousandth.*

**93.** $\sqrt{46.8}$

**94.** $\sqrt{39.5}$

**95.** $\sqrt[3]{-123.8}$

**96.** $\sqrt[4]{70.9}$

# ▤ APPLICATIONS

### BUSINESS AND ECONOMICS

**Total Income** **97.** A mine produced $600,000 of income during its first year. Each year thereafter, income increased by 20%. Find the total income produced in the first 5 years of the mine's life.

**Sinking Fund** **98.** In four years, Vern Heeren must pay a pledge of $5000 to his church's building fund. He wants to set up a sinking fund to accumulate that amount. What should each semiannual payment into the fund be at 8% compounded semiannually?

**Annuity** **99.** Georgette Dahl deposits $491 at the end of each quarter for 9 years. If the account pays 9.4% compounded quarterly, find the final amount in the account.

**Annuity** **100.** J. Euclid deposits $1526.38 at the end of each 6-month period in an account paying 7.6% compounded semiannually. How much will be in the account after 5 years?

**Amortization** **101.** Vicki Manchester borrows $20,000 from the bank to help her expand her business. She agrees to repay the money in equal payments at the end of each year for 9 years. Interest is at 8.9% compounded annually. Find the amount of each payment.

**Amortization** **102.** Ken Murrill wants to expand his pharmacy. To do this, he takes out a bank loan of $49,275 and agrees to repay it at 12.2% compounded monthly over 48 months. Find the amount of each payment necessary to amortize this loan.

**House Payments** *Find the monthly house payments for the following mortgages.*

**103.** $56,890 at 14.74% for 25 years

**104.** $77,110 at 16.45% for 30 years

# Tables

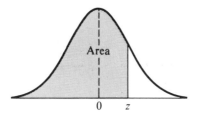

Area Under a Normal Curve to the Left of $z$, Where $z = \dfrac{x - \mu}{\sigma}$

| z | .00 | .01 | .02 | .03 | .04 | .05 | .06 | .07 | .08 | .09 |
|---|---|---|---|---|---|---|---|---|---|---|
| − 3.4 | .0003 | .0003 | .0003 | .0003 | .0003 | .0003 | .0003 | .0003 | .0003 | .0002 |
| − 3.3 | .0005 | .0005 | .0005 | .0004 | .0004 | .0004 | .0004 | .0004 | .0004 | .0003 |
| − 3.2 | .0007 | .0007 | .0006 | .0006 | .0006 | .0006 | .0006 | .0005 | .0005 | .0005 |
| − 3.1 | .0010 | .0009 | .0009 | .0009 | .0008 | .0008 | .0008 | .0008 | .0007 | .0007 |
| − 3.0 | .0013 | .0013 | .0013 | .0012 | .0012 | .0011 | .0011 | .0011 | .0010 | .0010 |
| − 2.9 | .0019 | .0018 | .0017 | .0017 | .0016 | .0016 | .0015 | .0015 | .0014 | .0014 |
| − 2.8 | .0026 | .0025 | .0024 | .0023 | .0023 | .0022 | .0021 | .0021 | .0020 | .0019 |
| − 2.7 | .0035 | .0034 | .0033 | .0032 | .0031 | .0030 | .0029 | .0028 | .0027 | .0026 |
| − 2.6 | .0047 | .0045 | .0044 | .0043 | .0041 | .0040 | .0039 | .0038 | .0037 | .0036 |
| − 2.5 | .0062 | .0060 | .0059 | .0057 | .0055 | .0054 | .0052 | .0051 | .0049 | .0048 |
| − 2.4 | .0082 | .0080 | .0078 | .0075 | .0073 | .0071 | .0069 | .0068 | .0066 | .0064 |
| − 2.3 | .0107 | .0104 | .0102 | .0099 | .0096 | .0094 | .0091 | .0089 | .0087 | .0084 |
| − 2.2 | .0139 | .0136 | .0132 | .0129 | .0125 | .0122 | .0119 | .0116 | .0113 | .0110 |
| − 2.1 | .0179 | .0174 | .0170 | .0166 | .0162 | .0158 | .0154 | .0150 | .0146 | .0143 |
| − 2.0 | .0228 | .0222 | .0217 | .0212 | .0207 | .0202 | .0197 | .0192 | .0188 | .0183 |
| − 1.9 | .0287 | .0281 | .0274 | .0268 | .0262 | .0256 | .0250 | .0244 | .0239 | .0233 |
| − 1.8 | .0359 | .0352 | .0344 | .0336 | .0329 | .0322 | .0314 | .0307 | .0301 | .0294 |
| − 1.7 | .0446 | .0436 | .0427 | .0418 | .0409 | .0401 | .0392 | .0384 | .0375 | .0367 |
| − 1.6 | .0548 | .0537 | .0526 | .0516 | .0505 | .0495 | .0485 | .0475 | .0465 | .0455 |
| − 1.5 | .0668 | .0655 | .0643 | .0630 | .0618 | .0606 | .0594 | .0582 | .0571 | .0559 |
| − 1.4 | .0808 | .0793 | .0778 | .0764 | .0749 | .0735 | .0722 | .0708 | .0694 | .0681 |
| − 1.3 | .0968 | .0951 | .0934 | .0918 | .0901 | .0885 | .0869 | .0853 | .0838 | .0823 |
| − 1.2 | .1151 | .1131 | .1112 | .1093 | .1075 | .1056 | .1038 | .1020 | .1003 | .0985 |
| − 1.1 | .1357 | .1335 | .1314 | .1292 | .1271 | .1251 | .1230 | .1210 | .1190 | .1170 |
| − 1.0 | .1587 | .1562 | .1539 | .1515 | .1492 | .1469 | .1446 | .1423 | .1401 | .1379 |

## Area Under a Normal Curve (continued)

| z | .00 | .01 | .02 | .03 | .04 | .05 | .06 | .07 | .08 | .09 |
|---|-----|-----|-----|-----|-----|-----|-----|-----|-----|-----|
| −.9 | .1841 | .1814 | .1788 | .1762 | .1736 | .1711 | .1685 | .1660 | .1635 | .1611 |
| −.8 | .2119 | .2090 | .2061 | .2033 | .2005 | .1977 | .1949 | .1922 | .1894 | .1867 |
| −.7 | .2420 | .2389 | .2358 | .2327 | .2296 | .2266 | .2236 | .2206 | .2177 | .2148 |
| −.6 | .2743 | .2709 | .2676 | .2643 | .2611 | .2578 | .2546 | .2514 | .2483 | .2451 |
| −.5 | .3085 | .3050 | .3015 | .2981 | .2946 | .2912 | .2877 | .2843 | .2810 | .2776 |
| −.4 | .3446 | .3409 | .3372 | .3336 | .3300 | .3264 | .3228 | .3192 | .3156 | .3121 |
| −.3 | .3821 | .3783 | .3745 | .3707 | .3669 | .3632 | .3594 | .3557 | .3520 | .3483 |
| −.2 | .4207 | .4168 | .4129 | .4090 | .4052 | .4013 | .3974 | .3936 | .3897 | .3859 |
| −.1 | .4602 | .4562 | .4522 | .4483 | .4443 | .4404 | .4364 | .4325 | .4286 | .4247 |
| −.0 | .5000 | .4960 | .4920 | .4880 | .4840 | .4801 | .4761 | .4721 | .4681 | .4641 |
| .0 | .5000 | .5040 | .5080 | .5120 | .5160 | .5199 | .5239 | .5279 | .5319 | .5359 |
| .1 | .5398 | .5438 | .5478 | .5517 | .5557 | .5596 | .5636 | .5675 | .5714 | .5753 |
| .2 | .5793 | .5832 | .5871 | .5910 | .5948 | .5987 | .6026 | .6064 | .6103 | .6141 |
| .3 | .6179 | .6217 | .6255 | .6293 | .6331 | .6368 | .6406 | .6443 | .6480 | .6517 |
| .4 | .6554 | .6591 | .6628 | .6664 | .6700 | .6736 | .6772 | .6808 | .6844 | .6879 |
| .5 | .6915 | .6950 | .6985 | .7019 | .7054 | .7088 | .7123 | .7157 | .7190 | .7224 |
| .6 | .7257 | .7291 | .7324 | .7357 | .7389 | .7422 | .7454 | .7486 | .7517 | .7549 |
| .7 | .7580 | .7611 | .7642 | .7673 | .7704 | .7734 | .7764 | .7794 | .7823 | .7852 |
| .8 | .7881 | .7910 | .7939 | .7967 | .7995 | .8023 | .8051 | .8078 | .8106 | .8133 |
| .9 | .8159 | .8186 | .8212 | .8238 | .8264 | .8289 | .8315 | .8340 | .8365 | .8389 |
| 1.0 | .8413 | .8438 | .8461 | .8485 | .8508 | .8531 | .8554 | .8577 | .8599 | .8621 |
| 1.1 | .8643 | .8665 | .8686 | .8708 | .8729 | .8749 | .8770 | .8790 | .8810 | .8830 |
| 1.2 | .8849 | .8869 | .8888 | .8907 | .8925 | .8944 | .8962 | .8980 | .8997 | .9015 |
| 1.3 | .9032 | .9049 | .9066 | .9082 | .9099 | .9115 | .9131 | .9147 | .9162 | .9177 |
| 1.4 | .9192 | .9207 | .9222 | .9236 | .9251 | .9265 | .9278 | .9292 | .9306 | .9319 |
| 1.5 | .9332 | .9345 | .9357 | .9370 | .9382 | .9394 | .9406 | .9418 | .9429 | .9441 |
| 1.6 | .9452 | .9463 | .9474 | .9484 | .9495 | .9505 | .9515 | .9525 | .9535 | .9545 |
| 1.7 | .9554 | .9564 | .9573 | .9582 | .9591 | .9599 | .9608 | .9616 | .9625 | .9633 |
| 1.8 | .9641 | .9649 | .9656 | .9664 | .9671 | .9678 | .9686 | .9693 | .9699 | .9706 |
| 1.9 | .9713 | .9719 | .9726 | .9732 | .9738 | .9744 | .9750 | .9756 | .9761 | .9767 |
| 2.0 | .9772 | .9778 | .9783 | .9788 | .9793 | .9798 | .9803 | .9808 | .9812 | .9817 |
| 2.1 | .9821 | .9826 | .9830 | .9834 | .9838 | .9842 | .9846 | .9850 | .9854 | .9857 |
| 2.2 | .9861 | .9864 | .9868 | .9871 | .9875 | .9878 | .9881 | .9884 | .9887 | .9890 |
| 2.3 | .9893 | .9896 | .9898 | .9901 | .9904 | .9906 | .9909 | .9911 | .9913 | .9916 |
| 2.4 | .9918 | .9920 | .9922 | .9925 | .9927 | .9929 | .9931 | .9932 | .9934 | .9936 |
| 2.5 | .9938 | .9940 | .9941 | .9943 | .9945 | .9946 | .9948 | .9949 | .9951 | .9952 |
| 2.6 | .9953 | .9955 | .9956 | .9957 | .9959 | .9960 | .9961 | .9962 | .9963 | .9964 |
| 2.7 | .9965 | .9966 | .9967 | .9968 | .9969 | .9970 | .9971 | .9972 | .9973 | .9974 |
| 2.8 | .9974 | .9975 | .9976 | .9977 | .9977 | .9978 | .9979 | .9979 | .9980 | .9981 |
| 2.9 | .9981 | .9982 | .9982 | .9983 | .9984 | .9984 | .9985 | .9985 | .9986 | .9986 |
| 3.0 | .9987 | .9987 | .9987 | .9988 | .9988 | .9989 | .9989 | .9989 | .9990 | .9990 |
| 3.1 | .9990 | .9991 | .9991 | .9991 | .9992 | .9992 | .9992 | .9992 | .9993 | .9993 |
| 3.2 | .9993 | .9993 | .9994 | .9994 | .9994 | .9994 | .9994 | .9995 | .9995 | .9995 |
| 3.3 | .9995 | .9995 | .9995 | .9996 | .9996 | .9996 | .9996 | .9996 | .9996 | .9997 |
| 3.4 | .9997 | .9997 | .9997 | .9997 | .9997 | .9997 | .9997 | .9997 | .9997 | .9998 |

# Compound Interest

$(1 + i)^n$

| n \ i | 1% | $1\frac{1}{2}$% | 2% | 3% | 4% | 5% | 6% | 8% |
|---|---|---|---|---|---|---|---|---|
| 1 | 1.01000 | 1.01500 | 1.02000 | 1.03000 | 1.04000 | 1.05000 | 1.06000 | 1.08000 |
| 2 | 1.02010 | 1.03023 | 1.04040 | 1.06090 | 1.08160 | 1.10250 | 1.12360 | 1.16640 |
| 3 | 1.03030 | 1.04568 | 1.06121 | 1.09273 | 1.12486 | 1.15763 | 1.19102 | 1.25971 |
| 4 | 1.04060 | 1.06136 | 1.08243 | 1.12551 | 1.16986 | 1.21551 | 1.26248 | 1.36049 |
| 5 | 1.05101 | 1.07728 | 1.10408 | 1.15927 | 1.21665 | 1.27628 | 1.33823 | 1.46933 |
| 6 | 1.06152 | 1.09344 | 1.12616 | 1.19405 | 1.26532 | 1.34010 | 1.41852 | 1.58687 |
| 7 | 1.07214 | 1.10984 | 1.14869 | 1.22987 | 1.31593 | 1.40710 | 1.50363 | 1.71382 |
| 8 | 1.08286 | 1.12649 | 1.17166 | 1.26677 | 1.36857 | 1.47746 | 1.59385 | 1.85093 |
| 9 | 1.09369 | 1.14339 | 1.19509 | 1.30477 | 1.42331 | 1.55133 | 1.68948 | 1.99900 |
| 10 | 1.10462 | 1.16054 | 1.21899 | 1.34392 | 1.48024 | 1.62889 | 1.79085 | 2.15892 |
| 11 | 1.11567 | 1.17795 | 1.24337 | 1.38423 | 1.53945 | 1.71034 | 1.89830 | 2.33164 |
| 12 | 1.12683 | 1.19562 | 1.26824 | 1.42576 | 1.60103 | 1.79586 | 2.01220 | 2.51817 |
| 13 | 1.13809 | 1.21355 | 1.29361 | 1.46853 | 1.66507 | 1.88565 | 2.13293 | 2.71962 |
| 14 | 1.14947 | 1.23176 | 1.31948 | 1.51259 | 1.73168 | 1.97993 | 2.26090 | 2.93719 |
| 15 | 1.16097 | 1.25023 | 1.34587 | 1.55797 | 1.80094 | 2.07893 | 2.39656 | 3.17217 |
| 16 | 1.17258 | 1.26899 | 1.37279 | 1.60471 | 1.87298 | 2.18287 | 2.54035 | 3.42594 |
| 17 | 1.18430 | 1.28802 | 1.40024 | 1.65285 | 1.94790 | 2.29202 | 2.69277 | 3.70002 |
| 18 | 1.19615 | 1.30734 | 1.42825 | 1.70243 | 2.02582 | 2.40662 | 2.85434 | 3.99602 |
| 19 | 1.20811 | 1.32695 | 1.45681 | 1.75351 | 2.10685 | 2.52695 | 3.02560 | 4.31570 |
| 20 | 1.22019 | 1.34686 | 1.48595 | 1.80611 | 2.19112 | 2.65330 | 3.20714 | 4.66096 |
| 21 | 1.23239 | 1.36706 | 1.51567 | 1.86029 | 2.27877 | 2.78596 | 3.39956 | 5.03383 |
| 22 | 1.24472 | 1.38756 | 1.54598 | 1.91610 | 2.36992 | 2.92526 | 3.60354 | 5.43654 |
| 23 | 1.25716 | 1.40838 | 1.57690 | 1.97359 | 2.46472 | 3.07152 | 3.81975 | 5.87146 |
| 24 | 1.26973 | 1.42950 | 1.60844 | 2.03279 | 2.56330 | 3.22510 | 4.04893 | 6.34118 |
| 25 | 1.28243 | 1.45095 | 1.64061 | 2.09378 | 2.66584 | 3.38635 | 4.29187 | 6.84848 |
| 26 | 1.29526 | 1.47271 | 1.67342 | 2.15659 | 2.77247 | 3.55567 | 4.54938 | 7.39635 |
| 27 | 1.30821 | 1.49480 | 1.70689 | 2.22129 | 2.88337 | 3.73346 | 4.82235 | 7.98806 |
| 28 | 1.32129 | 1.51722 | 1.74102 | 2.28793 | 2.99870 | 3.92013 | 5.11169 | 8.62711 |
| 29 | 1.33450 | 1.53998 | 1.77584 | 2.35657 | 3.11865 | 4.11614 | 5.41839 | 9.31727 |
| 30 | 1.34785 | 1.56308 | 1.81136 | 2.42726 | 3.24340 | 4.32194 | 5.74349 | 10.06266 |
| 31 | 1.36133 | 1.58653 | 1.84759 | 2.50008 | 3.37313 | 4.53804 | 6.08810 | 10.86767 |
| 32 | 1.37494 | 1.61032 | 1.88454 | 2.57508 | 3.50806 | 4.76494 | 6.45339 | 11.73708 |
| 33 | 1.38869 | 1.63448 | 1.92223 | 2.65234 | 3.64838 | 5.00319 | 6.84059 | 12.67605 |
| 34 | 1.40258 | 1.65900 | 1.96068 | 2.73191 | 3.79432 | 5.25335 | 7.25103 | 13.69013 |
| 35 | 1.41660 | 1.68388 | 1.99989 | 2.81386 | 3.94609 | 5.51602 | 7.68609 | 14.78534 |
| 36 | 1.43077 | 1.70914 | 2.03989 | 2.89828 | 4.10393 | 5.79182 | 8.14725 | 15.96817 |
| 37 | 1.44508 | 1.73478 | 2.08069 | 2.98523 | 4.26809 | 6.08141 | 8.63609 | 17.24563 |
| 38 | 1.45953 | 1.76080 | 2.12230 | 3.07478 | 4.43881 | 6.38548 | 9.15425 | 18.62528 |
| 39 | 1.47412 | 1.78721 | 2.16474 | 3.16703 | 4.61637 | 6.70475 | 9.70351 | 20.11530 |
| 40 | 1.48886 | 1.81402 | 2.20804 | 3.26204 | 4.80102 | 7.03999 | 10.28572 | 21.72452 |
| 41 | 1.50375 | 1.84123 | 2.25220 | 3.35990 | 4.99306 | 7.39199 | 10.90286 | 23.46248 |
| 42 | 1.51879 | 1.86885 | 2.29724 | 3.46070 | 5.19278 | 7.76159 | 11.55703 | 25.33948 |
| 43 | 1.53398 | 1.89688 | 2.34319 | 3.56452 | 5.40050 | 8.14967 | 12.25045 | 27.36664 |
| 44 | 1.54932 | 1.92533 | 2.39005 | 3.67145 | 5.61652 | 8.55715 | 12.98548 | 29.55597 |
| 45 | 1.56481 | 1.95421 | 2.43785 | 3.78160 | 5.84118 | 8.98501 | 13.76461 | 31.92045 |
| 46 | 1.58046 | 1.98353 | 2.48661 | 3.89504 | 6.07482 | 9.43426 | 14.59049 | 34.47409 |
| 47 | 1.59626 | 2.01328 | 2.53634 | 4.01190 | 6.31782 | 9.90597 | 15.46592 | 37.23201 |
| 48 | 1.61223 | 2.04348 | 2.58707 | 4.13225 | 6.57053 | 10.40127 | 16.39387 | 40.21057 |
| 49 | 1.62835 | 2.07413 | 2.63881 | 4.25622 | 6.83335 | 10.92133 | 17.37750 | 43.42742 |
| 50 | 1.64463 | 2.10524 | 2.69159 | 4.38391 | 7.10668 | 11.46740 | 18.42015 | 46.90161 |

Present Value

$$\frac{1}{(1 + i)^n}$$

| i / n | 1% | 1½% | 2% | 3% | 4% | 5% | 6% | 8% |
|---|---|---|---|---|---|---|---|---|
| 1 | .99010 | .98522 | .98039 | .97087 | .96154 | .95238 | .94340 | .92593 |
| 2 | .98030 | .97066 | .96117 | .94260 | .92456 | .90703 | .89000 | .85734 |
| 3 | .97059 | .95632 | .94232 | .91514 | .88900 | .86384 | .83962 | .79383 |
| 4 | .96098 | .94218 | .92385 | .88849 | .85480 | .82270 | .79209 | .73503 |
| 5 | .95147 | .92826 | .90573 | .86261 | .82193 | .78353 | .74726 | .68058 |
| 6 | .94205 | .91454 | .88797 | .83748 | .79031 | .74622 | .70496 | .63017 |
| 7 | .93272 | .90103 | .87056 | .81309 | .75992 | .71068 | .66506 | .58349 |
| 8 | .92348 | .88771 | .85349 | .78941 | .73069 | .67684 | .62741 | .54027 |
| 9 | .91434 | .87459 | .83676 | .76642 | .70259 | .64461 | .59190 | .50025 |
| 10 | .90529 | .86167 | .82035 | .74409 | .67556 | .61391 | .55839 | .46319 |
| 11 | .89632 | .84893 | .80426 | .72242 | .64958 | .58468 | .52679 | .42888 |
| 12 | .88745 | .83639 | .78849 | .70138 | .62460 | .55684 | .49697 | .39711 |
| 13 | .87866 | .82403 | .77303 | .68095 | .60057 | .53032 | .46884 | .36770 |
| 14 | .86996 | .81185 | .75788 | .66112 | .57748 | .50507 | .44230 | .34046 |
| 15 | .86135 | .79985 | .74301 | .64186 | .55526 | .48102 | .41727 | .31524 |
| 16 | .85282 | .78803 | .72845 | .62317 | .53391 | .45811 | .39365 | .29189 |
| 17 | .84438 | .77639 | .71416 | .60502 | .51337 | .43630 | .37136 | .27027 |
| 18 | .83602 | .76491 | .70016 | .58739 | .49363 | .41552 | .35034 | .25025 |
| 19 | .82774 | .75361 | .68643 | .57029 | .47464 | .39573 | .33051 | .23171 |
| 20 | .81954 | .74247 | .67297 | .55368 | .45639 | .37689 | .31180 | .21455 |
| 21 | .81143 | .73150 | .65978 | .53755 | .43883 | .35894 | .29416 | .19866 |
| 22 | .80340 | .72069 | .64684 | .52189 | .42196 | .34185 | .27751 | .18394 |
| 23 | .79544 | .71004 | .63416 | .50669 | .40573 | .32557 | .26180 | .17032 |
| 24 | .78757 | .69954 | .62172 | .49193 | .39012 | .31007 | .24698 | .15770 |
| 25 | .77977 | .68921 | .60953 | .47761 | .37512 | .29530 | .23300 | .14602 |
| 26 | .77205 | .67902 | .59758 | .46369 | .36069 | .28124 | .21981 | .13520 |
| 27 | .76440 | .66899 | .58586 | .45019 | .34682 | .26785 | .20737 | .12519 |
| 28 | .75684 | .65910 | .57437 | .43708 | .33348 | .25509 | .19563 | .11591 |
| 29 | .74934 | .64936 | .56311 | .42435 | .32065 | .24295 | .18456 | .10733 |
| 30 | .74192 | .63976 | .55207 | .41199 | .30832 | .23138 | .17411 | .09938 |
| 31 | .73458 | .63031 | .54125 | .39999 | .29646 | .22036 | .16425 | .09202 |
| 32 | .72730 | .62099 | .53063 | .38834 | .28506 | .20987 | .15496 | .08520 |
| 33 | .72010 | .61182 | .52023 | .37703 | .27409 | .19987 | .14619 | .07889 |
| 34 | .71297 | .60277 | .51003 | .36604 | .26355 | .19035 | .13791 | .07305 |
| 35 | .70591 | .59387 | .50003 | .35538 | .25342 | .18129 | .13011 | .06763 |
| 36 | .69892 | .58509 | .49022 | .34503 | .24367 | .17266 | .12274 | .06262 |
| 37 | .69200 | .57644 | .48061 | .33498 | .23430 | .16444 | .11579 | .05799 |
| 38 | .68515 | .56792 | .47119 | .32523 | .22529 | .15661 | .10924 | .05369 |
| 39 | .67837 | .55953 | .46195 | .31575 | .21662 | .14915 | .10306 | .04971 |
| 40 | .67165 | .55126 | .45289 | .30656 | .20829 | .14205 | .09722 | .04603 |
| 41 | .66500 | .54312 | .44401 | .29763 | .20028 | .13528 | .09172 | .04262 |
| 42 | .65842 | .53509 | .43530 | .28896 | .19257 | .12884 | .08653 | .03946 |
| 43 | .65190 | .52718 | .42677 | .28054 | .18517 | .12270 | .08163 | .03654 |
| 44 | .64545 | .51939 | .41840 | .27237 | .17805 | .11686 | .07701 | .03383 |
| 45 | .63905 | .51171 | .41020 | .26444 | .17120 | .11130 | .07265 | .03133 |
| 46 | .63273 | .50415 | .40215 | .25674 | .16461 | .10600 | .06854 | .02901 |
| 47 | .62646 | .49670 | .39427 | .24926 | .15828 | .10095 | .06466 | .02686 |
| 48 | .62026 | .48936 | .38654 | .24200 | .15219 | .09614 | .06100 | .02487 |
| 49 | .61412 | .48213 | .37896 | .23495 | .14634 | .09156 | .05755 | .02303 |
| 50 | .60804 | .47500 | .37153 | .22811 | .14071 | .08720 | .05429 | .02132 |

## Amount of an Annuity

$$s_{\overline{n}|i} = \frac{(1 + i)^n - 1}{i}$$

| $i$ / $n$ | 1% | $1\frac{1}{2}$% | 2% | 3% | 4% | 5% | 6% | 8% |
|---|---|---|---|---|---|---|---|---|
| 1 | 1.00000 | 1.00000 | 1.00000 | 1.00000 | 1.00000 | 1.00000 | 1.00000 | 1.00000 |
| 2 | 2.01000 | 2.01500 | 2.02000 | 2.03000 | 2.04000 | 2.05000 | 2.06000 | 2.08000 |
| 3 | 3.03010 | 3.04523 | 3.06040 | 3.09090 | 3.12160 | 3.15250 | 3.18360 | 3.24640 |
| 4 | 4.06040 | 4.09090 | 4.12161 | 4.18363 | 4.24646 | 4.31013 | 4.37462 | 4.50611 |
| 5 | 5.10101 | 5.15227 | 5.20404 | 5.30914 | 5.41632 | 5.52563 | 5.63709 | 5.86660 |
| 6 | 6.15202 | 6.22955 | 6.30812 | 6.46841 | 6.63298 | 6.80191 | 6.97532 | 7.33593 |
| 7 | 7.21354 | 7.32299 | 7.43428 | 7.66246 | 7.89829 | 8.14201 | 8.39384 | 8.92280 |
| 8 | 8.28567 | 8.43284 | 8.58297 | 8.89234 | 9.21423 | 9.54911 | 9.89747 | 10.63663 |
| 9 | 9.36853 | 9.55933 | 9.75463 | 10.15911 | 10.58280 | 11.02656 | 11.49132 | 12.48756 |
| 10 | 10.46221 | 10.70272 | 10.94972 | 11.46388 | 12.00611 | 12.57789 | 13.18079 | 14.48656 |
| 11 | 11.56683 | 11.86326 | 12.16872 | 12.80780 | 13.48635 | 14.20679 | 14.97164 | 16.64549 |
| 12 | 12.68250 | 13.04121 | 13.41209 | 14.19203 | 15.02581 | 15.91713 | 16.86994 | 18.97713 |
| 13 | 13.80933 | 14.23683 | 14.68033 | 15.61779 | 16.62684 | 17.71298 | 18.88214 | 21.49530 |
| 14 | 14.94742 | 15.45038 | 15.97394 | 17.08632 | 18.29191 | 19.59863 | 21.01507 | 24.21492 |
| 15 | 16.09690 | 16.68214 | 17.29342 | 18.59891 | 20.02359 | 21.57856 | 23.27597 | 27.15211 |
| 16 | 17.25786 | 17.93237 | 18.63929 | 20.15688 | 21.82453 | 23.65749 | 25.67253 | 30.32428 |
| 17 | 18.43044 | 19.20136 | 20.01207 | 21.76159 | 23.69751 | 25.84037 | 28.21288 | 33.75023 |
| 18 | 19.61475 | 20.48938 | 21.41231 | 23.41444 | 25.64541 | 28.13238 | 30.90565 | 37.45024 |
| 19 | 20.81090 | 21.79672 | 22.84056 | 25.11687 | 27.67123 | 30.53900 | 33.75999 | 41.44626 |
| 20 | 22.01900 | 23.12367 | 24.29737 | 26.87037 | 29.77808 | 33.06595 | 36.78559 | 45.76196 |
| 21 | 23.23919 | 24.47052 | 25.78332 | 28.67649 | 31.96920 | 35.71925 | 39.99273 | 50.42292 |
| 22 | 24.47159 | 25.83758 | 27.29898 | 30.53678 | 34.24797 | 38.50521 | 43.39229 | 55.45676 |
| 23 | 25.71630 | 27.22514 | 28.84496 | 32.45288 | 36.61789 | 41.43048 | 46.99583 | 60.89330 |
| 24 | 26.97346 | 28.63352 | 30.42186 | 34.42647 | 39.08260 | 44.50200 | 50.81558 | 66.76476 |
| 25 | 28.24320 | 30.06302 | 32.03030 | 36.45926 | 41.64591 | 47.72710 | 54.86451 | 73.10594 |
| 26 | 29.52563 | 31.51397 | 33.67091 | 38.55304 | 44.31174 | 51.11345 | 59.15638 | 79.95442 |
| 27 | 30.82089 | 32.98668 | 35.34432 | 40.70963 | 47.08421 | 54.66913 | 63.70577 | 87.35077 |
| 28 | 32.12910 | 34.48148 | 37.05121 | 42.93092 | 49.96758 | 58.40258 | 68.52811 | 95.33883 |
| 29 | 33.45039 | 35.99870 | 38.79223 | 45.21885 | 52.96629 | 62.32271 | 73.63980 | 103.96594 |
| 30 | 34.78489 | 37.53868 | 40.56808 | 47.57542 | 56.08494 | 66.43885 | 79.05819 | 113.28321 |
| 31 | 36.13274 | 39.10176 | 42.37944 | 50.00268 | 59.32834 | 70.76079 | 84.80168 | 123.34587 |
| 32 | 37.49407 | 40.68829 | 44.22703 | 52.50276 | 62.70147 | 75.29883 | 90.88978 | 134.21354 |
| 33 | 38.86901 | 42.29861 | 46.11157 | 55.07784 | 66.20953 | 80.06377 | 97.34316 | 145.95062 |
| 34 | 40.25770 | 43.93309 | 48.03380 | 57.73018 | 69.85791 | 85.06696 | 104.18375 | 158.62667 |
| 35 | 41.66028 | 45.59209 | 49.99448 | 60.46208 | 73.65222 | 90.32031 | 111.43478 | 172.31680 |
| 36 | 43.07688 | 47.27597 | 51.99437 | 63.27594 | 77.59831 | 95.83632 | 119.12087 | 187.10215 |
| 37 | 44.50765 | 48.98511 | 54.03425 | 66.17422 | 81.70225 | 101.62814 | 127.26812 | 203.07032 |
| 38 | 45.95272 | 50.71989 | 56.11494 | 69.15945 | 85.97034 | 107.70955 | 135.90421 | 220.31595 |
| 39 | 47.41225 | 52.48068 | 58.23724 | 72.23423 | 90.40915 | 114.09502 | 145.05846 | 238.94122 |
| 40 | 48.88637 | 54.26789 | 60.40198 | 75.40126 | 95.02552 | 120.79977 | 154.76197 | 259.05652 |
| 41 | 50.37524 | 56.08191 | 62.61002 | 78.66330 | 99.82654 | 127.83976 | 165.04768 | 280.78104 |
| 42 | 51.87899 | 57.92314 | 64.86222 | 82.02320 | 104.81960 | 135.23175 | 175.95054 | 304.24352 |
| 43 | 53.39778 | 59.79199 | 67.15947 | 85.48389 | 110.01238 | 142.99334 | 187.50758 | 329.58301 |
| 44 | 54.93176 | 61.68887 | 69.50266 | 89.04841 | 115.41288 | 151.14301 | 199.75803 | 356.94965 |
| 45 | 56.48107 | 63.61420 | 71.89271 | 92.71986 | 121.02939 | 159.70016 | 212.74351 | 386.50562 |
| 46 | 58.04589 | 65.56841 | 74.33056 | 96.50146 | 126.87057 | 168.68516 | 226.50812 | 418.42607 |
| 47 | 59.62634 | 67.55194 | 76.81718 | 100.39650 | 132.94539 | 178.11942 | 241.09861 | 452.90015 |
| 48 | 61.22261 | 69.56522 | 79.35352 | 104.40840 | 139.26321 | 188.02539 | 256.56453 | 490.13216 |
| 49 | 62.83483 | 71.60870 | 81.94059 | 108.54065 | 145.83373 | 198.42666 | 272.95840 | 530.34274 |
| 50 | 64.46318 | 73.68283 | 84.57940 | 112.79687 | 152.66708 | 209.34800 | 290.33590 | 573.77016 |

## Present Value of an Annuity

$$a_{\overline{n}|i} = \frac{1 - (1 + i)^{-n}}{i}$$

| $i$ / $n$ | 1% | $1\frac{1}{2}$% | 2% | 3% | 4% | 5% | 6% | 8% |
|---|---|---|---|---|---|---|---|---|
| 1 | .99010 | .98522 | .98039 | .97087 | .96154 | .95238 | .94340 | .92593 |
| 2 | 1.97040 | 1.95588 | 1.94156 | 1.91347 | 1.88609 | 1.85941 | 1.83339 | 1.78326 |
| 3 | 2.94099 | 2.91220 | 2.88388 | 2.82861 | 2.77509 | 2.72325 | 2.67301 | 2.57710 |
| 4 | 3.90197 | 3.85438 | 3.80773 | 3.71710 | 3.62990 | 3.54595 | 3.46511 | 3.31213 |
| 5 | 4.85343 | 4.78264 | 4.71346 | 4.57971 | 4.45182 | 4.32948 | 4.21236 | 3.99271 |
| 6 | 5.79548 | 5.69719 | 5.60143 | 5.41719 | 5.24214 | 5.07569 | 4.91732 | 4.62288 |
| 7 | 6.72819 | 6.59821 | 6.47199 | 6.23028 | 6.00205 | 5.78637 | 5.58238 | 5.20637 |
| 8 | 7.65168 | 7.48593 | 7.32548 | 7.01969 | 6.73274 | 6.46321 | 6.20979 | 5.74664 |
| 9 | 8.56602 | 8.36052 | 8.16224 | 7.78611 | 7.43533 | 7.10782 | 6.80169 | 6.24689 |
| 10 | 9.47130 | 9.22218 | 8.98259 | 8.53020 | 8.11090 | 7.72173 | 7.36009 | 6.71008 |
| 11 | 10.36763 | 10.07112 | 9.78685 | 9.25262 | 8.76048 | 8.30641 | 7.88687 | 7.13896 |
| 12 | 11.25508 | 10.90751 | 10.57534 | 9.95400 | 9.38507 | 8.86325 | 8.38384 | 7.53608 |
| 13 | 12.13374 | 11.73153 | 11.34837 | 10.63496 | 9.98565 | 9.39357 | 8.85268 | 7.90378 |
| 14 | 13.00370 | 12.54338 | 12.10625 | 11.29607 | 10.56312 | 9.89864 | 9.29498 | 8.24424 |
| 15 | 13.86505 | 13.34323 | 12.84926 | 11.93794 | 11.11839 | 10.37966 | 9.71225 | 8.55948 |
| 16 | 14.71787 | 14.13126 | 13.57771 | 12.56110 | 11.65230 | 10.83777 | 10.10590 | 8.85137 |
| 17 | 15.56225 | 14.90765 | 14.29187 | 13.16612 | 12.16567 | 11.27407 | 10.47726 | 9.12164 |
| 18 | 16.39827 | 15.67256 | 14.99203 | 13.75351 | 12.65930 | 11.68959 | 10.82760 | 9.37189 |
| 19 | 17.22601 | 16.42617 | 15.67846 | 14.32380 | 13.13394 | 12.08532 | 11.15812 | 9.60360 |
| 20 | 18.04555 | 17.16864 | 16.35143 | 14.87747 | 13.59033 | 12.46221 | 11.46992 | 9.81815 |
| 21 | 18.85698 | 17.90014 | 17.01121 | 15.41502 | 14.02916 | 12.82115 | 11.76408 | 10.01680 |
| 22 | 19.66038 | 18.62082 | 17.65805 | 15.93692 | 14.45112 | 13.16300 | 12.04158 | 10.20074 |
| 23 | 20.45582 | 19.33086 | 18.29220 | 16.44361 | 14.85684 | 13.48857 | 12.30338 | 10.37106 |
| 24 | 21.24339 | 20.03041 | 18.91393 | 16.93554 | 15.24696 | 13.79864 | 12.55036 | 10.52876 |
| 25 | 22.02316 | 20.71961 | 19.52346 | 17.41315 | 15.62208 | 14.09394 | 12.78336 | 10.67478 |
| 26 | 22.79520 | 21.39863 | 20.12104 | 17.87684 | 15.98277 | 14.37519 | 13.00317 | 10.80998 |
| 27 | 23.55961 | 22.06762 | 20.70690 | 18.32703 | 16.32959 | 14.64303 | 13.21053 | 10.93516 |
| 28 | 24.31644 | 22.72672 | 21.28127 | 18.76411 | 16.66306 | 14.89813 | 13.40616 | 11.05108 |
| 29 | 25.06579 | 23.37608 | 21.84438 | 19.18845 | 16.98371 | 15.14107 | 13.59072 | 11.15841 |
| 30 | 25.80771 | 24.01584 | 22.39646 | 19.60044 | 17.29203 | 15.37245 | 13.76483 | 11.25778 |
| 31 | 26.54229 | 24.64615 | 22.93770 | 20.00043 | 17.58849 | 15.59281 | 13.92909 | 11.34980 |
| 32 | 27.26959 | 25.26714 | 23.46833 | 20.38877 | 17.87355 | 15.80268 | 14.08404 | 11.43500 |
| 33 | 27.98969 | 25.87895 | 23.98856 | 20.76579 | 18.14765 | 16.00255 | 14.23023 | 11.51389 |
| 34 | 28.70267 | 26.48173 | 24.49859 | 21.13184 | 18.41120 | 16.19290 | 14.36814 | 11.58693 |
| 35 | 29.40858 | 27.07559 | 24.99862 | 21.48722 | 18.66461 | 16.37419 | 14.49825 | 11.65457 |
| 36 | 30.10751 | 27.66068 | 25.48884 | 21.83225 | 18.90828 | 16.54685 | 14.62099 | 11.71719 |
| 37 | 30.79951 | 28.23713 | 25.96945 | 22.16724 | 19.14258 | 16.71129 | 14.73678 | 11.77518 |
| 38 | 31.48466 | 28.80505 | 26.44064 | 22.49246 | 19.36786 | 16.86789 | 14.84602 | 11.82887 |
| 39 | 32.16303 | 29.36458 | 26.90259 | 22.80822 | 19.58448 | 17.01704 | 14.94907 | 11.87858 |
| 40 | 32.83469 | 29.91585 | 27.35548 | 23.11477 | 19.79277 | 17.15909 | 15.04630 | 11.92461 |
| 41 | 33.49969 | 30.45896 | 27.79949 | 23.41240 | 19.99305 | 17.29437 | 15.13802 | 11.96723 |
| 42 | 34.15811 | 30.99405 | 28.23479 | 23.70136 | 20.18563 | 17.42321 | 15.22454 | 12.00670 |
| 43 | 34.81001 | 31.52123 | 28.66156 | 23.98190 | 20.37079 | 17.54591 | 15.30617 | 12.04324 |
| 44 | 35.45545 | 32.04062 | 29.07996 | 24.25427 | 20.54884 | 17.66277 | 15.38318 | 12.07707 |
| 45 | 36.09451 | 32.55234 | 29.49016 | 24.51871 | 20.72004 | 17.77407 | 15.45583 | 12.10840 |
| 46 | 36.72724 | 33.05649 | 29.89231 | 24.77545 | 20.88465 | 17.88007 | 15.52437 | 12.13741 |
| 47 | 37.35370 | 33.55319 | 30.28658 | 25.02471 | 21.04294 | 17.98102 | 15.58903 | 12.16427 |
| 48 | 37.97396 | 34.04255 | 30.67312 | 25.26671 | 21.19513 | 18.07716 | 15.65003 | 12.18914 |
| 49 | 38.58808 | 34.52468 | 31.05208 | 25.50166 | 21.34147 | 18.16872 | 15.70757 | 12.21216 |
| 50 | 39.19612 | 34.99969 | 31.42361 | 25.72976 | 21.48218 | 18.25593 | 15.76186 | 12.23348 |

## Powers of e and Natural Logarithms

| x | $e^x$ | $e^{-x}$ | ln x | x | $e^x$ | $e^{-x}$ | ln x |
|---|-------|----------|------|---|-------|----------|------|
| .00 | 1.00000 | 1.00000 | | 4.0 | 54.5981 | .01832 | 1.3863 |
| .01 | 1.01005 | .99004 | | 4.1 | 60.3402 | .01657 | 1.4110 |
| .02 | 1.02020 | .98019 | −4.6052 | 4.2 | 66.6863 | .01500 | 1.4351 |
| .03 | 1.03045 | .97044 | −3.9120 | 4.3 | 73.6997 | .01357 | 1.4586 |
| .04 | 1.04081 | .96078 | −3.5066 | 4.4 | 81.4508 | .01228 | 1.4816 |
| .05 | 1.05127 | .95122 | −3.2189 | 4.5 | 90.0170 | .01111 | 1.5041 |
| .06 | 1.06183 | .94176 | −2.9957 | 4.6 | 99.4842 | .01005 | 1.5261 |
| .07 | 1.07250 | .93239 | −2.8134 | 4.7 | 109.947 | .00910 | 1.5476 |
| .08 | 1.08328 | .92311 | −2.6593 | 4.8 | 121.510 | .00823 | 1.5686 |
| .09 | 1.09417 | .91393 | −2.5257 | 4.9 | 134.290 | .00745 | 1.5892 |
| | | | −2.4079 | 5.0 | 148.413 | .00674 | 1.6094 |
| .10 | 1.10517 | .90483 | −2.3026 | 5.1 | 164.022 | .00610 | 1.6292 |
| .11 | 1.11628 | .89583 | −2.2073 | 5.2 | 181.272 | .00552 | 1.6487 |
| .12 | 1.12750 | .88692 | −2.1203 | 5.3 | 200.336 | .00499 | 1.6677 |
| .13 | 1.12883 | .87810 | −2.0402 | 5.4 | 221.406 | .00452 | 1.6864 |
| .14 | 1.15027 | .86936 | −1.9661 | 5.5 | 244.691 | .00445 | 1.7047 |
| .15 | 1.16183 | .86071 | −1.8971 | 5.6 | 270.426 | .00370 | 1.7228 |
| .16 | 1.17351 | .85214 | −1.8326 | 5.7 | 298.867 | .00335 | 1.7405 |
| .17 | 1.18530 | .84366 | −1.7720 | 5.8 | 330.299 | .00303 | 1.7579 |
| .18 | 1.19722 | .83527 | −1.7148 | 5.9 | 365.036 | .00274 | 1.7750 |
| .19 | 1.20925 | .82696 | −1.6607 | 6.0 | 403.428 | .00248 | 1.7918 |
| .2 | 1.22140 | .81873 | −1.6094 | 6.1 | 445.856 | .00224 | 1.8083 |
| .3 | 1.34985 | .74081 | −1.2040 | 6.2 | 492.748 | .00203 | 1.8245 |
| .4 | 1.49182 | .67032 | −.9163 | 6.3 | 544.570 | .00184 | 1.8405 |
| .5 | 1.64872 | .60653 | −.6931 | 6.4 | 601.843 | .00166 | 1.8563 |
| .6 | 1.82211 | .54881 | −.5108 | 6.5 | 665.139 | .00150 | 1.8718 |
| .7 | 2.01375 | .49658 | −.3567 | 6.6 | 735.093 | .00136 | 1.8871 |
| .8 | 2.22554 | .44932 | −.2231 | 6.7 | 812.403 | .00123 | 1.9021 |
| .9 | 2.45960 | .40656 | −.1054 | 6.8 | 897.844 | .00111 | 1.9169 |
| | | | | 6.9 | 992.271 | .00101 | 1.9315 |
| 1.0 | 2.71828 | .36787 | .0000 | 7.0 | 1096.63 | .00091 | 1.9459 |
| 1.1 | 3.00416 | .33287 | .0953 | 7.1 | 1211.96 | .00083 | 1.9601 |
| 1.2 | 3.32011 | .30119 | .1823 | 7.2 | 1339.43 | .00075 | 1.9741 |
| 1,3 | 3.66929 | .27253 | .2624 | 7.3 | 1480.29 | .00068 | 1.9879 |
| 1.4 | 4.05519 | .24659 | .3365 | 7.4 | 1635.98 | .00061 | 2.0015 |
| 1.5 | 4.48168 | .22313 | .4055 | 7.5 | 1808.03 | .00055 | 2.0149 |
| 1.6 | 4.95302 | .20189 | .4700 | 7.6 | 1998.19 | .00050 | 2.0281 |
| 1.7 | 5.47394 | .18268 | .5306 | 7.7 | 2208.34 | .00045 | 2.0412 |
| 1.8 | 6.04964 | .16529 | .5878 | 7.8 | 2440.59 | .00041 | 2.0541 |
| 1.9 | 6.68589 | .14956 | .6419 | 7.9 | 2697.27 | .00037 | 2.0669 |
| 2.0 | 7.38905 | .13533 | .6931 | 8.0 | 2980.94 | .00034 | 2.0794 |
| 2.1 | 8.16616 | .12245 | .7419 | 8.1 | 3294.45 | .00030 | 2.0919 |
| 2.2 | 9.02500 | .11080 | .7885 | 8.2 | 3640.94 | .00027 | 2.1041 |
| 2.3 | 9.97417 | .10025 | .8329 | 8.3 | 4023.86 | .00025 | 2.1163 |
| 2.4 | 11.0231 | .09071 | .8755 | 8.4 | 4447.05 | .00022 | 2.1282 |
| 2.5 | 12.1824 | .08208 | .9163 | 8.5 | 4914.75 | .00020 | 2.1401 |
| 2.6 | 13.4637 | .07427 | .9555 | 8.6 | 5431.65 | .00018 | 2.1518 |
| 2.7 | 14.8797 | .06720 | .9933 | 8.7 | 6002.90 | .00017 | 2.1633 |
| 2.8 | 16.4446 | .06081 | 1.0296 | 8.8 | 6634.23 | .00015 | 2.1748 |
| 2.9 | 18.1741 | .05502 | 1.0647 | 8.9 | 7331.96 | .00014 | 2.1861 |
| | | | | 9.0 | 8103.08 | .00012 | 2.1972 |
| 3.0 | 20.0855 | .04978 | 1.0986 | 9.1 | 8955.29 | .00011 | 2.2083 |
| 3.1 | 22.1979 | .04505 | 1.1314 | 9.2 | 9897.13 | .00010 | 2.2192 |
| 3.2 | 24.5325 | .04076 | 1.1632 | 9.3 | 10938.0 | .00009 | 2.2300 |
| 3.3 | 27.1126 | .03688 | 1.1939 | 9.4 | 12088.4 | .00008 | 2.2407 |
| 3.4 | 29.9641 | .03337 | 1.2238 | 9.5 | 13359.7 | .000075 | 2.2513 |
| 3.5 | 33.1154 | .03020 | 1.2528 | 9.6 | 14764.8 | .000068 | 2.2618 |
| 3.6 | 36.5982 | .02732 | 1.2809 | 9.7 | 16317.6 | .000061 | 2.2721 |
| 3.7 | 40.4473 | .02472 | 1.3083 | 9.8 | 18033.8 | .000055 | 2.2824 |
| 3.8 | 44.7012 | .02237 | 1.3350 | 9.9 | 19930.4 | .000050 | 2.2925 |
| 3.9 | 49.4024 | .02024 | 1.3610 | 10.0 | 22026.5 | .000045 | 2.3026 |

# Integrals

(C is an arbitrary constant; in all expressions involving ln x, it is assumed that $x > 0$.)

**1.** $\int x^n \, dx = \dfrac{1}{n+1} x^{n+1} + C \qquad$ (if $n \neq -1$)

**2.** $\int e^{kx} \, dx = \dfrac{1}{k} e^{kx} + C$

**3.** $\int \dfrac{a}{x} \, dx = a \ln |x| + C \qquad (a \neq 0)$

**4.** $\int \ln |ax| \, dx = x \, (\ln |ax| - 1) + C$

**5.** $\int \dfrac{1}{\sqrt{x^2 + a^2}} \, dx = \ln \left| \dfrac{x + \sqrt{x^2 + a^2}}{a} \right| + C$

**6.** $\int \dfrac{1}{\sqrt{x^2 - a^2}} \, dx = \ln \left| \dfrac{x + \sqrt{x^2 - a^2}}{a} \right| + C$

**7.** $\int \dfrac{1}{a^2 - x^2} \, dx = \dfrac{1}{2a} \cdot \ln \left| \dfrac{a + x}{a - x} \right| + C \qquad (x^2 \neq a^2)$

**8.** $\int \dfrac{1}{x^2 - a^2} \, dx = \dfrac{1}{2a} \cdot \ln \left| \dfrac{x - a}{x + a} \right| + C \qquad (x^2 \neq a^2)$

**9.** $\int \dfrac{1}{x\sqrt{a^2 - x^2}} \, dx = -\dfrac{1}{a} \cdot \ln \left| \dfrac{a + \sqrt{a^2 - x^2}}{x} \right| + C \qquad (0 < x < a)$

**10.** $\int \dfrac{1}{x\sqrt{a^2 + x^2}} \, dx = -\dfrac{1}{a} \cdot \ln \left| \dfrac{a + \sqrt{a^2 + x^2}}{x} \right| + C$

**11.** $\int \dfrac{x}{ax + b} \, dx = \dfrac{x}{a} - \dfrac{b}{a^2} \cdot \ln |ax + b| + C \qquad (a \neq 0)$

**12.** $\int \dfrac{x}{(ax + b)^2} \, dx = \dfrac{b}{a^2(ax + b)} + \dfrac{1}{a^2} \cdot \ln |ax + b| + C \qquad (a \neq 0)$

**13.** $\int \dfrac{1}{x(ax + b)} \, dx = \dfrac{1}{b} \cdot \ln \left| \dfrac{x}{ax + b} \right| + C \qquad (b \neq 0)$

**14.** $\int \dfrac{1}{x(ax + b)^2} \, dx = \dfrac{1}{b(ax + b)} + \dfrac{1}{b^2} \cdot \ln \left| \dfrac{x}{ax \cdot b} \right| + C \qquad (b \neq 0)$

**15.** $\int \sqrt{x^2 + a^2} \, dx = \dfrac{x}{2} \sqrt{x^2 + a^2} + \dfrac{a^2}{2} \cdot \ln \left| x + \sqrt{x^2 + a^2} \right| + C$

**16.** $\int x^n \cdot \ln x \, dx = x^{n+1} \left[ \dfrac{\ln |x|}{n + 1} - \dfrac{1}{(n + 1)^2} \right] + C \qquad (n \neq -1)$

**17.** $\int x^n e^{ax} \, dx = \dfrac{x^n e^{ax}}{a} - \dfrac{n}{a} \cdot \int x^{n-1} e^{ax} \, dx + C \qquad (a \neq 0)$

# Integrals Involving Trigonometric Functions

**18.** $\int \sin u \, du = -\cos u + C$

**19.** $\int \cos u \, du = \sin u + C$

**20.** $\int \sec^2 u \, du = \tan u + C$

**21.** $\int \csc^2 u \, du = -\cot u + C$

**22.** $\int \sec u \tan u \, du = \sec u + C$

**23.** $\int \csc u \cot u \, du = -\csc u + C$

**24.** $\int \tan u \, du = \ln |\sec u| + C$

**25.** $\int \cot u \, du = \ln |\sin u| + C$

**26.** $\int \sec u \, du = \ln |\sec u + \tan u| + C$

**27.** $\int \csc u \, du = \ln |\csc u - \cot u| + C$

**28.** $\int \dfrac{du}{\sqrt{a^2 - u^2}} = \text{Arcsin} \dfrac{u}{a} + C$

**29.** $\int \dfrac{du}{a^2 + u^2} = \dfrac{1}{a} \text{Arctan} \dfrac{u}{a} + C$

**30.** $\int \dfrac{du}{u\sqrt{u^2 - a^2}} = \dfrac{1}{a} \text{Arcsec} \dfrac{u}{a} + C$

**31.** $\int \sqrt{a^2 - u^2} \, du = \dfrac{u}{2} \sqrt{a^2 - u^2} + \dfrac{a^2}{2} \text{Arcsin} \dfrac{u}{a} + C$

**32.** $\int u^2 \sqrt{a^2 - u^2} \, du = \dfrac{u}{8} (2u^2 - a^2) \sqrt{a^2 - u^2} + \dfrac{a^4}{8} \text{Arcsin} \dfrac{u}{a} + C$

**33.** $\int \dfrac{\sqrt{a^2 - u^2}}{u^2} \, du = -\dfrac{1}{u} \sqrt{a^2 - u^2} - \text{Arcsin} \dfrac{u}{a} + C$

**34.** $\int \dfrac{u^2 \, du}{\sqrt{a^2 - u^2}} = -\dfrac{u}{2} \sqrt{a^2 - u^2} + \dfrac{a^2}{2} \text{Arcsin} \dfrac{u}{a} + C$

**35.** $\int \dfrac{\sqrt{u^2 - a^2}}{u} \, du = \sqrt{u^2 - a^2} - a \, \text{Arccos} \dfrac{a}{u} + C$

**36.** $\displaystyle\int \sin^n u\, du = -\frac{1}{n}\sin^{n-1} u\cos u + \frac{n-1}{aan}\int \sin^{n-2} u\, du$

**37.** $\displaystyle\int \cos^n u\, du = \frac{1}{n}\cos^{n-1} u\sin u + \frac{n-1}{n}\int \cos^{n-2} u\, du$

**38.** $\displaystyle\int \tan^n u\, du = \frac{1}{n-1}\tan^{n-1} u - \int \tan^{n-2} u\, du$

**39.** $\displaystyle\int \sec^n u\, du = \frac{1}{n-1}\tan u\sec^{n-2} u + \frac{n-2}{n-1}\int \sec^{n-2} u\, du$

**40.** $\displaystyle\int \sin au\sin bu\, du = \frac{\sin (a-b)u}{2(a-b)} - \frac{\sin (a+b)u}{2(a+b)} + C$

**41.** $\displaystyle\int \cos au\cos bu\, du = \frac{\sin (a-b)u}{2(a-b)} + \frac{\sin (a+b)u}{2(a+b)} + C$

**42.** $\displaystyle\int \sin au\cos bu\, du = -\frac{\cos(a-b)u}{2(a-b)} - \frac{\cos (a+b)u}{2(a+b)} + C$

**43.** $\displaystyle\int u\sin u\, du = \sin u - u\cos u + C$

**44.** $\displaystyle\int u^n\sin u\, du = -u^n\cos u + n\int u^{n-1}\cos u\, du$

**45.** $\displaystyle\int u\,\mathrm{Arccos}\, u\, du = \frac{2u^2-1}{4}\,\mathrm{Arccos}\, u - \frac{u\sqrt{1-u^2}}{4} + C$

**46.** $\displaystyle\int u\,\mathrm{Arcsin}\, u\, du = \frac{2u^2-1}{4}\,\mathrm{Arcsin}\, u + \frac{u\sqrt{1-u^2}}{4} + C$

**47.** $\displaystyle\int e^{au}\sin bu\, du = \frac{e^{au}}{a^2+b^2}(a\sin bu - b\cos bu) + C$

**48.** $\displaystyle\int e^{au}\cos bu\, du = \frac{e^{au}}{a^2+b^2}(a\cos bu + b\sin bu) + C$

# Trigonometric Functions

| Deg | Rad | Sin | Cos | Tan | Deg | Rad | Sin | Cos | Tan |
|---|---|---|---|---|---|---|---|---|---|
| 0 | .0000 | .0000 | 1.0000 | .0000 | 45 | .7854 | .7071 | .7071 | 1.0000 |
| 1 | .0175 | .0175 | .9998 | .0175 | 46 | .8029 | .7193 | .6947 | 1.0355 |
| 2 | .0349 | .0349 | .9994 | .0349 | 47 | .8203 | .7314 | .6820 | 1.0724 |
| 3 | .0524 | .0523 | .9986 | .0524 | 48 | .8378 | .7431 | .6691 | 1.1106 |
| 4 | .0698 | .0698 | .9976 | .0699 | 49 | .8552 | .7547 | .6561 | 1.1504 |
| 5 | .0873 | .0872 | .9962 | .0875 | 50 | .8727 | .7660 | .6428 | 1.1918 |
| 6 | .1047 | .1045 | .9945 | .1051 | 51 | .8901 | .7771 | .6293 | 1.2349 |
| 7 | .1222 | .1219 | .9925 | .1228 | 52 | .9076 | .7880 | .6157 | 1.2799 |
| 8 | .1396 | .1392 | .9903 | .1405 | 53 | .9250 | .7986 | .6018 | 1.3270 |
| 9 | .1571 | .1564 | .9877 | .1584 | 54 | .9425 | .8090 | .5878 | 1.3764 |
| 10 | .1745 | .1736 | .9848 | .1763 | 55 | .9599 | .8192 | .5736 | 1.4281 |
| 11 | .1920 | .1908 | .9816 | .1944 | 56 | .9774 | .8290 | .5592 | 1.4826 |
| 12 | .2094 | .2079 | .9781 | .2126 | 57 | .9948 | .8387 | .5446 | 1.5399 |
| 13 | .2269 | .2250 | .9744 | .2309 | 58 | 1.0123 | .8480 | .5299 | 1.6003 |
| 14 | .2443 | .2419 | .9703 | .2493 | 59 | 1.0297 | .8572 | .5150 | 1.6643 |
| 15 | .2618 | .2588 | .9659 | .2679 | 60 | 1.0472 | .8660 | .5000 | 1.7321 |
| 16 | .2793 | .2756 | .9613 | .2867 | 61 | 1.0647 | .8746 | .4848 | 1.8040 |
| 17 | .2967 | .2924 | .9563 | .3057 | 62 | 1.0821 | .8829 | .4695 | 1.8807 |
| 18 | .3142 | .3090 | .9511 | .3249 | 63 | 1.0996 | .8910 | .4540 | 1.9626 |
| 19 | .3316 | .3256 | .9455 | .3443 | 64 | 1.1170 | .8988 | .4384 | 2.0503 |
| 20 | .3491 | .3420 | .9397 | .3640 | 65 | 1.1345 | .9063 | .4226 | 2.1445 |
| 21 | .3665 | .3584 | .9336 | .3839 | 66 | 1.1519 | .9135 | .4067 | 2.2460 |
| 22 | .3840 | .3746 | .9272 | .4040 | 67 | 1.1694 | .9205 | .3907 | 2.3559 |
| 23 | .4014 | .3907 | .9205 | .4245 | 68 | 1.1868 | .9272 | .3746 | 2.4751 |
| 24 | .4189 | .4067 | .9135 | .4452 | 69 | 1.2043 | .9336 | .3584 | 2.6051 |
| 25 | .4363 | .4226 | .9063 | .4663 | 70 | 1.2217 | .9397 | .3420 | 2.7475 |
| 26 | .4538 | .4384 | .8988 | .4877 | 71 | 1.2392 | .9455 | .3256 | 2.9042 |
| 27 | .4712 | .4540 | .8910 | .5095 | 72 | 1.2566 | .9511 | .3090 | 3.0777 |
| 28 | .4887 | .4695 | .8829 | .5317 | 73 | 1.2741 | .9563 | .2924 | 3.2709 |
| 29 | .5061 | .4848 | .8746 | .5543 | 74 | 1.2915 | .9613 | .2756 | 3.4874 |
| 30 | .5236 | .5000 | .8660 | .5774 | 75 | 1.3090 | .9659 | .2588 | 3.7321 |
| 31 | .5411 | .5150 | .8572 | .6009 | 76 | 1.3265 | .9703 | .2419 | 4.0108 |
| 32 | .5585 | .5299 | .8480 | .6249 | 77 | 1.3439 | .9744 | .2250 | 4.3315 |
| 33 | .5760 | .5446 | .8387 | .6494 | 78 | 1.3614 | .9781 | .2079 | 4.7046 |
| 34 | .5934 | .5592 | .8290 | .6745 | 79 | 1.3788 | .9816 | .1908 | 5.1446 |
| 35 | .6109 | .5736 | .8192 | .7002 | 80 | 1.3963 | .9848 | .1736 | 5.6713 |
| 36 | .6283 | .5878 | .8090 | .7265 | 81 | 1.4137 | .9877 | .1564 | 6.3138 |
| 37 | .6458 | .6018 | .7986 | .7536 | 82 | 1.4312 | .9903 | .1392 | 7.1154 |
| 38 | .6632 | .6157 | .7880 | .7813 | 83 | 1.4486 | .9925 | .1219 | 8.1442 |
| 39 | .6807 | .6293 | .7771 | .8098 | 84 | 1.4661 | .9945 | .1045 | 9.5144 |
| 40 | .6981 | .6428 | .7660 | .8391 | 85 | 1.4835 | .9962 | .0872 | 11.4301 |
| 41 | .7156 | .6561 | .7547 | .8693 | 86 | 1.5010 | .9976 | .0698 | 14.3007 |
| 42 | .7330 | .6691 | .7431 | .9004 | 87 | 1.5184 | .9986 | .0523 | 19.0811 |
| 43 | .7505 | .6820 | .7314 | .9325 | 88 | 1.5359 | .9994 | .0349 | 28.6363 |
| 44 | .7679 | .6947 | .7193 | .9657 | 89 | 1.5533 | .9998 | .0175 | 57.2900 |
|  |  |  |  |  | 90 | 1.5708 | 1.0000 | .0000 | — |

# ANSWERS TO SELECTED EXERCISES

If you need further help with this course, you may want to obtain a copy of the *Student's Solutions Manual* that accompanies this textbook. This manual provides detailed step-by-step solutions to the odd-numbered exercises in the textbook and can help you study and understand the course material. Your college bookstore either has this manual or can order it for you.

> Note that answers to computer problems may vary slightly, because different computers and software may be used to solve the problems.

## REVIEW OF ALGEBRA    SECTION R.1    (PAGE 4)

**1.** $-x^2 + x + 9$    **2.** $-6y^2 + 3y + 10$    **3.** $-14q^2 + 11q - 14$    **4.** $9r^2 - 4r + 19$
**5.** $-.327x^2 - 2.805x - 1.458$    **6.** $-2.97r^2 - 8.083r + 7.81$    **7.** $-18m^3 - 27m^2 + 9m$    **8.** $12k^2 - 20k + 3$
**9.** $25r^2 + 5rs - 12s^2$    **10.** $18k^2 - 7kq - q^2$    **11.** $\frac{6}{25}y^2 + \frac{11}{40}yz + \frac{1}{16}z^2$    **12.** $\frac{15}{16}r^2 - \frac{7}{12}rs - \frac{2}{9}s^2$
**13.** $.0036x^2 - .04452x - .0918$    **14.** $4.34m^2 + 5.68m - 4.42$    **15.** $27p^3 - 1$    **16.** $6p^3 - 11p^2 + 14p - 5$
**17.** $8m^3 + 1$    **18.** $12k^4 + 21k^3 - 5k^2 + 3k + 2$    **19.** $m^2 + mn - 2n^2 - 2km + 5kn - 3k^2$
**20.** $2r^2 - 7rs + 3s^2 + 3rt - 4st + t^2$

## SECTION R.2    (PAGE 6)

**1.** $8a(a^2 - 2a + 3)$    **2.** $3y(y^2 + 8y + 3)$    **3.** $5p^2(5p^2 - 4pq + 20q^2)$    **4.** $10m^2(6m^2 - 12mn + 5n^2)$
**5.** $(m + 7)(m + 2)$    **6.** $(x + 5)(x - 1)$    **7.** $(z + 4)(z + 5)$    **8.** $(b - 7)(b - 1)$    **9.** $(a - 5b)(a - b)$
**10.** $(s - 5t)(s + 7t)$    **11.** $(y - 7z)(y + 3z)$    **12.** $6(a - 10)(a + 2)$    **13.** $3m(m + 3)(m + 1)$
**14.** $(2x + 1)(x - 3)$    **15.** $(3a + 7)(a + 1)$    **16.** $(2a - 5)(a - 6)$    **17.** $(5y + 2)(3y - 1)$
**18.** $(7m + 2n)(3m + n)$    **19.** $2a^2(4a - b)(3a + 2b)$    **20.** $4z^3(8z + 3a)(z - a)$    **21.** $(x + 8)(x - 8)$
**22.** $(3m + 5)(3m - 5)$    **23.** $(11a + 10)(11a - 10)$    **24.** Prime    **25.** $(z + 7y)^2$    **26.** $(m - 3n)^2$    **27.** $(3p - 4)^2$
**28.** $(a - 6)(a^2 + 6a + 36)$    **29.** $(2r - 3s)(4r^2 + 6rs + 9s^2)$    **30.** $(4m + 5)(16m^2 - 20m + 25)$

## SECTION R.3    (PAGE 12)

**1.** 12    **2.** $-2/7$    **3.** $-7/8$    **4.** $-1$    **5.** $-11/3, 7/3$    **6.** $-11/7, 19/7$    **7.** $-12, -4/3$    **8.** $-5, 3/4$
**9.** $-4, 5$    **10.** $-3, -2$    **11.** $-1, 6$    **12.** $-8, 3$    **13.** $-1, 3$    **14.** 4    **15.** $-2, 5/2,$    **16.** $-1/2, 4/3,$
**17.** 2, 5    **18.** $-4/3, 4/3$    **19.** $-4, 1/2$    **20.** 0, 4    **21.** $(5 + \sqrt{13})/6 \approx 1.434,$    $(5 - \sqrt{13})/6 \approx .232$
**22.** $(1 + \sqrt{33})/4 \approx 1.686,$    $(1 - \sqrt{33})/4 \approx -1.186$    **23.** $(-1 + \sqrt{5})/2 \approx .618,$    $(-1 - \sqrt{5})/2 \approx -1.618$
**24.** $5 + \sqrt{5} \approx 7.236,$    $5 - \sqrt{5} \approx 2.764$    **25.** $(-6 + \sqrt{26})/2 \approx -.450, (-6 - \sqrt{26})/2 \approx -5.550$
**26.** 1, 5/2    **27.** 1/2, 4/3    **28.** $-5, 2$    **29.** $-1, 5/2$    **30.** No real-number solutions
**31.** $(-1 + \sqrt{73})/6 \approx 1.257,$    $(-1 - \sqrt{73})/6 \approx -1.591$    **32.** $-1, 0$    **33.** 3    **34.** 12    **35.** $-12/5$
**36.** $-48/71$    **37.** $-59/6$    **38.** $-11/5$    **39.** No real-number solutions    **40.** $-5/2$    **41.** $-3, 4$
**42.** 2/3    **43.** 1    **44.** $(-13 - \sqrt{185})/4 \approx -6.650,$    $(-13 + \sqrt{185})/4 \approx .150$

## SECTION R.4    (PAGE 17)

**1.** $(-\infty, -1]$

**2.** $(-\infty, 1)$

**3.** $(-1, +\infty)$

**4.** $(-\infty, 1]$

**5.** $(1/5, +\infty)$

**6.** $(1/3, +\infty)$

**7.** $(-5, 6)$

**8.** $[7/3, 4]$

**9.** $[-11/2, 7/2]$

**10.** $[-1, 2]$

**11.** $[-17/7, +\infty)$

**12.** $(-\infty, 50/9]$

**13.** $(-2, 4)$

**14.** $(-\infty, -6] \cup [1, +\infty)$

**15.** $(1, 2)$

**16.** $(-\infty, -4) \cup (1/2, +\infty)$

**17.** $[1, 6]$

**18.** $[-3/2, 5]$

**19.** $(-\infty, -1/2) \cup (1/3, +\infty)$

**20.** $[-1/2, 2/5]$

**21.** $[-3, 1/2]$

**22.** $(-\infty, -2) \cup (5/3, +\infty)$ ·

**23.** $[-5, 5]$

**24.** $(-\infty, 0) \cup (16, +\infty)$

**25.** $(-5, 3]$    **26.** $(-\infty, -1) \cup (1, +\infty)$    **27.** $(-\infty, -2)$    **28.** $(-2, 3/2)$    **29.** $[-8, 5)$
**30.** $(-\infty, -3/2) \cup [-13/9, +\infty)$    **31.** $(-2, +\infty)$    **32.** $(-\infty, -1)$    **33.** $(-\infty, -1) \cup (-1/2, 1) \cup (2, +\infty)$
**34.** $(-4, -2) \cup (0, 2)$    **35.** $(1, 3/2]$    **36.** $(-\infty, -2) \cup (-2, 2) \cup [4, +\infty)$

## SECTION R.5    (PAGE 20)

**1.** $z/2$    **2.** $5p/2$    **3.** $8/9$    **4.** $3/(t - 3)$    **5.** $2(x + 2)/x$    **6.** $4(y + 2)$    **7.** $(m - 2)/(m + 3)$
**8.** $(r + 2)/(r + 4)$    **9.** $(x + 4)/(x + 1)$    **10.** $(z - 3)/(z + 2)$    **11.** $(2m + 3)/(4m + 3)$    **12.** $(2y + 1)/(y + 1)$
**13.** $3k/5$    **14.** $25p^2/9$    **15.** $6/(5p)$    **16.** $2$    **17.** $2/9$    **18.** $3/10$    **19.** $2(a + 4)/(a - 3)$    **20.** $2/(r + 2)$

**21.** $(k + 2)/(k + 3)$    **22.** $(m + 6)/(m + 3)$    **23.** $(m - 3)/(2m - 3)$    **24.** $(2n - 3)/(2n + 3)$    **25.** 1
**26.** $(6 + p)/(2p)$    **27.** $(8 - y)/(4y)$    **28.** $137/(30m)$    **29.** $(3m - 2)/[m(m - 1)]$    **30.** $(r - 12)/[r(r - 2)]$
**31.** $14/[3(a - 1)]$    **32.** $23/[20(k - 2)]$    **33.** $(7x + 9)/[(x - 3)(x + 1)(x + 2)]$    **34.** $y^2/[(y + 4)(y + 3)(y + 2)]$
**35.** $k(k - 13)/[(2k - 1)(k + 2)(k - 3)]$    **36.** $m(3m - 19)/[(3m - 2)(m + 3)(m - 4)]$

## ▰▰ SECTION R.6    (PAGE 23)

**1.** $1/64$    **2.** $1/81$    **3.** $1/216$    **4.** 1    **5.** 1    **6.** $3/4$    **7.** $-1/16$    **8.** $1/16$    **9.** $-1/9$    **10.** $1/9$
**11.** $25/64$    **12.** $216/343$    **13.** 8    **14.** 125    **15.** $49/4$    **16.** $27/64$    **17.** $1/7^4$    **18.** $1/3^6$    **19.** $1/2^3$
**20.** $1/6^2$    **21.** $4^3$    **22.** $8^5$    **23.** $1/10^8$    **24.** 5    **25.** $x^2$    **26.** $y^3$    **27.** $2^3k^3$    **28.** $1/(3z^7)$
**29.** $x^2/(2y)$    **30.** $m^3/5^4$    **31.** $a^3b^6$    **32.** $d^6/(2^2c^4)$    **33.** $1/6$    **34.** $-17/9$    **35.** $-13/66$    **36.** $81/26$
**37.** $35/18$    **38.** $213/200$    **39.** 9    **40.** 3    **41.** 4    **42.** 100    **43.** 4    **44.** $-25$    **45.** $2/3$    **46.** $4/3$
**47.** $1/32$    **48.** $1/5$    **49.** $4/3$    **50.** $1000/1331$    **51.** $2^2$    **52.** $27^{1/3}$    **53.** $4^2$    **54.** 1    **55.** $r$    **56.** $12^3/y^8$
**57.** $1/(2^2 \cdot 3k^{5/2})$ or $1/(12k^{5/2})$    **58.** $1/(2p^2)$    **59.** $a^{2/3}b^2$    **60.** $y/(x^{4/3}z^{1/2})$    **61.** $h^{1/3}t^{1/5}/k^{2/5}$    **62.** $m^3p/n$    **63.** 1
**64.** $m^{6/5}$    **65.** $x^{11/12}$    **66.** $-4a^2$    **67.** $16/y^{11/12}$    **68.** $9k^{1/3}$    **69.** $3x^3(x^2 - 1)^{-1/2}$
**70.** $5(5x + 2)^{-1/2}(45x^2 + 3x - 5)$    **71.** $(2x + 5)(x^2 - 4)^{-1/2}(4x^2 + 5x - 8)$    **72.** $(4x^2 + 1)(2x - 1)^{-1/2}(8x^2 - 1)$

## ▰▰ SECTION R.7    (PAGE 27)

**1.** 5    **2.** 6    **3.** $-5$    **4.** $5\sqrt{2}$    **5.** $20\sqrt{5}$    **6.** $4y^2\sqrt{2y}$    **7.** $7\sqrt{2}$    **8.** $9\sqrt{3}$    **9.** $2\sqrt{5}$    **10.** $\sqrt{2}$
**11.** $32\sqrt{3}$    **12.** $-2\sqrt{7}$    **13.** $2\sqrt[3]{2}$    **14.** $5\sqrt[3]{2}$    **15.** $7\sqrt[3]{3}$    **16.** $3\sqrt[3]{4}$    **17.** $xyz^2\sqrt{2x}$    **18.** $7rs^2t^5\sqrt{2r}$
**19.** $2zx^2y\sqrt[3]{2z^2x^2y}$    **20.** $x^2yz^2\sqrt[4]{y^3z^3}$    **21.** $ab\sqrt{ab}(b - 2a^2 + b^3)$    **22.** $p^2\sqrt{pq}(pq - q^4 + p^2)$    **23.** $5\sqrt{7}/7$
**24.** $-2\sqrt{3}/3$    **25.** $-\sqrt{3}/2$    **26.** $\sqrt{2}$    **27.** $-3(1 + \sqrt{5})/4$    **28.** $-5(2 + \sqrt{6})/2$    **29.** $-2(\sqrt{3} + \sqrt{2})$
**30.** $(\sqrt{10} - \sqrt{3})/7$    **31.** $(\sqrt{r} + \sqrt{3})/(r - 3)$    **32.** $5(\sqrt{m} + \sqrt{5})/(m - 5)$    **33.** $\sqrt{y} + \sqrt{5}$    **34.** $\sqrt{z} + \sqrt{11}$
**35.** $-2x - 2\sqrt{x(x + 1)} - 1$    **36.** $(p^2 + p + 2\sqrt{p(p^2 - 1)} - 1)/(-p^2 + p + 1)$    **37.** $-1/[2(1 - \sqrt{2})]$
**38.** $-2/[3(1 + \sqrt{3})]$    **39.** $x/(\sqrt{x} + x)$    **40.** $p/(\sqrt{p} - p)$    **41.** $-1/(2x - 2\sqrt{x(x + 1)} + 1)$
**42.** $(-p^2 + p + 1)/(p^2 + p - 2\sqrt{p(p^2 - 1)} - 1)$

## ▰▰ CHAPTER 1    SECTION 1.1    (PAGE 39)

**1.** Not a function    **3.** Function    **5.** Function    **7.** Not a function
**9.** $(-2, -3), (-1, -2), (0, -1), (1, 0), (2, 1),$    **11.** $(-2, 17), (-1, 13), (0, 9), (1, 5), (2, 1),$
$(3, 2)$; range: $\{-3, -2, -1, 0, 1, 2\}$        $(3, -3)$; range: $\{-3, 1, 5, 9, 13, 17\}$

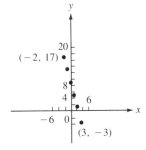

**13.** $(-2, 13)$, $(-1, 11)$, $(0, 9)$, $(1, 7)$, $(2, 5)$, $(3, 3)$; range: $\{3, 5, 7, 9, 11, 13\}$

**15.** $(-2, 3/2)$, $(-1, 2)$, $(0, 5/2)$, $(1, 3)$, $(2, 7/2)$, $(3, 4)$; range: $\{3/2, 2, 5/2, 3, 7/2, 4\}$

**17.** $(-2, 2)$, $(-1, 0)$, $(0, 0)$, $(1, 2)$, $(2, 6)$, $(3, 12)$; range: $\{0, 2, 6, 12\}$

**19.** $(-2, 4)$, $(-1, 1)$, $(0, 0)$, $(1, 1)$, $(2, 4)$, $(3, 9)$; range: $\{0, 1, 4, 9\}$

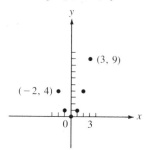

**21.** $(-2, 1)$, $(-1, 1/2)$, $(0, 1/3)$, $(1, 1/4)$, $(2, 1/5)$, $(3, 1/6)$; range: $\{1, 1/2, 1/3, 1/4, 1/5, 1/6\}$

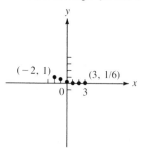

**23.** $(-2, -3)$, $(-1, -3/2)$, $(0, -3/5)$, $(1, 0)$, $(2, 3/7)$, $(3, 3/4)$; range: $\{-3, -3/2, -3/5, 0, 3/7, 3/4\}$

**25.** $(-\infty, 0)$

**27.** $[1, 2)$

**29.** $(-\infty, -9)$

**31.** $-4 < x < 3$   **33.** $x \le -1$   **35.** $-2 \le x < 6$   **37.** $x \le -4$ or $x \ge 4$   **39.** $(-\infty, +\infty)$   **41.** $(-\infty, +\infty)$
**43.** $[-4, 4]$   **45.** $[3, +\infty)$   **47.** $x \ne \pm 2$, or $(-\infty, -2) \cup (-2, 2) \cup (2, +\infty)$   **49.** $(-\infty, +\infty)$
**51.** $(-\infty, -1] \cup [5, +\infty)$   **53.** Domain: $[-5, 4]$; range: $[-2, 6]$   **55.** Domain: $(-\infty, +\infty)$; range: $(-\infty, 12]$
**57. (a)** 14   **(b)** $-7$   **(c)** $1/2$   **(d)** $3a + 2$   **(e)** $6/m + 2$   **59. (a)** 5   **(b)** $-23$   **(c)** $-7/4$   **(d)** $-a^2 + 5a + 1$
**(e)** $-4/m^2 + 10/m + 1$ or $(-4 + 10m + m^2)/m^2$   **61. (a)** $9/2$   **(b)** 1   **(c)** 0   **(d)** $(2a + 1)/(a - 2)$
**(e)** $(4 + m)/(2 - 2m)$   **63. (a)** 0   **(b)** 4   **(c)** 3   **(d)** 4   **65. (a)** $-3$   **(b)** $-2$   **(c)** $-1$   **(d)** 2   **67.** $6m - 20$

**69.** $r^2 + 2rh + h^2 - 2r - 2h + 5$     **71.** $9/q^2 - 6/q + 5$ or $(9 - 6q + 5q^2)/q^2$     **73.** Function     **75.** Not a function
**77.** Function     **79. (a)** $x^2 + 2xh + h^2 - 4$   **(b)** $2xh + h^2$   **(c)** $2x + h$     **81. (a)** $6x + 6h + 2$   **(b)** $6h$   **(c)** 6
**83. (a)** $1/(x + h)$   **(b)** $-h/[x(x + h)]$   **(c)** $-1/[x(x + h)]$     **85. (a)** \$11   **(b)** \$11   **(c)** \$18   **(d)** \$32   **(e)** \$32   **(f)** \$39
**(g)** \$39   **(h)** Continue the horizontal bars up and to the right.   **(i)** $x$, the number of days   **(j)** $S$, the cost of renting a saw

## ▰▰ SECTION 1.2   (PAGE 52)

**1.** 3/5     **3.** Not defined     **5.** 2     **7.** 5/9     **9.** Not defined     **11.** 2     **13.** .5785     **15.** $2x + y = 5$     **17.** $y = 1$
**19.** $x + 3y = 10$     **21.** $3x + 4y = 12$     **23.** $2x - 3y = 6$     **25.** $x = -6$     **27.** $5.081x + y = -4.69$

**29.**

**31.**

**33.**

**35.**

**37.**

**39.**

**41.**

**43.** $x + 3y = 11$     **45.** $x - y = 7$     **47.** $-2x + y = 4$
**49.** $2x - 3y = -6$     **51.** No     **59. (a)**
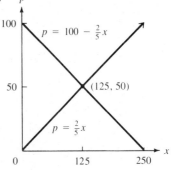

**(b)** 125 units   **(c)** 50

**61.** **(a)** $52   **(b)** $52   **(c)** $52
**(d)** $79   **(e)** $106
**(f)**

**(g)** Yes   **(h)** No

**63.** **(a)** $h = (8/3)t + 211/3$   **(b)** About 172 cm; about 190 cm
**(c)** About 45 cm
**65.** $m = 2.5$; $y = 2.5x - 70$

---

## ▦ SECTION 1.3   (PAGE 61)

**1.** If $C(x)$ is the cost of renting a saw for $x$ hours, then $C(x) = 12 + x$.   **3.** If $C(x)$ is the cost in cents of parking for $x$ half-hours, then $C(x) = 35x + 50$.   **5.** $C(x) = 30x + 100$   **7.** $C(x) = 25x + 1000$   **9.** $C(x) = 50x + 500$
**11.** $C(x) = 90x + 2500$   **13.** **(a)** 2600  **(b)** 2900  **(c)** 3200  **(d)** 2000  **(e)** 300   **15.** **(a)** $y = (800,000/7)x + 200,000$
**(b)** About $429,000   **(c)** About $1,230,000   **17.** **(a)** $97  **(b)** $97.097  **(c)** $.097 or 9.7¢  **(d)** $.097 or 9.7¢
**19.** **(a)** $100  **(b)** $36  **(c)** $24   **21.** 500 units; 30,000   **23.** Break-even point is 45 units; don't produce.
**25.** Break-even point is $-50$ units; impossible to make a profit here.   **27.** About $140 billion   **29.** **(a)** $y = .3x + 10.3$
**(b)** .3% per year; they are the same   **31.** **(a)** .3 cm  **(b)** .6 cm  **(c)** 1.5 cm  **(d)** 3.0 cm  **(e)** .03   **33.** **(a)** 13.95 min
**(b)** 26.7 min   **(c)** 52.2 min  **(d)** 103.2 min  **(e)** About 69 min  **(f)** About 104.5 min   **35.** **(a)** 37°C  **(b)** $-40$°F  **(c)** 68°F

---

## ▦ SECTION 1.4   (PAGE 68)

**1.**

**3.**

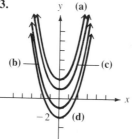

**5.** Vertex is (2, 0); axis is $x = 2$

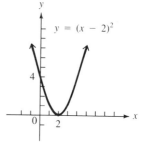

**7.** Vertex is $(-3, -4)$; axis is $x = -3$

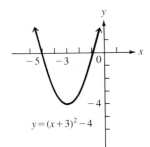

$y = (x+3)^2 - 4$

**9.** Vertex is $(-3, 2)$; axis is $x = -3$

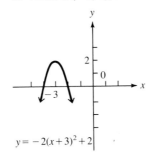

$y = -2(x+3)^2 + 2$

**11.** Vertex is $(-1, -3)$; axis is $x = -1$

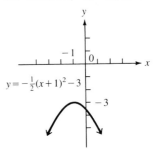

$y = -\frac{1}{2}(x+1)^2 - 3$

**13.** Vertex is $(1, 2)$; axis is $x = 1$

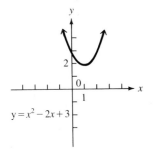

$y = x^2 - 2x + 3$

**15.** Vertex is $(-2, 6)$; axis is $x = -2$

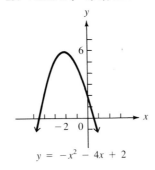

$y = -x^2 - 4x + 2$

**17.** Vertex is $(1, 3)$; axis is $x = 1$

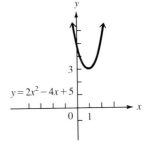

$y = 2x^2 - 4x + 5$

**19.** Vertex is $(3, 7)$; axis is $x = 3$

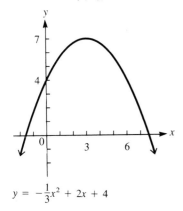

$y = -\frac{1}{3}x^2 + 2x + 4$

**21.** Vertex is $(-1, -1)$; axis is $x = -1$

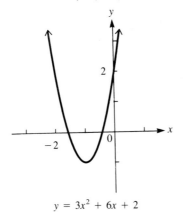

$y = 3x^2 + 6x + 2$

**23.**

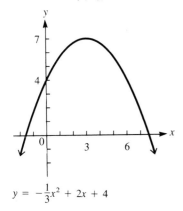

$y = .14x^2 + .56x - .3$

**25.**

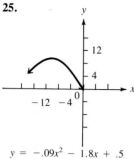

$y = -.09x^2 - 1.8x + .5$

**27.** $x = 1/2$
**29.** Maximum revenue is \$5625;
25 unsold seats

**31. (a)** $x(500 - x) = 500x - x^2$
    **(b)** $R(x)$

    **(c)** 250    **(d)** 62,500

**33. (a)** 60   **(b)** 70   **(c)** 90
    **(d)** 100   **(e)** 80   **(f)** 20
**39.** 10, 10

**35.** 16 ft; 2 sec

**41.** $10\sqrt{3}$ m ≈ 17.32 m

**37.** 80 ft by 160 ft

■■■ **SECTION 1.5**    **(PAGE 78)**

**1.**

$f(x) = (x + 1)^3$

**3.**

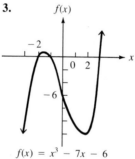

$f(x) = x^3 - 7x - 6$

**5.**

$f(x) = x^4 - 5x^2 + 6$

**7.**

$f(x) = 6x^3 + 11x^2 - x - 6$

**9.**

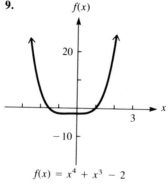

$f(x) = x^4 + x^3 - 2$

**11.**

$f(x) = 8x^4 - 2x^3 - 47x^2$
$- 52x - 15$

**13.**

$f(x) = x^3 + 3x^2 - 9x - 11$

**15.**

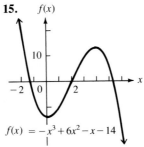

$f(x) = -x^3 + 6x^2 - x - 14$

**17.**

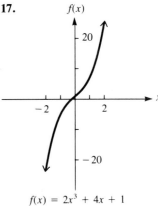

$f(x) = 2x^3 + 4x + 1$

**19.**

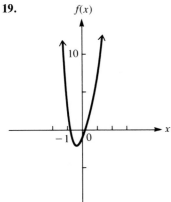

$f(x) = 2x^4 - 3x^3 + 4x^2 + 5x - 1$

**21.** $y = 0; x = 3$

$y = \dfrac{-4}{x-3}$

**23.** $y = 0; x = -3/2$

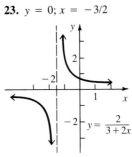

$y = \dfrac{2}{3+2x}$

**25.** $y = 3; x = 1$

$y = \dfrac{3x}{x-1}$

**27.** $y = 1; x = 4$

$y = \dfrac{x+1}{x-4}$

**29.** $y = -2/5; x = -4$

$y = \dfrac{1-2x}{5x+20}$

**31.** $y = -1/3; x = -2$

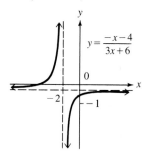

$y = \dfrac{-x-4}{3x+6}$

**33.** **(a)** $12.50; $10; $6.25; $4.76; $3.85
**(b)** Probably $(0, +\infty)$; doesn't seem reasonable to discuss the average cost per unit of no unit
**(c)**

$C(x) = \dfrac{500}{x+30}$

**35.** **(a)** $42.9 million   **(b)** $40 million
**(c)** $30 million   **(d)** $0 million
**(e)**

$y = \dfrac{60x - 6000}{x - 120}$

**37.** **(a)** $6700   **(b)** $15,600
**(c)** $26,800   **(d)** $60,300
**(e)** $127,300   **(f)** $328,300
**(g)** $663,300   **(h)** No
**(i)**

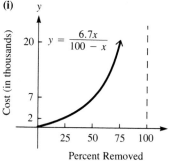

$y = \dfrac{6.7x}{100 - x}$

**39.**

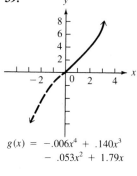

$g(x) = -.006x^4 + .140x^3 - .053x^2 + 1.79x$

**41.** **(a)** $A(x)$

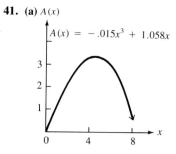

$A(x) = -.015x^3 + 1.058x$

**(b)** Between 4 and 5 hr, but closer to 5 hr
**(c)** From about 1 hr to about 7 hr

**43.** **(a)** $R = -1000$   **(b)** $G(R) = 1$
**(c)**

$G(R) = \dfrac{R}{1000 + R}$

**45.** **(a)** About $10,000
**(b)** About $20,000

**47.** Maximum of $-10.013$ when $x = .3$; minimum of $-11.328$ when $x = .8$

**49.** Maximum of 84 when $x = -2$; minimum of $-13$ when $x = -1$

**51.**

| $x$ | $-3$ | $-2.5$ | $-2$ | $-1.5$ | $-1$ | $-.5$ | $0$ | $.5$ | $1$ | $1.5$ |
|---|---|---|---|---|---|---|---|---|---|---|
| $f(x)$ | 7 | 9.125 | 9 | 7.375 | 5 | 2.625 | 1 | .875 | 3 | 8.125 |

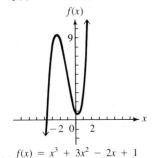

$f(x) = x^3 + 3x^2 - 2x + 1$

**53.**

| $x$ | $-6$ | $-5.5$ | $-5$ | $-4.5$ | $-4$ | $-3.5$ | $-3$ | $-2.5$ |
|---|---|---|---|---|---|---|---|---|
| $f(x)$ | $-2.769$ | $-2.988$ | $-3.333$ | $-3.951$ | $-5.333$ | $-10.889$ | 18 | 3.333 |

| $x$ | $-1.5$ | $-1$ | $-.5$ | $0$ | $.5$ | $1$ | $1.5$ | $2$ |
|---|---|---|---|---|---|---|---|---|
| $f(x)$ | .581 | .222 | .051 | 0 | .051 | .222 | .581 | .1333 |

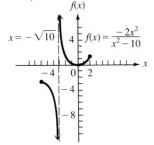

$x = -\sqrt{10}$   $f(x) = \dfrac{-2x^2}{x^2 - 10}$

**55.** No horizontal asymptote

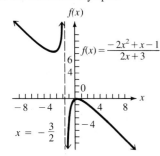

$f(x) = \dfrac{-2x^2 + x - 1}{2x + 3}$

$x = -\dfrac{3}{2}$

**57.** $y = 2$

$f(x) = \dfrac{2x^2 - 5}{x^2 - 1}$

# ▬▬ CHAPTER 1  REVIEW EXERCISES  (PAGE 83)

**1.** $(-3, -16/5)$, $(-2, -14/5)$, $(-1, -12/5)$, $(0, -2)$,
$(1, -8/5)$, $(2, -6/5)$, $(3, -4/5)$; range: $\{-16/5, -14/5,$
$-12/5, -2, -8/5, -6/5, -4/5\}$

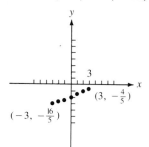

**3.** $(-3, 20)$, $(-2, 9)$, $(-1, 2)$, $(0, -1)$, $(1, 0)$, $(2, 5)$,
$(3, 14)$; range: $\{-1, 0, 2, 5, 9, 14, 20\}$

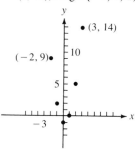

**5.** $(-3, 7)$, $(-2, 2)$, $(-1, -1)$, $(0, -2)$, $(1, -1)$, $(2, 2)$,
$(3, 7)$; range: $\{-2, -1, 2, 7\}$

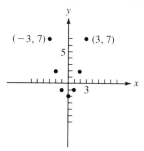

**7.** $(-3, 1/5)$, $(-2, 2/5)$, $(-1, 1)$, $(0, 2)$, $(1, 1)$, $(2, 2/5)$,
$(3, 1/5)$; range: $\{1/5, 2/5, 1, 2\}$

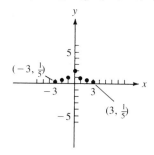

**9.** $(-3, -1)$, $(-2, -1)$, $(-1, -1)$, $(0, -1)$, $(1, -1)$,
$(2, -1)$, $(3, -1)$; range: $\{-1\}$

**11.** **(a)** 23  **(b)** $-9$  **(c)** $-17$  **(d)** $4r + 3$

**13.** **(a)** $-28$  **(b)** $-12$  **(c)** $-28$  **(d)** $-r^2 - 3$

**15.** **(a)** $-13$  **(b)** 3  **(c)** $-32$  **(d)** 22  **(e)** $-k^2 - 4k$
 **(f)** $-9m^2 + 12m$  **(g)** $-k^2 + 14k - 45$  **(h)** $12 - 5p$

**17.**

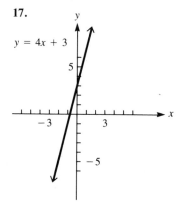

$y = 4x + 3$

**19.**

$3x - 5y = 15$

**21.**

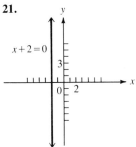

$x + 2 = 0$

**23.**

$y = 2x$

**25.** 1/3
**31.** Not defined
**37.** $x = -1$

**27.** $-2/11$
**33.** $2x - 3y = 13$
**39.** $5x - 8y = -40$

**29.** $-2/3$
**35.** $5x + 4y = 17$
**41.** $y = -5$

**43.**

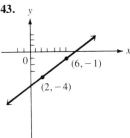

$(6, -1)$
$(2, -4)$

**45.**

$(-3, 4)$
$(-4, 1)$

**47.**

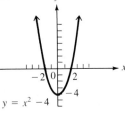

$y = x^2 - 4$

**49.**

$y = 3(x + 1)^2 - 5$

**51.**

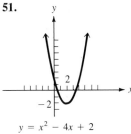

$y = x^2 - 4x + 2$

**53.**

$y = x^3 + 5$

**55.**

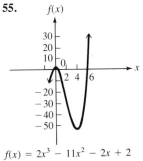

$f(x) = 2x^3 - 11x^2 - 2x + 2$

**57.**

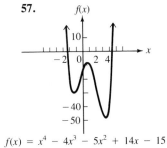

$f(x) = x^4 - 4x^3 - 5x^2 + 14x - 15$

**59.**

$y = \dfrac{8}{x}$

**61.**

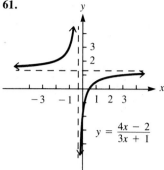

$$y = \frac{4x - 2}{3x + 1}$$

**63. (a)** 7/6; 9/2   **(b)** 2; 2   **(c)** 5/2; 1/2
**(d)** p                      **(e)** 15   **(f)** 2; 2

**65.** $C(x) = 30x + 60$
**67.** $C(x) = 46x + 120$
**69.** The third day; 104.2°F

**71. (a)** Approximately 1.2 yr and 9.8 yr
**(b)** $f(A) > g(A)$ for $2 < A < 9.8$;
$g(A) > f(A)$ for $A < 1.2$ or $A > 9.8$

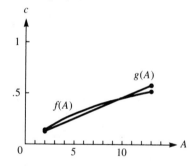

**(c)** At 5 yr and at 13 yr

## ▬▬ EXTENDED APPLICATION   (PAGE 87)

**1.** 4.8 million units

**2.**

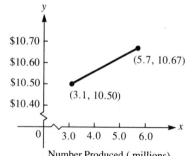

Number Produced ( millions)

**3.** In the interval under discussion (3.1 to 5.7 million units), the marginal cost always exceeds the selling price.

**4. (a)** 9.87; 10.22
**(b)**

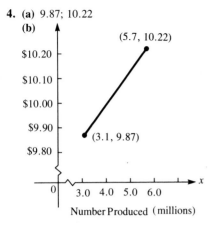

**(c)** .83 million units, which is not in the interval under discussion

## CHAPTER 2    SECTION 2.1    (PAGE 100)

**1.** 3    **3.** Does not exist    **5.** 0    **7.** −2    **9.** Does not exist    **11.** 10    **13.** Does not exist    **15.** 41    **17.** 9/7
**19.** −1    **21.** 6    **23.** −5    **25.** $\sqrt{5}$    **27.** Does not exist    **29.** 4    **31.** −1/9    **33.** 1/10    **35.** 1/(2$\sqrt{5}$)
**37.** 2x    **39.** 3/5    **41.** 1/2    **43.** 1/2    **45.** 0    **47.** 0    **49. (a)** $500    **(b)** Does not exist    **(c)** $1500    **(d)** $500
**(e)** $1500    **(f)** $1500    **(g)** 15    **51. (a)** 25/3    **(b)** 825/19    **(c)** 825/19    **(d)** 75    **53. (a)** .038    **(b)** .057    **(c)** .057    **(d)** 0
**55.** −4.178    **57.** .5    **59.** Within .00333

## SECTION 2.2    (PAGE 113)

**1.** 5    **3.** 8    **5.** 1/3    **7.** −1/3    **9.** 17    **11.** 25    **13.** 50    **15. (a)** 3    **(b)** 7/2    **(c)** 1    **(d)** 0    **(e)** −2
**(f)** −2    **(g)** −3    **(h)** −1    Sales are increasing, on the average, in (a)–(c); sales are not changing (on the average) in (d); sales
are decreasing (on the average) in (e)–(h).    **17. (a)** About 3 cents per minute per year    **(b)** About −2/3 cent per minute per
year    **(c)** About 1/5 cent per minute per year    **(d)** About 1 cent per minute per year    The average rate of change of costs is
increasing in (a), (c), and (d), and decreasing in (b).    **19. (a)** $500 per year    **(b)** $375 per year    **21.** 11    **23. (a)** −25
boxes per dollar    **(b)** −20 boxes per dollar    **(c)** −30 boxes per dollar    **25. (a)** −.3 word per minute    **(b)** −.9 word per
minute    **27. (a)** 7/3 ft per sec    **(b)** 1 ft per sec    **(c)** 2 ft per sec    **(d)** 1 ft per sec    **(e)** 11/6 ft per sec    **(f)** 5/4 ft per sec
Since the average velocity is positive in each case, the car is always moving in the same direction (forward).    **29. (a)** Total
distance    **(b)** Velocity

## SECTION 2.3    (PAGE 131)

**1.** 27    **3.** 1/8    **5.** 1/8    **7.** $y = 8x - 9$    **9.** $5x + 4y = 20$    **11.** $3y = 2x + 18$    **13.** 2    **15.** 1/5
**17.** 0    **19.** $f'(x) = -8x + 11$; −5; 11; 35    **21.** $f'(x) = 8$; 8; 8; 8    **23.** $f'(x) = 3x^2 + 3$; 15; 3; 30
**25.** $f'(x) = 2/x^2$; 1/2; does not exist; 2/9    **27.** $f'(x) = -4/(x - 1)^2$; −4; −4; −1/4    **29.** $f'(x) = 1/(2\sqrt{x})$; 1/(2$\sqrt{2}$); does
not exist; does not exist    **31.** 0    **33.** −6; 6    **35.** −3; 0; 2; 3; 5    **37. (a)** −4p + 4    **(b)** −36 is the approximate
decrease in demand for a small increase in price.    **39. (a)** $16 per table    **(b)** $16    **(c)** $15.998 or $16    **41. (a)** 30    **(b)** 20
**(c)** 10    **(d)** 0    **(e)** −10    **43.** 1000; 700; 250    **45.** −.0435    **47.** 4.269    **49.** −.1075

## ■ SECTION 2.4    (PAGE 146)

**1.** $f'(x) = 18x - 8$     **3.** $y' = 30x^2 - 18x + 6$     **5.** $y' = 4x^3 - 15x^2 + 18x$     **7.** $f'(x) = 9x^5 - 2x^{-.5}$ or $9x^5 - 2/x^{.5}$
**9.** $y' = -48x^{2.2} + 3.8x^{.9}$     **11.** $y' = 36t^{1/2} + 2t^{-1/2}$ or $36t^{1/2} + 2/t^{1/2}$     **13.** $y' = 4x^{-1/2} + (9/2)x^{-1/4}$ or $4/x^{1/2} + 9/(2x^{1/4})$
**15.** $g'(x) = -30x^{-6} + x^{-2}$ or $-30/x^6 + 1/x^2$     **17.** $y' = 8x^{-3} - 9x^{-4}$ or $8/x^3 - 9/x^4$
**19.** $y' = -20x^{-3} - 12x^{-5} - 6$ or $-20/x^3 - 12/x^5 - 6$     **21.** $f'(t) = -6t^{-2} + 16t^{-3}$ or $-6/t^2 + 16/t^3$
**23.** $y' = -36x^{-5} + 24x^{-4} - 2x^{-2}$ or $-36/x^5 + 24/x^4 - 2/x^2$     **25.** $f'(x) = -6x^{-3/2} - (3/2)x^{-1/2}$ or $-6/x^{3/2} - 3/(2x^{1/2})$
**27.** $p'(x) = 5x^{-3/2} - 12x^{-5/2}$ or $5/x^{3/2} - 12/x^{5/2}$     **29.** $y' = (-3/2)x^{-5/4}$ or $-3/(2x^{5/4})$     **31.** $y' = (-5/3)t^{-2/3}$ or $-5/(3t^{2/3})$
**33.** $-40x^{-6} + 36x^{-5}$ or $-40/x^6 + 36/x^5$     **35.** $(-9/2)x^{-3/2} - 3x^{-5/2}$ or $-9/(2x^{3/2}) - 3/x^{5/2}$     **37.** $-28$
**39.** 11/16     **41.** $-28$; $28x + y = 34$     **43.** 79/6     **45.** $-11/27$     **47. (a)** 30 **(b)** 0 **(c)** $-10$     **49.** $-10$
**51. (a)** \$0 **(b)** $-$\$1 **(c)** \$32 **(d)** \$207 **(e)** The profit will not change when sales are increased from 1000 to 1001; the profit
will decrease by \$1 when sales are increased from 2000 to 2001; the profit will increase by \$32 when sales are increased from 5000
to 5001; the profit will increase by \$207 when sales are increased from 10,000 to 10,001.     **53. (a)** $C'(x) = 2$
**(b)** $R'(x) = 6 - x/500$ **(c)** $P'(x) = 4 - x/500$ **(d)** $x = 2000$ **(e)** \$4000     **55. (a)** The blood sugar level is decreasing at a
rate of 4 points per unit of insulin. **(b)** The blood sugar level is decreasing at a rate of 10 points per unit of insulin.
**57. (a)** 264 **(b)** 510 **(c)** 97/4 or 24.25     **59. (a)** $v(t) = 6$ **(b)** 6; 6; 6     **61. (a)** $v(t) = 22t + 4$ **(b)** 4; 114; 224
**63. (a)** $v(t) = 12t^2 + 16t$ **(b)** 0; 380; 1360     **65. (a)** $-32$ ft per sec; $-64$ ft per sec **(b)** In 3 sec **(c)** $-96$ ft per sec

## ■ SECTION 2.5    (PAGE 155)

**1.** $y' = 4x + 3$     **3.** $f'(x) = 48x + 66$     **5.** $y' = 18x^2 - 6x + 4$     **7.** $y' = 9t^2 - 4t - 5$
**9.** $y' = 36x^3 + 21x^2 - 18x - 7$     **11.** $y' = 4(2x - 5)$ or $8x - 20$     **13.** $k'(t) = 4t(t^2 - 1)$ or $4t^3 - 4t$
**15.** $y' = (3/2)x^{1/2} + (1/2)x^{-1/2} + 2$ or $3x^{1/2}/2 + 1/(2\sqrt{x}) + 2$     **17.** $g'(x) = 10 + (3/2)x^{-1/2}$ or $10 + 3/(2\sqrt{x})$
**19.** $y' = -3/(2x - 1)^2$     **21.** $f'(x) = 53/(3x + 8)^2$     **23.** $y' = -6/(3x - 5)^2$     **25.** $y' = -17/(4 + t)^2$
**27.** $y' = (x^2 - 2x - 1)/(x - 1)^2$     **29.** $f'(t) = (-t^2 + 4t + 1)/(t^2 + 1)^2$     **31.** $y' = (-2x^2 - 6x - 1)/(2x^2 - 1)^2$
**33.** $g'(x) = (x^2 + 6x - 14)/(x + 3)^2$     **35.** $p'(t) = [-\sqrt{t}/2 - 1/(2\sqrt{t})]/(t - 1)^2$ or $(-t - 1)/[2\sqrt{t}(t - 1)^2]$
**37.** $y' = (5\sqrt{x}/2 - 3/\sqrt{x})/x$ or $(5x - 6)/(2x\sqrt{x})$     **39.** $y = -2x + 9$     **41. (a)** 108.80 **(b)** 198.40 **(c)** $9x + 18 + 8/x$
**(d)** $d\overline{C}/dx = 9 - 8/x^2$     **43. (a)** $G'(20) = -1/200$; go faster **(b)** $G'(40) = 1/400$; go slower     **45. (a)** $s'(x) = m/(m + nx)^2$
**(b)** $1/2560 \approx .000391$     **47. (a)** $-100$ **(b)** $-1/100 = -.01$

## ■ SECTION 2.6    (PAGE 165)

**1.** 1122     **3.** 97     **5.** $256k^2 + 48k + 2$     **7.** $24x + 4$; $24x + 35$     **9.** $-64x^3 + 2$; $-4x^3 + 8$     **11.** $1/x^2$; $1/x^2$
**13.** $\sqrt{8x^2 - 4}$; $8x + 10$     **15.** $x/(2 - 5x)$; $2(x - 5)$     **17.** $\sqrt{(x - 1)/x}$; $-1/\sqrt{x + 1}$
**19.** If $f(x) = x^{1/3}$ and $g(x) = 3x - 7$, then $y = f[g(x)]$.     **21.** If $f(x) = x^{1/2}$ and $g(x) = 9 - 4x$, then $y = f[g(x)]$.
**23.** If $f(x) = (x + 3)/(x - 3)$ and $g(x) = \sqrt{x}$, then $y = f[g(x)]$.
**25.** If $f(x) = x^2 + x + 5$ and $g(x) = x^{1/2} - 3$, then $y = f[g(x)]$.
**27.** $y' = 4(2x + 9)$     **29.** $f'(x) = 90(5x - 1)^2$     **31.** $k'(x) = -144x(12x^2 + 5)^2$     **33.** $y' = 36(2x + 5)(x^2 + 5x)^3$
**35.** $s'(t) = 36(2t + 5)^{1/2}$     **37.** $y' = -21(8x + 9)(4x^2 + 9x)^{1/2}/2$     **39.** $f'(t) = 16(4t + 7)^{-1/2}$ or $16/(4t + 7)^{1/2}$
**41.** $y' = -(2x + 4)(x^2 + 4x)^{-1/2}$ or $-(2x + 4)/(x^2 + 4x)^{1/2}$     **43.** $r'(t) = 16t(2t + 3) + 4(2t + 3)^2$ or $12(2t + 3)(2t + 1)$
**45.** $y' = 2(x + 2)(x - 1) + (x - 1)^2$ or $3(x - 1)(x + 1)$     **47.** $f'(x) = 10(x + 3)^2(x - 1) + 10(x - 1)^2(x + 3)$
or $20(x + 3)(x - 1)(x + 1)$     **49.** $y' = \dfrac{(5x + 1)^2}{(2x)^{1/2}} + 10(5x + 1)(2x)^{1/2}$ or $\dfrac{(5x + 1)(25x + 1)}{(2x)^{1/2}}$     **51.** $y' = -6(3x - 4)^{-3}$
or $\dfrac{-6}{(3x - 4)^3}$     **53.** $p'(t) = \dfrac{8(2t + 3)(t - 2)}{(4t - 1)^2}$     **55.** $y' = \dfrac{8(1 - x)}{(3x + 2)^3}$     **57.** $y' = \dfrac{x^{-1/2}(x^{1/2} - 1)^{-1/2}(3x^{1/2} - 1)}{4}$
or $\dfrac{3x^{1/2} - 1}{4x^{1/2}(x^{1/2} - 1)^{1/2}}$     **59.** $D(c) = \dfrac{-c^2 + 10c - 25}{25} + 500$     **61. (a)** \$101.22 **(b)** \$111.86 **(c)** \$117.59
**63. (a)** $-$\$1050 **(b)** $-$\$457.06     **65.** \$400 per additional worker     **67.** $18a^2 + 24a + 9$     **69.** $A[r(t)] = A(t) = 4\pi t^2$;
this function gives the area of the pollution in terms of the time since the pollutants were first emitted.
**71. (a)** $-.5$ **(b)** $-1/54 \approx -.02$ **(c)** $-.011$ **(d)** $-1/128 \approx -.008$

## SECTION 2.7    (PAGE 175)

**1.** Discontinuous at $-1$; derivative fails to exist at $-1$     **3.** Discontinuous at 1; derivative fails to exist at $-4$ and 1
**5.** Discontinuous at 0 and 2; derivative fails to exist at $-3$, 0, 2, 3, and 5     **7.** Yes; no     **9.** No; no; yes     **11.** Yes; no; no
**13.** Yes; no; yes     **15.** Yes; no; yes     **17.** Yes; yes; yes     **19.** No; yes; yes     **21.** Yes; yes; no     **23.** No; yes; yes
**25. (a)** $-4$ is not in the domain of $f[g(x)]$; yes; yes
   **(b)** Yes; yes; yes

**27. (a)** $96   **(b)** $150
   **(c)** $F(x)$

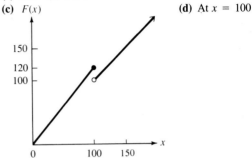

   **(d)** At $x = 100$

**29. (a)** $30   **(b)** $30   **(c)** $25   **(d)** $21.43
   **(e)** $22.50   **(f)** $30   **(g)** $25

**31.** At $t = m$

## CHAPTER 2   REVIEW EXERCISES   (PAGE 180)

**1.** 4     **3.** $-3$     **5.** 4     **7.** 17/3     **9.** 8     **11.** $-13$     **13.** 1/6     **15.** 1/5     **17.** 3/4     **19.** 1/4     **21.** $-3/2$
**23.** 30     **25.** 9/77     **27.** $y' = 4$     **29.** $y' = -3x^2 + 7$     **31.** $-2$; $y + 2x = -4$     **33.** $-3/4$; $3x + 4y = -9$
**35.** 38; $y = 38x + 22$     **37.** 3/4; $4y = 3x + 7$     **39.** $y' = 10x - 7$     **41.** $y' = 14x^{4/3}$
**43.** $f'(x) = -3x^{-4} + (1/2)x^{-1/2}$ or $-3/x^4 + 1/(2x^{1/2})$     **45.** $y' = 15t^4 + 12t^2 - 7$
**47.** $y' = 12x^{1/2} - 6x^{-1/2}$ or $12x^{1/2} - 6/x^{1/2}$     **49.** $g'(t) = -20t^{1/3} - 7t^{-2/3}$ or $-20t^{1/3} - 7/t^{2/3}$
**51.** $y' = 9x^{-3/4} - 45x^{-7/4}$ or $9/x^{3/4} - 45/x^{7/4}$     **53.** $k'(x) = 15/(x + 5)^2$     **55.** $y' = (2 - x + 2x^{1/2})/[2x^{1/2}(x + 2)^2]$
**57.** $y' = (x^2 - 2x)/(x - 1)^2$     **59.** $f'(x) = 12(3x - 2)^3$     **61.** $y' = 1/(2t - 5)^{1/2}$     **63.** $y' = 3(2x + 1)^2(8x + 1)$
**65.** $r'(t) = (-15t^2 + 52t - 7)(3t + 1)^4$     **67.** $y' = (x^2 - 4x + 4)(x - 2)^2 = 1$     **69.** $-1/[x^{1/2}(x^{1/2} - 1)^2]$
**71.** $(1 + 2t^{1/2})/[4t^{1/2}(t^{1/2} + t)^{1/2}]$     **73.** $-2/3$     **75.** No; yes; yes     **77.** Yes; no; yes; no     **79.** Yes; no; yes
**81.** Yes; yes; yes     **83.** Yes; yes; yes     **85.** Derivative fails to exist at $x_1$, $x_2$, $x_3$, and $x_4$; discontinuous at $x_2$ and $x_4$
**87. (a)** 315   **(b)** 1015   **(c)** 1515     **89. (a)** 55/3; sales will increase by 55 million dollars when 3 thousand more dollars are
spent on research   **(b)** 65/4; sales will increase by 65 million dollars when 4 thousand more dollars are spent on research   **(c)** 15;
sales will increase by 15 million dollars when 1 thousand more dollars are spent on research.     **91. (a)** $-9.5$; costs will decrease
by $950 for the next $100 spent on training   **(b)** $-2.375$; costs will decrease by $2375 for the next $1000 spent on training
**93. (a)** $150   **(b)** $187.50   **(c)** $189   **(d)** $C(x)$     **(e)** Discontinous at $x = $125

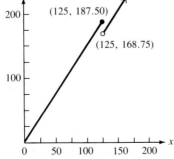

# ▰ CHAPTER 3   SECTION 3.1   (PAGE 195)

**1. (a)** $(1, +\infty)$  **(b)** $(-\infty, 1)$   **3. (a)** $(-\infty, -2)$  **(b)** $(-2, +\infty)$   **5. (a)** $(-\infty, -4), (-2, +\infty)$  **(b)** $(-4, -2)$

**7. (a)** $(-7, -4), (-2, +\infty)$  **(b)** $(-\infty, -7), (-4, -2)$

**9. (a)** $(-6, +\infty)$  **(b)** $(-\infty, -6)$   **11. (a)** $(-\infty, 3/2)$  **(b)** $(3/2, +\infty)$   **13. (a)** $(-\infty, -3), (4, +\infty)$

**(c)**   **(b)** $(-3, 4)$

**(c)**

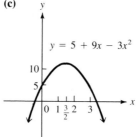

$f(x) = x^2 + 12x - 6$

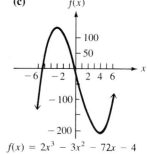

$f(x) = 2x^3 - 3x^2 - 72x - 4$

**15. (a)** $(-\infty, -3/2), (4, +\infty)$   **17. (a)** None  **(b)** $(-\infty, +\infty)$   **19. (a)** None

**(b)** $(-3/2, 4)$   **(c)**   **(b)** $(-\infty, -1), (-1, +\infty)$

**(c)**   **(c)**

$f(x) = 4x^3 - 15x^2 - 72x + 5$

$y = -3x + 6$

$f(x) = \dfrac{x + 2}{x + 1}$

**21. (a)** $(-4, +\infty)$  **(b)** $(-\infty, -4)$   **23. (a)** None  **(b)** $(1, +\infty)$   **25. (a)** $(0, +\infty)$  **(b)** $(-\infty, 0)$

**(c)**   **(c)**   **(c)**

$y = |x + 4|$

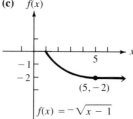

$(5, -2)$

$f(x) = -\sqrt{x - 1}$

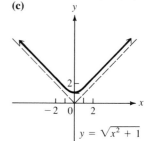

$y = \sqrt{x^2 + 1}$

**27. (a)** Nowhere  **(b)** Everywhere   **29.** $[0, 1125)$   **31.** After 10 days

**33. (a)** $(0, 3)$  **(b)** $(3, +\infty)$ (Remember: $x$ must be at least 0.)

# SECTION 3.2 (PAGE 208)

**1.** Relative minimum of $-4$ at 1    **3.** Relative maximum of 3 at $-2$    **5.** Relative maximum of 3 at $-4$; relative minimum of 1 at $-2$    **7.** Relative maximum of 3 at $-4$; relative minimum of $-2$ at $-7$ and $-2$    **9.** Relative minimum of $-44$ at $-6$    **11.** Relative maximum of 17 at $-3$    **13.** Relative maximum of $-8$ at $-3$; relative minimum of $-12$ at $-1$    **15.** Relative maximum of $827/96 - 1/4$; relative minimum of $-377/6$ at $-5$    **17.** Relative maximum of 57 at 2; relative minimum of 30 at 5    **19.** Relative maximum of $-4$ at 0; relative minimum of $-85$ at 3 and $-3$    **21.** Relative maximum of 0 at 8/5    **23.** Relative maximum of 1 at $-1$; relative minimum of 0 at 0    **25.** No relative extrema    **27.** Relative minimum of 0 at 0    **29.** Relative maximum of 0 at 1; relative minimum of 8 at 5    **31.** (2, 7)    **33.** (5/4, $-9/8$)    **35. (a)** 40  **(b)** 15  **(c)** 375    **37. (a)** 80  **(b)** 10  **(c)** 25    **39.** $x = 5$; $p = 275/6$    **41.** 4:44 P.M., 5:46 A.M.    **43.** 67 ft

# SECTION 3.3 (PAGE 216)

**1.** Absolute maximum at $x_3$; no absolute minimum    **3.** No absolute extrema    **5.** Absolute minimum at $x_1$; no absolute maximum    **7.** Absolute maximum at $x_1$; absolute minimum at $x_2$    **9.** Absolute maximum at 0; absolute minimum at $-3$    **11.** Absolute maximum at $-1$; absolute minimum at $-5$    **13.** Absolute maximum at $-2$; absolute minimum at 4    **15.** Absolute maximum at $-2$; absolute minimum at 3    **17.** Absolute maximum at 6; absolute minimum at $-4$ and 4    **19.** Absolute maximum at 0; absolute minimum at 2    **21.** Absolute maximum at 6; absolute minimum at 4    **23.** Absolute maximum at $\sqrt{2}$; absolute minimum at 0    **25.** Absolute maximum at $-3$ and 3, absolute minimum at 0    **27.** Absolute maximum at 0; absolute minimum at $-4$    **29.** Absolute maximum at 0; absolute minimum at $-1$ and 1    **31.** 1000 manuals; more than \$2.20    **33.** 6 months; 6%    **35.** 25; 16.1    **37.** The piece formed into a circle should have length $12\pi/(4 + \pi)$ ft, or about 5.28 ft.

# SECTION 3.4 (PAGE 233)

**1.** $f''(x) = 18x$; 0; 36; $-54$    **3.** $f''(x) = 36x^2 - 30x + 4$; 4; 88, 418    **5.** $f''(x) = 6$; 6; 6; 6    **7.** $f''(x) = 6(x + 4)$; 24; 36; 6    **9.** $f''(x) = 10/(x - 2)^3$; $-5/4$; $f''(2)$ does not exist; $-2/25$    **11.** $f''(x) = 2/(1 + x)^3$; 2; 2/27; $-1/4$    **13.** $f''(x) = -1/[4(x + 4)^{3/2}]$; $-1/32$; $-1/(4 \cdot 6^{3/2}) \approx -.0170$; $-1/4$    **15.** $f''(x) = (-6/5)x^{-7/5}$ or $-6/(5x^{7/5})$; $f''(0)$ does not exist; $-6/(5 \cdot 2^{7/5}) \approx -.4547$; $-6/[5 \cdot (-3)^{7/5}] \approx .2578$    **17.** $f'''(x) = -24x$; $f^{(4)}(x) = -24$    **19.** $f'''(x) = 240x^2 + 144x$; $f^{(4)}(x) = 480x + 144$    **21.** $f'''(x) = 18(x + 2)^{-4}$ or $18/(x + 2)^4$; $f^{(4)}(x) = -72(x + 2)^{-5}$ or $-72/(x + 2)^5$    **23.** $f'''(x) = -36(x - 2)^{-4}$ or $-36/(x - 2)^4$; $f^{(4)}(x) = 144(x - 2)^{-5}$ or $144/(x - 2)^5$    **25.** Concave upward on $(2, +\infty)$; concave downward on $(-\infty, 2)$; point of inflection at (2, 3)    **27.** Concave upward on $(-\infty, -1)$ and $(8, +\infty)$; concave downward on $(-1, 8)$; points of inflection at $(-1, 7)$ and (8, 6)    **29.** Concave upward on $(2, +\infty)$; concave downward on $(-\infty, 2)$; no points of inflection    **31.** Always concave upward; no points of inflection    **33.** Always concave downward; no points of inflection    **35.** Concave upward on $(-1, +\infty)$; concave downward on $(-\infty, -1)$; point of inflection at $(-1, 44)$    **37.** Concave upward on $(-\infty, 3/2)$; concave downward on $(3/2, +\infty)$; point of inflection at (3/2, 525/2)    **39.** Concave upward on $(5, +\infty)$; concave downward on $(-\infty, 5)$ no points of inflection    **41.** Concave upward on $(-10/3, +\infty)$; concave downward on $(-\infty, -10/3)$; point of inflection at $(-10/3, -250/27)$    **43.** Relative maximum at $-5$    **45.** Relative maximum at 0; relative minimum at 2/3    **47.** Relative maximum at 3; relative minimum at $-6$

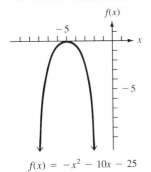

$f(x) = -x^2 - 10x - 25$

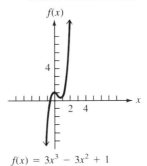

$f(x) = 3x^3 - 3x^2 + 1$

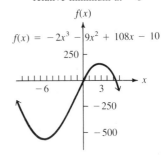

$f(x) = -2x^3 - 9x^2 + 108x - 10$

**49.** Relative maximum at $-5/3$;
relative minimum at $1/2$

$$f(x) = 2x^3 + \frac{7}{2}x^2 - 5x + 3$$

**51.** Relative minimum at $-3$

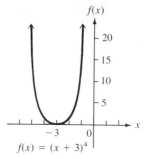

$$f(x) = (x + 3)^4$$

**53.** Relative maximum at 0;
relative minima at $-3$ and 3

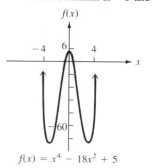

$$f(x) = x^4 - 18x^2 + 5$$

**55.** Relative maximum at $-1$;
relative minimum at 1

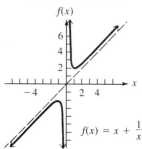

$$f(x) = x + \frac{1}{x}$$

**57.** Relative maximum at $-5$;
relative minimum at 5

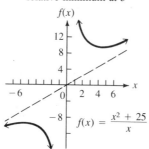

$$f(x) = \frac{x^2 + 25}{x}$$

**59.** No critical values;
no maximum or minimum

$$f(x) = \frac{x - 1}{x + 1}$$

**61.** **(a)** Between car phones and polaroid cameras    **(b)** Near black-and-white TVs    **(c)** Car phones and polaroid cameras; the rate of growth of sales will now decline    **(d)** Black-and-white TVs; the rate of decline of sales is starting to slow    **63.** $(22, 208)$
**65.** **(a)** At 4 hr    **(b)** 1160 million    **67.** **(a)** After 3 hr    **(b)** 2/9%
**69.** $v(t) = 16t + 4$; $a(t) = 16$; $v(0) = 16$ cm/sec; $v(4) = 68$ cm/sec; $a(0) = 16$ cm/sec$^2$; $a(4) = 16$ cm/sec$^2$
**71.** $v(t) = -15t^2 - 16t + 6$; $a(t) = -30t - 16$; $v(0) = 6$ cm/sec; $v(4) = -298$ cm/sec; $a(0) = -16$ cm/sec$^2$;
$a(4) = -136$ cm/sec$^2$    **73.** $v(t) = 6(3t + 4)^{-2}$ or $6/(3t + 4)^2$; $a(t) = -36(3t + 4)^{-3}$ or $-36/(3t + 4)^3$; $v(0) = 3/8$ cm/sec;
$v(4) = 3/128$ cm/sec; $a(0) = -9/16$ cm/sec$^2$; $a(4) = -9/1024$ cm/sec$^2$    **75.** **(a)** $-96$ ft/sec    **(b)** $-160$ ft/sec
**(c)** $-256$ ft/sec    **(d)** $-32$ ft/sec$^2$    **77.** **(a)** Increasing on $(0, 2)$ and $(4, 5)$; decreasing on $(-5, -.5)$ and $(2.5, 3.5)$
**(b)** Minima between $-.5$ and 0 and between 3.5 and 4; a maximum between 2 and 2.5    **(c)** Concave upward on $(-5, .5)$ and
$(3.5, 22)$; concave downward on $(1,3)$    **(d)** Inflection points between .5 and 1 and between 3 and 3.5
**79.** **(a)** Decreasing on $(-1, 1)$; increasing on $(1.2, 2)$    **(b)** Minimum between 1 and 1.2    **(c)** Concave upward on $(-1, 0)$ and
$(.8, 2)$; concave downward on $(.2, .6)$    **(d)** Inflection points between .6 and .8 and at 0

## ▨ SECTION 3.5   (PAGE 244)

**1.** **(a)** $y = 100 - x$    **(b)** $P = x(100 - x)$    **(c)** $P' = 100 - 2x$; $x = 50$    **(d)** 50 and 50    **(e)** $50 \cdot 50 = 2500$
**3.** 100; 100    **5.** 100; 50    **7.** 5; $-5$    **9.** **(a)** $1200 - 2x$    **(b)** $A(x) = 1200x - 2x^2$    **(c)** 300 m    **(d)** 180,000 sq m
**11.** 405,000 sq m    **13.** **(a)** $R(x) = 100,000x - 100x^2$    **(b)** 500    **(c)** 25,000,000 cents    **15.** **(a)** $\sqrt{3200} \approx 56.6$ mph
**(b)** \$45.24    **17.** 20,000 sq ft, with 200 ft on the \$3 sides and 100 ft on the \$6 sides    **19.** **(a)** $40 - 2x$    **(b)** $100 + 5x$
**(c)** $R(x) = 4000 - 10x^2$    **(d)** Now    **(e)** \$40 per tree    **21.** **(a)** 90    **(b)** \$405    **23.** 4 inches by 4 inches by 2 inches

**25.** 3 ft by 6 ft by 2 ft    **27.** 44.7 mph gives the minimum cost of $894    **29.** 1 mile from point $A$    **31.** 250 thousand

**33.** 56.25 thousand    **35.** 5.071 thousand    **37.** Point $P$ is $3\sqrt{7}/7 \approx 1.134$ miles from point $A$.    **39. (a)** $F = \dfrac{\pi(r - r_0)}{2ak}r^4;$

$r = \dfrac{4r_0}{5}$    **(b)** $\bar{v} = \dfrac{(r - r_0)r^2}{2ak}; \bar{v}$ is a maximum when $r = \dfrac{2r_0}{3}; \quad v(0) = \dfrac{(r - r_0)r^2}{ak}; v(0)$ is a maximum when $r = \dfrac{2r_0}{3}$

## ▰▰▰ SECTION 3.6    (PAGE 259)

**3.** 10    **5.** 10,000    **7.** 4899    **9. (a)** $E = p/(500 - p)$    **(b)** 12,500    **11. (a)** $E = 1$    **(b)** None    **13. (a)** $E = 2$;
elastic; a percentage increase in price will result in a greater percentage decrease in demand    **(b)** $E = 1/2$; inelastic; a percentage
change in price will result in a smaller percentage change in demand    **15. (a)** 9/8    **(b)** $x = 50/3; p = 5\sqrt{3}/3$    **(c)** 1

## ▰▰▰ SECTION 3.7    (PAGE 266)

**1.** $dy/dx = -4x/(3y)$    **3.** $dy/dx = -y/(y + x)$    **5.** $dy/dx = 2/y$    **7.** $dy/dx = -3y^2/(6xy - 4)$
**9.** $dy/dx = (-y - x)/x$    **11.** $dy/dx = (-6x - 4y)/(4x + y)$    **13.** $dy/dx = 3x^2/(2y)$    **15.** $dy/dx = y^2/x^2$
**17.** $dy/dx = -3x(2 + y)^2/2$    **19.** $dy/dx = -2xy/(x^2 + 3y^2)$    **21.** $dy/dx = -y^{1/2}/x^{1/2}$
**23.** $dy/dx = -y^{1/2}x^{-1/2}/(x^{1/2}y^{-1/2} + 2)$    **25.** $4y = 3x + 25$    **27.** $y = x + 2$    **29.** $24x + y = 57$    **31.** $x + 4y = 12$
**33. (a)** $3x + 4y = 50; -3x + 4y = -50$    **35.** $2y = x + 1$    **39.** $dv/du = -(2v + 1)^{1/2}/(2u^{1/2})$
**(b)**    **41. (a)** $dC/dx = (x^{3/2} + 25)/(Cx^{1/2})$; when $x = 5$ the approximate increase
in cost of an additional unit is .94

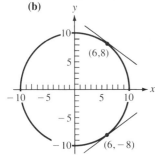

**(b)** $dR/dx = (180 - 36x)/R$; when $x = 5$ the approximate
change in revenue for a unit increase in sales is zero
**43.** $ds/dt = (-s + 6\sqrt{st})/(8s\sqrt{st} + t)$

## ▰▰▰ SECTION 3.8    (PAGE 272)

**1.** 440    **3.** $-15/2$    **5.** $-5/7$    **7.** 1/5    **9.** $200 per month    **11. (a)** Revenue is increasing at a rate of $180 per
day.    **(b)** Cost is increasing at a rate of $50 per day.    **(c)** Profit is increasing at a rate of $130 per day.    **13.** Demand is
decreasing at a rate of approximately 343 units over time.    **15.** $-.24$ mm per min    **17.** .067 mm per min    **19.** $-.370$
**21.** 7/6 ft per min    **23.** $16\pi$ sq ft per min    **25.** $50\pi$ cu in per min    **27.** 1/16 ft per min

## ▰▰▰ SECTION 3.9    (PAGE 281)

**1.** $dy = 12x\,dx$    **3.** $dy = (14x - 9)dx$    **5.** $dy = x^{-1/2}\,dx$    **7.** $dy = [-22/(x - 3)^2]dx$    **9.** $dy = (3x^2 - 1 + 4x)dx$
**11.** $dy = (x^{-2} + 6x^{-3})dx$ or $(1/x^2 + 6/x^3)dx$    **13.** $-2.6$    **15.** .1    **17.** .130    **19.** $-.023$    **21.** .24
**23.** $-.00444$    **25. (a)** $-34$ thousand pounds    **(b)** $-169.2$ thousand pounds    **27.** $-5.625$ housing starts
**29.** About $-$990,000    **31.** $21,600\pi$ cu in    **33. (a)** .347 million    **(b)** $-.022$ million    **35.** $1568\pi$ cu mm
**37.** $80\pi$ sq mm    **39.** $-7.2\pi$ cu cm    **41.** $\pm 1.224$ sq in    **43.** $\pm.116$ cu in

## CHAPTER 3 REVIEW EXERCISES (PAGE 284)

**1.** Increasing on $(5/2, +\infty)$; decreasing on $(-\infty, 5/2)$    **3.** Increasing on $(-4, 2/3)$; decreasing on $(-\infty, -4)$ and $(2/3, +\infty)$
**5.** Decreasing on $(-\infty, 4)$ and $(4, +\infty)$    **7.** Relative maximum of $-4$ at 2    **9.** Relative minimum of $-7$ at 2
**11.** Relative maximum of 101 at $-3$; relative minimum of $-24$ at 2    **13.** Relative maximum of $-151/54$ at 1/3; relative
minimum of $-27/8$ at $-1/2$    **15.** Relative maximum of 1 at 0; relative minima of $-2/3$ at $-1$ and $-29/3$ at 2
**17.** No extrema
**19.** Concave upward (on $-\infty, -1/12$); concave downward on
$(-1/12, +\infty)$; point of inflection at $(-1/12, 1007/216)$

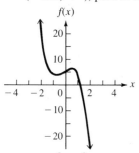

$$f(x) = -4x^3 - x^2 + 4x + 5$$

**21.** Always concave upward; no points of inflection

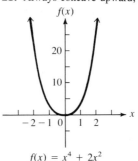

$$f(x) = x^4 + 2x^2$$

**23.** Concave upward on $(0, +\infty)$; concave downward on
$(-\infty, 0)$; no point of inflection

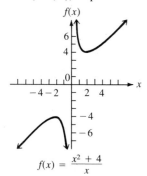

$$f(x) = \frac{x^2 + 4}{x}$$

**25.** Concave upward on $(-\infty, 3)$; concave downward on
$(3, +\infty)$; no point of inflection

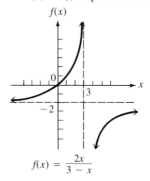

$$f(x) = \frac{2x}{3 - x}$$

**27.** $f''(x) = 36x^2 - 10$; 26; 314    **29.** $f''(x) = -68(2x + 3)^{-3}$ or $-68/(2x + 3)^3$; $-68/125$; 68/27
**31.** $f''(t) = (t^2 + 1)^{-3/2}$ or $1/(t^2 + 1)^{3/2}$; $1/2^{3/2} \approx .354$; $1/10^{3/2} \approx .032$    **33.** Absolute maximum of 29/4 at 5/2; absolute minima
of 5 at 1 and 4    **35.** Absolute maximum of 39 at $-3$; absolute minimum of $-319/27$ at 5/3
**37.** $dy/dx = (-4y - 2xy^3)/(3x^2y^2 + 4x)$    **39.** $dy/dx = (-4 - 9x^{3/2})/(24x^2y^2)$    **41.** $dy/dx = (2y - 2y^{1/2})/(4y^{1/2} - x + 9y)$
(This form of the answer to Exercise 41 was obtained by multiplying both sides of the given function by $x - 3y$.)
**43.** $dy/dx = [9(4y^2 - 3x)^{1/3} + 3]/(8y)$    **45.** $23x + 16y = 94$    **47.** 272    **49.** $-2$    **51.** $dy = (24x^2 - 4x)dx$
**53.** $dy = [-16/(2 + x)^2]dx$    **55.** .1    **57.** 25/2 and 25/2    **59. (a)** \$2000 **(b)** \$40,000    **61.** 2m by 4m by 4m
**63.** 3 inches    **65.** 8000    **67.** 80    **69.** 225 m by 450 m    **71.** $1/(4.8\pi) \approx .0663$ ft per min    **73.** $1.28\pi$ cu in or
about 4.021 cu in

■■■ **EXTENDED APPLICATION**    **(PAGE 288)**

**1.** $-C_1/m^2 + DC_3/2$    **2.** $m = \sqrt{2C_1/(DC_3)}$    **3.** About 3.33    **4.** $m^+ = 4$ and $m^- = 3$
**5.** $Z(m^+) = Z(4) = \$11,400$; $Z(m^-) = Z(3) = \$11,300$    **6.** 3 months; 9 trainees per batch

---

≡≡≡ **CHAPTER 4**    **SECTION 4.1**    **(PAGE 297)**

**1.**

**3.**

**5.**

**7.**

**9.**

**11.**

**13.**

**15.**
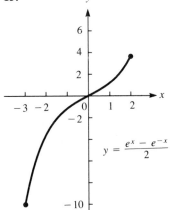

**17.** $-3$    **19.** 2    **21.** 3/2    **23.** 3/2    **25.** $-2$    **27.** $-5/4$    **29.** $-1/7$    **31.** 4, $-4$    **33.** 2    **35.** 0, 5/3
The answers to Exercises 37–43 were found on a calculator. Values from the table may differ slightly.
**37.** 12.18249    **39.** .87810    **41.** 2.44949    **43.** .79553    **45.** A sequence of points, one for each rational number
**47.** (a) \$4693.28   (b) \$4802.44   (c) \$4859.47   (d) \$4898.46    **49.** \$11,966.53
**51.** Choose the 8% investment, which would yield \$111.30 additional interest.
**53.** (a) 1.12, 1.25, 1.40, 1.57, 1.97, 2.21, 2.48, 2.77
    (b) $y$

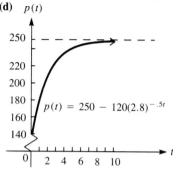

**55.** (a) About 207   (b) About 235   (c) About 249
    (d) $p(t)$

**57.** (a) $2.8 \times 10^{16}$, $3.7 \times 10^{8}$ cm$^3$   (b) $5.5 \times 10^{12}$,
    $3.25 \times 10^{4}$ cm$^3$   (c) $5.5 \times 10^{9}$, 32.5 cm$^3$

**59.** (a) About 1,120,000   (b) About 1,260,000   (c) About
    1,410,000   (d) About 1,590,000   (e) 2,000,000
    (f) 4,000,000   (g) 8,000,000   (h) 16,000,000

**61.** (a) 1,000,000   (b) About 1,410,000   (c) 2,000,000
    (d) 4,000,000
    (e) $P(t)$

$P(t) = 1,000,000(2^{.2t})$

(y-axis label: Population in millions)

---

■■■ SECTION 4.2    (PAGE 311)

**1.** $\log_2 8 = 3$    **3.** $\log_3 81 = 4$    **5.** $\log_{1/3} 9 = -2$    **7.** $2^7 = 128$    **9.** $25^{-1} = 1/25$    **11.** $10^4 = 10,000$    **13.** 2
**15.** 3    **17.** $-2$    **19.** $-2/3$
**21.** 1/9, 1/3, 1, 3, 27
**23.** $y$

$y = \log_4 x$

$y = \log_3 x$

**25.**

**27.**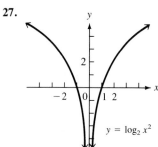

**29.** $\log_9 7 + \log_9 m$ **31.** $1 + \log_3 p - \log_3 5 - \log_3 k$ **33.** $\log_3 5 + (1/2)\log_3 2 - (1/4)\log_3 7$ **35.** $3a$
**37.** $a + 3c$ **39.** $2c + 3a + 1$ **41.** 2.9957 **43.** 6.6846 **45.** 7.4956 **47.** 10.9151 **49.** $-.7985$
**51.** 1.86 **53.** 9.35 **55.** 2.62 **57.** $x = 1/5$ **59.** $m = 3/2$ **61.** $y = 16$ **63.** $x = 8/5$ **65.** $x = 1$
**67.** $x = 1.47$ **69.** $k = 2.39$ **71.** $a = -.12$ **73.** $k = 1.36$ **77. (a)** 23.4 yr **(b)** 11.9 yr **(c)** 9.0 yr
**(d)** 23.3 yr; 12 yr; 9 yr **79.** 1 **81. (a)** 2.03 **(b)** 2.28 **(c)** 2.17 **(d)** 1.21
**83. (a)** About 240 **(b)** About 480 **(c)** About 760
**(d)** $F(t)$

**85.** 4.3 ml/min; 7.8 ml/min
**87. (a)** 20 **(b)** 30 **(c)** 50 **(d)** 60
**89. (a)** 3 **(b)** 6 **(c)** 8

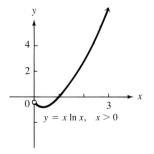

## SECTION 4.3 (PAGE 321)

**1.** $y' = 1/x$ **3.** $y' = -1/(3 - x)$ or $1/(x - 3)$ **5.** $y' = (4x - 7)/(2x^2 - 7x)$ **7.** $y' = 1/[2(x + 5)]$
**9.** $y' = 3(2x^2 + 5)/[x(x^2 + 5)]$ **11.** $y' = -3x/(x + 2) - 3\ln|x + 2|$ **13.** $y' = x + 2x \ln|x|$
**15.** $y' = [2x - 4(x + 3)\ln|x + 3|]/[x^3(x + 3)]$ **17.** $y' = (4x + 7 - 4x \ln|x|)/[x(4x + 7)^2]$
**19.** $y' = (6x \ln|x| - 3x)/(\ln|x|)^2$ **21.** $y' = 4(\ln|x + 1|)^3/(x + 1)$ **23.** $y' = 1/(x \ln|x|)$
**25.** $y' = 1/(x \ln 10)$ **27.** $y' = -1/[(\ln 10)(1 - x)]$ or $1/[(\ln 10)(x - 1)]$ **29.** $y' = 5/[2(\ln 5)(5x + 2)]$
**31.** $y' = 3(x + 1)/[(\ln 3)(x^2 + 2x)]$
**33.** Minimum of $-1/e \approx -.3679$ at $1/e \approx .3679$

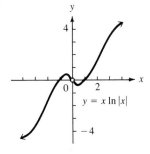

**35.** Minimum of $-1/e \approx -.3679$ at $1/e \approx .3679$; maximum of $1/e$ at $-1/e$

**37.** Minimum of 0 at $x = 1$ and $x = -1$

$y = (\ln |x|)^2$

**39.** Maximum of $1/e \approx .3679$ at $x = e \approx 2.718$

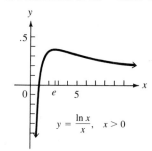

$y = \dfrac{\ln x}{x}, \quad x > 0$

**41.** As $x \to +\infty$, the slope approaches 0; as $x \to 0$, the slope becomes infinitely large.    **45. (a)** $\dfrac{dR}{dx} = 100 + \dfrac{50(\ln x - 1)}{(\ln x)^2}$

**(b)** \$112.48    **47. (a)** $R(x) = 100x - 10 x \ln x, \; 1 < x < 20{,}000$   **(b)** $\dfrac{dR}{dn} = 60(9 - \ln x)$   **(c)** About 360; hiring an

additional worker will produce an increase in revenue of about \$360    **49.** 26.9; 13.1    **51.**

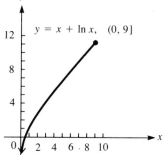

$y = x + \ln x, \quad (0, 9]$

---

### ■ SECTION 4.4   (PAGE 330)

**1.** $y' = 4e^{4x}$    **3.** $y' = 12e^{-2x}$    **5.** $y' = -16e^{2x}$    **7.** $y' = -16e^{x+1}$    **9.** $y' = 2xe^{x^2}$    **11.** $y' = 12xe^{2x^2}$
**13.** $y' = 16xe^{2x^2-4}$    **15.** $y' = xe^x + e^x = e^x(x + 1)$    **17.** $y' = 2(x - 3)(x - 2)e^{2x}$    **19.** $y' = e^{x^2}/x + 2xe^{x^2} \ln x$
**21.** $y' = (xe^x \ln x - e^x)/[x(\ln x)^2]$    **23.** $y' = (2xe^x - x^2e^x)/e^{2x} = x(2 - x)/e^x$    **25.** $y' = [x(e^x - e^{-x}) - (e^x + e^{-x})]/x^2$
**27.** $y' = -20{,}000e^{\cdot 4x}/(1 + 10e^{\cdot 4x})^2$    **29.** $y' = 8{,}000e^{-\cdot 2x}/(9 + 4e^{-\cdot 2x})^2$    **31.** $y' = 2(2x + e^{-x^2})(2 - 2xe^{-x^2})$
**33.** $y' = 5 \ln 8e^{5x \ln 8}$ or $y' = 5 (\ln 8)8^{5x}$    **35.** $y' = 6x(\ln 4)e^{(x^2+2)\ln 4}$ or $y' = 6x(\ln 4)4^{x^2+2}$
**37.** $y' = [(\ln 3)e^{\sqrt{x} \ln 3}]/\sqrt{x}$ or $y' = [(\ln 3)^{\sqrt{x}}]/\sqrt{x}$

**39.** Maximum of $1/e$ at $x = -1$

$y = -xe^x$

**41.** Minimum of 0 at $x = 0$;
maximum of $4/e^2 \approx .54$ at $x = 2$

$y = x^2e^{-x}$

**43.** Minimum of 2 at $x = 0$

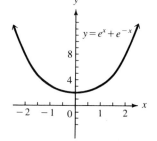

$y = e^x + e^{-x}$

**45. (a)** As $x \to -\infty$, the slope approaches 0. **(b)** As $x \to 0$, the slope approaches 1, the slope of a 45° line.    **47. (a)** 20   **(b)** 6
**49. (a)** 100%   **(b)** 94%   **(c)** 89%   **(d)** 83%   **(e)** $-3.045$   **(f)** $-2.865$   **(g)** The percent of these cars on the road is decreasing, but at a slower rate, as they age.    **51. (a)** .005   **(b)** .0007   **(c)** .000013   **(d)** $-.022$   **(e)** $-.0029$   **(f)** $-.000054$
**53. (a)** 36.8   **(b)** .00454   **(c)** Approximately 0   **(d)** $100e^{-.1N}$ is always positive since powers of $e$ are never negative. This means that repetition always makes a habit stronger.    **55.**  *y*

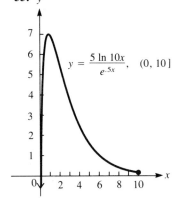

$y = \dfrac{5 \ln 10x}{e^{.5x}}, \quad (0, 10]$

---

![](SECTION 4.5)  **SECTION 4.5**    **(PAGE 342)**

**1. (a)** $10.94   **(b)** $11.27   **(c)** $11.62
**3.** The 10% investment compounded quarterly will yield $622.56 more than the other investment.
**5. (a)** 1000   **(b)** About 4500   **(c)** 4999.8   **(d)** At about 2 yr   **(e)** 5000   **(f)**  $S(x)$

$S(x) = 5000 - 4000e^{-x}$

**7. (a)** $y = 100e^{.11t}$   **(b)** About 15 months    **9. (a)** $y = 50{,}000e^{-.102t}$   **(b)** About 6.8 hr    **11. (a)** $E(t) = 10^6 e^{.0098t}$
**(b)** About $1.1 \times 10^6$   **(c)** About 41 min    **13. (a)** $y = 10{,}000e^{.091t}$   **(b)** About 7.6 yr
**15. (a)** 306   **(b)** 326   **(c)** 346   **(d)** After 1.6 months   **(e)** 350   **(f)**  $p(t)$

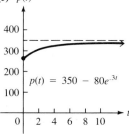

$p(t) = 350 - 80e^{.3t}$

**17.** **(a)** 0    **(b)** About 432    **(c)** About 497
**(d)** After about 1.6 days    **(e)** 500
**(f)** $P(x)$

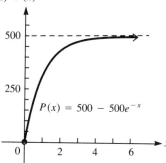

$P(x) = 500 - 500e^{-x}$

**19.** .23    **21.** About .01 g    **23.** About 37/500 or 7.4%
**25.** **(a)** $y = 10e^{.0095t}$    **(b)** 42.7° C    **27.** About 7400 yr
**29.** About 37,200 yr    **31.** 128.6°    **33.** About 30 min

## ▰▰ SECTION 4.6    (PAGE 352)

**1.** 5.11619%    **3.** 18.81%    **5.** 11.46213%    **7.** $1043.79    **9.** $4537.71    **11.** $5583.95    **13.** $40,720.81
**15.** $142,676.91    **17.** $6520.61    **19.** $481,482.93    **21.** 9.2025%    **23.** 6.16778%    **25.** **(a)** $257,107.67
**(b)** $49,892.33    **(c)** $68,189.54    **27.** $1759.12    **29.** $1813.70    **31.** **(a)** $9216.65; 4.6    **(b)** $6787.27; 3.4    **(c)** 1.35;
investment (a) will yield approximately 35% more after-tax dollars than investment (b)    **33.** $M$ is an increasing function of $t$.
**35.** 7.78757%    **37.** $2285.80    **39.** $1683.10    **41.** The conclusions are that the multiplier function is an increasing
function of $i$. The advantage of the IRA over a regular account widens as the interest rate $i$ increases and is particularly dramatic for
high income tax rates.

## ▰▰ CHAPTER 4   REVIEW EXERCISES   (PAGE 354)

**1.** $-1$    **3.** 2    **5.**

$y = 5^x$

**7.**

$y = \left(\dfrac{1}{5}\right)^x$

**9.** $\log_2 64 = 6$    **11.** $\ln 1.09417 = .09$    **13.** $2^5 = 32$    **15.** $e^{4.41763} = 82.9$    **17.** 4    **19.** 4/5    **21.** 3/2
**23.** 1.8245    **25.** 6.1800    **27.** $\log_5 (21 \, k^4)$    **29.** $\log_2 (x^2/m^3)$    **31.** $p = 1.416$    **33.** $m = -2.807$
**35.** $x = -3.305$    **37.** $m = 1.7547$    **39.** $y' = -12e^{2x}$    **41.** $y' = -6x^2 e^{-2x^3}$
**43.** $y' = 10xe^{2x} + 5e^{2x} = 5e^{2x}(2x + 1)$    **45.** $y' = 2x/(2 + x^2)$    **47.** $y' = (x - 3 - x \ln|3x|)/[x(x - 3)^2]$
**49.** $y' = [e^x (x + 1)(x^2 - 1)\ln|x^2 - 1| - 2x^2 e^x]/[(x^2 - 1)(\ln|x^2 - 1|)^2]$
**51.** $y' = 2(x^2 + e^x)(2x + e^x)$

**53.** Relative minimum of $-e^{-1} \approx -.368$ at $x = -1$

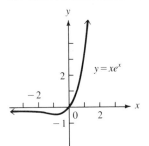

**55.** Relative minimum of $e^2 \approx 7.39$ at $x = 2$

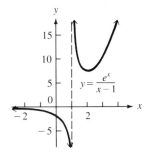

**57.** \$10,631.51     **59.** \$14,204.18     **61.** \$21,190.14     **63.** \$31,611.97     **65.** \$17,901.90     **67.** \$13,829.88
**69.** \$6245.97     **71.** \$6821.51     **73.** \$32,550.63     **75.** \$22,040.14     **77.** \$21,828.75     **79.** About 9.59%
**81. (a)** No  **(b)** \$21.57     **83.** About 7.7 m     **85. (a)** .0769277  **(b)** .0769231  **(c)** .0769231  **(d)** Never
**87. (a)** .8 m  **(b)** 1.2 m  **(c)** 1.5 m  **(d)** 1.8 m     **89. (a)** $y = 100e^{-.23t}$  **(b)** About 3 yr
**(e)** $h$

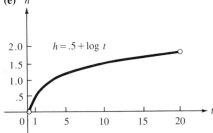

■■■ EXTENDED APPLICATION   (PAGE 360)

**1.** .0315     **2.** $\lambda y_0$ is about 118,000; the painting is a forgery.     **3.** $\lambda y_0 = 163,000$; forgery     **4.** $\lambda y_0 = 24,000$; cannot be a
modern forgery     **5.** $\lambda y_0 = 142,000$; forgery     **6.** $\lambda y_0 = 135,000$; forgery

---

■■■ **CHAPTER 5**     SECTION 5.1     (PAGE 371)

**1.** $2x^2 + C$     **3.** $5t^3/3 + C$     **5.** $6k + C$     **7.** $z^2 + 3z + C$     **9.** $x^3/3 + 3x^2 + C$     **11.** $t^3/3 - 2t^2 + 5t + C$
**13.** $z^4 + z^3 + z^2 - 6z + C$     **15.** $10z^{3/2}/3 + C$     **17.** $2u^{3/2}/3 + 2u^{5/2}/5 + C$     **19.** $6x^{5/2} + 4x^{3/2}/3 + C$
**21.** $4u^{5/2} - 4u^{7/2} + C$     **23.** $-1/z + C$     **25.** $-1/(2y^2) - 2y^{1/2} + C$     **27.** $9/t - 2 \ln|t| + C$     **29.** $e^{2t}/2 + C$
**31.** $-15e^{-.2x} + C$     **33.** $3 \ln|x| - 8e^{-.5x} + C$     **35.** $\ln|t| + 2t^3/3 + C$     **37.** $e^{2u}/2 + 2u^2 + C$
**39.** $x^3/3 + x^2 + x + C$     **41.** $6x^{7/6}/7 + 3x^{2/3}/2 + C$     **43.** $f(x) = 3x^{5/3}/5$     **45.** $C(x) = 2x^2 - 5x + 8$
**47.** $C(x) = .2x^3/3 + 5x^2/2 + 10$     **49.** $C(x) = 2x^{3/2}/3 - 8/3$     **51.** $C(x) = x^3/3 - x^2 + 3x + 6$
**53.** $C(x) = \ln|x| + x^2 + 7.45$     **55.** $P(x) = x^2 + 20x - 50$     **57. (a)** $f(t) = -e^{-.01t} + k$  **(b)** .095 unit
**59.** $v(t) = t^3/3 + t + 6$     **61.** $s(t) = -16t^2 + 6400$; 20 sec

## SECTION 5.2   (PAGE 379)

**1.** $2(2x + 3)^5/5 + C$   **3.** $-2(y - 2)^{-2} + C$   **5.** $-(2m + 1)^{-2}/2 + C$   **7.** $-(x^2 + 2x - 4)^{-3}/3 + C$
**9.** $(z^2 - 5)^{3/2}/3 + C$   **11.** $-2e^{2p} + C$   **13.** $e^{2x^3}/2 + C$   **15.** $e^{2t - t^2}/2 + C$   **17.** $-e^{1/z} + C$
**19.** $-8 \ln|1 + 3x|/3 + C$   **21.** $(\ln|2t + 1|)/2 + C$   **23.** $-(3v^2 + 2)^{-3}/18 + C$   **25.** $-(2x^2 - 4x)^{-1}/4 + C$
**27.** $[(1/r) + r]^2/2 + C$   **29.** $(x^3 + 3x)^{1/3} + C$   **31.** $(p + 1)^7/7 - (p + 1)^6/6 + C$
**33.** $2(5t - 1)^{5/2}/125 + 2(5t - 1)^{3/2}/75 + C$   **35.** $2(u - 1)^{3/2}/3 + 2(u - 1)^{1/2} + C$   **37.** $(x^2 + 12x)^{3/2}/3 + C$
**39.** $(\ln|t^2 + 2|)/2 + C$   **41.** $e^{2z^2}/4 + C$   **43.** $R(x) = (x^2 + 50)^3/3 + 137,919.33$   **45.** $p(x) = -e^{-x^2}/2 + .01$

## SECTION 5.3   (PAGE 389)

**1.** 18   **3.** 65   **5.** 20   **7.** 8   **9.** 56   **11.** 32; 38   **13.** 15; 31/2   **15.** 20; 30   **17.** 16; 14
**19.** 12.8; 27.2   **21.** 2.67; 2.08   **23.** (a) 3  (b) 3.5  (c) 4   **25.** About 10,000,000 cars   **27.** A concentration of
about 19 units   **29.** About 4   **31.** About 2004 ft   **33.** 13.9572   **35.** 1.28857

## SECTION 5.4   (PAGE 400)

**1.** $-6$   **3.** 3/2   **5.** 28/3   **7.** 13   **9.** 1/3   **11.** 76   **13.** 4/3   **15.** 112/25
**17.** $20e^{-.2} - 20e^{-.3} + 3 \ln 3 - 3 \ln 2 \approx 2.775$   **19.** $e^{10}/5 - e^5/5 - 1/2 \approx 4375.1$   **21.** $-3724/3 \approx -1241.33$
**23.** $447/7 \approx 63.857$   **25.** 42   **27.** 76   **29.** 24   **31.** 41/2   **33.** $e^2 - 3 + 1/e \approx 4.757$   **35.** 1   **37.** 16
**39.** $e^3 - 2e^2 + e \approx 8.026$   **45.** (a) \$22,000  (b) \$62,000  (c) 49.5 days   **47.** (a) \$8700  (b) \$9300   **49.** No
**51.** (a) 1.37 ft  (b) .32 ft   **53.** (a) 14.26  (b) 3.55   **55.** (a) $c(t) = 1.2e^{.04t}$  (b) $\int_0^{10} 1.2e^{.04t}\, dt$  (c) $30e^{.4} - 30 \approx$
14.75 billion  (d) About 12.8 yr  (e) About 14.4 yr   **57.** $(72/.014)(e^{.014T} - 1)$

## SECTION 5.5   (PAGE 414)

**1.** 15   **3.** 4   **5.** 26   **7.** 366.1667   **9.** 4/3   **11.** $5 + \ln 6 \approx 6.792$   **13.** $6 - e + 1/e \approx 3.650$
**15.** $e^2 - e - \ln 2 \approx 3.978$   **17.** (a) 8 yr  (b) About 148  (c) about 771   **19.** (a) 19.6 days  (b) \$176,792  (c) \$80,752
(d) \$96,040   **21.** 5733.33 cents or \$57.33   **23.** 83.33 cents or \$.83
**25.** (a) $p(x)$

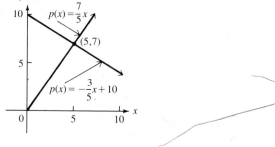

(b) $x = 5$  (c) 7.50  (d) 17.50

**27.** (a) .019; the lower 10% of the income producers earn
1.9% of the total income of the population  (b) .184; the
lower 40% of the income producers earn 18.4% of the
total income of the population  (c) .384; the lower 60%
of the income producers earn 38.4% of the total income
of the population  (d) .819; the lower 90% of the income
producers earn 81.9% of the total income of the population
(e) $I(x)$                            (f) .15

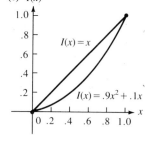

**29.** 12.83 (Simpson's rule)   **31.** .2585 (Simpson's rule)

## CHAPTER 5 REVIEW EXERCISES (PAGE 417)

**1.** $6x + C$    **3.** $x^2 + 3x + C$    **5.** $x^3/3 - 3x^2/2 + 2x + C$    **7.** $2x^{3/2} + C$    **9.** $2x^{3/2}/3 + 9x^{1/3} + C$
**11.** $2x^{-2} + C$    **13.** $-3e^{2x}/2 + C$    **15.** $2\ln|x - 1| + C$    **17.** $e^{3x^2}/6 + C$    **19.** $(3\ln|x^2 - 1|)/2 + C$
**21.** $2(x^2 - 3)^{3/2}/3 + C$    **23.** $-(x^3 + 5)^{-3}/9 + C$    **25.** $\ln|2x^2 - 5x| + C$    **27.** $-e^{-3x^4}/12 + C$
**29.** $2e^{-5x}/5 + C$    **31.** $64/3$    **33.** $1/6$    **35.** $20$    **37.** $24$    **39.** $12$    **41.** $72/25 \approx 2.88$
**43.** $2\ln 3$ or $\ln 9 \approx 2.1972$    **45.** $2e^4 - 2 \approx 107.1963$    **47.** $18$    **49.** $e^2 - 1 \approx 6.3891$
**51.** $C(x) = 10x - x^2 + 4$    **53.** $C(x) = (2x - 1)^{3/2} + 145$    **55.** (a) \$608   (b) \$644    **57.** 36,000
**59.** 2.5 yr; about \$99,000    **61.** 990

## EXTENDED APPLICATION (PAGE 422)

**1.** About 102 yr    **2.** About 98 yr    **3.** About 45.4 yr    **4.** About 90 yr

## CHAPTER 6   SECTION 6.1   (PAGE 429)

**1.** $xe^x - e^x + C$    **3.** $(-5xe^{-3x})/3 - (5e^{-3x})/9 + 3e^{-3x} + C$ or $(-5xe^{-3x})/3 + (22e^{-3x})/9 + C$
**5.** $-5e^{-1} + 3 \approx 1.1606$    **7.** $11\ln 2 - 3 \approx 4.6246$    **9.** $(x^2\ln x)/2 - x^2/4 + C$    **11.** $e^4 + e^2 \approx 61.9872$
**13.** $(x^2e^{2x})/2 - (xe^{2x})/2 + (e^{2x})/4 + C$    **15.** $243/8 - (3\sqrt[3]{2})/4 \approx 29.4301$    **17.** $4x^2\ln(5x) + 7x\ln(5x) - 2x^2 - 7x + C$
**19.** $[2x^2(x + 2)^{3/2}]/3 - [8x(x + 2)^{5/2}]/15 + [16(x + 2)^{7/2}]/105 + C$ or $(2/7)(x + 2)^{7/2} - (8/5)(x + 2)^{5/2} + (8/3)(x + 2)^{3/2} + C$
**21.** $.13077$    **23.** $-4\ln|(x + \sqrt{x^2 + 36})/6| + C$ or $-4\ln|x + \sqrt{x^2 + 36}| + C$
**25.** $\ln|(x - 3)/(x + 3)| + C$, $(x^2 > 9)$    **27.** $(4/3)\ln|(3 + \sqrt{9 - x^2})/x| + C$, $0 < x < 3$
**29.** $-2x/3 + 2\ln|3x + 1|/9 + C$    **31.** $(-2/15)\ln|x/(3x - 5)| + C$    **33.** $\ln|(2x - 1)/(2x + 1)| + C$
**35.** $-3\ln|(1 + \sqrt{1 - 9x^2})/(3x)| + C$    **37.** $2x - 3\ln|2x + 3| + C$
**39.** $1/[25(5x - 1)] - (\ln|5x - 1|)/25 + C$    **41.** About \$110,829    **43.** $15(5e^6 + 1)/2 \approx 15,136.08$ microbes

## SECTION 6.2 (PAGE 438)

**1.** (a) 2.7500   (b) 2.6667   (c) $8/3 \approx 2.6667$    **3.** (a) 1.6833   (b) 1.6222   (c) $\ln 5 \approx 1.6094$    **5.** (a) 16   (b) 14.6667
(c) $44/3 \approx 14.6667$    **7.** (a) .9436   (b) .8374   (c) $4/5 = .8$    **9.** (a) 9.3741   (b) 9.3004    **11.** (a) 5.9914   (b) 6.1672
(c) 6.2832; Simpson's rule is more accurate
**13.** (a) $f(x)$

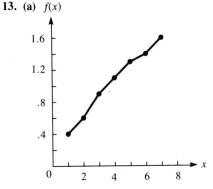

(b) 6.3   (c) 6.27

**15.** (a) 2.4759   (b) 2.3572
**17.** About 30 mcg/ml; this represents the total amount of drug available to the patient
**19.** About 8 mcg/ml; this represents the total effective amount of the drug available to the patient

**21. (a)** $f(x)$

**(b)** 71.5  **(c)** 69.0

(*Note:* The answers for Exercises 23–29 may vary depending on the computer and the software used.)
**23.** Trapezoidal: 12.603; Simpson: 12.6029
**25.** Trapezoidal: 9.83271; Simpson: 9.83377
**27.** Trapezoidal: 14.5192; Simpson: 14.5193
**29.** Trapezoidal: 3979.24; Simpson: 3979.24

## SECTION 6.3  (PAGE 447)

**1.** $8\pi/3$  **3.** $364\pi/3$  **5.** $386\pi/27$  **7.** $3\pi/2$  **9.** $18\pi$  **11.** $\pi(e^4 - 1)/2 \approx 84.19$  **13.** $\pi \ln 4 \approx 4.36$
**15.** $3124\pi/5$  **17.** $16\pi/15$  **19.** $4\pi/3$  **21.** $4\pi r^3/3$  **23.** $-57.67$  **25.** $31/9 \approx 3.44$  **27.** $e - 1 \approx 1.72$
**29. (a)** $2\pi k \int_0^R r(R^2 - r^2)dr$  **(b)** $\pi k R^4/2$  **31. (a)** 40  **(b)** 45  **(c)** 50

## SECTION 6.4  (PAGE 457)

**1. (a)** $5823.38  **(b)** $19,334.31  **3. (a)** $2911.69  **(b)** $9667.16  **5. (a)** $2637.47  **(b)** $8756.70  **7. (a)** $27,979.55
**(b)** $92,895.37  **9. (a)** $2.34  **(b)** $7.78  **11. (a)** $582.57  **(b)** $1934.20  **13. (a)** $3480.51  **(b)** $11,555.70
**15.** $74,565.94  **17.** $20,560.21; $86,778.41  **19.** $4175.52

## SECTION 6.5  (PAGE 462)

**1.** 1/2  **3.** Divergent  **5.** $-1$  **7.** 1000  **9.** 1  **11.** 3/5  **13.** 1  **15.** 4  **17.** Divergent  **19.** 1
**21.** Divergent  **23.** $(2 \ln 2.5)/21 \approx .087$  **25.** $4(\ln 2 - 1/2)/9 \approx .086$  **27.** Divergent  **29.** Divergent  **31.** 1
**33.** 0  **35.** $750,000  **37. (a)** $75,000  **(b)** $60,000  **39.** $30,000  **41.** 1250

## CHAPTER 6  REVIEW EXERCISES  (PAGE 463)

**1.** $[-2x(8 - x)^{5/2}]/5 - [4(8 - x)^{7/2}]/35 + C$  **3.** $xe^x - e^x + C$  **5.** $[(2x + 3)(\ln|2x + 3| - 1)]/2 + C$
**7.** $7(e^2 - 1)/4 \approx 11.181$  **9.** .4143  **11.** 3.983  **13.** 10.28  **15.** 1.459  **17.** $125\pi/6 \approx 65.45$
**19.** $\pi(e^4 - e^{-2})/2 \approx 85.55$  **21.** $406\pi/15 \approx 85.03$  **23.** 75.3982  **25.** 13/6  **27.** 1/2  **29.** $6/e \approx 2.207$
**31.** Divergent  **33.** 3  **35.** 10.55 million  **37.** $28,513.76  **39.** $402.64  **41.** $10,254.22  **43.** $464.49
**45.** $5715.89  **47.** $555,555.56  **49.** 2000 gal

## EXTENDED APPLICATION  (PAGE 467)

**1.** 10.65  **2.** 175.8  **3.** 441.46

## EXTENDED APPLICATION  (PAGE 471)

**1. (a)** 46  **(b)** 45.33  **2. (a)** .0002 liters per sec  **(b)** .0002 liters per sec  **3.** 34.08 ml per sec

# ⊟ CHAPTER 7    SECTION 7.1    (PAGE 481)

**1. (a)** 6   **(b)** $-8$   **(c)** $-20$   **(d)** 43    **3. (a)** $\sqrt{43}$   **(b)** 6   **(c)** $\sqrt{19}$   **(d)** $\sqrt{11}$

**5.**

**7.**

**9.**

**11.**

**13.**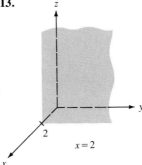

**15.** $\sqrt{3}$    **17.** $\sqrt{99}$ or $3\sqrt{11}$    **19.** $\sqrt{59}$    **21.** $(x+1)^2 + (y-4)^2 + (z-2)^2 = 9$
**23.** $(x-2)^2 + y^2 + (z+3)^2 = 16$    **25.** $(5, 3, -4)$; $r = 3$    **27.** $(-3,0,1)$; $r = \sqrt{12}$ or $2\sqrt{3}$    **29. (c)**    **31. (e)**
**33. (b)**    **35. (a)** $18x + 9h$    **(b)** $-6y - 3h$    **37. (a)** 1986.95 (rounded)   **(b)** 595 (rounded)   **(c)** 359,767.81 (rounded)
**39.** 1.197 (rounded)    **41.** $y = (500^{5/3})/x^{2/3} \approx 31{,}498/x^{2/3}$    **43. (a)** 7.85   **(b)** 4.02

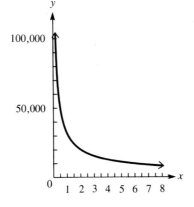

## SECTION 7.2 (PAGE 491)

**1. (a)** $24x - 8y$ **(b)** $-8x + 6y$ **(c)** 24 **(d)** $-20$ **3.** $f_x = -2y; f_y = -2x + 18y^2$; 2; 170
**5.** $f_x = 9x^2y^2; f_y = 6x^3y$; 36; $-1152$ **7.** $f_x = e^{x+y}; f_y = e^{x+y}; e^1$ or $e$; $e^{-1}$ or $1/e$
**9.** $f_x = -15e^{3x-4y}; f_y = 20e^{3x-4y}$; $-15e^{10}$; $20e^{-24}$
**11.** $f_x = (-x^4 - 2xy^2 - 3x^2y^3)/(x^3 - y^2)^2; f_y = (3x^3y^2 - y^4 + 2x^2y)/(x^3 - y^2)^2$; $-8/49$; $-1713/5329$
**13.** $f_x = 6xy^3/(1 + 3x^2y^3); f_y = 9x^2y^2/(1 + 3x^2y^3)$; 12/11; 1296/1297 **15.** $f_x = e^{x^2y}(2x^2y + 1); f_y = x^3e^{x^2y}$; $-7e^{-4}$; $-64e^{48}$
**17.** $f_{xx} = 36xy; f_{yy} = -18; f_{xy} = f_{yx} = 18x^2$ **19.** $R_{xx} = 8 + 24y^2; R_{yy} = -30xy + 24x^2; R_{xy} = R_{yx} = -15y^2 + 48yx$
**21.** $r_{xx} = -8y/(x + y)^3; r_{yy} = 8x/(x + y)^3; r_{xy} = r_{yx} = (4x - 4y)/(x + y)^3$ **23.** $z_{xx} = 0; z_{yy} = 4xe^y; z_{xy} = z_{yx} = 4e^y$
**25.** $r_{xx} = -1/(x + y)^2; r_{yy} = -1/(x + y)^2; r_{xy} = r_{yx} = -1/(x + y)^2$ **27.** $z_{xx} = 1/x; z_{yy} = -x/y^2; z_{xy} = z_{yx} = 1/y$
**29.** $x = -4; y = 2$ **31.** $x = 0, y = 0$, or $x = 3, y = 3$ **33.** $f_x = 2x; f_y = z; f_z = y + 4z^3; f_{yz} = 1$
**35.** $f_x = 6/(4z + 5); f_y = -5/(4z + 5); f_z = -4(6x - 5y)/(4z + 5)^2; f_{yz} = 20/(4z + 5)^2$
**37.** $f_x = (2x - 5z^2)/(x^2 - 5xz^2 + y^4); f_y = 4y^3/(x^2 - 5xz^2 + y^4); f_z = -10xz/(x^2 - 5xz^2 + y^4); f_{yz} = 40xy^3z/(x^2 - 5xz^2 + y^4)^2$
**39. (a)** 80 **(b)** 180 **(c)** 110 **(d)** 360 **41. (a)** $206,800
**(b)** $f_p = 132 - 2i - .02p; f_i = -2p$; the rate at which sales revenue is changing per unit of change in price ($f_p$) or interest rate ($f_i$)
**(c)** A sales revenue drop of $18,800 **43. (a)** 46.656 **(b)** $f_x(27, 64) = .6912$ and is the rate at which production is changing when
labor changes by 1 unit from 27 to 28 and capital remains constant; $f_y(27, 64) = .4374$ and is the rate at which production is
changing when capital changes by one unit from 64 to 65 and labor remains constant **(c)** Production would increase at a rate
of $f_x(x, y) = (1/3)x^{-4/3}[(1/3)x^{-1/3} + (2/3)y^{-1/3}]^{-4}$. **45.** $\partial z/\partial x = .4x^{-.6}y^{.6}; \partial z/\partial y = .6x^{.4}y^{-.4}$ **47. (a)** 168 **(b)** 5448
**(c)** 96 **(d)** 3882 **49. (a)** $.08585 W^{-.575}H^{.725}$ **(b)** .0112 **(c)** $.14645W^{.425}H^{-.275}$ **(d)** .783 **51. (a)** 4.125 1b
**(b)** $\partial f/\partial n = (1/4)n$, the rate of change of weight loss per unit change in workouts **(c)** An additional loss of 1/4 lb
**53. (a)** 2.04% **(b)** 1.26% **(c)** .05%, .003% **55. (a)** $-ae^{-ax}\left(\dfrac{N}{1 + be^{-kt}}\right)$ **(b)** $bkNe^{-(ax + kt)}/(1 + be^{-kt})^2$
**(c)** $a^2e^{-ax}\left(\dfrac{N}{1 + be^{-kt}}\right)$ **(d)** $\dfrac{(be^{-kt} - 1)bk^2Ne^{-(ax + kt)}}{(1 + be^{-kt})^3}$ **(e)** $\dfrac{-abkNe^{-(ax + kt)}}{(1 + be^{-kt})^2}$ **(f)** $\dfrac{-abkNe^{-(ax + kt)}}{(1 + be^{-kt})^2}$

## SECTION 7.3 (PAGE 502)

**1.** Saddle point at $(1, -1)$ **3.** Relative minimum at $(-1, -1/2)$ **5.** Relative minimum at $(-2, -2)$ **7.** Relative
minimum at $(15, -8)$ **9.** Relative maximum at $(2/3, 4/3)$ **11.** Saddle point at $(2, -2)$ **13.** Saddle point at $(1, 2)$
**15.** Saddle point at $(0, 0)$; relative minimum at $(4, 8)$ **17.** Saddle point at $(0, 0)$; relative minimum at $(9/2, 3/2)$ **19.** Saddle
point at $(0, 0)$ **21.** Relative maximum of 1 1/8 at $(-1, 1)$; saddle point at $(0, 0)$; (a) **23.** Relative minima of $-2$ 1/16 at
$(0, 1)$ and at $(0, -1)$; saddle point at $(0, 0)$; (b) **25.** Relative maxima of 1 1/16 at $(1, 0)$ and at $(-1, 0)$; relative minima of
$-15/16$ at $(0, 1)$ and $(0, -1)$; saddle points at $(-1, 1)$, $(1, -1)$, $(1, 1)$, and $(-1, -1)$; (e) **29.** 12 units of labor and $40 per
unit of goods produce maximum profit of $274,000. **31.** 12 lb of Super and 72 lb of Brand A **33.** Twelve units of
electrical tape and 25 units of packing tape give the minimum cost of $2237.

## SECTION 7.4 (PAGE 510)

**1.** $f(6, 6) = 72$ **3.** $f(4/3, 4/3) = 64/27$ **5.** $f(5, 3) = 28$ **7.** $f(20, 2) = 360$ **9.** $f(3/2, 3/2, 3) = 81/4$
**11.** 10, 10 **13.** $x = 6, y = 12$ **15.** 30, 30, 30 **17.** 60 ft by 60 ft **19.** Make 10 large, no small
**21.** 167 units of labor and 178 units of capital **23.** 50 m by 50 m **25.** $r = 5$ inches, $h = 10$ inches
**27.** 12.91 m by 12.91 m by 6.46 m

## ▰▰ SECTION 7.5   (PAGE 518)

**1.** $y' = 3.98x + 5.28$; 21.2, 33.14

**3.** $y' = 3.125x - 8.875$; 6.68; 13.08

**5. (a)** $y$

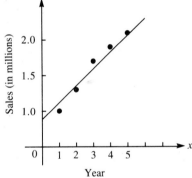

**(b)** $y' = .28x + .76$
**(c)** 2.44 million

**7.** $y' = 8.06x + 49.52$

**9. (a)** 26,920; 23,340; 19,770   **(b)** 29,370; 25,790; 22,210   **(c)** \$42
**11. (a)** $y' = 3.35x - 78.4$   **(b)** 123 lb   **(c)** 156 lb
**13. (a)**   $y$

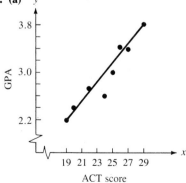

**(b)** $y' = .16x - .82$   **(c)** 3.6

## ■ SECTION 7.6 (PAGE 526)

**1.** $dz = 36x^3\,dx - 15y^2\,dy$ **3.** $dz = 2xy^3\,dx + (3x^2y^2 + 1)dy$ **5.** $dz = [-2y/(x - y)^2]dx + [2x/(x - y)^2]dy$
**7.** $dz = [y^{1/2}/x^{1/2} - 1/[2(x + y)^{1/2}]]\,dx + x^{1/2}/y^{1/2} - 1/[2(x + y)^{1/2}]]dy$
**9.** $dz = (3\sqrt{1 - 2y})dx + [-(3x + 2)/(1 - 2y)^{1/2}]dy$ **11.** $dz = [2x/(x^2 + 2y^4)]dx + [8y^3/(x^2 + 2y^4)]dy$
**13.** $dz = [y^2e^{x+y}(x + 1)]dx + [xye^{x+y}(y + 2)]dy$ **15.** $dz = (2x - y/x)dx + (-\ln x)dy$
**17.** $dw = (4x^3yz^2)dx + (x^4z^3)dy + (3x^4yz^2)dz$ **19.** $dw = (-6/x^2)dx + (12/y^2)dy + (18/z^2)dz$
**21.** $-.2$ **23.** $-.009$ **25.** $-.0394$ **27.** $-.00769$ **29.** $.11$ **31.** $-.335$ **33.** 19.16 cu cm
**35.** 71.2 cu in **37.** $-\$256.10$ **39.** .0769 units **41.** 6.096 cu cm **43.** 2252 matings **45.** $-.917$ units
**47.** 33.21 sq cm

## ■ SECTION 7.7 (PAGE 541)

**1.** $93y/4$ **3.** $3x + 609x^2/4 - 6$ **5.** $(1/9)(48 + y)^{3/2} - (1/9)(24 + y)^{3/2}$ **7.** $(1/3)(36 + 3y)^{3/2} - (1/3)(9 + 3y)^{3/2}$
**9.** $6 + 10y$ **11.** $\ln|75 + 10y| - \ln|27 + 6y|$ or $\ln|(75 + 10y)/(27 + 6y)|$ **13.** $(1/4)e^{x+4} - (1/4)e^{x-4}$
**15.** $(1/2)e^{25+9y} - (1/2)e^{9y}$ **17.** $(1/3)(16 + 3x)^{3/2} - (1/3)(3x)^{3/2}$ **19.** $(20y/3)(18 + 5y^2)^{1/2} - (20y/3)(9 + 5y^2)^{1/2}$
**21.** $279/8$ **23.** $179/2$ **25.** $(2/45)(39^{5/2} - 12^{5/2} - 7533)$ **27.** 21 **29.** 3716/45 **31.** $(\ln 2)^2$ **33.** $8 \ln 2 + 4$
**35.** 256 **37.** $(4/15)[33 - 2^{5/2} - 3^{5/2}]$ **39.** $-2 \ln 6/7$ **41.** $(1/2)(e^7 - e^6 - e^3 + e^2)$ **43.** 48 **45.** 4/3
**47.** $(2/15)(2^{5/2} - 2)$ **49.** $(1/4 \ln (17/8)$ **51.** $e^2 - 3$ **53.** 189/2 **55.** 97,632/105 **57.** 128/9 **59.** 27/8
**61.** $\ln 3$ **63.** 64/3 **65.** 116 **67.** 10/3 **69.** $7(e - 1)/3$ **71.** 16/3 **73.** 117/5 **75.** $4 \ln 2 - 2$
**77.** 13/3 **79.** $(e^7 - e^6 - e^5 + e^4)/2$ **81.** \$2583 **83.** \$933.33

## ■ CHAPTER 7 REVIEW EXERCISES (PAGE 545)

**1.** $-19; -255$ **3.** $-1; -5/2$ **5.** $\sqrt{43}$ **7.** $\sqrt{61}$
**9.**

**11.**

**13.**

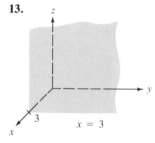

**15.** $(2, -3, 4); r = 3$ **17.** (a) $-10x + 7y$ (b) $-15$ (c) 7 **19.** $f_x = 27x^2y^2 - 5; f_y = 18x^3y$
**21.** $f_x = 4x/(4x^2 + y^2)^{1/2}; f_y = y/(4x^2 + y^2)^{1/2}$ **23.** $f_x = 2x \cdot e^{2y}; f_y = 2x^2 \cdot e^{2y}$ **25.** $f_x = 4x/(2x^2 + y^2); f_y = 2y/(2x^2 + y^2)$
**27.** $f_{xx} = 24xy^2; f_{xy} = 24x^2y - 8$ **29.** $f_{xx} = 8y/(x - 2y)^3; f_{xy} = (-4x - 8y)/(x - 2y)^3$ **31.** $f_{xx} = 2e^y; f_{xy} = 2xe^y$
**33.** $f_{xx} = (-2x^2y^2 - 4y)/(2 - x^2y)^2; f_{xy} = -4x/(2 - x^2y)^2$ **35.** Relative minimum at $(0, 1)$ **37.** Saddle point at $(0, 2)$
**39.** Saddle point at $(3, 1)$ **41.** Relative minimum at $(1, 1/2)$; saddle point at $(-1/3, 11/6)$ **43.** Extrema of 0 at $(0, 4)$ and
256/27 at $(8/3, 4/3)$ **45.** Extremum of 0 at $(0, 0)$ **47.** $x = 160/3, y = 80/3$ **49.** 5 inches by 5 inches by 5 inches
**51.** $dz = 18x^2\,dx - 20y\,dy$ **53.** $dz = [4y/(x + 2y)^2]dx + [-4x/(x + 2y)^2]dy$
**55.** $dz = [xye^{x-y}(x + 2)]dx + [x^2e^{x-y}(1 - y)]dy$ **57.** $dw = 5x^4\,dx + 4y^3\,dy - 3z^2\,dz$ **59.** 1.7 **61.** $64y^2/3 + 40$
**63.** $(1/9)[(30 + 3y)^{3/2} - (12 + 3y)^{3/2}]$ **65.** $12y - 16$ **67.** $(3/2)[(100 + 2y^2)^{1/2} - (2y^2)^{1/2}]$ **69.** 1232/9
**71.** $2[(42)^{5/2} - (24)^{5/2} - (39)^{5/2} + (21)^{5/2}]/135$ **73.** $2 \ln 2$ or $\ln 4$ **75.** 26 **77.** $(4/15)(782 - 8^{5/2})$ **79.** 1900
**81.** 1/2 **83.** 1/48 **85.** 3 **87.** (a) $\$(150 + \sqrt{10})$ (b) $\$(400 + \sqrt{10})$ (c) $\$(1200 + 2\sqrt{5})$
**89.** (a) $.6x^{-.4}y^{.4}$ or $.6y^{.4}/x^{.4}$ (b) $.4x^{.6}y^{-.6}$ or $.4x^{.6}/y^{.6}$ **91.** \$431 **93.** 7.92 cu cm **95.** 15.6 cu cm
**97.** (a) $y' = .97x + 31.5$ (b) About 216

# CHAPTER 8     SECTION 8.1   (PAGE 554)

**1.** $y = x^3/3 + C$    **3.** $y = -x^2 + x^3 + C$    **5.** $y = e^x + C$    **7.** $y = -2e^{-x} + C$    **9.** $y = -4x^3/9 + C$
**11.** $y = 3x^4/8 + C$    **13.** $y = x^3 - x^2 + 2$    **15.** $y = x^3 - 2x^2 + 2x + 8$    **17.** $y = \ln x + 10$
**19.** $y = -2xe^{-x} - 2e^{-x} + 44$    **21. (a)** $1011.75  **(b)** $1024.52  **(c)** No; if $x = 8$, the denominator becomes 0
**23.** $x = -2p^2 + C$    **25.** About 332,000

# SECTION 8.2   (PAGE 562)

**1.** $y^2 = x^2 + C$    **3.** $y = ke^{x^2}$    **5.** $y = ke^{(x^3 - x^2)}$    **7.** $y = Mx$    **9.** $y = Me^x + 5$    **11.** $y = -1/(e^x + C)$
**13.** $y^2 = 2x^3/3 + 9$    **15.** $y = e^{x^2 + 3x}$    **17.** $y = -5/(5 \ln x - 6)$    **19.** $y^2/2 - 3y = x^2 + x - 4$
**21.** $y = (e^x - 3)/(e^x - 2)$    **27. (a)** $155.24  **(b)** $161.55  **(c)** $164.62    **29.** About 11.6 yr    **31.** About 260
**33.** About 4.4 cc    **35.** About 387    **37. (a)** $k \approx .8$  **(b)** 11  **(c)** 55  **(d)** 2981    **39. (a)** $dy/dt = -.05y$
**(b)** $y = Me^{-.05t}$  **(c)** $y = 90e^{-.05t}$  **(d)** About 55 g

# SECTION 8.3   (PAGE 570)

**1.** $y = 5/2 + Ce^{-2x}$    **3.** $y = 3 + Ce^{-x^2/2}$    **5.** $y = x \ln x + Cx$    **7.** $y = -1/2 + Ce^{x^2/2}$    **9.** $y = x^2/4 + x + Cx^{-2}$
**11.** $y = -x^3/2 + Cx$    **13.** $y = e^x + 99e^{-x}$    **15.** $y = -1 + 11e^{(x^2-1)/2}$    **17.** $y = x^2/7 + 2560/(7x^5)$
**19.** $y = (3 + 197e^{4-x})/x$    **21. (a)** $y = c/(p + kce^{-cx})$  **(b)** $y = (cy_0)/[py_0 + (c - py_0)e^{-cx}]$  **(c)** $c/p$
**23.** $y = e^{-t} + 9999e^{.02t}$    **25.** $y = -50t - 2500 + 12{,}500e^{.02}$    **27. (a)** $t = 140e^{-1.3863t} + 40$  **(b)** 75°  **(c)** 40°

# SECTION 8.4   (PAGE 577)

**1.** 1.837    **3.** 4.315    **5.** 1.491    **7.** .087    **9.** $-.540; -.520$    **11.** 2.030; 2.042    **13.** 3.806; 4.759
**15.** $-.025; -.036$

**17.**

| $x_i$ | $y_i$ | $f(x_i)$ | $y_i - f(x_i)$ |
|---|---|---|---|
| 0 | 0 | 0 | 0 |
| .2 | 0 | .08772053 | $-.08772053$ |
| .4 | .11696071 | .22104189 | $-.10408118$ |
| .6 | .26432197 | .37954470 | $-.11522273$ |
| .8 | .43300850 | .55699066 | $-.12398216$ |
| 1.0 | .61867206 | .75000000 | $-.13132794$ |

**19.**

| $x_i$ | $y_i$ | $f(x_i)$ | $y_i - f(x_i)$ |
|---|---|---|---|
| 0 | 0 | 0 | 0 |
| .2 | .2 | .1812692 | .0187308 |
| .4 | .36 | .32967995 | .03032005 |
| .6 | .488 | .45118836 | .03681164 |
| .8 | .5904 | .55067104 | .03972896 |
| 1.0 | .67232 | .63212056 | .04019944 |

**21.**

**23.**

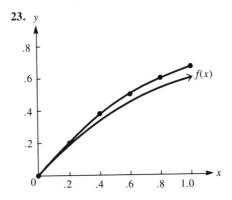

**25. (a)** $dy/dt = -3y + .02y^2$  **(b)** About 66 firms    **27.** About 20 species    **29.** About 7,000
(*Note:* The answers for Exercises 31 and 33 may vary depending on the computer and the software used.)
**31.** 4.047    **33.** 1.448

## SECTION 8.5    (PAGE 587)

**1.** $31,048.20    **3.** About 5.7 yr    **5. (a)** $3 \ln x - 2x + 2 \ln y - 3y = \ln 4 - 8$  **(b)** $x = 3/2, y = 2/3$
**7. (a)** $y = (24,995,000)/(4999 + e^{.25t})$  **(b)** 3671  **(c)** 91  **(d)** On the thirty-fourth day    **9. (a)** $y = (50e^{.18t})/(1 + .005e^{.18t})$
**(b)** In about 29 days    **11. (a)** $y = .005 + .015e^{-1.010t}$  **(b)** $y = .00273 + .00727e^{-1.1t}$
**13. (a)** $y = (5.55e^{.45t})/(1 + .111e^{.45t})$  **(b)** In about 6 days    **15. (a)** $y = 347e^{-4.24e^{-.1t}}$  **(b)** In about 5.5 days
**17. (a)** $y = [2(t + 100)^3 - 1,800,000]/(t + 100)^2$  **(b)** About 250 lb  **(c)** It continues to increase.    **19. (a)** $y = 20e^{-.02t}$
**(b)** About 6 lb  **(c)** It continues to decrease.    **21. (a)** $y = [.1(t + 100)^2 - 500]/(t + 100)$  **(b)** About 9.2 g

## CHAPTER 8  REVIEW EXERCISES    (PAGE 590)

**1.** $y = x^4/2 + 3x^2 + C$    **3.** $y = 4e^x + C$    **5.** $y = x^3/3 - 5x^2/2 + 1$    **7.** $y = -5e^{-x} - 5x + 22$
**9.** $y^2 = 3x^2 + 2x + C$    **11.** $y = (Cx^2 - 1)/2$    **13.** $y = 2e^{5x-x^2}$    **15.** $y^2 + 6y = 2x - 2x^2 + 352$
**17.** $y = 12/5 + Ce^{-5x}$    **19.** $y = 1 + Ce^{-x^2/6}$    **21.** $y = 1 + 7e^{-x^3/3}$    **23.** $y = 3/2 + (27e^{x^2})/2$    **25.** 2.138
**27.** 1.215; 1.223; $-.008$    **29. (a)** $5800 **(b)** $25,100    **31.** About 13.9 yr    **33.** About 219    **35.** 17.3 min
**37. (a)** $y = 489,300/(699 + e^{1.140t})$  **(b)** About 135  **(c)** After 5.7 wk    **39. (a)** $y = 100/(1 + 24e^{-.481x})$  **(b)** About 32
**41. (a)** 170.27 million  **(b)** 204.15 million  **(c)** 334.13 million  **(d)** There were different population changes over the 50-year
intervals, resulting in different $y$-values.    **43.** 213°

## EXTENDED APPLICATION    (PAGE 595)

**1. (a)** 2.4 yr  **(b)** 3.1 yr    **2. (a)** 28.2 yr  **(b)** 37.1 yr    **3. (a)** 173 yr  **(b)** 228 yr

## CHAPTER 9    SECTION 9.1    (PAGE 602)

**1.** Yes    **3.** Yes    **5.** No; $\int_0^3 4x^3 \, dx \neq 1$    **7.** No; $\int_{-2}^2 \frac{x^2}{16} \, dx \neq 1$    **9.** No; $f(x) < 0$ on $[-2, 0)$    **11.** $k = 3/14$
**13.** $k = 3/125$    **15.** $k = 2/9$    **17.** $k = 1/12$    **19.** $k = 3/26$    **21. (a)** .4226  **(b)** .2071  **(c)** .4082
**23. (a)** .3935  **(b)** .3834  **(c)** .3679    **25. (a)** .9975  **(b)** .0024    **27. (a)** .268  **(b)** .414  **(c)** .318
**29. (a)** $4/7 \approx .5714$  **(b)** $5/21 \approx .2381$    **31. (a)** .5  **(b)** .25  **(c)** .75

## SECTION 9.2    (PAGE 610)

**1.** $\mu = 5$; Var$(X) \approx 1.33$; $\sigma = 1.15$    **3.** $\mu = 14/3 \approx 4.67$; Var$(X) = .89$; $\sigma = .94$    **5.** $\mu = 2.83$;
Var$(X) = .59$; $\sigma = .77$    **7.** $\mu = 4/3 \approx 1.33$; Var$(X) = 2/9 = .22$; $\sigma = .47$    **9. (a)** 5.40  **(b)** 5.55  **(c)** 2.36
**(d)** .54  **(e)** .60    **11. (a)** $4/3 = 1.33$  **(b)** .22  **(c)** .47  **(d)** .56  **(e)** .63    **13. (a)** 5  **(b)** 0    **15. (a)** 4.83  **(b)** .055
**17. (a)** $\sqrt[4]{2} \approx 1.19$  **(b)** .18    **19. (a)** 310.3 hr  **(b)** 267 hr  **(c)** .206    **21. (a)** 2 mo  **(b)** 2 mo  **(c)** .632
**23. (a)** 2  **(b)** .89  **(c)** .62    **25. (a)** 3.194  **(b)** 1.265  **(c)** .0368

## ▄▄ SECTION 9.3   (PAGE 622)

**1.** (a) 4.4 cm   (b) .23 cm   (c) .29   **3.** (a) 33.33 yr   (b) 33.33 yr   (c) .23   **5.** (a) 1 day   (b) 1 day   (c) .23
**7.** 49.98%   **9.** 17.36%   **11.** 8.01%   **13.** 89.69%   **15.** $-1.28$   **17.** .92   **19.** $m = (b + a)/2$
**21.** (a) \$47,500   (b) .47   **23.** (a) $f(x) = .235e^{-.235x}$ on $[0, +\infty)$   (b) .095   **25.** (a) .1587   (b) .7698
**27.** (a) 28 days   (b) .375   **29.** (a) 1/2   (b) .49   **31.** 3.07 ft and 3.33 ft   **33.** 1.00002   **35.** 4.000506
**37.** (a) $\mu = 2.7416 \times 10^8 \approx 0$   (b) $\sigma = .999433 \approx 1$

## ▄▄ CHAPTER 9   REVIEW EXERCISES   (PAGE 625)

**1.** Probability density function   **3.** Probability density function   **5.** $k = 3/14$   **7.** (a) .828   (b) .536   (c) .364
**9.** $\mu = 6.5$; Var($X$) = 2.083; $\sigma = 1.443$   **11.** $\mu = 2.405$; Var($X$) = .759; $\sigma = .871$   **13.** (a) 7/12 $\approx$ .583
(b) .244   (c) .482   (d) .611   **15.** (a) 13.571   (b) 6.681   (c) .582   **17.** 6.3%   **19.** 77.77%   **21.** 99.38%
**23.** $-.81$   **25.** .406   **27.** (a) $f(x) = (e^{-x/8})/8$; $[0, +\infty)$   (b) 8   (c) 8   (d) .249   **29.** .1921   **31.** (a) 2.377g
(b) 1.533 g   (c) .851   **33.** (a) 21 inches   (b) .526   **35.** .1379

## ▄▄ CHAPTER 10   SECTION 10.1   (PAGE 641)

**1.** $\pi/3$   **3.** $5\pi/6$   **5.** $7\pi/6$   **7.** $13\pi/6$   **9.** 315°   **11.** 330°   **13.** 288°   **15.** 48°
(*Note:* In Exercises 17 and 19 we give the answers in the order sine, cosine, tangent, cotangent, secant, and cosecant.)
**17.** 4/5; $-3/5$; $-4/3$; $-3/4$; $-5/3$; 5/4   **19.** 4/5; 3/5; 4/3; 3/4; 5/3; 5/4   **21.** +; +; +; +; +; +
**23.** $-$; $-$; +; +; $-$; $-$   **25.** $\sqrt{3}/3$; $\sqrt{3}$; 2   **27.** $\sqrt{3}/2$; $\sqrt{3}/3$; $2\sqrt{3}/3$   **29.** $-1$; $-1$
**31.** $-\sqrt{3}/2$; $-2\sqrt{3}/3$   **33.** $\sqrt{3}/2$   **35.** 1   **37.** $2\sqrt{3}/3$   **39.** $-1$   **41.** $-\sqrt{3}/2$   **43.** $-\sqrt{2}$   **45.** $-\sqrt{3}$
**47.** 1/2   **49.** .6293   **51.** 7.1154   **53.** .3907   **55.** .1564
**57.**

**59.**

**61.**

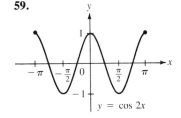

**63.**

**65. (a)** 1000    **(b)** 750    **(c)** 500    **(d)** 0    **(e)** 500

**(f)** $S(t)$

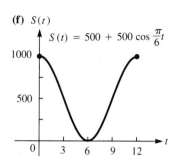

**67.** $2 \times 10^8$ m per sec
**69.** 120°
**71. (a)** 30° F    **(b)** 44° F    **(c)** 66° F    **(d)** 84° F    **(e)** 88° F

## SECTION 10.2    (PAGE 649)

**1.** $y' = 12 \cos 6x$    **3.** $y' = 108 \sec^2(9x + 1)$    **5.** $y' = -4 \cos^3 x \sin x$    **7.** $y' = 5 \tan^4 x \sec^2 x$
**9.** $y' = -5(4x \cos 4x + \sin 4x)$    **11.** $y' = (x \cos x - \sin x)/x^2$    **13.** $y' = 5e^{5x} \cos e^{5x}$    **15.** $y' = (\cos x)e^{\sin x}$
**17.** $y' = (2/x)\cos(\ln 4x^2)$    **19.** $y' = (2x \cos x^2)/\sin x^2$ or $2x \cot x^2$    **21.** $y' = 6 \cos x/(3 - 2 \sin x)^2$
**23.** $y' = \sqrt{\sin 3x}(\sin 3x \cos x - 3 \sin x \cos 3x)/[2\sqrt{\sin x}(\sin^2 3x)]$
**25. (a)** $R'(x) = -200\pi \sin 2\pi x$    **(b)** $-100\pi$    **(c)** 0    **(d)** $100\pi$

**27. (a)**

$y = \dfrac{\pi}{8} \cos 3\pi \left(t - \dfrac{1}{3}\right)$

**(b)** $v = dy/dt = -(3\pi^2/8) \sin 3\pi(t - 1/3)$
$a = d^2y/dt^2 = -(9\pi^3/8)\cos 3\pi(t - 1/3)$    **(d)** At $t = 1$,
acceleration is negative, arm is moving clockwise;
at $t = 4/3$, acceleration is positive, arm is moving
counterclockwise; at $t = 5/3$, acceleration is negative,
arm is moving clockwise.

## SECTION 10.3    (PAGE 654)

**1.** $(1/5)\sin 5x + C$    **3.** $5 \sin x - 2 \cos x + C$    **5.** $(-\cos x^2)/2 + C$    **7.** $-3 \tan 2x + C$    **9.** $(1/8)\sin^8 x + C$
**11.** $(2/3)\sin^{3/2} x + C$    **13.** $-\ln|1 + \cos x| + C$    **15.** $(1/6) \sin x^6 + C$    **17.** $-4 \ln|\cos (x/4)| + C$
**19.** $(1/3)\ln|\sin x^3| + C$    **21.** $-\cos e^x + C$    **23.** $1 - \sqrt{2}/2$    **25.** $-\ln 1/2$ or $\ln 2$    **27.** $\sqrt{3}/2 - 1$
**29.** $(-6/5)x \sin 5x - (6/25) \cos 5x + C$    **31.** $-8x \cos x + 8 \sin x + C$    **33.** $(-3/4)x \sin 8x - (3/32)\cos 8x + C$
**35.** 1/2    **37.** 6000

## SECTION 10.4    (PAGE 660)

**1.** $-\pi/3$    **3.** $\pi/4$    **5.** $-\pi/2$    **7.** $\pi/3$    **9.** $3\pi/4$    **11.** $-\pi/3$    **13.** $-8°$    **15.** 154°    **17.** 22°    **19.** 48°
**21.** $-42°$    **23.** 68°    **25.** $y' = 12/\sqrt{1 - 144x^2}$    **27.** $y' = 3/(1 + 9x^2)$    **29.** $y' = -2/[x\sqrt{x^2 - 4}]$
**31.** $y' = 1/[x(1 + [\ln|7x|]^2)]$    **33.** $y' = 1/[(1 + (x + 1)^2)\tan^{-1}(x + 1)]$    **35.** 18°    **37.** 15°
**39.** $(1/4)\tan^{-1} x^4 + C$    **41.** $\sin^{-1} e^x + C$    **43.** $(4/3)\sin^{-1} 3x + C$    **45.** $\tan^{-1} \ln|x| + C$

## SECTION 10.5　(PAGE 663)

**1.** 1　**3.** −1　**5.** 1　**7.** −2;2　**9.** 9; −9　**11.** 1/2; −1/2　**13.** About 1490　**15.** About 2610
**17.** About 5380　**19.** About 7010　**21.** About 1330
**23.** $f(t)$

**25.** $\sin \theta = s/L_2$　**27.** $\cot \theta = (L_0 - L_1)/s$　**29.** $k \cdot L_1/r_1^4$
**31.** $R = k + L_1/r_1^4 + k \cdot L_2/r_2^4$　**33.** $R' = k(s \csc^2 \theta)/r_1^4 + k(-s \cos \theta/\sin^2 \theta)/r_2^4$
**35.** $0 = k/r_1^4 - k \cos \theta/r_2^4$　**37.** $\cos \theta = 1/256 \approx .0039; \theta \approx 90°$　**39.** 340 cm
**41.** When $x = \pi/a$

## CHAPTER 10 REVIEW EXERCISES　(PAGE 666)

**1.** $\pi/2$　**3.** $7\pi/6$　**5.** $2\pi$　**7.** 1260°　**9.** 81°　**11.** 117°　**13.** $\sqrt{3}/2$　**15.** $\sqrt{3}/2$　**17.** $2\sqrt{3}/3$　**19.** 1/2
**21.** $-\sqrt{2}$　**23.** .7314　**25.** 6.314　**27.** .9945　**29.** .8290
**31.**

**33.**

**35.** $y' = -28 \cos 7x$　**37.** $y' = 8x \sec^2(4x^2 + 3)$　**39.** $y' = -18 \cos^5 x \sin x$　**41.** $y' = -20x^4 \csc^2(4x^5)$
**43.** $y' = -x^2 \sin x + 2x \cos x$　**45.** $y' = 2 \cos x/(\sin x + 1)^2$　**47.** $y' = [\sin x - x \cos x + 2 \cos x]/\sin^2 x$
**49.** $y' = -\tan x$　**51.** $-(1/2)\cos 2x + C$　**53.** $-(1/9) \ln|\cos 9x| + C$　**55.** $5 \tan x + C$
**57.** $-(1/6) \cos 3x^2 + C$　**59.** $-(2/3)(\cos x)^{3/2} + C$　**61.** $-(1/22) \ln|\cos 11x^2| + C$　**63.** $2(\sin x)^{7/2}/7 + C$
**65.** 1　**67.** $20\pi$　**69.** $-\pi/6$　**71.** 0　**73.** $-\pi/2$　**75.** $y' = -3/\sqrt{1 - 9x^2}$　**77.** $y' = 10x/(1 + 25x^4)$
**79.** $3 \sin^{-1} x + C$　**81.** $-(6/5) \tan^{-1} x^5 + C$　**83.** 105; 75

### ■■■ EXTENDED APPLICATION    (PAGE 671)

**1.** $i_0 e^{-5\alpha}$    **2.** $\sqrt{2}i_0/2$    **3. (a)** $[-\alpha i_0 \sin(t\pi/12)e^{-\alpha x/\sqrt{1-(9/16)\cos^2(t\pi/12)}}]/\sqrt{1-(9/16)\cos^2(t\pi/12)}$, which represents the rate of change of the light intensity with respect to change in distance below the surface while time remains constant
**(b)** $(\pi i^0/192)\cos(t\pi/12)e^{-\alpha x/\sqrt{1-(9/16)\cos^2(t\pi/12)}} \cdot (16 + 9[\sin^2(t\pi/12)]\alpha x[1-(9/16)\cos^2(t\pi/12)]^{-3/2})$, which represents the rate of change of the light intensity with respect to change in time while the distance below the surface remains constant.

### ■■■ CHAPTER 11    SECTION 11.1    (PAGE 676)

**1.** 2, 6, 18, 54    **3.** 1/2, 2, 8, 32    **5.** 3/2, 3, 6, 12, 24    **7.** $a_5 = 324$; $a_n = 4 \cdot 3^{n-1}$
**9.** $a_5 = -1875$; $a_n = -3(-5)^{n-1}$    **11.** $a_5 = 24$; $a_n = (3/2) \cdot 2^{n-1}$    **13.** $a_5 = -256$; $a_n = -(-4)^{n-1}$
**15.** $r = 2$; $a_n = 6 \cdot 2^{n-1}$    **17.** $r = 2$; $a_n = (3/4) \cdot 2^{n-1}$    **19.** Not geometric    **21.** $r = -2/3$; $a_n = (-5/8)(-2/3)^{n-1}$
**23.** 93    **25.** 33/4    **27.** 33    **29.** 860.95    **31.** 90    **33.** 170    **35.** 160    **37.** 255/4    **39. (a)** \$1681
**(b)** \$576    **41.** $2^{30}$ or \$1,073,741,824; $2^{31} - \$1$    **43.** 61%    **45.** About 95 times

### ■■■ SECTION 11.2    (PAGE 686)

(*Note:* Answers in this section may differ by a few cents, depending on whether or not calculators are used.)
**1.** \$1318.08    **3.** \$305,390.00    **5.** \$671,994.62    **7.** \$222,777.26    **9.** \$40,652.46    **11.** \$526.95    **13.** \$952.33
**15.** \$39,434.37    **17.** \$8045.29    **19.** \$1,367,773.96    **21.** \$103,796.60    **23.** \$85,594.80    **25.** \$476.90
**27.** \$11,942.55    **29.** \$1626.86    **31.** \$137,895.84    **33.** \$159.49    **35.** \$522.85    **37.** \$112,796.87
**39.** \$209,348.00    **41.** \$4765.40    **43. (a)** \$1200    **(b)** \$3511.58

**(c)**

| Payment Number | Amount of Deposit | Interest Earned | Total |
|---|---|---|---|
| 1 | \$3511.58 | \$0 | \$3511.58 |
| 2 | \$3511.58 | \$105.35 | \$7128.51 |
| 3 | \$3511.58 | \$213.86 | \$10,853.95 |
| 4 | \$3511.58 | \$325.62 | \$14,691.15 |
| 5 | \$3511.58 | \$440.73 | \$18,643.46 |
| 6 | \$3511.58 | \$559.30 | \$22,714.34 |
| 7 | \$3511.58 | \$681.43 | \$26,907.35 |
| 8 | \$3511.58 | \$807.22 | \$31,226.15 |
| 9 | \$3511.58 | \$936.78 | \$35,674.51 |
| 10 | \$3511.58 | \$1070.24 | \$40,256.33 |
| 11 | \$3511.58 | \$1207.69 | \$44,975.60 |
| 12 | \$3511.58 | \$1349.27 | \$49,836.45 |
| 13 | \$3511.58 | \$1495.09 | \$54,843.12 |
| 14 | \$3511.59 | \$1645.29 | \$60,000.00 |

**45.** \$12,493.78    **47.** \$158.00; \$1584    **49.** \$663.75    **51.** \$742.51    **53. (a)** \$4025.90    **(b)** \$2981.93

**55.**

| Payment Number | Amount of Payment | Interest for Period | Portion to Principal | Principal at End of Period |
|---|---|---|---|---|
| 0 | — | — | — | $4000 |
| 1 | $1207.68 | $320.00 | $887.68 | $3112.32 |
| 2 | $1207.68 | $248.99 | $958.69 | $2153.63 |
| 3 | $1207.68 | $172.29 | $1035.39 | $1118.24 |
| 4 | $1207.70 | $89.46 | $1118.24 | $0 |

**57.**

| Payment Number | Amount of Payment | Interest for Period | Portion to Principal | Principal at End of Period |
|---|---|---|---|---|
| 0 | — | — | — | $7184 |
| 1 | $211.03 | $107.76 | $103.27 | $7080.73 |
| 2 | $211.03 | $106.21 | $104.82 | $6975.91 |
| 3 | $211.03 | $104.64 | $106.39 | $6869.52 |
| 4 | $211.03 | $103.04 | $107.99 | $6761.53 |
| 5 | $211.03 | $101.42 | $109.61 | $6651.92 |
| 6 | $211.03 | $99.78 | $111.25 | $6540.67 |

## SECTION 11.3 (PAGE 698)

**1.** $1 - 2x + 2^2x^2/2! - 2^3x^3/3! + 2^4x^4/4!$ **3.** $e + ex + ex^2/2! + ex^3/3! + ex^4/4!$
**5.** $3 + x/6 - x^2/216 + x^3/3888 - 5x^4/279{,}936$ **7.** $-1 + x/3 + x^2/9 + 5x^3/81 + 10x^4/243$
**9.** $1 + x/4 - 3x^2/32 + 7x^3/128 - 77x^4/2048$ **11.** $-x - x^2/2 - x^3/3 - x^4/4$ **13.** $2x^2 - 2x^4$ **15.** $1 - x^2/2 + x^4/24$
**17.** $x - x^2 + x^3/2 - x^4/6$ **19.** $27 - 9x/2 + x^2/8 + x^3/432 + x^4/10{,}368$ **21.** $1 - x + x^2 - x^3 + x^4$ **23.** $x^2 - x^4/6$
**25.** $.9802$ **27.** $2.7456$ **29.** $2.9833$ **31.** $-1.0066$ **33.** $1.0074$ **35.** $-.0101$ **37.** $.0002$
**39.** $.99995$ or $1.0000$ **41.** $.401$ thousand dollars, or $401 **43.** $.905$ dollars **45.** **(a)** $1 + \lambda N + \lambda^2 N^2/2$
**(c)** $N = \sqrt{2k\lambda}/\lambda$ **47.** $.1500$ ml

## SECTION 11.4 (PAGE 705)

**1.** Converges to 24 **3.** $r = 2$; diverges **5.** Converges to 64/3 **7.** Converges to 1000/9 **9.** Converges to 3/2
**11.** Converges to 1/5 **13.** Converges to $e^2/(e + 1)$ **15.** $S_1 = 1, S_2 = 3/2, S_3 = 11/6, S_4 = 25/12, S_5 = 137/60$
**17.** $S_1 = 1/7, S_2 = 16/63, S_3 = 239/693, S_4 = 3800/9009, S_5 = 22{,}003/45{,}045$
**19.** $S_1 = 1/6, S_2 = 1/4, S_3 = 3/10, S_4 = 1/3, S_5 = 5/14$ **21.** $10{,}000/9 \approx 1111$ units **23.** 1600 rotations **25.** 12 m

## SECTION 11.5 (PAGE 716)

**1.** $5/2 + 5x/2^2 + 5x^2/2^3 + 5x^3/2^4 + \ldots + 5x^n/2^{n+1} + \ldots; (-2, 2)$
**3.** $8x - 8 \cdot 3x^2 + 8 \cdot 3^2x^3 - 8 \cdot 3^3x^4 + \ldots + (-1)^n \cdot 8 \cdot 3^nx^{n+1} + \ldots; (-1/3, 1/3)$
**5.** $x^2/4 + x^3/4^2 + x^4/4^3 + x^5/4^4 + \ldots + x^{n+2}/4^{n+1} + \ldots; (-4, 4)$
**7.** $4x - 4^2x^2/2 + 4^3x^3/3 - 4^4x^4/4 + \ldots + (-1)^n \cdot 4^{n+1}x^{n+1}/(n + 1) + \ldots; (-1/4, 1/4)$
**9.** $1 + 4x^2 + 4^2x^4/2! + 4^3x^6/3! + \ldots + 4^nx^{2n}/n! + \ldots; (-\infty, +\infty)$
**11.** $x^3 - x^4 + x^5/2! - x^6/3! + \ldots + (-1)^n \cdot x^{n+3}/n! + \ldots; (-\infty, +\infty)$
**13.** $2 - 2x^2 + 2x^4 - 2x^6 + \ldots + (-1)^n \cdot 2x^{2n} + \ldots; (-1, 1)$
**15.** $1 + x^2/2! + x^4/4! + x^6/6! + \ldots + x^{2n}/(2n)! + \ldots; (-\infty, +\infty)$
**17.** $2x^4 - 2^2x^8/2 + 2^3x^{12}/3 - 2^4x^{16}/4 + \ldots + (-1)^n \cdot 2^{n+1} \cdot x^{4n+4}/(n + 1) + \ldots; (-1/\sqrt[4]{2}, 1/\sqrt[4]{2}]$

**19.** $1 + x - x^2/2! - x^3/3! + x^4/4! + x^5/5! - x^6/6! - x^7/7! + . . .; (-\infty, +\infty)$
**21.** $1 - x^6/2! + x^{12}/4! - x^{18}/6! + . . . + (-1)^n x^{6n}/(2n)! + . . .; (-\infty, +\infty)$
**23.** $x^2 - x^6/3 + x^{10}/5 - x^{14}/7 + . . . + (-1)^n x^{4n+2}/(2n + 1) + . . .; [-1, 1]$
**25.** $1 + 2x + 2x^2 + 2x^3 + . . . + 2x^n + . . .$    **29.** .54499    **31.** .590789    **33.** .19146    **35.** .00999983
**37.** .9987503

## ▰ SECTION 11.6    (PAGE 722)

**1.** 4    **3.** 0    **5.** 1    **7.** Does not exist    **9.** 1    **11.** Does not exist    **13.** $1/2\sqrt{2})$    **15.** 1/4    **17.** 1/12
**19.** 9    **21.** 0    **23.** 1/8    **25.** 1/9    **27.** 1    **29.** Does not exist    **31.** 5    **33.** 1    **35.** 1    **37.** 3    **39.** 2
**41.** Does not exist    **43.** Does not exist    **45.** 0    **47.** 1/2

## ▰ SECTION 11.7    (PAGE 726)

**1.** 2.73    **3.** .33    **5.** .67    **7.** 2.33    **9.** $-2.20$    **11.** 3.10    **13.** $-2.00; .67; 3.50$    **15.** $-1; 3.67$
**17.** 1.414    **19.** 3.317    **21.** 15.811    **23.** 2.080    **25.** 4.642    **27.** Relative maximum at $-1.65$; relative minimum
at 3.65    **29.** Relative minima at $-.71$ and 1.77; relative maximum at 1.19    **31.** 6.13
**33.** (a) $f'(i) = [-1 + ni(1 + i)^{-n-1} + (1 + i)^{-n}]/i^2$
(b) $[Mi - Mi(1 + i)^{-n} - Pi^2]/[M(in(1 + i)^{-n-1} - 1 + (1 + i)^{-n})]$    (c) .01371229    (d) .01383273
**35.** $i_2 = .01036283; i_3 = .01036541$    **37.** 1.5538

## ▰ CHAPTER 11    REVIEW EXERCISES    (PAGE 728)

**1.** $a_5 = -64; a_n = -4(-2)^{n-1}$    **3.** $a_5 = 1; a_n = 81(1/3)^{n-1}$    **5.** 186    **7.** 121/3    **9.** 126    **11.** 363/2
**13.** $15,162.14    **15.** $199,870.32    **17.** $886.05    **19.** $3339.86    **21.** $2815.31    **23.** $34,357.52
**25.** $12,806.37    **27.** $2806.66

**29.**

| Payment Number | Amount of Payment | Interest for Period | Portion to Principal | Principal at End of Period |
|---|---|---|---|---|
| 0 | — | — | — | $5000 |
| 1 | $985.09 | $250.00 | $735.09 | $4264.91 |
| 2 | $985.09 | $213.25 | $771.84 | $3493.07 |
| 3 | $985.09 | $174.65 | $810.44 | $2682.63 |
| 4 | $985.09 | $134.13 | $850.96 | $1831.67 |
| 5 | $985.09 | $91.58 | $893.51 | $938.16 |
| 6 | $985.07 | $46.91 | $938.16 | $0 |

**31.** $e^2 - e^2 x + e^2 x^2/2! - e^2 x^3/3! + e^2 x^4/4!$    **33.** $1 + x/2 - x^2/8 + x^3/16 - 5x^4/128$
**35.** $\ln 2 - x/2 - x^2/8 - x^3/24 - x^4/64$    **37.** $1 - 2x^2 + 2x^4/3$    **39.** $1 + 2x/3 - x^2/9 + 4x^3/81 - 7x^4/243$
**41.** 7.3155    **43.** .9899    **45.** .7080    **47.** .9982    **49.** 1.0067    **51.** Converges to 27/5    **53.** $r = 3$; diverges
**55.** Converges to 1/3    **57.** $S_1 = 1; S_2 = 4/3; S_3 = 23/15; S_4 = 176/105; S_5 = 563/315$
**59.** $4/3 + 4x/3^2 + 4x^2/3^3 + 4x^3/3^4 + . . . + 4x^n/3^{n+1} + . . .; (-3, 3)$
**61.** $x^2 - x^3 + x^4 - x^5 + . . . + (-1)^n x^{n+2} + . . .; (-1, 1)$
**63.** $-2x - 2^2 x^2/2 - 2^3 x^3/3 - 2^4 x^4/4 - . . . - 2^{n+1} x^{n+1}/(n + 1) + . . .; [-1/2, 1/2)$
**65.** $1 - 2x^2 + 2^2 x^4/2! - 2^3 x^6/3! + . . . + (-1)^n 2^n x^{2n}/n! + . . .; (-\infty, +\infty)$
**67.** $2x^3 - 2 \cdot 3x^4 + 2 \cdot 3^2 x^5/2! - 2 \cdot 3^3 x^6/3! + . . . + (-1)^n \cdot 2 \cdot 3^n x^{n+3}/n! + . . .; (-\infty, +\infty)$

**69.** $1 - x^4/2! + x^8/4! - x^{12}/6! + \ldots + (-1)^n x^{4n}/(2n)! + \ldots; (-\infty, +\infty)$

**71.** $3x^2 - 3^3 \cdot x^6/3! + 3^5 \cdot x^{10}/5! - 3^7 \cdot x^{14}/7! + \ldots + (-1)^n \cdot 3^{2n+1} \cdot x^{4n+2}/(2n+1)! + \ldots; (-\infty, +\infty)$　　**73.** 7/4

**75.** Does not exist　　**77.** 5/7　　**79.** $-1/2$　　**81.** Does not exist　　**83.** 1/6　　**85.** Does not exist　　**87.** Does not exist

**89.** 4.73　　**91.** 2.65　　**93.** 6.841　　**95.** $-4.984$　　**97.** \$4,464,960　　**99.** \$27,320.71　　**101.** \$3322.43

**103.** \$717.21

# INDEX

## 5.2 SUBSTITUTION METHOD

| *Form of the Integral* | *Form of the Antiderivative* |
|---|---|

**1.** $\int [u(x)]^n \cdot u'(x)\, dx, \; n \neq -1$     $\dfrac{[u(x)]^{n+1}}{n+1} + C$

**2.** $\int e^{u(x)} \cdot u'(x)\, dx$     $e^{u(x)} + C$

**3.** $\int \dfrac{u'(x)\, dx}{u(x)}$     $\ln |u(x)| + C$

## 5.4 FUNDAMENTAL THEOREM OF CALCULUS

Let $f$ be continuous on the interval $[a, b]$, and let $F$ be *any* antiderivative of $f$. Then

$$\int_a^b f(x)\, dx = F(b) - F(a) = F(x)\Big|_a^b.$$

## 6.1 INTEGRATION BY PARTS

If $u$ and $v$ are differentiable functions, then

$$\int u\, dv = uv - \int v\, du.$$

## 6.2 TRAPEZOIDAL RULE

Let $f$ be a continuous function on $[a, b]$ and let $[a, b]$ be divided into $n$ equal subintervals by the points $a = x_0, x_1, x_2, \ldots, x_n = b$. Then, by the trapezoidal rule,

$$\int_a^b f(x)\, dx \approx \left(\frac{b-a}{n}\right)\left[\frac{1}{2} f(x_0) + f(x_1) + \cdots + f(x_{n-1}) + \frac{1}{2} f(x_n)\right].$$

## 6.2 SIMPSON'S RULE

Let $f$ be a continuous function on $[a, b]$ and let $[a, b]$ be divided into an even number $n$ of equal subintervals by the points $a = x_0, x_1, x_2, \ldots, x_n = b$. Then, by Simpson's rule,

$$\int_a^b f(x)\, dx \approx \frac{b-a}{3n}\, [f(x_0) + 4f(x_1) + 2f(x_2) + 4f(x_3) + \cdots + 2f(x_{n-2}) + 4f(x_{n-1}) + f(x_n)].$$

## 6.5 IMPROPER INTEGRALS

If $f$ is continuous on the indicated interval and if the indicated limits exist, then

$$\int_a^{+\infty} f(x)\, dx = \lim_{b \to +\infty} \int_a^b f(x)\, dx,$$

$$\int_{-\infty}^b f(x)\, dx = \lim_{a \to -\infty} \int_a^b f(x)\, dx,$$

$$\int_{-\infty}^{+\infty} f(x)\, dx = \int_{-\infty}^c f(x)\, dx + \int_c^{+\infty} f(x)\, dx,$$

for real numbers $a$, $b$, and $c$.

## 7.2 PARTIAL DERIVATIVES (INFORMAL DEFINITION)

The partial derivative of $f$ with respect to $x$ is the derivative of $f$ obtained by treating $x$ as a variable and $y$ as a constant.

The partial derivative of $f$ with respect to $y$ is the derivative of $f$ obtained by treating $y$ as a variable and $x$ as a constant.